GEODYNAMICS

Second Edition

First published in 1982, Don Turcotte and Jerry Schubert's *Geodynamics* became a classic textbook for several generations of students of geophysics and geology. In this second edition, the authors bring this classic text completely up-to-date. Important additions include a chapter on chemical geodynamics, an updated coverage of comparative planetology based on recent planetary missions, and a variety of other new topics.

Geodynamics provides the fundamentals necessary for an understanding of the workings of the solid Earth. The Earth is a heat engine, with the source of the heat the decay of radioactive elements and the cooling of the Earth from its initial accretion. The work output includes earthquakes, volcanic eruptions, and mountain building. *Geodynamics* comprehensively explains these concepts in the context of the role of mantle convection and plate tectonics. Observations such as the Earth's gravity field, surface heat flow, distribution of earthquakes, surface stresses and strains, and distribution of elements are discussed. The rheological behavior of the solid Earth, from an elastic solid to fracture to plastic deformation to fluid flow, is considered. Important inputs come from a comparison of the similarities and differences between the Earth, Venus, Mars, Mercury, and the Moon. An extensive set of student exercises is included.

This new edition of *Geodynamics* will once again prove to be a classic textbook for intermediate to advanced undergraduates and graduate students in geology, geophysics, and Earth science.

Donald L. Turcotte is Maxwell Upson Professor of Engineering, Department of Geological Sciences, Cornell University. In addition to this book, he is author or co-author of 3 books and 276 research papers, including *Fractals and Chaos in Geology and Geophysics* (Cambridge University Press, 1992 and 1997) and *Mantle Convection in the Earth and Planets* (with Gerald Schubert and Peter Olson; Cambridge University Press, 2001). Professor Turcotte is a Fellow of the American Geophysical Union, Honorary Fellow of the European Union of Geosciences, and Fellow of the Geological Society of America. He is the recipient of several medals, including the Day Medal of the Geological Society of America, the Wegener Medal of the European Union of Geosciences, the Whitten Medal of the American Geophysical Union, the Regents (New York State) Medal of Excellence, and Caltech's Distinguished Alumnus Award. Professor Turcotte is a member of the National Academy of Sciences and the American Academy of Arts and Sciences.

Gerald Schubert is a Professor in the Department of Earth and Space Sciences and the Institute of Geophysics and Planetary Physics at the University of California, Los Angeles. He is co-author with Donald Turcotte and Peter Olson of *Mantle Convection in the Earth and Planets* (Cambridge University Press, 2001), and author of over 400 research papers. He has participated in a number of NASA's planetary missions and has been on the editorial boards of many journals, including *Icarus, Journal of Geophysical Research, Geophysical Research Letters*, and *Annual Reviews of Earth and Planetary Sciences*. Professor Schubert is a Fellow of the American Geophysical Union and a recipient of the Union's James B. MacElwane medal. He is a member of the American Academy of Arts and Sciences.

GEODYNAMICS

Second Edition

DONALD L. TURCOTTE

Professor of Geological Sciences
Cornell University

GERALD SCHUBERT

Professor of Earth and Space Sciences
University of California, Los Angeles

CAMBRIDGE
UNIVERSITY PRESS

CAMBRIDGE UNIVERSITY PRESS
Cambridge, New York, Melbourne, Madrid, Cape Town, Singapore, São Paulo, Delhi

Cambridge University Press
32 Avenue of the Americas, New York, NY 10013-2473, USA

www.cambridge.org
Information on this title: www.cambridge.org/9780521661867

First edition published by John Wiley & Sons, Inc., 1982
Second edition published by Cambridge University Press 2002
Reprinted 2005, 2006, 2007

Printed in the United States of America

A catalog record for this publication is available from the British Library.

Library of Congress Cataloging in Publication Data

Turcotte, Donald Lawson.
Geodynamics / Donald L. Turcotte, Gerald Schubert. – 2nd ed
p.   cm.
Rev. ed. of: Geodynamics applications of continuum physics to geological
   problems, c1982.
Includes bibliographical references and index.
ISBN 0-521-66186-2 – ISBN 0-521-66624-4 (pb.)
1. Geodynamics. I. Schubert, Gerald. II. Turcotte, Donald Lawson. Geodynamics
   applications of continuum physics to geological problems, c1982. III. Title.
QE501 .T83   2001   2001025802

ISBN 978-0-521-66186-7 hardback
ISBN 978-0-521-66624-4 paperback

Contents

SEVEN. Rock Rheology   292

EIGHT. Faulting   339

NINE. Flows in Porous Media   374

Preface

This textbook deals with the fundamental physical processes necessary for an understanding of plate tectonics and a variety of geological phenomena. We believe that the appropriate title for this material is *geodynamics*. The contents of this textbook evolved from a series of courses given at Cornell University and UCLA to students with a wide range of backgrounds in geology, geophysics, physics, mathematics, chemistry, and engineering. The level of the students ranged from advanced undergraduate to graduate.

In all cases we present the material with a minimum of mathematical complexity. We have not introduced mathematical concepts unless they are essential to the understanding of physical principles. For example, our treatment of elasticity and fluid mechanics avoids the introduction or use of tensors. We do not believe that tensor notation is necessary for the understanding of these subjects or for most applications to geological problems. However, solving partial differential equations is an essential part of this textbook. Many geological problems involving heat conduction and solid and fluid mechanics require solutions of such classic partial differential equations as Laplace's equation, Poisson's equation, the biharmonic equation, and the diffusion equation. All these equations are derived from first principles in the geological contexts in which they are used. We provide elementary explanations for such important physical properties of matter as solid-state viscosity, thermal coefficient of expansion, specific heat, and permeability. Basic concepts involved in the studies of heat transfer, Newtonian and non-Newtonian fluid behavior, the bending of thin elastic plates, the mechan-

ical behavior of faults, and the interpretation of gravity anomalies are emphasized. Thus it is expected that the student will develop a thorough understanding of such fundamental physical laws as Hooke's law of elasticity, Fourier's law of heat conduction, and Darcy's law for fluid flow in porous media.

The problems are an integral part of this textbook. It is only through solving a substantial number of exercises that an adequate understanding of the underlying physical principles can be developed. Answers to selected problems are provided.

The first chapter reviews plate tectonics; its main purpose is to provide physics, chemistry, and engineering students with the geological background necessary to understand the applications considered throughout the rest of the textbook. We hope that the geology student can also benefit from this summary of numerous geological, seismological, and paleomagnetic observations. Since plate tectonics is a continuously evolving subject, this material may be subject to revision. Chapter 1 also briefly summarizes the geological and geophysical characteristics of the other planets and satellites of the solar system. Chapter 2 introduces the concepts of stress and strain and discusses the measurements of these quantities in the Earth's crust. Chapter 3 presents the basic principles of linear elasticity. The bending of thin elastic plates is emphasized and is applied to problems involving the bending of the Earth's lithosphere. Chapter 4 deals mainly with heat conduction and the application of this theory to temperatures in the continental crust and the continental and oceanic lithospheres. Heat transfer by convection is briefly discussed and applied to a determination of temperature in the Earth's mantle. Surface heat flow measurements are reviewed and interpreted in terms of the theory. The sources of the Earth's surface heat flow are discussed. Problems involving the solidification of magmas and extrusive lava flows are also treated. The basic principles involved in the interpretation of gravity measurements are given in Chapter 5. Fluid mechanics is studied in Chapter 6; problems involving mantle convection and postglacial rebound are emphasized. Chapter 7 deals with the rheology of rock or the manner in which it deforms or flows under applied forces. Fundamental processes are discussed from a microscopic point of view. The mechanical behavior of faults is discussed in Chapter 8 with particular attention being paid to observations of displacements along the San Andreas fault. Finally, Chapter 9 discusses the principles of fluid flow in porous media, a subject that finds application to hydrothermal circulations in the oceanic crust and in continental geothermal areas.

The contents of this textbook are intended to provide the material for a coherent one-year course. In order to accomplish this goal, some important aspects of geodynamics have had to be omitted. In particular, the fundamentals of seismology are not included. Thus the wave equation and its solutions are not discussed. Many seismic studies have provided important data relevant to geodynamic processes. Examples include (1) the radial distribution of density in the Earth as inferred from the radial profiles of seismic velocities, (2) important information on the locations of plate boundaries and the locations of descending plates at ocean trenches provided by accurate determinations of the epicenters of earthquakes, and (3) details of the structure of the continental crust obtained by seismic reflection profiling using artificially generated waves. An adequate treatment of seismology would have required a very considerable expansion of this textbook. Fortunately, there are a number of excellent textbooks on this subject.

A comprehensive study of the spatial and temporal variations of the Earth's magnetic field is also considered to be outside the scope of this textbook. A short discussion of the Earth's magnetic field relevant to paleomagnetic observations is given in Chapter 1. However, mechanisms for the generation of the Earth's magnetic field are not considered.

In writing this textbook, several difficult decisions had to be made. One was the choice of units; we use SI units throughout. This system of units is defined in Appendix 1. We feel there is a strong trend toward the use of SI units in both geology and geophysics. We recognize, however, that many cgs units are widely used. Examples include μcal cm^{-2} s^{-1} for heat flow, kilobar for stress, and milligal for gravity anomalies. For this reason we have often included the equivalent cgs unit in parentheses after the SI unit, for example, MPa (kbar). Another decision involved the referencing of original work. We do not believe that it is appropriate to include a large number of references in a basic textbook. We have credited those individuals making major contributions to the development of the theory of plate tectonics and continental drift in our brief discussion of the history of this subject in Chapter 1. We also provide references to data. At the end of each chapter a list of recommended reading is given. In many instances these are textbooks and reference books, but in some cases review papers are included. In each case the objective is to provide background material for the chapter or to extend its content.

Many of our colleagues have read all or parts of various drafts of this textbook. We acknowledge the contributions made by Jack Bird, Peter Bird, Muawia Barazangi, Allan Cox, Walter Elsasser, Robert Kay, Suzanne Kay, Mark Langseth, Bruce Marsh, Jay Melosh, John Rundle, Sean Solomon, David Stevenson, Ken Torrance, and David Yuen. We particularly wish to acknowledge the many contributions to our work made by Ron Oxburgh and the excellent manuscript preparation by Tanya Harter.

Preface to the Second Edition

As we prepared our revisions for this second edition of *Geodynamics* we were struck by the relatively few changes and additions that were required. The reason is clear: this textbook deals with fundamental physical processes that do not change. However, a number of new ideas and concepts have evolved and have been included where appropriate.

In revising the first chapter on plate tectonics we placed added emphasis on the concept of mantle plumes. In particular we discussed the association of plume heads with continental flood basalts. We extensively revised the sections on comparative planetology. We have learned new things about the Moon, and the giant impact hypothesis for its origin has won wide acceptance. For Venus, the Magellan mission has revolutionized our information about the planet. The high-resolution radar images, topography, and gravity data have provided new insights that emphasize the tremendous differences in structure and evolution between Venus and the Earth. Similarly, the Galileo mission has greatly enhanced our understanding of the Galilean satellites of Jupiter.

In Chapter 2 we introduce the crustal stretching model for the isostatic subsidence of sedimentary basins. This model provides a simple explanation for the formation of sedimentary basins. Space-based geodetic observations have revolutionized our understanding of surface strain fields associated with tectonics. We introduce the reader to satellite data obtained from the global positioning system (GPS) and synthetic aperture radar interferometry (INSAR). In Chapter 4 we introduce the plate cooling model for the thermal structure of the oceanic lithosphere as a complement to the half-space cooling model. We also present in this chapter the Culling model for the diffusive erosion and deposition of sediments. In Chapter 5 we show how geoid anomalies are directly related to the forces required to maintain topography.

In Chapter 6 we combine a pipe-flow model with a Stokes-flow model in order to determine the structure and strength of plume heads and plume tails. The relationship between hotspot swells and the associated plume flux is also introduced. In addition to the steady-state boundary-layer model for the structure of mantle convection cells, we introduce a transient boundary-layer model for the stability of the lithosphere.

Finally, we conclude the book with a new Chapter 10 on chemical geodynamics. The concept of chemical geodynamics has evolved since the first edition was written. The object is to utilize geochemical data, particularly the isotope systematics of basalts, to infer mantle dynamics. Questions addressed include the homogeneity of the mantle, the fate of subducted lithosphere, and whether whole mantle convection or layered mantle convection is occurring.

The use of SI units is now firmly entrenched in geology and geophysics, and we use these units throughout the book. Since *Geodynamics* is meant to be a textbook, large numbers of references are inappropriate. However, we have included key references and references to sources of data in addition to recommended collateral reading.

In addition to the colleagues who we acknowledge in the preface to the first edition, we would like to add Claude Allègre, Louise Kellogg, David Kohlstedt, Bruce Malamud, Mark Parmentier, and David Sandwell. We also acknowledge the excellent manuscript preparation by Stacey Shirk and Judith Hohl, and figure preparation by Richard Sadakane.

ONE

Plate Tectonics

1-1 Introduction

Plate tectonics is a model in which the outer shell of the Earth is divided into a number of thin, rigid plates that are in relative motion with respect to one another. The relative velocities of the plates are of the order of a few tens of millimeters per year. A large fraction of all earthquakes, volcanic eruptions, and mountain building occurs at plate boundaries. The distribution of the major surface plates is illustrated in Figure 1–1.

The plates are made up of relatively cool rocks and have an average thickness of about 100 km. The plates are being continually created and consumed. At ocean ridges adjacent plates diverge from each other in a process known as *seafloor spreading*. As the adjacent plates diverge, hot mantle rock ascends to fill the gap. The hot, solid mantle rock behaves like a fluid because of solid-state creep processes. As the hot mantle rock cools, it becomes rigid and accretes to the plates, creating new plate area. For this reason ocean ridges are also known as *accreting plate boundaries*. The accretionary process is symmetric to a first approximation so that the rates of plate formation on the two sides of a ridge are approximately equal. The rate of plate formation on one side of an ocean ridge defines a half-spreading velocity u. The two plates spread with a relative velocity of $2u$. The global system of ocean ridges is denoted by the heavy dark lines in Figure 1–1.

Because the surface area of the Earth is essentially constant, there must be a complementary process of plate consumption. This occurs at ocean trenches. The surface plates bend and descend into the interior of the Earth in a process known as *subduction*. At an ocean trench the two adjacent plates converge, and one descends beneath the other. For this reason ocean trenches are also known as *convergent plate boundaries*. The worldwide distribution of trenches is shown in Figure 1–1 by the lines with triangular symbols, which point in the direction of subduction.

A cross-sectional view of the creation and consumption of a typical plate is illustrated in Figure 1–2. That part of the Earth's interior that comprises the plates is referred to as the *lithosphere*. The rocks that make up the lithosphere are relatively cool and rigid; as a result the interiors of the plates do not deform significantly as they move about the surface of the Earth. As the plates move away from ocean ridges, they cool and thicken. The solid rocks beneath the lithosphere are sufficiently hot to be able to deform freely; these rocks comprise the *asthenosphere*, which lies below the lithosphere. The lithosphere slides over the asthenosphere with relatively little resistance.

As the rocks of the lithosphere become cooler, their density increases because of thermal contraction. As a result the lithosphere becomes gravitationally unstable with respect to the hot asthenosphere beneath. At the ocean trench the lithosphere bends and sinks into the interior of the Earth because of this negative buoyancy. The downward gravitational body force on the descending lithosphere plays an important role in driving plate tectonics. The lithosphere acts as an elastic plate that transmits large elastic stresses without significant deformation. Thus the gravitational body force can be transmitted directly to the surface plate and this force pulls the plate toward the trench. This body force is known as *trench pull*. Major faults separate descending lithospheres from adjacent overlying lithospheres. These faults are the sites of most great earthquakes. Examples are the Chilean earthquake in 1960 and the Alaskan earthquake in 1964. These are the largest earthquakes that have occurred since modern seismographs have been available. The locations of the descending lithospheres can be accurately determined from the earthquakes occurring in the cold, brittle rocks of the lithospheres. These planar zones of earthquakes associated with subduction are known as *Wadati–Benioff zones*.

Lines of active volcanoes lie parallel to almost all ocean trenches. These volcanoes occur about 125 km above the descending lithosphere. At least a fraction of the magmas that form these volcanoes are produced

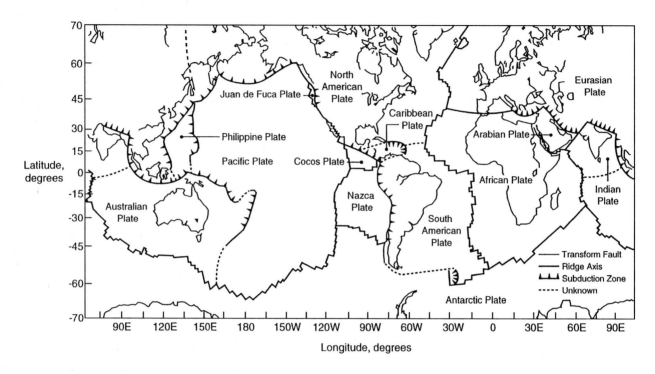

1–1 Distribution of the major plates. The ocean ridge axis (accretional plate margins), subduction zones (convergent plate margins), and transform faults that make up the plate boundaries are shown.

near the upper boundary of the descending lithosphere and rise some 125 km to the surface. If these volcanoes stand on the seafloor, they form an *island arc*, as typified by the Aleutian Islands in the North Pacific. If the trench lies adjacent to a continent, the volcanoes grow from the land surface. This is the case in the western

1–2 Accretion of a lithospheric plate at an ocean ridge and its subduction at an ocean trench. The asthenosphere, which lies beneath the lithosphere, is shown along with the line of volcanic centers associated with subduction.

United States, where a volcanic line extends from Mt. Baker in the north to Mt. Shasta in the south. Mt. St. Helens, the site of a violent eruption in 1980, forms a part of this volcanic line. These volcanoes are the sites of a large fraction of the most explosive and violent volcanic eruptions. The eruption of Mt. Pinatubo in the Philippines in 1991, the most violent eruption of the 20th century, is another example. A typical subduction zone volcano is illustrated in Figure 1–3.

The Earth's surface is divided into continents and oceans. The oceans have an average depth of about 4 km, and the continents rise above sea level. The reason for this difference in elevation is the difference in

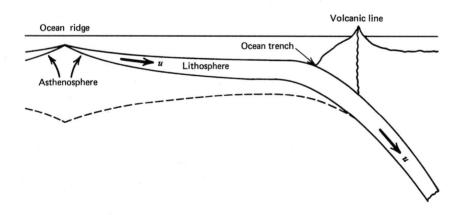

1-3 Izalco volcano in El Salvador, an example of a subduction zone volcano (NOAA–NGDC Howell Williams).

the thickness of the crust. Crustal rocks have a different composition from that of the mantle rocks beneath and are less dense. The crustal rocks are therefore gravitationally stable with respect to the heavier mantle rocks. There is usually a well-defined boundary, the *Moho* or Mohorovičić discontinuity, between the crust and mantle. A typical thickness for *oceanic crust* is 6 km; *continental crust* is about 35 km thick. Although oceanic crust is gravitationally stable, it is sufficiently thin so that it does not significantly impede the subduction of the gravitationally unstable oceanic lithosphere. The oceanic lithosphere is continually cycled as it is accreted at ocean ridges and subducted at ocean trenches. Because of this cycling the average age of the ocean floor is about 10^8 years (100 Ma).

On the other hand, the continental crust is sufficiently thick and gravitationally stable so that it is not subducted at an ocean trench. In some cases the denser lower continental crust, along with the underlying gravitationally unstable continental mantle lithosphere, can be recycled into the Earth's interior in a process known as *delamination*. However, the light rocks of the upper continental crust remain in the continents. For this reason the rocks of the continental crust, with an average age of about 10^9 years (1 Ga), are much older than the rocks of the oceanic crust. As the lithospheric plates move across the surface of the Earth, they carry the continents with them. The relative motion of continents is referred to as *continental drift*.

Much of the historical development leading to plate tectonics concerned the validity of the hypothesis of continental drift: that the relative positions of continents change during geologic time. The similarity in shape between the west coast of Africa and the east coast of South America was noted as early as 1620 by Francis Bacon. This "fit" has led many authors to speculate on how these two continents might have been attached. A detailed exposition of the hypothesis of continental drift was put forward by Frank B. Taylor (1910). The hypothesis was further developed by Alfred Wegener beginning in 1912 and summarized in his book *The Origin of Continents and Oceans* (Wegener, 1946). As a meteorologist, Wegener was particularly interested in the observation that glaciation had occurred in equatorial regions at the same time that tropical conditions prevailed at high latitudes. This observation in itself could be explained by *polar wander*, a shift of the rotational axis without other surface deformation. However, Wegener also set forth many of the qualitative arguments that the continents had formerly been attached. In addition to the observed fit of continental margins, these arguments included the correspondence of geological provinces, continuity of structural features such as relict mountain ranges, and the correspondence of fossil types. Wegener argued that a single supercontinent, Pangaea, had formerly existed. He suggested that tidal forces or forces associated with the rotation of the Earth were responsible for the breakup of this continent and the subsequent continental drift.

Further and more detailed qualitative arguments favoring continental drift were presented by Alexander du Toit, particularly in his book *Our Wandering Continents* (du Toit, 1937). Du Toit argued that instead of a single supercontinent, there had formerly been a northern continent, Laurasia, and a southern continent, Gondwanaland, separated by the Tethys Ocean.

During the 1950s extensive exploration of the seafloor led to an improved understanding of the worldwide range of mountains on the seafloor known as mid-ocean ridges. Harry Hess (1962) hypothesized that the seafloor was created at the axis of a ridge and moved away from the ridge to form an ocean in a process now referred to as *seafloor spreading*. This process explains the similarity in shape between continental margins. As a continent breaks apart, a new ocean ridge forms. The ocean floor created is formed symmetrically at this ocean ridge, creating a new ocean. This is how the

Atlantic Ocean was formed; the mid-Atlantic ridge where the ocean formed now bisects the ocean.

It should be realized, however, that the concept of continental drift won general acceptance by Earth scientists only in the period between 1967 and 1970. Although convincing qualitative, primarily geological, arguments had been put forward to support continental drift, almost all Earth scientists and, in particular, almost all geophysicists had opposed the hypothesis. Their opposition was mainly based on arguments concerning the rigidity of the mantle and the lack of an adequate driving mechanism.

The propagation of seismic shear waves showed beyond any doubt that the mantle was a solid. An essential question was how horizontal displacements of thousands of kilometers could be accommodated by solid rock. The fluidlike behavior of the Earth's mantle had been established in a general way by gravity studies carried out in the latter part of the nineteenth century. Measurements showed that mountain ranges had low-density roots. The lower density of the roots provides a negative relative mass that nearly equals the positive mass of the mountains. This behavior could be explained by the principle of *hydrostatic equilibrium* if the mantle behaved as a fluid. Mountain ranges appear to behave similarly to blocks of wood floating on water.

The fluid behavior of the mantle was established quantitatively by N. A. Haskell (1935). Studies of the elevation of beach terraces in Scandinavia showed that the Earth's surface was still rebounding from the load of the ice during the last ice age. By treating the mantle as a viscous fluid with a viscosity of 10^{20} Pa s, Haskell was able to explain the present uplift of Scandinavia. Although this is a very large viscosity (water has a viscosity of 10^{-3} Pa s), it leads to a fluid behavior for the mantle during long intervals of geologic time.

In the 1950s theoretical studies had established several mechanisms for the very slow creep of crystalline materials. This creep results in a fluid behavior. Robert B. Gordon (1965) showed that solid-state creep quantitatively explained the viscosity determined from observations of postglacial rebound. At temperatures that are a substantial fraction of the melt temperature, thermally activated creep processes allow mantle rock to flow at low stress levels on time scales greater than 10^4 years. The rigid lithosphere includes rock that is sufficiently cold to preclude creep on these long time scales.

The creep of mantle rock was not a surprise to scientists who had studied the widely recognized flow of ice in glaciers. Ice is also a crystalline solid, and gravitational body forces in glaciers cause ice to flow because its temperature is near its melt temperature. Similarly, mantle rocks in the Earth's interior are near their melt temperatures and flow in response to gravitational body forces.

Forces must act on the lithosphere in order to make the plates move. Wegener suggested that either tidal forces or forces associated with the rotation of the Earth caused the motion responsible for continental drift. However, in the 1920s Sir Harold Jeffreys, as summarized in his book *The Earth* (Jeffreys, 1924), showed that these forces were insufficient. Some other mechanism had to be found to drive the motion of the plates. Any reasonable mechanism must also have sufficient energy available to provide the energy being dissipated in earthquakes, volcanoes, and mountain building. Arthur Holmes (1931) hypothesized that thermal convection was capable of driving mantle convection and continental drift. If a fluid is heated from below, or from within, and is cooled from above in the presence of a gravitational field, it becomes gravitationally unstable, and thermal convection can occur. The hot mantle rocks at depth are gravitationally unstable with respect to the colder, more dense rocks in the lithosphere. The result is thermal convection in which the colder rocks descend into the mantle and the hotter rocks ascend toward the surface. The ascent of mantle material at ocean ridges and the descent of the lithosphere into the mantle at ocean trenches are parts of this process. The Earth's mantle is being heated by the decay of the radioactive isotopes uranium 235 (^{235}U), uranium 238 (^{238}U), thorium 232 (^{232}Th), and potassium 40 (^{40}K). The volumetric heating from these isotopes and the *secular cooling* of the Earth drive mantle convection. The heat generated by the radioactive isotopes decreases with time as they decay. Two billion years ago the heat generated was about twice the present value. Because the amount of heat generated is less today, the vigor of the mantle convection required today to extract the heat is also less. The vigor of mantle convection depends on the mantle viscosity. Less vigorous mantle convection implies a lower viscosity. But the mantle viscosity is a strong function of mantle temperature; a lower mantle viscosity implies a cooler mantle. Thus as mantle convection becomes less vigorous, the mantle cools; this is secular cooling. As a result, about 80% of the heat lost from the interior of the Earth is from the decay of the radioactive isotopes and about 20% is due to the cooling of the Earth (secular cooling).

During the 1960s independent observations supporting continental drift came from paleomagnetic studies. When magmas solidify and cool, their iron component is magnetized by the Earth's magnetic field. This remanent magnetization provides a fossil record of the orientation of the magnetic field at that time. Studies of the orientation of this field can be used to determine the movement of the rock relative to the Earth's magnetic poles since the rock's formation. Rocks in a single surface plate that have not been deformed locally show the same position for the Earth's magnetic poles. Keith Runcorn (1956) showed that rocks in North America and Europe gave different positions for the magnetic poles. He concluded that the differences were the result of continental drift between the two continents.

Paleomagnetic studies also showed that the Earth's magnetic field has been subject to episodic reversals. Observations of the magnetic field over the oceans indicated a regular striped pattern of *magnetic anomalies* (regions of magnetic field above and below the average field value) lying parallel to the ocean ridges. Frederick Vine and Drummond Matthews (1963) correlated the locations of the edges of the striped pattern of magnetic anomalies with the times of magnetic field reversals and were able to obtain quantitative values for the rate of seafloor spreading. These observations have provided the basis for accurately determining the relative velocities at which adjacent plates move with respect to each other.

By the late 1960s the framework for a comprehensive understanding of the geological phenomena and processes of continental drift had been built. The basic hypothesis of plate tectonics was given by Jason Morgan (1968). The concept of a mosaic of rigid plates in relative motion with respect to one another was a natural consequence of thermal convection in the mantle. A substantial fraction of all earthquakes, volcanoes, and mountain building can be attributed to the interactions among the lithospheric plates at their boundaries (Isacks et al., 1968). Continental drift is an inherent part of plate tectonics. The continents are carried with the plates as they move about the surface of the Earth.

PROBLEM 1-1 If the area of the oceanic crust is 3.2×10^8 km^2 and new seafloor is now being created at the rate of 2.8 km^2 yr^{-1}, what is the mean age of the oceanic crust? Assume that the rate of seafloor creation has been constant in the past.

1-2 The Lithosphere

An essential feature of plate tectonics is that only the outer shell of the Earth, the lithosphere, remains rigid during intervals of geologic time. Because of their low temperature, rocks in the lithosphere do not significantly deform on time scales of up to 10^9 years. The rocks beneath the lithosphere are sufficiently hot so that solid-state creep can occur. This creep leads to a fluidlike behavior on geologic time scales. In response to forces, the rock beneath the lithosphere flows like a fluid.

The lower boundary of the lithosphere is defined to be an isotherm (surface of constant temperature). A typical value is approximately 1600 K. Rocks lying above this isotherm are sufficiently cool to behave rigidly, whereas rocks below this isotherm are sufficiently hot to readily deform. Beneath the ocean basins the lithosphere has a thickness of about 100 km; beneath the continents the thickness is about twice this value. Because the thickness of the lithosphere is only 2 to 4% of the radius of the Earth, the lithosphere is a thin shell. This shell is broken up into a number of plates that are in relative motion with respect to one another. The rigidity of the lithosphere ensures, however, that the interiors of the plates do not deform significantly.

The rigidity of the lithosphere allows the plates to transmit elastic stresses during geologic intervals. The plates act as stress guides. Stresses that are applied at the boundaries of a plate can be transmitted throughout the interior of the plate. The ability of the plates to transmit stress over large distances has important implications with regard to the driving mechanism of plate tectonics.

The rigidity of the lithosphere also allows it to bend when subjected to a load. An example is the load applied by a volcanic island. The load of the Hawaiian Islands causes the lithosphere to bend downward around the load, resulting in a region of deeper water around the islands. The elastic bending of the lithosphere under vertical loads can also explain the structure of ocean trenches and some sedimentary basins.

However, the entire lithosphere is not effective in transmitting elastic stresses. Only about the upper half of it is sufficiently rigid so that elastic stresses are not relaxed on time scales of 10^9 years. This fraction of the lithosphere is referred to as the *elastic lithosphere*. Solid-state creep processes relax stresses in the lower, hotter part of the lithosphere. However, this part of the lithosphere remains a coherent part of the plates. A detailed discussion of the difference between the thermal and elastic lithospheres is given in Section 7–10.

1-3 Accreting Plate Boundaries

Lithospheric plates are created at ocean ridges. The two plates on either side of an ocean ridge move away from each other with near constant velocities of a few tens of millimeters per year. As the two plates diverge, hot mantle rock flows upward to fill the gap. The upwelling mantle rock cools by conductive heat loss to the surface. The cooling rock accretes to the base of the spreading plates, becoming part of them; the structure of an accreting plate boundary is illustrated in Figure 1–4.

As the plates move away from the ocean ridge, they continue to cool and the lithosphere thickens. The elevation of the ocean ridge as a function of distance from the ridge crest can be explained in terms of the temperature distribution in the lithosphere. As the lithosphere cools, it becomes more dense; as a result it sinks downward into the underlying mantle rock. The topographic elevation of the ridge is due to the greater buoyancy of

the thinner, hotter lithosphere near the axis of accretion at the ridge crest. The elevation of the ocean ridge also provides a body force that causes the plates to move away from the ridge crest. A component of the gravitational body force on the elevated lithosphere drives the lithosphere away from the accretional boundary; it is one of the important forces driving the plates. This force on the lithosphere is known as *ridge push* and is a form of *gravitational sliding*.

The volume occupied by the ocean ridge displaces seawater. Rates of seafloor spreading vary in time. When rates of seafloor spreading are high, ridge volume is high, and seawater is displaced. The result is an increase in the global sea level. Variations in the rates of seafloor spreading are the primary cause for changes in sea level on geological time scales. In the Cretaceous ($\approx$80 Ma) the rate of seafloor spreading was about 30% greater than at present and sea level was about 200 m higher than today. One result was that a substantial fraction of the continental interiors was covered by shallow seas.

Ocean ridges are the sites of a large fraction of the Earth's volcanism. Because almost all the ridge system is under water, only a small part of this volcanism can be readily observed. The details of the volcanic processes at ocean ridges have been revealed by exploration using submersible vehicles. Ridge volcanism can also be seen in Iceland, where the oceanic crust is sufficiently thick so that the ridge crest rises above sea level. The volcanism at ocean ridges is caused by *pressure-release melting*. As the two adjacent plates move apart, hot

1–4 An accreting plate margin at an ocean ridge.

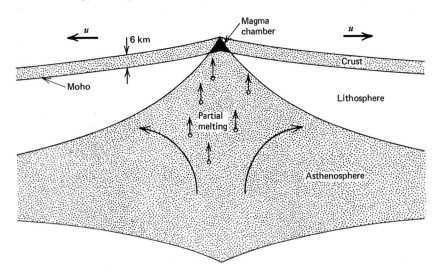

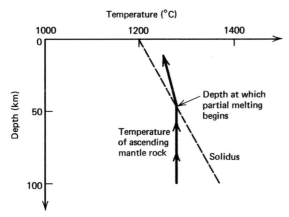

1–5 The process of pressure-release melting is illustrated. Melting occurs because the nearly isothermal ascending mantle rock encounters pressures low enough so that the associated solidus temperatures are below the rock temperatures.

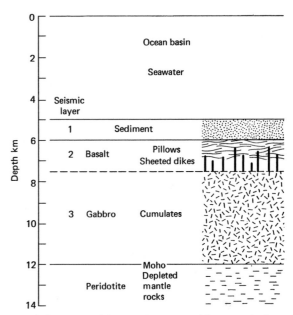

1–6 Typical structure of the oceanic crust, overlying ocean basin, and underlying depleted mantle rock.

mantle rock ascends to fill the gap. The temperature of the ascending rock is nearly constant, but its pressure decreases. The pressure p of rock in the mantle is given by the simple hydrostatic equation

$$p = \rho g y, \qquad (1\text{–}1)$$

where ρ is the density of the mantle rock, g is the acceleration of gravity, and y is the depth. The solidus temperature (the temperature at which the rock first melts) decreases with decreasing pressure. When the temperature of the ascending mantle rock equals the solidus temperature, melting occurs, as illustrated in Figure 1–5. The ascending mantle rock contains a low-melting-point, basaltic component. This component melts to form the oceanic crust.

PROBLEM 1-2 At what depth will ascending mantle rock with a temperature of 1600 K melt if the equation for the solidus temperature T is

$$T(K) = 1500 + 0.12\,p\,(MPa).$$

Assume $\rho = 3300 \text{ kg m}^{-3}$, $g = 10 \text{ m s}^{-2}$, and the mantle rock ascends at constant temperature.

The *magma* (melted rock) produced by partial melting beneath an ocean ridge is lighter than the residual mantle rock, and buoyancy forces drive it upward to the surface in the vicinity of the ridge crest. Magma chambers form, heat is lost to the seafloor, and this magma solidifies to form the oceanic crust. In some localities

slices of oceanic crust and underlying mantle have been brought to the surface. These are known as *ophiolites*; they occur in such locations as Cyprus, Newfoundland, Oman, and New Guinea. Field studies of ophiolites have provided a detailed understanding of the oceanic crust and underlying mantle. Typical oceanic crust is illustrated in Figure 1–6. The crust is divided into layers 1, 2, and 3, which were originally associated with different seismic velocities but subsequently identified compositionally. Layer 1 is composed of sediments that are deposited on the volcanic rocks of layers 2 and 3. The thickness of sediments increases with distance from the ridge crest; a typical thickness is 1 km. Layers 2 and 3 are composed of basaltic rocks of nearly uniform composition. A typical composition of an ocean basalt is given in Table 1–1. The basalt is composed primarily of two rock-forming minerals, plagioclase feldspar and pyroxene. The plagioclase feldspar is 50 to 85% anorthite ($CaAl_2Si_2O_8$) component and 15 to 50% albite ($NaAlSi_3O_8$) component. The principal pyroxene is rich in the diopside ($CaMgSi_2O_6$) component. Layer 2 of the oceanic crust is composed of extrusive volcanic flows that have interacted with the seawater to form pillow lavas and intrusive flows primarily in the form of sheeted dikes. A typical thickness for layer 2 is 1.5 km. Layer 3 is made up of gabbros and related

TABLE 1–1 **Typical Compositions of Important Rock Types**

	Granite	Diorite	Clastic Sediments	Continental Crust	Basalt	Harzburgite	"Pyrolite"	Chondrite
SiO_2	70.8	57.6	70.4	61.7	50.3	45.3	46.1	33.3
Al_2O_3	14.6	16.9	14.3	15.8	16.5	1.8	4.3	2.4
Fe_2O_3	1.6	3.2	—	—	—	—	—	—
FeO	1.8	4.5	5.3	6.4	8.5	8.1	8.2	35.5
MgO	0.9	4.2	2.3	3.6	8.3	43.6	37.6	23.5
CaO	2.0	6.8	2.0	5.4	12.3	1.2	3.1	2.3
Na_2O	3.5	3.4	1.8	3.3	2.6	—	0.4	1.1
K_2O	4.2	3.4	3.0	2.5	0.2	—	0.03	—
TiO_2	0.4	0.9	0.7	0.8	1.2	—	0.2	—

cumulate rocks that crystallized directly from the magma chamber. Gabbros are coarse-grained basalts; the larger grain size is due to slower cooling rates at greater depths. The thickness of layer 3 is typically 4.5 km.

Studies of ophiolites show that oceanic crust is underlain primarily by a peridotite called harzburgite. A typical composition of a harzburgite is given in Table 1–1. This peridotite is primarily composed of olivine and orthopyroxene. The olivine consists of about 90% forsterite component (Mg_2SiO_4) and about 10% fayalite component (Fe_2SiO_4). The orthopyroxene is less abundant and consists primarily of the enstatite component ($MgSiO_3$). Relative to basalt, harzburgite contains lower concentrations of calcium and aluminum and much higher concentrations of magnesium. The basalt of the oceanic crust with a density of 2900 kg m^{-3} is gravitationally stable with respect to the underlying peridotite with a density of 3300 kg m^{-3}. The harzburgite has a greater melting temperature ($\simeq$500 K higher) than basalt and is therefore more refractory.

Field studies of ophiolites indicate that the harzburgite did not crystallize from a melt. Instead, it is the crystalline residue left after partial melting produced the basalt. The process by which partial melting produces the basaltic oceanic crust, leaving a refractory residuum of peridotite, is an example of igneous *fractionation*.

Molten basalts are less dense than the solid, refractory harzburgite and ascend to the base of the oceanic crust because of their buoyancy. At the base of the crust they form a magma chamber. Since the forces driving plate tectonics act on the oceanic lithosphere, they produce a fluid-driven fracture at the ridge crest. The molten basalt flows through this fracture, draining the magma chamber and resulting in surface flows. These surface flows interact with the seawater to generate pillow basalts. When the magma chamber is drained, the residual molten basalt in the fracture solidifies to form a dike. The solidified rock in the dike prevents further migration of molten basalt, the magma chamber refills, and the process repeats. A typical thickness of a dike in the vertical sheeted dike complex is 1 m.

Other direct evidence for the composition of the mantle comes from *xenoliths* that are carried to the surface in various volcanic flows. Xenoliths are solid rocks that are entrained in erupting magmas. Xenoliths of mantle peridotites are found in some basaltic flows in Hawaii and elsewhere. Mantle xenoliths are also carried to the Earth's surface in kimberlitic eruptions. These are violent eruptions that form the kimberlite pipes where diamonds are found.

It is concluded that the composition of the upper mantle is such that basalts can be fractionated leaving harzburgite as a residuum. One model composition for the parent undepleted mantle rock is called *pyrolite* and its chemical composition is given in Table 1–1. In order to produce the basaltic oceanic crust, about 20% partial melting of pyrolite must occur. Incompatible elements such as the heat-producing elements uranium, thorium, and potassium do not fit into the crystal structures of the principal minerals of the residual harzburgite; they are therefore partitioned into the basaltic magma during partial melting.

Support for a pyrolite composition of the mantle also comes from studies of meteorites. A pyrolite composition of the mantle follows if it is hypothesized that the Earth was formed by the accretion of parental material similar to Type 1 carbonaceous chondritic meteorites. An average composition for a Type 1 carbonaceous chondrite is given in Table 1–1. In order to generate a pyrolite composition for the mantle, it is necessary to remove an appropriate amount of iron to form the core as well as some volatile elements such as potassium.

A 20% fractionation of pyrolite to form the basaltic ocean crust and a residual harzburgite mantle explains the major element chemistry of these components. The basalts generated over a large fraction of the ocean ridge system have near-uniform compositions in both major and trace elements. This is evidence that the parental mantle rock from which the basalt is fractionated also has a near-uniform composition. However, both the basalts of normal ocean crust and their parental mantle rock are systematically depleted in incompatible elements compared with the model chondritic abundances. The missing incompatible elements are found to reside in the continental crust.

Seismic studies have been used to determine the thickness of the oceanic crust on a worldwide basis. The thickness of the basaltic oceanic crust has a nearly constant value of about 6 km throughout much of the area of the oceans. Exceptions are regions of abnormally shallow bathymetry such as the North Atlantic near Iceland, where the oceanic crust may be as thick as 25 km. The near-constant thickness of the basaltic oceanic crust places an important constraint on mechanisms of partial melting beneath the ridge crest. If the basalt of the oceanic crust represents a 20% partial melt, the thickness of depleted mantle beneath the oceanic crust is about 24 km. However, this depletion is gradational so the degree of depletion decreases with depth.

1–4 Subduction

As the oceanic lithosphere moves away from an ocean ridge, it cools, thickens, and becomes more dense because of thermal contraction. Even though the basaltic rocks of the oceanic crust are lighter than the underlying mantle rocks, the colder subcrustal rocks in the lithosphere become sufficiently dense to make old oceanic lithosphere heavy enough to be gravitationally unstable with respect to the hot mantle rocks immediately underlying the lithosphere. As a result of this gravitational instability the oceanic lithosphere founders and begins to sink into the interior of the Earth at ocean trenches. As the lithosphere descends into the mantle, it encounters increasingly dense rocks. However, the rocks of the lithosphere also become increasingly dense as a result of the increase of pressure with depth (mantle rocks are compressible), and they continue to be heavier than the adjacent mantle rocks as they descend into the mantle so long as they remain colder than the surrounding mantle rocks at any depth. Phase changes in the descending lithosphere and adjacent mantle and compositional variations with depth in the ambient mantle may complicate this simple picture of thermally induced gravitational instability. Generally speaking, however, the descending lithosphere continues to subduct as long as it remains denser than the immediately adjacent mantle rocks at any depth. The subduction of the oceanic lithosphere at an ocean trench is illustrated schematically in Figure 1–7.

The negative buoyancy of the dense rocks of the descending lithosphere results in a downward body force. Because the lithosphere behaves elastically, it can transmit stresses and acts as a stress guide. The body force acting on the descending plate is transmitted to the surface plate, which is pulled toward the ocean trench. This is one of the important forces driving plate tectonics and continental drift. It is known as *slab pull*.

Prior to subduction the lithosphere begins to bend downward. The convex curvature of the seafloor defines the seaward side of the ocean trench. The oceanic lithosphere bends continuously and maintains its structural integrity as it passes through the subduction zone. Studies of elastic bending at subduction zones are in good agreement with the morphology of some subduction zones seaward of the trench axis (see Section 3–17). However, there are clearly significant deviations from a simple elastic rheology. Some trenches exhibit a sharp "hinge" near the trench axis and this has been attributed to an elastic–perfectly plastic rheology (see Section 7–11).

As a result of the bending of the lithosphere, the near-surface rocks are placed in tension, and block faulting often results. This block faulting allows some of the overlying sediments to be entrained in the upper part of the basaltic crust. Some of these sediments

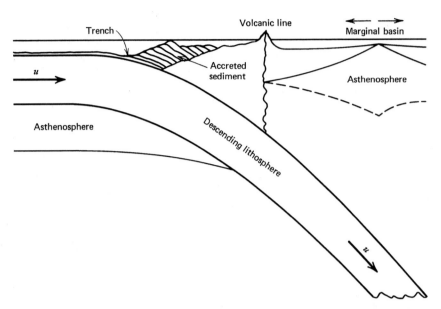

1-7 Subduction of oceanic lithosphere at an ocean trench. Sediments forming layer 1 of the oceanic crust are scraped off at the ocean trench to form the accretionary prism of sediments. The volcanic line associated with subduction and the marginal basin sometimes associated with subduction are also illustrated.

are then subducted along with the basaltic rocks of the oceanic crust, but the remainder of the sediments are scraped off at the base of the trench. These sediments form an *accretionary prism* (Figure 1–7) that defines the landward side of many ocean trenches. Mass balances show that only a fraction of the sediments that make up layer 1 of the oceanic crust are incorporated into accretionary prisms. Since these sediments are derived by the erosion of the continents, the subduction of sediments is a mechanism for subducting continental crust and returning it to the mantle.

The arclike structure of many ocean trenches (see Figure 1–1) can be qualitatively understood by the *ping-pong ball analogy*. If a ping-pong ball is indented, the indented portion will have the same curvature as the original ball, that is, it will lie on the surface of an imaginary sphere with the same radius as the ball, as illustrated in Figure 1–8. The lithosphere as it bends downward might also be expected to behave as a flexible but inextensible thin spherical shell. In this case the angle of dip α of the lithosphere at the trench can be related to the radius of curvature of the island arc. A cross section of the subduction zone is shown in Figure 1–8*b*.

The triangles OAB, BAC, and BAD are similar right triangles so that the angle subtended by the indented section of the sphere at the center of the Earth is equal to the angle of dip. The radius of curvature of the indented section, defined as the great circle distance BQ, is thus $a\alpha/2$, where a is the radius of the Earth. The radius of curvature of the arc of the Aleutian trench is about 2200 km. Taking $a = 6371$ km, we find that $\alpha = 39.6°$. The angle of dip of the descending lithosphere along much of the Aleutian trench is near 45°. Although the ping-pong ball analogy provides a framework for understanding the arclike structure of some trenches, it should be emphasized that other trenches do not have an arclike form and have radii of curvature that are in poor agreement with this relationship. Interactions of the descending lithosphere with an adjacent continent may cause the descending lithosphere to deform so that the ping-pong ball analogy would not be valid.

Ocean trenches are the sites of many of the largest earthquakes. These earthquakes occur on the fault zone separating the descending lithosphere from the overlying lithosphere. Great earthquakes, such as the 1960 Chilean earthquake and the 1964 Alaskan earthquake, accommodate about 20 m of downdip motion of the oceanic lithosphere and have lengths of about 350 km along the trench. A large fraction of the relative displacement between the descending lithosphere and the overlying mantle wedge appears to be accommodated by great earthquakes of this type. A typical velocity of

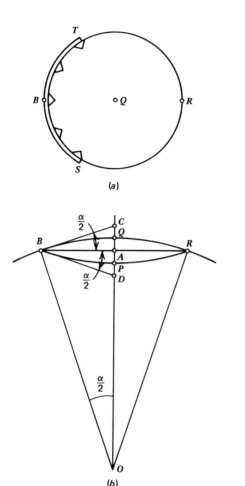

1-8 The ping-pong ball analogy for the arc structure of an ocean trench. (a) Top view showing subduction along a trench extending from S to T. The trench is part of a small circle centered at Q. (b) Cross section of indented section. BQR is the original sphere, that is, the surface of the Earth. BPR is the indented sphere, that is, the subducted lithosphere. The angle of subduction α is CBD. O is the center of the Earth.

subduction is 0.1 m yr^{-1} so that a great earthquake with a displacement of 20 m would be expected to occur at intervals of about 200 years.

Earthquakes within the cold subducted lithosphere extend to depths of about 660 km. The locations of these earthquakes delineate the structure of the descending plate and are known as the *Wadati-Benioff zone*. The shapes of the upper boundaries of several descending lithospheres are given in Figure 1–9. The positions of the trenches and the volcanic lines are also shown. Many subducted lithospheres have an angle of dip near 45°. In the New Hebrides the dip is signifi-

cantly larger, and in Peru and North Chile the angle of dip is small.

The lithosphere appears to bend continuously as it enters an ocean trench and then appears to straighten out and descend at a near-constant dip angle. A feature of some subduction zones is paired belts of deep seismicity. The earthquakes in the upper seismic zone, near the upper boundary of the descending lithosphere, are associated with compression. The earthquakes within the descending lithosphere are associated with tension. These double seismic zones are attributed to the "unbending," i.e., straightening out, of the descending lithosphere. The double seismic zones are further evidence of the rigidity of the subducted lithosphere. They are also indicative of the forces on the subducted lithosphere that are straightening it out so that it descends at a typical angle of 45°.

Since the gravitational body force on the subducted lithosphere is downward, it would be expected that the subduction dip angle would be 90°. In fact, as shown in Figure 1–9, the typical dip angle for a subduction zone is near 45°. One explanation is that the oceanic lithosphere is "foundering" and the trench is migrating oceanward. In this case the dip angle is determined by the flow kinematics. While this explanation is satisfactory in some cases, it has not been established that all slab dips can be explained by the kinematics of mantle flows. An alternative explanation is that the subducted slab is supported by the induced flow above the slab. The descending lithosphere induces a corner flow in the mantle wedge above it, and the pressure forces associated with this corner flow result in a dip angle near 45° (see Section 6–11).

One of the key questions in plate tectonics is the fate of the descending plates. Earthquakes terminate at a depth of about 660 km, but termination of seismicity does not imply cessation of subduction. This is the depth of a major seismic discontinuity associated with the solid–solid phase change from spinel to perovskite and magnesiowüstite; this phase change could act to deter penetration of the descending lithosphere. In some cases seismic activity spreads out at this depth, and in some cases it does not. Shallow subduction earthquakes generally indicate extensional stresses whereas the deeper earthquakes indicate compressional stresses. This is also an indication of a resistance to subduction. Seismic velocities in the cold descending

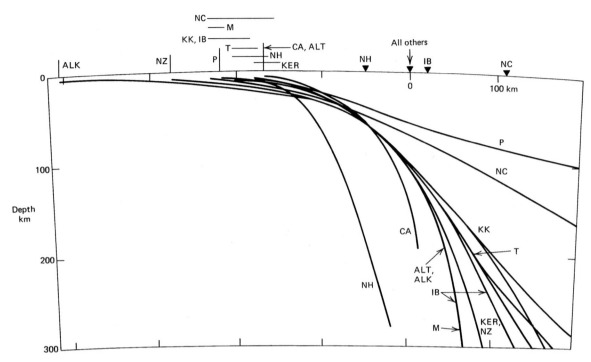

1-9 The shapes of the upper boundaries of descending lithospheres at several oceanic trenches based on the distributions of earthquakes. The names of the trenches are abbreviated for clarity (NH = New Hebrides, CA = Central America, ALT = Aleutian, ALK = Alaska, M = Mariana, IB = Izu-Bonin, KER = Kermadec, NZ = New Zealand, T = Tonga, KK = Kurile-Kamchatka, NC = North Chile, P = Peru). The locations of the volcanic lines are shown by the solid triangles. The locations of the trenches are shown either as a vertical line or as a horizontal line if the trench–volcanic line separation is variable (Isacks and Barazangi, 1977).

may result in an island arc or they may occur on the continental crust (Figure 1–10). The volcanoes lie 125 to 175 km above the descending plate, as illustrated in Figure 1–9.

It is far from obvious why volcanism is associated with subduction. The descending lithosphere is cold compared with the surrounding mantle, and thus it

lithosphere are significantly higher than in the surrounding hot mantle. Systematic studies of the distribution of seismic velocities in the mantle are known as *mantle tomography*. These studies have provided examples of the descending plate penetrating the 660-km depth.

The fate of the descending plate has important implications regarding mantle convection. Since plates descend into the *lower mantle*, beneath a depth of 660 km, some form of *whole mantle* convection is required. The entire upper and at least a significant fraction of the lower mantle must take part in the plate tectonic cycle. Although there may be a resistance to convection at a depth of 660 km, it is clear that the plate tectonic cycle is not restricted to the *upper mantle* above 660 km.

Volcanism is also associated with subduction. A line of regularly spaced volcanoes closely parallels the trend of the ocean trench in almost all cases. These volcanics

1-10 Eruption of ash and steam from Mount St. Helens, Washington, on April 3, 1980. Mount St. Helens is part of a volcanic chain, the Cascades, produced by subduction of the Juan de Fuca plate beneath the western margin of the North American plate (Washington Department of Natural Resources).

should act as a heat sink rather than as a heat source. Because the flow is downward, magma cannot be produced by pressure-release melting. One source of heat is frictional dissipation on the fault zone between the descending lithosphere and the overlying mantle. However, there are several problems with generating island-arc magmas by frictional heating. When rocks are cold, frictional stresses can be high, and significant heating can occur. However, when the rocks become hot, the stresses are small, and it appears to be impossible to produce significant melting simply by frictional heating.

It has been suggested that interactions between the descending slab and the induced flow in the overlying mantle wedge can result in sufficient heating of the descending oceanic crust to produce melting. However, thermal models of the subduction zone show that there is great difficulty in producing enough heat to generate the observed volcanism. The subducted cold lithospheric slab is a very large heat sink and strongly depresses the isotherms above the slab. It has also been argued that water released from the heating of hydrated minerals in the subducted oceanic crust can contribute to melting by depressing the solidus of the crustal rocks and adjacent mantle wedge rocks. However, the bulk of the volcanic rocks at island arcs have near-basaltic compositions and erupt at temperatures very similar to eruption temperatures at accretional margins. Studies of the petrology of island-arc magmas indicate that they are primarily the result of the partial melting of rocks in the mantle wedge above the descending lithosphere. Nevertheless, geochemical evidence indicates that partial melting of subducted sediments and oceanic crust does play an important role in island-arc volcanism. Isotopic studies have shown conclusively that subducted sediments participate in the melting process. Also, the locations of the surface volcanic lines have a direct geometrical relationship to the geometry of subduction. In some cases two adjacent slab segments subduct at different angles, and an offset occurs in the volcanic line; for the shallower dipping slab, the volcanic line is farther from the trench keeping the depth to the slab beneath the volcanic line nearly constant.

Processes associated with the subducted oceanic crust clearly trigger subduction zone volcanism. However, the bulk of the volcanism is directly associated with the melting of the mantle wedge in a way similar to the melting beneath an accretional plate margin. A possible explanation is that "fluids" from the descending oceanic crust induce melting and create sufficient buoyancy in the partially melted mantle wedge rock to generate an ascending flow and enhance melting through pressure release. This process may be three-dimensional with ascending diapirs associated with individual volcanic centers.

In some trench systems a secondary accretionary plate margin lies behind the volcanic line, as illustrated in Figure 1–7. This *back-arc spreading* is very similar to the seafloor spreading that is occurring at ocean ridges. The composition and structure of the ocean crust that is being created are nearly identical. Back-arc spreading creates *marginal basins* such as the Sea of Japan. A number of explanations have been given for back-arc spreading. One hypothesis is that the descending lithosphere induces a secondary convection cell, as illustrated in Figure 1–11a. An alternative hypothesis is that the ocean trench migrates away from an adjacent continent because of the "foundering" of the descending lithosphere. Back-arc spreading is required to fill the gap, as illustrated in Figure 1–11b. If the adjacent continent is being driven up against the trench, as in South America, marginal basins do not develop. If the adjacent continent is stationary, as in the western Pacific, the foundering of the lithosphere leads to a series of marginal basins as the trench migrates seaward. There is observational evidence that back-arc spreading centers are initiated at volcanic lines. Heating of the lithosphere at the volcanic line apparently weakens it sufficiently so that it fails under tensional stresses.

PROBLEM 1–3 If we assume that the current rate of subduction, $0.09 \text{ m}^2 \text{ s}^{-1}$, has been applicable in the past, what thickness of sediments would have to have been subducted in the last 3 Gyr if the mass of subducted sediments is equal to one-half the present mass of the continents? Assume the density of the continents ρ_c is 2700 kg m^{-3}, the density of the sediments ρ_s is 2400 kg m^{-3}, the continental area A_c is $1.9 \times 10^8 \text{ km}^2$, and the mean continental thickness h_c is 35 km.

1-5 Transform Faults

In some cases the rigid plates slide past each other along *transform faults*. The ocean ridge system is not

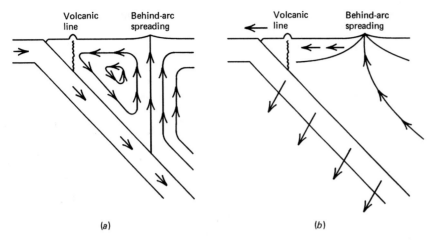

Volcanic line   Behind-arc spreading

Volcanic line   Behind-arc spreading

(a)   (b)

1–11 Models for the formation of marginal basins. (*a*) Secondary mantle convection induced by the descending lithosphere. (*b*) Ascending convection generated by the foundering of the descending lithosphere and the seaward migration of the trench.

a continuous accretional margin; rather, it is a series of ridge segments offset by transform faults. The ridge segments lie nearly perpendicular to the spreading direction, whereas the transform faults lie parallel to the spreading direction. This structure is illustrated in Figure 1–12*a*. The orthogonal ridge–transform system has been reproduced in the laboratory using wax that

1–12 (*a*) Segments of an ocean ridge offset by a transform fault. (*b*) Cross section along a transform fault.

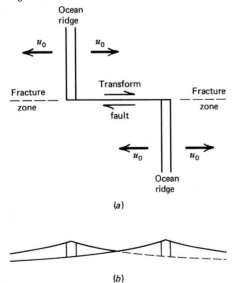

Ocean ridge

u_0 u_0

Fracture zone   Transform   Fracture zone

fault

u_0 u_0

Ocean ridge

(a)

(b)

solidifies at the surface. Even with this analogy, the basic physics generating the orthogonal pattern is not understood. The relative velocity across a transform fault is twice the spreading velocity. This relative velocity results in seismicity (earthquakes) on the transform fault between the adjacent ridge sections. There is also differential vertical motion on transform faults. As the seafloor spreads away from a ridge crest, it also subsides. Since the adjacent points on each side of a transform fault usually lie at different distances from the ridge crest where the crust was formed, the rates of subsidence on the two sides differ. A cross section along a transform fault is given in Figure 1–12*b*. The extensions of the transform faults into the adjacent plates are known as *fracture zones*. These fracture zones are often deep valleys in the seafloor. An ocean ridge segment that is not perpendicular to the spreading direction appears to be unstable and transforms to the orthogonal pattern.

A transform fault that connects two ridge segments is known as a *ridge–ridge transform*. Transform faults can also connect two segments of an ocean trench. In some cases one end of a transform fault terminates in a *triple junction* of three surface plates. An example is the San Andreas fault in California; the San Andreas accommodates lateral sliding between the Pacific and North American plates.

1–6 Hotspots and Mantle Plumes

Hotspots are anomalous areas of surface volcanism that cannot be directly associated with plate tectonic processes. Many hotspots lie well within the interiors of

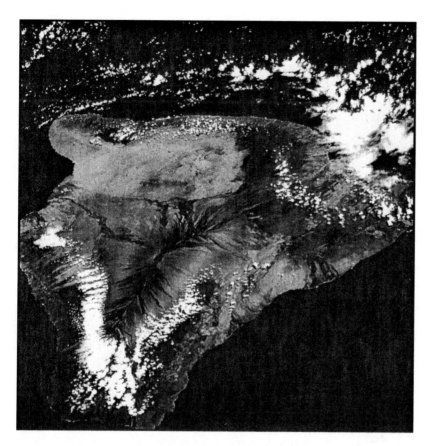

1-13 Satellite photograph of the island of Hawaii. The island is dominated by the active volcano Mauna Loa near its center (NASA STS61A-50-0057).

plates; an example is the volcanism of the Hawaiian Islands (Figure 1–13). Other hotspots lie at or near an ocean ridge, an example is the volcanism that forms Iceland. Much more voluminous than normal ocean ridge volcanism; this volcanism resulted in a thick oceanic crust and the elevation of Iceland above sea level.

In many cases hotspots lie at the end of well-defined lines of volcanic edifices or volcanic ridges. These are known as *hotspot tracks*. The hotspot track associated with the Hawaiian hotspot is the Hawaiian–Emperor island–seamount chain that extends across the Pacific plate to the Aleutian Islands.

There is little agreement on the total number of hotspots. The positions of thirty hotspots are given in Table 1–2, and twenty of the most prominent hotspots are shown in Figure 1–14. Also shown in this figure are some of the hotspot tracks. Some compilations of hotspots list as many as 120 (see Figure 1–15). The def-

inition of a hotspot tends to be quite subjective, particularly with regard to volcanism on or adjacent to plate boundaries. Hotspots occur both in the oceans and on the continents. They do not appear to be uniformly distributed over the Earth's surface. There are numerous hotspots in Africa and relatively few in South America, North America, Europe, and Asia.

Jason Morgan (1971) attributed hotspot volcanism to a global array of deep mantle plumes. Mantle plumes are quasi-cylindrical concentrated upwellings of hot mantle rock and they represent a basic form of mantle convection. Pressure-release melting in the hot ascending plume rock produces the basaltic volcanism associated with most hotspots. The hypothesis of fixed mantle plumes impinging on the base of the moving lithospheric plates explains the origin of hotspot tracks (see Figure 1–16).

The prototype example of a hotspot track is the Hawaiian–Emperor chain of volcanic islands and seamounts illustrated in Figure 1–17. The associated hotspot volcanism has resulted in a nearly continuous

TABLE 1-2 Hotspot Locations

Hotspot	Overlying Plate	Latitude (Degrees)	Longitude (Degrees)
Hawaii	Pacific	20	−157
Samoa	Pacific	−13	−173
St. Helena	Africa	−14	−6
Bermuda	N. America	33	−67
Cape Verde	Africa	14	−20
Pitcairn	Pacific	−26	−132
MacDonald	Pacific	−30	−140
Marquesas	Pacific	−10	−138
Tahiti	Pacific	−17	−151
Easter	Pac-Naz	−27	−110
Reunion	Indian	−20	55
Yellowstone	N. America	43	−111
Galapagos	Nazca	0	−92
Juan Fernandez	Nazca	−34	−83
Ethiopia	Africa	8	37
Ascencion	S. Am–Afr	−8	−14
Afar	Africa	10	43
Azores	Eurasia	39	−28
Iceland	N. Am–Eur	65	−20
Madeira	Africa	32	−18
Canary	Africa	28	−17
Hoggar	Ind–Ant	−49	69
Bouvet	Afr–Ant	−54	2
Pr. Edward	Afr–Ant	−45	50
Eifel	Eurasia	48	8
San Felix	Nazca	−24	−82
Tibesti	Africa	18	22
Trinadade	S. America	−20	−30
Tristan	S. Am–Afr	−36	−13

Source: After Crough and Jurdy (1980).

volcanic ridge that extends some 4000 km from near the Aleutian Islands to the very active Kilauea volcano on the island of Hawaii. There is a remarkably uniform age progression, with the age of each volcanic shield increasing systematically with distance from Kilauea. Directly measured ages and ages inferred from seafloor magnetic anomalies are given in Figure 1–17. These ages are given as a function of distance from Kilauea in Figure 1–18 and they correlate very well with a propagation rate of 90 mm yr^{-1} across the Pacific plate.

A striking feature of this track is the bend that separates the near-linear trend of the Emperor chain from the near-linear trend of the Hawaiian chain. The bend in the track occurred at about 43 Ma when there was an abrupt shift in the motion of the Pacific plate. This shift was part of a global reorientation of plate motions over a span of a few million years. This shift has been attributed to the continental collision between India and Asia, which impeded the northward motion of the Indian plate.

Many hotspots are associated with linear tracks as indicated in Figure 1–14. When the relative motions of the plates are removed the hotspots appear to be nearly fixed with respect to each other. However, they are certainly not precisely fixed. Systematic studies have shown that the relative motion among hotspots amounts to a few mm yr^{-1}. These results are consistent with plumes that ascend through a mantle in which horizontal velocities are about an order of magnitude smaller than the plate velocities.

Many hotspots are also associated with topographic swells. Hotspot swells are regional topographic highs with widths of about 1000 km and anomalous elevations of up to 3 km. The hotspot swell associated with the Hawaiian hotspot is illustrated in Figure 1–19. The swell is roughly parabolic in form and extends upstream from the active hotspot. The excess elevation associated with the swell decays rather slowly down the track of the hotspot. Hotspot swells are attributed to the interaction between the ascending hot mantle rock in the plume and the lithospheric plate upon which the plume impinges.

The volcanic rocks produced at most hotspots are primarily basalt. In terms of overall composition, the rocks are generally similar to the basaltic rocks produced at ocean ridges. It appears that these volcanic rocks are also produced by about 20% partial melting of mantle rocks with a pyrolite composition. However, the concentrations of incompatible elements and isotopic ratios differ from those of normal mid-ocean ridge basalts. Whereas the mid-ocean ridge basalts are nearly uniformly depleted in incompatible elements, the concentrations of these elements in hotspot basalts have considerable variation. Some volcanoes produce basalts that are depleted, some produce basalts that have near chondritic ratios, and some volcanoes produce basalts that are enriched in the incompatible elements. These differences will be discussed in some detail in Chapter 10.

The earthquakes of the Wadati–Benioff zone define the geometry of the subducted oceanic lithosphere.

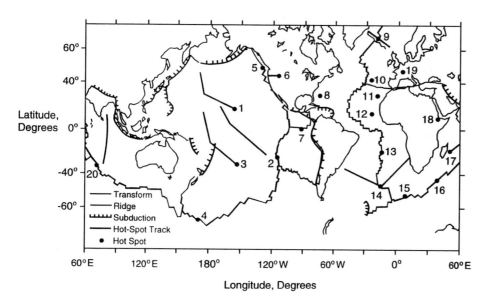

1-14 Hotspot and hotspot track locations: 1, Hawaii (Hawaiian–Emperor Seamount Chain); 2, Easter (Tuomoto–Line Island Chain); 3, MacDonald Seamount (Austral–Gilbert–Marshall Island Chain); 4, Bellany Island; 5, Cobb Seamount (Juan de Fuca Ridge); 6, Yellowstone (Snake River Plain–Columbia Plateau); 7, Galapagos Islands; 8, Bermuda; 9, Iceland; 10, Azores; 11, Canary Islands; 12, Cape Verde Islands; 13, St. Helena; 14, Tristan de Cunha (Rio Grande Ridge (w), Walvis Ridge (e)); 15, Bouvet Island; 16, Prince Edward Island; 17, Reunion Island (Mauritius Plateau, Chagos–Lacadive Ridge); 18, Afar; 19, Eifel; 20, Kerguelen Plateau (Ninety-East Ridge).

No seismicity is associated with mantle plumes, and little direct observational evidence exists of their structure and origin. Thus we must depend on analytical, numerical, and laboratory studies for information. These studies indicate that plumes originate in a lower hot thermal boundary layer either at the base of the mantle (the D″-layer of seismology) or at an interface in the lower mantle between an upper convecting mantle layer and an isolated lower mantle layer. Plumes result from the gravitational instability of the hot lower thermal boundary layer just as the subducted lithosphere results from the gravitational instability of the cold, surface thermal boundary layer, the lithosphere.

Numerical and laboratory studies of the initiation of plumes show a leading diapir or *plume head* followed by a thin cylindrical conduit or *plume tail* that connects the diapir to the source region. An example from a laboratory experiment is given in Figure 1–20. Confirmation of this basic model comes from the association of massive flood basalts with plume heads.

There is convincing observational evidence that flood basalt eruptions mark the initiation of hotspot tracks. As specific examples, the hotspot tracks of the currently active Reunion, Iceland, Tristan da Cunha, and Prince Edward hotspots originate, respectively, in the Deccan, Tertiary North Atlantic, Parana, and Karoo flood basalt provinces.

The association of the Reunion hotspot with the Deccan flood basalt province is illustrated in Figure 1–21. Pressure-release melting in the plume head as it approached and impinged on the lithosphere can explain the eruption of the Deccan traps in India with a volume of basaltic magma in excess of 1.5×10^6 km^3 in a time interval of less than 1 Myr. Since then, Reunion hotspot volcanism has been nearly continuous for 60 Myr with an average eruption rate of 0.02 km^3 yr^{-1}. As the Indian plate moved northward the hotspot track formed the Chagos–Laccadive Ridge. The hotspot track is then offset by seafloor spreading on the central Indian Ridge and forms the Mascarene Ridge on the Indian plate that connects to the currently active volcanism of the Reunion Islands.

1-7 Continents

As described in the previous sections, the development of plate tectonics primarily involves the ocean basins, yet the vast majority of geological data comes from the continents. There is essentially no evidence for plate tectonics in the continents, and this is certainly one

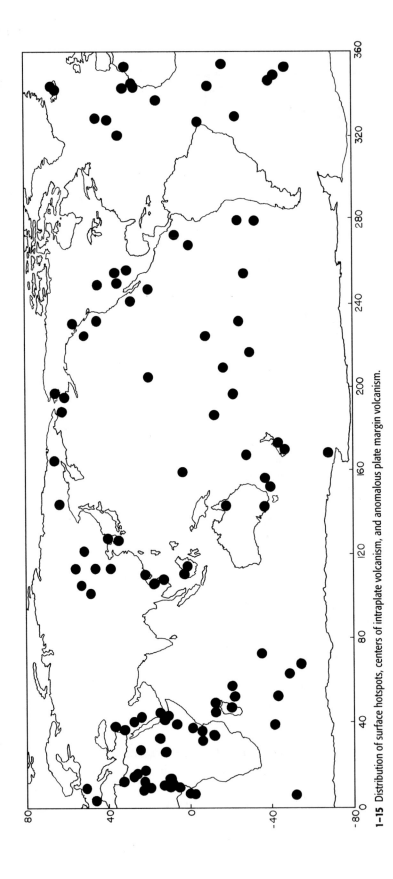

1-15 Distribution of surface hotspots, centers of intraplate volcanism, and anomalous plate margin volcanism.

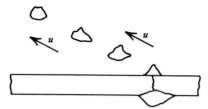

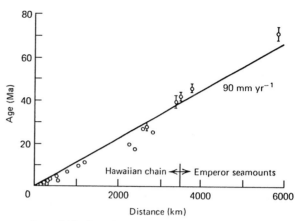

1-16 Formation of an island–seamount chain by the movement of a lithospheric plate over a melting anomaly in the upper mantle.

reason why few geologists were willing to accept the arguments in favor of continental drift and mantle convection for so long. The near surface rocks of the continental crust are much older than the rocks of the oceanic crust. They also have a more silicic composition. The continents include not only the area above sea level but also the continental shelves. It is difficult to provide an absolute definition of the division between oceanic and continental crust. In most cases it is appropriate to define the transition as occurring at an ocean depth of 3 km. The area of the continents, including the margins, is about 1.9×10^8 km^2, or 37% of the surface of the Earth.

1-18 Ages of islands and seamounts in the Hawaiian–Emperor chain as a function of distance from the currently active Kilauea volcano. The straight line gives a constant rate of propagation across the Pacific plate of 90 mm yr^{-1}.

The rocks that make up the continental crust are, in bulk, more silicic and therefore less dense than the basaltic rocks of the oceanic crust. Also, the continental crust with a mean thickness of about 40 km is considerably thicker than the oceanic crust. These two effects make the continental lithosphere gravitationally stable and prevent it from being subducted. Although continental crust cannot be destroyed by subduction, it can be recycled indirectly by delamination. The mantle portion of the continental lithosphere is sufficiently cold

1-17 Age progression of the Hawaiian–Emperor seamount and island chain (Molnar and Stock, 1987). Dated seamounts and islands are shown in normal print and dates inferred from magnetic anomalies are shown in bold print.

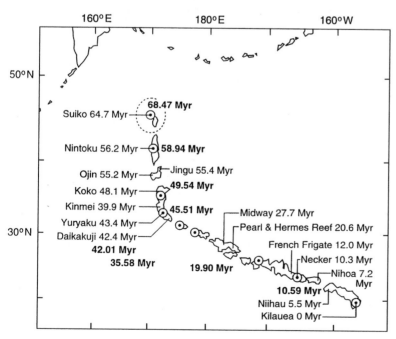

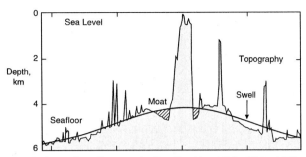

1-19 Bathymetric profile across the Hawaiian ridge at Oahu (Watts, 1976).

lamination is an efficient mechanism for the removal of continental lithosphere. Continental crust can also be recycled into the mantle by the subduction of sediments. Although there is evidence of the recycling of the continental crust, it is much less efficient than the recycling of oceanic crust by the plate tectonic cycle. The result is that the continental crust is nearly a factor of 10 older than oceanic crust. Continental crust older than 1 billion years is common, and some is older than 3 billion years.

Determining the relative age of continental rocks has been an important aspect of the historical development of geology. The early classification of the age of rocks was based on the fossils found in sedimentary rocks. By studying the evolution of the species involved, and their relative positions in the *stratigraphic column*, an uncalibrated, relative time scale was developed. The divisions of the time scale were associated with sedimentary *unconformities*. These are recognized as discontinuities in the sedimentation process, where adjacent strata often contain dissimilar fossils. These unconformities occur worldwide.

It is now recognized that major unconformities correspond with times of low sea level. During these periods *erosion* occurred over a large fraction of the continents, causing gaps in the sedimentary record. During periods of high sea level much of the area of the continents was covered with shallow seas, and sediments were deposited. The causes of the periods of high and low sea levels are not fully understood. Clearly, ice ages can cause periods of low sea level. Also, on a longer time scale, variations in the volume of the ocean ridge system can change the sea level.

Quantitative measurements of the concentrations of radioactive isotopes and their daughter products in rocks have provided an absolute geological time scale. The science of dating rocks by radioisotopic techniques is known as *geochronology*. Geochronological methods will be discussed in Section 10–2.

The radiometrically calibrated geological time scale is given in Table 1–3. Note that the Precambrian period, during which fossils were not available for classification purposes, represents 88% of the Earth's history.

Erosion and sedimentation play an important role in shaping the surface of the continents. Mountain ranges that are built by plate tectonic processes are eroded to near sea level in a few million years. Any areas of the

and dense to be gravitationally unstable. Thus it is possible for the lower part of the continental lithosphere, including the lower continental crust, to delaminate and sink into the lower mantle. This is partial subduction or delamination. It has been suggested that delamination is occurring in continental collision zones such as the Himalayas and the Alps and behind subduction zones such as in the Altiplano in Peru. There are a number of continental areas in which the mantle lithosphere is absent. One example is the western United States. Crustal doubling such as in Tibet has also been attributed to the absence of mantle lithosphere beneath Asia. De-

1-20 Photograph of a low-density, low-viscosity glucose fluid plume ascending in a high-density, high-viscosity glucose fluid (Olson and Singer, 1985).

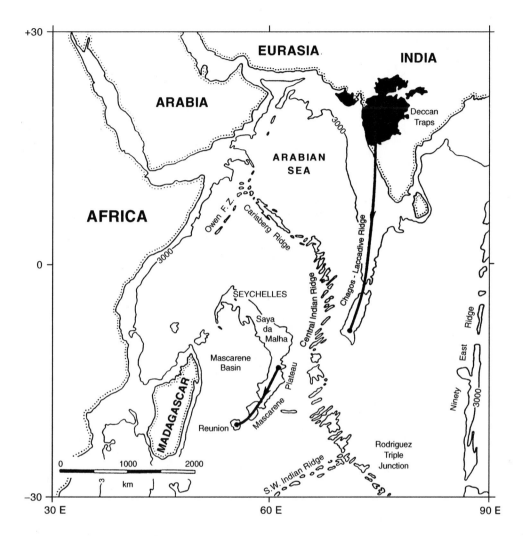

1–21 The relationship between the Reunion hotspot track and the Deccan flood basalts (White and McKenzie, 1989).

continents that are depressed below sea level are filled with these sediments to form sedimentary basins. The base of a sedimentary basin is referred to as the *basement*. Because the ages of basement rocks are not well known on a worldwide basis, it is difficult to specify a mean age for continental rocks. Regions of the continents where Precambrian metamorphic and igneous rocks are exposed are known as *continental shields*. Detailed studies of Precambrian terrains indicate that the plate tectonic processes that are occurring today have been going on for at least 3 billion years.

It is relatively easy to estimate the composition of the upper continental crust, but it is difficult to esti-

mate the composition of the crust as a whole. Direct evidence for the composition of the lower continental crust comes from surface exposures of high-grade metamorphic rocks and lower crustal xenoliths transported to the surface in diatremes and magma flows. Indirect evidence of the composition of the lower crust comes from comparisons between seismic velocities and laboratory studies of relevant minerals. An estimate of the bulk composition of the continental crust is given in Table 1–1. It is compared with the mean composition of clastic sediments (representative of the upper continental crust) and with a typical basalt composition. Estimates of the mean composition of the continental crust are clearly more basic (less silicic) than the composition of the upper continental crust, but they do not approach a basaltic composition.

TABLE 1-3 **Geologic Time Scale**

Age (Ma)		Period	Era	Eon
0				
0.01		Holocene	Quaternary	
	Upper		Quaternary	
0.13	Middle	Pleistocene	Quaternary	
0.8	Lower		Quaternary	
1.8				
	Upper	Pliocene	Neogene	
3.6	Lower		Neogene	
5.3	Upper		Neogene	
11.2	Middle	Miocene	Neogene	
16.4	Lower		Neogene	Cenozoic
23.8				Cenozoic
	Upper	Oligocene	Tertiary / Paleogene	
28.5	Lower			
33.7	Upper			
37.0	Middle	Eocene	Paleogene	
49.0	Lower			
54.8				
	Upper	Paleocene		
61.0	Lower			
65.0				
	Upper	Cretaceous		
98.9	Lower			
142.0				
	Upper			
159.4	Middle	Jurassic	Mesozoic	
180.1	Lower			
205.7				Phanerozoic
	Upper			
227.4	Middle	Triassic		
241.7	Lower			
248.2				
	Upper	Permian		
256.0	Lower			
290.0				
	Upper	Carboniferous		
323.0	Lower			
354.0	Upper			
370.0	Middle	Devonian	Paleozoic	
391.0	Lower			
417.0				
443.0		Silurian		
	Upper	Ordovician		
470.0	Lower			
495.0	Upper			
505.0	Middle	Cambrian		
518.0	Lower			
545.0				
2500		Precambrian	Proterozoic	
4550			Archean	

An important question is: How is continental crust formed? One hypothesis is that it is formed from partial melts of the mantle. But as we have discussed, mantle melts have near-basaltic compositions. Thus, if this were the case, the mean composition of the continental crust would also be basaltic. However, as seen in Table 1–1, the mean continental crust is considerably more silicic than the composition of basalts. A preferred hypothesis for the generation of the continental crust consists of three steps: 1) Basaltic volcanism from the mantle associated with island-arc volcanics, continental rifts, and hotspots is responsible for the formation of the continental crust. 2) Intracrustal melting and high-temperature metamorphism are responsible for the differentiation of the continental crust so that the upper crust is more silicic and the lower crust is more basic. Basaltic magmas from the mantle that intrude into a basaltic continental crust in the presence of water can produce the granitic rocks associated with the bulk continental crust. 3) Delamination of substantial quantities of continental lithosphere including the mantle and lower crust returns a fraction of the more basic lower crust to the mantle. The residuum, composed primarily of the upper crust, thus becomes more silicic and forms the present continental crust.

1–8 **Paleomagnetism and the Motion of the Plates**

Although qualitative geological arguments had long favored the continental drift theory, it remained for paleomagnetic studies to provide quantitative confirmation. Paleomagnetism is the study of the Earth's past magnetic field from the records preserved in magnetized rocks. The silicate minerals making up the bulk of a rock are either *paramagnetic* (olivine, pyroxene, garnet, amphiboles) or *diamagnetic* (quartz, feldspar) and are incapable of acquiring a *permanent magnetization*. However, rocks containing small amounts of *ferromagnetic*, or more accurately *ferrimagnetic*, minerals, that is, iron oxides such as magnetite Fe_3O_4 and hematite Fe_2O_3 and iron sulfides such as pyrrhotite $Fe_{1-y}S$, can acquire a weak permanent magnetism when they are formed. The *fossil magnetism* in a rock is referred to as *natural remanent magnetism* (*NRM*).

A rock can acquire NRM in several ways. When a mineral is heated above its *Curie temperature*, all magnetism is lost. For magnetite the Curie temperature is 851 K. When a rock containing ferromagnetic minerals is cooled to a temperature below the Curie temperature, known as the *blocking temperature*, in the presence of a magnetic field, it can acquire a remanent magnetism. This is known as *thermoremanent magnetism (TRM)*. In some cases magnetic minerals are formed by chemical processes at low temperatures. As a grain of a ferromagnetic mineral grows, it reaches a size where it becomes magnetically stable. If this occurs in an applied magnetic field, a *chemical remanent magnetism (CRM)* may be acquired. A sedimentary rock may also acquire a remanent magnetism during its formation. As small particles of ferromagnetic minerals fall through water in the presence of a magnetic field, their *magnetic moments* become partially aligned with the ambient magnetic field; the result is that the sedimentary rock that is formed with these particles present has a *depositional remanent magnetism (DRM)*.

Rocks may also acquire magnetism after they are formed. This type of magnetism may usually be removed by subjecting the rock to alternating magnetic fields or by heating the rock to a substantial fraction of the Curie temperature. After it has been confirmed that the magnetism in a rock is in fact the remanent magnetism acquired at the time of its formation, the orientation or direction of the remanent field is determined. This is normally expressed in terms of the *declination D* or *magnetic azimuth*, which is the angle between geographic north and the magnetic field direction measured positive clockwise (0 to 360°), and the *inclination I*, which is the angle between the horizontal and the field direction measured positive downward (−90 to +90°) (Figure 1–22).

In addition to declination and inclination, the complete specification of a remanent magnetic field requires the determination of its magnitude B. The SI unit of B is the tesla or weber m^{-2}. Figure 1–22 clearly shows that the horizontal B_H and vertical B_V components of the magnetic field are related to the magnitude of the field and the inclination by

$$B_H = B \cos I \tag{1-2}$$
$$B_V = B \sin I. \tag{1-3}$$

The horizontal field can be further resolved into a northward component B_{HN} and an eastward component B_{HE} given by

$$B_{HN} = B \cos I \cos D \tag{1-4}$$
$$B_{HE} = B \cos I \sin D. \tag{1-5}$$

The present-day magnetic field of the Earth can be reasonably approximated as a *dipole magnetic field*, the form of which is sketched in Figure 1–23. The horizontal and vertical components of the Earth's dipole magnetic field, B_θ and B_r, at its surface, assuming that the Earth is a sphere of radius a, are given by

$$B_\theta = \frac{\mu_0 m}{4\pi a^3} \sin \theta_m \tag{1-6}$$
$$B_r = \frac{\mu_0 m}{2\pi a^3} \cos \theta_m, \tag{1-7}$$

1-22 Declination and inclination of the magnetic field.

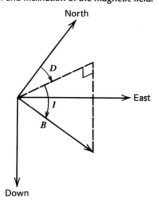

1-23 The Earth's dipole magnetic field.

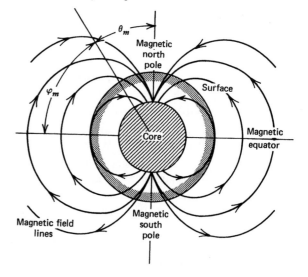

where μ_0 is the permeability of free space ($\mu_0 = 4\pi \times 10^{-7}$ tesla m A^{-1}), m is the dipole moment (A m^2), and θ_m is the magnetic colatitude (magnetic latitude $\phi_m = \pi/2 - \theta_m$) (see Figure 1–23). The magnetic poles are the positions where the dipole field lines are vertical. At the north magnetic pole ($\theta_m = 0$, $\phi_m = \pi/2$), $B_\theta = 0$, $B_r = \mu_0 m/2\pi a^3$, the inclination is $\pi/2$ rad or 90°, and the field is directed into the Earth. At the south magnetic pole ($\theta_m = \pi$, $\phi_m = -\pi/2$), $B_\theta = 0$, $B_r = -\mu_0 m/2\pi a^3$, the inclination is $-\pi/2$ rad or $-90°$, and the field is directed out from the Earth. The magnetic field lines of the Earth's present dipole magnetic field leave at the south magnetic pole and enter at the north magnetic pole (Figure 1–23). At the magnetic equator ($\theta_m = \pi/2$, $\phi_m = 0$), $B_r = 0$, $B_\theta = \mu_0 m/4\pi a^3$, the field lines are horizontal, and the inclination is zero. The angle of inclination of the dipole magnetic field is given by

$$\tan I = \frac{B_r}{B_\theta}, \tag{1-8}$$

and its magnitude B can be written

$$B = \left(B_r^2 + B_\theta^2\right)^{1/2}. \tag{1-9}$$

By substituting for B_r and B_θ from Equations (1–6) and (1–7), we can rewrite these expressions for I and B as

$$\tan I = 2\cot\theta_m = 2\tan\phi_m \tag{1-10}$$

$$\begin{aligned}
B &= \frac{\mu_0 m}{4\pi a^3}(\sin^2\theta_m + 4\cos^2\theta_m)^{1/2} \\
&= \frac{\mu_0 m}{4\pi a^3}(1 + 3\cos^2\theta_m)^{1/2} \\
&= \frac{\mu_0 m}{4\pi a^3}(1 + 3\sin^2\phi_m)^{1/2}. \tag{1-11}
\end{aligned}$$

The Earth's magnetic field is only approximately a dipole. The present locations (latitude and longitude) of the magnetic poles are 73°N, 100°W and 68°S, 143°E. The magnetic poles of the dipole field that is the best fit to the Earth's field are at 79°N, 70°W and 79°S, 110°E. Thus the axis of the dipole field makes an angle of about 11° with the Earth's rotational axis. The moment of the dipole field is $m = 7.94 \times 10^{22}$ A m^2, and the surface magnetic field at the magnetic equator is $B_\theta = 3.07 \times 10^{-5}$ teslas. Maps of the magnitude, declination, and inclination of the present magnetic field of the Earth are presented in Figure 1–24.

PROBLEM 1–4 Assume that the Earth's magnetic field is a dipole. What is the maximum intensity of the field at the core–mantle boundary?

PROBLEM 1–5 Assume that the Earth's magnetic field is a dipole. At what distance above the Earth's surface is the magnitude of the field one-half of its value at the surface?

If a dipole field is a reasonable approximation of the Earth's magnetic field throughout geologic time, a paleomagnetic measurement of declination and inclination can be used to locate the magnetic pole position at the time the rock acquired its magnetization. Suppose that the paleomagnetic measurement is carried out at a north latitude ϕ and an east longitude ψ, as in Figure 1–25. From the definition of declination it is clear that the paleomagnetic north pole lies an angular distance θ_m along a great circle making an angle D with the meridian through the measurement point. Geographic north, paleomagnetic north, and the measurement point define a spherical triangle with sides $\pi/2 - \phi$, θ_m, and $\pi/2 - \phi_p$, where ϕ_p is the latitude of the paleomagnetic pole. The triangle contains the included angle D. Using a result from spherical trigonometry, we can write

$$\begin{aligned}
\cos\left(\frac{\pi}{2} - \phi_p\right) &= \cos\left(\frac{\pi}{2} - \phi\right)\cos\theta_m \\
&\quad + \sin\left(\frac{\pi}{2} - \phi\right)\sin\theta_m \cos D. \tag{1-12}
\end{aligned}$$

This can be simplified by noting that $\cos(\pi/2 - \phi_p) = \sin\phi_p$, $\cos(\pi/2 - \phi) = \sin\phi$, and $\sin(\pi/2 - \phi) = \cos\phi$. The result is

$$\sin\phi_p = \sin\phi \cos\theta_m + \cos\phi \sin\theta_m \cos D. \tag{1-13}$$

The magnetic colatitude θ_m can be determined from Equations (1–8) and (1–10). The angle between the meridians passing through the measurement point and the paleomagnetic north pole is $\psi_p - \psi$, where ψ_p is the east longitude of the paleomagnetic pole. A second spherical trigonometric formula allows us to write

$$\frac{\sin(\psi_p - \psi)}{\sin\theta_m} = \frac{\sin D}{\sin(\pi/2 - \phi_p)} = \frac{\sin D}{\cos\phi_p} \tag{1-14}$$

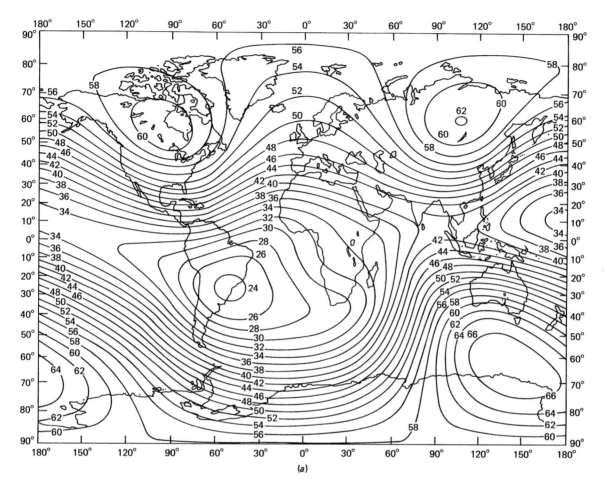

1-24 Present-day magnetic field of the Earth. (*a*) Magnitude, μT. Continued on pp. 26–7.

or

$$\sin(\psi_p - \psi) = \frac{\sin \theta_m \sin D}{\cos \phi_p}, \qquad (1-15)$$

if $\cos \theta_m > \sin \phi \sin \phi_p$. If $\cos \theta_m < \sin \phi \sin \phi_p$, Equation (1–15) must be replaced by

$$\sin(\pi + \psi - \psi_p) = \frac{\sin \theta_m \sin D}{\cos \phi_p}. \qquad (1-16)$$

Paleomagnetic measurements are useful only if the orientation of the sample has remained fixed with respect to the rest of the geological province, since the sample was magnetized. Usually the absence of subsequent deformation can be established with some certainty and the reliability of the measurement established.

PROBLEM 1-6 The measured declination and inclination of the paleomagnetic field in Upper Triassic rocks at 41.5°N and 72.7°W are $D = 18°$ and $I = 12°$. Determine the paleomagnetic pole position.

PROBLEM 1-7 The measured declination and inclination of the paleomagnetic field in Oligocene rocks at 51°N and 14.7°E are $D = 200°$ and $I = -63°$. Determine the paleomagnetic pole position.

PROBLEM 1-8 The measured declination and inclination of the paleomagnetic field in Lower Cretaceous rocks at 45.5°N and 73°W are $D = 154°$ and $I = -58°$. Determine the paleomagnetic pole position.

Paleomagnetic measurements can indicate the position of the magnetic pole as a function of time for rocks of different ages. However, before discussing these results, we should note that one of the early conclusions of paleomagnetic measurements was that the

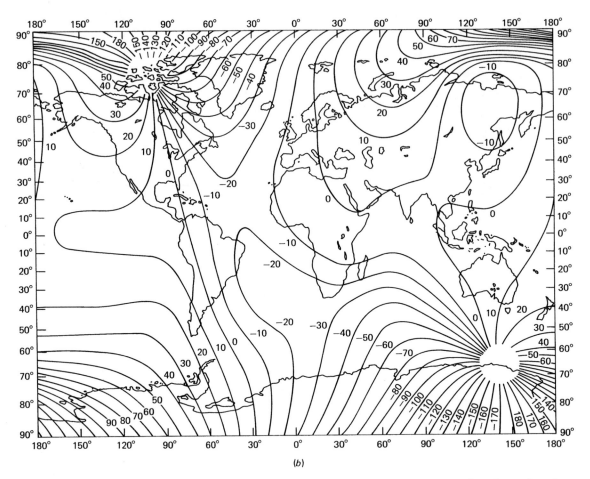

1–24 (cont.) (*b*) Declination, deg.

Earth's magnetic field has been subject to periodic *reversals* in which the north magnetic pole became the south magnetic pole and vice versa. This was apparent from the reversed orientations of the remanent magnetic field in a series of rocks of different ages from the same locality. A summary of dated rocks with *normal and reversed polarities* for the last 5 Ma is given in Figure 1–26. Measurements indicate that for the past 720,000 years the magnetic field has been in its present (normal) orientation; this magnetic time period is referred to as the Brunhes epoch. Between 0.72 and 2.45 Ma, there was a period known as the Matuyama epoch during which the orientation of the field was predominantly reversed. Periods of normal polarity for the last 170 Ma are given in Table 1–4.

The mechanism for magnetic field reversals is not known. In fact, the way in which the Earth's magnetic field is generated is only qualitatively understood. It is well established from seismology that the outer core of the Earth is primarily composed of liquid iron. Presumably, electric currents in the highly electrically conducting liquid iron generate the Earth's magnetic field. However, the currents that create the magnetic field are themselves driven by motions of the conducting liquid in the presence of the magnetic field. The field generation mechanism requires the presence of the field itself. The process by which fluid motions maintain the magnetic field against its tendency to decay because of ohmic dissipation is known as *regenerative dynamo action.*

An energy source is required to overcome the resistive losses. Possible energy sources are the decay of radioactive elements in the core, the cooling of the core, the latent heat release upon solidification of the inner core, and the gravitational energy release that accompanies solidification of the inner core. The last energy source exists because the outer core contains an alloying element lighter than iron. The light element

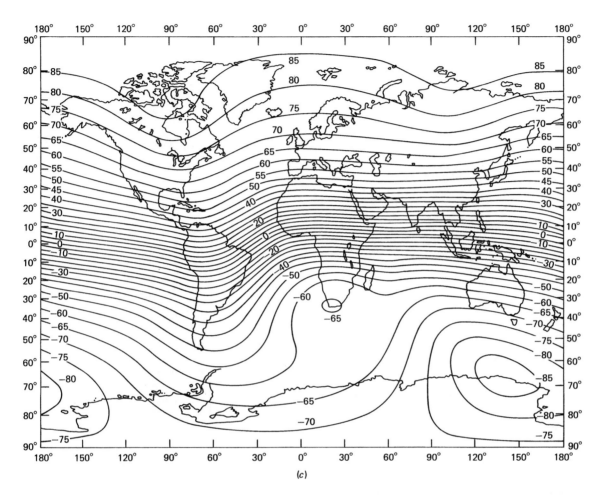

1–24 (cont.) (c) Inclination, deg.

does not enter the inner core when solidification occurs at the inner core–outer core boundary. As a result, growth of the inner core concentrates the light element in the outer core, causing outer-core liquid to become increasingly lighter with time. This releases gravitational potential energy in the same way that separation of the entire core did early in the Earth's evolution.

One or more of these energy sources drives the thermal or chemical convective motions of the highly conducting liquid iron that result in a self-excited dynamo; however, detailed theories of the process are not available. Self-excited mechanical dynamos built in the laboratory exhibit random reversals of the resulting field. Presumably, the dynamo in the Earth's core is subject to random fluctuations that aperiodically lead to field reversals.

It is believed that the rotation of the Earth has an important influence on the generation of the field. We have already noted that the Earth's present dipole axis is nearly aligned with its axis of rotation. It is implicitly assumed in the use of paleomagnetic measurements that the magnetic poles and the geographic poles coincide. A measurement of a paleomagnetic pole can then be used to deduce the motion of the plate on which the measurement was made.

Many paleomagnetic measurements have been made. Data are divided into geological periods and into continental areas that appear to have remained a single unit over the periods considered. Average pole positions are given in Table 1–5. If no relative motion occurred among the continental blocks, all measurements during a particular period should give the same pole position. Clearly, as can be seen from Table 1–5, this is not the case. If a sequence of pole positions for a particular continental area is plotted, it should

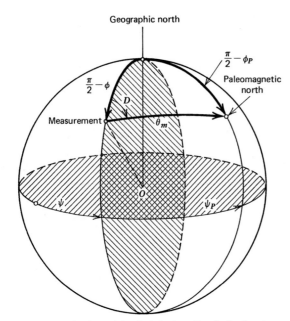

1–25 Geometry for determining the latitude and longitude of a paleomagnetic field.

form a continuous path terminating close to the present position of the magnetic pole; this is known as the *polar wandering path* for the magnetic pole. A polar wandering path of a plate can be used to determine the absolute position of that plate relative to the geographic poles. The relation between the polar wandering paths of two adjacent plates can be used to determine relative velocities between the plates. The polar wandering paths for North America and Europe are shown in Figure 1–27. The systematic divergence of the paths over the past several hundred million years was one of the first pieces of quantitative evidence that continental drift was occurring. Unfortunately the considerable scatter in paleomagnetic measurements makes it difficult to obtain reliable data. Much of this scatter can be attributed to deviations of the magnetic poles from the geographic poles.

The magnitude of the magnetic field at the Earth's surface varies both in space and in time. The spatial variations are known as *magnetic anomalies*. In the continents, regions of high magnetic field, that is, positive magnetic anomalies, are usually associated with concentrations of magnetic minerals in the Earth's crust. Regional surveys of the magnetic field are an important method of exploration for economic deposits of minerals.

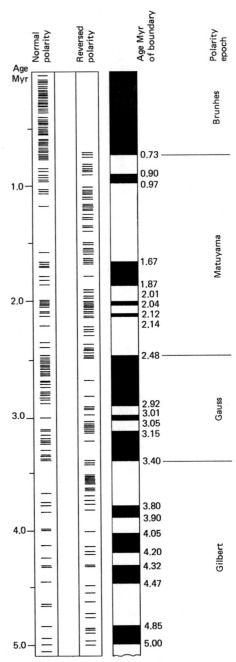

1–26 Measurements of the polarity of the Earth's magnetic field for the last 5 million years. Each short line indicates a dated polarity determination from a volcanic rock. The shaded periods are intervals of predominantly normal polarity.

TABLE 1–4 Ages in Ma of Periods of Normal Polarity of the Earth's Magnetic Field for the Last 170 Ma

Anomaly	Normal Polarity Interval		Anomaly	Normal Polarity Interval		Anomaly	Normal Polarity Interval	
1.1	0.00	0.72	6.7	22.90	23.05	M1	118.70	121.81
1.2	0.91	0.97	6.8	23.25	23.38	M2	122.25	123.03
2.1	1.65	1.88	6.9	23.62	23.78	M4	125.36	126.46
2.2	2.06	2.09	7.1	25.01	25.11	M6	127.05	127.21
2.3	2.45	2.91	7.2	25.17	25.45	M7	127.34	127.52
2.4	2.98	3.07	7.3	25.84	26.01	M8	127.97	128.33
2.5	3.17	3.40	8.1	26.29	26.37	M9	128.60	128.91
3.1	3.87	3.99	8.2	26.44	27.13	M10.1	129.43	129.82
3.2	4.12	4.26	9.1	27.52	28.07	M10.2	130.19	130.57
3.3	4.41	4.48	9.2	28.12	28.51	M10.3	130.63	131.00
3.4	4.79	5.08	10.1	29.00	29.29	M10.4	131.02	131.36
3.5	5.69	5.96	10.2	29.35	29.58	M11.1	131.65	132.53
3.6	6.04	6.33	11.1	30.42	30.77	M11.2	133.03	133.08
3.7	6.66	6.79	11.2	30.82	31.21	M11.3	133.50	134.31
4.1	7.01	7.10	12	31.60	32.01	M12.1	134.42	134.75
4.2	7.17	7.56	13.1	34.26	34.44	M12.2	135.56	135.66
4.3	7.62	7.66	13.2	34.50	34.82	M12.3	135.88	136.24
4.4	8.02	8.29	15.1	36.12	36.32	M13	136.37	136.64
4.5	8.48	8.54	15.2	36.35	36.54	M14	137.10	137.39
4.6	8.78	8.83	15.3	36.93	37.16	M15	138.30	139.01
5.1	8.91	9.09	16.1	37.31	37.58	M16	139.58	141.20
5.2	9.14	9.48	16.2	37.63	38.01	M17	141.85	142.27
5.3	9.49	9.80	17.1	38.28	39.13	M18	143.76	144.33
5.4	9.83	10.13	17.2	39.20	39.39	M19.1	144.75	144.88
5.5	10.15	10.43	17.3	39.45	39.77	M19.2	144.96	145.98
5.6	10.57	10.63	18.1	39.94	40.36	M20.1	146.44	146.75
5.7	11.11	11.18	18.2	40.43	40.83	M20.2	146.81	147.47
5.8	11.71	11.90	18.3	40.90	41.31	M21	148.33	149.42
5.9	12.05	12.34	19	42.14	42.57	M22.1	149.89	151.46
5.10	12.68	12.71	20	43.13	44.57	M22.2	151.51	151.56
5.11	12.79	12.84	21	47.01	48.51	M22.3	151.61	151.69
5.12	13.04	13.21	22	50.03	50.66	M22.4	152.53	152.66
5.13	13.40	13.64	23.1	51.85	52.08	M23.1	152.84	153.21
5.14	13.87	14.24	23.2	52.13	52.83	M23.2	153.49	153.52
5.15	14.35	14.79	23.3	53.15	53.20	M24.1	154.15	154.48
5.16	14.98	15.07	24.1	53.39	53.69	M24.2	154.85	154.88
5.17	15.23	15.35	24.2	54.05	54.65	M24.3	155.08	155.21
5.18	16.27	16.55	25	57.19	57.80	M24.4	155.48	155.84
5.19	16.59	16.75	26	58.78	59.33	M25.1	156.00	156.29
5.20	16.82	16.99	27	61.65	62.17	M25.2	156.55	156.70
5.21	17.55	17.87	28	62.94	63.78	M25.3	156.78	156.88
5.22	18.07	18.09	29	64.16	64.85	M25.4	156.96	157.10
5.23	18.50	19.00	30	65.43	67.14	M26.1	157.20	157.30
6.1	19.26	20.23	31	67.23	68.13	M26.2	157.38	157.46
6.2	20.52	20.74	32.1	70.14	70.42	M26.3	157.53	157.61
6.3	20.97	21.37	32.2	70.69	72.35	M26.4	157.66	157.85
6.4	21.60	21.75	32.3	72.77	72.82	M27	158.01	158.21
6.5	21.93	22.03	33	73.12	79.09	M28	158.37	158.66
6.6	22.23	22.60	34	84.00	118.00	M29	158.87	159.80
						J-QZ	160.33	169.00

Source: Harland et al. (1990).

TABLE 1-5 Position of the North Magnetic Pole in Different Geological Periods as Determined by Paleomagnetic Studies

		North America	Europe	Russian Platform	Siberian Platform	Africa	South America	Australia	India
Tertiary	U	87N, 140E	80N, 157E	78N, 191E	66N, 234E	87N, 152E	82N, 62E	77N, 275E	
	L	85N, 197E	75N, 151E	68N, 192E	57N, 152E	85N, 186E		70N, 306E	
Cretaceous		64N, 187E	86N, 0E	66N, 166E	77N, 176E	61N, 260E	78N, 236E	53N, 329E	22N, 295E
Jurassic		76N, 142E	36N, 50E	65N, 138E		65N, 262E	84N, 256E	48N, 331E	
Triassic		62N, 100E	45N, 143E	51N, 154E	47N, 151E		80N, 71E		20N, 308E
Permian		46N, 117E	45N, 160E	44N, 162E		27N, 269E	60N, 180E		7S, 304E
Carboniferous	U	37N, 126E	38N, 161E	43N, 168E	34N, 144E	46N, 220E		46N, 315E	26S, 312E
	L			22N, 168E		26N, 206E	43N, 151E	73N, 34E	
Devonian		29N, 123E	0N, 136E	36N, 162E	28N, 151E			72N, 174E	
Silurian				28N, 149E	24N, 139E			54N, 91E	
Ordovician		28N, 192E	10N, 176E		25S, 131E	24S, 165E	11S, 143E	2N, 188E	
Cambrian		7N, 140E	22N, 167E	8N, 189E	36S, 127E				28N, 212E

Source: After M. W. McElhinny (1973).

Similar magnetic surveys over the oceans have shown a pattern of striped magnetic anomalies, that is, elongated continuous zones of positive magnetic anomalies some tens of kilometers wide separated from one another by zones of negative magnetic anomalies.

1-27 Polar wandering paths for North America and Europe. Numbers give time before present in millions of years.

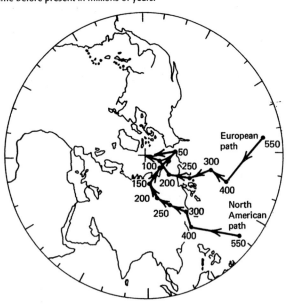

The zones of striped magnetic anomalies generally lie parallel to ocean ridges and are symmetric with respect to the ridge crest. A typical pattern adjacent to the mid-Atlantic ridge is shown in Figure 1–28. A typical magnetic anomaly profile perpendicular to the East Pacific Rise is given in Figure 1–29. The magnitude of any individual anomaly is a few hundred nanoteslas, or about 1% of the Earth's dipole field at the surface. The magnetic anomalies are attributed to thermal remanent magnetism in the basaltic oceanic crust. As the volcanic rocks of the oceanic crust cool through the magnetic blocking temperature near the ocean ridge, a thermal remanent magnetism is acquired in the direction of the Earth's magnetic field. This magnetization of the oceanic crust produces the magnetic anomalies as a consequence of the episodic reversals in the Earth's magnetic field. Ocean floor created in the last 720,000 years has been magnetized in the direction of the Earth's present magnetic field, leading to a positive magnetic anomaly (see Figure 1–26). However, ocean floor created between 2.45 and 0.72 Ma was primarily magnetized in the direction of the reversed field. This magnetization is opposite to the present Earth's field and therefore subtracts from it, leading to a zone of low field or a negative magnetic anomaly, as illustrated in Figure 1–29b. The conclusion is that the stripes

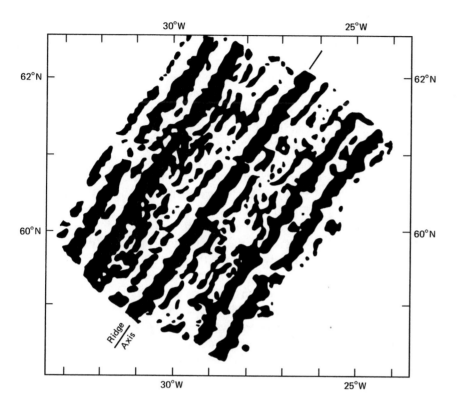

1–28 Striped pattern of magnetic anomalies parallel to the Mid-Atlantic ridge (Heirtzler et al., 1966).

1–29 (a) Magnetic anomaly profile perpendicular to the East Pacific Rise (52°S, 118°W). (b) Induced magnetization in the oceanic crust due to episodic reversals of the Earth's magnetic field. (c) Correlation of the positions x of the magnetic anomalies with t of field reversals to give the velocity u of seafloor spreading.

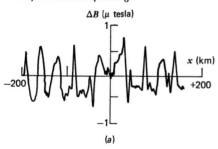

(a)

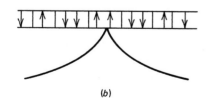

(b)

of seafloor with positive magnetic anomalies were created during periods of normal polarity of the Earth's magnetic field and stripes of the seafloor with negative magnetic anomalies were created during periods of reversed polarity of the Earth's magnetic field.

Since the dates of the field reversals are known independently from geochronological studies, the widths of the magnetic stripes can be used to determine the velocity of seafloor spreading. For the example given in Figure 1–29a, the distance from the ridge crest to the edge of each anomaly is plotted against the time

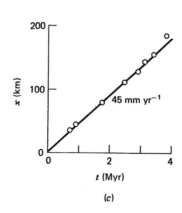

(c)

of known field reversal in Figure 1–29c. The result is nearly a straight line, the slope of which is the velocity of seafloor spreading, 45 mm yr^{-1} in this case. Velocities of seafloor spreading (half-spreading rates) range upward to about 100 mm yr^{-1}.

PROBLEM 1–9 Determine the velocity of seafloor spreading on the East Pacific Rise from the magnetic anomaly profile given in Figure 1–30a.

PROBLEM 1–10 Determine the velocity of seafloor spreading on the South East Indian Rise from the magnetic anomaly profile given in Figure 1–30b.

Shipboard magnetometers have been used to obtain maps of magnetic anomalies over a large fraction of the world's oceans. Striped patterns of magnetic anomalies have not been obtained near the paleomagnetic equator. At the magnetic equator the magnetic field is horizontal, and the magnetization of the ferromagnetic minerals in the oceanic crust does not produce a significant surface magnetic anomaly. The maps of magnetic anomalies have been used to determine the age of a large fraction of the ocean floor (Figure 1–31). This distribution of ages has been confirmed by the *Deep Sea Drilling Project* (*DSDP*). The deep-sea capability of the drilling ship *Glomar Challenger* made it

possible to drill a large number of cored holes through the sedimentary cover and into the underlying basaltic oceanic crust. If we hypothesize that the age of the oldest sediments in the sedimentary sequence adjacent to the volcanic crust, as determined from studies of fossils, corresponds to the age of the volcanic rocks, then we can determine the age of the seafloor. This has been done for a number of DSDP holes, and the results have been compared with the age of the seafloor inferred from studies of the magnetic anomalies in Figure 1–32. The excellent agreement is striking confirmation of the magnetic method for determining the age of the seafloor.

Because the surface area of the Earth remains essentially constant, the velocities of seafloor spreading at ocean ridges can be related to velocities of subduction at ocean trenches. As a result the relative velocities among the rigid plates can be determined. The ten major plates are illustrated in Figure 1–1. The relative motion between two adjacent plates can be obtained using Euler's theorem. This theorem states that any line on the surface of a sphere can be translated to any other position and orientation on the sphere by a single rotation about a suitably chosen axis passing through the center of the sphere. In terms of the Earth this means that a rigid surface plate can be shifted to a new position by a rotation about a uniquely defined axis. The point where this axis intersects the surface of the Earth is known as a *pole of rotation*. This is illustrated in Figure 1–33, where plate B is rotating counterclockwise with respect to plate A. Ridge segments lie on lines of longitude emanating from the pole of rotation P. Transform faults lie on small circles with their centers at the pole of rotation.

The relative motion between two adjacent plates is completely specified when the latitude and longitude of the pole of rotation together with the angular velocity of rotation ω are given. The location of the pole of rotation can be determined from the orientations of ridge crests, magnetic lineaments, and transform faults. The angular velocity of rotation can be obtained from the seafloor-spreading velocities determined from widths of the magnetic lineaments and the requirement that surface area must be preserved.

The latitudes and longitudes of the poles of rotation for relative motions among ten plates are given in Table 1–6. The angular velocities of rotation are also

1–30 Typical profiles of the magnetic anomaly pattern (a) perpendicular to the East Pacific Rise at 61°S and 151°W and (b) perpendicular to the South East Indian Rise at 54°S and 142°E.

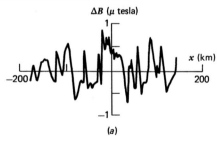

(a)

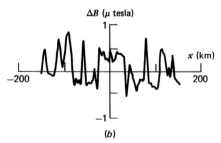

(b)

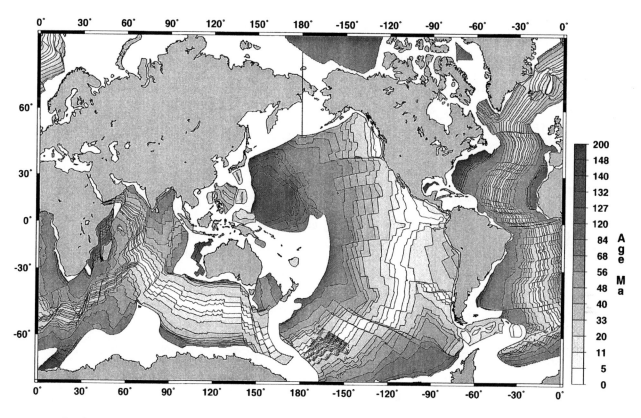

1-31 Map of seafloor ages (Muller et al., 1997).

1-32 Correlation of the ages of the oldest sediments in DSDP holes with the predicted ages of the oceanic crust based on seafloor magnetic anomalies.

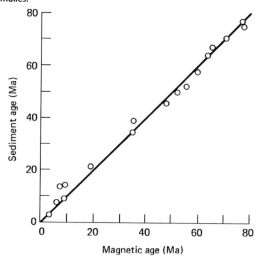

1-33 Plate B is moving counterclockwise relative to plate A. The motion is defined by the angular velocity ω about the pole of rotation P. Double lines are ridge segments, and arrows denote directions of motion on transform faults.

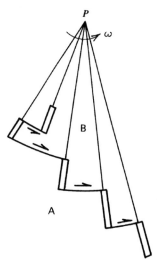

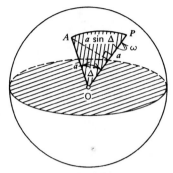

TABLE 1–6 Pole Positions and Rates of Rotation for Relative Motion Between Adjacent Surface Plates*			
Plates	**Lat. (N)**	**Long. (E)**	**ω (deg/Myr)**
EU–NA	62.4	135.8	0.21
AF–NA	78.8	38.3	0.24
AF–EU	21.0	−20.6	0.12
NA–SA	16.3	−58.1	0.15
AF–SA	62.5	−39.4	0.31
AN–SA	86.4	−40.7	0.26
NA–CA	−74.3	−26.1	0.10
CA–SA	50.0	−65.3	0.18
NA–PA	48.7	−78.2	0.75
CO–PA	36.8	−108.6	2.00
CO–NA	27.9	−120.7	1.36
CO–NZ	4.8	−124.3	0.91
NZ–PA	55.6	−90.1	1.36
NZ–AN	40.5	−95.9	0.52
NZ–SA	56.0	−94.0	0.72
AN–PA	64.3	−84.0	0.87
PA–AU	−60.1	−178.3	1.07
EU–PA	61.1	−85.8	0.86
CO–CA	24.1	−119.4	1.31
NZ–CA	56.2	−104.6	0.55
AU–AN	13.2	38.2	0.65
AF–AN	5.6	−39.2	0.13
AU–AF	12.4	49.8	0.63
AU–IN	−5.6	77.1	0.30
IN–AF	23.6	28.5	0.41
AR–AF	24.1	24.0	0.40
IN–EU	24.4	17.7	0.51
AR–EU	24.6	13.7	0.50
AU–EU	15.1	40.5	0.69
IN–AR	3.0	91.5	0.03

* Global plate motion model NUVEL-1A. The first plate moves counterclockwise relative to the second plate. Abbreviations: PA, Pacific; NA, North America; SA, South America; AF, Africa; CO, Cocos; NZ, Nazca; EU, Eurasia; AN, Antarctica; AR, Arabia; IN, India; AU, Australia; CA, Caribbean. See Figure 1–1 for plate geometries.
Source: DeMets et al. (1994).

1–34 Geometry for the determination of the relative plate velocity at point A on the boundary between two plates in terms of the rate of rotation ω about pole P.

longitude ψ' of the point on the plate boundary by the same spherical trigonometry formula used in Equation (1–12). By referring to Figure 1–35 we can write

$$\cos \Delta = \cos \theta \cos \theta' + \sin \theta \sin \theta' \cos(\psi - \psi'). \tag{1–18}$$

The surface distance s between points A and P is

$$s = a\Delta, \tag{1–19}$$

with Δ in radians. This relation along with Equation (1–18) can be used to determine the distance between two points on the surface of the Earth given the latitudes and longitudes of the points. Using Equations (1–17) and (1–18), one can find the relative velocity between two plates, at any point on the boundary between the plates, once the latitude and longitude of the point on the boundary have been specified.

1–35 Geometry for determining the angle between point A on a plate boundary and a pole of rotation.

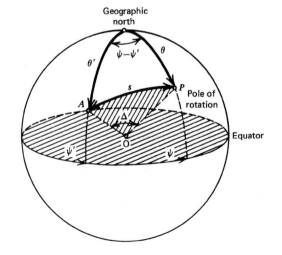

given. The relative velocity u between plates at any plate boundary is

$$u = \omega a \sin \Delta, \tag{1–17}$$

where a is the radius of the Earth and Δ is the angle subtended at the center of the Earth by the pole of rotation P and point A on the plate boundary (see Figure 1–34). Note that ω is in radians per unit time. The angle Δ can be related to the colatitude θ and east longitude ψ of the pole of rotation and the colatitude θ' and east

As a specific example let us determine the relative velocity across the San Andreas fault at San Francisco (37.8°N, 122°W). We assume that the entire relative velocity between the rigid Pacific and North American plates is accommodated on this fault. From Table 1–6 we find $\theta = 90° - 48.7° = 41.3°$ and $\psi = -78.2°$. Since $\theta' = 52.2°$ and $\psi' = 238°$, we find from Equation (1–18) that $\Delta = 33.6°$; with $\omega = 0.75°$ Myr^{-1}, we find from Equation (1–17) that the relative velocity across the fault is 46 mm yr^{-1}.

PROBLEM 1-11 Determine the declination and inclination of the Earth's magnetic field at Boston ($\phi = 42.5°$, $\psi = -71°$). Use the dipole approximation to the field, but do not assume that the geographic and magnetic poles coincide.

PROBLEM 1-12 Determine the declination and inclination of the Earth's magnetic field at Chicago ($\phi = 41.8°$, $\psi = -87.5°$). Use the dipole approximation to the field, but do not assume that the geographic and magnetic poles coincide.

PROBLEM 1-13 What are the surface distances between the Earth's magnetic poles and geographic poles?

PROBLEM 1-14 What is the surface distance between the Earth's magnetic poles and the best-fit dipole poles?

PROBLEM 1-15 Plot the distance between the paleomagnetic poles obtained from North American and European rocks as a function of time, and discuss the results.

PROBLEM 1-16 Plot the distance between the paleomagnetic poles obtained from the Russian and Siberian Platform rocks as a function of time, and discuss the results.

PROBLEM 1-17 What is the spreading rate between the North American and Eurasian plates in Iceland (65°N, 20°W)?

PROBLEM 1-18 What is the relative plate velocity between the Nazca and South American plates at Lima, Peru (12°S, 77°W)?

PROBLEM 1-19 What is the relative plate velocity between the Indian and Eurasian plates in the Himalayas (30°N, 81°E)?

1-9 Triple Junctions

A plate boundary can end only by intersecting another plate boundary; this intersection is a *triple junction*. Since there are three types of plate boundaries – ridges, trenches, and transform faults – there are in principle ten types of triple junctions. However, some of these triple junction cannot exist. An example is a triple junction of three transform faults. The required condition for the existence of a triple junction is that the three vector velocities defining relative motions between plate pairs at a triple junction must form a closed triangle. For many types of triple junctions this condition requires a particular orientation of the plate boundaries.

As a specific example let us consider the ridge–ridge–ridge (RRR) triple junction illustrated in Figure 1–36a. The ridge between plates A and B lies in the north-south direction (an azimuth with respect to the triple junction of 0°). Since the relative velocity across a ridge is perpendicular to the ridge, the vector velocity of plate B relative to plate A, $\mathbf{u}_{BA}$, has an azimuth, measured clockwise from north, of 90°; we assume that the magnitude is $u_{BA} = 100$ mm yr^{-1}. The ridge between plates B and C has an azimuth of 110° relative to the triple junction. The vector velocity of plate C relative to plate B, $\mathbf{u}_{CB}$, therefore has an azimuth of 200°; we assume that the magnitude $u_{CB} = 80$ mm yr^{-1}.

1-36 (*a*) Schematic of a ridge–ridge–ridge (RRR) triple junction of plates A, B, and C. (*b*) Vector velocities for relative motion between the plates.

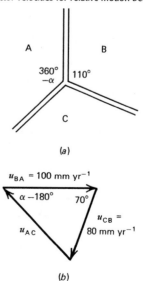

The problem is to find the azimuth of the ridge between plates A and C, α, and the azimuth and magnitude of the relative velocity $\mathbf{u}_{AC}$.

The velocity condition for all triple junctions requires that

$$\mathbf{u}_{BA} + \mathbf{u}_{CB} + \mathbf{u}_{AC} = 0. \qquad (1\text{-}20)$$

This is illustrated in Figure 1–36b. In order to determine the magnitude of the velocity u_{AC} we use the law of cosines:

$$u_{AC} = (100^2 + 80^2 - 2 \cdot 100 \cdot 80 \cdot \cos 70°)^{1/2}$$
$$= 104.5 \text{ mm yr}^{-1}. \qquad (1\text{-}21)$$

The angle α is then determined using the law of sines:

$$\sin(\alpha - 180°) = \frac{80}{104.5} \sin 70° = 0.7518 = -\sin\alpha,$$
$$\alpha = 228.7°. \qquad (1\text{-}22)$$

The azimuth of the ridge is 228.7°, and the azimuth of $\mathbf{u}_{AC}$ is 318.7°. An example of an RRR triple junction is the intersection of the Nazca, Cocos, and Pacific plates (see Figure 1–1).

PROBLEM 1–20 Consider an RRR triple junction of plates A, B, and C. The ridge between plates A and B lies in a north–south direction (an azimuth of 0° with respect to the triple junction) and has a relative velocity of 60 mm yr^{-1}. The ridge between plates B and C has an azimuth of 120° with respect to the triple junction, and the ridge between plates A and C has an azimuth of 270° with respect to the triple junction. Determine the azimuths and magnitudes of the relative velocities between plates B and C and C and A.

We next consider a trench–trench–trench (TTT) triple junction. In general this type of triple junction cannot exist. A geometry that is acceptable is illustrated in Figure 1–37a. Both plates A and B are being subducted beneath plate C along a single north–south trench. Plate A is also being subducted beneath plate B along a trench that has an azimuth of 135° with respect to the triple junction. Since *oblique subduction* can occur, the relative velocities between plates where subduction is occurring need not be perpendicular to the trench. We assume that the velocity of plate A relative to plate B has a magnitude $u_{AB} = 50$ mm yr^{-1}

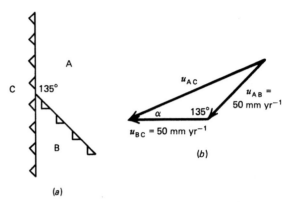

1–37 (*a*) Illustration of a trench–trench–trench (TTT) triple junction of plates A, B, and C. (*b*) Vector velocities for relative motion between the plates.

and an azimuth of 225°. We also assume that the relative velocity of plate B with respect to plate C has a magnitude $u_{BC} = 50$ mm yr^{-1} and an azimuth of 270°. Applying the law of cosines to the velocity triangle of Figure 1–37b, we find

$$u_{AC} = (50^2 + 50^2 - 2 \cdot 50 \cdot 50 \cdot \cos 135°)^{1/2}$$
$$= 92.4 \text{ mm yr}^{-1}. \qquad (1\text{-}23)$$

The angle α in Figure 1–37b is determined from the law of sines:

$$\sin\alpha = \frac{50}{92.4} \sin 135° = 0.383, \quad \alpha = 22.5°, \quad (1\text{-}24)$$

so that the azimuth of $\mathbf{u}_{AC}$ is 247.5°. The velocity at which subduction is occurring is $u_{AC} \cos\alpha = 85.4$ mm yr^{-1}, and the velocity of migration of the triple junction along the north–south trench is $u_{AC} \sin\alpha = 35.4$ mm yr^{-1}. An example of a TTT triple junction is the intersection of the Eurasian, Pacific, and Philippine plates (see Figure 1–1).

PROBLEM 1–21 Show that a triple junction of three transform faults cannot exist.

PROBLEM 1–22 Consider the TTT triple junction illustrated in Figure 1–38. This triple junction is acceptable because the relative velocity between plates C and A, $\mathbf{u}_{CA}$, is parallel to the trench in which plate B is being subducted beneath plate C. The trench between plates C and B has an azimuth of 180° so that $\mathbf{u}_{CA}$ has an azimuth of 0°; assume that $u_{CA} = 50$ mm yr^{-1}. Also assume that the azimuth and

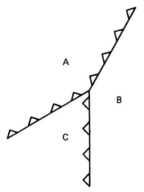

1-38 Another TTT triple junction.

magnitude of $\mathbf{u}_{BA}$ are 315° and 60 mm yr^{-1}. Determine the azimuth and magnitude of $\mathbf{u}_{BC}$.

As our final example we consider a ridge–trench–fault (RTF) triple junction. This is another type of triple junction that cannot generally exist. An acceptable geometry is illustrated in Figure 1–39a; the trench and the transform fault are aligned in the north–south direction. Plate C is being subducted beneath plate B; plate A is sliding past plate B on a transform fault. The velocity of plate B relative to plate A has a magnitude $u_{BA} = 50$ mm yr^{-1} and an azimuth of 180° (the orientation of the fault requires an azimuth of either 0° or 180°). The ridge has an azimuth of 225° with respect to the triple junction. This constrains the relative velocity between plates A and C to have an azimuth of 315°; we assume that $u_{AC} = 40$ mm yr^{-1}. Applying the law of

cosines to the velocity triangle in Figure 1–39b we get

$$u_{CB} = (50^2 + 40^2 - 2 \cdot 40 \cdot 50 \cos 45°)^{1/2}$$
$$= 35.7 \text{ mm yr}^{-1}, \tag{1-25}$$

and from the law of sines we find

$$\sin \alpha = \frac{40}{35.7} \sin 45° = 0.79, \quad \alpha = 52.4°. \tag{1-26}$$

The rate at which the ridge is migrating northward along the trench–transform boundary is $u_{CB} \cos \alpha + u_{AC} \cos 45° = 50.1$ mm yr^{-1}. An example of an RTF triple junction is the intersection of the Pacific, North American, and Cocos plates (see Figure 1–1).

It should be emphasized that the relative plate motions given in Table 1–6 are only instantaneously valid. As plates evolve, their poles of rotation migrate, and their angular velocities change. Plate boundaries and triple junctions must also evolve. One result is that a plate boundary may cease to be active or new plate boundaries and triple junctions may form. Another consequence is that plate boundaries may become broad zones of diffuse deformation. The western United States is an example of such a zone; the deformation associated with the interaction of the Pacific, Juan de Fuca, and North American plates extends from the Colorado Front in Wyoming, Colorado, and New Mexico, to the Pacific Coast (see Section 1–13).

PROBLEM 1-23 Consider the TTR triple junction illustrated in Figure 1–40. A ridge with an azimuth of 135° relative to the triple junction is migrating along a north–south trench. If the azimuth and magnitude of

1-40 A TTR triple junction.

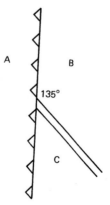

1-39 (a) A trench–ridge–fault (TRF) triple junction of plates A, B, and C. (b) Vector velocities for the relative motions between the plates.

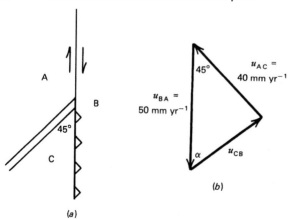

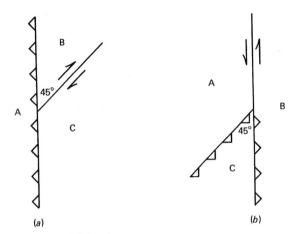

1–41 Two TTF triple junctions.

$\mathbf{u}_{BA}$ are 270° and 50 mm yr^{-1} and $u_{CB} = 40$ mm yr^{-1}, determine the azimuth and magnitude of $\mathbf{u}_{CA}$. Also determine the direction and rate of migration of the ridge relative to plate A.

PROBLEM 1–24 Consider the TTF triple junction illustrated in Figure 1–41a. A right-lateral transform fault has an azimuth of 45° with respect to the triple junction that is migrating along a north–south trench. If the azimuth and magnitude of $\mathbf{u}_{BA}$ are 270° and 50 mm yr^{-1} and $u_{CB} = 50$ mm yr^{-1}, determine the azimuth and magnitude of $\mathbf{u}_{CA}$. Also determine the direction and rate of migration of the fault along the trench.

PROBLEM 1–25 Consider the TTF triple junction illustrated in Figure 1–41b. A left-lateral transform fault has an azimuth of 0°, and two trenches have azimuths of 180° and 225°. If the azimuth and magnitude of $\mathbf{u}_{CB}$ are 90° and 10 mm yr^{-1} and $u_{AB} = 50$ mm yr^{-1}, determine the azimuth and magnitude of $\mathbf{u}_{AC}$.

1–10 The Wilson Cycle

J. Tuzo Wilson (1966) proposed that continental drift is cyclic. In particular he proposed that oceans open and close cyclically; this concept is now known as the Wilson cycle and was based on the opening and closing of the Atlantic Ocean. The Wilson cycle, in its simplest form, is illustrated in Figure 1–42.

The first step in the Wilson cycle, illustrated in Figure 1–42a, is the breakup of a continent. This occurs on continental rift zones. The first stage of the splitting process is the formation of a *rift valley*. When a continent starts to fracture under tensional stresses, a rift valley is formed. The central block of the rift valley, known as a *graben*, subsides, as shown in Figure 1–42a, and the edges of the adjacent blocks are uplifted. The faults that occur on the sides of the down-dropped central graben are known as *normal faults*. Displacements on the normal faults accommodate horizontal extension. Examples of rift valleys that may be in the first stage of continental splitting include the East African rift system and the Rio Grande (river valley) rift. There is ample evidence in the geological record, however, that some rift valleys never evolve into an ocean. The splitting process may be aborted. Once the formation of the rift valley relieves the tensional stresses, no further horizontal extension may occur.

The Red Sea and the Gulf of Aden are rift valleys that have progressed to the formation of accreting plate margins. Together with the East African rift they define a *three-armed pattern* that can be seen in the satellite photograph in Figure 1–43. If all the rifts of a three-armed system develop into accreting plate margins, an RRR triple junction is formed. In many cases, however, only two arms develop into accreting margins, and the third becomes a relict rift zone in the continent. This third arm is referred to as a *failed arm*. An example of a failed arm is the Benue rift on the western margin of Africa shown in Figure 1–53. The other two arms of this system became part of the early mid-Atlantic ridge at which the Atlantic Ocean formed. The failed arm of the system eventually became filled with sediments; the sediment-filled fossil rift is known as an *aulacogen*.

The second stage of continent splitting is the formation of a seafloor-spreading center, or ocean ridge. This is illustrated in Figure 1–42b. The normal faults associated with the margins of the rift valley now form the margins of a new ocean. Upwelling hot mantle rock partially melts to form new ocean crust and the first stages of an ocean ridge. An example of an ocean at this early stage of development is the Red Sea (Figure 1–43). As seafloor spreading continues at the spreading center, an ocean is formed. Because the creation of new seafloor at an ocean ridge is very nearly a symmetric process, the ocean ridge bisects the newly created ocean. This is illustrated in Figure 1–42c. An example is the Atlantic Ocean. The margins of the opening ocean are known

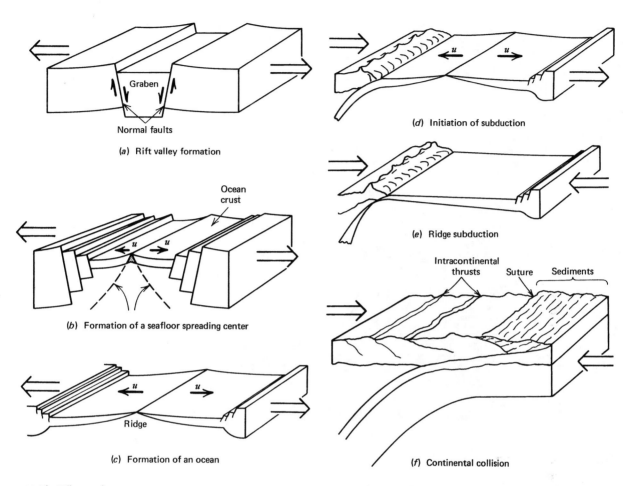

(a) Rift valley formation

(b) Formation of a seafloor spreading center

(c) Formation of an ocean

(d) Initiation of subduction

(e) Ridge subduction

(f) Continental collision

1–42 The Wilson cycle.

as *passive continental margins* in contrast to the *active continental margins* where subduction is occurring.

As the seafloor at the continental margin grows older, the lithosphere becomes thicker and more dense. Eventually the lithosphere becomes sufficiently unstable so that it founders and an ocean trench develops and subduction begins. This is illustrated in Figure 1–42*d*. Trenches apparently form immediately adjacent to one of the continents. This is the site of the oldest, coldest, and most unstable oceanic lithosphere. Also, the continental margin is inherently a zone of weakness. As the ocean basin adjacent to a continent grows older, it continues to subside relative to the continent. This differential subsidence is accommodated on the normal faults associated with the continental margin. These normal faults are zones of weakness, and they may play a key role in the formation of new ocean trenches, when

a passive continental margin is converted to an active continental margin.

If the rate of subduction is greater than the rate of seafloor spreading, the size of the ocean will decrease. Eventually the ocean ridge itself will be subducted (see Figure 1–42*e*). Ridge subduction is occurring along the west coast of North America. The remanents of the Juan de Fuca ridge form the boundary between the Juan de Fuca plate and the Pacific plate (Figure 1–1). The northern part of this ridge was subducted beneath the Aleutian trench. Other parts of the ridge were subducted off the west coast of California. In these cases, the subduction led to the transformation of the convergent plate boundaries between the North American plate and the Juan de Fuca plate (also known as the Farallon plate) to the present transform fault boundaries between the North American and Pacific plates.

After ridge subduction, the remainder of the oceanic plate will be subducted and the continents will collide

1–43 Satellite photograph of the Red Sea (NASA STS040-078-088).

(Figure 1–42*f*). The implications of a continental collision are discussed in the next section.

Evidence of the past motions of the continents comes from many sources. The distribution of magnetic lineations on the seafloor can be used to reconstruct the positions of the continents for about the last 150 Ma. Because there is very little seafloor older than this, reconstructions prior to about 150 Ma are primarily based on paleomagnetic measurements in continental rocks. Many other sources of information contribute to paleoreconstructions. Dated orogenic events provide information on the locations of ocean trenches and continental collision zones. The spatial distributions of fossils, glaciations, and morphological features provide additional latitude control.

Continental reconstructions for the last 170 Ma are given in Figure 1–44. Reassembly of the continents clearly resembles the construction of a jigsaw puzzle.

Not only does South America fit against Africa, but Australia can be fit together with Southeast Asia and Antarctica as well as Greenland with North America and Europe. Continental reconstructions can be extended even farther back in time, but the uncertainties become large.

At 170 Ma the supercontinent Pangaea was being rifted to form the northern continent Laurasia (composed of North America, Europe, and Asia) and the southern continent Gondwanaland composed of South America, Africa, India, Australia, and parts of Antarctica and Southeast Asia. Between these continents the Tethys Ocean was being formed. Between 170 and 100 Ma the central Atlantic Ocean began to form as North America rotated away from Africa. Simultaneously the Tethys Ocean was closing. Between 100 and 50 Ma the Atlantic Ocean continued to form and the Indian Ocean formed as Australia and Antarctica rotated away from Africa. The Tethys Ocean continued to close. In the last 50 Ma the Atlantic has continued to open, India has collided with Eurasia, and Australia

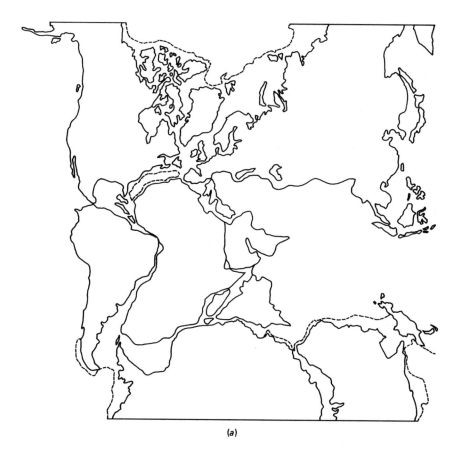

1-44 Continental reconstructions. (*a*) 170 Ma. Continued on pp. 42–3. (Smith et al., 1974)

has moved northward relative to Antarctica. Today the Mediterranean Sea, Black Sea, and Caspian Sea are the only relics of the Tethys Ocean.

1-11 Continental Collisions

The collision of two continents when an ocean closes is a major cause of mountain building. At present a continental collision is occurring along a large fraction of the southern boundary of the Eurasian plate. The style of this collision varies considerably from west to east. The mountain building associated with a continental collision is referred to as an *orogeny*. The region where mountain building is occurring is referred to as an *orogenic zone*. The collision between the Eurasian and African plates has resulted in the formation of the Alps; this is a relatively subdued continental collision and the Alpine orogenic zone is relatively narrow. One model for this collision is illustrated in the cross section given in Figure 1–45. A flake of the upper continental crust of the Eurasian plate has overridden the continen-

tal crust of the African plate. The forces associated with the southward dipping subduction of the Eurasian plate has driven the upper Eurasian crust several hundred kilometers over the African crust. The lower Eurasian crust has been delaminated and is being subducted into the mantle with the Eurasian lithosphere. The underlying African crust is exposed through the overlying upper Eurasian crust at several points in the Alps. The splitting of the Eurasian crust at a depth of about 15 km requires an intracrustal *decollement*. This type of splitting is often observed in the geological record and is attributed to a soft crustal rheology at intermediate crustal depths.

The continental collision between the Eurasian and the Indian plates has resulted in a much broader orogenic zone that extends throughout much of China. This orogenic zone is illustrated in Figure 1–46. The collision has resulted in the formation of the Himalayas,

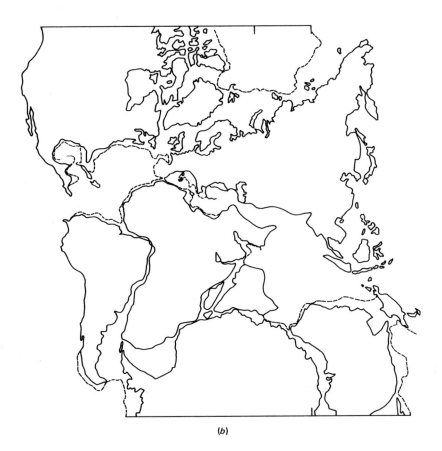

(b)

1–44 (cont.) (b) 100 Ma. (Smith et al., 1974)

the highest and most extensive mountain range in the world. A satellite photograph of the Himalayas looking to the northwest is given in Figure 1–47. Imbedded within the Himalayas is the Indus *suture*, the actual boundary between the Indian plate and the Eurasian plate. The Tibetan plateau is a broad region of elevated topography with extensive faulting but very little recent volcanism. Faulting extends throughout much of China. A substantial fraction of the largest historical earthquakes has occurred on these faults and in many cases the death toll has been very high. Reports claimed that there were 655,000 deaths during the Tangshan earthquake of July 28, 1976. The largest reported death toll in an earthquake was the 800,000 deaths attributed to the Shensi earthquake on January 23, 1556.

It is necessary to explain why this orogenic zone is so broad and why the orogeny is principally on the Eurasian plate with relatively little deformation on the Indian plate. One explanation for the asymmetric deformation is that the Eurasian lithosphere in Tibet and China was thin and weak prior to the collision. This area may have resembled the present western United States which has a weak and easily deformable lithosphere. A simplified model for this continental collision is given in Figure 1–48. The Indian continental crust and lithosphere have been thrust beneath the Eurasian crust across the entire width of the Tibetan plateau.

Continental collisions can produce large amounts of horizontal strain. It is estimated that the original continental crust in the Himalayas has been shortened by 300 km or more. Strain in the crust is accommodated by both brittle and ductile mechanisms. The brittle upper crust can be compressed and thickened by displacements on a series of *thrust faults* that form a *thrust belt* such as that illustrated in Figure 1–49; each of the upthrust blocks forms a mountain range. Sedimentary basins often form over the downthrust blocks. In the Wyoming thrust belt these sedimentary basins are the sites of major oil fields. Crustal thickening and

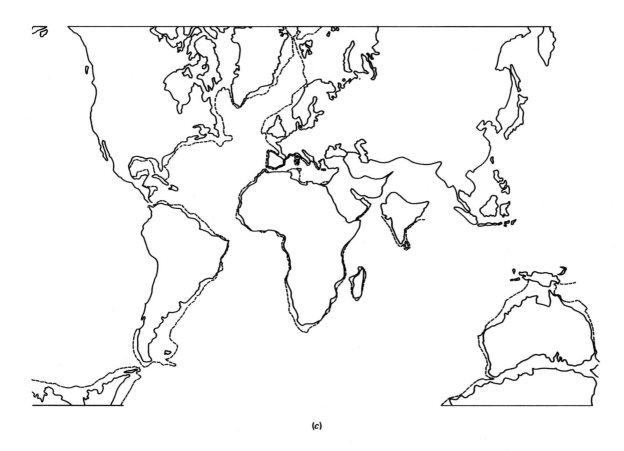

(c)

1-44 (cont.) (c) 50 Ma. (Smith et al., 1974)

1-45 Cross section of the Alpine orogenic zone (after Schmid et al., 1997).

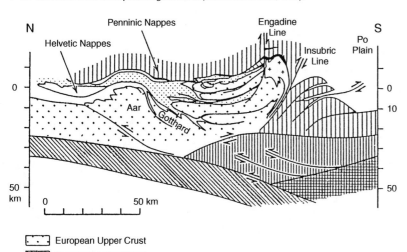

European Upper Crust
European Lower Crust
Adriatic Upper Crust
Adriatic Lower Crust

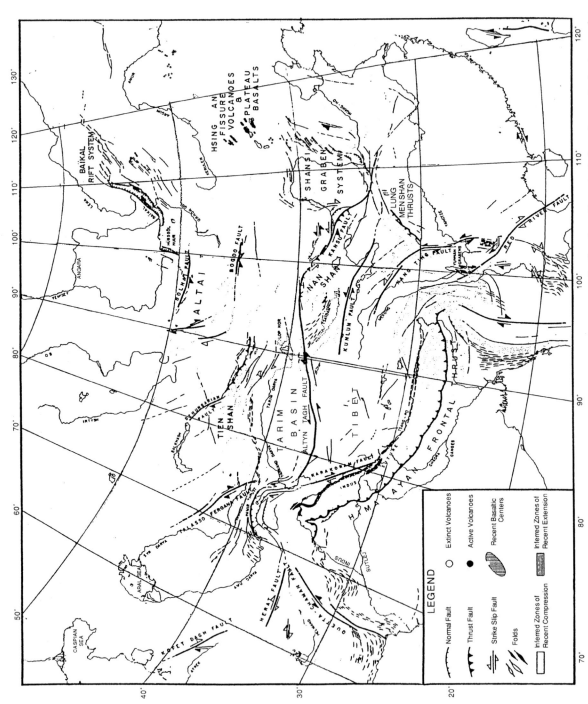

1–46 Illustration of the orogenic zone resulting from the continental collision between the Eurasian and Indian plates. The major faults and zones of volcanism are shown. The Indus suture is the probable boundary between the plates (after Tapponier and Molnar, 1977).

1–47 Satellite photograph of the Himalayas and the Tibetan plateau (NASA STS 41G-120-0022).

shortening resulting from thrusting during the collision of India and Asia are illustrated in Figure 1–48. In some cases the entire brittle part of the continental crust is thrust over the adjacent continental crust as a *thrust sheet*. Evidence indicates that a thrust sheet in the southern Appalachian Mountains extends over hun-

1–48 A schematic cross section of the Himalaya and the southern Tibetan plateau showing underthrusting of the Indian continental crust and lithosphere beneath the Eurasian crust. MBT, Main Boundary Thrust; MCT, Main Central Thrust; STD, South Tibetan Detachment (after Zhao and Nelson, 1993).

dreds of kilometers. This structure is associated with the continental collision that occurred when the proto-Atlantic ocean closed at about 250 Ma.

The crust can also be compressed by ductile deformation, one result of which is *folding*. The convex upward or top of a fold is known as an *anticline*, and the concave upward or bottom of a fold is known as a *syncline*. On a large scale these are known as *anticlinoria* and *synclinoria*. Folding is illustrated in Figure 1–50. When a region of large-scale folding is eroded, the easily eroded strata form valleys, and the resistant strata form ridges. This type of valley and ridge topography is found in Pennsylvania and West Virginia (Figure 1–51). The ridges are primarily sandstone, and

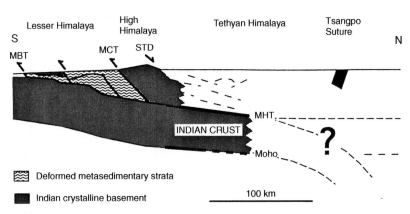

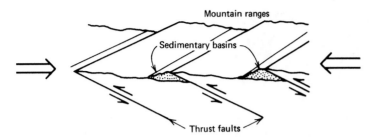

1–49 Horizontal compression resulting in continental collision and a series of thrust faults. Each uplifted block creates a mountain range and each downdropped block a sedimentary basin.

the valleys are the result of the erosion of shales. The large-scale folding in this area was also the result of the continental collision that occurred at about 250 Ma. An extreme amount of deformation occurs in the formation of *nappes*. A nappe may be either a thrust sheet or a *recumbent fold*, that is, a fold whose limbs are almost parallel and roughly horizontal (Figure 1–45).

The effects of continental collisions appear to vary widely. The collision between India and Asia is responsible not only for the Himalaya Mountains but also for tectonics and seismicity throughout China. In contrast, the Alpine orogeny in Europe is narrowly confined. There is also observational evidence that the continental collision that resulted in the formation of the Appalachian Mountains was relatively mild. This difference in collisional orogenies may be attributed to the characteristics of the orogenic zone prior to collision. China may have resembled the western United States; that is, its lithosphere may have been very thin prior to collision, and it may therefore have been easily deformed during the collision.

1–50 Large-scale folding resulting from horizontal compression. The easily eroded strata form valleys, and the resistive strata form ridges.

1–12 Volcanism and Heat Flow

As we have previously discussed, volcanism is associated with both accretionary plate margins and subduction zones. The worldwide distribution of active volcanoes is shown in Figure 1–52. Clearly, most volcanoes are associated with plate margins. Continuous volcanism occurs along the worldwide ocean ridge system, where it forms the 6-km-thick basaltic oceanic crust. Only a small fraction of this volcanism is included in the distribution of active volcanoes given in Figure 1–52; the remainder occurs on the seafloor, where it cannot be readily observed. The linear chains of active volcanoes associated with ocean trenches are clearly illustrated in Figure 1–52. However, significant gaps in the chains do occur even when active subduction is going on. Examples are in Peru and central Chile.

While a large fraction of the Earth's volcanism can be directly attributed to plate boundaries, there are many exceptions. The most obvious example is the volcanism of the Hawaiian Islands. This intraplate volcanism occurs near the center of the Pacific plate. As discussed in Section 1–6 centers of intraplate volcanism are referred to as hotspots. The locations of thirty hotspots are given in Table 1–2 and the distribution of twenty hotspots is shown on the map in Figure 1–14. These are both intraplate hotspots and hotspots located on or near oceanic ridges. One example of a hotspot on an oceanic ridge is Iceland, where very high rates of

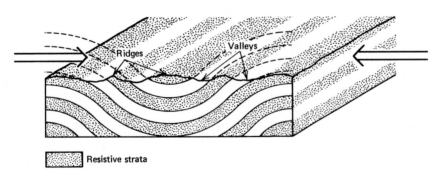

Resistive strata

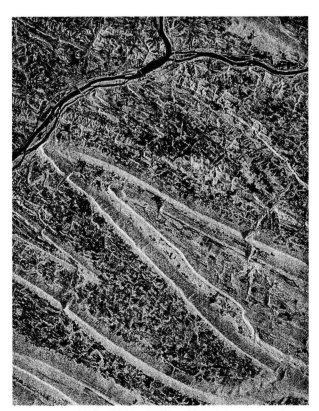

1-51 Space shuttle radar image of the Appalachian foldbelt in Pennsylvania. The more resistant strata such as sandstones form the narrow ridges (NASA PIA01306).

volcanism have produced anomalously thick oceanic crust. Other examples are the Azores and Galapagos Islands, where anomalous volcanism has produced groups of islands near an ocean ridge.

In many ways hotspot volcanics are notable for their differences rather than their similarities. We will now discuss in somewhat more detail the hotspot volcanics of Africa and western North America. The distribution of volcanic rocks in Africa that are younger than 26 million years is shown in Figure 1–53. Active volcanism is occurring throughout much of Africa. The East African rift system is a nearly linear feature extending southwest several thousand kilometers from its junction with the Red Sea and the Gulf of Aden. The rift is characterized by tensional tectonics and horizontal extension. As discussed previously, the rift may represent the first stage of continental rupture.

The East African rift is also characterized by near-circular regions of elevated topography referred to as *swells*. The relationship of these features to the rifting

process is uncertain. They may be associated with convective plumes in the mantle. An alternative hypothesis for continental rifts is that they are the direct result of tensional stresses in the lithosphere. The tensional stresses cause the continental lithosphere to rupture, leading to volcanism and uplift.

Other areas of extensive volcanism in Africa are the Tibesti area in northeast Chad and Hoggar to the west. In addition to volcanism these areas are associated with crustal swells; however, they do not appear to be associated with any linear features. The Haruj volcanics north of Tibesti are not associated with any apparent crustal elevation. Adjacent to the continental margin in the Gulf of Guinea is the Cameroon line of recent volcanics. This is a series of active volcanic centers that pass from oceanic onto continental crust. Although this is a linear chain of volcanic centers, the linear progression in time is not well defined. Farther to the north on the continental margin of Africa lie the Canary Islands. In this group of volcanic islands, volcanism has been centered for a long period.

It is evident that very diverse types of intraplate volcanism occur in Africa. It is not clear whether all this volcanism can be attributed to a single mechanism. If mantle plumes are responsible for this volcanism, how many are required? Why does the volcanism in Africa have such a variety of forms?

The western United States is another area of extensive volcanism. The distribution of recent surface volcanic rocks (with ages of less than 7 million years) is given in Figure 1–54. Because the San Andreas fault in California is recognized as a major plate boundary between the Pacific and North American plates, the volcanism of this area may be classified as being plate-margin-related; however, the volcanism extends more than 1500 km from the plate margin.

Yellowstone National Park in the northwest corner of Wyoming is the center of extensive recent volcanism that occurs at the end of a track of volcanism that extends along the Snake River plain. For this reason the Yellowstone area is classified as a hotspot (see Figure 1–14), and it is thus a possible site of a mantle plume. The ages of surface volcanic rocks on the Snake River plain are given as a function of the distance from Yellowstone in Figure 1–55. The ages of the oldest volcanic rocks tend to increase with distance from Yellowstone; however, young volcanic rocks occur

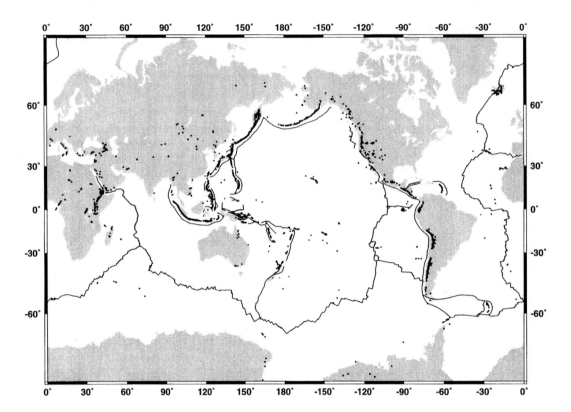

1–52 Distribution of active volcanoes in the Quaternary.

along much of the length of the Snake River plain. It is difficult to associate these young volcanics, which extend over a distance of some 500 km, with a single melting anomaly beneath Yellowstone. Also, it is clear from Figure 1–54 that very young volcanics extend throughout the western United States. Some of these volcanics form the volcanic line associated with subduction of the Juan de Fuca plate beneath Washington, Oregon, and northern California. But much volcanism remains unexplained. Small amounts of volcanism appear to be associated with the tensional tectonics of the Rio Grande rift in New Mexico and the Basin and Range province in Nevada and Arizona. Further discussion of this is given in the next section.

Variations in surface heat flow can also be correlated with the distribution of plates. On a worldwide basis the average surface heat flow is about 87 mW m^{-2}. For rocks with an average thermal conductivity this corresponds to an increase in temperature with depth of 25 K km^{-1}. The heat being lost to the surface of the Earth can be attributed to the heat produced by the decay of the radioactive isotopes and to the cooling of the Earth.

Plate margins and other areas where volcanism occurs are generally characterized by surface heat flows much higher than the average value just given. The high heat flows may be the result of a thin lithosphere or, in some cases, the migration of magma through a relatively thick lithosphere. The cooling of the oceanic lithosphere as it spreads from an oceanic ridge leads to a systematic decrease with age in the heat flux to the seafloor.

The occurrence of hot springs is also strongly correlated with volcanism. In continental areas with no volcanism, the temperatures of springs seldom exceed 293 K. Most boiling hot springs and geysers (Figure 1–56) are directly associated with the cooling of a magma body at a relatively shallow depth. The circulation of heated ground water near a cooling intrusion accelerates the solidification of the intrusion and plays an important role in the emplacement of ore deposits. Minerals dissolve in the hot water; when the water boils or is cooled, the minerals precipitate to form ore deposits. Hydrothermal circulation of seawater in the

1-53 Distribution of volcanic rocks in Africa. Dark areas are surface volcanic rocks with ages less than 26 Ma. Also shown are some of the tectonic provinces and areas of crustal doming.

oceanic crust also is believed to play a significant role in removing the heat at ocean ridges and in concentrating minerals on the seafloor and in the oceanic crust. Exploration with the deep submersible *Alvin* has provided actual observations of hot water from the crust venting directly to the ocean. Submarine thermal springs have been discovered on the Galápagos rift and the East Pacific rise crest. Spectacular submarine hot springs with exit water temperatures near 700 K have also been discovered at the latter location.

1-13 Seismicity and the State of Stress in the Lithosphere

Just as in the case of volcanoes, the occurrences of earthquakes strongly correlate with plate boundaries.

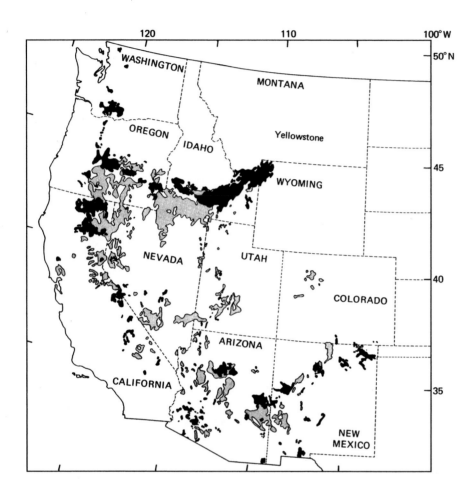

1-54 Distribution of recent volcanic rocks in the western United States. Dark areas are rocks younger than 1.5 Ma; shaded areas are rocks with ages between 1.5 and 7 Ma.

1-55 Ages of volcanic rocks in the Snake River plain as a function of the distance from Yellowstone Caldera.

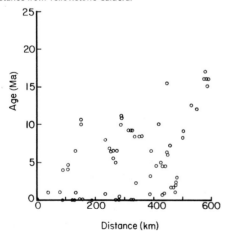

1-56 Old Faithful Geyser in Yellowstone National Park, Wyoming (J. R. Stacy 692, U.S. Geological Survey).

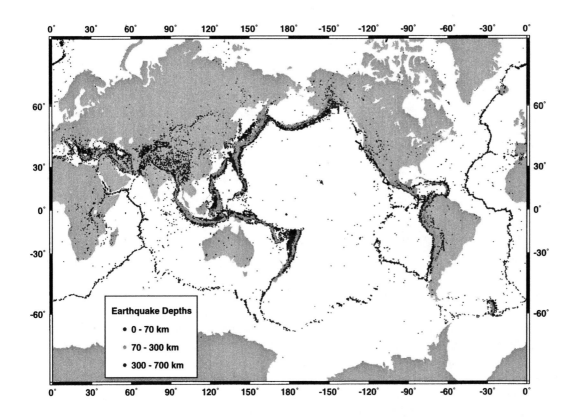

1-57 Global distribution of earthquakes with magnitudes greater than 5.1 for the period 1964–1995 (Engdahl et al., 1998).

1-58 View along the San Andreas fault in Choia Valley. Note the streams have been offset by the right-lateral displacement on the fault (R. E. Wallace 304, U.S. Geological Survey).

The worldwide distribution of seismicity is illustrated in Figure 1–57. Earthquakes occur on all types of plate boundaries; however, subduction zones and areas of continental collision are noted for their very large earthquakes. Large earthquakes also occur in plate interiors but with a much lower frequency.

Earthquakes are associated with displacements on preexisting faults. A typical displacement in a very large earthquake is 10 m. If the relative velocity across a plate boundary is 50 mm yr^{-1}, it would take 200 years to accumulate this displacement. Large earthquakes at subduction zones and major transform faults such as the San Andreas reoccur in about this period. Since regular displacements do not have to be accommodated in plate interiors, the period between major intraplate earthquakes is much longer.

The near-surface expressions of major faults are broad zones of fractured rock with a width of a kilometer or more (Figure 1–58). Smaller faults may have zones of *fault gouge* with widths of a few centimeters

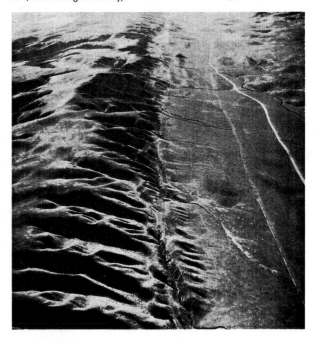

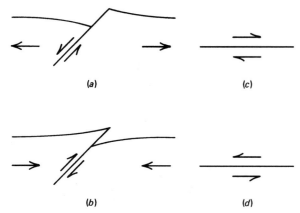

1–59 Cross sections of (*a*) a normal fault and (*b*) a thrust fault and top views of (*c*) right-lateral and (*d*) left-lateral strike-slip faults.

or less. Small faults grade down to rock fractures across which there is no offset displacement. The total offset across major faults may be hundreds of kilometers. A fault zone is a zone of weakness. When the regional stress level becomes sufficiently large, the fault ruptures and an earthquake occurs. There is extensive geological evidence that faults become reactivated. Large stresses can reactivate faults that have been inactive for tens or hundreds of millions of years.

The direction of the displacement on a fault can be used to determine the state of stress responsible for the displacement. Since voids cannot be created in the Earth's deep interior, displacements on faults are parallel to the fault surface. If a region is in a state of tensional stress, *normal faulting* will occur, as illustrated in Figure 1–59*a*. If a region is in a state of compressional stress, *thrust faulting* will occur, as illustrated in Figure 1–59*b*.

If a region is in a state of shear, *strike–slip* faulting will occur, as illustrated in Figures 1–59*c* and 1–59*d*. If, to an observer standing on one side of the fault, the motion on the other side of the fault is to the left, the fault is a *left-lateral* or *sinistral* strike–slip fault. If the motion on the other side of the fault is to the right, it is a *right-lateral* or *dextral* strike–slip fault. The displacement during many earthquakes combines the horizontal displacement associated with strike–slip faulting and the vertical displacement associated with either normal or thrust faulting.

As discussed previously, the lithosphere can transmit stress over large distances. There are several sources for the stress in the lithosphere. One source is the body forces that drive the motion of the surface plates. These include the negative buoyancy on the descending plate at a subduction zone and the gravitational sliding of a plate off an ocean ridge. Changes of temperature lead to *thermal stresses*. When the temperature increases or decreases, rock expands or contracts. The expansion or contraction can cause very large stresses. Erosion and sedimentation also cause a buildup of stress through the addition or removal of surface loads. Glaciation and deglaciation act similarly. Because the Earth is not a perfect sphere but rather a spheroid with polar flattening and an equatorial bulge, plates must deform as they change latitude. This deformation leads to *membrane stresses* in the lithosphere. Plate interactions such as continental collisions are an important source of stress. Large displacements of the cool, near-surface rocks often occur in these zones. If these deformations occur on faults, high stress levels and major earthquakes can be expected. The state of stress in the lithosphere is the result of all these factors and other contributions.

As a specific example of seismicity and stress we again turn to the western United States. The distribution of seismicity in this region is given in Figure 1–60. Also included in the figure are the relative velocities between plates and the directions of lithospheric stress inferred from displacements on faults. The Juan de Fuca plate is being formed on the Juan de Fuca ridge with a half-spreading rate of 29 mm yr^{-1}. The seismicity on a transform fault offsetting two segments of the ridge is clearly illustrated. Because the lithosphere is thin at the ridge and the rocks are hot and weak, relatively little seismicity is associated with the spreading center at the ridge crest. The Juan de Fuca plate is being subducted at a rate of about 15 mm yr^{-1} at a trench along the Oregon and Washington coasts. The seismicity in Oregon and Washington associated with this subduction is also relatively weak. This may be due to *aseismic slip* on the fault zone between the descending oceanic plate and the overlying continental lithosphere. In this case the relative displacement is accommodated without the buildup of the large stresses required for extensive seismicity. An alternative explanation is that the accumulated displacement was relieved in a great earthquake several hundred years ago and insufficient displacement has accumulated to cause high stresses. The historic record of earthquakes in

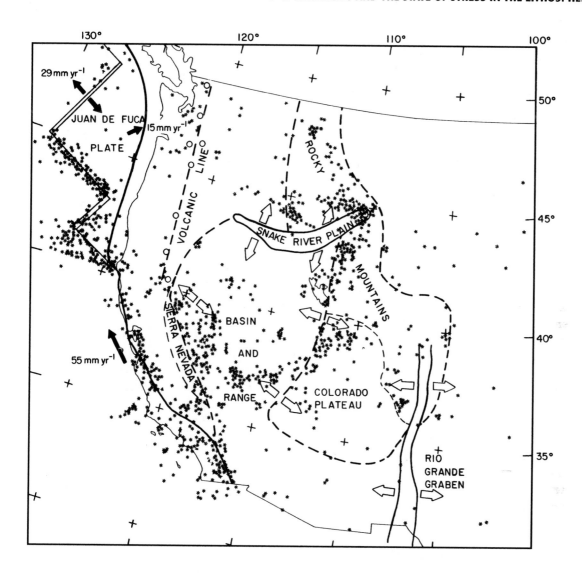

1-60 Distribution of seismicity in the geological provinces of the western United States (stars). Solid arrows give relative plate velocities; open arrows give stress directions inferred from seismic focal mechanism studies.

the western United States is relatively short, and since the subduction velocity is quite small, the recurrence period would be expected to be of the order of 500 years. Although the extensive seismicity usually associated with active subduction is absent in the Pacific Northwest, a well-defined line of active volcanoes lies parallel to the trench. The volcanoes extend from Mount Baker in Washington to Mount Shasta in northern California. These volcanoes have had violent eruptions throughout the recent geological past.

An eruption about 6000 years ago removed the upper 2 km of Mount Mazama, creating Crater Lake in Oregon. The spectacular eruption of Mount St. Helens, Washington (Figure 1–10), on May 18, 1980, blew out its entire north flank, a volume of about 6 km^3.

The velocity between the Pacific and North American plates is 47 mm yr^{-1} in California. A large fraction of this is accommodated by displacements on the San Andreas fault. In the north the fault terminates in a fault–fault–trench (FFT) triple junction with the Juan de Fuca plate. In the south the fault terminates in a series of small spreading centers (ocean ridges) extending down the Gulf of California. Along much of the fault, displacements are almost entirely right-lateral strike–slip. However, north of Los Angeles the fault bends, introducing a thrusting component. Motion on thrust faults in this region is responsible for the formation of a series of mountain ranges known as the Transverse Ranges.

A great earthquake occurred on the northern section of the San Andreas fault in 1906; the average displacement was 4 m. A great earthquake occurred on the southern section of the San Andreas fault in 1857; the average displacement was 7 m. A detailed discussion of the San Andreas fault is given in Section 8–8.

It is clear that the displacements on accreting plate margins, subduction zones, and transform faults cannot explain the entire distribution of seismicity in the western United States. Major earthquakes occur throughout the region. Rapid mountain building is associated with the Rocky Mountains and the Sierra Nevada. The Basin and Range province is a region of extensive normal faulting. The presence of many graben structures is evidence of crustal extension due to tensional stresses. The asthenosphere rises to the base of the continental crust in this region, and the lithosphere is thin and weak. Considerable volcanism occurs throughout the province. The Rio Grande rift, which marks the eastern boundary of this area of volcanism, seismicity, and mountain building, is also an extensional feature. The stress directions shown in Figure 1–60 indicate the entire western United States appears to be extending because of tensional stresses. Although there is no comprehensive understanding of this area, it is likely that the seismicity, volcanism, and mountain building are the result of complex interactions of the Pacific, North American, and Juan de Fuca plates that are deforming the entire region. It is likely that there is a geometrical incompatibility between the strike–slip motion on the San Andreas fault and the time-dependent relative displacement between the Pacific and North American plates. As a result the western part of the North American plate is being deformed.

China is another region of extensive tectonics. It is the site of extensive seismicity and mountain building. Deformation associated with the continental collision between India and Asia extends several thousands of kilometers north of the suture zone.

Seismicity can also occur in plate interiors. An example is New Madrid, Missouri, where three very large earthquakes struck in 1811 and 1812. A significant number of small earthquakes occur in this region at the present time. It should not be surprising that earthquakes occur in plate interiors, since the elastic lithosphere can transmit large stresses. These intraplate earthquakes are likely to occur where the elastic properties of the plate change and stress concentrates.

1–14 The Driving Mechanism

Plate tectonics provides a general framework for understanding the worldwide distribution of seismicity, volcanism, and mountain building. These phenomena are largely associated with plate interactions at plate margins. However, an explanation must also be given for the relative motions of the plates. The basic mechanism responsible for plate tectonics must provide the energy for the earthquakes, volcanism, and mountain building. The only source of energy of sufficient magnitude is heat from the interior of the Earth. This heat is the result of the radioactive decay of the uranium isotopes ^{238}U and ^{235}U, the thorium isotope ^{232}Th, and the potassium isotope ^{40}K as well as the cooling of the Earth. An accurate estimate of the heat lost from the interior of the Earth can be obtained from measurements of the surface heat flow. The energy associated with seismicity, volcanism, and mountain building is about 1% of the heat flow to the surface.

Heat can be converted to directed motion by *thermal convection*. Consider a horizontal fluid layer in a gravitational field that is heated from within and cooled from above. The cool fluid near the upper boundary is heavier than the hotter fluid in the layer. Buoyancy forces cause the cool fluid to sink, and it is replaced by hot rising fluid. Laboratory experiments show that under appropriate conditions two-dimensional convection cells develop, as illustrated in Figure 1–61. A thin *thermal boundary layer* of cool fluid forms adjacent to the upper boundary of the layer. Thermal boundary layers from two adjacent cells merge and separate from the upper boundary to form a *cool descending plume*. The negative buoyancy of the cool descending plume drives the flow. The thin thermal boundary layer is directly analogous to the lithosphere. The separation of the thermal boundary layers to form the cool descending plume is analogous to subduction. The buoyancy body force on the cool descending plume is analogous to the body force on the descending lithosphere.

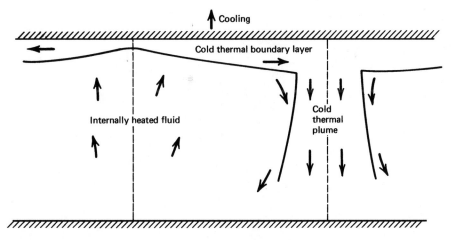

1-61 Boundary layer model for two-dimensional thermal convection in a fluid layer heated from within and cooled from above.

Ascending mantle plumes can also be associated with thermal convection. There is expected to be a hot thermal boundary layer at the base of the convecting mantle. Buoyancy forces on the low-density mantle rock would be expected to form hot ascending plumes either at the base of the mantle (the D″-layer of seismology) or at an interface in the lower mantle between an upper convecting layer and an isolated lower layer.

The fluidlike behavior of the Earth's crystalline mantle can be explained by solid-state creep processes. At low stress levels and temperatures approaching the rock solidus, the dominant creep process is the diffusion of ions and vacancies through the crystal lattice. This diffusion relieves an applied stress and results in strain. The strain rate is proportional to the stress, resulting in a *Newtonian fluid behavior*. At higher stress levels creep is the result of the movement of *dislocations* through the crystal lattice. In this case, the strain rate is proportional to the third power of the stress, resulting in a *non-Newtonian fluid behavior*. Both diffusion creep and dislocation creep are thermally activated; that is, the strain rates are inversely proportional to the exponential of the inverse absolute temperature.

The strain rate in the mantle is so small that it cannot be reproduced in the laboratory. However, extrapolations of laboratory measurements give fluid properties consistent with thermal convection in the mantle. Independent information on the fluid behavior of the mantle comes from studies of glacial loading and unloading.

When an ice sheet forms, its weight forces mantle rock to flow from beneath it, and the Earth's surface subsides. When the ice sheet melts, the mantle rock returns, and the Earth's surface rebounds. It takes about 10,000 years for the rebound to take place. Data on the rate of rebound come from dated, elevated beaches. These data have been used to obtain a viscosity for the mantle of about 10^{21} Pa s. Although this viscosity is large, it allows a fluid behavior of the mantle on geological time scales.

1-15 Comparative Planetology

Space missions have provided extensive information on the other planets and the planetary satellites of the solar system. It appears that plate tectonics is unique to the Earth. The Moon and Mercury have continuous lithospheres whose surfaces have been shaped largely by impacts and volcanic processes. Although impact cratering and volcanism have also been prevalent on Mars, its surface has also been modified by its atmosphere and the flow of a surface fluid, presumably water. Cloud-covered Venus has yielded its secrets to the eyes of Earth-based and spacecraft radar systems. Cratering and volcanism have extensively modified its surface, but there is no direct evidence of plate tectonic features such as extensive ridge or trench systems. The Galilean satellites of Jupiter have been shown to have very diverse features including very active volcanism on Io. The surface of Ganymede shows impact craters and tectonic structures resulting from dynamical processes in an underlying predominantly ice lithosphere. Callisto is a heavily cratered object about the same size as Ganymede, but there is no sign that its surface has been altered by internal activity. Europa is mainly a rocky object, somewhat smaller than the Moon, with a relatively thin outer shell of water that is ice at the surface but may be liquid at depth.

The surface of Io has been recently formed by a style of volcanism apparently unique to that body. Io is the only body in the solar system, other than the Earth, on which we have observed active volcanism; this satellite is the most volcanically active body in the solar system. Lithospheric plate evolution has destroyed much of the evidence of the early evolution of the Earth by continuously creating new surface rocks and returning old surface rocks to the mantle. The pervasive volcanism of Io has had the same effect by blanketing the surface with recently formed lavas. However, bodies such as Mercury, the Moon, Callisto, and the satellites of Mars preserve the early records of their evolutions in their cratered surfaces and thereby provide information on the early history of the solar system. Some of the physical properties of the terrestrial-like bodies of the solar system are summarized in Appendix 2, Section C.

1-16 The Moon

Early telescopic observations showed that the near side of the Moon is composed of two types of surfaces: topographically low, dark areas-referred to as *maria* (or seas) and topographically elevated, light areas known as *highlands* (Figure 1–62). The highlands are more heavily cratered and are therefore presumed to be older because the flux of meteories is known to have decreased with time. Because of its synchronous rotation with respect to the Earth, the far side of the Moon was first observed from spacecraft in lunar orbit. Highland terrain dominates the far side of the Moon; there are no extensive maria on the farside lunar surface.

The first manned landing on the Moon took place on July 20, 1969. This Apollo 11 mission as well as the subsequent Apollo 12, 15, and 17 missions landed on the lunar maria. Chemical studies of the rocks returned on these missions showed that the maria are composed of basaltic rocks similar in major element chemistry to the basalts of the oceanic crust. Radioactive dating of these rocks gives ages of 3.16 to 3.9 Ga.

1–62 Full hemispheric image of the Moon taken by the Galileo spacecraft on December 7, 1992, on its way to explore the Jupiter system in 1995–97. The distinct bright ray crater at the bottom of the image is the Tycho impact basin. The dark areas are basaltic-rock-filled impact basins: Oceanus Procellarum (on the left), Mare Imbrium (center left), Mare Serenitatis and Mare Tranquillitatis (center), and Mare Crisium (near the right edge) (NASA Image PIA00405).

The Apollo 14 and 16 missions returned samples from the lunar highlands. These rocks have a much more complex chemical history than the mare rocks. They have been extensively shocked and melted by meteorite bombardment. Detailed chemical studies have shown, however, that these rocks are highly fractionated igneous rocks. Radioactive dating of the highland rocks indicates that they crystallized about 4.5 Ga ago, close to the estimated age of the solar system.

The evolution of the Moon can be divided into three phases: (1) highlands formation, (2) mare formation, and (3) surface quiescence. The highlands of the Moon formed early in its evolution, one hypothesis being that they crystallized from a global *magma ocean*. It is difficult to date the crystallization of the highlands exactly, but it certainly terminated by 4.0 Ga and probably before 4.4 Ga. The terminal bombardment between 3.8 and 4.0 Ga resulted in the excavation of many large, deep basins. These basins, particularly on the nearside, were subsequently filled by mare basaltic volcanism. The lunar maria constitute some 17% of the surface area.

One of the major discoveries of the Apollo missions was that the Moon is made up of a variety of igneous rock types that differ widely in both their chemistry and mineral composition. The major differences between the lunar maria and highlands indicate large-scale chemical differentiation of the Moon. Early recognition of the fact that the highlands are composed mostly of plagioclase, a relatively light mineral, led to the suggestion that this mineral represents crystal flotation at the top of a deep magma ocean. The basic argument for a "magma ocean" is the need for a mechanism to float a plagioclase-rich crust, while denser minerals such as olivine and pyroxene sink. As the Moon formed, its outer portions consisted of a layer of molten silicate magma in which plagioclase floated and accumulated into the first stable lunar crust.

The solidification of the magma ocean must have occurred in about 100 Myr after the formation of the solar system because of the age of the returned lunar samples. Seismic studies carried out on the Apollo missions showed that the lunar crust has a thickness between 60 and 100 km. Seismic velocities and the mean density of the Moon indicate that the lunar mantle is composed primarily of peridotite similar to the Earth's mantle. It is hypothesized that the lunar crust represents about a 20% partial melt fraction of a primitive lunar mantle with a composition similar to pyrolite. It is expected that there is a layer of depleted mantle rock beneath the lunar crust with a thickness of about 300–500 km.

Subsequent to the solidification of the magma ocean, the morphology of the lunar surface was strongly affected by collisions with the remaining planetesimals and large meteorites. These collisions created large basins; the largest of the colliding bodies created the Imbrium basin, an event that has been dated at 3.86 Ga. A period of volcanism lasting 1 Gyr then filled the floors of these preexisting impact basins with the dark basaltic rocks that form the lunar maria. This volcanism terminated at about 3 Ga. Since then the lunar surface has remained virtually unaltered.

All of the smooth dark regions visible on the Moon's nearside consist of basaltic rocks that partly or completely fill the multiring mare basins. Nearly all of the basalts occur on the nearside. A significant time interval elapsed between the formation of a large mare basin by impact and its subsequent filling with basaltic magma

flows to form a dark lunar maria. Current information dates the Imbrium basin at 3.86 Ga, but the lavas that fill it date at about 3.3 Ga. The primary landforms resulting from lunar basaltic volcanism are vast, smooth plains, indicating low lava viscosities and high eruption rates. Major basaltic eruptions lasted a minimum of 800 million years, i.e., from 3.9 to 3.1 Ga. On the basis of low crater densities on some formations, minor eruptions could have continued until as recently as 2 Ga.

Although lunar rocks are similar to igneous rocks on the Earth, there are significant differences between the two bodies. Unlike the Earth, the Moon does not have a large iron core. The Moon may have a small iron core, but its radius is constrained by the measured values of lunar mass, radius, and moment of inertia to have a value less than about 350 km. Since the mean density of the Moon is only 3340 kg m^{-3}, the missing iron cannot be distributed through the lunar mantle. It is therefore concluded that the Moon is deficient in metallic iron relative to the Earth. The Moon also has fewer volatile elements than the Earth; there is no evidence of a significant presence of water during the evolution of the Moon.

Magnetic field measurements were made by small satellites left in lunar orbit by the Apollo 15 and 16 missions. Although localized regions of magnetized rock were detected by the subsatellites, no global lunar magnetic field could be measured. A lunar magnetic dipole moment can be no larger than 10^{16} A m^2. This is nearly seven orders of magnitude smaller than the Earth's dipole moment. The absence of a present-day global lunar magnetic field is presumably due to the absence of an active dynamo in the Moon. This may indicate that the Moon has no core; on the other hand, a small lunar core could have cooled, or solidified, sufficiently so that convective motions in it are no longer possible. It has been suggested that the localized areas of remanent lunar magnetism were magnetized in the ambient field of an ancient lunar dynamo.

The Moon is the only body other than the Earth for which we have in situ determinations of the surface heat flux. Two lunar heat flow measurements have been made, one on the Apollo 15 mission and the other on Apollo 17. The measured heat flow values are 20 mW m^{-2} and 16 mW m^{-2}. Although these two determinations may not be representative of the average lunar heat flow, the values are consistent with the Earth's surface heat loss if the differences in the sizes of the planets are accounted for.

The lunar gravity field is known quite well from the radio tracking of the many spacecraft that have been placed in lunar orbit. A map of the Moon's gravity field is shown in a later chapter (Figure 5–13). The lunar maria are sites of positive gravity anomalies, or excess concentrations of mass known as *mascons*. These surface loads appear to be supported by the lunar lithosphere, an observation that implies that the Moon's lithosphere is thicker and therefore stronger than the Earth's. The Earth's lithosphere is not thick enough to support large excess loads – mountains, for example – with the consequence that such loads tend to depress the lithosphere and subside. Since the maria were formed by 3 Ga, the Moon's lithosphere must have thickened sufficiently by then to support the mascons.

The Moon's motion about the Earth is prograde; that is, it rotates in the same sense as the rotation of the planets about the sun. In its present *prograde orbit* the tidal interactions between the Earth and the Moon cause the separation between the bodies to increase; in the past the Moon was closer to the Earth. Extrapolation of the present rate of *tidal dissipation* back in time would bring the Moon to within a few Earth radii of the Earth between 1.5 and 3 Ga. Since there is little evidence to support a close approach of the two bodies during this period, it is presumed that the rate of tidal dissipation in the past has been lower than at present. Nevertheless, it is highly likely that the Moon has been considerably closer to the Earth than it is today.

Theories for the origin of the Moon have been debated for more than a century. The classic theories claim (1) that the Moon formed as a separate planet and was captured by the Earth, (2) that the Moon was originally part of the Earth and that the Earth broke into two parts, and (3) that the Earth and moon formed as a *binary planet*. None of these theories has been able to satisfy all the major constraints on lunar origin, which include the large prograde angular momentum of the Earth–Moon system relative to the other planets and the Moon's depletion in volatile elements and iron compared with the cosmic (chondritic) abundances. Another theory proposes that the Moon formed by accreting from a disc of ejecta orbiting the Earth after the impact of a Mars-size body with the Earth. The giant impact origin of the Moon has gained widespread support because it does not violate any of the major observational constraints on lunar origin. One of the major consequences of the giant impact hypothesis of lunar origin is a hot, partially molten (or perhaps completely molten) Moon upon accretion from the circumterrestrial ejecta disk.

1–17 Mercury

Although it is the smallest of the terrestrial planets, Mercury is the densest (Appendix 2, Section C). If the planet has the cosmic abundance of heavy elements, then its large density requires that Mercury is 60% to 70% Fe by mass. With the iron concentrated in a central core, Mercury could best be described as a ball of iron surrounded by a thin silicate shell.

In photographs obtained by the Mariner 10 spacecraft during 1974 and 1975 (Figure 1–63), portions of Mercury's surface strongly resemble the heavily cratered lunar highlands. In addition, there are large areas of relatively smooth terrain and a number of *ringed basins* believed to be impact structures. The largest of these is the 1300-km-diameter Caloris basin,

1–63 Hemispheric image of Mercury acquired by the Mariner 10 spacecraft on March 24, 1974 (NASA Image PIA00437).

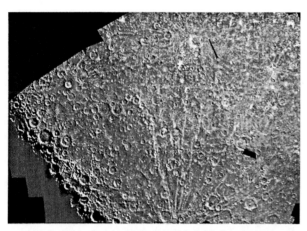

1–64 Photomosaic of Mariner 10 images of the Michelangelo Quadrangle H-12 on Mercury. In addition to the numerous impact craters, ejecta deposits are seen as bright lines or rays radiating outward from some young craters. Several large lobate scarps are visible in the lower left side of the image (NASA Image PIA02237).

which is similar to the Imbrium and Orientale basins on the Moon. The Caloris basin is covered with a relatively smooth plains type of material, perhaps similar to the lunar maria, having many fewer craters than the heavily cratered terrain. Areas of relatively smooth terrain known as *intercrater plains* are also found interspaced between the basins and craters. *Lobate scarps*, probably curved fault scarps, which are several kilometers high and extend for hundreds of kilometers across Mercury's surface, have no lunar counterpart (Figure 1–64). These scarps are suggestive of thrust faults resulting from crustal shortening and compression.

Several hypotheses have been advanced to explain the compressional surface features on Mercury. The first hypothesis concerns tidal despining. Early in its evolution Mercury may have had a rapid rotation. If the planet was hot it would have had a near hydrostatic shape with considerable polar flattening and an equatorial bulge. As the planet cooled, a global lithosphere developed with considerable rigidity and ellipticity. However, tidal interactions with the sun gradually slowed the rotation of the planet. The rigidity of the lithosphere preserved a fossil ellipticity associated with the early rapid rotation but as a result large lithospheric stresses developed. The resultant compressional stresses in the equatorial region are one explanation for the observed compressional features. An alterative explanation is that they were caused by the

formation and/or solidification of the large iron core on Mercury. Mercury's high mean density of 5440 kg m^{-3}, almost equal to the Earth's, is attributed to a large iron core with a 500 to 600 km thick cover of silicate rocks. One explanation for the high mean density is that a massive collision blasted off a large fraction of an early mantle of larger size.

Magnetic field measurements by Mariner 10 showed that Mercury has an intrinsic global magnetic field. Because of the limited amount of data, there are large uncertainties in the inferred value of Mercury's magnetic dipole moment. Most estimates lie in the range of 2 to $5 \times 10^{19} \text{ A m}^2$, or about 5×10^{-4} of the Earth's magnetic field strength. Although a magnetized crust cannot be ruled out as a source of this field, it seems more likely that it originates by dynamo action in a liquid part of Mercury's core.

Because of the similarities in the surfaces of Mercury and the Moon, their evolutions must have been similar in several respects. Separation of the iron and silicates in Mercury and crustal differentiation must have occurred very early in its history because the planet's surface preserves an ancient record of heavy bombardment similar to the lunar highlands. The filling of the Caloris basin must have occurred subsequent to the termination of this severe cratering phase because the basin material is relatively free of craters. The lobate scarps must also have formed at the end of or subsequent to the early phase of severe bombardment because they sometimes pass through and deform old craters (Figure 1–64). The scarps may be a consequence of the cooling and contraction of the core, and if so, they are the only surface features that distinguish Mercury with its large core from the Moon with only a very small core or none at all.

1–18 Mars

The first detailed photographs of the Martian surface were returned from the flybys of Mariner 4 (1965) and Mariners 6 and 7 (1969). These photographs showed a heavily cratered surface much like that of the Moon. However, the Mariner 9 (1971) photographs revealed that the earlier spacecraft had photographed only a single type of terrain on a planet of great geological diversity. There are volcanoes that dwarf the largest volcanic structures on Earth, a huge canyon complex comparable to the East African rift system, meandering channels with multiple braided features and stream-lined islands, sand dunes, and polar caps. The richness and variety of Martian geologic forms was not fully re-alized prior to the pictures returned by the Viking 1 and 2 orbiters and landers (1976) and the Mars orbiter camera aboard the Mars Global Surveyor (1999). The surface of Mars is characterized by a wide variety of volcanic and tectonic landforms. However, there is no evidence of the global system of ridges and trenches that are characteristic of active plate tectonics. Thus, it is concluded that Mars does not have active plate tectonics.

The most striking global feature of the Martian sur-face is its hemispheric asymmetry. Much of the south-ern hemisphere of Mars is covered by densely cratered highlands, whereas most of the northern hemisphere is made up of lightly cratered plains. The heavily cratered terrain in the southern hemisphere is probably the rem-nant of the postaccretionary surface of the planet, and the younger northern plains are probably volcanic in origin.

The southern highlands cover more than 60% of the surface of Mars. Most of the highland terrain consists of ancient densely cratered rock (largely impact breccias) formed early in the planet's history when impact rates were high. Extensive lava flows have covered large areas within the highlands. The large, roughly circular basins of Argyre and Hellas (Figure 1–65) are located in the southern hemisphere and are generally believed to be impact basins similar to the mare basins on the Moon. The Hellas basin has a rim diameter of ~2300 km and is one of the largest impact structures in the solar system. It is the dominant surface feature of the Martian southern highlands. The Argyre basin has a diameter in excess of 1500 km. Volcanic plains cover much of the northern lowlands of Mars. These volcanic plains are similar to the volcanic plains that dominate other planetary surfaces, e.g., Venus, and they are much less cratered than the southern highlands.

The approximately hemispherical dichotomy is gen-erally held to be an ancient, first-order feature of the Martian crust. The dichotomy has been ascribed vari-ously to a very long-wavelength mantle convective planform, to subcrustal erosion due to mantle convec-tion, to post-accretional core formation, to one large impact, and to several impact events. Sleep (1994) has proposed that the lowland crust was formed in an episode of seafloor spreading on Mars. He hypothe-sized a hemispheric subduction event that destroyed the original primitive crust in the northern hemisphere, and proposed a well-defined sequence of seafloor-spreading events that created the northern volcanic plains.

One of the major volcanotectonic features on Mars is the Tharsis region, which is a large elevated region composed of relatively young volcanics. The horizontal scale is some 3000 km with the elevation rising about 10 km above the mean surface elevation. The region ex-hibits a complex history of episodic tectonism, closely associated with volcanism on local and regional scales. The entire Tharsis uplift appears to be the result of extensive volcanism.

Three immense volcanic shields (Arsia, Pavonis, and Ascraeus Montes) form the Tharsis Montes, a linear chain of volcanoes extending northeastward across the Tharsis rise (Figures 1–65 and 1–66). These three shields have gentle slopes of a few degrees (the upper slopes are commonly steeper than the lower slopes), wide calderas, and flank vents. The shields appear to be the result of basaltic flows and are similar to the intraplate shield volcanoes of the Hawaiian Islands. The Martian shield volcanoes rise 10 to 18 km above the Tharsis rise and attain elevations of 18 to 26 km. Along the Tharsis axial trend, volcanoes stretch from Arsia Mons to near Tempe Patera, some 4000 km. Lava flows that were erupted from the Tharsis Montes and surrounding vents cover nearly 7×10^6 km^2.

Olympus Mons (Figures 1–65 and 1–66) is a shield volcano nearly 600 km in diameter and over 26 km high, the tallest mountain in the solar system. Flows on the flanks of the volcano and adjacent volcanic plains that were erupted from fissures east of the volcano are among the youngest flows on Mars. The extreme height of the Martian volcanoes can be attributed to the low surface gravity and the lack of relative motion between the lithosphere and the magma source. The presence of shield volcanoes on Mars and their absence on the Moon can be attributed to differences in the viscosities of the erupted lavas. A significant gravity anomaly is associated with the Tharsis uplift. This gravity anomaly can be explained if the volcanic construct is partially supported by the elastic lithosphere on Mars. Because Mars is smaller than the Earth, it would be expected to cool more efficiently – it has a larger surface area

1–65 Composite images of the two hemispheres of Mars. Upper left is the "eastern" hemisphere. The hemispheric dichotomy between the young, smooth, low-lying northern plains and the heavily cratered, old, southern highlands is clearly illustrated. The dark circular Hellas basin in the south is accepted to be an impact structure. Lower right is the "western" hemisphere. The three giant shield volcanoes that form the linear Tharsis Montes chain lie near the equator. Olympus Mons, the tallest mountain in the solar system, lies to the northwest of this chain. To the east the Valles Marineris canyon system is seen (NASA Image PIA02040).

to volume ratio – and has a thicker lithosphere, other factors being the same. This additional thickness and the smaller radius give the elastic lithosphere on Mars a much greater rigidity.

Another major tectonic feature on Mars is an enormous canyon system, Valles Marineris, extending eastward from Tharsis for about 4500 km (Figure 1–67; see also Figure 1–68). Individual canyons are up to 200 km wide and several kilometers deep. In the central section (Figure 1–68), the system is about 600 km wide and over 7 km deep. The Valles Marineris system might be a complex set of fractures in the Martian crust caused by the large topographic bulge containing the Tharsis volcanic region. The system is roughly radial to this bulge, as are other prominent fractures.

Numerous channels are widely distributed over the Martian surface. They display a variety of morphologic forms, including braiding and stream-lined islands, strongly suggestive of formation by flowing water (Figure 1–69). If water did flow on the surface of Mars some time in the past, the water may have originated by the melting of subsurface ice. This is supported by the association of the apparent sources of many

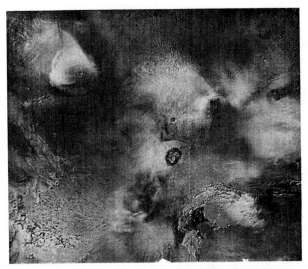

1–66 Image mosaic of the shield volcanoes in the Tharsis region of Mars obtained on a single Martian afternoon by the Mars orbiter camera on board the Mars Global Surveyor. Olympus Mons is the large shield in the upper left corner. Arsia Mons, Pavonis Mons, and Ascraeus Mons lie on a line trending SW–NE near the center of the image. The great canyon system, Valles Marineris, can be seen in the lower right corner (NASA Image PIA02049).

channels with so-called *chaotic terrain*: areas of large, irregular blocks probably formed by collapse following the removal of a subsurface material such as *ground ice*. Chaotic terrain is visible in the photomosaic in Figure 1–67 at the eastern end of the Vallis Marineris system; a broad collapsed area extends outward from Capri Chasma. Many of the north-trending channels in this area appear to originate in this chaotic terrain. Martian

1–67 Mars Global Surveyor image of the Valles Marineris canyon system on Mars (NASA Image PIA00422).

channels give the impression of having been formed by episodic flooding of large areas, as might be expected from the sudden release of large amounts of subsurface water. Possible terrestrial analogs to these channels are the scablands of the Columbia plateau in the United States and the Sandur plains in Iceland, both of which formed by the sudden release of large quantities of glacial meltwater. The existence of a Martian ground ice is also indicated by the unusual forms of some crater ejecta. Figure 1–70 shows a *lobate ejecta* flow surrounding an impact crater on Mars. The unique appearance of the ejecta pattern suggests the incorporation of large amounts of water into the ejecta, as would occur if the impact penetrated a ground-ice-rich subsurface. In addition to the small amount of water currently present in the thin CO_2 atmosphere of Mars, the planet presently contains water in the form of ice in its permanent or residual polar caps, which underlie the seasonal CO_2 ice caps.

Although processes associated with liquid flow may have been active only in the past, the present surface of Mars is being actively modified by *atmospheric erosion and deposition*. It is hardly surprising, in view of the perennial dust storms that blanket the planet, that windblown sand effectively alters the present surface of Mars. Figure 1–71 is a photograph of a large *dune field* on Mars. Winds are an effective means of transporting material over the Martian surface; there are layered deposits in the polar regions that are believed to be accumulations of material carried by the atmosphere from other regions of the planet.

The mean density of 3950 kg m^{-3} and the relatively small moment of inertia of Mars are evidence that Mars

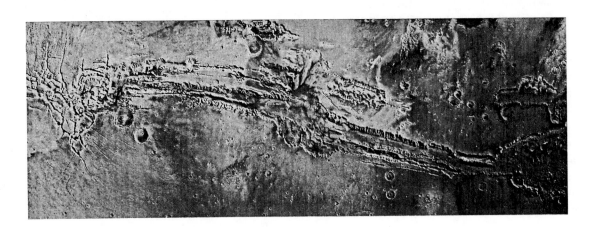

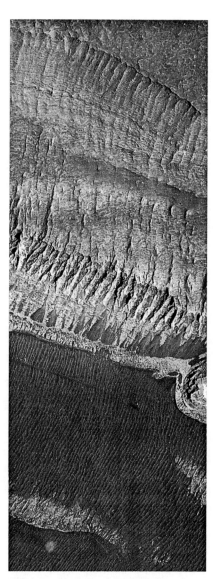

1–68 Mars Global Surveyor image showing the layered structure in the walls of a mesa in southern Melas Chasma in Valles Marineris. This image covers an area 3 km wide and 8.2 km long. Erosion by landslides has exposed tens of layers several meters in thickness and has created the dark fan-shaped deposits seen near the center of the image. The floor of the canyon is dark and is covered by many parallel ridges and grooves (lower third of the image) (NASA Image PIA02398).

a global intrinsic magnetic field. Early magnetic field measurements from the Mars 2, 3, and 5 spacecraft were interpreted to imply that Mars had a small magnetic field with a dipole moment 3×10^{-4} times the Earth's magnetic dipole moment. Data from the Mars Global Surveyor have settled the question of the existence of a global Martian magnetic field – there is none,

1–69 Mars Global Surveyor image showing a branching valley "network" in an ancient cratered terrain. This image covers an area of 11.5 by 27.4 km. The eroded valleys are bright and taken as evidence that Mars had liquid water running across its surface about 4 billion years ago (NASA Image PIA01499).

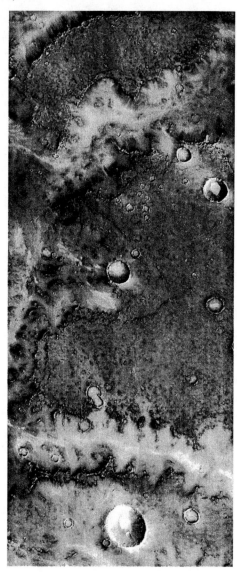

has a metallic core. The size of the core depends on assumptions about its composition, whether the core is Fe–FeS, for example; model values of core radius vary between 0.4 and 0.6 of the Martian radius. Even though Mars has a metallic core, it does not have

1-70 This Mars Global Surveyor image shows two small impact craters with dark ejecta deposits that were blown out of the craters during the impacts. The large crater has a diameter of about 89 m and the smaller crater about 36 m. The ejecta is darker than the surrounding substrate because the impacts broke through the upper, brighter surface material and penetrated to a layer of darker material beneath (NASA Image PIA01683).

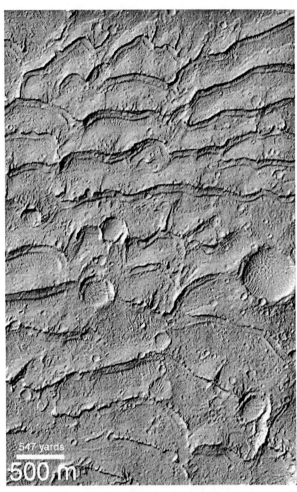

1-71 Mars Global Surveyor image of a sand dune field. The number of impact craters in the image indicate that the dunes are quite ancient (NASA Image PIA02359).

but the crust of Mars has strong concentrations of remanent magnetism implying that Mars had a global magnetic field in the past. Crustal magnetization on Mars is mainly confined to the ancient highlands of the southern hemisphere and it is largely organized into east-west-trending linear features of alternating polarity extending over distances as large as 2000 km. The magnetization features are reminiscent of the magnetic stripes on the Earth's seafloor, and suggest the possibility of a plate tectonic regime with seafloor spreading early in the history of Mars. The absence of crustal magnetism near large impact basins such as Hellas and Argyre implies that the early Martian dynamo ceased to operate before about 4 Ga. The major evidence for an initially hot and differentiated Mars is the acceptance of Mars as a parent body of the SNC meteorites. This is a class of meteorites found on Earth that

apparently escaped from the Martian gravity field after one or more large impacts. The radiometric ages for SNC meteorites are about 4.6 Ga, the U/Pb isotopic composition of SNC meteorites require core formation at about 4.6 Ga, and the old age (≥ 4 Ga) of the southern highlands suggests early crustal differentiation. Other evidence for a hot early Mars includes water-carved features on the Martian surface suggesting early outgassing and an early atmosphere.

1-19 Phobos and Deimos

The two satellites of Mars, Phobos and Deimos, are very small, irregularly shaped objects. Little was known of these bodies until the Mariner 9, Viking, and the Mars Global Surveyor missions provided detailed

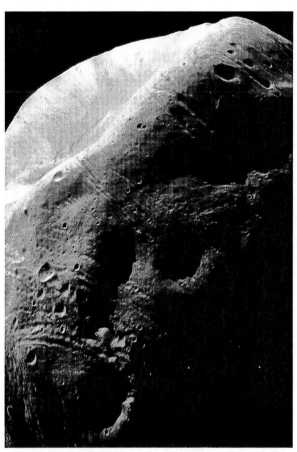

1-72 This image of Phobos, the inner and larger of the two moons of Mars, was taken by the Mars Global Surveyor on August 19, 1998. It shows a close-up of the largest crater on Phobos, Stickney, 10 km in diameter, nearly half the size of the entire body. Crossing at and near the rim of Stickney are shallow, elongated depressions which may be fractures that resulted from the impact that created Stickney (NASA Image PIA01333).

photographs of them (Figure 1–72). Roughly speaking, they are triaxial ellipsoids about 10 to 20 km across. Their surfaces are heavily cratered, but not identical in appearance. In particular, Phobos has a system of long linear depressions or grooves each of which is about 100 to 200 m wide and 10 to 20 m deep (Figure 1–72). There are no grooves on Deimos. The grooves on Phobos are probably related to fractures caused by a large impact, perhaps the one responsible for the Stickney crater (Figure 1–72). There are no craters on Deimos as large as Stickney; this may explain the absence of grooves on Deimos. The low mean density of Phobos, 2000 kg m^{-3}, and its *reflectance spectrum* suggest that it is made of a dark gray material similar to Types I or II *carbonaceous chondrite* meteorites.

1-20 Venus

In terms of size and density Venus is the planet that is most similar to the Earth. An obvious working hypothesis would be that the tectonics of Venus would be similar to the tectonics of the Earth and there would be plate tectonics. We now know that this is not the case and that mantle convection on Venus has a different surface expression than mantle convection on the Earth.

The cloud cover on Venus has prevented optical observations. However, Pioneer Venus radar, Earth-based radar observations, Venera 15–16 orbital imaging radar, and the Magellan radar images have provided clear views of the surface of Venus. These views, along with topography and gravity data, indicate that Earth and Venus are in fact quite different. On Earth the global oceanic rift system and the arcuate ocean trenches are the primary surface manifestations of plate tectonics. The almost total absence of these features on Venus has led to the conclusion that active plate tectonics is not occurring on that planet at this time. Clearly, any comprehensive understanding of tectonism and volcanism on Venus requires an understanding of how heat is transported in the absence of plate tectonics.

There are other ways in which Venus is strikingly different. It rotates in a retrograde sense with a period of 243 days; a Venusian day is 117 Earth days. Venus has a massive, mostly CO_2 atmosphere with a surface temperature of about 750 K and a surface pressure of nearly 10 MPa. Sulfuric acid clouds blanket the planet and prevent us from directly viewing the surface. Because of its earthlike size and mass, Venus most likely has a metallic core similar to Earth's. However, magnetic field measurements by the Pioneer Venus orbiter during 1979 and 1980 revealed that Venus does not have an intrinsic global magnetic field; these observations determined that if Venus had an intrinsic magnetic field, its dipole moment would have to be much less than 10^{19} A m^2.

Studies of the surface of Venus during the Magellan mission have provided a wealth of data on its tectonic and volcanic processes. The radar images of the surface are complemented by global topography and gravity data. The planet is remarkably smooth; 64% of the surface comprises a plains province with elevation differences of 2 km or less; highland areas stand as much as 10 km above the plains but they constitute only

about 5% of the surface; lowlands are 2 to 3 km below the plains and occupy the remaining 31% of the surface. Figure 1–73 shows the distribution of these topographic provinces. Although local elevation extremes on Venus and Earth are roughly comparable, global topographic variations are much smaller on Venus; the planet's surface is at a nearly uniform level.

There are tectonic features on Venus that resemble major tectonic features on the Earth. Beta Regio has many of the features of a continental rift on Earth. It has a domal structure with a diameter of about 2000 km and a swell amplitude of about 2 km. It has a well-defined central rift valley with a depth of 1–2 km and there is some evidence of a three-armed planform (aulacogen). It is dominated by two shieldlike features, Theia Mons and Rhea Mons, which rise about 4 km above the mean level. The U.S.S.R. Venera 9 and 10 spacecraft, which landed just east of Beta Regio, measured a basaltic composition and a density of 2800 kg m^{-3} for the surface rocks. These observations substantiate the identification of Theia Mons and Rhea Mons as shield volcanoes. Atla, Eistla, and Bell Regiones have rift zone characteristics similar to Beta Regio.

Most of the highlands on Venus are concentrated into two main continental-sized areas: Ishtar Terra, the size of Australia, in the northern hemisphere, and Aphrodite Terra, about the size of Africa, near the

1–73 Composite radar images of the two hemispheres of Venus. The left hemispheric view is centered at 0°E longitude. The light region near the north pole is Maxwell Montes, the highest region on Venus. The circular structure near the center is Heng-o Corona. The light stippled region south of this is Alpha Regio. The right hemispheric view is centered at 180°E longitude. The bright equatorial region just south of the equator on the left is Aphrodite Terra. The large circular feature just south of this is Artemis Corona (NASA Image PIA00157).

equator (Figures 1–73 and 1–74). Aphrodite Terra, with a length of some 1500 km, is reminiscent of major continental collision zones on Earth, such as the mountain belt that extends from the Alps to the Himalayas. Ishtar Terra is a region of elevated topography with a horizontal scale of 2000–3000 km. A major feature is Lakshmi Planum which is an elevated plateau similar to Tibet with a mean elevation of about 4 km. This plateau is surrounded by linear mountain belts. Akna, Danu, Freyja, and Maxwell Montes, reaching elevations of 10 km, are similar in scale and elevation to the Himalayas.

The gravitational anomalies associated with topographic planetary features further constrain their origin. Gravity anomalies obtained from tracking Pioneer Venus provided further major surprises. Unlike on the Earth, gravity anomalies correlate with high topography on Venus. Large positive gravity anomalies are directly associated with Beta Regio and eastern Aphrodite Terra.

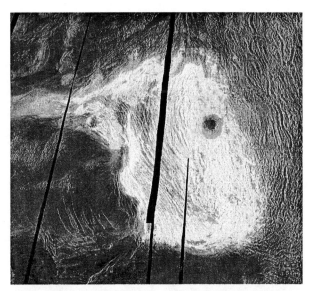

1-74 Magellan image of Maxwell Montes, the highest mountain on Venus, rising almost 11 km above the mean planetary radius. The prominent circular structure is Cleopatra, an impact basin with a diameter of about 100 km and a depth of 2.5 km (NASA Image PIA00149).

One of the most important observational constraints on the geodynamics of Venus comes from studies of impact cratering on the surface (Figure 1–75). Some 840 impact craters have been identified from Magellan images with diameters ranging from 2 to 280 km. The

1-75 Magellan radar image of three large impact craters in the Lavinia region of Venus. The craters range in diameter from 37 km to 50 km. The bright areas are rough (radar-bright) ejecta (NASA Image PIA00086).

distribution of craters on Venus cannot be distinguished from a random distribution. Unlike the Moon and Mars, older and younger terrains cannot be identified. The surface of Venus appears to be of a near-uniform age. Correlations of this impact flux with craters on the Moon, the Earth, and Mars indicate a mean surface age of 0.5 ± 0.3 Ga. Another important observation is that 52% of the craters are slightly fractured and only 4.5% are embayed by lava flows. These observations led Schaber et al. (1992) to hypothesize that a global volcanic resurfacing event had occurred at about 500 Ma and that relatively little surface volcanism has occurred since. Further statistical tests have shown that a large fraction of the surface of Venus ($\approx$80–90%) was covered by fresh volcanic flows during a period of 10–50 Myr. It is well established that the geologic evolution of Venus is far more catastrophic than the Earth's.

Other major features unique to Venus are the coronae. These are quasi-circular features, 100–2600 km in diameter, with raised interiors and elevated rims, often with annular troughs. It has been suggested that the perimeters of several large coronae on Venus, specifically Artemis (Figure 1–76), Latona, and Eithinoha, resemble terrestrial subduction zones in both planform and topography. Artemis chasma has a radius of curvature similar to that of the South Sandwich subduction zone on the Earth. Large coronae may be incipient circular subduction zones. The foundering lithosphere is replaced by ascending hot mantle in a manner similar to back-arc spreading on the Earth.

1-21 The Galilean Satellites

The innermost satellites of Jupiter, in order of distance from the planet, are Amalthea, Io, Europa, Ganymede, and Callisto. The latter four were discovered by Galileo in 1610 and are collectively referred to as the *Galilean satellites*. Amalthea was discovered by Barnard in 1892. They all have nearly circular prograde orbits lying almost exactly in Jupiter's equatorial plane. Our knowledge of the Galilean satellites increased considerably as a consequence of the flybys of Voyagers 1 and 2 on March 5, 1979 and July 9, 1979, respectively, and the Galileo mission (1995–2000) has yielded a further enormous jump in our knowledge of these bodies. We now know as much about the surfaces and interiors of the Galilean satellites as we do about some of the terrestrial

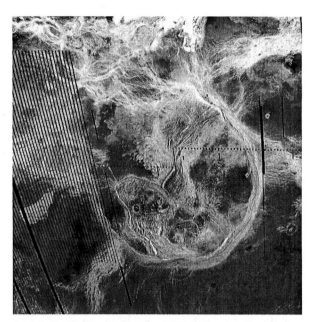

1–76 Composite Magellan radar image of Artemis corona. The near circular trough of the Artemis chasma has a diameter of 2100 km. The concentric features outside the chasma are attributed to normal faulting associated with lithospheric flexure similar to that occurring seaward of subduction zones on the Earth. The geometry of Artemis corona is generally similar to the Aleutian island arc and has been associated with an aborted subduction zone (NASA Image PIA00101).

1–77 High-resolution image of Jupiter's moon Io acquired by the Galileo spacecraft on July 3, 1999. The surface is covered by volcanic centers, many of them active (NASA Image PIA02308).

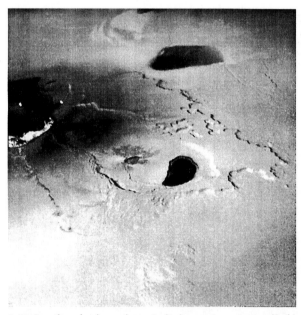

1–78 An active volcanic eruption on Jupiter's moon Io was captured in this image taken on February 22, 2000, by the Galileo spacecraft. This picture is about 250 km across. The eruption is occurring at Tvashtar Catena, a chain of giant volcanic calderas (NASA Image PIA02550).

planets in our inner solar system. These satellites are very different from one another and from the terrestrial planets; many of the physical processes occurring in their interiors and on their surfaces are unique to these bodies.

From Appendix 2, Section D, it can be seen that Io is only slightly larger and more massive than the Moon. Its similarity to the Moon extends no further, however; the Voyager and Galileo missions showed Io to be the most volcanically active body in the solar system. During the flybys of both the Voyager and Galileo spacecraft numerous active volcanic plumes were observed, some extending to heights of hundreds of kilometers above the surface. Io (Figures 1–77 and 1–78) displays a great diversity of color and *albedo*; *spectral reflectance* data suggest that its surface is dominated by sulfur-rich materials that account for the variety of colors – orange, red, white, black, and brown. Io's volcanism is predominantly silicate-based as on the terrestrial planets though sulfur-based volcanism

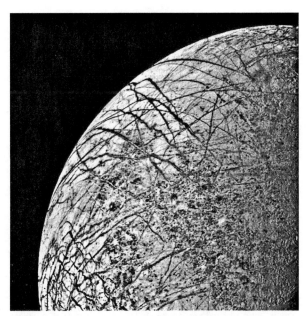

1-79 Near hemispheric image of Jupiter's satellite Europa taken by the Voyager 2 spacecraft on July 9, 1979. The linear crack-like features are clearly illustrated as well as the darker mottled regions (NASA Image PIA01523).

also occurs. The circular features on Io (Figure 1–77) are caldera-like depressions (Figure 1–78); some have diameters as large as 200 km. There are no recognizable impact craters on the satellite, although the flux of impacting objects in the early Jovian system is believed to be as large as it was around the terrestrial planets. Io's surface is geologically very young, the silicate and sulfur lavas having only recently resurfaced the planet. Relatively few of Io's calderas are associated with structures of significant positive relief. Thus they are quite unlike the calderas of the Hawaiian volcanoes or the Tharsis volcanoes on Mars. There are isolated mountains with considerable relief on Io (~10 km), but their exact height as well as their mode of origin is uncertain.

The source of heat for Io's volcanism is *tidal dissipation*. The gravitational interaction of Io with Europa and Ganymede forces Io into an orbit with higher eccentricity than it would have if it were circling Jupiter by itself. The resulting tidal flexing of Io in Jupiter's gravity field dissipates very large quantities of heat; Io's surface heat flow exceeds the global heat loss from the Earth by a factor of 3 or more. Tidal dissipation is insignificant as

a heat source for the terrestrial planets that are heated mainly by the decay of radioactive elements. However, the special circumstances of Io's orbit around a massive planet makes tidal heating an unusually effective heat source for Io.

Io's density and moment of inertia constrain its internal structure. The satellite has a large metallic core whose exact size is uncertain because we do not know the composition of the core. The core radius could be as large as about half of Io's radius and is surrounded by a silicate mantle. Io's extensive volcanism suggests that the satellite has a crust and a partially molten asthenosphere. Much of the tidal dissipative heating probably occurs in this asthenosphere. Io is known to be in hydrostatic equilibrium under the action of the Jovian tidal forces and its rotation. It is not known if Io has its own magnetic field.

Europa is only slightly smaller and less massive than the Moon (Appendix 2, Section D), but it also looks quite different from our satellite. Figure 1–79 is a Voyager 2 picture of Europa that shows the surface to consist of two major terrain types: a uniformly bright terrain crossed by numerous dark linear markings and a somewhat darker mottled terrain. Relatively few impact craters exist on Europa indicating that the surface is geologically young. The linear markings are ridges and fractures; they have little or no vertical relief. They extend over distances as large as thousands of kilometers and vary in width from several kilometers to about 100 km. Europa's density and moment of inertia indicate that, although it is composed mainly of silicates, it must contain a large fraction (about 20% by mass) of water. The water is believed to be in a surface layer about 100 km thick surrounding a silicate mantle and metallic core. The water layer may be completely frozen or it may consist of ice above liquid. Infrared spectra of Europa and its high albedo indicate that the surface is covered with water ice or frost. High-resolution Galileo pictures show features such as ice rafts that have rotated and separated from each other over an underlying soft ice layer or an internal liquid ocean. The relative absence of craters on Europa may have resulted from the freezing of a competent ice layer only after the termination of the early phase of severe bombardment or it may be due to geologically recent resurfacing of the satellite; the global fracture pattern may be a consequence of tidal stresses and nonsynchronous rotation of Europa's outer shell of ice. The surfaces of Europa and, as we shall see, Ganymede and Callisto are shaped by processes occurring in a predominantly ice shell. Although large ice-covered regions of the Earth give us some clues about what surface features to expect, the icy Galilean satellites provide a unique example of surfaces shaped by global-scale ice tectonic processes at extremely low temperatures (the surface temperatures of the Galilean satellites are about 150 K). The geologist studying Io must be mainly a volcanologist; the geologist investigating Europa, Ganymede, and Callisto, on the other hand, must be mainly a glaciologist! If there is an internal ocean on Europa, the satellite must then be considered a possible site for extra-terrestrial life. Some tidal heating of Europa is necessary to prevent the freezing of an internal liquid water ocean.

Ganymede and Callisto, the icy Galilean satellites, are about the size of Mercury (Appendix 2, Sections C and D). Their low mean densities, less than $2000\,kg\,m^{-3}$, indicate that they are composed of silicates with very substantial amounts of water. The fraction of water contained in these bodies depends on the density of the silicates; as much as 50% of the satellites could be water. Multiple flybys of Ganymede and Callisto by the Galileo spacecraft have provided us with knowledge of the satellites' densities and moments of inertia from which we can infer the internal structures of the bodies. Ganymede is fully differentiated into a metallic core, silicate mantle, and thick (many hundreds of kilometers) outer ice shell. In contrast, Callisto is only partially differentiated. Most of the satellite consists of a primordial ice–rock mixture; only in the outer few hundred kilometers can the ice and rock have separated.

The Galileo spacecraft found that Ganymede has its own magnetic field while Callisto does not. Ganymede is the only moon in our solar system known to have an intrinsic global magnetic field at present. Ganymede's magnetic field is so large that the field must be generated by dynamo action in a liquid metallic core just as a dynamo in the Earth's outer core produces our magnetic field. Observations of Ganymede by the Galileo spacecraft provide strong support for the complete differentiation of the satellite and for the existence of a liquid metallic core in its interior.

A major unsolved question is why Ganymede is fully differentiated while Callisto is only slightly differentiated; both bodies are about the same size and are made up of about the same proportions of ice and rock. One possible explanation is that Ganymede was tidally heated in the past whereas Callisto was not. The appearances of Ganymede and Callisto are consistent with a differentiated interior for Ganymede and a relatively primordial interior for Callisto.

Unlike Europa and Io, Ganymede has numerous impact craters. Two major terrain types have been identified: relatively dark heavily cratered terrain and lighter grooved terrain. The former has a crater density comparable with that of the lunar highlands and other ancient cratered surfaces of the inner planets. Although the grooved terrain contains fewer craters, it nonetheless has a crater density comparable with the oldest lunar maria and Martian plains units. Bright-rayed impact craters are abundant on both types of terrain. Almost all the grooved terrain is a mosaic of sets of grooves; groove systems are 10 to 100 km wide and 10 to 1000 km long. Individual grooves are a few hundred meters deep. The craters on Ganymede display a variety of morphologic forms. Craters several hundred kilometers in diameter are found only as subdued scars on the oldest parts of Ganymede's surface. These presumably degraded impact craters appear today as circular bright patches without rims or central depressions; they have been described as *crater palimpsests*. Creep in a predominantly ice surface is probably responsible for the severe degradation of the large craters. Craters in the younger grooved terrain are generally better preserved that those in the older heavily cratered areas. There is no major relief on Ganymede; that is, there are no large mountains or basins. Galileo observations of Ganymede confirm that tectonism has been a major factor in shaping the satellite's surface. Tectonic activity on Ganymede is in accord with its differentiated interior.

REFERENCES

Crough, S. T., and D. M. Jurdy (1980), Subducted lithosphere, hotspots, and the geoid, *Earth Planet. Sci. Let.* **48**, 15–22.

DeMets, C., R. G. Gordon, D. F. Argus, and S. Stein (1994), Effect of recent revisions to the geomagnetic reversal time scale on estimates of current plate motions, *Geophys. Res. Let.* **21**, 2191–2194.

du Toit, A. (1937), *Our Wandering Continents* (Oliver and Boyd, Edinburgh).

Engdahl, E. R., R. van der Hilst, and R. Buland (1998), Global teleseismic earthquake relocation with improved travel times and procedures for depth determination, *Bull. Seis. Soc. Am.* **88**, 722–743.

Gordon, R. B. (1965), Diffusion creep in the Earth's mantle, *J. Geophys. Res.* **70**, 2413–2418.

Harland, W. B., R. L. Armstrong, A. V. Cox, L. E. Craig, A. G. Smith, and D. G. Smith (1990). *A Geologic Time Scale 1989* (Cambridge University Press, Cambridge).

Haskell, N. A. (1935), The motion of a viscous fluid under a surface load, *Physics* **6**, 265–269.

Heirtzler, J. R., X. LePichon, and J. G. Baron (1966), Magnetic anomalies over the Reykjanes Ridge, *Deep Sea Res.* **13**, 427–443.

Hess, H. (1962), History of ocean basins, in *Petrologic Studies: A Volume in Honour of A. E. Buddington*, A. E. J. Engle, ed., pp. 599–620 (Geological Society of America, Boulder).

Holmes, A. (1931), Radioactivity and Earth movement XVIII, *Trans. Geol. Soc. Glasgow* **18**, 559–606.

Isacks, B. L., and M. Barazangi (1977), Geometry of Benioff zones: Lateral segmentation and downwards bending of the subducted lithosphere, in *Island Arcs, Deep Sea Trenches, and Back-Arc Basins*, pp. 99–114 (American Geophysical Union, Washington, D.C.).

Isacks, B. L., J. Oliver, and L. R. Sykes (1968), Seismology and the new global tectonics, *J. Geophys. Res.* **73**, 5855–5899.

Jeffreys, H. (1924), *The Earth* (Cambridge University Press, Cambridge).

McElhinny, M. W. (1973), *Paleomagnetism and Plate Tectonis* (Cambridge University Press, Cambridge).

Molnar, P., and J. Stock (1987), Relative motions of hotspots in the Pacific, Atlantic and Indian Oceans since the late Cretaceous, *Nature* **327**, 587–591.

Morgan, W. J. (1968), Rises, trenches, great faults, and crustal blocks, *J. Geophys. Res.* **73**, 1959–1982.

Morgan, W. J. (1971), Convection plumes in the lower mantle, *Nature* **230**, 42–43.

Muller, R. D., W. R. Roest, J. Y. Royer, L. M. Gahagan, and others (1997), Digital isochrons of the world's ocean floor, *J. Geophys. Res.* **102**, 3211–3214.

Olson, P., and H. Singer (1985), Creeping plumes, *J. Fluid Mech.* **158**, 511–531.

Runcorn, K. (1956), Paleomagnetic comparisons between Europe and North America, *Proc. Geol. Assoc. Canada* **8**, 77–85.

Schaber, G. G., R. G. Strom, H. J. Moore, L. A. Soderblom, R. L. Kirk, D. J. Chadwick, D. D. Dawson, L. R. Gaddis, J. M. Boyce, and J. Russell (1992), Geology and distribution of impact craters on Venus: What are they telling us? *J. Geophys. Res.* **97**, 13,257–13,301.

Schmid, S. M., O. A. Pfiffner, G. Schönborn, N. Froitzheim, and E. Kissling (1997), Integrated cross section and tectonic evolution of the Alps along the eastern traverse, in *Deep Structure of the Swiss Alps: Results of NRP 20*, O. A. Pfiffner, P. Lehner, P. Heitzmann, S. Mueller, and A. Steck, eds., pp. 289–304 (Birkhäuser, Cambridge, Mass.).

Sleep, N. H. (1994), Martian plate tectonics, *J. Geophys. Res.* **99**, 5639–5655.

Smith, A. G., J. C. Briden, and G. E. Drewry (1974), Phanerozoic world maps, in *Organisms and Continents through Time*, N. F. Hughes, ed., pp. 1–42 (The Paleontological Association, London).

Tapponier, P., and P. Molnar (1977), Active faulting and tectonics in China, *J. Geophys. Res.* **82**, 2905–2930.

Taylor, F. B. (1910), Bearing of the Tertiary mountain belt on the origin of the Earth's plan, *Bull. Geol. Soc. Am.* **21**, 179–226.

Vine, F., and D. Matthews (1963), Magnetic anomalies over ocean ridges, *Nature* **199**, 947.

Watts, A. B. (1976), Gravity and bathymetry in the central Pacific Ocean, *J. Geophys. Res.* **81**, 1533–1553.

Wegener, A. (1946), *The Origin of Continents and Oceans*, 4th Ed. (Dover, New York).

White, R., and D. McKenzie (1989), Magmatism at rift zones: The generation of volcanic continental margins and flood basalts, *J. Geophys. Res.* **94**, 7685–7729.

Wilson, J. T. (1966), Did the Atlantic close and then reopen? *Nature* **211**, 676–681.

Zhao, W., and K. D. Nelson (1993), Deep seismic reflection evidence for continental underthrusting beneath southern Tibet, *Nature* **366**, 557–559.

COLLATERAL READING

Continents Adrift, Readings from Scientific American (W. H. Freeman and Company, San Francisco, 1972), 172 pages.

A collection of 15 papers on plate tectonics originally published in Scientific American. *The papers are divided into three major sections dealing with the Earth's interior, seafloor spreading and continental drift, and consequences and examples of continental drift. Each section is preceded by a brief introduction by J. T. Wilson.*

Decker, R., and B. Decker, *Volcanoes* (W. H. Freeman and Company, San Francisco, 1981), 244 pages.

An introductory discussion of volcanoes, how they work, and how they are produced by plate tectonic processes.

Jacobs, J. A., *Reversals of the Earth's Magnetic Field*, 2nd Ed. (Cambridge University Press, Cambridge, 1994), 346 pages.

A detailed discussion of observations and theory associated with the reversals of the Earth's magnetic fields.

LePichon, X., J. Francheteau, and J. Bonnin, *Plate Tectonics* (Elsevier, Amsterdam, 1973), 300 pages.

One of the first textbooks on plate tectonics. There are major chapters on the rheology of the mantle, kinematics of plate movements, and physical processes at accreting and consuming plate boundaries.

Lowrie, W., *Fundamentals of Geophysics* (Cambridge University Press, Cambridge, 1997), 354 pages.

This is a comprehensive treatment of basic geophysics at a moderately advanced level. Topics include gravity, seismology, geomagnetism, and geodynamics.

McDonald, G. A., *Volcanoes* (Prentice Hall, Englewood Cliffs, NJ, 1972), 510 pages.

A largely descriptive and in-depth discussion of the physical aspects of volcanology. Major chapters include volcanic rocks and magmas, lava flows, volcanic eruptions and edifices, craters, calderas, fumaroles, hot springs, and geysers.

McElhinny, M. W., *Paleomagnetism and Plate Tectonics* (Cambridge, London, 1973), 358 pages.

Seven chapters deal with geomagnetism, rock magnetism, experimental methods in paleomagnetism, reversals of the Earth's field, seafloor spreading and plate tectonics, apparent polar wandering, and paleomagnetic poles.

Merrill, R. T., M. W. McElhinny, and P. L. McFadden, *The Magnetic Field of the Earth* (Academic Press, San Diego, 1996), 531 pages.

A comprehensive discussion of all aspects of the Earth's magnetic field. Topics include the present geomagnetic field, paleomagnetism, reversals, and dynamo theory.

Press, F., and R. Siever, *Earth* (W. H. Freeman and Company, San Francisco, 1974), 945 pages.

An introductory textbook on Earth science. The book is divided into three major sections dealing with the geological history of the Earth and its surface and interior.

Ringwood, A. E., *Composition and Petrology of the Earth's Mantle* (McGraw-Hill, New York, 1975), 618 pages.

An advanced textbook that combines observational data from natural petrology with experimental results on phase equilibria of natural rock systems at high temperature and pressure to discuss the composition and petrology of the upper mantle–crust system. There are also chapters discussing the lower mantle and the origin and evolution of the Earth.

Williams, H., and A. R. McBirney, *Volcanology* (Freeman, Cooper and Company, San Francisco, 1979), 397 pages.

An advanced textbook with chapters on the physical nature of magmas, generation, rise, and storage of magma, eruptive mechanisms, lava flows, pyroclastic flows, fissure eruptions, oceanic volcanism, and hydrothermal phenomena.

Wyllie, P. J., *The Dynamic Earth: Textbook in Geosciences* (John Wiley, New York, 1971), 416 pages.

An advanced textbook designed mainly for graduate students in geology and geochemistry. Chapters deal with the structure, composition, mineralogy, and petrology of the crust and mantle, mantle phase transitions, magma generation, plate tectonics, and the Earth's interior.

TWO

Stress and Strain in Solids

2-1 Introduction

Plate tectonics is a consequence of the gravitational body forces acting on the solid mantle and crust. Gravitational forces result in an increase of pressure with depth in the Earth; rocks must support the weight of the overburden that increases with depth. A static equilibrium with pressure increasing with depth is not possible, however, because there are horizontal variations in the gravitational body forces in the Earth's interior. These are caused by horizontal variations in density associated with horizontal differences in temperature. The horizontal thermal contrasts are in turn the inevitable consequence of the heat release by radioactivity in the rocks of the mantle and crust. The horizontal variations of the gravitational body force produce the differential stresses that drive the relative motions associated with plate tectonics.

One of the main purposes of this chapter is to introduce the fundamental concepts needed for a quantitative understanding of stresses in the solid Earth. *Stresses* are forces per unit area that are transmitted through a material by interatomic force fields. Stresses that are transmitted perpendicular to a surface are *normal stresses*; those that are transmitted parallel to a surface are *shear stresses*. The mean value of the normal stresses is the pressure. We will describe the techniques presently used to measure the state of stress in the Earth's crust and discuss the results of those measurements.

Stress in an elastic solid results in *strain* or deformation of the solid. The simplest example of strain is the decrease in volume accompanying an increase in pressure due to the *compressibility* of a solid. *Normal*

strain is defined as the ratio of the change in length of a solid to its original length. The *shear strain* is defined as one-half of the decrease in a right angle in a solid when it is deformed. The surface of the Earth is continually being strained by tectonic processes. These changes in strain can be measured directly by geodetic techniques. This chapter also discusses the basic concepts required for a quantitative understanding of strain and changes in strain in the solid Earth.

2-2 Body Forces and Surface Forces

The forces on an element of a solid are of two types: body forces and surface forces. Body forces act throughout the volume of the solid. The magnitude of the body force on an element is thus directly proportional to its volume or mass. An example is the downward force of gravity, that is, the weight of an element, which is the product of its mass and the acceleration of gravity g. Since density ρ is mass per unit volume, the gravitational body force on an element is also the product of ρg and the element's volume. Thus the downward gravitational body force is g per unit mass and ρg per unit volume.

The densities of some common rocks are listed in Appendix 2, Section E. The densities of rocks depend on the pressure; the values given are zero-pressure densities. Under the high pressures encountered deep in the mantle, rocks are as much as 50% denser than the zero-pressure values. The variation of density with depth in the Earth is discussed in Chapter 4. Typical mantle rocks have zero-pressure densities of 3250 kg m^{-3}. Basalt and gabbro, which are the principal constituents of the oceanic crust, have densities near 2950 kg m^{-3}. Continental igneous rocks such as granite and diorite are significantly lighter with densities of 2650 to 2800 kg m^{-3}. Sedimentary rocks are generally the lightest and have the largest variations in density, in large part because of variations in porosity and water content in the rocks.

Surface forces act on the surface area bounding an element of volume. They arise from interatomic forces exerted by material on one side of the surface onto material on the opposite side. The magnitude of the surface force is directly proportional to the area of the surface on which it acts. It also depends on the orientation of the surface. As an example, consider the force that must act at the base of the column of rock at a

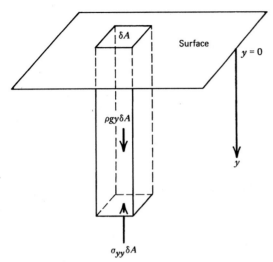

2–1 Body and surface forces acting on a vertical column of rock.

depth y beneath the surface to support the weight of the column, as illustrated in Figure 2–1. The weight of the column of cross-sectional area δA, is $\rho g y \delta A$. This weight must be balanced by an upward surface force $\sigma_{yy} \delta A$ distributed on the horizontal surface of area δA at depth y. We are assuming that no vertical forces are acting on the lateral surfaces of the column and that the density ρ is constant; σ_{yy} is thus the surface force per unit area acting perpendicular to a horizontal surface, that is, stress. Since the forces on the column of rock must be equal if the column is in equilibrium, we find

$$\sigma_{yy} = \rho g y. \tag{2–1}$$

The normal force per unit area on horizontal planes increases linearly with depth. The normal stress due to the weight of the overlying rock or *overburden* is known as the *lithostatic stress* or pressure.

To find a typical value for stress in the lithosphere, let us determine the lithostatic stress on a horizontal plane at the base of the continental crust. Assume that the crust is 35 km thick and that its mean density is

2750 kg m^{-3}; from Equation (2–1) we find that

$$\sigma_{yy} = 2750 \text{ kg m}^{-3} \times 10 \text{ m s}^{-2} \times 3.5 \times 10^4 \text{ m}$$
$$= 9.625 \times 10^8 \text{ Pa} = 962.5 \text{ MPa}.$$

The SI unit for pressure or stress is the pascal (Pa). Pressures and stresses in the Earth are normally given in megapascals (Mpa); 1 megapascal = 10^6 pascals.

Think of continents as blocks of wood floating on a sea of mantle rock, as illustrated in Figure 2–2. The mean density of the continent, say $\rho_c = 2750$ kg m^{-3}, is less than the mean upper mantle density, say $\rho_m = 3300$ kg m^{-3}, so that the continent "floats." *Archimedes' principle* applies to continents; they are buoyed up by a force equal to the weight of mantle rock displaced. At the base of the continent $\sigma_{yy} = \rho_c g h$, where ρ_c is the density of the continent and h is its thickness. At this depth in the mantle, σ_{yy} is $\rho_m g b$, where ρ_m is the mantle density and b is the depth in the mantle to which the continent "sinks." Another statement of Archimedes' principle, also known as *hydrostatic equilibrium*, is that these stresses are equal. Therefore we find

$$\rho_c h = \rho_m b. \tag{2–2}$$

The height of the continent above the surrounding mantle is

$$h - b = h - \frac{\rho_c}{\rho_m}h = h\left(1 - \frac{\rho_c}{\rho_m}\right). \tag{2–3}$$

Using the values given earlier for the densities and the thickness of the continental crust $h = 35$ km, we find from Equation (2–3) that $h - b = 5.8$ km. This analysis is only approximately valid for determining the depth of the oceans relative to the continents, since we have neglected the contribution of the seawater and the oceanic crust. The application of hydrostatic equilibrium to the continental crust is known as *isostasy*; it is discussed in more detail in Chapter 5.

PROBLEM 2–1 An average thickness of the oceanic crust is 6 km. Its density is 2900 kg m^{-3}. This is

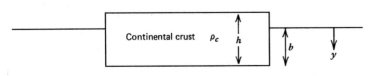

2–2 A continental block "floating" on the fluid mantle.

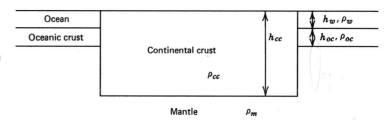

2–3 Isostasy of the continental crust relative to an ocean basin.

overlain by 5 km of water ($\rho_w = 1000$ kg m^{-3}) in a typical ocean basin. Determine the normal force per unit area on a horizontal plane at the base of the oceanic crust due to the weight of the crust and the overlying water.

PROBLEM 2–2 A mountain range has an elevation of 5 km. Assuming that $\rho_m = 3300$ kg m^{-3}, $\rho_c = 2800$ kg m^{-3}, and that the reference or normal continental crust has a thickness of 35 km, determine the thickness of the continental crust beneath the mountain range. Assume that hydrostatic equilibrium is applicable.

PROBLEM 2–3 There is observational evidence from the continents that the sea level in the Cretaceous was 200 m higher than today. After a few thousand years, however, the seawater is in isostatic equilibrium with the ocean basins. What was the corresponding increase in the depth of the ocean basins? Take $\rho_w = 1000$ kg m^{-3} and the density of the displaced mantle to be $\rho_m = 3300$ kg m^{-3}.

A more realistic model for the depth of the ocean basins is illustrated in Figure 2–3. The continental crust has a thickness h_{cc} and a density ρ_{cc}; its upper surface is at sea level. The oceanic crust is covered with water of depth h_w and density ρ_w. The oceanic crust has a thickness h_{oc} and density ρ_{oc}. The mantle density is ρ_m. Application of the principle of isostasy to the base of the continental crust gives

$$\rho_{cc}h_{cc} = \rho_w h_w + \rho_{oc}h_{oc} + \rho_m(h_{cc} - h_w - h_{oc}). \tag{2-4}$$

The depth of the ocean basin relative to the continent is given by

$$h_w = \frac{(\rho_m - \rho_{cc})}{(\rho_m - \rho_w)}h_{cc} - \frac{(\rho_m - \rho_{oc})}{(\rho_m - \rho_w)}h_{oc}. \tag{2-5}$$

Taking $h_{cc} = 35$ km, $h_{oc} = 6$ km, $\rho_m = 3300$ kg m^{-3}, $\rho_w = 1000$ kg m^{-3}, $\rho_{cc} = 2800$ kg m^{-3}, and $\rho_{oc} = 2900$ kg m^{-3}, we find $h_w = 6.6$ km.

Subsidence of the surface of the continental crust often results in the formation of a *sedimentary basin*. Assume that the surface of the continental crust is initially at sea level and, as it subsides, sediments are deposited so that the surface of the sediments remains at sea level. One cause of the subsidence is the thinning of the continental crust. As the crust is thinned, isostasy requires that the surface subside. A simple model for this subsidence applicable to some sedimentary basins is the crustal stretching model (McKenzie, 1978). This two-dimensional model is illustrated in Figure 2–4. A section of continental crust with an initial width w_0 is stretched to a final width w_b. The stretching factor α is defined by

$$\alpha = \frac{w_b}{w_0}. \tag{2-6}$$

2–4 Illustration of the crustal stretching model for the formation of a sedimentary basin. A section of continental crust of initial width w_0, illustrated in (a), is stretched by a stretching factor $\alpha = 4$ to a final width w_b to form the sedimentary basin illustrated in (b).

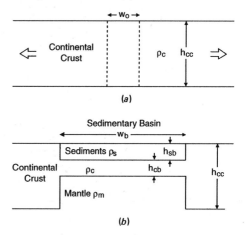

In order to conserve the volume of the stretched continental crust we assume a constant crustal density ρ_{cc} and require that

$$w_b h_{cb} = w_0 h_{cc}, \tag{2-7}$$

where h_{cc} is the initial thickness of the continental crust and h_{cb} is the final thickness of the stretched crust. The combination of Equations (2–6) and (2–7) gives

$$h_{cb} = \frac{h_{cc}}{\alpha}. \tag{2-8}$$

The surface of this stretched continental crust subsides and is assumed to be covered with sediments of density $\rho_s (\rho_s < \rho_{cc})$ to sea level. The sediments have a thickness h_{sb} and the lower boundary of the sediments is referred to as *basement*. Application of the principle of isostasy to the base of the reference continental crust gives

$$\rho_{cc} h_{cc} = \rho_s h_{sb} + \rho_{cc} h_{cb} + \rho_m (h_{cc} - h_{sb} - h_{cb}). \tag{2-9}$$

The combination of Equations (2–8) and (2–9) gives the thickness of the sedimentary basin in terms of the stretching factor as

$$h_{sb} = h_{cc} \left(\frac{\rho_m - \rho_{cc}}{\rho_m - \rho_s} \right) \left(1 - \frac{1}{\alpha} \right). \tag{2-10}$$

The thickness of the sedimentary basin is given as a function of the stretching factor in Figure 2–5 for $h_{cc} = 35$ km, $\rho_m = 3300$ kg m^{-3}, $\rho_{cc} = 2800$ kg m^{-3}, and $\rho_s =$

2500 kg m^{-3}. The maximum thickness of the sedimentary basin for an infinite stretching factor is $h_{sb} = 22$ km.

PROBLEM 2–4 A sedimentary basin has a thickness of 4 km. Assuming that the crustal stretching model is applicable and that $h_{cc} = 35$ km, $\rho_m = 3300$ kg m^{-3}, $\rho_{cc} = 2750$ kg m^{-3}, and $\rho_s = 2550$ kg m^{-3}, determine the stretching factor.

PROBLEM 2–5 A sedimentary basin has a thickness of 7 km. Assuming that the crustal stretching model is applicable and that $h_{cc} = 35$ km, $\rho_m = 3300$ kg m^{-3}, $\rho_{cc} = 2700$ kg m^{-3}, and $\rho_s = 2450$ kg m^3, determine the stretching factor.

PROBLEM 2–6 A simple model for a continental mountain belt is the crustal compression model illustrated in Figure 2–6. A section of the continental crust of width w_0 is compressed to a width w_{mb}. The compression factor β is defined by

$$\beta = \frac{w_0}{w_{mb}}. \tag{2-11}$$

Show that the height of the mountain belt h is given by

$$h = h_{cc} \frac{(\rho_m - \rho_{cc})}{\rho_m} (\beta - 1). \tag{2-12}$$

2–6 Illustration of the crustal compression model for a mountain belt. A section of continental crust of width w_0, shown in (*a*), is compressed by compression factor $\beta = 2$ to form a mountain belt as shown in (*b*).

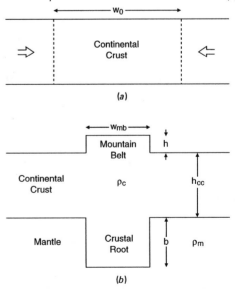

2–5 Thickness of a sedimentary basin h_{sb} as a function of the crustal stretching factor α.

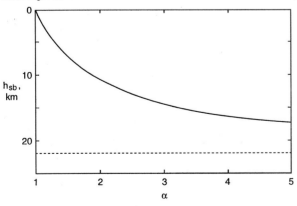

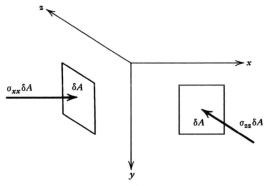

2–7 Horizontal surface forces acting on vertical planes.

Assuming $\beta = 2$, $h_{cc} = 35$ km, $\rho_m = 3300$ kg m^{-3}, and $\rho_{cc} = 2800$ kg m^{-3}, determine the height of the mountain belt h and the thickness of the crustal root b.

Just as there are normal surface forces per unit area on horizontal planes in the Earth, there are also normal surface forces per unit area on vertical planes, as sketched in Figure 2–7. The horizontal normal stress components σ_{xx} and σ_{zz} can include large-scale tectonic forces, in which case $\sigma_{xx} \neq \sigma_{zz} \neq \sigma_{yy}$. On the other hand, there are many instances in which rock was heated to sufficiently high temperatures or was sufficiently weak initially so that the three stresses σ_{xx}, σ_{zz}, and σ_{yy} are equal to the weight of the overburden; that is,

$$p_L \equiv \sigma_{xx} = \sigma_{zz} = \sigma_{yy} = \rho g y. \qquad (2\text{--}13)$$

When the three normal stresses are equal, they are defined to be the pressure. The balance between pressure and the weight of the overburden is known as a *lithostatic state of stress*. It is completely equivalent to the hydrostatic state of stress in a motionless body of fluid wherein pressure forces are exerted equally in all directions and pressure increases proportionately with depth.

We will now show that the continental block illustrated in Figure 2–2 cannot simply be in a lithostatic state of stress. The force balance on the continental block is illustrated in Figure 2–8. A horizontal force is acting on the edge of the block F_m. We assume that this force is due to the lithostatic pressure in the mantle rock of density ρ_m. The vertical distribution of this pressure is given in Figure 2–9. The horizontal force F_m is obtained by integrating the lithostatic pressure:

$$F_m = \int_0^b p_L \, dy = \rho_m g \int_0^b y \, dy = \frac{1}{2} \rho_m g b^2. \qquad (2\text{--}14)$$

This force is per unit width of the block so that it has dimensions of force per unit length. The total force per unit width is proportional to the area under the stress distribution given in Figure 2–9.

We next determine the horizontal force per unit width acting at a typical cross section in the continental block F_c. We assume that the horizontal normal stress acting in the continent σ_{xx} is made up of two parts, the lithostatic contribution $\rho_c g y$ and a constant tectonic contribution $\Delta \sigma_{xx}$,

$$\sigma_{xx} = \rho_c g y + \Delta \sigma_{xx}. \qquad (2\text{--}15)$$

The tectonic contribution is also known as the *deviatoric stress*. The horizontal force F_c is obtained by integrating the horizontal normal stress

$$F_c = \int_0^h \sigma_{xx} \, dy = \int_0^h (\rho_c g y + \Delta \sigma_{xx}) \, dy$$
$$= \frac{1}{2} \rho_c g h^2 + \Delta \sigma_{xx} h. \qquad (2\text{--}16)$$

In order to maintain a static balance, the two forces F_c and F_m must be equal. Using Equations (2–2), (2–14),

2–9 The area under the stress versus depth profile is proportional to the total horizontal force on a vertical plane.

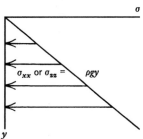

2–8 Force balance on a section of continental block.

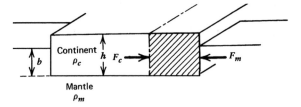

and (2–16), we obtain

$$\Delta\sigma_{xx} = \frac{1}{2}\frac{\rho_m g b^2}{h} - \frac{1}{2}\rho_c g h = -\frac{1}{2}\rho_c g h\left(1 - \frac{\rho_c}{\rho_m}\right).$$
(2–17)

A horizontal *tensile stress* is required to maintain the integrity of the continental block. The horizontal tensile stress is a force per unit area acting on vertical planes and tending to pull on such planes. A *compressive stress* is a normal force per unit area tending to push on a plane. We consider compressive stresses positive and tensile stresses negative, a convention generally adopted in the geological literature. This is opposite to the sign convention used in most elasticity textbooks in which positive stress is tensional. Taking $h = 35$ km, $\rho_m = 3300$ kg m^{-3}, and $\rho_c = 2750$ kg m^{-3}, we find from Equation (2–17) that $\Delta\sigma_{xx} = -80.2$ MPa. Typical values for deviatoric stresses in the continents are of the order of 10 to 100 MPa.

PROBLEM 2–7 Consider a continental block to have a thickness of 70 km corresponding to a major mountain range. If the continent has a density of 2800 kg m^{-3} and the mantle a density of 3300 kg m^{-3}, determine the tensional stress in the continental block.

PROBLEM 2–8 Determine the deviatoric stress in the continent for the oceanic–continental structure in Figure 2–3 by proceeding as follows. Show that the pressure as a function of depth in the continental crust p_c is

$$p_c = \rho_{cc} g y,$$
(2–18)

and that the pressures in the water, in the oceanic crust, and in the mantle beneath the oceanic crust are

$$\begin{aligned}
p_0 &= \rho_w g y \qquad 0 \le y \le h_w \\
&= \rho_w g h_w + \rho_{oc} g(y - h_w) \qquad h_w \le y \le h_w + h_{oc} \\
&= \rho_w g h_w + \rho_{oc} g h_{oc} + \rho_m g(y - h_w - h_{oc}) \\
&\qquad\qquad h_w + h_{oc} \le y \le h_{cc}.
\end{aligned}$$
(2–19)

Find the net difference in the hydrostatic pressure force between the continental and the oceanic crusts F by integrating the pressures over a depth equal to

the thickness of the continental crust. The result is

$$\begin{aligned}
F = g[&h_w h_{cc}(\rho_m - \rho_w) + h_{oc} h_{cc}(\rho_m - \rho_{oc}) \\
&- h_w h_{oc}(\rho_m - \rho_{oc}) - \tfrac{1}{2}h_w^2(\rho_m - \rho_w) \\
&- \tfrac{1}{2}h_{oc}^2(\rho_m - \rho_{oc}) - \tfrac{1}{2}h_{cc}^2(\rho_m - \rho_{cc})].
\end{aligned}$$
(2–20)

Calculate F for $h_w = 5$ km, $\rho_w = 1000$ kg m^{-3}, $h_{oc} = 7$ km, $\rho_{oc} = 2900$ kg m^{-3}, $\rho_{cc} = 2800$ kg m^{-3}, and $\rho_m = 3300$ kg m^{-3}. Find h_{cc} from Equation (2–5). If the elastic stresses required to balance this force are distributed over a depth equal to h_{cc}, determine the stress. If the stresses are exerted in the continental crust, are they tensional or compressional? If they act in the oceanic lithosphere, are they tensional or compressional?

Surface forces can act parallel as well as perpendicular to a surface. An example is provided by the forces acting on the area element δA lying in the plane of a strike–slip fault, as illustrated in Figure 2–10. The normal compressive force $\sigma_{xx}\delta A$ acting on the fault face is the consequence of the weight of the overburden and the tectonic forces tending to press the two sides of the fault together. The tangential or shear force on the element $\sigma_{xz}\delta A$ opposes the tectonic forces driving the left-lateral motion on the fault. This shear force is the result of the frictional resistance to motion on the fault. The quantity σ_{xz} is the tangential surface force per unit area or the *shear stress*. The first subscript refers to the direction normal to the surface element and the second subscript to the direction of the shear force.

Another example of the resistive force due to a shear stress is the emplacement of a *thrust sheet*. In zones of

2–10 Normal and tangential surface forces on an area element in the fault plane of a strike–slip fault.

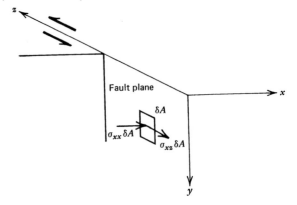

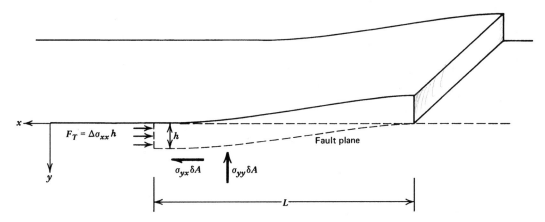

continental collision a thin sheet of crystalline rock is often *overthrust* upon adjacent continental rocks on a low-angle thrust fault. This process is illustrated in Figure 2–11, where the thrust sheet has been emplaced from the left as a consequence of horizontal tectonic forces. Neglecting the influence of gravity, which is considered in Section 8–4, we can write the total horizontal tectonic force F_T due to a horizontal tectonic stress $\Delta\sigma_{xx}$ as

$$F_T = \Delta\sigma_{xx}h, \tag{2–21}$$

where h is the thickness of the thrust sheet and F_T is a force per unit width of the sheet. This tectonic driving force is resisted by the shear stress σ_{yx} acting on the base of the thrust sheet. The total resisting shear force per unit width F_R is

$$F_R = \sigma_{yx}L, \tag{2–22}$$

where L is the length of the thrust sheet.

In many cases it is appropriate to relate the shear stress resisting the sliding of one surface over another to the normal force pressing the surfaces together. Empirically we often observe that these stresses are proportional to one another so that

$$\sigma_{yx} = f\sigma_{yy}, \tag{2–23}$$

where σ_{yy} is the vertical normal stress acting on the base of the thrust sheet and f, the constant of proportionality, is known as the *coefficient of friction*. Assuming that σ_{yy} has the lithostatic value

$$\sigma_{yy} = \rho_c g h, \tag{2–24}$$

and equating the driving tectonic force F_T to the

resisting shear force, we find that

$$\Delta\sigma_{xx} = f\rho_c g L. \tag{2–25}$$

This is the tectonic stress required to emplace a thrust sheet of length L. Taking a typical value for the tectonic stress to be $\Delta\sigma_{xx} = 100$ MPa and assuming a thrust sheet length $L = 100$ km and $\rho_c = 2750$ kg m^{-3}, we find that the required coefficient of friction is $f = 0.036$. The existence of long thrust sheets implies low values for the coefficient of friction.

PROBLEM 2–9 Assume that the friction law given in Equation (2–23) is applicable to the strike–slip fault illustrated in Figure 2–10 with $f = 0.3$. Also assume that the normal stress σ_{xx} is lithostatic with $\rho_c = 2750$ kg m^{-3}. If the fault is 10 km deep, what is the force (per unit length of fault) resisting motion on the fault? What is the mean tectonic shear stress over this depth $\bar{\sigma}_{zx}$ required to overcome this frictional resistance?

PROBLEM 2–10 Consider a block of rock with a height of 1 m and horizontal dimensions of 2 m. The density of the rock is 2750 kg m^{-3}. If the coefficient of friction is 0.8, what force is required to push the rock on a horizontal surface?

PROBLEM 2–11 Consider a rock mass resting on an inclined bedding plane as shown in Figure 2–12. By balancing the forces acting on the block parallel to the inclined plane, show that the tangential force per unit area $\sigma_{x'y'}$ on the plane supporting the block is

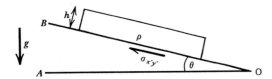

2-12 Gravitational sliding of a rock mass.

$\rho g h \sin\theta$ (ρ is the density and h is the thickness of the block). Show that the sliding condition is

$$\theta = \tan^{-1} f. \qquad (2\text{--}26)$$

PROBLEM 2-12 The pressure p_h of fluids (water) in the pores of rocks reduces the effective normal stress pressing the surfaces together along a fault. Modify Equation (2–25) to incorporate this effect.

2-3 Stress in Two Dimensions

In the previous section we were concerned primarily with stresses on the surface of a material. However, stress components can be defined at any point in a material. In order to illustrate this point, it is appropriate to consider a small rectangular element with dimensions $\delta x, \delta y,$ and δz defined in accordance with a cartesian x, y, z coordinate system, as illustrated in Figure 2–13. In this section we will consider a two-dimensional state of stress; the state is two-dimensional in the sense that there are no surface forces in the z direction and none

2-13 Surface forces acting on a small rectangular element in a two-dimensional state of stress.

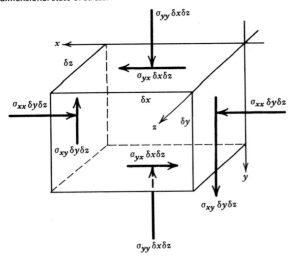

of the surface forces shown vary in the z direction. The *normal stresses* are σ_{xx} and σ_{yy}, and the *shear stresses* are σ_{xy} and σ_{yx}. The notation adopted in labeling the stress components allows immediate identification of the associated surface forces. The second subscript on σ gives the direction of the force, and the first subscript gives the direction of the normal to the surface on which the force acts.

The tangential or shear stresses σ_{xy} and σ_{yx} have associated surface forces that tend to rotate the element in Figure 2–13 about the z axis. The moment exerted by the surface force $\sigma_{xy}\delta y\delta z$ is the product of the force and the moment arm δx; that is, it is $\sigma_{xy}\delta x\delta y\delta z$. This couple is counteracted by the moment $\sigma_{yx}\delta x\delta y\delta z$ exerted by the surface force $\sigma_{yx}\delta x\delta z$ with a moment arm δy. Because the element cannot rotate if it is in equilibrium,

$$\sigma_{xy} = \sigma_{yx}. \qquad (2\text{--}27)$$

Thus the shear stresses are symmetric in that their value is independent of the order of the subscripts. Three independent components of stress $\sigma_{xx}, \sigma_{yy},$ and σ_{xy} must be specified in order to prescribe the two-dimensional state of stress.

The state of stress is dependent on the orientation of the coordinate system. We will now determine the three components of stress in a coordinate system x', y' inclined at an angle θ with respect to the x, y coordinate system as illustrated in Figure 2–14a. To determine the normal stress, we carry out a static force balance on the triangular element OAB illustrated in Figure 2–14b. The sides of the triangle lie in the $x, y,$ and y' directions. We first write a force balance in the y direction. The force in the y direction on face AO is

$$\sigma_{yy} AO,$$

and the force in the y direction on face OB is

$$\sigma_{xy} OB.$$

The force in the y direction on face AB is

$$-\sigma_{x'x'} AB \sin\theta - \sigma_{x'y'} AB \cos\theta.$$

The sum of these forces must be zero for the triangular element OAB to be in equilibrium. This gives

$$(\sigma_{x'x'} \sin\theta + \sigma_{x'y'} \cos\theta)AB = \sigma_{yy} AO + \sigma_{xy} OB. \qquad (2\text{--}28)$$

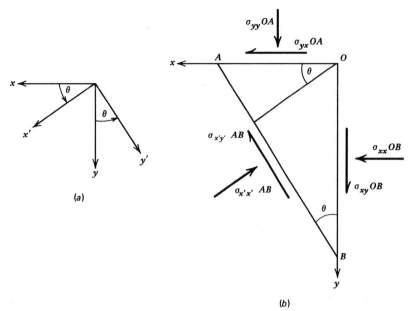

2–14 Transformation of stresses from the x, y coordinate system to the x', y' coordinate system. (a) Illustration of the coordinate systems. (b) Triangular element (with sides in the x, y, and y' directions) on which a static force balance is carried out.

However, the sides of triangle OAB are in the ratios

$$\frac{AO}{AB} = \sin\theta \qquad \frac{OB}{AB} = \cos\theta, \qquad (2\text{–}29)$$

so that

$$\sigma_{x'x'}\sin\theta + \sigma_{x'y'}\cos\theta = \sigma_{yy}\sin\theta + \sigma_{xy}\cos\theta. \qquad (2\text{–}30)$$

We next write a force balance in the x direction. The force in the x direction on face AO is

$$\sigma_{yx}AO,$$

and the force in the x direction on face OB is

$$\sigma_{xx}OB.$$

The force in the x direction on face AB is

$$-\sigma_{x'x'}AB\cos\theta + \sigma_{x'y'}AB\sin\theta.$$

Upon equating the sum of these forces to zero, we get

$$(\sigma_{x'x'}\cos\theta - \sigma_{x'y'}\sin\theta)AB = \sigma_{yx}AO + \sigma_{xx}OB. \qquad (2\text{–}31)$$

With the values of AO/AB and OB/AB as given in Equation (2–29), we find

$$\sigma_{x'x'}\cos\theta - \sigma_{x'y'}\sin\theta = \sigma_{yx}\sin\theta + \sigma_{xx}\cos\theta. \qquad (2\text{–}32)$$

We multiply Equation (2–30) by $\sin\theta$, multiply Equation (2–32) by $\cos\theta$, and add the results to obtain

$$\sigma_{x'x'}(\cos^2\theta + \sin^2\theta) = \sigma_{xx}\cos^2\theta + \sigma_{yy}\sin^2\theta$$
$$+ \sigma_{xy}\sin\theta\cos\theta$$
$$+ \sigma_{yx}\sin\theta\cos\theta. \qquad (2\text{–}33)$$

This can be further simplified by using

$$\cos^2\theta + \sin^2\theta = 1 \qquad (2\text{–}34)$$
$$\sigma_{xy} = \sigma_{yx} \qquad (2\text{–}35)$$
$$2\sin\theta\cos\theta = \sin 2\theta. \qquad (2\text{–}36)$$

The result is

$$\sigma_{x'x'} = \sigma_{xx}\cos^2\theta + \sigma_{yy}\sin^2\theta + \sigma_{xy}\sin 2\theta. \qquad (2\text{–}37)$$

By multiplying Equation (2–30) by $\cos\theta$ and subtracting the product of Equation (2–32) with $\sin\theta$, we find

$$\sigma_{x'y'}(\sin^2\theta + \cos^2\theta) = \sigma_{yy}\sin\theta\cos\theta + \sigma_{xy}\cos^2\theta$$
$$- \sigma_{xx}\sin\theta\cos\theta - \sigma_{yx}\sin^2\theta. \qquad (2\text{–}38)$$

By using the trigonometric relations already quoted, the symmetry of the shear stresses σ_{xy} and σ_{yx} and

$$\cos 2\theta = \cos^2\theta - \sin^2\theta, \qquad (2\text{–}39)$$

we can rewrite Equation (2–38) in the simpler form:

$$\sigma_{x'y'} = \tfrac{1}{2}(\sigma_{yy} - \sigma_{xx})\sin 2\theta + \sigma_{xy}\cos 2\theta. \qquad (2\text{–}40)$$

Equations (2–37) and (2–40) provide formulas for obtaining the normal and shear stresses on arbitrarily oriented elements of area in terms of σ_{xx}, σ_{yy}, and σ_{xy} (or σ_{yx}). Thus these three stress components completely specify the state of two-dimensional stress in a solid. When $\theta = 0$, the equations show that $\sigma_{x'x'}$ is σ_{xx} and $\sigma_{x'y'}$ is σ_{xy}, as required.

PROBLEM 2–13 Show that

$$\sigma_{y'y'} = \sigma_{xx}\sin^2\theta + \sigma_{yy}\cos^2\theta - \sigma_{xy}\sin 2\theta. \qquad (2\text{–}41)$$

PROBLEM 2–14 The state of stress at a point on a fault plane is $\sigma_{yy} = 150$ MPa, $\sigma_{xx} = 200$ MPa, and $\sigma_{xy} = 0$ (y is depth and the x axis points westward). What are the normal stress and the tangential stress on the fault plane if the fault strikes N–S and dips 35° to the west?

For any arbitrary two-dimensional state of stress $\sigma_{xx}, \sigma_{yy}, \sigma_{xy}$, it is possible to find a surface oriented in such a manner that no shear forces are exerted on the surface. We need simply set $\sigma_{x'y'}$ in Equation (2–40) to zero and solve for θ. Stress $\sigma_{x'y'}$ is zero if

$$\tan 2\theta = \frac{2\sigma_{xy}}{\sigma_{xx} - \sigma_{yy}}. \qquad (2\text{–}42)$$

The direction θ defined by Equation (2–42) is known as a *principal axis of stress*. If θ is a principal axis direction, then so is $\theta + \pi/2$ because $\tan 2\theta = \tan[2(\theta + \pi/2)]$; this can be seen as follows:

$$\tan\left[2\left(\theta + \frac{\pi}{2}\right)\right] = \tan(2\theta + \pi) = \frac{\tan 2\theta + \tan \pi}{1 - \tan 2\theta \tan \pi}$$
$$= \tan 2\theta. \qquad (2\text{–}43)$$

The last step is true because $\tan \pi = 0$. The coordinate axes defined by the orthogonal principal axis directions are called the principal axes. There are no shear stresses on area elements oriented perpendicular to the principal axes.

The normal stresses in the principal axis coordinate system are known as *principal stresses*. To solve for the principal stresses σ_1 and σ_2, substitute Equation (2–42) into the expression for $\sigma_{x'x'}$, Equation (2–37). Before

making the substitution, rewrite Equation (2–37) using the identities

$$\sin^2\theta = \frac{1 - \cos 2\theta}{2} \qquad (2\text{–}44)$$

$$\cos^2\theta = \frac{1 + \cos 2\theta}{2}. \qquad (2\text{–}45)$$

Equation (2–37) becomes

$$\sigma_{x'x'} = \frac{\sigma_{xx} + \sigma_{yy}}{2} + \frac{1}{2}\cos 2\theta(\sigma_{xx} - \sigma_{yy}) + \sigma_{xy}\sin 2\theta \qquad (2\text{–}46)$$

or

$$\sigma_{x'x'} = \frac{\sigma_{xx} + \sigma_{yy}}{2} + \frac{1}{2}\cos 2\theta(\sigma_{xx} - \sigma_{yy} + 2\sigma_{xy}\tan 2\theta). \qquad (2\text{–}47)$$

The determination of the principal stresses from Equation (2–47) requires an expression for $\cos 2\theta$ as well as for $\tan 2\theta$. The value of $\cos 2\theta$ can be obtained from the expression for $\tan 2\theta$ using

$$\tan^2 2\theta = \frac{\sin^2 2\theta}{\cos^2 2\theta} = \frac{1 - \cos^2 2\theta}{\cos^2 2\theta}, \qquad (2\text{–}48)$$

which can be rewritten as

$$\cos 2\theta = \frac{1}{(1 + \tan^2 2\theta)^{1/2}}. \qquad (2\text{–}49)$$

By substituting Equation (2–42) into Equation (2–49), one finds

$$\cos 2\theta = \frac{\sigma_{xx} - \sigma_{yy}}{\pm\left(4\sigma_{xy}^2 + (\sigma_{xx} - \sigma_{yy})^2\right)^{1/2}}. \qquad (2\text{–}50)$$

Upon substituting the expressions above for $\tan 2\theta$ and $\cos 2\theta$ into Equation (2–47), we get

$$\sigma_{1,2} = \frac{\sigma_{xx} + \sigma_{yy}}{2} \pm \left\{\frac{(\sigma_{xx} - \sigma_{yy})^2}{4} + \sigma_{xy}^2\right\}^{1/2}. \qquad (2\text{–}51)$$

Instead of specifying σ_{xx}, σ_{yy}, and σ_{xy}, we can describe the state of stress at a point in a solid by giving the orientation of the principal axes and the values of the principal stresses.

In deriving these formulas for the orientation of the principal axes and the magnitudes of the principal stresses, we have tacitly assumed $\sigma_{xx} - \sigma_{yy} \neq 0$. If $\sigma_{xx} = \sigma_{yy}$, then $\sigma_{x'y'} = \sigma_{xy}\cos 2\theta$, and the principal axes have angles of $\pm 45°$, assuming $\sigma_{xy} \neq 0$. If $\sigma_{xy} = 0$, the principal stresses are σ_{xx} and σ_{yy}. If $\sigma_{xx} = \sigma_{yy}$ and $\sigma_{xy} \neq 0$, the principal stresses are

$$\frac{(\sigma_{xx} + \sigma_{yy})}{2} \pm \sigma_{xy} = \sigma_{xx} \pm \sigma_{xy}. \qquad (2\text{-}52)$$

It is often convenient to have formulas for the normal and shear stresses for an arbitrarily oriented coordinate system in terms of the principal stresses and the angle of the coordinate system with respect to the principal axes. To derive such formulas, consider the x, y axes in Figure 2–14 to be principal axes so that $\sigma_1 = \sigma_{xx}$, $\sigma_2 = \sigma_{yy}$, and $\sigma_{xy} = 0$. The stresses $\sigma_{x'x'}$, $\sigma_{x'y'}$, and $\sigma_{y'y'}$ are then given as

$$\sigma_{x'x'} = \sigma_1 \cos^2 \theta + \sigma_2 \sin^2 \theta$$
$$= \frac{\sigma_1 + \sigma_2}{2} + \frac{(\sigma_1 - \sigma_2)}{2} \cos 2\theta \qquad (2\text{-}53)$$

$$\sigma_{x'y'} = -\tfrac{1}{2}(\sigma_1 - \sigma_2) \sin 2\theta \qquad (2\text{-}54)$$

$$\sigma_{y'y'} = \sigma_1 \sin^2 \theta + \sigma_2 \cos^2 \theta$$
$$= \frac{\sigma_1 + \sigma_2}{2} - \frac{(\sigma_1 - \sigma_2)}{2} \cos 2\theta. \qquad (2\text{-}55)$$

At this point, there is no particular reason to retain the primes on the coordinate axes. We can simplify future applications of Equations (2–53) to (2–55) by identifying the x', y' coordinate axes as "new" x, y coordinate axes. Therefore, if θ is considered to be the angle between the direction of σ_1 and the x direction (direction of σ_{xx}), we can write

$$\sigma_{xx} = \frac{\sigma_1 + \sigma_2}{2} + \frac{(\sigma_1 - \sigma_2)}{2} \cos 2\theta \qquad (2\text{-}56)$$

$$\sigma_{xy} = -\tfrac{1}{2}(\sigma_1 - \sigma_2) \sin 2\theta \qquad (2\text{-}57)$$

$$\sigma_{yy} = \frac{\sigma_1 + \sigma_2}{2} - \frac{(\sigma_1 - \sigma_2)}{2} \cos 2\theta. \qquad (2\text{-}58)$$

PROBLEM 2–15 Show that the sum of the normal stresses on any two orthogonal planes is a constant. Evaluate the constant.

PROBLEM 2–16 Show that the maximum and minimum normal stresses act on planes that are at right angles to each other.

By differentiating Equation (2–40) with respect to θ and equating the resulting expression to zero, we can find the angle at which the shear stress $\sigma_{x'y'}$ is a maximum; the angle is given by

$$\tan 2\theta = \frac{(\sigma_{yy} - \sigma_{xx})}{2\sigma_{xy}}. \qquad (2\text{-}59)$$

A comparison of Equations (2–42) and (2–59) shows that $\tan 2\theta$ for the principal axis orientation and $\tan 2\theta$ for the maximum shear stress orientation are negative reciprocals. Thus the angles 2θ differ by 90° and the axes that maximize the shear stress lie at 45° to the principal axes. The maximum value of the shear stress can thus be found by letting $\theta = \pi/4$ in Equation (2–57). One gets

$$(\sigma_{xy})_{\max} = \tfrac{1}{2}(\sigma_1 - \sigma_2). \qquad (2\text{-}60)$$

The maximum shear stress is half the difference of the principal stresses. It is also obvious from Equation (2–57) that $(\sigma_{xy})_{\max}$ is exerted on a surface whose normal is at 45° to the principal axes.

2-4 Stress in Three Dimensions

In three dimensions we require additional stress components to specify the surface forces per unit area on surfaces of arbitrary orientation. Figure 2–15 shows the surface forces per unit area, that is, the stresses, on the faces of a small rectangular parallelepiped. There

2–15 Stress components on the faces of a small rectangular parallelepiped.

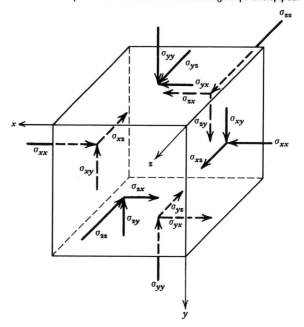

are nine components of stress required to describe the surface forces per unit area on the faces of the element. σ_{xx}, σ_{yy}, and σ_{zz} are the normal stresses, and σ_{xy}, σ_{yx}, σ_{xz}, σ_{zx}, σ_{yz}, and σ_{zy} are shear stresses. If the parallelepiped is not to rotate about any of its axes, then $\sigma_{xy} = \sigma_{yx}$, $\sigma_{xz} = \sigma_{zx}$, and $\sigma_{yz} = \sigma_{zy}$. Only six of the stress components are independent.

The transformation of coordinates to principal axes can also be carried out in three dimensions. Three orthogonal axes can always be chosen such that all shear stress components are zero. The normal stresses on planes perpendicular to these axes are the principal stresses, usually denoted as σ_1, σ_2, and σ_3. By convention these are chosen such that $\sigma_1 \geq \sigma_2 \geq \sigma_3$. Therefore, σ_1 is the maximum principal stress, σ_3 is the minimum principal stress, and σ_2 is the intermediate principal stress. The state of stress at a point in a solid is completely specified by giving σ_{xx}, σ_{yy}, σ_{zz}, σ_{xy}, σ_{yz}, and σ_{xz} or the orientation of the principal axes and the values of the principal stresses.

Clearly two or even three of the principal stresses may be equal. When all three are equal, the state of stress is *isotropic* and the principal stresses can be identified as the pressure $p = \sigma_1 = \sigma_2 = \sigma_3$. In any coordinate system the normal stresses are equal to the pressure, and there are no shear stresses. Any set of orthogonal axes qualifies as a principal axis coordinate system. This is referred to as a *hydrostatic state of stress*. The *lithostatic state of stress* is a hydrostatic state in which the stress increases proportionately with depth at a rate controlled by the density of the rock. When the three principal stresses are not equal, the pressure is defined to be their mean:

$$p = \tfrac{1}{3}(\sigma_1 + \sigma_2 + \sigma_3). \tag{2-61}$$

The pressure is invariant to the choice of coordinate system, that is, to the orientation of the coordinate axes, so that it is equal to the mean of the normal stresses in any coordinate system:

$$p = \tfrac{1}{3}(\sigma_{xx} + \sigma_{yy} + \sigma_{zz}). \tag{2-62}$$

Recall that we have taken normal stress to be positive for compression so that it has the same sign as the pressure.

In studying stress in the Earth, it is often convenient to subtract the mean stress, that is, the pressure, from the normal stress components. Accordingly, we define *deviatoric normal stresses* by

$$\sigma'_{xx} = \sigma_{xx} - p \qquad \sigma'_{yy} = \sigma_{yy} - p \qquad \sigma'_{zz} = \sigma_{zz} - p$$
$$\sigma'_{xy} = \sigma_{xy} \qquad \sigma'_{xz} = \sigma_{xz} \qquad \sigma'_{yz} = \sigma_{yz}, \tag{2-63}$$

where primes refer to the deviatoric stresses. By definition, the average of the normal deviatoric stresses is zero. Similarly *deviatoric principal stresses* can be defined as

$$\sigma'_1 = \sigma_1 - p \qquad \sigma'_2 = \sigma_2 - p \qquad \sigma'_3 = \sigma_3 - p, \tag{2-64}$$

and their average is zero.

We can determine the orientation of the plane on which the shear stress is a maximum, just as we did in the case of two-dimensional stress. The direction of the normal to this plane bisects the angle between the directions of the maximum and minimum principal stresses. The largest possible value of the shear stress is $(\sigma_1 - \sigma_3)/2$.

2–5 Pressures in the Deep Interiors of Planets

Because rocks can readily deform on geologic time scales at the high temperatures encountered deep in planetary interiors, it is a good approximation for many purposes to consider the planets to be in a hydrostatic state of stress completely described by the dependence of pressure p on radius r. Pressure must increase with depth because the weight of the material above any radius r increases as r decreases. The situation is completely analogous to the lithostatic state of stress already discussed. By differentiating Equation (2–13) with respect to y, we find that the rate of increase of pressure, or lithostatic stress, with depth is ρg. In spherical coordinates, with spherical symmetry, the rate of decrease of pressure with radius is given by

$$\frac{dp}{dr} = -\rho g. \tag{2-65}$$

In calculating the lithostatic stress near the surface of a planet, it is adequate to consider g to be constant. However, deep in a planet g is a function of radius, as shown in Figure 2–16. In addition, ρ is also generally a function of r. The gravitational acceleration $g(r)$ for a

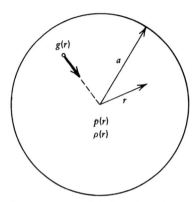

2–16 Spherically symmetric model of a planet for the purpose of calculating $p(r)$.

spherically symmetric body is given by

$$g(r) = \frac{GM(r)}{r^2}, \tag{2–66}$$

where G is the *universal gravitational constant* and $M(r)$ is the mass inside radius r.

$$M(r) = \int_0^r 4\pi r'^2 \rho(r') \, dr'. \tag{2–67}$$

A further discussion of planetary gravity is given in Chapter 5. Given a model of the density inside a planet, that is, given the form of $\rho(r)$, one can integrate Equation (2–67) to obtain $M(r)$; $g(r)$ follows from Equation (2–66). Equation (2–65) can then be integrated to solve for $p(r)$. In general, however, $\rho(r)$ is a function of $p(r)$; so an equation of state is required.

For a small planet, such as the Moon, the procedure is particularly straightforward, since ρ can be considered a constant; $M(r)$ is then $\frac{4}{3}\pi\rho r^3$, and the acceleration of gravity is

$$g(r) = \frac{4}{3}\pi\rho Gr. \tag{2–68}$$

The equation for p is

$$\frac{dp}{dr} = -\frac{4}{3}\pi\rho^2 Gr, \tag{2–69}$$

which upon integration gives

$$p = -\frac{2}{3}\pi\rho^2 Gr^2 + c. \tag{2–70}$$

The constant of integration c can be evaluated by equating the pressure to zero at the surface of the body $r = a$. One obtains

$$p = \frac{2}{3}\pi\rho^2 G(a^2 - r^2). \tag{2–71}$$

Pressure is a quadratic function of radius in a small constant-density planet.

PROBLEM 2-17 Determine the pressure at the center of the Moon. Assume $\rho = 3300 \text{ kg m}^{-3}$ and $a = 1738$ km. What is the variation of g with radius in the Moon?

PROBLEM 2-18 Consider a simple two-layer model of a planet consisting of a core of density ρ_c and radius b surrounded by a mantle of density ρ_m and thickness $a - b$. Show that the gravitational acceleration as a function of radius is given by

$$g(r) = \frac{4}{3}\pi\rho_c Gr \qquad 0 \le r \le b$$
$$= \frac{4}{3}\pi G[r\rho_m + b^3(\rho_c - \rho_m)/r^2] \qquad b \le r \le a. \tag{2–72}$$

and that the pressure as a function of radius is given by

$$p(r) = \frac{4}{3}\pi\rho_m Gb^3(\rho_c - \rho_m)\left(\frac{1}{r} - \frac{1}{a}\right)$$
$$+ \frac{2}{3}\pi G\rho_m^2(a^2 - r^2) \qquad b \le r \le a$$
$$= \frac{2}{3}\pi G\rho_c^2(b^2 - r^2) + \frac{2}{3}\pi G\rho_m^2(a^2 - b^2)$$
$$+ \frac{4}{3}\pi\rho_m Gb^3(\rho_c - \rho_m)\left(\frac{1}{b} - \frac{1}{a}\right)$$
$$0 \le r \le b. \tag{2–73}$$

Apply this model to the Earth. Assume $\rho_m = 4000$ kg m^{-3}, $b = 3486$ km, $a = 6371$ km. Calculate ρ_c given that the total mass of the Earth is 5.97×10^{24} kg. What are the pressures at the center of the Earth and at the core–mantle boundary? What is the acceleration of gravity at $r = b$?

2-6 Stress Measurement

The direct measurement of stress is an important source of information on the state of stress in the lithosphere. At shallow depths, the state of stress is strongly affected by the presence of faults and joints, and stress measurements near the surface yield little useful information on tectonic stresses in the lithosphere. At sufficiently large depths, the lithostatic pressure closes these zones of weakness, allowing stresses to be transmitted across them. Stress measurements made at depth are thus directly interpretable in terms of large-scale

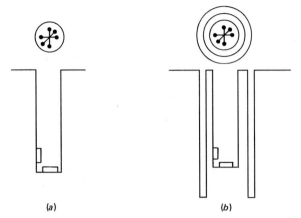

(a)   (b)

2-17 Schematic of overcoring technique for stress measurements. *(a)* A hole is drilled, and four strain gauges are installed: one on the side wall to measure σ_{yy} and three on the base to measure σ_{xx}, σ_{xz}, and σ_{zz}. It is assumed that the drilling of the hole has not affected the ambient state of stress. *(b)* The second annular hole is drilled. It is assumed that this annular hole completely relieves the initial stresses.

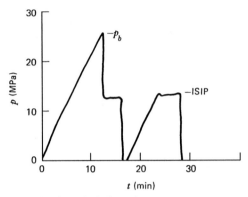

2-18 Pressure log during hydrofracturing.

tectonic stresses. Stress measurements at depth are carried out in mines and in deep boreholes. The two principal methods of making in situ stress measurements are *overcoring* and *hydrofracturing*.

The first step in overcoring is to drill a hole in rock that is free of faults and joints. *Strain (deformation) gauges* can be installed in three perpendicular directions on the base of the hole and on the side, as illustrated in Figure 2–17*a* (strain or deformation in response to stress is discussed quantitatively in the following section). Alternatively two holes are drilled at right angles, and strain gauges are installed on the bases of the two holes. We assume that the stress in the direction of the strain gauge is not affected by the drilling of the hole. The hole is then overcored; that is, an annular hole with radius larger than the original hole is drilled, as shown in Figure 2–17*b*. We assume that the overcoring completely relieves the stresses in the isolated block of rock to which the strain gauges have been attached. The displacements on the strain gauges can then be used to determine the original state of stress. An important limitation of this method is that the length of the hole used is limited to about 1 m. To make overcoring measurements at greater depths, it is necessary to drill the holes in mines.

The second method of direct stress measurement is hydrofracturing. In this method a section of a borehole that is free of fractures or other porosities is isolated

using inflatable packers. The isolated section is then pressurized by pumping fluid into it, and the pressure of the fluid is monitored. The pressure is increased until a fracture occurs. The fluid pressure at which the fracture occurs is referred to as the *breakdown pressure p_b*. A typical pressure–time history during hydrofracturing is illustrated in Figure 2–18. If the pump is shut off immediately and the hydraulic circuit kept closed, an *instantaneous shut-in pressure (ISIP)* is recorded, as illustrated in Figure 2–18. This is the pressure that is just sufficient to hold the fracture open. If the fluid pressure is dropped and then increased, the maximum pressure is the ISIP when the fracture is once again opened.

Several assumptions are implicit in the interpretation of the pressure record obtained during hydrofracturing. The first is that the resultant fracture is in a vertical plane. The second is that the rock fractures in pure tension so that the stress perpendicular to the fracture is the minimum horizontal principal stress. With these assumptions the magnitude of the minimum horizontal principal stress is equal to the ISIP. Using theories for the fracture of rock, the maximum horizontal principal stress can be deduced from p_b and ISIP, but with considerably less accuracy.

Measurements of the minimum horizontal stress σ_{min} as a function of depth in the Cajon Pass borehole in California are given in Figure 2–19. This borehole is adjacent to the San Andreas fault in southern California and was drilled to a depth of 3.5 km. A series of ISIP measurements were carried out using both hydrofractures and preexisting fractures, and it is assumed that these gave σ_{min}.

In general, measurements of the vertical component of stress indicate that it is nearly equal to the weight of

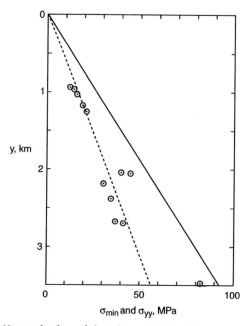

2-19 Measured values of the minimum horizontal stress σ_{min} as a function of depth y in the Cajon Pass borehole in California (Zoback and Healy, 1992). Also included in this figure are the vertical component of stress σ_{yy} shown by the solid line (assumed equal to the lithostatic pressure) and 0.6 σ_{yy} shown by the dashed line.

the overburden, that is, the lithostatic pressure. Using Equation (2–1) with $\rho g = 26.5$ MPa km^{-1} the vertical component of stress σ_{yy} is also given as a function of depth in Figure 2–19. The measured stresses correlate reasonably well with 0.6 σ_{yy}.

Another technique used to determine the orientation of crustal stresses is the observation of wellbore breakouts. Wellbore breakouts are the result of localized failure around a borehole in response to horizontal compression. Compression produces spallation zones along the wellbore at the azimuth of minimum principal stress where the circumferential compressive stress is a maximum. The spallation zones can be used to infer the directions of the horizontal principal stresses (Gough and Bell, 1981).

Observations of wellbore breakouts can be obtained from borehole televiewer data. The borehole televiewer is an ultrasonic well-logging tool which can image the orientation and distribution of fractures as well as the orientation of stress-induced wellbore breakouts.

PROBLEM 2-19 An overcoring stress measurement in a mine at a depth of 1.5 km gives normal stresses

TABLE 2-1 Stress Measurements at 200 m Depth vs. Distance from the San Andreas Fault

Distance from Fault (km)	Maximum Principal Stress (MPa)	Minimum Principal Stress (MPa)
2	9	8
4	14	8
22	18	8
34	22	11

of 62 MPa in the N–S direction, 48 MPa in the E–W direction, and 51 MPa in the NE–SW direction. Determine the magnitudes and directions of the principal stresses.

PROBLEM 2-20 The measured horizontal principal stresses at a depth of 200 m are given in Table 2–1 as a function of distance from the San Andreas fault. What are the values of maximum shear stress at each distance?

2-7 Basic Ideas about Strain

Stresses cause solids to deform; that is, the stresses produce changes in the distances separating neighboring small elements of the solid. In the discussion that follows we describe the ways in which this deformation can occur. Implicit in our discussion is the assumption that the deformations are small.

Figure 2–20 shows a small element of the solid in the shape of a rectangular parallelepiped. Prior to deformation it has sides δx, δy, and δz. The element may be deformed by changing the dimensions of its sides while maintaining its shape in the form of a rectangular parallelepiped. After deformation, the sides of the element are $\delta x - \varepsilon_{xx}\delta x$, $\delta y - \varepsilon_{yy}\delta y$, and $\delta z - \varepsilon_{zz}\delta z$. The quantities ε_{xx}, ε_{yy}, and ε_{zz} are normal components of *strain*; ε_{xx} is the change in length of the side parallel to the x axis divided by the original length of the side, and ε_{yy} and ε_{zz} are similar fractional changes in the lengths of the sides originally parallel to the y and z axes, respectively. The normal components of strain ε_{xx}, ε_{yy}, and ε_{zz} are assumed, by convention, to be positive if the deformation shortens the length of a side. This is consistent with the convention that treats compressive stresses as positive.

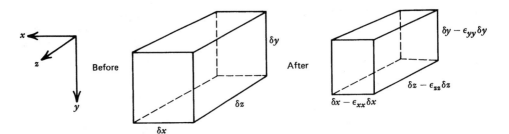

If the deformation of the element in Figure 2–20 is so small that squares and higher order products of the strain components can be neglected in computing the change in volume of the element, the fractional change in volume (volume change divided by original volume) is $\varepsilon_{xx} + \varepsilon_{yy} + \varepsilon_{zz}$. This quantity is known as the *dilatation* Δ; it is positive if the volume of the element is decreased by compression.

PROBLEM 2–21 Uplift and subsidence of large areas are also accompanied by horizontal or lateral strain because of the curvature of the Earth's surface. Show that the lateral strain ε accompanying an uplift Δy is given by

$$\varepsilon = \frac{\Delta y}{R}, \tag{2–74}$$

where R is the radius of the Earth.

PROBLEM 2–22 The *porosity* ϕ of a rock is defined as its void volume per unit total volume. If all the pore spaces could be closed, for example, by subjecting the rock to a sufficiently large pressure, what would be the dilatation? For loose sand ϕ is about 40%, and for oil sands it is usually in the range of 10 to 20%. Table 2–2 gives the porosities of several rocks.

The strain components of a small element of solid can be related to the *displacement* of the element. In

2–20 A deformation that changes the dimensions of a rectangular parallelepiped but not its shape.

order to simplify the derivation of this relationship, we consider the two-dimensional example in Figure 2–21. Prior to deformation, the rectangular element occupies the position *pqrs*. After deformation, the element is in the position *p′q′r′s′*. It is assumed to retain a rectangular shape. The coordinates of the corner *p* before strain are *x* and *y*; after strain the corner is displaced to the location denoted by *p′* with coordinates *x′*, *y′*. The displacement of the corner *p* as a result of the strain or deformation is

$$w_x(x, y) = x - x' \tag{2–75}$$

in the *x* direction and

$$w_y(x, y) = y - y' \tag{2–76}$$

in the *y* direction. Displacements in the negative *x* and *y* directions are considered positive to agree with the

2–21 Distortion of the rectangular element *pqrs* into the rectangular element *p′q′r′s′*.

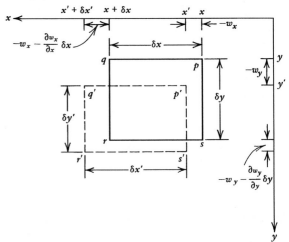

TABLE 2-2 **Rock Porosities**	
Rock	**Porosity (%)**
Hasmark dolomite	3.5
Marianna limestone	13.0
Berea sandstone	18.2
Muddy shale	4.7
Repetto siltstone	5.6

sign convention in which positive strains imply a contraction. Corner q at $x + \delta x, y$ is displaced to position q' with coordinates $x' + \delta x', y'$ as a result of the deformation. Its displacement in the x direction is

$$w_x(x + \delta x, y) = x + \delta x - (x' + \delta x'). \quad (2\text{–}77)$$

Similarly, the displacement of corner s in the y direction $w_y(x, y + \delta y)$ is given by the difference in the y coordinates of s' and s

$$w_y(x, y + \delta y) = y + \delta y - (y' + \delta y'). \quad (2\text{–}78)$$

In writing Equations (2–77) and (2–78), we have assumed that the strains $(\delta x - \delta x')/\delta x$ and $(\delta y - \delta y')/\delta y$ are small.

Since δx and δy are infinitesimal, we can expand $w_x(x + \delta x, y)$ and $w_y(x, y + \delta y)$ as

$$w_x(x + \delta x, y) = w_x(x, y) + \frac{\partial w_x}{\partial x}\delta x \quad (2\text{–}79)$$

$$w_y(x, y + \delta y) = w_y(x, y) + \frac{\partial w_y}{\partial y}\delta y. \quad (2\text{–}80)$$

Substitution of Equation (2–79) into Equation (2–77) and subtraction of Equation (2–75) yield

$$\delta x = \delta x' + \frac{\partial w_x}{\partial x}\delta x. \quad (2\text{–}81)$$

Similarly, substitution of Equation (2–80) into Equation (2–78) and subtraction of Equation (2–76) yield

$$\delta y = \delta y' + \frac{\partial w_y}{\partial y}\delta y. \quad (2\text{–}82)$$

From the definitions of the strain components and Equations (2–81) and (2–82) we find

$$\varepsilon_{xx} \equiv \frac{\delta x - \delta x'}{\delta x} = \frac{\partial w_x}{\partial x} \quad (2\text{–}83)$$

$$\varepsilon_{yy} \equiv \frac{\delta y - \delta y'}{\delta y} = \frac{\partial w_y}{\partial y}. \quad (2\text{–}84)$$

In *three-dimensional strain*, the third strain component ε_{zz} is clearly given by

$$\varepsilon_{zz} = \frac{\delta z - \delta z'}{\delta z} = \frac{\partial w_z}{\partial z}. \quad (2\text{–}85)$$

The components of strain in the x, y, and z directions are proportional to the derivatives of the associated

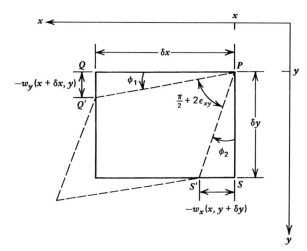

2-22 Distortion of a rectangle into a parallelogram by a strain field involving shear.

displacements in the respective directions. The dilatation Δ is given by

$$\Delta = \frac{\partial w_x}{\partial x} + \frac{\partial w_y}{\partial y} + \frac{\partial w_z}{\partial z}. \quad (2\text{–}86)$$

We have so far considered strains or deformations that do not alter the right angles between line elements that are mutually perpendicular in the unstrained state. Shear strains, however, can distort the shapes of small elements. For example, Figure 2–22 shows a rectangular element in two dimensions that has been distorted into a parallelogram. As illustrated in this figure, the shear strain ε_{xy} is defined to be one-half of the decrease in the angle SPQ

$$\varepsilon_{xy} \equiv -\tfrac{1}{2}(\phi_1 + \phi_2), \quad (2\text{–}87)$$

where ϕ_1 and ϕ_2 are the angles through which the sides of the original rectangular element are rotated. The sign convention adopted here makes ε_{xy} negative if the original right angle is altered to an acute angle. As in the case of stress, the shear strain is symmetric so that $\varepsilon_{yx} = \varepsilon_{xy}$. Figure 2–22 shows that the angles ϕ_1 and ϕ_2 are related to the displacements by

$$\tan \phi_1 = \frac{-w_y(x + \delta x, y)}{\delta x} = \phi_1 \quad (2\text{–}88)$$

$$\tan \phi_2 = \frac{-w_x(x, y + \delta y)}{\delta y} = \phi_2. \quad (2\text{–}89)$$

In Equations (2–88) and (2–89), we assume that the

rotations are infinitesimal so that the tangents of the angles are very nearly equal to the angles themselves.

We can express $w_y(x + \delta x, y)$ and $w_x(x, y + \delta y)$ in terms of the spatial derivatives of the displacements according to

$$w_y(x + \delta x, y) = \frac{\partial w_y}{\partial x}\delta x \qquad (2\text{–}90)$$

$$w_x(x, y + \delta y) = \frac{\partial w_x}{\partial y}\delta y, \qquad (2\text{–}91)$$

Where, for simplicity, we assume $w_x(x, y) = 0$ and $w_y(x, y) = 0$. Substitution of Equations (2–90) and (2–91) into Equations (2–88) and (2–89) and further substitution of the resulting expressions for ϕ_1 and ϕ_2 into Equation (2–87) yield

$$\varepsilon_{xy} = \frac{1}{2}\left(\frac{\partial w_y}{\partial x} + \frac{\partial w_x}{\partial y}\right) \qquad (2\text{–}92)$$

as the relation between shear strain and the spatial derivatives of displacements. In the engineering literature, $\gamma_{xy} = 2\varepsilon_{xy}$ is often used. Care should be exercised in dealing with these quantities.

Shear strain can also lead to a solid-body rotation of the element if $\phi_1 \neq \phi_2$. The *solid-body rotation* ω_z is defined by the relation

$$\omega_z = -\frac{1}{2}(\phi_1 - \phi_2). \qquad (2\text{–}93)$$

Substitution of Equations (2–88) and (2–89) into Equation (2–93) gives

$$\omega_z = \frac{1}{2}\left(\frac{\partial w_y}{\partial x} - \frac{\partial w_x}{\partial y}\right). \qquad (2\text{–}94)$$

The rotation of any element can be resolved in terms of the shear strain and the solid-body rotation. From Equations (2–87) and (2–93), the angle ϕ_1 through which a line element parallel to the x axis is rotated is

$$\phi_1 = -(\varepsilon_{xy} + \omega_z), \qquad (2\text{–}95)$$

and the angle ϕ_2 through which a line element in the y direction is rotated is

$$\phi_2 = \omega_z - \varepsilon_{xy}. \qquad (2\text{–}96)$$

Thus, in the absence of solid-body rotation, ε_{xy} is the clockwise angle through which a line element in the x direction is rotated. It is also the counterclockwise angle through which a line element in the y direction is rotated.

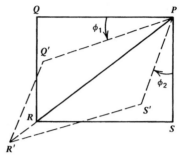

(a) Pure shear strain ($\phi_1 = \phi_2$)

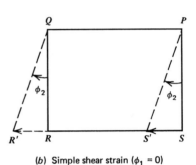

(b) Simple shear strain ($\phi_1 = 0$)

2–23 Sketch of (a) pure shear strain that involves no solid-body rotation of elements and (b) simple shear strain that includes such rotation.

If the amount of solid-body rotation is zero, the distortion is known as *pure shear*. In this case, illustrated in Figure 2–23a,

$$\phi_1 = \phi_2 \qquad (2\text{–}97)$$

$$\frac{\partial w_y}{\partial x} = \frac{\partial w_x}{\partial y} \qquad (2\text{–}98)$$

and the shear strain is

$$\varepsilon_{xy} = \frac{\partial w_x}{\partial y} = \frac{\partial w_y}{\partial x}. \qquad (2\text{–}99)$$

The case of *simple shear*, shown in Figure 2–23b, combines solid-body rotation and shear in such a manner that

$$\phi_1 = \frac{\partial w_y}{\partial x} = 0. \qquad (2\text{–}100)$$

From Equation (2–94), the amount of solid-body rotation is

$$\omega_z = -\frac{1}{2}\frac{\partial w_x}{\partial y}, \qquad (2\text{–}101)$$

and the shear strain is

$$\varepsilon_{xy} = \frac{1}{2} \frac{\partial w_x}{\partial y}. \tag{2-102}$$

Simple shear is often associated with strike–slip faulting.

The equations given for two-dimensional strains and solid-body rotation about one axis can be readily generalized to three dimensions. A pure shear strain in the xz plane has an associated shear strain component given by

$$\varepsilon_{xz} = \varepsilon_{zx} = \frac{1}{2} \left(\frac{\partial w_z}{\partial x} + \frac{\partial w_x}{\partial z} \right) \tag{2-103}$$

and a pure shear strain in the yz plane corresponds to

$$\varepsilon_{yz} = \varepsilon_{zy} = \frac{1}{2} \left(\frac{\partial w_z}{\partial y} + \frac{\partial w_y}{\partial z} \right). \tag{2-104}$$

A solid-body rotation about the x axis ω_x is related to displacement derivatives by

$$\omega_x = \frac{1}{2} \left(\frac{\partial w_z}{\partial y} - \frac{\partial w_y}{\partial z} \right). \tag{2-105}$$

Similarly, a solid-body rotation about the y axis is

$$\omega_y = \frac{1}{2} \left(\frac{\partial w_x}{\partial z} - \frac{\partial w_z}{\partial x} \right). \tag{2-106}$$

The strain components ε_{xx}, ε_{yy}, ε_{zz}, ε_{xy}, ε_{xz}, and ε_{yz} are sufficient to describe the general infinitesimal deformation of solid elements subjected to stresses. The solid-body rotations ω_x, ω_y, and ω_z do not alter distances between neighboring elements of a solid and, therefore, do not involve stresses. Accordingly, the strain components and their associated stresses are of primary concern to us in subsequent chapters.

Just as it was important to know the stresses on area elements whose normals make arbitrary angles with respect to x, y axes, so it is essential to know the fractional changes in length and the rotation angles of arbitrarily inclined line elements. For simplicity we consider the two-dimensional case. We wish to determine the strains in the x', y' coordinate system, which is inclined at an angle θ with respect to the x, y coordinate system, as shown in Figure 2–24a. As a result of the strain field ε_{xx}, ε_{yy}, ε_{xy} and the solid-body rotation ω_z, the line elements PR and PQ experience changes in length and rotations. Line element PR is parallel to the x' axis, and PQ is parallel to the y' axis. The extension in length of PR divided by the original length $\delta x'$ is the strain

component $-\varepsilon_{x'x'}$; the counterclockwise angle of rotation of PR is the angle $\phi_1' = -\varepsilon_{x'y'} - \omega_{z'}$. This is illustrated in Figure 2–24b. The extension in length of PQ divided by the original length $\delta y'$ is the strain component $-\varepsilon_{y'y'}$; the clockwise rotation of PQ is the angle $\phi_2' = \omega_{z'} - \varepsilon_{x'y'}$. This is shown in Figure 2–24c.

We first determine the strain component $-\varepsilon_{x'x'}$. The displacement of R to R' in Figure 2–24b is the net result of the combined elongations and rotations of δx and δy. The x component of the displacement of R' relative to R arises from the elongation of δx in the x direction, $-\varepsilon_{xx}\delta x$, and the rotation of δy through the clockwise angle ϕ_2. The latter contribution to the displacement is $\phi_2\delta y$, which, according to Equation (2–96), is $(\omega_z - \varepsilon_{xy})\delta y$. Thus the total x component of the displacement of R' with respect to R is

$$-\varepsilon_{xx}\delta x + (\omega_z - \varepsilon_{xy})\delta y.$$

The y component of the displacement of R' with respect to R is the sum of the elongation of δy, $-\varepsilon_{yy}\delta y$, and the contribution from the rotation of δx, which, with Equation (2–95), is $\phi_1\delta x = -(\varepsilon_{xy} + \omega_z)\delta x$. Thus the total y component of displacement of R' with respect to R is

$$-\varepsilon_{yy}\delta y - (\varepsilon_{xy} + \omega_z)\delta x.$$

For small strains, the change in length of PR is the sum of the x component of RR' projected on the line PR,

$$[-\varepsilon_{xx}\delta x + (\omega_z - \varepsilon_{xy})\delta y] \cos\theta,$$

and the y component of RR' projected on the line PR,

$$[-\varepsilon_{yy}\delta y - (\varepsilon_{xy} + \omega_z)\delta x] \sin\theta.$$

The strain component $\varepsilon_{x'x'}$ is thus

$$-\varepsilon_{x'x'} = \frac{[-\varepsilon_{xx}\delta x + (\omega_z - \varepsilon_{xy})\delta y]\cos\theta}{\delta x'}$$
$$+ \frac{[-\varepsilon_{yy}\delta y - (\varepsilon_{xy} + \omega_z)\delta x]\sin\theta}{\delta x'}. \tag{2-107}$$

Since

$$\frac{\delta x}{\delta x'} = \cos\theta \qquad \frac{\delta y}{\delta x'} = \sin\theta \tag{2-108}$$

Equation (2–107) can be rewritten as

$$\varepsilon_{x'x'} = \varepsilon_{xx}\cos^2\theta + \varepsilon_{yy}\sin^2\theta + 2\varepsilon_{xy}\sin\theta\cos\theta. \tag{2-109}$$

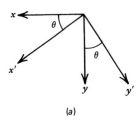

(a)

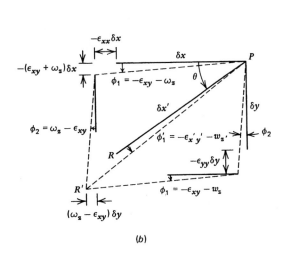

(b)

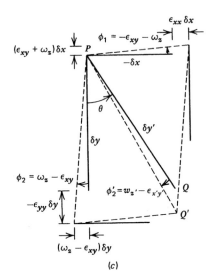

(c)

2-24 *(a)* The transformation of coordinates *x, y* through an angle θ to *x', y'*. *(b)* The transformation of the strain components onto the *x'* axis. *(c)* The transformation of the strain components onto the *y'* axis.

Using Equation (2–36), we can further rewrite Equation (2–109) as

$$\varepsilon_{x'x'} = \varepsilon_{xx} \cos^2 \theta + \varepsilon_{yy} \sin^2 \theta + \varepsilon_{xy} \sin 2\theta. \quad (2\text{--}110)$$

This has the same form as the transformation of the normal stress given in Equation (2–37).

We next determine the strain component $-\varepsilon_{y'y'}$. As can be seen in Figure 2–24c, the component of the displacement of Q' with respect to Q in the negative *x* direction is the sum of the elongation of δx, $-\varepsilon_{xx}\delta x$, and the contribution from the rotation of δy, $-\phi_2 \delta y = -(\omega_z - \varepsilon_{xy})\delta y$, that is,

$$-\varepsilon_{xx}\delta x - (\omega_z - \varepsilon_{xy})\delta y.$$

The *y* component of the displacement of Q' with respect to Q is the sum of the elongation of δy, $-\varepsilon_{yy}\delta y$, and the contribution due to the rotation of δx, $-\phi_1 \delta x =$

$(\varepsilon_{xy} + \omega_z)\delta x$, that is,

$$-\varepsilon_{yy}\delta y + (\omega_z + \varepsilon_{xy})\delta x.$$

After projection of these displacements onto the line PQ, the strain component $\varepsilon_{y'y'}$ can be written as

$$-\varepsilon_{y'y'} = \frac{-[\varepsilon_{xx}\delta x + (\omega_z - \varepsilon_{xy})\delta y]\sin\theta}{\delta y'}$$
$$+ \frac{[-\varepsilon_{yy}\delta y + (\omega_z + \varepsilon_{xy})\delta x]\cos\theta}{\delta y'}. \quad (2\text{--}111)$$

Since

$$\frac{\delta x}{\delta y'} = \sin\theta \qquad \frac{\delta y}{\delta y'} = \cos\theta, \quad (2\text{--}112)$$

Equation (2–111) can be put in the form

$$\varepsilon_{y'y'} = \varepsilon_{xx} \sin^2 \theta + \varepsilon_{yy} \cos^2 \theta - 2\varepsilon_{xy} \sin\theta\cos\theta.$$
$$(2\text{--}113)$$

By substituting Equation (2–36) into Equation (2–113), we get

$$\varepsilon_{y'y'} = \varepsilon_{xx} \sin^2 \theta + \varepsilon_{yy} \cos^2 \theta - \varepsilon_{xy} \sin 2\theta. \quad (2\text{--}114)$$

PROBLEM 2–23 Derive Equation (2–114) from Equation (2–110) by using the substitution $\theta' = \theta + \pi/2$. Why can this be done?

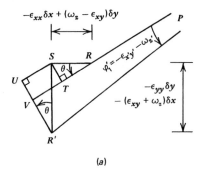

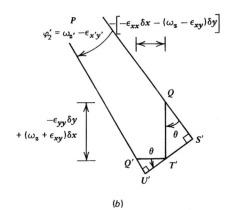

2-25 Geometrical determination of (a) ϕ_1' and (b) ϕ_2'.

We now turn to the determination of the shear strain, $\varepsilon_{x'y'}$, and the solid-body rotation $\omega_{z'}$ in the new coordinate system. We first determine the angle $\phi_1' = -\varepsilon_{x'y'} - \omega_{z'}$ from the geometrical relationships shown in Figure 2–25a. For sufficiently small strain, ϕ_1' is given by

$$\phi_1' = -\varepsilon_{x'y'} - \omega_{z'} = \frac{R'V}{\delta x'}. \qquad (2-115)$$

From Figure 2–25a we can see that

$$R'V = R'U - VU = R'U - TS, \qquad (2-116)$$

and

$$R'U = [-\varepsilon_{yy}\delta y - (\varepsilon_{xy} + \omega_z)\delta x]\cos\theta \qquad (2-117)$$
$$TS = [-\varepsilon_{xx}\delta x + (\omega_z - \varepsilon_{xy})\delta y]\sin\theta. \qquad (2-118)$$

By combining Equations (2–108) and (2–115) with (2–118), we obtain

$$\varepsilon_{x'y'} + \omega_{z'} = (\varepsilon_{yy} - \varepsilon_{xx})\sin\theta\cos\theta$$
$$+ \varepsilon_{xy}(\cos^2\theta - \sin^2\theta) + \omega_z. \qquad (2-119)$$

The angle ϕ_2' can be found from the geometrical relationships shown in Figure 2–25b; it is given by

$$\phi_2' = \omega_{z'} - \varepsilon_{x'y'} = \frac{U'S'}{\delta y'}. \qquad (2-120)$$

From Figure 2–25b it is seen that

$$U'S' = U'T' + T'S', \qquad (2-121)$$

and

$$U'T' = -[\varepsilon_{xx}\delta x - (\omega_z - \varepsilon_{xy})\delta y]\cos\theta \qquad (2-122)$$
$$T'S' = [-\varepsilon_{yy}\delta y + (\omega_z + \varepsilon_{xy})\delta x]\sin\theta. \qquad (2-123)$$

By combining Equations (2–112) and (2–120) with (2–123), we obtain

$$\omega_{z'} - \varepsilon_{x'y'} = (\varepsilon_{xx} - \varepsilon_{yy})\sin\theta\cos\theta$$
$$- \varepsilon_{xy}(\cos^2\theta - \sin^2\theta) + \omega_z. \qquad (2-124)$$

By adding and subtracting Equations (2–119) and (2–124), we can find separate equations for $\omega_{z'}$ and $\varepsilon_{x'y'}$:

$$\omega_{z'} = \omega_z \qquad (2-125)$$
$$\varepsilon_{x'y'} = (\varepsilon_{yy} - \varepsilon_{xx})\sin\theta\cos\theta + \varepsilon_{xy}(\cos^2\theta - \sin^2\theta). \qquad (2-126)$$

The solid-body rotation is invariant to the coordinate transformation, as expected, because it represents a rotation of an element without deformation. By introducing Equations (2–36) and (2–39) into Equation (2–126), we obtain

$$\varepsilon_{x'y'} = \tfrac{1}{2}(\varepsilon_{yy} - \varepsilon_{xx})\sin 2\theta + \varepsilon_{xy}\cos 2\theta. \qquad (2-127)$$

This has the same form as the transformation of the shear stress given in Equation (2–40).

Just as there are principal axes of stress in a solid, there are *principal axes of strain*. In the principal strain axis coordinate system, shear strain components are zero. Setting $\varepsilon_{x'y'} = 0$ in Equation (2–127) gives the direction of one of the principal axes of strain as

$$\tan 2\theta = \frac{2\varepsilon_{xy}}{\varepsilon_{xx} - \varepsilon_{yy}}. \qquad (2-128)$$

We have already shown, in connection with principal stress axes, that if θ is a principal axis direction, so is

$\theta + \pi/2$. The fractional changes in length along the directions of the principal strain axes are the *principal strains*. With θ given by Equation (2–128), Equation (2–110) determines the principal strain $\varepsilon_1 = \varepsilon_{x'x'}$. The principal strain ε_2 is identified with $\varepsilon_{y'y'}$. By a procedure analogous to the one used in deriving Equation (2–51) we find

$$\varepsilon_{1,2} = \tfrac{1}{2}(\varepsilon_{xx} + \varepsilon_{yy}) \pm \left\{\varepsilon_{xy}^2 + \tfrac{1}{4}(\varepsilon_{xx} - \varepsilon_{yy})^2\right\}^{1/2}.$$

$$(2\text{--}129)$$

It is convenient to have formulas for the normal and shear strains at an angle θ with respect to the ε_1 principal strain axis. Taking $\varepsilon_{xy} = 0$, $\varepsilon_{xx} = \varepsilon_1$, and $\varepsilon_{yy} = \varepsilon_2$ in Equations (2–109) and (2–127), we obtain

$$\varepsilon_{xx} = \varepsilon_1 \cos^2\theta + \varepsilon_2 \sin^2\theta \qquad (2\text{--}130)$$

$$\varepsilon_{xy} = -\tfrac{1}{2}(\varepsilon_1 - \varepsilon_2)\sin 2\theta. \qquad (2\text{--}131)$$

PROBLEM 2–24 Show that the principal strains are the minimum and the maximum fractional changes in length.

PROBLEM 2–25 Show that the maximum shear strain is given by $\tfrac{1}{2}(\varepsilon_1 - \varepsilon_2)$. What is the direction in which the shear strain is maximum?

Principal axes of strain can also be found for arbitrary three-dimensional strain fields. With respect to these axes all shear strain components are zero. The normal strains along these axes are the principal strains ε_1, ε_2, and ε_3. One can introduce the concept of *deviatoric strain* in analogy to deviatoric stress by referring the strain components to a state of *isotropic strain* equal to the average normal strain e. In three dimensions

$$e \equiv \tfrac{1}{3}(\varepsilon_{xx} + \varepsilon_{yy} + \varepsilon_{zz}) = \tfrac{1}{3}\Delta. \qquad (2\text{--}132)$$

The average normal strain and the dilatation are invariant to the choice of coordinate axes. The deviatoric strain components, denoted by primes, are

$$\varepsilon_{xx}' = \varepsilon_{xx} - e \qquad \varepsilon_{yy}' = \varepsilon_{yy} - e \qquad \varepsilon_{zz}' = \varepsilon_{zz} - e$$

$$\varepsilon_{xy}' = \varepsilon_{xy} \qquad \varepsilon_{xz}' = \varepsilon_{xz} \qquad \varepsilon_{yz}' = \varepsilon_{yz}. \qquad (2\text{--}133)$$

2–8 Strain Measurements

Strain or deformation at the Earth's surface is often a consequence of large-scale tectonic forces. Thus the measurement of surface strain can provide important information on fundamental geodynamic processes. For example, in order to understand the mechanical behavior of faults, it is essential to determine the distribution of the *coseismic surface strain* as a function of distance from the fault, a problem we discuss further in Chapter 8. Because surface strains are generally very small, sophisticated distance-measuring techniques are usually required to determine them. However, there are instances in which surface displacements are so large that they can be easily measured. An example is the surface offset on a fault when a great earthquake occurs; offsets of 10 m and more have been recorded. Tree lines, roads, railroad tracks, pipelines, fences, and the like can be used to make such measurements. Figure 2–26 shows a fence offset by 3 m during the 1906 earthquake on the San Andreas fault in California. Measured surface offsets resulting from this earthquake are summarized in Figure 2–27. Although there is considerable scatter in the data, an offset of about 4 m was observed along much of the fault break. The scatter of the data illustrates one of the principal problems in measuring surface strain. The Earth's crust is not a continuum material with uniform properties. Changes in rock

2–26 A fence offset by 3 m on the ranch of E. R. Strain, Marin County, California, as a result of slip along the San Andreas fault during the great 1906 earthquake (G. K. Gilbert 3028, U.S. Geological Survey.)

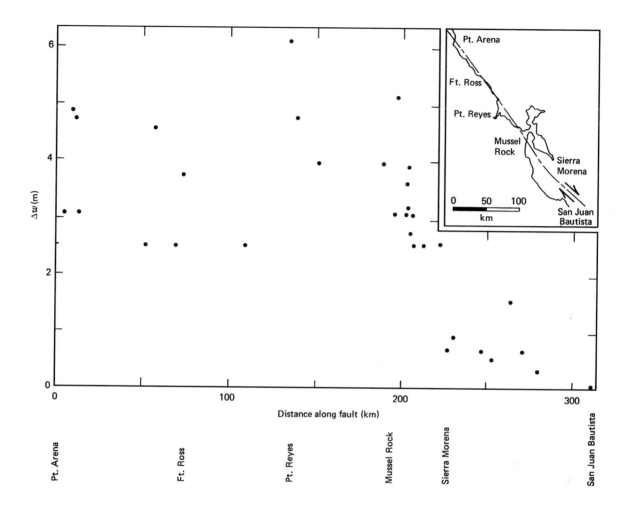

2-27 Observed surface offsets on the San Andreas fault resulting from the 1906 earthquake (Thatcher, 1975).

type, the presence of thick soil layers, and offsets on branching or secondary faults all contribute to the variations in the measured offsets.

The strain field associated with the 1906 earthquake can be estimated from the data in Figure 2–27. Since the San Andreas fault is a strike–slip fault, we assume that the strain field is a simple shear and that it extends 40 km from the fault. The distance that the cyclic strain field extends from the fault is considered in detail in Chapter 8. The value of 40 km is subject to considerable uncertainty. The mean displacement of 4 m across the fault during the earthquake is made up of 2 m displacements on opposite sides of the fault. The shear strain ε_{xz} can thus be estimated from Equation (2–102) as

$$\varepsilon_{xz} = \frac{1}{2} \frac{\partial w_x}{\partial z} \approx \frac{1}{2} \frac{2m}{40,000 \text{ m}} = 2.5 \times 10^{-5}. \quad (2-134)$$

If great earthquakes recur about every 100 years along the San Andreas fault, the *rate of shear strain* accumulation on the fault $\dot{\varepsilon}_{xz}$ is

$$\dot{\varepsilon}_{xz} = \frac{2.5 \times 10^{-5}}{100 \text{ yr}} = 0.25 \times 10^{-6} \text{ yr}^{-1}. \quad (2-135)$$

As we have already noted, surface strains of the magnitude calculated in Equation (2–134) are difficult to measure; they require extremely accurate determinations of distances. This has been the main concern of *geodesy* for several centuries. The traditional end product of geodetic surveys is the topographic map, constructed from the elevations of a network of *benchmarks*. Benchmarks are spaced over much of the United States at intervals of a few kilometers and

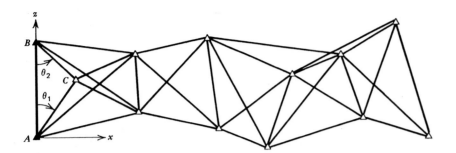

2–28 Illustration of triangulation. The x, z position of monument C can be determined from the line length AB and the angles θ_1 and θ_2. The positions of the other monuments can be similarly determined.

ground surveys are used to establish accurate benchmark elevations in a *geodetic network*. Geodetic networks are systematically resurveyed to determine the changes in elevation between benchmarks. Vertical displacements of benchmarks on the order of 10 to 100 cm are often found. In many instances, these displacements can be associated with subsidence due to the removal of ground water. However, in other cases they must be attributed to tectonic causes. Extensive geodetic measurements have been made along the San Andreas fault, and these are used to illustrate the concepts discussed in this section. Because the San Andreas fault is a strike–slip fault, the displacements associated with slip along the fault are predominantly horizontal. Thus we restrict our attention to the horizontal components of strain $\varepsilon_{xx}, \varepsilon_{xz}$, and ε_{zz}. Horizontal strains are obtained by measuring the positions of fixed *monuments*.

Historically, the standard method for determining the positions of monuments is *triangulation*, as illustrated in Figure 2–28. Assume that the absolute positions of the reference points A and B have been determined. The location of monument C can be found from the line length AB and the two angles θ_1 and θ_2 as follows. Applying the law of sines to triangle ABC produces

$$\frac{(AB)}{\sin(\pi - \theta_1 - \theta_2)} = \frac{(AC)}{\sin \theta_2}. \tag{2–136}$$

If we assume that point A defines the origin of the x, z coordinate system and that AB lies on the z axis, the coordinates of $C(x_c, z_c)$ are given by

$$x_c = (AC) \sin \theta_1 \qquad z_c = (AC) \cos \theta_1. \tag{2–137}$$

By solving Equation (2–136) for (AC) and substituting into Equation (2–137), we obtain

$$x_c = \frac{(AB) \sin \theta_1 \sin \theta_2}{\sin(\pi - \theta_1 - \theta_2)} \tag{2–138}$$

$$z_c = \frac{(AB) \cos \theta_1 \sin \theta_2}{\sin(\pi - \theta_1 - \theta_2)}. \tag{2–139}$$

The locations of the other monuments in the triangulation network can be similarly determined. The use of redundant triangles, as shown in Figure 2–28, improves the accuracy of the results.

PROBLEM 2–26 The coordinates x_A, z_A and x_B, z_B of monuments A and B shown in Figure 2–29 are assumed known. Determine the coordinates x_C, z_C of monument C in terms of the coordinates of monuments A and B and the angles θ_1 and θ_2.

The angles required for triangulation are obtained using a *theodolite*. The accuracy to which an angle can be determined is 0.3 to 1.0 second of arc, implying errors in distance determination of about 3 in 10^6. A typical maximum length over which a measurement is made is 50 km. The accuracy of triangulation observations is equivalent to about 10 years of shear strain accumulation on the San Andreas fault; see Equation (2–135). Therefore, considerable redundancy in a network is required to obtain meaningful results.

2–29 Sketch for Problem 2–26.

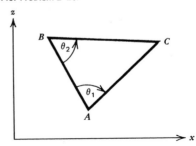

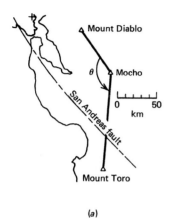

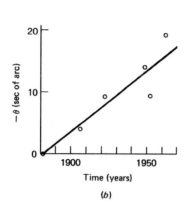

(a)

(b)

2-30 *(a)* The three monuments at Mount Diablo, Mocho, and Mount Toro are part of a primary triangulation network that spans the San Andreas fault south of San Francisco. *(b)* Observed changes in the angle θ between the monuments since 1882 (Savage and Burford, 1973).

An example of triangulation observations is given in Figure 2–30. Mount Diablo, Mocho, and Mount Toro are three monuments in a primary triangulation network that spans the San Andreas fault south of San Francisco (Figure 2–30a). The changes in the angle θ between these monuments in a series of surveys since 1882 are given in Figure 2–30b. Mount Toro lies 30 km southwest of the San Andreas fault, and Mocho lies 60 km northeast of the fault. If it is assumed that these monuments lie outside the zone of strain accumulation and release associated with great earthquakes, the relative motion across the San Andreas fault can be obtained from these observations. A reasonable fit to the data is $d\theta/dt = -0.192$ sec of arc yr^{-1}.

The length of the line between Mocho and Mount Toro is 125 km, and it crosses the San Andreas fault at

an angle of 45°. The calculated relative velocity across the fault is thus

$$u = \frac{125 \times 10^6 \times 0.192}{3600 \times 57.3 \times \sin 45°} = 41 \text{ mm yr}^{-1}. \quad (2\text{–}140)$$

This value is in quite good agreement with the predicted relative velocity of 46 mm yr^{-1} from plate tectonics (see Section 1–8).

PROBLEM 2–27 Figure 2–31 shows three monuments on Mount Diablo, Sonoma Mountain, and Farallon lighthouse and the change in the included angle θ relative to the 1855 measurement. Assuming that these three monuments lie outside the zone of strain accumulation and release on the San Andreas fault, determine the relative velocity across the fault.

2-31 A triangulation net across the San Andreas fault *(a)* and the measured angle θ since 1855 *(b)*.

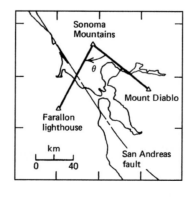

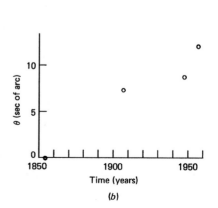

(a)

(b)

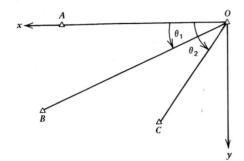

2-32 Sketch for Problem 2-28.

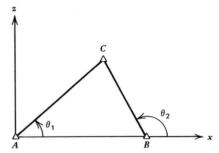

2-33 Illustration of how strain measurements between three monuments *A*, *B*, and *C* can be used to determine the strain field ε_{xx}, ε_{zz}, and ε_{xz}.

PROBLEM 2-28 Triangulation measurements at monument 0 give the time rate of change of θ_1, $\dot{\theta}_1$ and the time rate of change of θ_2, $\dot{\theta}_2$ (Figure 2–32). Show that

$$\dot{\varepsilon}_{xy} = \frac{1}{2} \frac{(\dot{\theta}_2 \sec\theta_2 \csc\theta_2 - \dot{\theta}_1 \sec\theta_1 \csc\theta_1)}{(\tan\theta_2 - \tan\theta_1)} \quad (2\text{–}141)$$

and

$$\dot{\varepsilon}_{yy} - \dot{\varepsilon}_{xx} = \frac{(\dot{\theta}_2 \csc^2\theta_2 - \dot{\theta}_1 \csc^2\theta_1)}{(\operatorname{ctn}\theta_1 - \operatorname{ctn}\theta_2)}, \quad (2\text{–}142)$$

where $\dot{\varepsilon}_{xx} = d\varepsilon_{xx}/dt$, and so on.

As we have shown, the accuracy of triangulation measurements is generally insufficient to obtain useful data on strain accumulation. Fortunately electro-optical distance-measuring instruments greatly improve the accuracy of strain measurements. However, they also greatly increase the expense. To make a distance measurement, a geodolite is placed on one monument and a reflector on the second monument. The geodolite emits a modulated laser beam that is reflected back to the instrument from the reflector. A comparison of the modulated phases of the emitted and returned beams determines the length of the optical path between the monuments as an unknown number of whole modulation lengths plus a precisely determined fractional modulation length. The unknown number of whole modulation lengths is determined by carrying out measurements at successively lower modulation frequencies.

Distances between measured monuments are typically 10 km, and the inherent accuracy of the geodimeter is about 1 mm. Therefore an accuracy of 1 part in 10^7 can in principle be achieved. This is about an order of magnitude better than triangulation measurement accuracy. In practice, however, the accuracy of distance determinations is limited by variations in refractivity

along the atmospheric path. In order to obtain accuracies approaching 1 part in 10^7, it is necessary to determine the temperature and humidity along the path. This is usually done by flying a suitably equipped airplane or helicopter along the path while the distance measurement is being carried out. The atmospheric pressure at the two terminal monuments is also required.

By carrying out measurements at three different wavelengths or frequencies, devices with multiwavelength capabilities have eliminated the need for meteorological observations along the optical path. This approach reduces costs and improves accuracy to a few parts in 10^8.

If it is assumed that the three monuments *A*, *B*, and *C* in Figure 2–33 are in a uniform strain field, measurements of the rates of change in the three line lengths $\dot{\varepsilon}_{xx} = -\Delta AB/AB$, $\dot{\varepsilon}_{x'x'} = -\Delta AC/AC$, and $\dot{\varepsilon}_{x''x''} = -\Delta BC/BC$ and the angles θ_1 and θ_2 give the entire rate of strain field $\dot{\varepsilon}_{xx}$, $\dot{\varepsilon}_{zz}$, and $\dot{\varepsilon}_{xz}$. From Equation (2–109) we have

$$\dot{\varepsilon}_{x'x'} = \dot{\varepsilon}_{xx} \cos^2\theta_1 + \dot{\varepsilon}_{zz} \sin^2\theta_1 + 2\dot{\varepsilon}_{xz} \sin\theta_1 \cos\theta_1 \quad (2\text{–}143)$$

$$\dot{\varepsilon}_{x''x''} = \dot{\varepsilon}_{xx} \cos^2\theta_2 + \dot{\varepsilon}_{zz} \sin^2\theta_2 + 2\dot{\varepsilon}_{xz} \sin\theta_2 \cos\theta_2. \quad (2\text{–}144)$$

These equations can be solved for $\dot{\varepsilon}_{zz}$ and $\dot{\varepsilon}_{xz}$; we find

$$\dot{\varepsilon}_{zz} = \frac{\dot{\varepsilon}_{xx}(\operatorname{ctn}\theta_1 - \operatorname{ctn}\theta_2) - \dot{\varepsilon}_{x'x'} \sec\theta_1 \csc\theta_1}{\tan\theta_2 - \tan\theta_1}$$
$$+ \frac{\dot{\varepsilon}_{x''x''} \sec\theta_2 \csc\theta_2}{\tan\theta_2 - \tan\theta_1} \quad (2\text{–}145)$$

$$\dot{\varepsilon}_{xz} = \frac{\dot{\varepsilon}_{xx}(\operatorname{ctn}^2\theta_1 - \operatorname{ctn}^2\theta_2) - \dot{\varepsilon}_{x'x'} \csc^2\theta_1}{2(\operatorname{ctn}\theta_2 - \operatorname{ctn}\theta_1)}$$
$$+ \frac{\dot{\varepsilon}_{x''x''} \csc^2\theta_2}{2(\operatorname{ctn}\theta_2 - \operatorname{ctn}\theta_1)}. \quad (2\text{–}146)$$

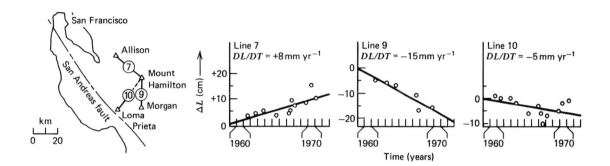

The results can be transformed into any other coordinate system using Equations (2–109), (2–114), and (2–127).

As an example of the direct measurement of strain accumulation, consider the data given in Figure 2–34. We assume that the three lines measured are in a uniform strain field. We further assume that line 7 (length 28 km) defines the x coordinate so that $\dot{\varepsilon}_{xx} = 8/(28 \times 10^6)$ yr^{-1} = 0.29×10^{-6} yr^{-1} and that line 10 (length 31 km) defines the z coordinate so that $\dot{\varepsilon}_{zz} = -5/(31 \times 10^6)$ yr^{-1} = -0.16×10^{-6} yr^{-1}. The angle between lines 9 and 10 is 30°. The rate of strain on line 9 (length 25 km) is $\dot{\varepsilon}_{x'x'} = -1.5/(25 \times 10^6)$ yr^{-1} = -0.6×10^{-6} yr^{-1}. The rate of shear strain $\dot{\varepsilon}_{xz}$ can be determined by inverting Equation (2–109):

$$\dot{\varepsilon}_{xz} = \tfrac{1}{2}(\dot{\varepsilon}_{x'x'} \sec\theta \csc\theta - \dot{\varepsilon}_{xx}\operatorname{ctn}\theta - \dot{\varepsilon}_{zz}\tan\theta).$$

$$(2\text{–}147)$$

With $\theta = 120°$ and the previously determined values of $\dot{\varepsilon}_{x'x'}$, $\dot{\varepsilon}_{xx}$, and $\dot{\varepsilon}_{zz}$ we obtain

$$\begin{aligned}\dot{\varepsilon}_{xz} &= \tfrac{1}{2}(-0.6 \times 1.15 \times 2 - 0.29 \times 1.73 \\ &\quad + 0.16 \times 0.58)\times 10^{-6}\ \text{yr}^{-1} \\ &= -0.90 \times 10^{-6}\ \text{yr}^{-1}.\end{aligned}$$

$$(2\text{–}148)$$

From Equation (2–128) the direction of one of the principal strain rate axes relative to the x axis is $\theta = -38°$. Assuming that line 9 trends N–S, the directions of the principal strain rate axes are 22° W of N and 22° N of E. These can be compared with the principal axis directions expected for a simple shear strain model of strain accumulation in this region; since the San Andreas trends 45° W of N in this area, the expected directions are north and east.

The values of the principal strain rates from Equation (2–129) are $\dot{\varepsilon}_{1,2} = 0.993 \times 10^{-6}$ yr^{-1},

2-34 Line length changes between the monument at Mt. Hamilton and the monuments at Allison (line 7), Loma Prieta (line 10), and Morgan (line 9) obtained between 1961 and 1971 using a geodimeter (Savage and Burford, 1973).

-0.863×10^{-6} yr^{-1}. Assuming $\dot{\varepsilon}_{xz} = \dot{\varepsilon}_1 = -\dot{\varepsilon}_2 = 0.93 \times 10^{-6}$ yr^{-1} (the average of the above two values) and that simple shear is occurring uniformly for a distance d from the fault, d can be determined from the shear strain rate and the relative velocity $u = 46$ mm yr^{-1} according to

$$d = \frac{u}{4\dot{\varepsilon}_{xz}} = \frac{46 \times 10^{-3}}{4 \times 0.93 \times 10^{-6}} = 12.4\ \text{km}. \qquad (2\text{–}149)$$

With the uniform strain assumption, the strain accumulation would be limited to a region closer to the fault than the geodetic net considered.

PROBLEM 2–29 Given in Figure 2–35 are the line lengths between the monument at Diablo and the monuments at Hills, Skyline, and Sunol obtained between 1970 and 1978 using a geodimeter. Assuming a uniform strain field, determine $\dot{\varepsilon}_{xx}$, $\dot{\varepsilon}_{yy}$, and $\dot{\varepsilon}_{xy}$. Take the Sunol–Diablo line to define the y coordinate. Discuss the results in terms of strain accumulation on the San Andreas fault, which can be assumed to trend at 45° with respect to the Sunol–Diablo line (Savage and Prescott, 1978).

Advances in space geodesy have revolutionized geodetic investigations of tectonic motions. Studies carried out in the 1980s utilized *satellite laser ranging* (SLR) and *very long baseline interferometry* (VLBI). SLR measures distances from a ground station to various satellites using an electro-optical instrument similar to the geodolite previously described. Signals are reflected from the satellite and the position of the station is determined relative to the Earth's center of mass. VLBI uses

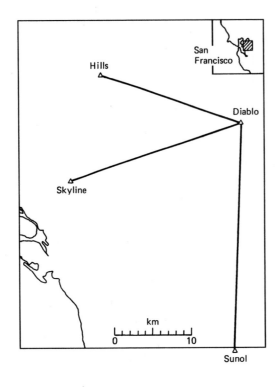

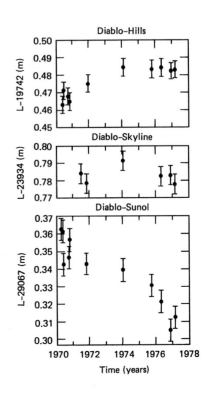

2–35 Geodetic net and measurements for use in Problem 2–29. The line length is L.

interstellar emissions from quasars to obtain interferometric patterns that determine an absolute position of a station. This technique can also be used for studies of the motion of Earth relative to the stars. Because the signals pass through the denser part of the atmosphere at an oblique angle, attenuation problems associated with water vapor are greatly reduced. The success of these techniques demonstrated that space-based geodetic systems could provide absolute positions on the surface of the Earth with a subcentimeter accuracy. However, both techniques have serious limitations due to their use of large stationary or mobile antennas that are bulky and expensive.

These difficulties were overcome when the *global positioning system* (GPS) became fully operational in the early 1990s. GPS consists of some 24 satellites that interact with ground-based receivers to provide accurate surface positions. GPS was introduced by the Department of Defense as a global navigation system with an accuracy of meters (Hofmann-Wallenhof et al., 1997). However, surface instruments were developed that use the carrier signals from the active GPS satellites in an interferometric mode to determine differential positions between surface benchmarks with a

subcentimeter accuracy (Larson, 1996). The great advantage of GPS is the low cost and availability of the instruments so that large numbers of surface observations can be made.

One of the first accomplishments of space geodesy was the confirmation that the plate tectonic velocities given in Section 1–8 are also valid on a year-to-year basis. As a specific example, we show in Figure 2–36 the relative displacements between the Yaragadee station (Perth, Australia) and the Maui station (Hawaiian Islands). These SLR observations give a relative velocity $u_{YM} = -90 \pm 5$ mm yr^{-1}.

We next compare this value with the value predicted by the plate motions given in Table 1–6. We first obtain the motion of the Maui station ($\theta' = 90° - 20.7° = 69.3°$, $\psi' = 203.7°$) relative to the fixed Australian plate. From Table 1–6 we find $\theta = 90° + 60.1° = 150.1°$, $\psi = -178.3°$, and $\omega = 1.07$ deg Myr^{-1} (0.0187 rad Myr^{-1}). Using Equation (1–18) we find that the angle $\Delta_{PM} = 82.76°$. Substitution into Equation (1–17) gives

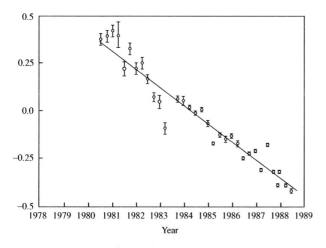

2-36 The geodetic time series for the change in distance along a great circle path between the Yaragadee station (Perth, Australia) and the Maui station (Hawaiian Islands). The distance changes are in meters. The data were obtained using satellite laser ranging (SLR) (Smith et al., 1990). The straight line correlation gives a velocity $u_{ym} = -90 \pm 5$ mm yr^{-1}.

2-37 Observed velocity vectors for geodetic stations in southern California obtained from a combined GPS and VLBI data set (Feigl et al., 1993). The velocities are given relative to a fixed Pacific plate. Error ellipses and major faults are also illustrated.

$u_{MP} = 118$ mm yr^{-1}. This is the velocity of the Maui station relative to the fixed Australian plate; this velocity is perpendicular to the great circle path passing through the pole of rotation and the Maui station and is in the counterclockwise direction.

The measured relative velocity between the Yaragadee and Maui stations, $u_{YM} = -90 \pm 5$ mm yr^{-1}, is in the direction of the great circle between the two stations (the negative sign indicates a convergence). The angle β between the two great circle paths YM and MP must be determined and the plate motion velocity must be resolved onto the YM great circle direction. From Figure 1–35 we see that we can determine the angle β using Equation (1–18) with the result

$$\cos \beta = \frac{\cos \Delta_{PY} - \cos \Delta_{PM} \cos \Delta_{YM}}{\sin \Delta_{PM} \sin \Delta_{YM}}, \qquad (2\text{–}150)$$

where Δ_{PY} is the angle subtended at the center of the Earth by the pole of rotation P and the Yaragadee station Y ($\theta'' = 90° + 29° = 119°$, $\psi'' = 115.3°$) and Δ_{YM} is the angle subtended by the Maui station M and the Yaragadee station Y. From Equation (1–18) we find $\Delta_{PY} = 53.6°$ and $\Delta_{YM} = 98.5°$. Substitution of these values into Equation (2–150) gives $\beta = 51.4°$. The relative velocity between the Yaragadee and Maui stations

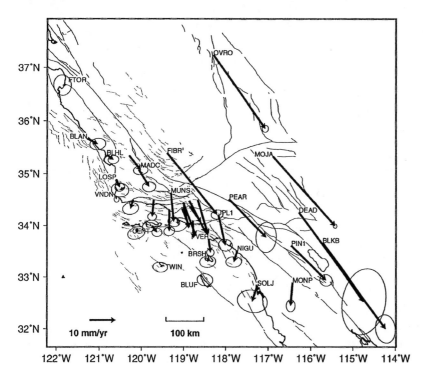

u_{YM} is related to the velocity of the Maui station relative to the Australian plate u_{MP} by

$$u_{YM} = u_{MP}\cos(90° - \beta) = u_{MP}\sin\beta. \qquad (2\text{–}151)$$

Taking $\beta = 51.4°$ and $u_{MP} = -118\,\text{mm yr}^{-1}$, Equation (2–151) gives $u_{YM} = -92\,\text{mm yr}^{-1}$. This is in excellent agreement with the observed value $u_{YM} = -90 \pm 5\,\text{mm yr}^{-1}$.

PROBLEM 2-30 Based on SLR observations, the relative velocity between the Greenbelt, USA (39°N, 283.2°E), and Weitzell, Germany (49.1°N, 12.9°E), stations is $18 \pm 4\,\text{mm yr}^{-1}$. Determine the expected relative velocities based on the plate motion data given in Table 1–6.

PROBLEM 2-31 Based on SLR observations, the relative velocity between the Simosato, Japan (33.5°N, 136°E), and the Maui stations is $-62 \pm 6\,\text{mm yr}^{-1}$. Determine the expected relative velocities based on the plate motion data given in Table 1–6.

PROBLEM 2-32 Based on SLR observations, the relative velocity between the Easter Island (27.1°S, 250.6°E) and the Arequipa, Peru (16.5°S, 288.5°E), stations is $-62 \pm 7\,\text{mm yr}^{-1}$. Determine the expected relative velocities based on the plate motion data given in Table 1–6.

The low cost and mobility of GPS systems allow detailed determinations of tectonic strain in active plate boundary regions. An example for central and southern California is shown in Figure 2–37 which gives the observed velocity vectors of geodetic stations obtained from a combined GPS and VLBI data set (Feigl et al., 1993). The velocities are given relative to a fixed Pacific plate. The velocity vectors of the OVRO (Owens Valley) and MOJA (Mojave) stations are representative of the relative southwest motion of the North American plate with respect to the Pacific plate. The virtual stationarity of the VNDN (Vandenberg AFB) and the BLUF (San Clemente Island) stations indicate their attachment to a rigid Pacific plate. The intermediate motion of the JPL (Pasadena) station represents the complex displacement field within the Los Angeles basin.

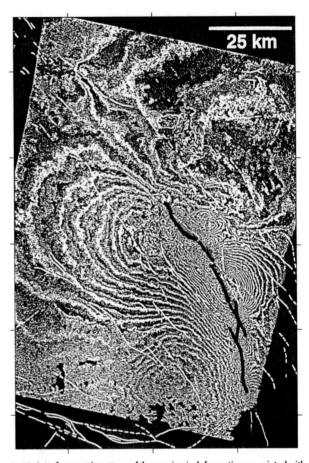

2-38 Interferometric pattern of the coseismic deformation associated with the magnitude 7.3 Landers, California, earthquake on June 28, 1992 (Price and Sandwell, 1998). The dark lines represent surface ruptures associated with the earthquake and the white lines represent other known faults in the region. Each interferometric fringe corresponds to a displacement of 28 mm.

PROBLEM 2-33 The displacement of the OVRO (Owens Valley) station is $20.1\,\text{mm yr}^{-1}$ to the east and $-28.0\,\text{mm yr}^{-1}$ to the north. Assuming the San Andreas fault to be pure strike slip, and that this displacement is associated only with motion on this fault, determine the mean slip velocity on the fault and its orientation.

PROBLEM 2-34 The displacement of the MOJA (Mojave) station is $23.9\,\text{mm yr}^{-1}$ to the east and $-26.6\,\text{mm yr}^{-1}$ to the north. Assuming the San Andreas fault to be pure strike slip and that this displacement is associated only with motion on this fault, determine the mean slip velocity on the fault and its orientation.

Synthetic aperture radar interferometry (INSAR) from satellites has opened a new era in geodetic observations. A synthetic aperture radar (SAR) image is obtained using radar backscatter returns from the Earth's surface. If the Earth's surface deforms between two SAR image acquisitions, a radar interferogram can be obtained to quantify the deformation. The point-wise product of the first image with the second image produces a fringe pattern associated with the phase differences between the two images. Each fringe represents a phase change of 2π radians.

An example of INSAR interferometry is given in Figure 2–38 (Price and Sandwell, 1998). This is the pattern of images associated with the magnitude 7.3 Landers, California, earthquake which occurred on June 28, 1992, and ruptured nearly 100 km of previously unmapped faults in the Mojave Desert, California. The maximum measured surface displacement was 5.1 m. The images were acquired by the ERS-1 satellite on April 24 and August 7, 1992. The satellite was at an altitude of 785 km and the radar images were collected along ray paths pointed west at an average angle of 23° from the vertical. Each fringe corresponds to 28 mm (half the 56-mm wavelength of the ERS-1 SAR). The distribution of coseismic deformation shown in Figure 2–38 will be considered in detail in Chapter 8.

REFERENCES

Feigl, K. L., et al. (1993), Space geodetic measurement of crustal deformation in central and southern California, 1984–1992, *J. Geophys. Res.* **98,** 21,677–21,712.

Gough, D. J., and J. S. Bell (1981), Stress orientations from borehole well fractures with examples from Colorado, east Texas, and northern Canada, *Can. J. Earth Sci.* **19,** 1,358–1,370.

Hofmann-Wallenhof, B., H. Lichtenegger, and J. Collins (1997), *Global Positioning System*, 4th Ed. (Springer, Vienna), 389p.

Larson, K. M. (1996), Geodesy, *Prog. Astron. Aeronaut.* **164,** 539–557.

McKenzie, D. (1978), Some remarks on the development of sedimentary basins, *Earth Planet. Sci. Let.* **40,** 25–32.

Price, E. J., and D. T. Sandwell (1998), Small-scale deformations associated with the 1992 Landers, California, earthquake mapped by synthetic aperture radar interferometry phase gradients, *J. Geophys. Res.* **103,** 27,001–27,016.

Savage, J. C., and R. O. Burford (1973), Geodetic determination of relative plate motion in central California, *J. Geophys. Res.* **78,** 832–845.

Savage, J. C., and W. H. Prescott (1978), Geodolite measurements near the Briones Hills, California, earthquake swarm of January 8, 1977, *Seis. Soc. Am. Bull.* **68,** 175–180.

Smith, D. E., et al. (1990), Tectonic motion and deformation from satellite laser ranging to LAGEOS, *J. Geophys. Res.* **95,** 22,013–22,041.

Thatcher, W. (1975), Strain accumulation and release mechanism of the 1906 San Francisco earthquake, *J. Geophys. Res.* **80,** 4,862–4,872.

Zoback, M. D., and J. H. Healy (1992), In situ stress measurements to 3.5 km depth in the Cajon Pass scientific research borehole: Implications for the mechanics of crustal faulting, *J. Geophys. Res.* **97,** 5,039–5,057.

COLLATERAL READING

Bomford, G., *Geodesy* (Oxford University Press, London, 1962), 561 pages.

An in-depth discussion of geodetic measurement techniques, methods of analysis, and implications of gravity observations for the figure of the Earth, crustal structure, and the state of stress in the crust.

Heiskanen, W. A., and H. Moritz, *Physical Geodesy* (W. H. Freeman and Company, San Francisco, 1967), 364 pages.

A graduate level textbook in geodesy. The contents includes chapters on potential theory, the gravity field of the Earth, gravimetric methods, astrogeodetic methods, and statistical and mathematical approaches in determinations of the Earth's figure.

Jaeger, J. C., and N. G. W. Cook, *Fundamentals of Rock Mechanics* (Chapman and Hall, London, 1976), 585 pages.

An advanced textbook presenting the mathematical and experimental foundations of the mechanical behavior of rock. There are chapters on stress and strain, friction, elasticity, rock strength, laboratory testing, ductile behavior, fluid flow in rocks, fracture, state of stress underground, measurements of underground stresses, mining and engineering applications, and geological and geophysical applications.

Jeffreys, H., *The Earth, Its Origin, History and Physical Constitution* (Cambridge University Press, Cambridge, 1962), 438 pages.

    A classic textbook on the physics of the solid Earth, which includes discussions of stress, strain, elasticity, mechanical behavior of rocks, seismology, gravity, and stress differences in the Earth.

Timoshenko, S., and J. N. Goodier, *Theory of Elasticity* (McGraw-Hill, New York, 1970), 567 pages.

    Fundamentals of the mathematical theory of elasticity with engineering applications. There are major chapters on plane stress and plane strain, bending of beams, two-dimensional stress problems in rectangular, polar, and curvilinear coordinates, solutions by the method of complex variables, three-dimensional stress-strain problems, torsion, bending of bars, thermal stresses, wave propagation, and finite-difference solutions. About half the chapters include student exercises.

THREE

Elasticity and Flexure

3-1 Introduction

In the previous chapter we introduced the concepts of stress and strain. For many solids it is appropriate to relate stress to strain through the laws of elasticity. *Elastic materials* deform when a force is applied and return to their original shape when the force is removed. Almost all solid materials, including essentially all rocks at relatively low temperatures and pressures, behave elastically when the applied forces are not too large. In addition, the elastic strain of many rocks is linearly proportional to the applied stress. The equations of linear elasticity are greatly simplified if the material is *isotropic*, that is, if its elastic properties are independent of direction. Although some metamorphic rocks with strong foliations are not strictly isotropic, the isotropic approximation is usually satisfactory for the earth's crust and mantle.

At high stress levels, or at temperatures that are a significant fraction of the rock solidus, deviations from elastic behavior occur. At low temperatures and confining pressures, rocks are brittle solids, and large deviatoric stresses cause fracture. As rocks are buried more deeply in the earth, they are subjected to increasingly large confining pressures due to the increasing weight of the overburden. When the confining pressure on the rock approaches its brittle failure strength, it deforms plastically. *Plastic deformation* is a continuous, irreversible deformation without fracture. If the applied force causing plastic deformation is removed, some fraction of the deformation remains. We consider plastic deformation in Section 7-11. As discussed in Chapter 1, hot mantle rocks behave as a fluid on

geological time scales; that is, they continuously deform under an applied force.

Given that rocks behave quite differently in response to applied forces, depending on conditions of temperature and pressure, it is important to determine what fraction of the rocks of the crust and upper mantle behave elastically on geological time scales. One of the fundamental postulates of plate tectonics is that the surface plates constituting the lithosphere do not deform significantly on geological time scales. Several observations directly confirm this postulate. We know that the transform faults connecting offset segments of the oceanic ridge system are responsible for the major linear fracture zones in the ocean. That these fracture zones remain linear and at constant separation is direct evidence that the oceanic lithosphere does not deform on a time scale of 10^8 years. Similar evidence comes from the linearity of the magnetic lineaments of the seafloor (see Section 1-8).

There is yet other direct evidence of the elastic behavior of the lithosphere on geological time scales. Although erosion destroys mountain ranges on a time scale of 10^6 to 10^7 years, many geological structures in the continental crust have ages greater than 10^9 years. The very existence of these structures is evidence of the elastic behavior of the lithosphere. If the rocks of the crust behaved as a fluid on geological time scales, the gravitational body force would have erased these structures. As an example, pour a very viscous substance such as molasses onto the bottom of a flat pan. If the fluid is sufficiently viscous and is poured quickly enough, a structure resembling a mountain forms (see Figure 3–1a). However, over time, the fluid will eventually cover the bottom of the pan to a uniform depth (see Figure 3–1b). The gravitational body force causes the fluid to flow so as to minimize the gravitational potential energy.

A number of geological phenomena allow the long-term elastic behavior of the lithosphere to be studied quantitatively. In several instances the lithosphere bends under surface loads. Direct evidence of this bending comes from the Hawaiian Islands and many other island chains, individual islands, and seamounts. There is also observational evidence of the elastic bending of the oceanic lithosphere at ocean trenches and of the continental lithosphere at sedimentary basins – the

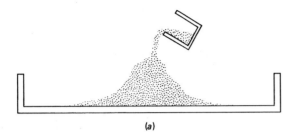

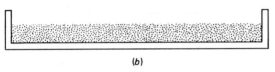

3–1 (a) Structure formed immediately after rapidly pouring a very viscous fluid into a container. (b) Final shape of the fluid after a long time has elapsed.

Michigan basin, for example. We will make quantitative comparisons of the theoretically predicted elastic deformations of these structures with the observational data in later sections of this chapter.

One important reason for studying the elastic behavior of the lithosphere is to determine the state of stress in the lithosphere. This stress distribution is responsible for the occurrence of earthquakes. Earthquakes are direct evidence of high stress levels in the lithosphere. An earthquake relieves accumulated strain in the lithosphere. The presence of mountains is also evidence of high stress levels. Elastic stresses must balance the gravitational body forces on mountains. Because of their elastic behavior, surface plates can transmit stresses over large horizontal distances.

3–2 Linear Elasticity

A linear, isotropic, elastic solid is one in which stresses are linearly proportional to strains and mechanical properties have no preferred orientations. The principal axes of stress and strain coincide in such a medium, and the connection between stress and strain can be conveniently written in this coordinate system as

$$\sigma_1 = (\lambda + 2G)\varepsilon_1 + \lambda\varepsilon_2 + \lambda\varepsilon_3 \tag{3–1}$$

$$\sigma_2 = \lambda\varepsilon_1 + (\lambda + 2G)\varepsilon_2 + \lambda\varepsilon_3 \tag{3–2}$$

$$\sigma_3 = \lambda\varepsilon_1 + \lambda\varepsilon_2 + (\lambda + 2G)\varepsilon_3, \tag{3–3}$$

where the material properties λ and G are known as *Lamé parameters*; G is also known as the *modulus of*

rigidity. The material properties are such that a principal strain component ε produces a stress $(\lambda + 2G)\varepsilon$ in the same direction and stresses $\lambda\varepsilon$ in mutually perpendicular directions.

Equations (3–1) to (3–3) can be written in the inverse form as

$$\varepsilon_1 = \frac{1}{E}\sigma_1 - \frac{\nu}{E}\sigma_2 - \frac{\nu}{E}\sigma_3 \tag{3–4}$$

$$\varepsilon_2 = -\frac{\nu}{E}\sigma_1 + \frac{1}{E}\sigma_2 - \frac{\nu}{E}\sigma_3 \tag{3–5}$$

$$\varepsilon_3 = -\frac{\nu}{E}\sigma_1 - \frac{\nu}{E}\sigma_2 + \frac{1}{E}\sigma_3, \tag{3–6}$$

and E and ν are material properties known as *Young's modulus* and *Poisson's ratio*, respectively. A principal stress component σ produces a strain σ/E in the same direction and strains $(-\nu\sigma/E)$ in mutually orthogonal directions.

The elastic behavior of a material can be characterized by specifying either λ and G or E and ν; the sets of parameters are not independent. Analytic formulas expressing λ and G in terms of E and ν, and vice versa, are obtained in the following sections. Values of E, G, and ν for various rocks are given in Section E of Appendix 2. Young's modulus of rocks varies from about 10 to 100 GPa, and Poisson's ratio varies between 0.1 and 0.4. The elastic properties of the earth's mantle and core can be obtained from seismic velocities and the density distribution. The elastic properties E, G, and ν inferred from a typical seismically derived earth model are given in Section F of Appendix 2. The absence of shear waves in the outer core ($G = 0$) is taken as conclusive evidence that the outer core is a liquid. In the outer core ν has the value 0.5, which we will see is appropriate to an incompressible fluid.

The behavior of linear solids is more readily illustrated if we consider idealized situations where several of the stress and strain components vanish. These can then be applied to important geological problems.

3–3 Uniaxial Stress

In a state of *uniaxial stress* only one of the principal stresses, σ_1 say, is nonzero. Under this circumstance Equations (3–2) and (3–3), with $\sigma_2 = \sigma_3 = 0$, give

$$\varepsilon_2 = \varepsilon_3 = \frac{-\lambda}{2(\lambda + G)}\varepsilon_1. \tag{3–7}$$

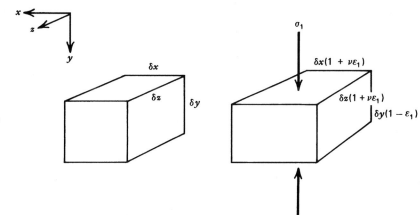

3–2 Deformation under uniaxial stress.

Not only does the stress σ_1 produce a strain ε_1, but it changes the linear dimensions of elements aligned perpendicular to the axis of stress. If σ_1 is a compression, then ε_1 is a decrease in length, and both ε_2 and ε_3 are increases in length. The element in Figure 3–2 has been shortened in the y direction, but its cross section in the xz plane has expanded.

Using Equations (3–4) to (3–6), we can also write

$$\varepsilon_2 = \varepsilon_3 = -\frac{\nu}{E}\sigma_1 = -\nu\varepsilon_1. \tag{3-8}$$

By comparing Equations (3–7) and (3–8), we see that

$$\nu = \frac{\lambda}{2(\lambda + G)}. \tag{3-9}$$

From Equations (3–1) and (3–7) we find

$$\sigma_1 = \frac{G(3\lambda + 2G)}{(\lambda + G)}\varepsilon_1, \tag{3-10}$$

which, with the help of Equation (3–8), identifies Young's modulus as

$$E = \frac{G(3\lambda + 2G)}{(\lambda + G)}. \tag{3-11}$$

Equations (3–9) and (3–11) can be inverted to yield the following formulas for G and λ in terms of E and ν

$$G = \frac{E}{2(1 + \nu)} \tag{3-12}$$

$$\lambda = \frac{E\nu}{(1 + \nu)(1 - 2\nu)}. \tag{3-13}$$

The relation between stress and strain in uniaxial compression or tension from Equation (3–8),

$$\sigma_1 = E\varepsilon_1, \tag{3-14}$$

is also known as *Hooke's law*. A linear elastic solid is said to exhibit Hookean behavior. Uniaxial compression testing in the laboratory is one of the simplest methods of determining the elastic properties of rocks. Figure 3–3 shows the data from such a test on a cylindrical sample of quartzite. The rock deforms approximately elastically until the applied stress exceeds the compressive strength of the rock, at which point failure occurs. Compressive strengths of rocks are hundreds to thousands of megapascals. As we discussed in the previous chapter, a typical tectonic stress is 10 MPa. With $E = 70$ GPa, this yields a typical tectonic strain in uniaxial stress of 1.4×10^{-4}.

The *dilatation* Δ or fractional volume change in uniaxial compression is, according to Equation (3–8),

$$\Delta = \varepsilon_1 + \varepsilon_2 + \varepsilon_3 = \varepsilon_1(1 - 2\nu). \tag{3-15}$$

The decrease in volume due to contraction in the direction of compressive stress is offset by an increase in volume due to expansion in the orthogonal directions. Equation (3–15) allows us to determine Poisson's ratio for an *incompressible material*, which cannot undergo a net change in volume. In order for Δ to equal zero in uniaxial compression, ν must equal 1/2. Under uniaxial compression, an incompressible material contracts in the direction of applied stress but expands exactly half as much in each of the perpendicular directions.

There are some circumstances in which the formulas of uniaxial compression can be applied to calculate the strains in rocks. Consider, for example, a rectangular column of height h that is free to expand or contract

3-3 Stress–strain curves for quartzite in uni-axial compression (Bieniawski, 1967).

in the horizontal; that is, it is laterally unconstrained. By this we mean that the horizontal stresses are zero ($\sigma_2 = \sigma_3 = 0$). Then the vertical stress σ_1 at a distance y from the top of the column of rock is given by the weight of the column,

$$\sigma_1 = \rho g y. \tag{3-16}$$

The vertical strain as a function of the distance y from the top is

$$\varepsilon_1 = \frac{\rho g y}{E}. \tag{3-17}$$

The slab contracts in the vertical by an amount

$$\delta h = \int_0^h \varepsilon_1 \, dy = \frac{\rho g}{E} \int_0^h y \, dy = \frac{\rho g h^2}{2E}. \tag{3-18}$$

3-4 Uniaxial Strain

The state of *uniaxial strain* corresponds to only one nonzero component of principal strain, ε_1 say. With $\varepsilon_2 = \varepsilon_3 = 0$, Equations (3–1) to (3–3) give

$$\sigma_1 = (\lambda + 2G)\varepsilon_1 \tag{3-19}$$

$$\sigma_2 = \sigma_3 = \lambda \varepsilon_1 = \frac{\lambda}{(\lambda + 2G)} \sigma_1. \tag{3-20}$$

Equations (3–4) to (3–6) simplify to

$$\sigma_2 = \sigma_3 = \frac{\nu}{(1 - \nu)} \sigma_1 \tag{3-21}$$

$$\sigma_1 = \frac{(1 - \nu) E \varepsilon_1}{(1 + \nu)(1 - 2\nu)}. \tag{3-22}$$

By comparing Equations (3–19) to (3–22), one can also derive the relations already given between λ, G and ν, E.

The equations of uniaxial strain can be used to determine the change in stress due to sedimentation or erosion. We first consider sedimentation and assume that an initial surface is covered by h km of sediments of density ρ, as shown in Figure 3–4. We also assume

3-4 Stresses on a surface covered by sediments of thickness h.

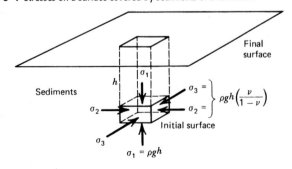

that the base of the new sedimentary basin is laterally confined so that the equations of uniaxial strain are applicable. The two horizontal components of strain are zero, $\varepsilon_2 = \varepsilon_3 = 0$. The vertical principal stress on the initial surface σ_1 is given by the weight of the overburden

$$\sigma_1 = \rho g h. \tag{3-23}$$

From Equation (3–21) the horizontal normal stresses are given by

$$\sigma_2 = \sigma_3 = \frac{\nu}{(1-\nu)} \rho g h. \tag{3-24}$$

The horizontal stresses are also compressive, but they are smaller than the vertical stress.

It is of interest to determine the deviatoric stresses after sedimentation. The pressure at depth h as defined by Equation (2–61) is

$$p = \frac{1}{3}(\sigma_1 + \sigma_2 + \sigma_3) = \frac{(1+\nu)}{3(1-\nu)} \rho g h. \tag{3-25}$$

The deviatoric stresses are then determined from Equations (2–63) with the result

$$\sigma_1' = \sigma_1 - p = \frac{2(1-2\nu)}{3(1-\nu)} \rho g h \tag{3-26}$$

$$\sigma_2' = \sigma_2 - p = \sigma_3' = \sigma_3 - p = -\frac{(1-2\nu)}{3(1-\nu)} \rho g h. \tag{3-27}$$

The horizontal deviatoric stress is tensional. For $\nu = 0.25$ the horizontal deviatoric stress is 2/9 of the lithostatic stress. With $\rho = 3000 \text{ kg m}^{-3}$ and $h = 2 \text{ km}$ the horizontal deviatoric stress is -13.3 MPa. This stress is of the same order as measured surface stresses.

We next consider erosion. If the initial state of stress before erosion is that given above, erosion will result in the state of stress that existed before sedimentation occurred. The processes of sedimentation and erosion are reversible. However, in many cases the initial state of stress prior to erosion is lithostatic. Therefore at a depth h the principal stresses are

$$\sigma_1 = \sigma_2 = \sigma_3 = \rho g h. \tag{3-28}$$

After the erosion of h km of overburden the vertical stress at the surface is $\bar{\sigma}_1 = 0$ (an overbar denotes a stress after erosion). The change in vertical stress $\Delta\sigma_1 = \bar{\sigma}_1 - \sigma_1$ is $-\rho g h$. If only ε_1 is nonzero,

Equation (3–21) gives

$$\Delta\sigma_2 = \Delta\sigma_3 = \left(\frac{\nu}{1-\nu}\right)\Delta\sigma_1. \tag{3-29}$$

The horizontal surface stresses after erosion $\bar{\sigma}_2$ and $\bar{\sigma}_3$ are consequently given by

$$\bar{\sigma}_2 = \bar{\sigma}_3 = \sigma_2 + \Delta\sigma_2 = \rho g h - \frac{\nu}{(1-\nu)} \rho g h$$

$$= \left(\frac{1-2\nu}{1-\nu}\right)\rho g h. \tag{3-30}$$

If $h = 5$ km, $\nu = 0.25$, and $\rho = 3000 \text{ kg m}^{-3}$, we find from Equation (3–30) that $\bar{\sigma}_2 = \bar{\sigma}_3 = 100$ MPa. Erosion can result in large surface compressive stresses due simply to the elastic behavior of the rock. This mechanism is one explanation for the widespread occurrence of near-surface compressive stresses in the continents.

PROBLEM 3-1 Determine the surface stress after the erosion of 10 km of granite. Assume that the initial state of stress is lithostatic and that $\rho = 2700 \text{ kg m}^{-3}$ and $\nu = 0.25$.

PROBLEM 3-2 An unstressed surface is covered with sediments with a density of 2500 kg m^{-3} to a depth of 5 km. If the surface is laterally constrained and has a Poisson's ratio of 0.25, what are the three components of stress at the original surface?

PROBLEM 3-3 A horizontal stress σ_1 may be accompanied by stress in other directions. If it is assumed that there is no displacement in the other horizontal direction and zero stress in the vertical, find the stress σ_2 in the other horizontal direction and the strain ε_3 in the vertical direction.

PROBLEM 3-4 Assume that the earth is unconstrained in one lateral direction ($\sigma_2 = \sigma_3$) and is constrained in the other ($\varepsilon_1 = 0$). Determine ε_2 and σ_1 when y kilometers of rock of density ρ are eroded away. Assume that the initial state of stress was lithostatic.

3-5 Plane Stress

The state of *plane stress* exists when there is only one zero component of principal stress; that is, $\sigma_3 = 0$, $\sigma_1 \neq 0$, $\sigma_2 \neq 0$. The situation is sketched in Figure 3–5, which shows a thin plate loaded on its edges. The strain

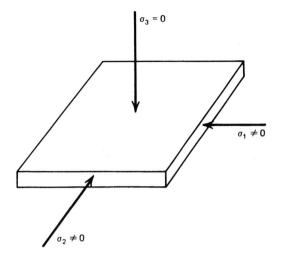

3–5 Plane stress.

components according to Equations (3–4) to (3–6) are

$$\varepsilon_1 = \frac{1}{E}(\sigma_1 - v\sigma_2) \tag{3–31}$$

$$\varepsilon_2 = \frac{1}{E}(\sigma_2 - v\sigma_1) \tag{3–32}$$

$$\varepsilon_3 = \frac{-v}{E}(\sigma_1 + \sigma_2). \tag{3–33}$$

The geometry of Figure 3–5 suggests that the plane stress formulas may be applicable to horizontal tectonic stresses in the lithosphere. Let us assume that in addition to the lithostatic stresses there are equal horizontal components of principal stress $\Delta\sigma_1 = \Delta\sigma_2$. According to Equations (3–31) to (3–33), the horizontal tectonic stresses produce the strains

$$\varepsilon_1 = \varepsilon_2 = \frac{(1-v)}{E}\Delta\sigma_1 \tag{3–34}$$

$$\varepsilon_3 = \frac{-2v}{E}\Delta\sigma_1. \tag{3–35}$$

If the horizontal tectonic stresses are compressive, vertical columns of lithosphere of initial thickness h_L, horizontal area A, and density ρ will undergo a decrease in area and an increase in thickness. The mass in a column will remain constant, however. Therefore we can write

$$\delta(\rho A h_L) = 0. \tag{3–36}$$

The weight per unit area at the base of the column $\rho g h_L$

will increase, as can be seen from

$$\begin{aligned}
\delta(\rho g h_L) &= \delta\left(\rho g h_L A \cdot \frac{1}{A}\right) \\
&= \frac{1}{A}\delta(\rho g h_L A) + \rho g h_L A\,\delta\left(\frac{1}{A}\right) \\
&= \rho g h_L A\left(-\frac{1}{A^2}\right)\delta A = \rho g h_L\left(-\frac{\delta A}{A}\right).
\end{aligned} \tag{3–37}$$

The term $\delta(\rho g h_L A)/A$ is zero from Equation (3–36); $\delta(\rho g h_L)$ is positive because $-\delta A/A$ is a positive quantity given by

$$-\frac{\delta A}{A} = \varepsilon_1 + \varepsilon_2 = \frac{2(1-v)}{E}\Delta\sigma_1. \tag{3–38}$$

The increase in the weight per unit area at the base of the lithospheric column gives the increase in the vertical principal stress $\Delta\sigma_3$. By combining Equations (3–37) and (3–38), we get

$$\Delta\sigma_3 = \frac{2(1-v)\rho g h_L}{E}\Delta\sigma_1 \tag{3–39}$$

or

$$\frac{\Delta\sigma_3}{\Delta\sigma_1} = \frac{2(1-v)\rho g h_L}{E}. \tag{3–40}$$

Taking $v = 0.25$, $E = 100$ GPa, $\rho = 3000$ kg m^{-3}, $g = 10$ m s^{-2}, and $h_L = 100$ km as typical values for the lithosphere, we find that $\Delta\sigma_3/\Delta\sigma_1 = 0.045$. Because the change in the vertical principal stress is small compared with the applied horizontal principal stresses, we conclude that the plane stress assumption is valid for the earth's lithosphere.

PROBLEM 3–5 Triaxial compression tests are a common laboratory technique for determining elastic properties and strengths of rocks at various pressures p and temperatures. Figure 3–6 is a schematic of the experimental method. A cylindrical rock specimen is loaded axially by a compressive stress σ_1. The sample is also uniformly compressed laterally by stresses $\sigma_2 = \sigma_3 < \sigma_1$.

Show that

$$\varepsilon_2 = \varepsilon_3$$

and

$$\sigma_1 - \sigma_2 = 2G(\varepsilon_1 - \varepsilon_2).$$

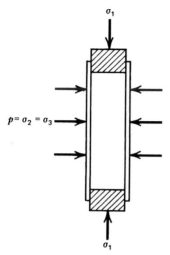

3-6 Sketch of a triaxial compression test on a cylindrical rock sample.

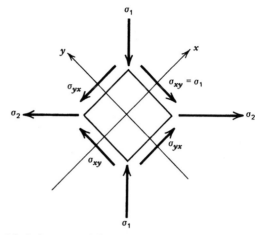

3-8 Principal stresses and shear stresses in the case of pure shear.

Thus if the measured stress difference $\sigma_1 - \sigma_2$ is plotted against the measured strain difference $\varepsilon_1 - \varepsilon_2$, the slope of the line determines $2G$.

From Equation (3–6) it is obvious that

$$\sigma_3 = \nu(\sigma_1 + \sigma_2). \qquad (3\text{--}44)$$

This can be used together with Equations (3–4) and (3–5) to find

$$\varepsilon_1 = \frac{(1 + \nu)}{E}\{\sigma_1(1 - \nu) - \nu\sigma_2\} \qquad (3\text{--}45)$$

$$\varepsilon_2 = \frac{(1 + \nu)}{E}\{\sigma_2(1 - \nu) - \nu\sigma_1\}. \qquad (3\text{--}46)$$

3-6 Plane Strain

In the case of *plane strain*, $\varepsilon_3 = 0$, for example, and ε_1 and ε_2 are nonzero. Figure 3–7 illustrates a plane strain situation. A long bar is rigidly confined between supports so that it cannot expand or contract parallel to its length. In addition, the stresses σ_1 and σ_2 are applied uniformly along the length of the bar.

Equations (3–1) to (3–3) reduce to

$$\sigma_1 = (\lambda + 2G)\varepsilon_1 + \lambda\varepsilon_2 \qquad (3\text{--}41)$$

$$\sigma_2 = \lambda\varepsilon_1 + (\lambda + 2G)\varepsilon_2 \qquad (3\text{--}42)$$

$$\sigma_3 = \lambda(\varepsilon_1 + \varepsilon_2). \qquad (3\text{--}43)$$

3-7 Pure Shear and Simple Shear

The state of stress associated with pure shear is illustrated in Figure 3–8. Pure shear is a special case of plane stress. One example of pure shear is $\sigma_3 = 0$ and $\sigma_1 = -\sigma_2$. From Equations (2–56) to (2–58) with $\theta = -45°$ (compare Figures 2–14 and 3–8), we find that

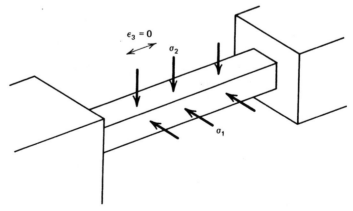

3-7 An example of plane strain.

$\sigma_{xx} = \sigma_{yy} = 0$ and $\sigma_{xy} = \sigma_1$. In this coordinate system only the shear stress is nonzero. From Equations (3–31) and (3–32) we find that

$$\varepsilon_1 = \frac{(1+v)}{E}\sigma_1 = \frac{(1+v)}{E}\sigma_{xy} = -\varepsilon_2, \qquad (3\text{–}47)$$

and from Equations (2–130) and (2–131) with $\theta = -45°$ we get $\varepsilon_{xx} = \varepsilon_{yy} = 0$ and $\varepsilon_{xy} = \varepsilon_1$. Equation (3–47) then gives

$$\sigma_{xy} = \frac{E}{1+v}\varepsilon_{xy}. \qquad (3\text{–}48)$$

By introducing the modulus of rigidity from Equation (3–12), we can write the shear stress as

$$\sigma_{xy} = 2G\varepsilon_{xy}, \qquad (3\text{–}49)$$

which explains why the modulus of rigidity is also known as the shear modulus. (Note: In terms of $\gamma_{xy} \equiv 2\varepsilon_{xy}$, $\sigma_{xy} = G\gamma_{xy}$.) These results are valid for both pure shear and simple shear because the two states differ by a solid-body rotation that does not affect the state of stress.

Simple shear is generally associated with displacements on a strike–slip fault such as the San Andreas in California. In Equation (2–134) we concluded that the shear strain associated with the 1906 San Francisco earthquake was 2.5×10^{-5}. With $G = 30$ GPa, Equation (3–49) gives the related shear stress as 1.5 MPa. This is a very small stress drop to be associated with a great earthquake. However, for the stress drop to have been larger, the width of the zone of strain accumulation would have had to have been even smaller. If the stress had been 15 MPa, the width of the zone of strain accumulation would have had to have been 4 km on each side of the fault. We will return to this problem in Chapter 8.

PROBLEM 3-6 Show that Equation (3–49) can also be derived by assuming plane strain.

3-8 Isotropic Stress

If all the principal stresses are equal $\sigma_1 = \sigma_2 = \sigma_3 \equiv p$, then the state of stress is isotropic, and the principal stresses are equal to the pressure. The principal strains in a solid subjected to isotropic stresses are also equal $\varepsilon_1 = \varepsilon_2 = \varepsilon_3 = \frac{1}{3}\Delta$; each component of strain is equal

to one-third of the dilatation. By adding Equations (3–1) to (3–3), we find

$$p = \left(\frac{3\lambda + 2G}{3}\right)\Delta \equiv K\Delta \equiv \frac{1}{\beta}\Delta. \qquad (3\text{–}50)$$

The quantity K is the *bulk modulus*, and its reciprocal is β, the *compressibility*. The ratio of p to the bulk modulus gives the fractional volume change that occurs under isotropic compression.

Because the mass of a solid element with volume V and density ρ must be conserved, any change in volume δV of the element must be accompanied by a change in its density $\delta\rho$. The fractional change in density can be related to the fractional change in volume, the dilatation, by rearranging the equation of mass conservation

$$\delta(\rho V) = 0, \qquad (3\text{–}51)$$

which gives

$$\rho\delta V + V\delta\rho = 0 \qquad (3\text{–}52)$$

or

$$\frac{-\delta V}{V} = \Delta = \frac{\delta\rho}{\rho}. \qquad (3\text{–}53)$$

Equation (3–53) of course assumes Δ to be small. The combination of Equations (3–50) and (3–53) gives

$$\delta\rho = \rho\beta p. \qquad (3\text{–}54)$$

This relationship can be used to determine the increase in density with depth in the earth.

Using Equations (3–11) to (3–13), we can rewrite the formula for K given in Equation (3–50) as

$$K = \frac{1}{\beta} = \frac{E}{3(1-2v)}. \qquad (3\text{–}55)$$

Thus as v tends toward $1/2$, that is, as a material becomes more and more incompressible, its bulk modulus tends to infinity.

3-9 Two-Dimensional Bending or Flexure of Plates

We have already discussed how plate tectonics implies that the near-surface rocks are rigid and therefore behave elastically on geological time scales. The thin elastic surface plates constitute the lithosphere, which floats on the relatively fluid mantle beneath. The plates are

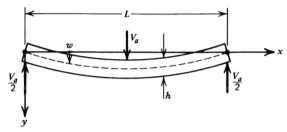

3–9 A thin plate of length L and thickness h pinned at its ends and bending under and applied load V_a.

subject to a variety of loads – volcanoes, seamounts, for example – that force the lithosphere to bend under their weights. By relating the observed *flexure* or *bending* of the lithosphere to known surface loads, we can deduce the elastic properties and thicknesses of the plates. In what follows, we first develop the theory of plate bending in response to applied forces and torques. The theory can also be used to understand fold trains in mountain belts by modeling the folds as deformations of elastic plates subject to horizontal compressive forces. Other geologic applications also can be made. For example, we will apply the theory to model the upwarping of strata overlying igneous intrusions (Section 3–12).

A simple example of plate bending is shown in Figure 3–9. A plate of thickness h and width L is pinned at its ends and bends under the load of a line force V_a (N m^{-1}) applied at its center. The plate is infinitely long in the z direction. A vertical, static force balance and the symmetry of the situation require that equal vertical line forces $V_a/2$ be applied at the supports. The plate is assumed to be thin compared with its width, $h \ll L$, and the vertical deflection of the plate w is taken to be small, $w \ll L$. The latter assumption is necessary to justify the use of linear elastic theory. The two-dimensional bending of plates is also referred to

as *cylindrical bending* because the plate takes the form of a segment of a cylinder.

The deflection of a plate can be determined by requiring it to be in equilibrium under the action of all the forces and torques exerted on it. The forces and torques on a small section of the plate between horizontal locations x and $x + dx$ are shown in Figure 3–10. A downward force per unit area $q(x)$ is exerted on the plate by whatever distributed load the plate is required to support. Thus, the downward load, per unit length in the z direction, between x and $x + dx$ is $q(x)\,dx$. A net shear force V, per unit length in the z direction, acts on the cross section of the plate normal to the plane of the figure; it is the resultant of all the shear stresses integrated over that cross-sectional area of the plate. A horizontal force P, per unit length in the z direction, is applied to the plate; it is assumed that P is independent of x. The net *bending moment M*, per unit length in the z direction, acting on a cross section of the plate is the integrated effect of the moments exerted by the normal stresses σ_{xx}, also known as the *fiber stresses*, on the cross section. We relate M to the fiber stresses in the plate later in the discussion. All quantities in Figure 3–10 are considered positive when they have the sense shown in the figure. At location x along the plate the shear force is V, the bending moment is M, and the deflection is w; at $x + dx$, the shear force is $V + dV$, the bending moment is $M + dM$, and the deflection is $w + dw$. It is to be emphasized that V, M, and P are per unit length in the z direction.

A force balance in the vertical direction on the element between x and $x + dx$ yields

$$q(x)\,dx + dV = 0 \tag{3–56}$$

3–10 Forces and torques on a small section of a deflecting plate.

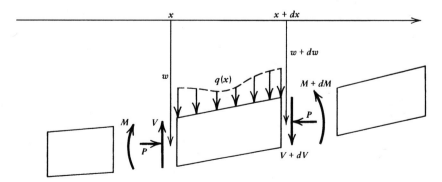

or

$$\frac{dV}{dx} = -q. \tag{3-57}$$

The moments M and $M + dM$ combine to give a net counterclockwise torque dM on the element. The forces V and $V + dV$ are separated by a distance dx (an infinitesimal *moment arm*) and exert a net torque $V dx$ on the element in a clockwise sense. (The change in V in going from x to $x + dx$ can be ignored in calculating the moment due to the shear forces.) The horizontal forces P exert a net counterclockwise torque $-P dw$ on the element through their associated moment arm $-dw$. (Note that dw is negative in going from x to $x + dx$.) A balance of all the torques gives

$$dM - P dw = V dx \tag{3-58}$$

or

$$\frac{dM}{dx} = V + P\frac{dw}{dx}. \tag{3-59}$$

We can eliminate the shear force on a vertical cross section of the plate V from Equation (3–59) by differentiating the equation with respect to x and substituting from Equation (3–57). One obtains

$$\frac{d^2 M}{dx^2} = -q + P\frac{d^2 w}{dx^2}. \tag{3-60}$$

Equation (3–60) can be converted into a differential equation for the deflection w if the bending moment M can be related to the deflection; we will see that M is inversely proportional to the *local radius of curvature of the plate R* and that R^{-1} is $-d^2w/dx^2$.

To relate M to the curvature of the plate, we proceed as follows. If the plate is deflected downward, as in Figure 3–11, the upper half of the plate is contracted, and the longitudinal stress σ_{xx} is positive; the lower part

of the plate is extended, and σ_{xx} is negative. The fiber stress σ_{xx} is zero on the midplane $y = 0$, which is a *neutral unstrained surface*. The net effect of these stresses is to exert a counterclockwise bending moment on the cross section of the plate. The curvature of the plate has, of course, been exaggerated in Figure 3–11 so that x is essentially horizontal. The force on an element of the plate's cross section of thickness dy is $\sigma_{xx} dy$. This force exerts a torque about the midpoint of the plate given by $\sigma_{xx} y \, dy$. If we integrate this torque over the cross section of the plate, we obtain the bending moment

$$M = \int_{-h/2}^{h/2} \sigma_{xx} y \, dy, \tag{3-61}$$

where h is the thickness of the plate.

The bending stress σ_{xx} is accompanied by longitudinal strain ε_{xx} that is positive (contraction) in the upper half of the plate and negative (extension) in the lower half. There is no strain in the direction perpendicular to the xy plane because the plate is infinite in this direction and *the bending is two-dimensional or cylindrical*; that is, $\varepsilon_{zz} = 0$. There is also zero stress normal to the surface of the plate; that is, $\sigma_{yy} = 0$. Because the plate is thin, we can take $\sigma_{yy} = 0$ throughout. Thus plate bending is an example of plane stress, and we can use Equations (3–31) and (3–32) to relate the stresses and strains; that is,

$$\varepsilon_{xx} = \frac{1}{E}(\sigma_{xx} - \nu\sigma_{zz}) \tag{3-62}$$

$$\varepsilon_{zz} = \frac{1}{E}(\sigma_{zz} - \nu\sigma_{xx}). \tag{3-63}$$

In writing these equations, we have identified the principal strains $\varepsilon_1, \varepsilon_2$ with $\varepsilon_{xx}, \varepsilon_{zz}$ and the principal stresses σ_1, σ_2 with σ_{xx}, σ_{zz}. With $\varepsilon_{zz} = 0$, Equations (3–62) and (3–63) give

$$\sigma_{xx} = \frac{E}{(1 - \nu^2)}\varepsilon_{xx}. \tag{3-64}$$

Equation (3–61) for the bending moment can be rewritten, using Equation (3–64), as

$$M = \frac{E}{(1 - \nu^2)}\int_{-h/2}^{h/2} \varepsilon_{xx} y \, dy. \tag{3-65}$$

The longitudinal strain ε_{xx} depends on the distance from the midplane of the plate y and the local radius of

3–11 The normal stresses on a cross section of a thin curved elastic plate.

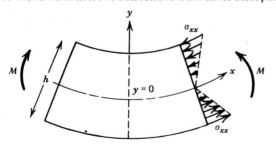

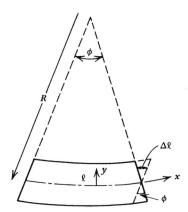

3–12 Longitudinal extension and contraction at a distance y from the midplane of the plate.

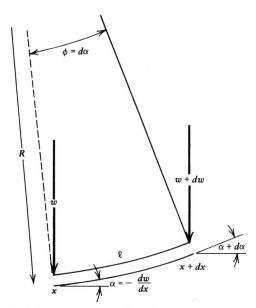

3–13 Sketch illustrating the geometrical relations in plate bending.

curvature of the plate R. Figure 3–12 shows a bent section of the plate originally of length l (l is infinitesimal). The length of the section measured along the midplane remains l. The small angle ϕ is l/R in radians. The geometry of Figure 3–12 shows that the change in length of the section Δl at a distance y from the midplane is

$$\Delta l = -y\phi = -y\frac{1}{R}, \tag{3–66}$$

where the minus sign is included because there is contraction when y is positive. Thus the strain is

$$\varepsilon_{xx} = -\frac{\Delta l}{l} = \frac{y}{R}. \tag{3–67}$$

Implicit in this relation is the assumption that plane sections of the plate remain plane.

The local radius of curvature R is determined by the change in slope of the plate midplane with horizontal distance. The geometry is shown in Figure 3–13. If w is small, $-dw/dx$, the slope of the midplane, is also the angular deflection of the plate from the horizontal α. The small angle ϕ in Figure 3–13 is simply the change in α, that is, $d\alpha$, in the small distance l or dx. Thus

$$\phi = d\alpha = \frac{d\alpha}{dx}dx = \frac{d}{dx}\left(-\frac{dw}{dx}\right)dx = -\frac{d^2w}{dx^2}\,dx, \tag{3–68}$$

and we find

$$\frac{1}{R} = \frac{\phi}{l} \approx \frac{\phi}{dx} = -\frac{d^2w}{dx^2}. \tag{3–69}$$

Finally, the strain is given by

$$\varepsilon_{xx} = -y\frac{d^2w}{dx^2}, \tag{3–70}$$

and the bending moment can be written

$$\begin{aligned}
M &= \frac{-E}{(1-\nu^2)}\frac{d^2w}{dx^2}\int_{-h/2}^{h/2} y^2\,dy \\
&= \frac{-E}{(1-\nu^2)}\frac{d^2w}{dx^2}\left(\frac{y^3}{3}\right)_{-h/2}^{h/2} \\
&= \frac{-Eh^3}{12(1-\nu^2)}\frac{d^2w}{dx^2}. \tag{3–71}
\end{aligned}$$

The coefficient of $-d^2w/dx^2$ on the right side of Equation (3–71) is called the *flexural rigidity* D of the plate

$$D \equiv \frac{Eh^3}{12(1-\nu^2)}. \tag{3–72}$$

According to Equations (3–69), (3–71), and (3–72), the bending moment is the flexural rigidity of the plate divided by its curvature

$$M = -D\frac{d^2w}{dx^2} = \frac{D}{R}. \tag{3–73}$$

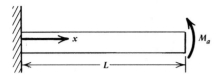

3–14 An embedded plate subject to an applied torque.

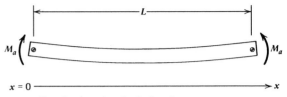

3–16 Bending of a plate pinned at both ends.

Upon substituting the second derivative of Equation (3–73) into Equation (3–60), we obtain the general equation for the deflection of the plate

$$D\frac{d^4w}{dx^4} = q(x) - P\frac{d^2w}{dx^2}. \qquad (3\text{–}74)$$

We next solve Equation (3–74) for plate deflection in a number of simple cases and apply the results to the deformation of crustal strata and to the bending of the lithosphere.

3–10 Bending of Plates under Applied Moments and Vertical Loads

Consider a plate *embedded* at one end and subject to an applied torque M_a at the other, as shown in Figure 3–14. Assume for simplicity that the plate is weightless. With $q = 0$, Equation (3–57) shows that the shear stress on a section of the plate V must be a constant. In fact, $V = 0$, since there is no applied force acting on the plate. This can easily be seen by considering a force balance on a section of the plate, as shown in Figure 3–15. Since $P = 0$ and since we have established $V = 0$, Equation (3–59) requires that $M = $ constant. The constant must be M_a, the applied torque, as shown by a *moment balance* on an arbitrary section of the plate (Figure 3–15).

To determine the deflection of the plate, we could integrate Equation (3–74) with $q = P = 0$. However, since we already know $M \equiv M_a$, it is simpler to integrate Equation (3–73), the twice integrated form of the fourth-order differential equation. The boundary conditions are $w = 0$ at $x = 0$ and $dw/dx = 0$ at $x = 0$. These boundary conditions at the left end of the plate clarify what is meant by an embedded plate; the embed-

ded end of the plate cannot be displaced, and its slope must be zero. The integral of Equation (3–73) subject to these boundary conditions is

$$w = \frac{-M_a x^2}{2D}. \qquad (3\text{–}75)$$

The bent plate has the shape of a parabola. w is negative according to the convention we established if M is positive; that is, the plate is deflected upward.

PROBLEM 3–7 What is the displacement of a plate pinned at both ends ($w = 0$ at $x = 0, L$) with equal and opposite bending moments applied at the ends? The problem is illustrated in Figure 3–16.

As a second example we consider the bending of a plate embedded at its left end and subjected to a concentrated force V_a at its right end, as illustrated in Figure 3–17. In this situation, $q = 0$, except at the point $x = L$, and Equation (3–57) gives $V = $ constant. The constant must be V_a, as shown by the vertical force balance on the plate sketched in Figure 3–18. With P also equal to zero, Equation (3–59) for the bending moment simplifies to

$$\frac{dM}{dx} = V_a. \qquad (3\text{–}76)$$

This equation can be integrated to yield

$$M = V_a x + \text{constant}, \qquad (3\text{–}77)$$

and the constant can be evaluated by noting that there is no applied torque at the end $x = L$; that is, $M = 0$ at

3–17 An embedded plate subjected to a concentrated load.

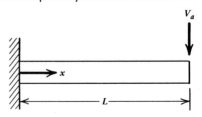

3–15 Force and torque balance on a section of the plate in Figure 3–14.

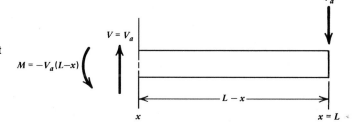

3-18 Forces and torques on a section of a plate loaded at its right end by a force V_a.

$x = L$. Thus we obtain

$$M = V_a(x - L). \tag{3-78}$$

The bending moment changes linearly from $-V_a L$ at the embedded end to zero at the free end. A simple torque balance on the section of the plate shown in Figure 3–18 leads to Equation (3–78), since M must balance the torque of the applied force V_a acting with moment arm $L - x$.

The displacement can be determined by integrating Equation (3–74), which simplifies to

$$\frac{d^4 w}{dx^4} = 0, \tag{3-79}$$

when $q = P = 0$. The integral of Equation (3–79) is

$$\frac{d^3 w}{dx^3} = \text{constant}. \tag{3-80}$$

The constant can be evaluated by differentiating Equation (3–73) with respect to x and substituting for dM/dx from Equation (3–76). The result is

$$\frac{d^3 w}{dx^3} = -\frac{V_a}{D}. \tag{3-81}$$

A second-order differential equation for w can be obtained by integrating Equation (3–81) and evaluating the constant of integration with the boundary condition $d^2 w/dx^2 = 0$ at $x = L$. Alternatively, the same equation can be arrived at by substituting for M from Equation (3–78) into Equation (3–73)

$$\frac{d^2 w}{dx^2} = -\frac{V_a}{D}(x - L). \tag{3-82}$$

This equation may be integrated twice more subject to the standard boundary conditions $w = dw/dx = 0$ at

$x = 0$. One finds

$$w = \frac{V_a x^2}{2D}\left(L - \frac{x}{3}\right). \tag{3-83}$$

PROBLEM 3-8 Determine the displacement of a plate of length L pinned at its ends with a concentrated load V_a applied at its center. This problem is illustrated in Figure 3–9.

As a third and final example, we consider the bending of a plate embedded at one end and subjected to a uniform loading $q(x) = \text{constant}$, as illustrated in Figure 3–19. Equation (3–74), with $P = 0$, becomes

$$\frac{d^4 w}{dx^4} = \frac{q}{D}. \tag{3-84}$$

We need four boundary conditions to integrate Equation (3–84). Two of them are the standard conditions $w = dw/dx = 0$ at the left end $x = 0$. A third boundary condition is the same as the one used in the previous example, namely, $d^2 w/dx^2 = 0$ at $x = L$, because there is no external torque applied at the right end of the plate – see Equation (3–73). The fourth boundary condition follows from Equation (3–59) with $P = 0$. Because there is no applied concentrated load at $x = L$, V must vanish there, as must dM/dx and from Equation (3–73), $d^3 w/dx^3$. After some algebra, one finds the solution

$$w = \frac{q x^2}{D}\left(\frac{x^2}{24} - \frac{Lx}{6} + \frac{L^2}{4}\right). \tag{3-85}$$

3-19 A uniformly loaded plate embedded at one end.

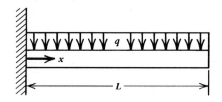

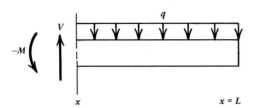

3–20 Section of a uniformly loaded plate.

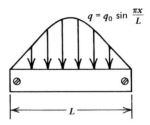

3–21 A freely supported plate loaded sinusoidally.

The shear force at $x = 0$ is $-D(d^3w/dx^3)_{x=0}$. From Equation (3–85) this is qL, a result that also follows from a consideration of the overall vertical equilibrium of the plate because qL is the total loading. The shear stress on the section $x = 0$ is qL/h. The bending moment on the section $x = 0$ is $-D(d^2w/dx^2)_{x=0}$ or $-qL^2/2$. The maximum bending or fiber stress, $\sigma_{xx}^{max} = \sigma_{xx}$ at $y = -h/2$, is given, from Equations (3–85), (3–64), and (3–70), by

$$\sigma_{xx}^{max} = \frac{E}{(1-v^2)}\frac{h}{2}\frac{d^2w}{dx^2} = \frac{6}{h^2}D\frac{d^2w}{dx^2} = -\frac{6M}{h^2}.$$

$$(3\text{–}86)$$

At $x = 0$, σ_{xx}^{max} is $3qL^2/h^2$. The ratio of the shear stress to the maximum bending stress at $x = 0$ is $h/3L$, a rather small quantity for a thin plate. It is implicit in the analysis of the bending of thin plates that shear stresses in the plates are small compared with the bending stresses.

PROBLEM 3–9 Calculate V and M by carrying out force and torque balances on the section of the uniformly loaded plate shown in Figure 3–20.

PROBLEM 3–10 A granite plate with $\rho = 2700$ kg m^{-3} is embedded at one end. If $L = 10$ m and $h = 1/4$ m, what is the maximum bending stress and the shear stress at the base?

PROBLEM 3–11 Determine the displacement of a plate that is embedded at the end $x = 0$ and has a uniform loading q from $x = L/2$ to $x = L$.

PROBLEM 3–12 Determine the deflection of a plate of length L that is embedded at $x = 0$ and has equal loads V_a applied at $x = L/2$ and at $x = L$.

PROBLEM 3–13 Find the deflection of a uniformly loaded beam pinned at the ends, $x = 0, L$. Where is the maximum bending moment? What is the maximum bending stress?

PROBLEM 3–14 A granite plate freely supported at its ends spans a gorge 20 m wide. How thick does the plate have to be if granite fails in tension at 20 MPa? Assume $\rho = 2700$ kg m^{-3}.

PROBLEM 3–15 Determine the deflection of a freely supported plate, that is, a plate pinned at its ends, of length L and flexural rigidity D subject to a sinusoidal load $q_a = q_0 \sin \pi x/L$, as shown in Figure 3–21.

3–11 Buckling of a Plate under a Horizontal Load

When an elastic plate is subjected to a horizontal force P, as shown in Figure 3–22a, the plate can *buckle*, as illustrated in Figure 3–22b, if the applied force is sufficiently large. Fold trains in mountain belts are believed

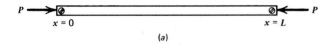

(a)

3–22 Plate buckling under a horizontal force.

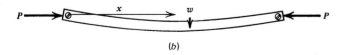

(b)

to result from the warping of strata under horizontal compression. We will therefore consider the simplest example of plate buckling under horizontal compression to determine the minimum force required for buckling to occur and the form, that is, the *wavelength*, of the resulting deflection. In a subsequent section we will carry out a similar calculation to determine if the lithosphere can be expected to buckle under horizontal tectonic compression.

We consider a plate pinned at both ends and subjected to a horizontal force P, as shown in Figure 3–22. The deflection of the plate is governed by Equation (3–74) with $q = 0$:

$$D\frac{d^4w}{dx^4} + P\frac{d^2w}{dx^2} = 0. \tag{3-87}$$

This can be integrated twice to give

$$D\frac{d^2w}{dx^2} + Pw = c_1 x + c_2. \tag{3-88}$$

However, we require that w is zero at $x = 0$, L and that $d^2w/dx^2 = 0$ at $x = 0$, L, since there are no applied torques at the ends. These boundary conditions require that $c_1 = c_2 = 0$, and Equation (3–88) reduces to

$$D\frac{d^2w}{dx^2} + Pw = 0. \tag{3-89}$$

Equation (3–89) has the general solution

$$w = c_1 \sin\left(\frac{P}{D}\right)^{1/2} x + c_2 \cos\left(\frac{P}{D}\right)^{1/2} x, \tag{3-90}$$

where c_1 and c_2 are constants of integration. Because w is equal to zero at $x = 0$, c_2 must be zero, and

$$w = c_1 \sin\left(\frac{P}{D}\right)^{1/2} x. \tag{3-91}$$

But w must also vanish at $x = L$, which implies that if $c_1 \neq 0$, then

$$\sin\left(\frac{P}{D}\right)^{1/2} L = 0. \tag{3-92}$$

Thus $(P/D)^{1/2} L$ must be an integer multiple of π,

$$\left(\frac{P}{D}\right)^{1/2} L = n\pi \qquad n = 1, 2, 3, \ldots \tag{3-93}$$

Solving this equation for P, we get

$$P = \frac{n^2\pi^2}{L^2} D. \tag{3-94}$$

Equation (3–94) defines a series of values of P for which nonzero solutions for w exist. The smallest such value is for $n = 1$ when P is given by

$$P = P_c = \frac{\pi^2}{L^2} D. \tag{3-95}$$

This is the minimum buckling load for the plate. If P is smaller than this *critical value*, known as an *eigenvalue*, the plate will not deflect under the applied load; that is, $c_1 = 0$ or $w = 0$. When P has the value given by Equation (3–95), the plate buckles or deflects under the horizontal load. At the onset of deflection the plate assumes the shape of a half sine curve

$$w = c_1 \sin\left(\frac{P}{D}\right)^{1/2} x$$
$$= c_1 \sin\frac{\pi x}{L}. \tag{3-96}$$

The amplitude of the deflection cannot be determined by the linear analysis carried out here. Nonlinear effects fix the magnitude of the deformation.

The application of plate flexure theory to fold trains in mountain belts requires somewhat more complex models than considered here. Although a number of effects must be incorporated to approximate reality more closely, one of the most important is the influence of the medium surrounding a folded stratum. The rocks above and below a folded layer exert forces on the layer that influence the form (wavelength) of the folds and the critical horizontal force necessary to initiate buckling.

3–12 Deformation of Strata Overlying an Igneous Intrusion

A *laccolith* is a sill-like igneous intrusion in the form of a round lens-shaped body much wider than it is thick. Laccoliths are formed by magma that is intruded along bedding planes of flat, layered rocks at pressures so high that the magma raises the overburden and deforms it into a domelike shape. If the flow of magma is along a crack, a two-dimensional laccolith can be formed. Our analysis is restricted to this case. A photograph of a laccolithic mountain is given in Figure 3–23 along with a sketch of our model.

(a)

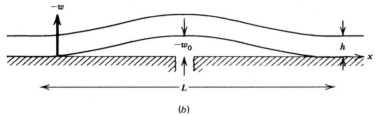

(b)

The overburden or elastic plate of thickness h is bent upward by the pressure p of the magma that will form the laccolith upon solidification. The loading of the plate $q(x)$ is the part of the upward pressure force p in excess of the lithostatic pressure $\rho g h$:

$$q = -p + \rho g h. \tag{3–97}$$

This problem is very similar to the one illustrated in Figure 3–19. In both cases the loading is uniform so that Equation (3–84) is applicable. We take $x = 0$ at the center of the laccolith. The required boundary conditions are $w = dw/dx = 0$ at $x = \pm L/2$. The solution of Equation (3–84) that satisfies these boundary conditions is obtained after some algebra in the form

$$w = -\frac{(p - \rho g h)}{24D}\left(x^4 - \frac{L^2 x^2}{2} + \frac{L^4}{16}\right). \tag{3–98}$$

Note that because of the symmetry of the problem the coefficients of x and x^3 must be zero. The maximum deflection at the center of the laccolith, $x = 0$, is

$$w_0 = -\frac{(p - \rho g h)L^4}{384D}. \tag{3–99}$$

3–23 (a) A laccolith in Red and White Mountain, Colorado. The overlying sedimentary rocks have been eroded (University of Colorado, Boulder). (b) A two-dimensional model for a laccolith.

In terms of its maximum value, the deflection is given by

$$w = w_0\left(1 - 8\frac{x^2}{L^2} + 16\frac{x^4}{L^4}\right). \tag{3–100}$$

PROBLEM 3–16 Show that the cross-sectional area of a two-dimensional laccolith is given by $(p - \rho g h)L^5 / 720D$.

PROBLEM 3–17 Determine the bending moment in the overburden above the idealized two-dimensional laccolith as a function of x. Where is M a maximum? What is the value of M_{max}?

PROBLEM 3–18 Calculate the fiber stress in the stratum overlying the two-dimensional laccolith as a function of y (distance from the centerline of the layer) and x. If dikes tend to form where tension is greatest in the base of the stratum forming the roof of a laccolith, where would you expect dikes to occur for the two-dimensional laccolith?

3-13 **Application to the Earth's Lithosphere**

When applying Equation (3–74) to determine the downward deflection of the earth's lithosphere due to an applied load, we must be careful to include in $q(x)$ the hydrostatic restoring force caused by the effective replacement of mantle rocks in a vertical column by material of smaller density. In the case of the oceanic lithosphere, water fills in "the space vacated" by mantle rocks moved out of the way by the deflected lithosphere. In the case of the continental lithosphere, the rocks of the thick continental crust serve as the fill. Figure 3–24a illustrates the oceanic case. The upper part of the figure shows a lithospheric plate of thickness h and density ρ_m floating on a "fluid" mantle also of density ρ_m. Water of density ρ_w and thickness h_w overlies the oceanic lithosphere. Suppose that an applied load deflects the lithosphere downward a distance w and that water fills in the space above the plate, as shown in the bottom part of Figure 3–24a. The weight per unit area of a vertical column extending from the base of the deflected lithosphere to the surface is

$$\rho_w g(h_w + w) + \rho_m g h.$$

The pressure at a depth $h_w + h + w$ in the surrounding mantle where there is no plate deflection is

$$\rho_w g h_w + \rho_m g(h + w).$$

3-24 Models for calculating the hydrostatic restoring force on lithospheric plates deflected by an applied load q_a. (a) Oceanic case. (b) Continental case.

Thus there is an upward hydrostatic force per unit area equal to

$$\rho_w g h_w + \rho_m g(h + w) - \{\rho_w g(h_w + w) + \rho_m g h\}$$
$$= (\rho_m - \rho_w)g w \tag{3-101}$$

tending to restore the deflected lithosphere to its original configuration. The hydrostatic restoring force per unit area is equivalent to the force that results from replacing mantle rock of thickness w and density ρ_m by water of thickness w and density ρ_w. The net force per unit area acting on the lithospheric plate is therefore

$$q = q_a - (\rho_m - \rho_w)g w, \tag{3-102}$$

where q_a is the applied load at the upper surface of the lithosphere. Equation (3–74) for the deflection of the elastic oceanic lithosphere becomes

$$D\frac{d^4 w}{dx^4} + P\frac{d^2 w}{dx^2} + (\rho_m - \rho_w)g w = q_a(x). \tag{3-103}$$

Figure 3–24b illustrates the continental case. The upper part of the figure shows the continental crust of thickness h_c and density ρ_c separated by the Moho from the rest of the lithosphere of thickness h and density ρ_m. The entire continental lithosphere lies on top of a fluid mantle of density ρ_m. The lower part of Figure 3–24b shows the plate deflected downward a distance w by an applied load such as excess topography. The Moho, being a part of the lithosphere, is also deflected downward a distance w. The space vacated by the deflected lithosphere is filled in by crustal rocks. The crust

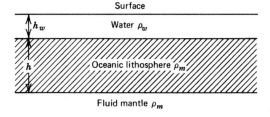

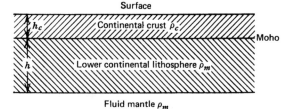

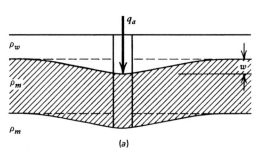

(a)   (b)

beneath the load is effectively thickened by the amount w by which the Moho is depressed. The weight per unit area of a vertical column extending from the base of the deflected plate to the surface is

$$\rho_c g(h_c + w) + \rho_m g h.$$

The pressure at a depth $h_c + h + w$ in the surrounding mantle far from the deflected plate is

$$\rho_c g h_c + \rho_m g(h + w).$$

The difference between these two quantities is the upward hydrostatic restoring force per unit area

$$\rho_c g h_c + \rho_m g(h + w) - \{\rho_c g(h_c + w) + \rho_m g h\}$$
$$= (\rho_m - \rho_c)g w. \qquad (3\text{–}104)$$

The restoring force is equivalent to the force that results from replacing mantle rock by crustal rock in a layer of thickness w. The net force per unit area acting on the elastic continental lithosphere is therefore

$$q = q_a - (\rho_m - \rho_c)g w. \qquad (3\text{–}105)$$

Equation (3–74) for the deflection of the plate becomes

$$D\frac{d^4 w}{dx^4} + P\frac{d^2 w}{dx^2} + (\rho_m - \rho_c)g w = q_a(x). \qquad (3\text{–}106)$$

We are now in a position to determine the elastic deflection of the lithosphere and the accompanying internal stresses (shear and bending) for different loading situations.

3–14 Periodic Loading

How does the positive load of a mountain or the negative load of a valley deflect the lithosphere? To answer this question, we determine the response of the lithosphere to a periodic load. We assume that the elevation of the topography is given by

$$h = h_0 \sin 2\pi \frac{x}{\lambda}, \qquad (3\text{–}107)$$

where h is the topographic height and λ is its wavelength. Positive h corresponds to ridges and negative h to valleys. Since the amplitude of the topography is small compared with the thickness of the elastic lithosphere, the influence of the topography on this thickness can be neglected. The load on the lithosphere

corresponding to the topography given by Equation (3–107) is

$$q_a(x) = \rho_c g h_0 \sin 2\pi \frac{x}{\lambda} \qquad (3\text{–}108)$$

where ρ_c is the density of the crustal rocks associated with the height variation. The equation for the deflection of the lithosphere is obtained by substituting this expression for $q_a(x)$ into Equation (3–106) and setting $P = 0$ to obtain

$$D\frac{d^4 w}{dx^4} + (\rho_m - \rho_c)g w = \rho_c g h_0 \sin 2\pi \frac{x}{\lambda}. \qquad (3\text{–}109)$$

Because the loading is periodic in x, the response or deflection of the lithosphere will also vary sinusoidally in x with the same wavelength as the topography. Thus we assume a solution of the form

$$w = w_0 \sin 2\pi \frac{x}{\lambda}. \qquad (3\text{–}110)$$

By substituting Equation (3–110) into Equation (3–109), we determine the amplitude of the deflection of the lithosphere to be

$$w_0 = \frac{h_0}{\dfrac{\rho_m}{\rho_c} - 1 + \dfrac{D}{\rho_c g}\left(\dfrac{2\pi}{\lambda}\right)^4}. \qquad (3\text{–}111)$$

The quantity $(D/\rho_c g)^{1/4}$ has the dimensions of a length. It is proportional to the natural wavelength for the flexure of the lithosphere.

If the wavelength of the topography is sufficiently short, that is, if

$$\lambda \ll 2\pi\left(\frac{D}{\rho_c g}\right)^{1/4}, \qquad (3\text{–}112)$$

then the denominator of Equation (3–111) is much larger than unity, and

$$w_0 \ll h_0. \qquad (3\text{–}113)$$

Short-wavelength topography causes virtually no deformation of the lithosphere. The lithosphere is infinitely rigid for loads of this scale. This case is illustrated in Figure 3–25a. If the wavelength of the topography is

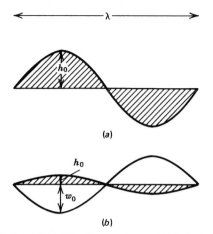

(a)

(b)

3-25 Deflection of the lithosphere under a periodic load. (a) Short-wavelength loading with no deflection of the lithosphere. (b) Long-wavelength loading with isostatic deflection of the lithosphere.

sufficiently long, that is, if

$$\lambda \gg 2\pi \left(\frac{D}{\rho_c g} \right)^{1/4}, \tag{3–114}$$

then Equation (3–111) gives

$$w = w_{0\infty} = \frac{\rho_c h_0}{(\rho_m - \rho_c)}. \tag{3–115}$$

This is the isostatic result obtained in Equation (2–3). For topography of sufficiently long wavelength, the lithosphere has no rigidity and the topography is fully compensated; that is, it is in hydrostatic equilibrium.

The degree of compensation C of the topographic load is the ratio of the deflection of the lithosphere to its maximum or hydrostatic deflection

$$C = \frac{w_0}{w_{0\infty}}. \tag{3–116}$$

Upon substituting Equations (3–111) and (3–115) into the equation for C, we obtain

$$C = \frac{(\rho_m - \rho_c)}{\rho_m - \rho_c + \dfrac{D}{g}\left(\dfrac{2\pi}{\lambda}\right)^4}. \tag{3–117}$$

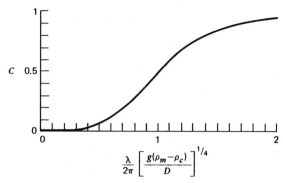

3-26 Dependence of the degree of compensation on the nondimensional wavelength of periodic topography.

This dependence is illustrated in Figure 3–26. For a lithosphere with elastic thickness 25 km, $E = 70$ GPa, $\nu = 0.25$, $\rho_m = 3300$ kg m^{-3}, and $\rho_c = 2800$ kg m^{-3} we find that topography is 50% compensated ($C = 0.5$) if its wavelength is $\lambda = 420$ km. Topography with a shorter wavelength is substantially supported by the rigidity of the lithosphere; topography with a longer wavelength is only weakly supported.

3-15 Stability of the Earth's Lithosphere under an End Load

We have already seen how a plate pinned at its ends can buckle if an applied horizontal load exceeds the critical value given by Equation (3–95). Let us investigate the stability of the lithosphere when it is subjected to a horizontal force P. We will see that when P exceeds a critical value, an infinitely long plate ($L \to \infty$) will become unstable and deflect into the sinusoidal shape shown in Figure 3–27.

The equation for the deflection of the plate is obtained by setting $q_a = 0$ in Equation (3–103):

$$D\frac{d^4w}{dx^4} + P\frac{d^2w}{dx^2} + (\rho_m - \rho_w)gw = 0. \tag{3–118}$$

This equation can be satisfied by a sinusoidal deflection

3-27 Buckling of an infinitely long plate under an applied horizontal load with a hydrostatic restoring force.

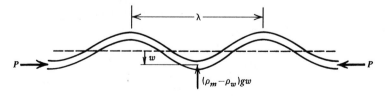

of the plate as given in Equation (3–110) if

$$D\left(\frac{2\pi}{\lambda}\right)^4 - P\left(\frac{2\pi}{\lambda}\right)^2 + (\rho_m - \rho_w)g = 0, \quad (3\text{–}119)$$

a result of directly substituting Equation (3–110) into Equation (3–118). Equation (3–119) is a quadratic equation for the square of the wavelength of the sinusoid λ. Its solution is

$$\left(\frac{2\pi}{\lambda}\right)^2 = \frac{P \pm [P^2 - 4(\rho_m - \rho_w)gD]^{1/2}}{2D}. \quad (3\text{–}120)$$

Because the wavelength of the deformed lithosphere must be real, there can only be a solution if P exceeds the critical value

$$P_c = \{4Dg(\rho_m - \rho_w)\}^{1/2}. \quad (3\text{–}121)$$

P_c is the minimum value for P for which the initially horizontal lithosphere will become unstable and acquire the sinusoidal shape. If $P < P_c$, the horizontal lithosphere is stable and will not buckle under the end load.

The eigenvalue P_c can also be written

$$P_c = \left(\frac{Eh^3(\rho_m - \rho_w)g}{3(1 - \nu^2)}\right)^{1/2} = \sigma_c h, \quad (3\text{–}122)$$

where σ_c is the critical stress associated with the force P_c. Solving Equation (3–122) for the critical stress we find

$$\sigma_c = \left(\frac{Eh(\rho_m - \rho_w)g}{3(1 - \nu^2)}\right)^{1/2}. \quad (3\text{–}123)$$

The wavelength of the instability that occurs when $P = P_c$ is given by Equation (3–120):

$$\lambda_c = 2\pi\left(\frac{2D}{P_c}\right)^{1/2} = 2\pi\left(\frac{D}{g(\rho_m - \rho_w)}\right)^{1/4}$$

$$= 2\pi\left(\frac{Eh^3}{12(1 - \nu^2)(\rho_m - \rho_w)g}\right)^{1/4}. \quad (3\text{–}124)$$

We wish to determine whether buckling of the lithosphere can lead to the formation of a series of synclines and anticlines. We consider an elastic lithosphere with a thickness of 50 km. Taking $E = 100$ GPa, $\nu = 0.25$, $\rho_m = 3300$ kg m^{-3}, and $\rho_w = 1000$ kg m^{-3}, we find from Equation (3–123) that $\sigma_c = 6.4$ GPa. A 50-km-thick elastic lithosphere can support a horizontal compres-

sive stress of 6.4 GPa without buckling. Because of the very large stress required, we conclude that such buckling does not occur. The lithosphere fails, presumably by the development of a fault, before buckling can take place. In general, horizontal forces have a small influence on the bending behavior of the lithosphere. For this reason we neglect them in the lithosphere bending studies to follow.

Horizontal forces are generally inadequate to buckle the lithosphere because of its large elastic thickness. However, the same conclusion may not apply to much thinner elastic layers, such as elastic sedimentary strata embedded between strata that behave as fluids and highly thinned lithosphere in regions of high heat flow. To evaluate the influence of horizontal forces on the bending of such thin layers, we take $h = 1$ km and the other parameters as before and find from Equation (3–123) that $\sigma_c = 900$ MPa. From Equation (3–124) we obtain $\lambda_c = 28$ km. We conclude that the buckling of thin elastic layers may contribute to the formation of folded structures in the earth's crust.

3–16 Bending of the Elastic Lithosphere under the Loads of Island Chains

Volcanic islands provide loads that cause the lithosphere to bend. The Hawaiian ridge is a line of volcanic islands and seamounts that extends thousands of kilometers across the Pacific. These volcanic rocks provide a linear load that has a width of about 150 km and an average amplitude of about 100 MPa. The bathymetric profile across the Hawaiian archipelago shown in Figure 3–28 reveals a depression, the Hawaiian Deep, immediately adjacent to the ridge and an outer *peripheral bulge* or upwarp.

To model the deflection of the lithosphere under linear loading, let us consider the behavior of a plate under a line load V_0 applied at $x = 0$, as shown in Figure 3–29. Since the applied load is zero except at $x = 0$, we

3–28 A bathymetric profile across the Hawaiian archipelago.

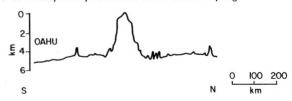

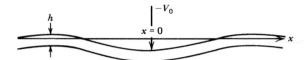

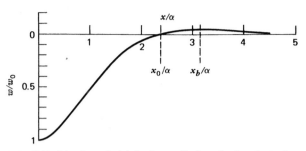

3-29 Deflection of the elastic lithosphere under a line load.

take $q_a(x) = 0$ and $P = 0$ in Equation (3–103) and solve

$$D\frac{d^4w}{dx^4} + (\rho_m - \rho_w)gw = 0. \tag{3-125}$$

The general solution of Equation (3–125) is

$$w = e^{x/\alpha}\left(c_1 \cos\frac{x}{\alpha} + c_2 \sin\frac{x}{\alpha}\right)$$
$$+ e^{-x/\alpha}\left(c_3 \cos\frac{x}{\alpha} + c_4 \sin\frac{x}{\alpha}\right), \tag{3-126}$$

where the constants $c_1, c_2, c_3,$ and c_4 are determined by the boundary conditions and

$$\alpha = \left[\frac{4D}{(\rho_m - \rho_w)g}\right]^{1/4}. \tag{3-127}$$

The parameter α is known as the *flexural parameter*.

Because there is symmetry about $x = 0$, we need only determine w for $x \geq 0$. We require that $w \to 0$ as $x \to \infty$ and that $dw/dx = 0$ at $x = 0$. Clearly, c_1 and c_2 must be zero and $c_3 = c_4$. Equation (3–126) becomes

$$w = c_3 e^{-x/\alpha}\left(\cos\frac{x}{\alpha} + \sin\frac{x}{\alpha}\right) \qquad x \geq 0. \tag{3-128}$$

The constant c_3 is proportional to the magnitude of the applied line load V_0. From Equation (3–81) we have

$$\frac{1}{2}V_0 = D\frac{d^3w}{dx^3}(x = 0) = \frac{4Dc_3}{\alpha^3}. \tag{3-129}$$

(Half the plate supports half the load applied at $x = 0$. Note also that a downward force on the left end of the plate is negative according to the sign convention illustrated in Figure 3–10.) Substituting for c_3 from Equation (3–129) into Equation (3–128), we obtain

$$w = \frac{V_0\alpha^3}{8D}e^{-x/\alpha}\left(\cos\frac{x}{\alpha} + \sin\frac{x}{\alpha}\right) \qquad x \geq 0. \tag{3-130}$$

The maximum amplitude of the deflection at $x = 0$ is given by

$$w_0 = \frac{V_0\alpha^3}{8D}. \tag{3-131}$$

3-30 Half of the theoretical deflection profile for a floating elastic plate supporting a line load.

In terms of w_0, the deflection of the plate is

$$w = w_0 e^{-x/\alpha}\left(\cos\frac{x}{\alpha} + \sin\frac{x}{\alpha}\right). \tag{3-132}$$

This profile is given in Figure 3–30.

The deflection of the lithosphere under a line load is characterized by a well-defined arch or forebulge. The half-width of the depression, x_0, is given by

$$x_0 = \alpha \tan^{-1}(-1) = \frac{3\pi}{4}\alpha. \tag{3-133}$$

The distance from the line load to the maximum amplitude of the forebulge, x_b, is obtained by determining where the slope of the profile is zero. Upon differentiating Equation (3–132) and setting the result to zero

$$\frac{dw}{dx} = -\frac{2w_0}{\alpha}e^{-x/\alpha}\sin\frac{x}{\alpha} = 0, \tag{3-134}$$

we find

$$x_b = \alpha \sin^{-1} 0 = \pi\alpha. \tag{3-135}$$

The height of the forebulge w_b is obtained by substituting this value of x_b into Equation (3–132):

$$w_b = -w_0 e^{-\pi} = -0.0432w_0. \tag{3-136}$$

The amplitude of the forebulge is quite small compared with the depression of the lithosphere under the line load.

This analysis for the line load is only approximately valid for the Hawaiian Islands, since the island load is distributed over a width of about 150 km. However, the distance from the center of the load to the crest of the arch can be used to estimate the thickness of the elastic lithosphere if we assume that it is equal to x_b.

A representative value of x_b for the Hawaiian archipelago is 250 km; with $x_b = 250$ km, Equation (3–135) gives a flexural parameter $\alpha = 80$ km. For $\rho_m - \rho_w = 2300$ kg m^{-3} and $g = 10$ m s^{-2} Equation (3–127) gives $D = 2.4 \times 10^{23}$ N m. Taking $E = 70$ GPa and $\nu = 0.25$, we find from Equation (3–72) that the thickness of the elastic lithosphere is $h = 34$ km.

PROBLEM 3–19 (a) Consider a lithospheric plate under a line load. Show that the absolute value of the bending moment is a maximum at

$$x_m = \alpha \cos^{-1} 0 = \frac{\pi}{2}\alpha \qquad (3–137)$$

and that its value is

$$M_m = -\frac{2Dw_0}{\alpha^2}e^{-\pi/2} = -0.416\frac{Dw_0}{\alpha^2}. \qquad (3–138)$$

(b) Refraction studes show that the Moho is depressed about 10 km beneath the center of the Hawaiian Islands. Assuming that this is the value of w_0 and that $h = 34$ km, $E = 70$ GPa, $\nu = 0.25$, $\rho_m - \rho_w = 2300$ kg m^{-3}, and $g = 10$ m s^{-2}, determine the maximum bending stress in the lithosphere.

Since volcanism along the Hawaiian ridge has weakened the lithosphere, it may not be able to sustain large bending moments beneath the load. In this case we should consider a model in which the lithosphere is fractured along the line of the ridge. Let us accordingly determine the deflection of a semi-infinite elastic plate floating on a fluid half-space and subjected to a line load $V_0/2$ at its end, as sketched in Figure 3–31. The deflection is given by Equation (3–126), with the constants of integration yet to be determined. Since the plate extends from $x = 0$ to $x = \infty$ and we require $w \to 0$ as $x \to \infty$, c_1 and c_2 must again be zero. We have assumed that no external torque is applied to the end $x = 0$. From Equation (3–73) we can conclude that $d^2w/dx^2 = 0$ at $x = 0$. This boundary condition requires that $c_4 = 0$. Finally, by equating the shear on

3-31 Deflection of a broken elastic lithosphere under a line load.

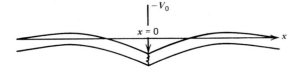

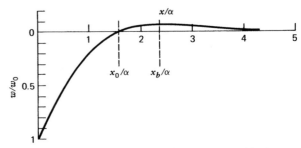

3-32 The deflection of the elastic lithosphere under an end load.

the end $x = 0$ to the applied line load, we find

$$\frac{1}{2}V_0 = D\frac{d^3w}{dx^3}(x = 0) = \frac{2Dc_3}{\alpha^3}. \qquad (3–139)$$

With the value of c_3 from Equation (3–139) and $c_1 = c_2 = c_4 = 0$, Equation (3–126) gives

$$w = \frac{V_0\alpha^3}{4D}e^{-x/\alpha}\cos\frac{x}{\alpha}. \qquad (3–140)$$

The maximum amplitude of the deflection at $x = 0$ is

$$w_0 = \frac{V_0\alpha^3}{4D}. \qquad (3–141)$$

For the same load, the deflection amplitude of a broken lithosphere is twice as great as it is for a lithosphere without a break. By substituting Equation (3–141) into Equation (3–140), we can write

$$w = w_0 e^{-x/\alpha}\cos\frac{x}{\alpha}. \qquad (3–142)$$

This profile is given in Figure 3–32.

The half-width of the depression and the position and amplitude of the forebulge are given by

$$x_0 = \frac{\pi}{2}\alpha \qquad (3–143)$$

$$x_b = \frac{3\pi}{4}\alpha \qquad (3–144)$$

$$w_b = w_0 e^{-3\pi/4}\cos\frac{3\pi}{4} = -0.0670w_0. \qquad (3–145)$$

The amplitude of the forebulge for the broken lithosphere model, although still small compared with the deflection of the lithosphere under the load, is considerably larger than the forebulge amplitude of an unbroken lithosphere supporting the same load.

We again evaluate the model results for the deflection of the lithosphere caused by the Hawaiian Islands.

With $x_b = 250$ km, we find from Equation (3–144) that $\alpha = 106$ km. This result, together with $\rho_m - \rho_w = 2300$ kg m^{-3}, g = 10 m s^{-2}, $E = 70$ GPa, and $\nu = 0.25$, gives $D = 7.26 \times 10^{23}$ Nm and $h = 49$ km. The thickness of a broken lithosphere turns out to be about 50% greater than the thickness of an unbroken lithosphere.

PROBLEM 3–20 (*a*) Consider a lithospheric plate under an end load. Show that the absolute value of the bending moment is a maximum at

$$x_m = \alpha \tan^{-1} 1 = \frac{\pi}{4} \alpha, \tag{3–146}$$

and that its value is

$$M_m = -\frac{2Dw_0}{\alpha^2} e^{-\pi/4} \sin \frac{\pi}{4} = -0.644 \frac{Dw_0}{\alpha^2}. \tag{3–147}$$

(*b*) Refraction studies show that the Moho is depressed about 10 km beneath the center of the Hawaiian Islands. Assuming that this is the value of w_0 and that $h = 49$ km, $E = 70$ GPa, $\nu = 0.25$, $\rho_m - \rho_w = 2300$ kg m^{-3}, and $g = 10$ m s^{-2}, determine the maximum bending stress in the lithosphere.

3–17 Bending of the Elastic Lithosphere at an Ocean Trench

Another example of the flexure of the oceanic elastic lithosphere is to be found at ocean trenches. Prior to subduction, considerable bending of the elastic lithosphere occurs. The bent lithosphere defines the oceanward side of the trench. To model this behavior, we will consider an elastic plate acted upon by an end load V_0 and a bending moment M_0, as illustrated in Figure 3–33.

The deflection of the plate is governed by Equation (3–125), and once again the general solution is given

3–33 Bending of the lithosphere at an ocean trench due to an applied vertical load and bending moment.

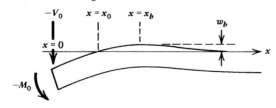

by Equation (3–126). We require $w \to 0$ as $x \to \infty$ so that $c_1 = c_2 = 0$ and

$$w = e^{-x/\alpha} \left(c_3 \cos \frac{x}{\alpha} + c_4 \sin \frac{x}{\alpha} \right). \tag{3–148}$$

At $x = 0$, the bending moment is $-M_0$; from Equation (3–73) we obtain

$$c_4 = \frac{-M_0 \alpha^2}{2D}. \tag{3–149}$$

Also, at $x = 0$, the shear force is $-V_0$; from Equations (3–59) and (3–73) we find

$$c_3 = (V_0 \alpha + M_0) \frac{\alpha^2}{2D}. \tag{3–150}$$

The equation for w can now be written as

$$w = \frac{\alpha^2 e^{-x/\alpha}}{2D} \left\{ -M_0 \sin \frac{x}{\alpha} + (V_0 \alpha + M_0) \cos \frac{x}{\alpha} \right\}. \tag{3–151}$$

Equation (3–151) reduces to Equation (3–140) in the case $M_0 = 0$. Note that the line load here is V_0; it was $V_0/2$ in Equation (3–140).

The elastic deflection of the oceanic lithosphere in terms of the vertical force and bending moment at the ocean trench axis is given by Equation (3–151). The vertical force and bending moment are the result of the gravitational body force acting on the descending plate. Unfortunately, V_0 and M_0 cannot be determined directly. Quantities that can be measured directly are the height of the forebulge w_b and the half-width of the forebulge $x_b - x_0$, as illustrated in Figure 3–33. We therefore express the trench profile in terms of these parameters. We can determine x_0 by setting $w = 0$:

$$\tan \frac{x_0}{\alpha} = 1 + \frac{\alpha V_0}{M_0}. \tag{3–152}$$

Similarly, we can determine x_b by setting $dw/dx = 0$:

$$\tan \frac{x_b}{\alpha} = -1 - \frac{2M_0}{\alpha V_0}. \tag{3–153}$$

The height of the forebulge is obtained by substituting this value of x_b into Equation (3–151):

$$w_b = \frac{\alpha^2}{2D} e^{-x_b/\alpha} \left[-M_0 \sin \frac{x_b}{\alpha} + (M_0 + V_0 \alpha) \cos \frac{x_b}{\alpha} \right]. \tag{3–154}$$

From Equations (3–152) and (3–154) we find

$$\tan\left(\frac{x_b - x_0}{\alpha}\right) = \frac{\sin\left(\frac{x_b}{\alpha} - \frac{x_0}{\alpha}\right)}{\cos\left(\frac{x_b}{\alpha} - \frac{x_0}{\alpha}\right)}$$

$$= \frac{\sin\frac{x_b}{\alpha}\cos\frac{x_0}{\alpha} - \cos\frac{x_b}{\alpha}\sin\frac{x_0}{\alpha}}{\cos\frac{x_b}{\alpha}\cos\frac{x_0}{\alpha} + \sin\frac{x_b}{\alpha}\sin\frac{x_0}{\alpha}}$$

$$= \frac{\tan\frac{x_b}{\alpha} - \tan\frac{x_0}{\alpha}}{1 + \tan\frac{x_b}{\alpha}\tan\frac{x_0}{\alpha}} = 1 \qquad (3\text{–}155)$$

and

$$x_b - x_0 = \frac{\pi}{4}\alpha. \qquad (3\text{–}156)$$

This half-width is a direct measure of the flexural parameter and, therefore, of the flexural rigidity and thickness of the elastic lithosphere.

By using Equation (3–152), we can rewrite Equation (3–151) for the deflection of the lithosphere as

$$w = \frac{\alpha^2 M_0}{2D} e^{-x/\alpha}\left(-\sin\frac{x}{\alpha} + \tan\frac{x_0}{\alpha}\cos\frac{x}{\alpha}\right)$$

$$= \frac{\alpha^2 M_0}{2D} e^{-[(x-x_0)/\alpha]-x_0/\alpha}$$

$$\times \left\{\frac{\sin\frac{x_0}{\alpha}\cos\frac{x}{\alpha} - \cos\frac{x_0}{\alpha}\sin\frac{x}{\alpha}}{\cos\frac{x_0}{\alpha}}\right\}$$

$$= -\frac{\alpha^2 M_0}{2D} e^{-[(x-x_0)/\alpha]}e^{-x_0/\alpha}\frac{\sin\left(\frac{x-x_0}{\alpha}\right)}{\cos\left(\frac{x_0}{\alpha}\right)}.$$

$$(3\text{–}157)$$

The height of the forebulge is thus given by

$$w_b = -\frac{\alpha^2 M_0}{2D} e^{-[(x_b-x_0)/\alpha]}e^{-x_0/\alpha}\frac{\sin\left(\frac{x_b-x_0}{\alpha}\right)}{\cos\left(\frac{x_0}{\alpha}\right)}.$$

$$(3\text{–}158)$$

Upon dividing Equation (3–157) by Equation (3–158) and eliminating α using Equation (3–156), we obtain

$$\frac{w}{w_b} = \frac{\exp\left[-\frac{\pi}{4}\left(\frac{x-x_0}{x_b-x_0}\right)\right]}{\exp\left(-\frac{\pi}{4}\right)}\frac{\sin\left[\frac{\pi}{4}\left(\frac{x-x_0}{x_b-x_0}\right)\right]}{\sin\frac{\pi}{4}}$$

$$= \sqrt{2}e^{\pi/4}\exp\left[-\frac{\pi}{4}\left(\frac{x-x_0}{x_b-x_0}\right)\right]\sin\left[\frac{\pi}{4}\left(\frac{x-x_0}{x_b-x_0}\right)\right].$$

$$(3\text{–}159)$$

The plot of w/w_b vs. $(x-x_0)/(x_b/x_0)$ shown in Figure 3–34a defines a *universal flexure profile*. The

profile is valid for any two-dimensional elastic flexure of the lithosphere under end loading.

We can solve for the bending moment in terms of $(x-x_0)/(x_b-x_0)$ by substituting Equation (3–159) into Equation (3–73)

$$M = \frac{\sqrt{2}\pi^2 e^{\pi/4}}{8}\frac{Dw_b}{(x_b-x_0)^2}\cos\left[\frac{\pi(x-x_0)}{4(x_b-x_0)}\right]$$

$$\times \exp\left[-\frac{\pi(x-x_0)}{4(x_b-x_0)}\right]. \qquad (3\text{–}160)$$

The dependence of $M(x_b-x_0)^2/Dw_b$ on $(x-x_0)/(x_b-x_0)$ is shown in Figure 3–34b. The bending moment is a maximum at $(x-x_0)/(x_b-x_0) = -1$. The shear force can be determined from Equations (3–59) and (3–160) to be

$$V = -\frac{\sqrt{2}\pi^3 e^{\pi/4}}{32}\frac{Dw_b}{(x_b-x_0)^3}\left[\cos\left\{\frac{\pi(x-x_0)}{4(x_b-x_0)}\right\}\right.$$

$$\left. + \sin\left\{\frac{\pi(x-x_0)}{4(x_b-x_0)}\right\}\right]\exp\left[-\frac{\pi(x-x_0)}{4(x_b-x_0)}\right].$$

$$(3\text{–}161)$$

The dimensionless shear force $V(x_b-x_0)^3/Dw_b$ is plotted vs. $(x-x_0)/(x_b-x_0)$ in Figure 3–34c. The shear force is zero at $(x-x_0)/(x_b-x_0) = -1$.

The universal flexure profile is compared with an observed bathymetric profile across the Mariana trench in Figure 3–35. In making the comparison, we take $x_b = 55$ km and $w_b = 500$ m ($x_0 = 0$). From Equation (3–156) we find that $\alpha = 70$ km. With $\rho_m - \rho_w = 2300$ kg m^{-3} and $g = 10$ m s^{-2}, Equation (3–127) gives $D = 1.4 \times 10^{23}$ N m. From Equation (3–72) with $E = 70$ GPa and $\nu = 0.25$ we find that the thickness of the elastic lithosphere is 28 km. This value is in quite good agreement with the thickness of the oceanic elastic lithosphere obtained by considering island loads. The largest bending stress is 900 MPa, and it occurs 20 km seaward of the trench axis. This is a very large deviatoric stress, and it is doubtful that the near-surface rocks have sufficient strength in tension. However, the yield stress of the mantle is likely to approach this value at depth where the lithostatic pressure is high.

Although the trench bathymetric profile given in Figure 3–35 appears to exhibit elastic flexure, other trench profiles exhibit an excessively large curvature near the point of the predicted maximum bending moment. This is discussed in Chapter 7, where we

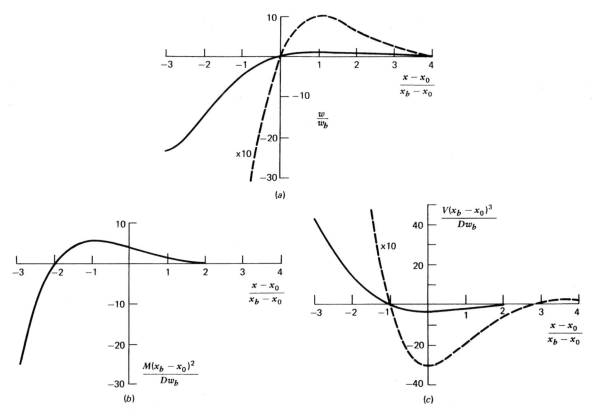

(a)

(b)

(c)

3–34 Universal solution for the deflection of an elastic lithosphere under a vertical end load and bending moment. (a) Dependence of the nondimensional displacement w/w_b on the nondimensional position $(x - x_0)/(x_b - x_0)$. The profile is also shown at an amplification of 10 to 1 to more clearly show the structure of the forebulge. (b) The dimensionless bending moment versus $(x - x_0)/(x_b - x_0)$. (c) The dimensionless vertical shear force as a function $(x - x_0)/(x_b - x_0)$.

3–35 Comparison of a bathymetric profile across the Mariana trench (solid line) with the universal lithospheric deflection profile given by Equation (3–159) (dashed line); $x_b = 55$ km and $w_b = 0.5$ km.

associate this excess curvature with the plastic failure of the lithosphere.

3–18 Flexure and the Structure of Sedimentary Basins

Lithospheric flexure is also associated with the structure of many sedimentary basins. A sedimentary basin is a region where the earth's surface has been depressed and the resulting depression has been filled by sediments. Typical sedimentary basins have depths up to 5 km, although some are as deep as 15 km. Because

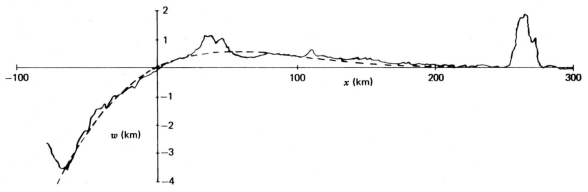

sedimentary basins contain reservoirs of petroleum, their structures have been studied in detail using seismic reflection profiling and well logs.

Some sedimentary basins are bounded by near-vertical faults along which the subsidence has occurred. Others, however, have a smooth basement, and the subsidence is associated with the flexure of the elastic lithosphere. The horizontal dimensions of these sedimentary basins, about 400 to 1000 km, reflect the magnitude of the flexural parameter based on sediments of density ρ_s replacing mantle rock of density ρ_m, $\alpha = [4D/(\rho_m - \rho_s)g]^{1/4}$.

Some sedimentary basins have a nearly two-dimensional structure. They are caused by the loading of a linear mountain belt and are known as *foreland basins*. Examples are the series of sedimentary basins lying east of the Andes in South America and the Appalachian basin in the eastern United States lying west of the Appalachian Mountains. Depth contours of the basement beneath the Appalachian basin are given in Figure 3–36a. A basement profile is shown in Figure 3–36b. The depth w is the depth below sea level, and the coordinate $-x$ is measured from the point where basement rocks are exposed at the surface.

It is appropriate to model the structure of the Appalachian basin as a two-dimensional lithospheric plate under a linear end load. Thus the universal flexure profile given in Equation (3–159) is directly applicable. In order to fit the basement profile given in Figure 3–36b we take $x_b = 122$ km and $w_b = 290$ m ($x_0 = 0$). Since the forebulge has been destroyed by erosion this choice of parameters is not unique. They can be varied somewhat, and a reasonable fit can still be obtained. However, these values are near the center of the acceptable range. From Equation (3–156) we find that they correspond to $\alpha = 155$ km. As we have already noted, the flexural rigidity must be based on the density difference between the mantle and the sediments $\rho_m - \rho_s$. With $\rho_m - \rho_s = 700$ kg m^{-3} and $g = 10$ m s^{-2} we find $D = 10^{24}$ N m. From Equation (3–72) with $E = 70$ GPa and $\nu = 0.25$ we find that the thickness of the elastic continental lithosphere is $h = 54$ km. This is somewhat larger than the values we obtained for the thickness of the elastic oceanic lithosphere. Flexure studies of other sedimentary basins give similar values of elastic thickness.

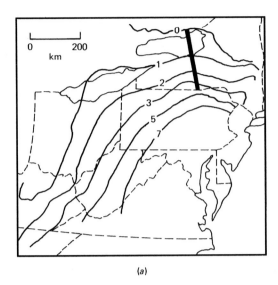

(a)

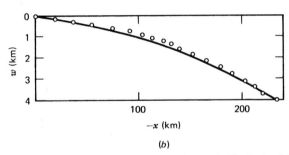

(b)

3–36 (a) Contours of basement (in km) in the Appalachian basin of the eastern United States. Data are from well logs and seismic reflection studies. (b) The data points are the depths of basement below sea level as a function of the distance from the point where basement rocks are exposed at the surface along the profile given by the heavy line in (a). The solid line is the universal flexure profile from Equation (3–159) with $x_b = 122$ km and $w_b = 290$ m ($x_0 = 0$).

PROBLEM 3–21 An ocean basin has a depth of 5.5 km. If it is filled to sea level with sediments of density 2600 kg m^{-3}, what is the maximum depth of the resulting sedimentary basin? Assume $\rho_m = 3300$ kg m^{-3}.

PROBLEM 3–22 The Amazon River basin in Brazil has a width of 400 km. Assuming that the basin is caused by a line load at its center and that the elastic lithosphere is not broken, determine the corresponding thickness of the elastic lithosphere. Assume $E = 70$ GPa, $\nu = 0.25$, and $\rho_m - \rho_s = 700$ kg m^{-3}.

REFERENCE

Bieniawski, Z. T. (1967), Mechanism of brittle fracture of rock: Part II. Experimental studies, *Int. J. Rock. Mech. Min. Sci.*, **4**, 407–423.

COLLATERAL READING

Eringen, A. C., *Mechanics of Continua* (John Wiley, New York, 1967), 502 pages.

A comprehensive treatment of the mechanics of continua at a relatively sophisticated level. The basic concepts of strain, stress, flow, thermodynamics, and constitutive equations are introduced. Applications are made to elasticity, fluid dynamics, thermoplasticity, and viscoelasticity.

Fung, Y. C., *Foundations of Solid Mechanics* (Prentice-Hall, Englewood Cliffs, NJ, 1965), 525 pages.

A graduate-level textbook on the mechanics of solids. The text is mainly concerned with the classical theory of elasticity, thermodynamics of solids, thermoelasticity, viscoelasticity, plasticity, and finite deformation theory. The book begins with an introductory chapter on elastic and viscoelastic behavior. Cartesian tensors are then introduced and used in the discussions of stress, strain, and the conservation laws. Subsequent chapters deal with linear elasticity, solutions of elastic problems by potentials, two-dimensional problems, energy theorems, Saint-Venant's principle, Hamilton's principle, wave propagation, elasticity and thermodynamics, thermoelasticity, viscoelasticity, and finite strain theory. Problems for the student are included.

Jaeger, J. C., *Elasticity, Fracture, and Flow*, 3rd edition (Methuen, London, 1969), 268 pages.

A monograph on the mathematical foundations of elasticity, plasticity, viscosity, and rheology. Chapter 1 develops the analysis of stress and strain with emphasis on Mohr's representations. Chapter 2 discusses stress–strain relations for elasticity, viscosity, and plasticity, and criteria for fracture and yield. Chapter 3 derives the equations of motion and equilibrium. Chapters 4 and 5 deal with stresses in the earth's crust, rock mechanics, and applications to structural geology.

Jaeger, J. C., and N. G. W. Cook, *Fundamentals of Rock Mechanics* (Chapman and Hall, London, 1976), 585 pages.

See collateral reading list for Chapter 2.

Kraus, H., *Thin Elastic Shells* (John Wiley, New York, 1967).

An extensive mathematical treatment of the deformation of thin elastic shells. It includes three chapters on the theory of thin elastic shells, four chapters on static analysis, two chapters on dynamic analysis, and two chapters on numerical methods.

Muskhelishvili, N. I., *Some Basic Problems of the Mathematical Theory of Elasticity*, 4th edition (P. Noordhoff, Groningen, 1963), 718 pages.

This treatise on the mathematical theory of elasticity is divided into seven major parts. Part 1 deals with the fundamental equations of the mechanics of an elastic body. It includes separate chapters on analyses of stress and strain, relation between stress and strain, the equilibrium equations of an elastic body, and the fundamental boundary value problems of static elasticity. Part 2 treats planar problems whose solutions are obtained with the aid of the stress function and its complex representation. The technique of conformal mapping is introduced. Part 3 develops the Fourier series approach to the solution of planar problems, while Parts 4 and 5 make use of Cauchy integrals. Part 6 presents solutions for special planar geometries and Part 7 deals with the extension, torsion and bending of bars.

Novozhilov, V. V., *Thin Shell Theory* (P. Noordhoff, Groningen, 1964), 377 pages.

A mathematical analysis of stresses and strains in thin shells using linear elasticity theory. There are four chapters on the general theory of thin elastic shells, the membrane theory of shells, cylindrical shells, and shells of revolution.

Timoshenko, S., and J. N. Goodier, *Theory of Elasticity*, (McGraw-Hill, New York, 1970), 567 pages.

See collateral reading list for Chapter 2.

Timoshenko, S., and D. H. Young, *Elements of Strength of Materials*, 5th edition (Van Nostrand, Princeton, NJ, 1968), 377 pages.

An undergraduate engineering textbook with an extensive treatment of the bending of beams and elastic stability. Problems with solutions are included.

FOUR

Heat Transfer

4-1 Introduction

In the previous chapter we studied the elastic behavior of the outer shell of the Earth. Our studies of the bending of the lithosphere have shown that a near-surface region with a thickness of 25 to 50 km behaves elastically on geological time scales. Seismic studies have shown that the entire mantle of the Earth to a depth of 2885 km is a solid because it transmits shear waves. In order to understand the presence of a thin elastic shell, it is necessary to allow for variations in the rheology of the solid rock as a function of depth. Although the behavior of the near-surface rocks is predominantly elastic, the deeper rocks must exhibit a fluid or creep behavior on geological time scales in order to relax the stresses. The fluid behavior of mantle rock also results in mantle convection and the associated movement of the surface plates.

We know from both laboratory and theoretical studies that the rheology of solids is primarily a function of temperature. Therefore, to understand the mechanical behavior of the Earth, we must understand its thermal structure. The rheology of mantle rocks is directly related to the temperature as a function of depth. This, in turn, is dependent on the rate at which heat can be lost from the interior to the surface. There are three mechanisms for the transfer of heat: *conduction, convection,* and *radiation*. Conductive heat transfer occurs through a medium via the net effect of molecular collisions. It is a diffusive process wherein molecules transmit their kinetic energy to other molecules by colliding with them. Heat is conducted through a medium in which there is a spatial variation in the temperature. Convective heat transport is associated with the motion of a medium.

If a hot fluid flows into a cold region, it will heat the region; similarly, if a cold fluid flows into a hot region, it will cool it. Electromagnetic radiation can also transport heat. An example is the radiant energy from the Sun. In the Earth, radiative heat transport is only important on a small scale and its influence can be absorbed into the definition of the thermal conductivity.

As the discussion of this chapter shows, both conduction and convection are important heat transport mechanisms in the Earth. The temperature distribution in the continental crust and lithosphere is governed mainly by the conductive heat loss to the surface of heat that is generated internally by the decay of radioactive isotopes in the rocks and heat that flows upward from the subcontinental mantle. The loss of the Earth's internal heat through the oceanic crust and lithosphere is controlled largely by conduction, although convective heat transport by water circulating through the basaltic crustal rocks is also important, especially near ridges. Intrusive igneous bodies cool by both conduction and the convective effects of circulating groundwater. The heating of buried sediments and the adjustment of subsurface temperatures to effects of surface erosion and glaciation occur via the process of conduction. Convection plays the dominant role in the transport of heat from the Earth's deep mantle and in controlling the temperature of its interior.

This chapter discusses mainly heat conduction and its application to geological situations. Because convective heat transfer involves fluid motions, we will postpone a detailed discussion of this subject to Chapter 6, where we will develop the fundamentals of fluid mechanics. However, the consequences of convective heat transport are incorporated into our discussion of the Earth's temperature toward the end of this chapter.

4-2 Fourier's Law of Heat Conduction

The basic relation for conductive heat transport is *Fourier's law*, which states that the heat flux q, or the flow of heat per unit area and per unit time, at a point in a medium is directly proportional to the temperature gradient at the point. In one dimension, Fourier's law takes the form

$$q = -k\frac{dT}{dy} \qquad (4-1)$$

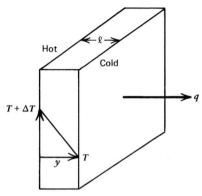

4-1 Heat transfer through a slab.

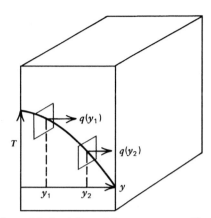

4-2 Heat flux and the local slope of the temperature profile when $T(y)$ has nonzero curvature.

where k is the coefficient of thermal conductivity and y is the coordinate in the direction of the temperature variation. The minus sign appears in Equation (4–1) since heat flows in the direction of decreasing temperature. With $dT/dy > 0$, T increases in the positive y direction, so that heat must flow in the negative y direction.

Figure 4–1 is a simple example of how Fourier's law can be used to give the heat flux through a slab of material of thickness l across which a temperature difference ΔT is maintained. In this case, the temperature gradient is

$$\frac{dT}{dy} = -\frac{\Delta T}{l}, \qquad (4\text{–}2)$$

and the heat flux, from Fourier's law, is

$$q = \frac{k \Delta T}{l}. \qquad (4\text{–}3)$$

Fourier's law applies even when the temperature distribution is not linear, as sketched in Figure 4–2. In this case, the local slope of the temperature profile must be used in Fourier's law, and for constant k the heat flux is a function of y, $q = q(y)$. We will see that curvature in a temperature profile implies either the occurrence of sources or sinks of heat or time dependence.

4-3 Measuring the Earth's Surface Heat Flux

The average heat flux at the Earth's surface provides important information on the amount of heat being produced in the Earth and the temperature distribution in its interior. In the 1800s it was recognized that the temperature in caves and mines increases with depth. Typical values for this increase are $dT/dy = 20$ to 30 K km^{-1}. Since the thermal conductivities of near-

surface rocks are usually in the range 2 to $3 \text{ W m}^{-1} \text{ K}^{-1}$, the heat flow to the surface of the Earth implied by these temperature gradients is, according to Equation (4–1), 40 to 90 mW m^{-2}. It is standard practice to take the upward surface heat flow to be a positive quantity, even though from Equation (4–1), with y measured positive downward, it has a negative value.

Although temperature measurements in caves and mines give approximate values for the near-surface thermal gradient, accurate measurements of the thermal gradient in continental areas require deep drill holes. Deep measurements are necessary because climatic variations in the Earth's surface temperature, particularly those due to ice ages, influence the temperatures in the near-surface rocks. These effects are considered quantitatively in Section 4–14. In order to reach the steady-state thermal structure, holes must be drilled deeper than about 300 m.

The thermal gradient is measured by lowering a thermistor (an accurate electronic thermometer) down the drill hole. Care must be exercised to prevent the circulation of drilling fluid during drilling from affecting the measured gradient. This can be done in either of two ways. Measurements can be made at the bottom of the drill hole during drilling. The drilling fluid does not have time to change the temperature at the bottom of the hole. Alternatively, the temperature log of the hole (the measurement of the temperature as a function of depth) can be carried out some time after drilling has ceased. It usually takes 1 to 2 years for a drill hole to equilibrate to the ambient geothermal gradient. Drill holes are invariably filled with groundwater. It is the temperature of this water that is measured by the thermistor. As long as the water is not flowing, its temperature equilibrates with that of the surrounding rock. However, many drill holes cross aquifers (underground channels of porous rock in which water flows), with the result that water will flow up or down the drill hole if it is not lined and will affect the temperature distribution.

From Fourier's law it is clear that the determination of the heat flux requires a measurement of both the temperature gradient and the thermal conductivity of the rock. The thermal conductivity of rocks can be determined in the laboratory by subjecting samples cut from drill holes to known heat fluxes and measuring the temperature drops across them. Figure 4–3 is a schematic of one way in which this can be done.

The rock sample of thermal conductivity k_r is placed between material – brass, for example – of thermal conductivity k_b. Thermocouples measure the temperatures of the hot and cold ends of the metal, T_H and T_c, respectively, and the temperatures at the surfaces of the metal adjoining the rock section, T_1 and T_2. The contact between the rock and metal (air perhaps) involves an unknown thermal resistance to the flow of heat. Since the same heat must be conducted through the rock and metal in steady state, Fourier's law (4–1) can be used to determine k_r in terms of the measurable quantities T_H, T_c, T_1, T_2, d, l and the conductivity k_b. Thermal conductivities of a variety of rocks are given in Section E of Appendix 2.

PROBLEM 4–1 For the situation sketched in Figure 4–3 and discussed above, show that the thermal conductivity of the rock sample can be determined from the equation

$$\frac{T_1 - T_2}{T_H - T_1} = \frac{k_b}{k_r}\frac{d}{l} + \frac{2\delta k_b}{l k_c} \tag{4–4}$$

The thermal resistance of the contacts are accounted for by associating a thermal conductivity k_c and a thickness δ with each contact. By making measurements on rock samples of different thicknesses and plotting $(T_1 - T_2)/(T_H - T_1)$ vs. d, one can determine k_r from the slope of the resulting straight line without knowing either δ or k_c.

We just discussed the determination of the surface heat flow in the continents. The heat flow can also be measured on the ocean floor. A large fraction of the seafloor is covered by a layer of soft sediments. A needlelike probe carrying a series of thermistors is dropped from a ship and penetrates the sediments. Typically the probe has a length of 3 m. The near-surface heat flow in the oceanic crust is almost a constant because climatic variations do not change the temperature of the seawater in the deep oceans. This water is buffered at a temperature between 1 and 2°C, the temperature at which the density of the seawater is a maximum. The variation is due to changes in salinity. In many cases, however, the near-surface heat flow in the sediments is influenced by the hydrothermal convection of seawater through the sediments and basaltic crustal rocks.

The thermal conductivity of the sediments can be determined using a heater in the heat-flow probe. The record of the increase in probe temperature with time after the heater is turned on can be interpreted to give the thermal conductivity of the sediments, as discussed later in this chapter.

PROBLEM 4–2 Temperatures at the interfaces between sedimentary layers of different rock types as determined from a well log are given in Table 4–1. The measured thermal conductivity of each layer is also given. Determine the heat flow through each layer and the mean value of the heat flow.

4–3 Laboratory device for measuring the thermal conductivity of a rock sample.

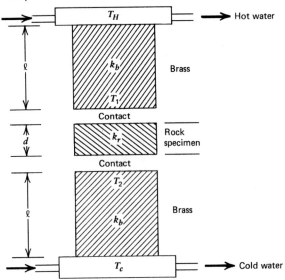

TABLE 4-1 Temperatures Between Layers of Rock Types			
Depth (m)	Temp. (°C)	Rock Type	k (Wm^{-1} K^{-1})
380	18.362		
		Sandstone	3.2
402	18.871		
		Shale	1.7
412	19.330		
		Sandstone	5.3
465	20.446		
		Salt	6.1
475	20.580		
		Sandstone	3.4
510	21.331		
		Shale	1.9
515	21.510		

4-4 The Earth's Surface Heat Flow

Tens of thousands of heat flow measurements have been made both in the continents and the oceans. Because the oceanic crust participates in the plate tectonic cycle and the continental crust does not, we can consider these regions separately.

The mean heat flow for all continents is $65 \pm 1.6 \, \text{mW m}^{-2}$. Regions of high heat flow in the continents are generally restricted to active volcanic areas. Examples are the lines of volcanoes associated with ocean trenches – the Andes, for example – and regions of tensional tectonics such as the western United States. The areas of high heat flow associated with volcanic lines are generally quite small and do not contribute significantly to the mean heat flow. Similarly, areas of tensional tectonics are quite small on a global basis. Broad regions of continental tectonics, such as the collision zone extending from the Alps through the Himalayas, have near-normal surface heat flows. Therefore, regions of active tectonics and mountain building make a relatively small contribution to the mean continental heat flow.

In stable continental areas, the surface heat flow has a strong correlation with the surface concentrations of the radioactive, heat-producing isotopes. This correlation, which is discussed in detail in Section 4-8, is illustrated in Figure 4-11. Approximately one-half of the surface heat flow in the continents can be at-tributed to the heat production from the radioactive isotopes of uranium, thorium, and potassium in the continental crust. Surface heat flow systematically decreases with the age of the surface rocks in stable continental areas. Similarly, the concentration of the radioactive isotopes in the surface rocks also decreases with the age of the rocks. This decrease is attributed to the progressive effects of erosion that remove the near-surface rocks with the largest concentrations of the heat-producing isotopes. The conclusion is that the decrease in surface heat flow with age in stable continental areas can be primarily attributed to the decrease in the crustal concentrations of the heat-producing isotopes.

The mean measured heat flow for all the oceans is $101 \pm 2.2 \, \text{mW m}^{-2}$. The concentration of the heat-producing isotopes in the oceanic crust is about one order of magnitude less than it is in the continental crust. Also, the oceanic crust is about a factor of 5 thinner than the continental crust. Therefore, the contribution of heat production by the radioactive isotopes in the oceanic crust to the surface heat flow is negligible ($\sim 2\%$).

The most striking feature of heat flow measurements in the oceans is the systematic dependence of the surface heat flow on the age of the seafloor. This can be understood as a consequence of the gradual cooling of the oceanic lithosphere as it moves away from the mid-ocean ridge. This process is analyzed in detail in Section 4-16, where it is shown that conductive cooling of the initially hot oceanic mantle can explain quantitatively the observed heat flow–age relation. The dependence of the oceanic heat flow measurements on age is given in Figure 4-25.

The total heat flow from the interior of the Earth Q can be obtained by multiplying the area of the continents by the mean continental heat flow and adding the product of the oceanic area and the mean oceanic heat flow. The continents, including the continental margins, have an area $A_c = 2 \times 10^8 \, \text{km}^2$. Multiplying this by the mean observed continental heat flow, $65 \, \text{mW m}^{-2}$, we get the total heat flow from the continents to be $Q_c = 1.30 \times 10^{13} \, \text{W}$. Similarly, taking the oceans, including the marginal basins, to have an area $A_o = 3.1 \times 10^8 \, \text{km}^2$ and a mean observed heat flow of $101 \, \text{mW m}^{-2}$, we find that the total heat flow from the oceans is $Q_o = 3.13 \times 10^{13} \, \text{W}$. Adding the heat flow through the continents and the oceans, we find that the total surface heat flow is $Q = 4.43 \times 10^{13}$ W. Dividing by the Earth's surface area $A = 5.1 \times 10^8$ km^2, we get 87 mW m^{-2} for the corresponding mean surface heat flow.

4–5 Heat Generation by the Decay of Radioactive Elements

A substantial part of the heat lost through the Earth's surface undoubtedly originates in the interior of the Earth by the decay of the radioactive elements uranium, thorium, and potassium. Some part of the surface heat loss must come from the overall cooling of the Earth through geologic time. An upper limit to the concentration of radioactive elements in the Earth can be derived by attributing all the surface heat loss to the radioactive heat generation. The mean heat generation per unit mass H is then given by

$$H = \frac{Q}{M}. \qquad (4\text{–}5)$$

If we take $M = 5.97 \times 10^{24}$ kg, the mass of the Earth, and $Q = 4.43 \times 10^{13}$ W, we find $H = 7.42 \times 10^{-12}$ W kg^{-1}. However, on the basis of geochemical studies, we can argue that the core cannot contain a significant fraction of the heat-producing elements. In this case, the mass in Equation (4–5) should be the mass of the mantle, $M = 4.0 \times 10^{24}$ kg and $H = 11.1 \times 10^{-12}$ W kg^{-1}.

A further reduction must be made in the value of H appropriate to the mantle because a substantial fraction of the heat lost from the continents originates in the highly concentrated radioactive isotopes of the continental crust. Although the mean continental heat flux of 65 mW m^{-2} is known with some certainty, we are uncertain as to the fraction that can be attributed to the heat-producing elements. Based on estimates of the mean concentrations of these elements in the continental crust, we attribute 37 mW m^{-2} to the heat-producing elements. The remaining 28 mW m^{-2} is attributed to basal heating of the continental lithosphere by mantle convection. This heat is then conducted through the mantle portion of the continental lithosphere to the base of the continental crust. Radiogenic heat production in the continental crust corresponds to a total heat flow of 7.4×10^{12} W, or 17% of the total surface heat flow. Reduction of the mantle heat production by this amount gives $H = 9.22 \times 10^{-12}$ W kg^{-1}.

A further correction to the radiogenic heat production in the mantle must be made because of the *secular cooling* of the Earth. Only a fraction of the present-day surface heat flow can be attributed to the decay of radioactive isotopes presently in the mantle. Because the radioactive isotopes decay into stable isotopes, heat production due to radioactive decay is decreasing with time. For example, we will show that the heat production 3 billion years ago was about twice as great as it is today. Since less heat is being generated in the Earth through time, less heat is also being convected to the surface. Thus, the vigor of the mantle convection system is decreasing with the age of the Earth. Because the strength of convection is dependent on viscosity and the viscosity of the mantle is a sensitive function of its temperature, a decrease in the heat flux with time leads to a decrease in the mean mantle temperature. This cooling of the Earth in turn contributes to the surface heat flow. We will consider this problem in some detail in Section 7–8 and conclude that about 80% of the present-day surface heat flow can be attributed to the decay of radioactive isotopes presently in the Earth and about 20% comes from the cooling of the Earth. We can thus reduce the present-day mantle heat production accordingly so that our preferred value is $H = 7.38 \times 10^{-12}$ W kg^{-1}.

Radioactive heating of the mantle and crust is attributed to the decay of the uranium isotopes ^{235}U and ^{238}U, the thorium isotope ^{232}Th, and the potassium isotope ^{40}K. The rates of heat production and the half-lives $\tau_{1/2}$ of these isotopes are given in Table 4–2. At the present time natural uranium is composed of 99.28% by weight ^{238}U and 0.71% ^{235}U. Natural thorium is 100% ^{232}Th. Natural potassium is composed of 0.0119% ^{40}K. The present rates of heat production of natural uranium and potassium are also given in Table 4–2.

The ratios of potassium to uranium and thorium to uranium are nearly constant in a wide range of terrestrial rocks. Based on these observed ratios we take $C_0^K/C_0^U = 10^4$ and $C_0^{Th}/C_0^U = 4$, where C_0^K, C_0^{Th}, and C_0^U are the present mass concentrations of potassium, thorium, and uranium, respectively. The total present-day production H_0 is related to the heat generation rates of the individual radioactive elements by

$$H_0 = C_0^U \left(H^U + \frac{C_0^{Th}}{C_0^U} H^{Th} + \frac{C_0^K}{C_0^U} H^K \right). \qquad (4\text{–}6)$$

TABLE 4–2 **Rates of Heat Release H and Half-Lives $\tau_{1/2}$ of the Important Radioactive Isotopes in the Earth's Interior**

Isotope	H (W kg^{-1})	$\tau_{1/2}$ (yr)	Concentration C (kg kg^{-1})
^{238}U	9.46×10^{-5}	4.47×10^{9}	30.8×10^{-9}
^{235}U	5.69×10^{-4}	7.04×10^{8}	0.22×10^{-9}
U	9.81×10^{-5}		31.0×10^{-9}
^{232}Th	2.64×10^{-5}	1.40×10^{10}	124×10^{-9}
^{40}K	2.92×10^{-5}	1.25×10^{9}	36.9×10^{-9}
K	3.48×10^{-9}		31.0×10^{-5}

Note: Heat release is based on the present mean mantle concentrations of the heat-producing elements.

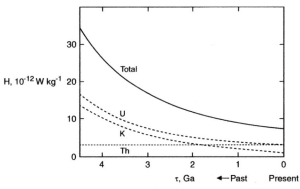

4–4 Mean mantle heat production rates due to the decay of the radioactive isotopes of U, Th, and K as functions of time measured back from the present.

Taking $H_0 = 7.38 \times 10^{-12}$ W kg^{-1} and the other parameters as given above and in Table 4–2, we find that $C_0^U = 3.1 \times 10^{-8}$ kg kg^{-1} or 31 ppb (parts per billion by weight). These preferred values for the mean mantle concentrations of heat-producing elements are also given in Table 4–2.

The mean heat production rate of the mantle in the past can be related to the present heat production rate using the half-lives of the radioactive isotopes. The concentration C of a radioactive isotope at time t measured backward from the present is related to the present concentration C_0 and the half-life of the isotope $\tau_{1/2}$ by

$$C = C_0 \exp\left(\frac{t \ln 2}{\tau_{1/2}}\right). \qquad (4\text{–}7)$$

Thus, the past mean mantle heat production rate is given by

$$\begin{aligned}
H = {}& 0.9928 C_0^U H^{U^{238}} \exp\left(\frac{t \ln 2}{\tau_{1/2}^{U^{238}}}\right) \\
& + 0.0071 C_0^U H^{U^{235}} \exp\left(\frac{t \ln 2}{\tau_{1/2}^{U^{235}}}\right) \\
& + C_0^{Th} H^{Th} \exp\left(\frac{t \ln 2}{\tau_{1/2}^{Th}}\right) \\
& + 1.19 \times 10^{-4} C_0^K H^{K^{40}} \exp\left(\frac{t \ln 2}{\tau_{1/2}^{K^{40}}}\right).
\end{aligned}$$

$$(4\text{–}8)$$

The rate of mean mantle heat production based on Equation (4–8) and parameter values in Table 4–2 is

plotted as a function of time before the present in Figure 4–4. The past contributions of the individual radioactive elements are also shown. We see that the rate of heat production 3×10^9 yr ago was about twice the present value. Today heat is produced primarily by ^{238}U and ^{232}Th, but in the distant past ^{235}U and ^{40}K were the dominant isotopes because of their shorter half-lives.

The concentrations of the heat-producing elements in surface rocks vary considerably. Some typical values are given in Table 4–3. The mantle values from Table 4–2 are included for reference. Partial melting at ocean ridges depletes mantle rock of incompatible elements such as uranium, thorium, and potassium. These incompatible elements are concentrated in the basaltic partial melt fraction. As a result, the oceanic crust is enriched in these elements by about a factor of 4 relative to the fertile mantle. Peridotites that have been depleted in the incompatible elements are sometimes found on the surface of the Earth. A typical example of the small concentrations of the heat-producing elements in a "depleted" peridotite is given in Table 4–3. Processes that lead to the formation of the continental crust, such as the volcanism associated with ocean trenches, further differentiate the incompatible elements. The concentrations of the heat-producing elements in a typical continental rock such as a granite are quite variable, but in general they are an order of magnitude greater than in tholeiitic basalts. Representative values of concentrations in granite are given in Table 4–3.

It is generally accepted that the chondritic class of meteorites is representative of primitive mantle

TABLE 4–3 Typical Concentrations of the Heat-Producing Elements in Several Rock Types and the Average Concentrations in Chondritic Meteorites

Rock Type	U (ppm)	Concentration Th (ppm)	K (%)
Reference undepleted (fertile) mantle	0.031	0.124	0.031
"Depleted" peridotites	0.001	0.004	0.003
Tholeiitic basalt	0.07	0.19	0.088
Granite	4.7	20	4.2
Shale	3.7	12	2.7
Average continental crust	1.42	5.6	1.43
Chondritic meteorites	0.008	0.029	0.056

material. The average concentrations of the heat-producing elements in chondritic meteorites are listed in Table 4–3. The concentrations of uranium and thorium are about a factor of 4 less than our mean mantle values, and the concentration of potassium is about a factor of 2 larger. The factor of 8 difference in the ratio C_0^K / C_0^U is believed to represent a fundamental difference in elemental abundances between the Earth's mantle and chondritic meteorites.

PROBLEM 4–3 Determine the present mean mantle concentrations of the heat-producing elements if the present value for the mean mantle heat production is 7.38×10^{-12} W kg^{-1} and $C_0^K / C_0^U = 6 \times 10^4$ and $C_0^{Th} / C_0^U = 4$.

PROBLEM 4–4 Determine the rates of heat production for the rocks listed in Table 4–3.

PROBLEM 4–5 The measured concentrations of the heat-producing elements in a rock are $C^U = 3.2$ ppb, $C^{Th} = 11.7$ ppb, and $C^K = 2.6\%$. Determine the rate of heat generation per unit mass in the rock.

4–6 One-Dimensional Steady Heat Conduction with Volumetric Heat Production

Heat conduction theory enables us to determine the distribution of temperature in a region given information about the temperatures or heat fluxes on the boundaries of the region and the sources of heat production in the region. In general, we can also use the theory to determine time variations in the temperature distribution. We first develop the theory for the simple situation in which heat is transferred in one direction only and there are no time variations (steady state) in the temperature or heat flow. The basic equation of conductive heat transfer theory is a mathematical statement of conservation of energy; the equation can be derived as follows.

Consider a slab of infinitesimal thickness δy, as sketched in Figure 4–5. The heat flux out of the slab $q(y + \delta y)$ crosses the face of the slab located at $y + \delta y$, and the heat flux into the slab $q(y)$ crosses the face located at y. The net heat flow out of the slab, per unit time and per unit area of the slab's face, is

$$q(y + \delta y) - q(y).$$

Since δy is infinitesimal, we can expand $q(y + \delta y)$ in a

4–5 Heat flow into $q(y)$ and out of $q(y + \delta y)$ a thin slab of thickness δy producing heat internally at the rate of H per unit mass.

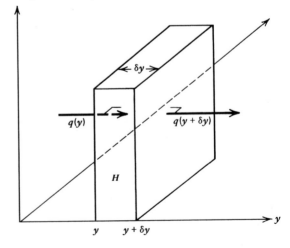

Taylor series as

$$q(y + \delta y) = q(y) + \delta y \frac{dq}{dy} + \cdots. \qquad (4\text{–}9)$$

Thus we find

$$q(y + \delta y) - q(y) = \delta y \frac{dq}{dy} = \delta y \frac{d}{dy}\left[-k\left(\frac{dT}{dy}\right)\right]$$
$$= \delta y \left[-k\left(\frac{d^2 T}{dy^2}\right)\right], \qquad (4\text{–}10)$$

where we have used Equation (4–1) (Fourier's law) for q, and we have assumed a constant thermal conductivity. The right side of Equation (4–10) is the net heat flow out of a slab of thickness δy, per unit time and per unit area. It is nonzero only when there is curvature in the temperature profile.

If there is a nonzero net heat flow per unit area out of the thin slab, as given by Equation (4–10), this heat flow must be supplied, in steady state, by heat generated internally in the slab. If H is the heat production rate per unit mass, the amount of heat generated in the slab per unit time and per unit area of the slab face is

$$\rho H \delta y, \qquad (4\text{–}11)$$

where ρ is the density of the slab. By equating (4–10) and (4–11), one obtains

$$0 = k \frac{d^2 T}{dy^2} + \rho H. \qquad (4\text{–}12)$$

This equation can be integrated to determine temperature as a function of position y once the region of interest and appropriate boundary conditions have been specified.

Assume that the medium is a half-space with the surface at $y = 0$ (see Figure 4–6). The coordinate y increases with distance into the half-space; thus y is a depth coordinate. One possible set of boundary conditions for Equation (4–12) is the specification of both temperature and heat flux at the surface. Thus we require the temperature T to be T_0 at $y = 0$ and the heat flux at the surface q to be $-q_0$. The reason for the difference in sign between q and q_0 is that q is positive in the direction of positive y, that is, downward, while q_0 is assumed to be positive upward (we anticipate application to the Earth for which the surface heat flux is indeed upward).

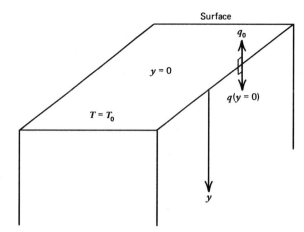

4–6 Geometry and boundary conditions for integration of Equation (4–12).

One integration of Equation (4–12) gives

$$\rho H y = -k \frac{dT}{dy} + c_1 = q + c_1, \qquad (4\text{–}13)$$

where c_1 is a constant of integration. Since $q = -q_0$ on $y = 0$, we find

$$c_1 = q_0 \qquad (4\text{–}14)$$

and

$$\rho H y = -k \frac{dT}{dy} + q_0. \qquad (4\text{–}15)$$

Integration of Equation (4–15) results in

$$\rho H \frac{y^2}{2} = -kT + q_0 y + c_2, \qquad (4\text{–}16)$$

where c_2 is another constant of integration. Since $T = T_0$ on $y = 0$, we find

$$c_2 = kT_0. $$

and

$$T = T_0 + \frac{q_0}{k} y - \frac{\rho H}{2k} y^2. \qquad (4\text{–}17)$$

PROBLEM 4–6 Consider a geological situation in which the subsurface is layered, with *bedding planes* making an angle θ with the horizontal surface, as shown in Figure 4–7a. Suppose that the thermal conductivity for heat conduction parallel to BC is k_1 and the conductivity for heat transport parallel to AB

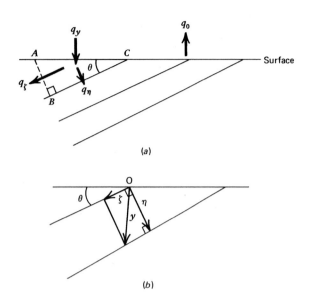

(a)

(b)

4–7 Geometry for Problem 4–6.

is k_3. Though the bedding planes are inclined to the horizontal, isotherms are nevertheless horizontal. Show that the upward surface heat flow is given by

$$q_0 = \{k_1 + (k_3 - k_1)\cos^2\theta\}\frac{\partial T}{\partial y}. \qquad (4\text{–}18)$$

HINT: Introduce coordinates ζ and η as shown in Figure 4–7b. Note that

$$y = \eta\cos\theta + \zeta\sin\theta \qquad (4\text{–}19)$$

and

$$\frac{\partial T}{\partial \eta} = \cos\theta\frac{\partial T}{\partial y} \qquad \frac{\partial T}{\partial \zeta} = \sin\theta\frac{\partial T}{\partial y}. \qquad (4\text{–}20)$$

Write a steady-state heat balance for the triangle ABC in Figure 4–7a. Use Fourier's law of heat conduction to evaluate q_ζ and q_η. Note $q_0 = -q_y$. Assume no heat sources.

4–7 A Conduction Temperature Profile for the Mantle

We can use Equation (4–17) to determine the temperature as a function of depth in the Earth, that is, the *geotherm*, assuming heat is transported by conduction. The depth profile of the temperature is given in Figure 4–8, assuming $T_0 = 0°C$, $q_0 = 70$ mW m^{-2}, $\rho = 3300$ kg m^{-3}, $H = 7.38 \times 10^{-12}$ W kg^{-1}, and $k = 4$ W m^{-1} K^{-1}. Also included in Figure 4–8 are the *liquidus* and *solidus* of basalt and the solidus of peridotite. Basalt is the low-melting-temperature fraction of the mantle. When the temperature of the mantle exceeds the basalt solidus, this fraction starts to melt, resulting in volcanism. This is the cause of the extensive basaltic volcanism that forms the oceanic crust. When the temperature reaches the basalt liquidus, this fraction is entirely melted, leaving a high-melting-temperature residuum that is primarily composed of the mineral olivine. When the mantle temperature reaches the olivine solidus, the remainder of the mantle rock melts. The ability of seismic shear waves to propagate through the mantle indicates that substantial melting does not occur. The conclusion is that this conduction analysis does not predict the temperature in the Earth's mantle.

In an attempt to assess the failure of the conductive mantle geotherm to model the Earth, one may ask whether the near-surface concentration of radioactive elements in crustal rocks can modify the analysis. (The

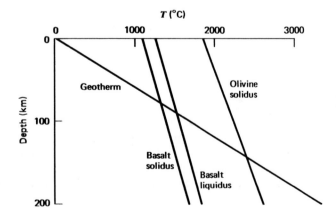

4–8 Temperature as a function of depth within the Earth assuming heat transport is by conduction (conduction geotherm). Also included are the solidus and liquidus of basalt and the solidus of peridotite (olivine).

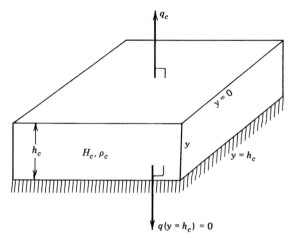

4-9 Heat flow through the top of a slab containing internal heat sources. No heat flows through the bottom of the slab.

partial melting processes that lead to the formation of the crust concentrate the radioactive elements.) The only way in which this could have an effect is through a reduction in the amount of the surface heat flow q_0 attributed to mantle heat sources. Thus we must assess the contribution of crustal radioactivity to surface heat flow. It is appropriate to do this for the oceanic crust because the suboceanic mantle geotherm dominates the temperature distribution of the mantle.

To determine the contribution q_c to the surface heat flow of a layer of crust of thickness h_c and heat production per unit mass H_c, we proceed as indicated in Figure 4–9. Equation (4–13) applies to this case also, with $\rho = \rho_c$ and $H = H_c$ (subscript c refers to the crust),

$$\rho_c H_c y = -k\frac{dT}{dy} + c_1 = q + c_1. \qquad (4\text{–}21)$$

To evaluate c_1, we note that $q = -q_c$ on $y = 0$ and

$$c_1 = q_c.$$

The heat flux in the slab satisfies

$$q + q_c = \rho_c H_c y. \qquad (4\text{–}22)$$

But $q = 0$ at $y = h_c$ because we have assumed that no heat enters the bottom of the slab (the appropriate boundary condition if we want to determine the heat flowing out the top of the slab due only to radioactive isotopes contained in it). Thus we find

$$q_c = \rho_c H_c h_c. \qquad (4\text{–}23)$$

The oceanic crust is primarily composed of basalts. Thus we take $\rho_c = 2900$ kg m^{-3}, $h_c = 6$ km, and $H_c = 2.6 \times 10^{-11}$ W kg^{-1}. (The radiogenic heat production rate per unit mass of basalts was calculated in Problem 4–4.) From Equation (4–23) the resultant contribution to the surface heat flow is $q_c = 0.45$ mW m^{-2}; this is a small fraction of the mean oceanic heat flow, which is about 100 mW m^{-2}. The conclusion is that heat production in the oceanic crust does not make a significant contribution to the oceanic surface heat flow. Therefore, an alternative explanation must be found for the failure of the simple conduction profile to model the suboceanic mantle geotherm. In later sections we show that heat flow due to mantle convection invalidates the conduction results.

4-8 Continental Geotherms

Whereas conductive temperature profiles fail to describe the mantle geotherm, they successfully model the geotherm in the continental crust and lithosphere, where the dominant thermal processes are radiogenic heat production and conductive heat transport to the surface. Because of the great age of the continental lithosphere, time-dependent effects can, in general, be neglected.

The surface rocks in continental areas have considerably larger concentrations of radioactive elements than the rocks that make up the oceanic crust. Although the surface rocks have a wide range of heat production, a typical value for a granite is $H_c = 9.6 \times 10^{-10}$ W kg^{-1} (H for granite was calculated in Problem 4–4). Taking $h_c = 35$ km and $\rho_c = 2700$ kg m^{-3}, one finds that the heat flow from Equation (4–23) is $q_c = 91$ mW m^{-2}. Since this value is considerably larger than the mean surface heat flow in continental areas (65 mW m^{-2}), we conclude that the concentration of the radioactive elements decreases with depth in the continental crust.

For reasons that we will shortly discuss in some detail it is appropriate to assume that the heat production due to the radioactive elements decreases exponentially with depth,

$$H = H_0 e^{-y/h_r}. \qquad (4\text{–}24)$$

Thus H_0 is the surface ($y = 0$) radiogenic heat production rate per unit mass, and h_r is a length scale for the decrease in H with depth. At the depth $y = h_r$, H is $1/e$ of its surface value. Substitution of

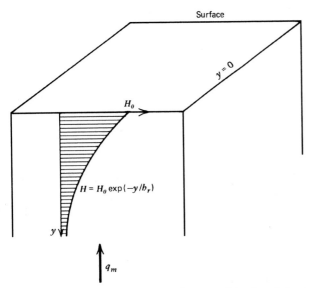

4–10 Model of the continental crust with exponential radiogenic heat source distribution.

Equation (4–24) into the equation of energy conservation (4–12) yields the differential equation governing the temperature distribution in the model of the continental crust:

$$0 = k\frac{d^2 T}{dy^2} + \rho H_0 e^{-y/h_r}. \tag{4–25}$$

Beneath the near-surface layer of heat-producing elements we assume that the upward heat flow at great depth is q_m; that is, $q \to -q_m$ as $y \to \infty$. This model for heat production in the continental crust is sketched in Figure 4–10.

An integration of Equation (4–25) yields

$$c_1 = k\frac{dT}{dy} - \rho H_0 h_r e^{-y/h_r} = -q - \rho H_0 h_r e^{-y/h_r}. \tag{4–26}$$

The constant of integration c_1 can be determined from the boundary condition on the heat flux at great depth, that is, from the mantle heat flux to the base of the lithosphere

$$c_1 = q_m. \tag{4–27}$$

Thus the heat flux at any depth is

$$q = -q_m - \rho H_0 h_r e^{-y/h_r}. \tag{4–28}$$

The surface heat flow $q_0 = -q(y = 0)$ is obtained by setting $y = 0$ with the result

$$q_0 = q_m + \rho h_r H_0. \tag{4–29}$$

With an exponential depth dependence of radioactivity, the surface heat flow is a linear function of the surface radioactive heat production rate.

In order to test the validity of the linear heat flow–heat production relation (4–29), determinations of the radiogenic heat production in surface rocks have been carried out for areas where surface heat flow measurements have been made. Several regional correlations are given in Figure 4–11. In each case a linear correlation appears to fit the data quite well. The corresponding length scale h_r is the slope of the best-fit straight line and the mantle (reduced) heat flow q_m is the vertical intercept of the line. For the Sierra Nevada data we have $q_m = 17$ mW m^{-2} and $h_r = 10$ km; for the eastern United States data we have $q_m = 33$ mW m^{-2} and $h_r = 7.5$ km; for the Norway and Sweden data,

4–11 Dependence of surface heat flow q_0 on the radiogenic heat production per unit volume in surface rock ρH_0 in selected geological provinces: Sierra Nevada (solid squares and very long dashed line), eastern U.S. (solid circles and intermediate dashed line), Norway and Sweden (open circles and solid line), eastern Canadian shield (open squares and short dashed line). In each case the data are fit with the linear relationship Equation (4–29).

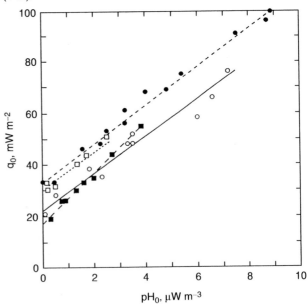

$q_m = 22$ mW m^{-2} and $h_r = 7.2$ km; and for the eastern Canadian shield data, $q_m = 30.5$ mW m^{-2} and $h_r = 7.1$ km. In all cases the length scale h_r is near 10 km. The values of the mantle or reduced heat flow q_m are reasonably consistent with the mean basal heating of the continental lithosphere $q_m = 28$ mW m^{-2} given in Section 4–5.

Thus a model of the continental crust with exponentially decreasing radioactivity can explain the linear surface heat flow–surface radioactivity relation. The exercises to follow show that the exponential radioactivity distribution is not unique in its ability to model the linear q_0 versus ρH_0 relation; other dependences of H on depth that confine radioactivity near the surface are consistent with observations. However, the exponential distribution is the only one that preserves the linear q_0 versus ρH_0 relation under differential erosion, a strong argument to support its relevance to the continental crust. The exponential depth dependence is also consistent with magmatic and hydrothermal differentiation processes, although a detailed understanding of these processes in the continental crust is not available.

PROBLEM 4–7 Table 4–4 gives a series of surface heat flow and heat production measurements in the Sierra Nevada Mountains in California. Determine the reduced heat flow q_m and the scale depth h_r.

PROBLEM 4–8 Consider one-dimensional steady-state heat conduction in a half-space with heat production that decreases exponentially with depth. The surface heat flow–heat production relation is $q_0 = q_m + \rho H_0 h_r$. What is the heat flow–heat production relation at depth $y = h^*$? Let q^* and H^* be the upward heat flux and heat production at $y = h^*$.

TABLE 4–4 Surface Heat Flow and Heat Production Data for the Sierra Nevada Mountains

q_0 (mW m^{-2})	ρH_0 (μW m^{-3})	q_0 (mW m^{-2})	ρH_0 (μW m^{-3})
18	0.3	31	1.5
25	0.8	34	2.0
25	0.9	42	2.6
29	1.3	54	3.7

PROBLEM 4–9 Assume that the radioactive elements in the Earth are uniformly distributed through a near-surface layer. The surface heat flow is 70 mW m^{-2}, and there is no heat flow into the base of the layer. If $k = 4$ W m^{-1} K^{-1}, $T_0 = 0°C$, and the temperature at the base of the layer is 1200°C, determine the thickness of the layer and the volumetric heat production.

PROBLEM 4–10 Consider one-dimensional steady-state heat conduction in a half-space. The heat sources are restricted to a surface layer of thickness b; their concentration decreases linearly with depth so that $H = H_0$ at the surface $y = 0$ and $H = 0$ at the depth $y = b$. For $y > b$, $H = 0$ and there is a constant upward heat flux q_m. What is the q_0 (upward surface heat flow)–H_0 relation? Determine the temperature profile as a function of y.

PROBLEM 4–11 The exponential depth dependence of heat production is preferred because it is self-preserving upon erosion. However, many alternative models can be prescribed. Consider a two-layer model with $H = H_1$ and $k = k_1$ for $0 \le y \le h_1$, and $H = H_2$ and $k = k_2$ for $h_1 \le y \le h_2$. For $y > h_2$, $H = 0$ and the upward heat flux is q_m. Determine the surface heat flow and temperature at $y = h_2$ for $\rho_1 = 2600$ kg m^{-3}, $\rho_2 = 3000$ kg m^{-3}, $k_1 = k_2 = 2.4$ W m^{-1} K^{-1}, $h_1 = 8$ km, $h_2 = 40$ km, $\rho_1 H_1 = 2 \mu W$ m^{-3}, $\rho_2 H_2 = 0.36 \mu W$ m^{-3}, $T_0 = 0 °C$, and $q_m = 28$ mW m^{-2}.

A further integration of Equation (4–28) using Equation (4–1) and the boundary condition $T = T_0$ at $y = 0$ gives

$$T = T_0 + \frac{q_m y}{k} + \frac{\rho H_0 h_r^2}{k}\left(1 - e^{-y/h_r}\right). \qquad (4\text{–}30)$$

or, alternatively, using Equation (4–29), we obtain

$$T = T_0 + \frac{q_m y}{k} + \frac{(q_0 - q_m)h_r}{k}\left(1 - e^{-y/h_r}\right). \qquad (4\text{–}31)$$

Figure 4–12 is a plot of a typical geotherm in the continental crust computed from Equation (4–31) with $T_0 = 10°C$, $q_0 = 56.5$ mW m^{-2}, $q_m = 30$ mW m^{-2}, $h_r = 10$ km, and $k = 3.35$ W m^{-1} K^{-1}.

PROBLEM 4–12 An alternative model for the continental crust is to assume that in addition to the exponentially decreasing near-surface radioactivity there is

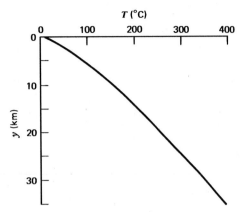

4–12 A typical geotherm in the continental crust.

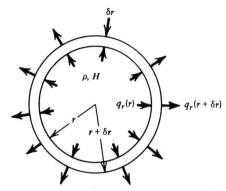

4–13 Heat flow into and out of a thin spherical shell with internal heat generation.

also a constant concentration of radioactivity H_0 to the depth h_c. Show that the crustal geotherm for this model is given by

$$T = T_0 + \frac{\rho H_0 h_r^2}{k}\left(1 - e^{-y/h_r}\right) - \frac{\rho H_c y^2}{2k}$$
$$+ \frac{(q_m + \rho H_c h_c)}{k} y \quad \text{for } 0 \leq y \leq h_c, \quad (4\text{–}32)$$

$$T = T_0 + \frac{\rho H_c h_c^2}{2k} + \frac{\rho H_0 h_r^2}{k} + \frac{q_m y}{k} \quad \text{for } y \geq h_c. \tag{4–33}$$

4–9 Radial Heat Conduction in a Sphere or Spherical Shell

We next consider the steady radial conduction of heat in a sphere or spherical shell with volumetric heat production. The temperature distributions in thick planetary lithospheres, such as the lithospheres of the Moon and Mars, are properly described by solutions of the heat conduction equation in spherical geometry. The effects of spherical geometry are not so important for the Earth's lithosphere, which is quite thin compared with the Earth's radius. However, on a small body like the Moon, the lithosphere may be a substantial fraction of the planet's radius. To describe heat conduction in spherical geometry, we must derive an energy balance equation.

Consider a spherical shell of thickness δr and inner radius r, as sketched in Figure 4–13. Assume that the conductive transport of heat occurs in a spherically symmetric manner. The total heat flow out of the shell

through its outer surface is

$$4\pi(r + \delta r)^2 q_r(r + \delta r),$$

and the total heat flow into the shell at its inner surface is

$$4\pi r^2 q_r(r).$$

The subscript r on the heat flux q indicates that the flow of heat is radial. Since δr is infinitesimal, we can expand $q_r(r + \delta r)$ in a Taylor series as

$$q_r(r + \delta r) = q_r(r) + \delta r \frac{dq_r}{dr} + \cdots. \tag{4–34}$$

Thus neglecting powers of δr, the net heat flow out of the spherical shell is given by

$$4\pi[(r + \delta r)^2 q_r(r + \delta r) - r^2 q_r(r)]$$
$$= 4\pi r^2 \left(\frac{2}{r} q_r + \frac{dq_r}{dr}\right)\delta r. \tag{4–35}$$

If the net heat flow from the shell is nonzero, then, by conservation of energy, this flow of heat must be supplied by heat generated internally in the shell (in steady state). With the rate of heat production per unit mass H, the total rate at which heat is produced in the spherical shell is

$$4\pi r^2 \rho H \delta r,$$

$4\pi r^2 \delta r$ being the approximate expression for the volume of the shell. By equating the rate of heat production to the net heat flow out of the spherical shell, Equation (4–35), we get

$$\frac{dq_r}{dr} + \frac{2q_r}{r} = \rho H. \tag{4–36}$$

The heat balance Equation (4–36) can be converted into an equation for the temperature by relating the radial heat flux q_r to the radial temperature gradient dT/dr. Fourier's law still applies in spherical geometry,

$$q_r = -k\frac{dT}{dr}. \tag{4-37}$$

Upon substituting Equation (4–37) into Equation (4–36), we find

$$0 = k\left(\frac{d^2 T}{dr^2} + \frac{2}{r}\frac{dT}{dr}\right) + \rho H \tag{4-38}$$

or

$$0 = k\frac{1}{r^2}\frac{d}{dr}\left(r^2\frac{dT}{dr}\right) + \rho H. \tag{4-39}$$

By twice integrating Equation (4–39), one obtains a general expression for the temperature in a sphere or spherical shell with internal heat production and in steady state:

$$T = -\frac{\rho H}{6k}r^2 + \frac{c_1}{r} + c_2. \tag{4-40}$$

The constants of integration c_1 and c_2 depend on the boundary conditions of a particular problem. As an example, we determine the temperature distribution in a sphere of radius a that has a uniform internal rate of heat production. The boundary condition is that the outer surface of the sphere has a temperature T_0. In order to have a finite temperature at the center of the sphere, we must set $c_1 = 0$. To satisfy the temperature boundary condition at the surface of the sphere, we require

$$c_2 = T_0 + \frac{\rho H a^2}{6k}. \tag{4-41}$$

The temperature in the sphere is therefore given by

$$T = T_0 + \frac{\rho H}{6k}(a^2 - r^2). \tag{4-42}$$

From Equation (4–37), the surface heat flux q_0 at $r = a$ is given by

$$q_0 = \frac{1}{3}\rho H a. \tag{4-43}$$

Equation (4–43) is a statement of conservation of energy that applies no matter what the mode of internal heat transfer in the sphere is. The temperature distribution in the sphere is shown in Figure 4–14.

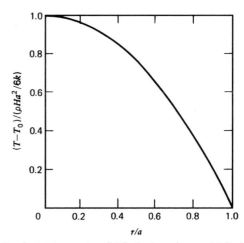

4-14 Steady-state temperature distribution in a sphere containing internal heat sources.

PROBLEM 4-13 Derive the equation $q_0 = \rho H a/3$ for a sphere with uniform volumetric heating and density by making a simple overall steady-state heat balance.

PROBLEM 4-14 What would the central temperature of the Earth be if it were modeled by a sphere with uniform volumetric heating? Take $q_0 = 70$ mW m^{-2}, $k = 4$ W m^{-1} K^{-1}, and $T_0 = 300$ K.

PROBLEM 4-15 Derive an expression for the temperature at the center of a planet of radius a with uniform density ρ and internal heat generation H. Heat transfer in the planet is by conduction only in the lithosphere, which extends from $r = b$ to $r = a$. For $0 \le r \le b$ heat transfer is by convection, which maintains the temperature gradient dT/dr constant at the adiabatic value $-\Gamma$. The surface temperature is T_0. To solve for $T(r)$, you need to assume that T and the heat flux are continuous at $r = b$.

PROBLEM 4-16 It is assumed that a constant density planetary body of radius a has a core of radius b. There is uniform heat production in the core but no heat production outside the core. Determine the temperature at the center of the body in terms of a, b, k, T_0 (the surface temperature), and q_0 (the surface heat flow).

4-10 Temperatures in the Moon

The Moon is a relatively small planetary body so it is a good approximation to assume that its density is constant. If we also assume that the Moon is in a steady-state thermal balance and that the mean heat production is the same as the value we derived for the Earth's mantle, that is, $H = 7.38 \times 10^{-12}$ W kg^{-1}, we can determine the surface heat flow on the Moon using Equation (4–43). With $\rho = 3300$ kg m^{-3} and $a = 1738$ km we find that $q_0 = 14.1$ mW m^{-2}. The mean of two lunar heat flow measurements on Apollos 15 and 17 is $q_s = 18$ mW m^{-2}. This approximate agreement suggests that the mean lunar abundances of the radioactive isotopes are near those of the Earth. The difference may be partially attributable to the cooling of the Moon.

Assuming that the conduction solution is applicable and that the Moon has a uniform distribution of radioactivity, the maximum temperature at the center of the Moon can be obtained from Equation (4–42) with the result $T_{max} = 3904$ K, assuming $k = 3.3$ W m^{-1} K^{-1} and that the surface temperature is $T_0 = 250$ K. This conduction solution indicates that a substantial fraction of the interior of the Moon is totally melted. However, the limited seismic results from the Apollo missions suggest that a sizable liquid core in the Moon is unlikely. Thus, either the conductive solution is not valid or the radioactive isotopes are not distributed uniformly throughout the Moon. There should be some upward concentration of radioactive isotopes in the relatively thick lunar highland crust (60 km) by analogy with the upward concentration of radioactive isotopes in the Earth's continental crust.

PROBLEM 4–17 Determine the steady-state conduction temperature profile for a spherical model of the Moon in which all the radioactivity is confined to an outer shell whose radii are b and a (a is the lunar radius). In the outer shell H is uniform.

4–11 Steady Two- and Three-Dimensional Heat Conduction

Obviously, not all heat conduction problems of geologic interest can be solved by assuming that heat is transported in one direction only. In this section, we generalize the heat conduction equation to account for heat transfer in two dimensions. The further generalization to three dimensions will be obvious and stated without proof. The first step is to write an appropriate energy conservation equation. If heat can be con-

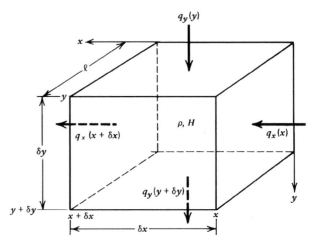

4–15 Heat flow into and out of a rectangular element.

ducted in both the x and y directions, we must consider the heat balance on a small rectangular element with dimensions δx and δy, as illustrated in Figure 4–15.

The heat flux in the x direction is q_x, and in the y direction it is q_y. The rate at which heat flows into the element in the y direction is $q_y(y)\delta xl$, where l is an arbitrary length in the third direction (in two-dimensional heat conduction we assume that nothing varies in the third dimension). Similarly, heat flows into the element in the x direction at the rate $q_x(x)\delta yl$. The heat flow rates out of the element are $q_y(y+\delta y)\delta xl$ and $q_x(x+\delta x)\delta yl$. The net heat flow rate out of the element is

$$\{q_x(x+\delta x) - q_x(x)\}\delta yl + \{q_y(y+\delta y) - q_y(y)\}\delta xl$$

$$= \frac{\partial q_x}{\partial x}\delta x\delta yl + \frac{\partial q_y}{\partial y}\delta x\delta yl = \left(\frac{\partial q_x}{\partial x} + \frac{\partial q_y}{\partial y}\right)\delta x\delta yl.$$

$$(4\text{–}44)$$

Taylor series expansions have been used for q_x $(x+\delta x)$ and $q_y(y+\delta y)$ to simplify the expression in Equation (4–44). Partial derivative symbols appear in Equation (4–44) because q_x can depend on both x and y; similarly q_y can be a function of both x and y.

In steady state, a nonzero value of the right side of Equation (4–44) requires that heat be produced internally in the rectangular element. The rate of heat generation in the element is $\rho H(\delta x\delta yl)$; equating this to the right side of Equation (4–44) yields

$$\frac{\partial q_x}{\partial x} + \frac{\partial q_y}{\partial y} = \rho H. \qquad (4\text{–}45)$$

Clearly, if we had heat conduction in three dimensions, Equation (4–45) would be replaced by

$$\frac{\partial q_x}{\partial x} + \frac{\partial q_y}{\partial y} + \frac{\partial q_z}{\partial z} = \rho H. \tag{4-46}$$

Fourier's law of heat conduction relates the heat flow in any direction to the temperature gradient in that direction. If we assume that the thermal conductivity of the rock is isotropic, that is, the rock conducts heat equally readily in any direction, Fourier's law can be written

$$q_x = -k\frac{\partial T}{\partial x} \tag{4-47}$$

$$q_y = -k\frac{\partial T}{\partial y}. \tag{4-48}$$

Upon substitution of Equations (4–47) and (4–48) into Equation (4–45), we obtain

$$-k\left(\frac{\partial^2 T}{\partial x^2} + \frac{\partial^2 T}{\partial y^2}\right) = \rho H. \tag{4-49}$$

Generalizing this to three-dimensional heat conduction gives

$$-k\left(\frac{\partial^2 T}{\partial x^2} + \frac{\partial^2 T}{\partial y^2} + \frac{\partial^2 T}{\partial z^2}\right) = \rho H. \tag{4-50}$$

If there are no internal heat sources, the temperature satisfies

$$\frac{\partial^2 T}{\partial x^2} + \frac{\partial^2 T}{\partial y^2} = 0. \tag{4-51}$$

Equation (4–51) is known as *Laplace's equation*. In three dimensions, Laplace's equation is

$$\frac{\partial^2 T}{\partial x^2} + \frac{\partial^2 T}{\partial y^2} + \frac{\partial^2 T}{\partial z^2} = 0. \tag{4-52}$$

Laplace's equation is encountered in many other fields, including fluid flow, diffusion, and magnetostatics.

PROBLEM 4–18 If the medium conducting heat is anisotropic, separate thermal conductivities must be used for heat transfer in the x and y directions, k_x and k_y, respectively. What equation replaces Equation (4–49) for determining the temperature distribution in such a medium?

4-12 Subsurface Temperature Due to Periodic Surface Temperature and Topography

As an example of a two-dimensional heat conduction problem, we solve for the temperatures beneath the surface in a region where there are lateral variations in surface temperature. Such surface temperature variations can arise as a result of topographic undulations and the altitude dependence of temperature in the Earth's atmosphere. Horizontal surface temperature variations also occur at the boundaries between land and bodies of water, such as lakes or seas. A knowledge of how surface temperature variations affect subsurface temperature is important for the correct interpretation of borehole temperature measurements in terms of surface heat flow.

Consider again a semi-infinite half-space in the region $y \geq 0$. The surface is defined by the plane $y = 0$. For simplicity, we assume that the surface temperature T_s is a periodic function of x (horizontal distance) given by

$$T_s = T_0 + \Delta T \cos\frac{2\pi x}{\lambda}, \tag{4-53}$$

where λ is the wavelength of the spatial temperature variation. The situation is sketched in Figure 4–16. We also assume that there are no radiogenic heat sources in the half-space, since our interest here is in the nature of the subsurface temperature variations caused by the periodic surface temperature. To determine the temperature distribution, we must solve Equation (4–51) with the boundary condition given by Equation (4–53).

We assume that the method of separation of variables is appropriate; that is,

$$T(x, y) = T_0 + X(x)Y(y). \tag{4-54}$$

4-16 Temperature in a half-space whose surface temperature varies periodically with distance.

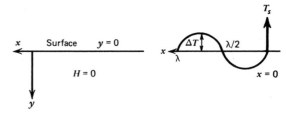

In order to satisfy the surface boundary condition, we must have

$$X(x) = \cos \frac{2\pi x}{\lambda}; \tag{4-55}$$

that is, the horizontal variations in temperature are the same at all depths. When Equations (4–54) and (4–55) are substituted into Equation (4–51), we obtain

$$0 = -\frac{4\pi^2}{\lambda^2} Y + \frac{d^2 Y}{dy^2}, \tag{4-56}$$

which is an ordinary differential equation for Y. The general solution of this equation is

$$Y(y) = c_1 e^{-2\pi y/\lambda} + c_2 e^{2\pi y/\lambda}, \tag{4-57}$$

where c_1 and c_2 are the constants of integration. Since the temperature must be finite as $y \to \infty$, we must require that $c_2 = 0$. To satisfy the boundary condition given in Equation (4–53), it is necessary that $c_1 = \Delta T$. The solution for the temperature distribution in the half-space is

$$T(x, y) = T_0 + \Delta T \cos \frac{2\pi x}{\lambda} e^{-2\pi y/\lambda}. \tag{4-58}$$

The temperature disturbance introduced by the surface temperature variation decays exponentially with depth in a distance proportional to the horizontal wavelength of the surface temperature variation.

The energy equation is linear in the temperature. Therefore, solutions to the equation can be added, and the result is still a solution of the energy equation. This is known as the *principle of superposition*. If the temperature in the continental crust is given by Equation (4–30) but the surface temperature has a periodic variation given by Equation (4–53), the temperature distribution in the crust is obtained by adding Equations (4–30) and (4–58):

$$T = T_0 + \frac{q_m y}{k} + \frac{\rho H_0 h_r^2}{k}\left(1 - e^{-y/h_r}\right)$$
$$+ \Delta T \cos \frac{2\pi x}{\lambda} e^{-2\pi y/\lambda}. \tag{4-59}$$

This result satisfies the applicable energy equation (4–49) and the required surface boundary condition (4–53).

The analysis in this section can also be used to determine the effect of small amplitude, periodic topography on the near-surface temperature distribution. This problem is illustrated in Figure 4–17. The topography is given by the relation

$$h = h_0 \cos \frac{2\pi x}{\lambda} \tag{4-60}$$

We assume that the atmosphere has a vertical temperature gradient β so that the surface temperature T_s is given by

$$T_s = T_0 + \beta y \qquad y = h. \tag{4-61}$$

A typical value for β is 6.5 K km^{-1}.

To apply these results, we must project the surface temperature values that are known on $y = h$ onto $y = 0$. This is because the temperature given by Equation (4–59) is written in terms of ΔT, the amplitude of the periodic temperature variation on $y = 0$; see Equation (4–53). Because the topography is shallow, this can be accomplished with just the first term of a Taylor series expansion:

$$T(y = 0) = T(y = h) - \left(\frac{\partial T}{\partial y}\right)_{y=0} h. \tag{4-62}$$

The temperature on $y = h$ is given by Equation (4–61), and the temperature gradient $(\partial T/\partial y)_{y=0}$ is given to sufficient accuracy by the value of the gradient in the absence of topography because h is small. From

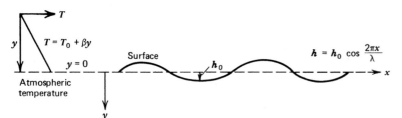

4-17 An undulating surface topography results in surface temperature variations that extend downward.

Equation (4–29) we can write

$$\left(\frac{\partial T}{\partial y}\right)_{y=0} = \frac{q_0}{k} = \frac{q_m + \rho h_r H_0}{k}. \tag{4–63}$$

The result of substituting Equations (4–61) and (4–63) into Equation (4–62) is

$$T(y = 0) = T_0 + \beta h - \frac{(q_m + \rho h_r H_0)}{k} h$$
$$= T_0 + \left\{\beta - \left(\frac{q_m + \rho h_r H_0}{k}\right)\right\} h_0 \cos \frac{2\pi x}{\lambda}. \tag{4–64}$$

Comparison of Equation (4–53) and (4–64) shows that

$$\Delta T = \left(\beta - \frac{q_m}{k} - \frac{\rho H_0 h_r}{k}\right) h_0. \tag{4–65}$$

Finally, substitution of Equation (4–65) into Equation (4–59) gives

$$T = T_0 + \frac{q_m y}{k} + \frac{\rho H_0 h_r^2}{k}\left(1 - e^{-y/h_r}\right)$$
$$+ \left(\beta - \frac{q_m}{k} - \frac{\rho H_0 h_r}{k}\right) h_0 \cos \frac{2\pi x}{\lambda} e^{-2\pi y/\lambda}, \tag{4–66}$$

for the temperature distribution in the continental crust with periodic topography.

PROBLEM 4–19 If a spatially periodic surface temperature variation has a wavelength of 1 km, at what depth is the horizontal variation 1% of that at the surface?

PROBLEM 4–20 A mountain range can be represented as a periodic topography with a wavelength of 100 km and an amplitude of 1.2 km. Heat flow in a valley is measured to be 46 mW m^{-2}. If the atmospheric gradient is 6.5 K km^{-1} and $k = 2.5$ Wm^{-1} K^{-1}, determine what the heat flow would have been without topography; that is, make a topographic correction.

4–13 One-Dimensional, Time-Dependent Heat Conduction

Many of the important geological problems involving heat conduction are time dependent. Examples that we consider later are the cooling of intrusive igneous bodies, the cooling of the oceanic lithosphere, erosion or sedimentation effects on temperature, and others. Volumetric heat production usually plays a minor role in

these phenomena, and we accordingly assume $H = 0$. In addition, it is adequate to consider heat conduction in one direction only.

If there are not heat sources in the medium, a net heat flow out of the slab illustrated in Figure 4–5 must reduce its temperature. The specific heat c of the medium is the energy required to raise the temperature of a unit mass of material by one degree. Thus, an element of the slab of thickness δy and unit cross-sectional area requires an energy flow per unit time given by

$$\rho c \frac{\partial T}{\partial t} \delta y$$

to maintain a temperature change at the rate $\partial T/\partial t$ ($\rho \delta y$ is the slab mass per unit cross-sectional area and $\rho c \delta y$ is the slab's heat capacity per unit cross-sectional area). Thus we can equate the right side of Equation (4–10) with $-\delta y \, \rho c \partial T/\partial t$, since a net heat flow out of the slab leads to a decrease in slab temperature

$$\rho c \frac{\partial T}{\partial t} = k \frac{\partial^2 T}{\partial y^2}. \tag{4–67}$$

Equation (4–67) is the basic equation governing the time and spatial variations of the temperature when heat is transferred in one dimension by conduction. Partial derivatives are required because T is a function of both time and space. We can rewrite Equation (4–67) in the form

$$\frac{\partial T}{\partial t} = \kappa \frac{\partial^2 T}{\partial y^2}, \tag{4–68}$$

where κ, the thermal diffusivity, is

$$\kappa = \frac{k}{\rho c}. \tag{4–69}$$

Note that κ has units of length^2/time such as square meters per second. If temperature changes occur with a characteristic time interval τ, they will propagate a distance on the order of $\sqrt{\kappa \tau}$. Similarly, a time l^2/κ is required for temperature changes to propagate a distance l. Such simple considerations can be used to obtain useful estimates of thermal effects. We now proceed to solve Equation (4–68) for a number of situations of geological and geophysical interest.

PROBLEM 4–21 Derive the time-dependent heat conduction equation for a situation in which heat transport occurs radially toward or away from a line of

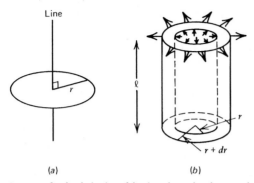

4–18 Geometry for the derivation of the time-dependent heat conduction equation in cylindrical coordinates.

infinite length. The heat flux q and the temperature T depend only on the perpendicular distance from the line r and time t (see Figure 4–18a).

HINT: Write an energy balance for a cylindrical shell of length l, inner radius r, and outer radius $r + \delta r$. The heat flows occur over the entire lateral surfaces of the cylindrical shell, as sketched in Figure 4–18b. Fourier's law of heat conduction in the form $q = -k(\partial T/\partial r)$ applies. The answer is

$$\frac{\partial T}{\partial t} = \frac{\kappa}{r}\frac{\partial}{\partial r}\left(r\frac{\partial T}{\partial r}\right). \qquad (4\text{–}70)$$

PROBLEM 4–22 Derive the time-dependent heat conduction equation appropriate to the situation in which heat transport is always radially toward or away from a point. Equation (4–35) gives the net heat flow out of a thin spherical shell. This must be equated to the time rate of change of temperature of the shell times the heat capacity of the shell. Fourier's law in the form of Equation (4–37) applies. The answer is

$$\frac{\partial T}{\partial t} = \frac{\kappa}{r}\frac{\partial^2}{\partial r^2}(rT). \qquad (4\text{–}71)$$

PROBLEM 4–23 Using the relation $\tau = l^2/\kappa$ and taking $\kappa = 1\ \mathrm{mm^2\ s^{-1}}$, determine the characteristic times for the conductive cooling of the Earth, Moon, Mars, Venus, and Mercury. What are the implications of these estimates?

PROBLEM 4–24 If the mean surface heat flow on the Earth ($\bar{q}_0 = 87\ \mathrm{mW\ m^{-2}}$) were attributed entirely to the cooling of the Earth, what would be the mean rate of cooling? (Take $\bar{c} = 1\ \mathrm{kJ\ kg^{-1}\ K^{-1}}$.)

PROBLEM 4–25 If the mean surface heat flow on the Moon ($\bar{q}_0 = 18\ \mathrm{mW\ m^{-2}}$) were attributed entirely to the cooling of the Moon, what would be the mean rate of cooling (Take $\bar{c} = 1\ \mathrm{kJ\ kg^{-1}\ K^{-1}} q_0$).

4–14 Periodic Heating of a Semi-Infinite Half-Space: Diurnal and Seasonal Changes in Subsurface Temperature

The surface temperature of the Earth regularly changes with time because of day–night variations and the changes of season. On a longer time scale, it changes because of the quasi-periodic nature of glaciations, for example. In this section we use the one-dimensional, time-dependent heat conduction equation to determine how these time-periodic surface temperature changes affect temperatures below the surface.

Again consider a semi-infinite half-space in the region $y \geq 0$ whose surface is defined by the plane $y = 0$. We assume that the surface temperature is a periodic function of time

$$T_s = T_0 + \Delta T \cos \omega t. \qquad (4\text{–}72)$$

The circular frequency ω is related to the frequency f by

$$\omega = 2\pi f. \qquad (4\text{–}73)$$

In addition, the period of the temperature fluctuations τ is

$$\tau = \frac{1}{f} = \frac{2\pi}{\omega}. \qquad (4\text{–}74)$$

If this represents the daily variation of the surface temperature, then $\tau = 1$ day, $f = 1.157 \times 10^{-5}\ \mathrm{s^{-1}}$, and $\omega = 7.272 \times 10^{-5}\ \mathrm{rad\ s^{-1}}$. We also assume that $T \to T_0$ as $y \to \infty$; that is, very far beneath the surface, the temperature is the average surface temperature.

To solve Equation (4–68) with this surface boundary condition, we use the method of separation of variables:

$$T(y, t) = T_0 + Y(y)T'(t). \qquad (4\text{–}75)$$

Because the surface temperature is time-periodic, we can assume that the subsurface temperature also varies periodically with time at the same frequency. However, it is not correct to assume that $T'(t)$ is simply $\cos \omega t$, as that would imply that the subsurface temperatures are in exact time phase with the surface temperature. In other words, if $T'(t)$ were $\cos \omega t$, the maximum and minimum temperatures would be reached at the same times, independent of depth. In fact, we will see that the temperature changes at different depths are not in phase; the maximum temperature at any depth lags behind the maximum temperature at the surface, for example, because of the finite amount of time required for the temperature maximum to diffuse downward. The difference in phase between temperature variations at the surface and those at depth can be accounted for by using both $\cos \omega t$ and $\sin \omega t$ dependences for $T'(t)$. We generalize Equation (4–75) to

$$T(y, t) = T_0 + Y_1(y) \cos \omega t + Y_2(y) \sin \omega t. \quad (4\text{--}76)$$

By substituting Equation (4–76) into Equation (4–68), we find

$$-\omega Y_1 = \kappa \frac{d^2 Y_2}{dy^2} \qquad \omega Y_2 = \kappa \frac{d^2 Y_1}{dy^2}. \quad (4\text{--}77)$$

These are two coupled ordinary differential equations for the unknowns Y_1 and Y_2. We can solve the first of these equations for Y_1 and substitute into the second equation to obtain

$$\frac{d^4 Y_2}{dy^4} + \frac{\omega^2}{\kappa^2} Y_2 = 0. \quad (4\text{--}78)$$

Alternatively, we could have solved the second of Equations (4–77) for Y_2 and substituted into the first equation. Had we done so, we would have found that Y_1 satisfies the same fourth-order ordinary differential equation as does Y_2.

A standard technique for solving ordinary differential equations with constant coefficients, of which Equation (4–78) is an example, is to assume a solution of the form

$$Y_2 = c e^{\alpha y}. \quad (4\text{--}79)$$

If Equation (4–79) is to satisfy Equation (4–78), then

$$\alpha^4 + \frac{\omega^2}{\kappa^2} = 0 \quad (4\text{--}80)$$

or

$$\alpha = \pm \left(\frac{1 \pm i}{\sqrt{2}} \right) \sqrt{\frac{\omega}{\kappa}}, \quad (4\text{--}81)$$

where i is the square root of -1. Because four values of α satisfy Equation (4–80), the general solution for

Y_2 (or Y_1) must be written

$$\begin{aligned} Y_2 = {} & c_1 \exp \left(\frac{(1+i)}{\sqrt{2}} \sqrt{\frac{\omega}{\kappa}} y \right) \\ & + c_2 \exp \left(\frac{(1-i)}{\sqrt{2}} \sqrt{\frac{\omega}{\kappa}} y \right) \\ & + c_3 \exp \left(\frac{-(1+i)}{\sqrt{2}} \sqrt{\frac{\omega}{\kappa}} y \right) \\ & + c_4 \exp \left(\frac{-(1-i)}{\sqrt{2}} \sqrt{\frac{\omega}{\kappa}} y \right). \quad (4\text{--}82) \end{aligned}$$

Because the temperature fluctuations must decay with depth, the constants c_1 and c_2 are zero, and Y_2 takes the form

$$\begin{aligned} Y_2 = {} & \exp \left(-y \sqrt{\frac{\omega}{2\kappa}} \right) \left[c_3 \exp \left(-iy \sqrt{\frac{\omega}{2\kappa}} \right) \right. \\ & \left. + c_4 \exp \left(iy \sqrt{\frac{\omega}{2\kappa}} \right) \right]. \quad (4\text{--}83) \end{aligned}$$

It is convenient to rewrite the solution for Y_2 as

$$Y_2 = \exp \left(-y \sqrt{\frac{\omega}{2\kappa}} \right) \left(b_1 \cos \sqrt{\frac{\omega}{2\kappa}} y + b_2 \sin \sqrt{\frac{\omega}{2\kappa}} y \right), \quad (4\text{--}84)$$

where b_1 and b_2 are constants that can be related to c_3 and c_4, although it is unnecessary to do so. The transition from Equation (4–83) to Equation (4–84) is possible because the trigonometric functions $\sin x$ and $\cos x$ can be written in terms of the exponentials e^{ix} and e^{-ix}, and vice versa. The unknown function Y_1 has a similar form

$$Y_1 = \exp \left(-y \sqrt{\frac{\omega}{2\kappa}} \right) \left(b_3 \cos \sqrt{\frac{\omega}{2\kappa}} y + b_4 \sin \sqrt{\frac{\omega}{2\kappa}} y \right). \quad (4\text{--}85)$$

The remaining constants of integration can be determined as follows. If Y_1 and Y_2 are to satisfy Equations (4–77), then

$$b_2 = b_3 \quad \text{and} \quad b_1 = -b_4. \quad (4\text{--}86)$$

Also, the surface temperature must be of the form (4–72), which requires

$$b_1 = 0 \quad \text{and} \quad b_3 = \Delta T. \quad (4\text{--}87)$$

Thus, the temperature variation in the half-space due to a time-periodic surface temperature is

$$T = T_0 + \Delta T \exp\left(-y\sqrt{\frac{\omega}{2\kappa}}\right)$$
$$\times (\cos)\left(\omega t \cos y\sqrt{\frac{\omega}{2\kappa}} + \sin \omega t \sin y\sqrt{\frac{\omega}{2\kappa}}\right),$$
$$(4\text{-}88)$$

$$T = T_0 + \Delta T \exp\left(-y\sqrt{\frac{\omega}{2\kappa}}\right)\cos\left(\omega t - y\sqrt{\frac{\omega}{2\kappa}}\right).$$
$$(4\text{-}89)$$

Equation (4–89) shows that the amplitude of the time-dependent temperature fluctuation decreases exponentially with depth. This fluctuation decreases to $1/e$ of its surface value in a skin depth d_ω given by

$$d_\omega = \left(\frac{2\kappa}{\omega}\right)^{1/2}.$$
$$(4\text{-}90)$$

For the daily variation of temperature, the frequency is $\omega = 7.27 \times 10^{-5}$ rad s^{-1}. With $\kappa = 1$ mm^2 s^{-1}, the skin depth for diurnal temperature changes from Equation (4–90) is 0.17 m. Except for a factor of $\pi^{-1/2}$, the skin depth for the penetration of the surface temperature variation is just what one would have estimated on the basis of dimensional arguments, that is, $\sqrt{\kappa\tau}(\tau = $ period). Because skin depth is inversely proportional to the square root of frequency, it is clear that the more rapid the fluctuation in temperature, the less it penetrates beneath the surface.

The argument of the trigonometric factor in Equation (4–89) shows that the phase difference ϕ between temperature fluctuations at the surface and those at depth y is

$$\phi = y\sqrt{\frac{\omega}{2\kappa}}.$$
$$(4\text{-}91)$$

If the depth y is the skin depth, the fluctuations are out of phase by 1 radian (57.3°). Figure 4–19 illustrates how the amplitudes of the temperature variations decay with depth and how the phases of the fluctuations shift with depth.

PROBLEM 4–26 Assume that the yearly temperature variation is periodic. What is the skin depth? At what depth is the temperature 180° out of phase with the surface variation? Assume $\kappa = 1$ mm^2 s^{-1}.

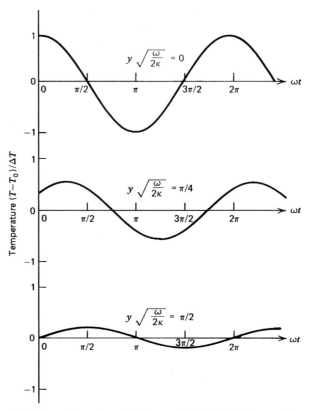

4–19 Phase shift and amplitude decay with depth of a time-periodic surface temperature variation.

PROBLEM 4–27 Assume that the temperature effects of glaciations can be represented by a periodic surface temperature with a period of 10^4 yr. If it is desired to drill a hole to a depth that the temperature effect of the glaciations is 5% of the surface value, how deep must the hole be drilled? Assume $\kappa = 1$ mm^2 s^{-1}.

PROBLEM 4–28 Estimate the depth to which frost penetrates in the ground at a latitude where the annual surface temperatures vary between −5 and 25°C. Assume that the water content of the ground is sufficiently small so that the latent heat can be ignored on freezing and thawing. Assume κ for the soil is 0.8 mm^2 s^{-1}.

PROBLEM 4–29 Estimate the effects of variations in bottom water temperature on measurements of oceanic heat flow by using the model of a semi-infinite half-space subjected to periodic surface temperature fluctuations. Such water temperature variations at a specific location on the ocean floor can be due to, for example, the transport of water with variable temperature past the site by deep ocean currents. Find the amplitude of water temperature variations that cause surface heat flux variations of $40 \, \text{mW m}^{-2}$ above and below the mean on a time scale of 1 day. Assume that the thermal conductivity of sediments is $0.8 \, \text{W m}^{-1} \, \text{K}^{-1}$ and the sediment thermal diffusivity is $0.2 \, \text{mm}^2 \, \text{s}^{-1}$.

PROBLEM 4–30 Consider a semi-infinite half-space $(y \geq 0)$ whose surface temperature is given by Equation (4–72). At what values of ωt is the surface heat flow zero?

4–15 Instantaneous Heating or Cooling of a Semi-Infinite Half-Space

A number of important geological problems can be modeled by the instantaneous heating or cooling of a semi-infinite half-space. In the middle of the nineteenth century Lord Kelvin used this solution to estimate the age of the Earth. He assumed that the surface heat flow resulted from the cooling of an initially hot Earth and concluded that the age of the Earth was about 65 million years. We now know that this estimate was in error for two reasons – the presence of radioactive isotopes in the mantle and solid-state thermal convection in the mantle.

In many cases magma flows through preexisting joints or cracks. When the flow commences, the wall rock is subjected to a sudden increase in temperature. Heat flows from the hot magma into the cold country rock, thus increasing its temperature. The temperature of the wall rock as a function of time can be obtained by solving the one-dimensional, time-dependent heat conduction equation for a semi-infinite half-space, initially at a uniform temperature, whose surface is suddenly brought to a different temperature at time $t = 0$ and maintained at this new temperature for later times.

This solution can also be used to determine the thermal structure of the oceanic lithosphere. At the crest of an ocean ridge, hot mantle rock is subjected to a cold surface temperature. As the seafloor spreads away from the ridge crest, the near-surface rocks lose heat to the cold seawater. The cooling near-surface rocks form the rigid oceanic lithosphere.

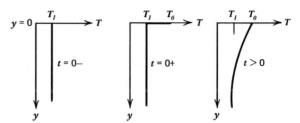

4–20 Heating of a semi-infinite half-space by a sudden increase in surface temperature.

We now obtain the solution to Equation (4–68) in a semi-infinite half-space defined by $y > 0$ whose surface is given an instantaneous change in temperature. Initially at $t = 0$, the half-space has a temperature T_1; for $t > 0$, the surface $y = 0$ is maintained at a constant temperature T_0. As a result, heat is transferred into the half-space if $T_0 > T_1$, and the temperature increases. If $T_1 > T_0$, the half-space cools, and its temperature decreases. The situation is sketched in Figure 4–20 for the case $T_0 > T_1$.

The temperature distribution in the rock is the solution of Equation (4–68) subject to the conditions

$$
\begin{aligned}
T &= T_1 \quad \text{at} \quad t = 0, \quad y > 0 \\
T &= T_0 \quad \text{at} \quad y = 0 \quad t > 0 \\
T &\to T_1 \quad \text{as} \quad y \to \infty \quad t > 0.
\end{aligned}
$$
(4–92)

The problem posed by Equations (4–68) and (4–92) is a familiar one in the theory of partial differential equations. It can be solved in a rather straightforward way using an approach known as *similarity*. First, it is convenient to introduce the dimensionless temperature ratio θ

$$
\theta = \frac{T - T_1}{T_0 - T_1}
$$
(4–93)

as a new unknown. The equation for θ is identical with the one for T,

$$
\frac{\partial \theta}{\partial t} = \kappa \frac{\partial^2 \theta}{\partial y^2},
$$
(4–94)

but the conditions on θ are simpler

$$
\begin{aligned}
\theta(y, 0) &= 0 \\
\theta(0, t) &= 1 \\
\theta(\infty, t) &= 0.
\end{aligned}
$$
(4–95)

The similarity approach to determining θ is based on the idea that the only length scale in the problem, that is, the only quantity that has the dimensions of length other than y itself, is $\sqrt{\kappa t}$, the characteristic *thermal diffusion distance* (recall that the diffusivity κ has dimensions of length^2/time). It is reasonable to suppose that, in this circumstance, θ is not a function of t and y separately, but rather it is a function of the dimensionless ratio

$$\eta = \frac{y}{2\sqrt{\kappa t}}. \tag{4–96}$$

The factor of 2 is introduced to simplify the subsequent results. It is not only reasonable that θ should depend only on η, but a theorem in dimensional analysis shows that this must be the case.

The dimensionless parameter η is known as the *similarity variable*. The solutions at different times are "similar" to each other in the sense that the spatial dependence at one time can be obtained from the spatial dependence at a different time by stretching the coordinate y by the square root of the ratio of the times. We will see from the solution to this problem that the characteristic thermal diffusion length is the distance over which the effects of a sudden, localized change in temperature can be felt after a time t has elapsed from the onset of the change.

The equations (4–94) and (4–95) must be rewritten in terms of η. This requires that we determine the partial derivatives of θ with respect to t and y in terms of derivatives with respect to η. This can be accomplished using the chain rule for differentiation as follows:

$$\frac{\partial \theta}{\partial t} = \frac{d\theta}{d\eta}\frac{\partial \eta}{\partial t} = \frac{d\theta}{d\eta}\left(-\frac{1}{4}\frac{y}{\sqrt{\kappa t}}\frac{1}{t}\right) = \frac{d\theta}{d\eta}\left(-\frac{1}{2}\frac{\eta}{t}\right) \tag{4–97}$$

$$\frac{\partial \theta}{\partial y} = \frac{d\theta}{d\eta}\frac{\partial \eta}{\partial y} = \frac{d\theta}{d\eta}\frac{1}{2\sqrt{\kappa t}} \tag{4–98}$$

$$\frac{\partial^2 \theta}{\partial y^2} = \frac{1}{2\sqrt{\kappa t}}\frac{d^2\theta}{d\eta^2}\frac{\partial \eta}{\partial y} = \frac{1}{4}\frac{1}{\kappa t}\frac{d^2\theta}{d\eta^2}. \tag{4–99}$$

Equation (4–94) becomes

$$-\eta\frac{d\theta}{d\eta} = \frac{1}{2}\frac{d^2\theta}{d\eta^2}. \tag{4–100}$$

The boundary conditions are easy to deal with; $y = 0$ maps into $\eta = 0$ and both $y = \infty$ and $t = 0$ map into

$\eta = \infty$. Thus the conditions (4–95) reduce to

$$\theta(\infty) = 0$$
$$\theta(0) = 1. \tag{4–101}$$

The fact that the introduction of the similarity variable reduces the partial differential equation (4–94) to an ordinary differential equation (4–100) with respect to η and reduces the separate conditions in t and y to consistent conditions involving η alone, is a posteriori proof of the validity of the approach.

Equation (4–100) can be integrated by letting

$$\phi = \frac{d\theta}{d\eta}. \tag{4–102}$$

Rewriting Equation (4–100), we obtain

$$-\eta\phi = \frac{1}{2}\frac{d\phi}{d\eta} \tag{4–103}$$

or

$$-\eta\,d\eta = \frac{1}{2}\frac{d\phi}{\phi}. \tag{4–104}$$

Integration of Equation (4–104) is straightforward:

$$-\eta^2 = \ln\phi - \ln c_1, \tag{4–105}$$

where $-\ln c_1$ is the constant of integration. It follows that

$$\phi = c_1 e^{-\eta^2} = \frac{d\theta}{d\eta}. \tag{4–106}$$

Upon integrating Equation (4–106), we get

$$\theta = c_1 \int_0^\eta e^{-\eta'^2}\,d\eta' + 1, \tag{4–107}$$

where η' is a dummy variable of integration and the condition $\theta(0) = 1$ was used to evaluate the second constant of integration. Since $\theta(\infty) = 0$, we must have

$$0 = c_1 \int_0^\infty e^{-\eta'^2}\,d\eta' + 1. \tag{4–108}$$

The integral in Equation (4–108) is well known:

$$\int_0^\infty e^{-\eta'^2}\,d\eta' = \frac{\sqrt{\pi}}{2}. \tag{4–109}$$

Thus the constant c_1 is $-2/\sqrt{\pi}$ and

$$\theta = 1 - \frac{2}{\sqrt{\pi}}\int_0^\eta e^{-\eta'^2}\,d\eta'. \tag{4–110}$$

The function defined by the integral in Equation (4–110) occurs so often in solutions of physical problems that it is given a special name, the *error function* $erf(\eta)$

$$\text{erf}(\eta) \equiv \frac{2}{\sqrt{\pi}} \int_0^{\eta} e^{-\eta'^2} d\eta'. \qquad (4\text{--}111)$$

Thus we can rewrite θ as

$$\theta = 1 - \text{erf}(\eta) = \text{erfc}\,\eta \qquad (4\text{--}112)$$

where $erfc(\eta)$ is the *complementary error function*. Values of the error function and the complementary error function are listed in Table 4–5. The functions are also shown in Figure 4–21.

The solution for the temperature as a function of time t and distance y is Equation (4–112). It can be written in terms of the original variables as

$$\frac{T - T_1}{T_0 - T_1} = \text{erfc}\frac{y}{2\sqrt{\kappa t}}. \qquad (4\text{--}113)$$

At $y = 0$, the complementary error function is 1 and $T = T_0$. As $y \to \infty$ or $t = 0$, erfc is 0 and $T = T_1$. The general solution for θ or $(T - T_1)/(T_0 - T_1)$ is shown as erfc η in Figure 4–21.

The near-surface region in which there is a significant temperature change is referred to as a *thermal boundary layer*. The thickness of the thermal boundary layer requires an arbitrary definition, since the temperature T approaches the initial rock temperature T_1 asymptotically. We define the thickness of the boundary layer y_T as the distance to where $\theta = 0.1$. This distance changes with time as the half-space heats up or cools off. The condition $\theta = 0.1$ defines a unique value of the similarity variable η_T, however. From Equation (4–112) and Table (4–5) we obtain

$$\eta_T = \text{erfc}^{-1}0.1 = 1.16 \qquad (4\text{--}114)$$

and from Equation (4–96) we get

$$y_T = 2\eta_T\sqrt{\kappa t} = 2.32\sqrt{\kappa t}. \qquad (4\text{--}115)$$

The thickness of the thermal boundary layer is 2.32 times the characteristic thermal diffusion distance $\sqrt{\kappa t}$.

PROBLEM 4–31 Derive an expression for the thickness of the thermal boundary layer if we define it to be the distance to where $\theta = 0.01$.

TABLE 4–5 **The Error Function and the Complementary Error Function**

η	**erf η**	**erfc η**
0	0	1.0
0.02	0.022565	0.977435
0.04	0.045111	0.954889
0.06	0.067622	0.932378
0.08	0.090078	0.909922
0.10	0.112463	0.887537
0.15	0.167996	0.832004
0.20	0.222703	0.777297
0.25	0.276326	0.723674
0.30	0.328627	0.671373
0.35	0.379382	0.620618
0.40	0.428392	0.571608
0.45	0.475482	0.524518
0.50	0.520500	0.479500
0.55	0.563323	0.436677
0.60	0.603856	0.396144
0.65	0.642029	0.357971
0.70	0.677801	0.322199
0.75	0.711156	0.288844
0.80	0.742101	0.257899
0.85	0.770668	0.229332
0.90	0.796908	0.203092
0.95	0.820891	0.179109
1.0	0.842701	0.157299
1.1	0.880205	0.119795
1.2	0.910314	0.089686
1.3	0.934008	0.065992
1.4	0.952285	0.047715
1.5	0.966105	0.033895
1.6	0.976348	0.023652
1.7	0.983790	0.016210
1.8	0.989091	0.010909
1.9	0.992790	0.007210
2.0	0.995322	0.004678
2.2	0.998137	0.001863
2.4	0.999311	0.000689
2.6	0.999764	0.000236
2.8	0.999925	0.000075
3.0	0.999978	0.000022

PROBLEM 4–32 If the surface temperature is increased 10 K, how long is it before the temperature increases 2 K at a depth of 1 m ($\kappa = 1$ mm^2 s^{-1})?

The heat flux at the surface $y = 0$ is given by differentiating Equation (4–113) according to Fourier's law

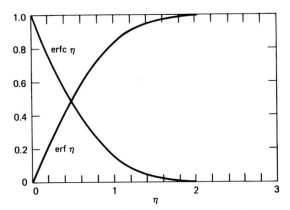

4-21 The error function and the complementary error function.

and evaluating the result at $y = 0$ such that

$$
\begin{aligned}
q &= -k\left(\frac{\partial T}{\partial y}\right)_{y=0} \\
&= -k(T_0 - T_1)\frac{\partial}{\partial y}\left(\text{erfc}\frac{y}{2\sqrt{\kappa t}}\right)_{y=0} \\
&= k(T_0 - T_1)\frac{\partial}{\partial y}\left(\text{erf}\frac{y}{2\sqrt{\kappa t}}\right)_{y=0} \\
&= \frac{k(T_0 - T_1)}{2\sqrt{\kappa t}}\frac{d}{d\eta}(\text{erf}\,\eta)_{\eta=0} \\
&= \frac{k(T_0 - T_1)}{2\sqrt{\kappa t}}\left(\frac{2}{\sqrt{\pi}}e^{-\eta^2}\right)_{\eta=0} = \frac{k(T_0 - T_1)}{\sqrt{\pi \kappa t}}.
\end{aligned}
$$
$$(4\text{--}116)$$

The surface heat flux q is infinite at $t = 0$ because of the sudden application of the temperature T_0 at $t = 0$. However, q decreases with time, and the total heat into the semi-infinite half-space up to any time, Q, is finite; it is given by the integral of Equation (4–116) from $t = 0$ to t

$$
Q = \int_0^t q\,dt' = \frac{2k(T_0 - T_1)}{\sqrt{\kappa \pi}}\sqrt{t}. \qquad (4\text{--}117)
$$

Except for the factor $\pi^{-1/2}$ the heat flux into the rock is k times the temperature difference $(T_0 - T_1)$ divided by the thermal diffusion length $\sqrt{\kappa t}$.

In the mid-1800s William Thompson, later Lord Kelvin, used the theory for the conductive cooling of a semi-infinite half-space to estimate the age of the Earth. He hypothesized that the Earth was formed at a uniform high temperature T_1 and that its surface was sub-sequently maintained at the low temperature T_0. He assumed that a thin near-surface boundary layer developed as the Earth cooled. Since the boundary layer would be thin compared with the radius of the Earth, he reasoned that the one-dimensional model developed above could be applied. From Equation (4–116) he concluded that the age of the Earth t_0 was given by

$$
t_0 = \frac{(T_1 - T_0)^2}{\pi \kappa (\partial T/\partial y)_0^2}, \qquad (4\text{--}118)
$$

where $(\partial T/\partial y)_0$ is the present near-surface thermal gradient. With $(\partial T/\partial y)_0 = 25$ K km^{-1}, $T_1 - T_0 = 2000$ K, and $\kappa = 1$ mm^2 s^{-1}, the age of the Earth from Equation (4–118) is $t_0 = 65$ million years. It was not until radioactivity was discovered about 1900 that this estimate was seriously questioned.

PROBLEM 4-33 One way of determining the effects of erosion on subsurface temperatures is to consider the instantaneous removal of a thickness l of ground. Prior to the removal $T = T_0 + \beta y$, where y is the depth, β is the geothermal gradient, and T_0 is the surface temperature. After removal, the new surface is maintained at temperature T_0. Show that the subsurface temperature after the removal of the surface layer is given by

$$
T = T_0 + \beta y + \beta l\,\text{erf}\left(\frac{y}{2\sqrt{\kappa t}}\right).
$$

How is the surface heat flow affected by the removal of surface material?

PROBLEM 4-34 Determine the effect of a glacial epoch on the surface geothermal gradient as follows. At the start of the glacial epoch $t = -\tau$, the subsurface temperature is $T_0 + \beta y$. The surface is $y = 0$, and y increases downward. During the period of glaciation the surface temperature drops to $T_0 - \Delta T_0$. At the end of the glacial period, $t = 0$, the surface temperature again rises to T_0. Find the subsurface temperature $T(y, t)$ and the surface heat flow for $t > -\tau$. If the last glaciation began at 13,000 year BP and ended 8000 year BP and $\Delta T_0 = 20$ K ($\kappa = 1$ mm^2 s^{-1}, $k = 3.3$ W m^{-1} K^{-1}), determine the effect on the present surface heat flow.

HINT: Use the idea of superposition to combine the elementary solutions to the heat conduction equation in such a way as to develop the solution of this problem without having to solve a differential equation again.

PROBLEM 4-35 One technique for measuring the thermal conductivity of sediments involves the insertion of a very thin cylinder, or needle, heated by an internal heater wire at a known and constant rate, into the sediments. A small thermistor inside the needle measures the rise of temperature T with time t. After the heater has been on for a short time, measurements of T show a linear growth with $\ln t$,

$$T = c_1 \ln t + c_2.$$

The sediment conductivity can be deduced from the slope of a T versus $\ln t$ plot, c_1, with the aid of a theoretical formula you can derive as follows. Consider the temperature field due to an infinite line source that emits Q units of heat per unit time and per unit length for times $t > 0$ in an infinite medium initially at temperature T_0. Determine $T(r, t)$ by solving Equation (4–70) subject to the appropriate initial and boundary conditions.

HINT: A similarity solution with the similarity variable $\eta = r^2/4\kappa t$ works. In fact, the solution is

$$T - T_0 = \frac{Q}{4\pi k} \int_\eta^\infty \frac{e^{-\eta'}}{\eta'} d\eta'. \tag{4–119}$$

The integral $\int_\eta^\infty (1/\eta')e^{-\eta'} d\eta'$ is known as the *exponential integral* $E_1(\eta)$. Thus $T - T_0 = (Q/4\pi k)E_1(\eta)$. The function $E_1(\eta)$ can be evaluated numerically and tabulated, just as the error function. (Values of $E_1(\eta)$ are given in Table 8–4.) Furthermore, it can be shown that for η sufficiently small (t large enough)

$$E_1(\eta) = -\gamma - \ln \eta + \cdots, \tag{4–120}$$

where γ is Euler's constant $0.5772156649\ldots$. Thus, after a sufficiently long time,

$$
\begin{aligned}
T - T_0 &= \frac{Q}{4\pi k}(-\gamma - \ln \eta + \cdots) \\
&= \frac{-Q\gamma}{4\pi k} - \frac{Q}{4\pi k} \ln \frac{r^2}{4\kappa t} + \cdots \\
&= \frac{-Q\gamma}{4\pi k} - \frac{Q}{4\pi k} \ln \frac{r^2}{4\kappa} + \frac{Q}{4\pi k} \ln t + \cdots.
\end{aligned}
\tag{4–121}
$$

The measured slope c_1 is thus $Q/4\pi k$, and, with Q known, k can be determined.

PROBLEM 4-36 Displacements along faults can bring rock masses with different temperatures into sudden contact. Thrust sheets result in the emplacement of buried crustal rocks above rocks that were previously at the surface. The transform faults that offset ocean ridge segments juxtapose oceanic lithospheres of different ages. Consider therefore how temperature varies with time and position when two semi-infinite half-spaces initially at temperatures $T_-(y < 0)$ and $T_+(y > 0)$ are placed adjacent to each other along $y = 0$ at time $t = 0$. Show that T is given by

$$T = \frac{(T_+ + T_-)}{2} + \frac{(T_+ - T_-)}{2} \text{erf}\left(\frac{y}{2\sqrt{\kappa t}}\right). \tag{4–122}$$

Consider also how temperature varies with time and depth for a situation in which the initial temperature distribution in a half-space ($y > 0$) is $T = T_1$ for $0 < y < b$ and $T = T_2$ for $y > b$. Assume that the surface $y = 0$ is maintained at $T = T_0$ for $t > 0$ and that $T \to T_2$ as $y \to \infty$ for $t > 0$. Show that $T(y, t)$ is given by

$$T = (T_1 - T_0)\text{erf}\left(\frac{y}{2\sqrt{\kappa t}}\right) + \frac{(T_2 - T_1)}{2}$$

$$\times \left\{ \text{erf}\frac{(y - b)}{2\sqrt{\kappa t}} + \text{erf}\frac{(y + b)}{2\sqrt{\kappa t}} \right\} + T_0. \tag{4–123}$$

4-16 Cooling of the Oceanic Lithosphere

As we have already noted, the solution developed in Section 4–15 is also relevant to the cooling of the oceanic lithosphere adjacent to a mid-ocean ridge. In Chapter 1 we discussed how the mid-ocean ridge system is associated with ascending mantle convection. The surface plates on either side of the ridge move horizontally with a velocity u, as illustrated in Figure 4–22. The plates are created from the hot mantle rock that is flowing upward beneath the ridge. This rock is cooled by the seawater and forms the rigid plates that move away from the ridge. Since the oceanic lithosphere is the surface plate that moves rigidly over the deeper mantle, it can be identified with the part of the upper mantle whose temperature is less than some value below which mantle rocks do not readily deform over geologic time. High-temperature deformation of rocks in the

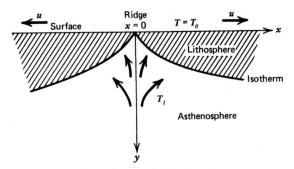

4–22 Schematic of the cooling oceanic lithosphere.

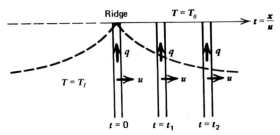

4–23 Vertical columns of mantle and lithosphere moving horizontally away from the ridge and cooling vertically to the surface ($t_2 > t_1 > 0$).

laboratory indicates that this temperature is about 1600 K. Thus we can think of the lithosphere as the region between the surface and a particular isotherm, as shown in the figure. The depth to this isotherm increases with the age of the lithosphere; that is, the lithosphere thickens as it moves farther from the ridge, since it has more time to cool. We refer to the age of the lithosphere as the amount of time t required to reach the distance x from the ridge (because of symmetry we consider x positive); $t = x/u$.

The temperature of the rock at the ridge crest $x = 0$ and beneath the plate is T_1. The seawater cools the surface to the temperature T_0. Thus, a column of mantle is initially at temperature T_1, and its surface is suddenly brought to the temperature T_0. As the column moves away from the ridge, its surface temperature is maintained at T_0, and it gradually cools. This problem is identical to the sudden cooling of a half-space, treated in Section 4–15, if we neglect horizontal heat conduction compared with vertical heat conduction. This is a good approximation as long as the lithosphere is thin. With horizontal heat conduction neglected, heat conduction is vertical in columns of mantle and lithosphere, as it is in the half-space problem. Although a thin column may not resemble a semi-infinite half-space, the essential feature both have in common that makes the cooling problem identical for both is the vertical heat conduction. Figure 4–23 illustrates columns of mantle moving laterally away from the ridge and cooling to the surface.

To adapt the half-space sudden cooling solution to the oceanic lithosphere cooling situation, let $t = x/u$, and rewrite Equation (4–113) as

$$\frac{T_1 - T}{T_1 - T_0} = \operatorname{erfc}\left(\frac{y}{2\sqrt{\kappa x/u}}\right). \qquad (4\text{–}124)$$

This can be further rearranged as

$$\frac{T_1 - T}{T_1 - T_0} = 1 - \frac{T - T_0}{T_1 - T_0} = 1 - \operatorname{erf}\left(\frac{y}{2\sqrt{\kappa x/u}}\right)$$

and

$$\frac{T - T_0}{T_1 - T_0} = \operatorname{erf}\left(\frac{y}{2\sqrt{\kappa x/u}}\right). \qquad (4\text{–}125)$$

According to Equation (4–125) the surface temperature is T_0, since $\operatorname{erf}(0) = 0$ and $T \to T_1$ as the depth $y \to \infty$, since $\operatorname{erf}(\infty) = 1$. Figure 4–24 shows the isotherms beneath the ocean surface as a function of the age of the seafloor for $T_1 - T_0 = 1300$ K, and $\kappa = 1$ mm^2 s^{-1}. The isotherms in Figure 4–24 have the shape of parabolas. The thickness of the oceanic lithosphere y_L can be

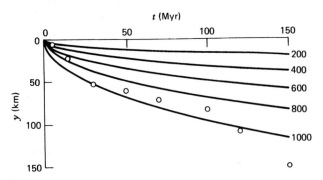

4–24 The solid lines are isotherms, $T - T_0$ (K), in the oceanic lithosphere from Equation (4–125). The data points are the thicknesses of the oceanic lithosphere in the Pacific determined from studies of Rayleigh wave dispersion data (Leeds et al., 1974).

obtained directly from Equation (4–115) by replacing t with x/u:

$$y_L = 2.32(\kappa t)^{1/2} = 2.32\left(\frac{\kappa x}{u}\right)^{1/2}. \qquad (4\text{–}126)$$

With $\kappa = 1 \text{ mm}^2 \text{ s}^{-1}$ the thickness of the lithosphere at an age of 80 Myr is 116 km. It should be emphasized that the thickness given in Equation (4–126) is arbitrary in that it corresponds to $(T - T_0)/(T_1 - T_0) = 0.9$. Also included in Figure 4–24 are thicknesses of the oceanic lithosphere in the Pacific obtained from studies of Rayleigh wave dispersion.

The surface heat flux q_0 as a function of age and distance from the ridge crest is given by Equation (4–116)

$$q_0 = \frac{k(T_1 - T_0)}{\sqrt{\pi \kappa t}} = k(T_1 - T_0)\left(\frac{u}{\pi \kappa x}\right)^{1/2}. \qquad (4\text{–}127)$$

This is the surface heat flow predicted by the half-space cooling model.

Many measurements of the surface heat flow in the oceans have been carried out and there is considerable scatter in the results. A major cause of this scatter is hydrothermal circulations through the oceanic crust. The heat loss due to these circulations causes observed heat flows to be systematically low. Lister et al. (1990) considered only measurements in thick sedimentary cover that blocked hydrothermal circulations. Their values of surface heat flow are given in Figure 4–25 as a function of the age of the seafloor. The results, for the half-space cooling model from Equation (4–127) are compared with the observations taking $k = 3.3 \text{ W m}^{-1} \text{ K}^{-1}$ and the other parameter values as above. Quite good agreement is found at younger ages but the data appear to lie above the theoretical prediction for older ages. This discrepancy will be discussed in detail in later sections.

The cumulative area of the ocean floor A as a function of age, that is, the area of the seafloor with ages less than a specified value, is given in Figure 4–26. The mean age of the seafloor is 60.4 Myr. Also included in Figure 4–26 is the cumulative area versus age for a model seafloor that has been produced at a rate $dA/dt = 0.0815 \text{ m}^2 \text{ s}^{-1}$ and subducted at an age τ of 120.8 Myr (dashed line). This is the average rate of seafloor accretion over this time. It should be noted that the present

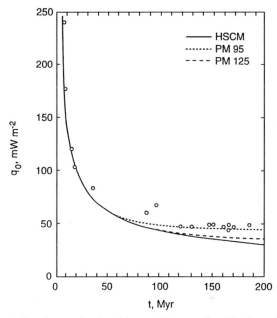

4–25 Heat flow as a function of the age of the ocean floor. The data points are from sediment covered regions of the Atlantic and Pacific Oceans (Lister et al., 1990). Comparisons are made with the half-space cooling model (HSCM) from Equation (4–127) and the plate model from Equation (4–133) with $y_{L0} = 95$ km (PM 95) and with $y_{L0} = 125$ km (PM 125).

rate of seafloor accretion is about $0.090 \text{ m}^2 \text{ s}^{-1}$; very close to the long-term average value.

For a constant rate of seafloor production and for subduction at an age τ, the mean oceanic heat flow $\bar{q}_0$ is

$$\bar{q}_0 = \frac{1}{\tau}\int_0^\tau q_0 \, dt = \frac{1}{\tau}\int_0^\tau \frac{k(T_1 - T_0)}{\sqrt{\pi \kappa t}} \, dt = \frac{2k(T_1 - T_0)}{\sqrt{\pi \kappa \tau}}. \qquad (4\text{–}128)$$

Taking $\tau = 120.8$ Myr and the other parameters as above, we find that the mean oceanic heat flow is $\bar{q}_0 = 78.5 \text{ mW m}^{-2}$. This is in reasonable agreement with the mean value of oceanic heat flow measurements (101 mW m^{-2}) given in Section 4–4. This agreement is somewhat fortuitous since the data are not evenly distributed with respect to the age of the seafloor; oceanic heat flow data are also biased toward areas of the seafloor that are well covered by sediments. Nevertheless, we can conclude that a substantial fraction of the heat lost from the interior of the Earth is directly attributable to the cooling of the oceanic lithosphere. An oceanic geotherm corresponding to the mean age of

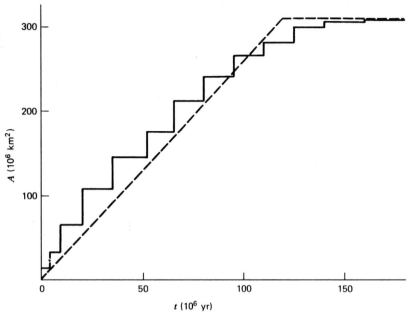

4–26 Cumulative area of seafloor A as a function of age t (the area of seafloor with ages younger than a given age) (solid lines). The dashed line is a cumulative area function for a model seafloor produced at a constant rate of 0.0815 m² s⁻¹, and subducted at an age of 120.8 Myr.

60.4 Myr as obtained from Equation (4–125) is given in Figure 4–27.

So far we have discussed only the oceanic lithosphere. We can also apply the one-dimensional cooling model to the continental lithosphere and compare the surface heat flow predicted by the model with heat flow measurements. The heat flow values that are relevant to this purpose are the reduced heat flows defined in Section 4–8. Recall that the reduced heat flux is the mantle contribution to the surface heat flow; it is deduced from the correlation of the surface heat flow with the surface concentration of heat-producing radioactive isotopes (see Equation (4–29) and Figure 4–11). Reduced heat flux values for several geological provinces are plotted against the ages of the provinces in Figure 4–28. If the mantle heat flow in continental areas were due to the conductive cooling of the lithosphere, q_m would be given by Equation (4–127) with t the age of the continental crust. The prediction of this equation for $k = 3.3$ W m⁻¹ K⁻¹, $\kappa = 1$ mm² s⁻¹, and $T_1 - T_0 = 1300$ K is also shown in Figure 4–28. Clearly, the values of mantle heat flow deduced from observations lie considerably above the conductive cooling prediction for the older provinces. The measured

4–27 Mean oceanic geotherm determined from Equation (4–125) with $t = 60.4$ Myr.

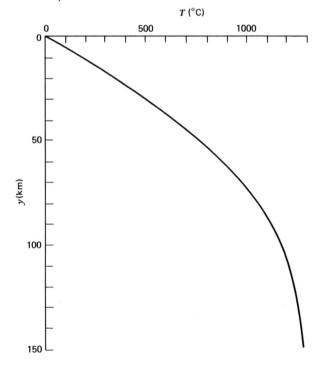

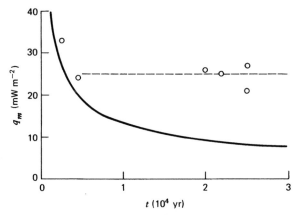

4–28 Dependence of the mantle heat flow on age for several continental geological provinces. The solid line is the predicted heat flow for a conductively cooling continental lithosphere from Equation (4–127), and the dashed line is a constant mantle heat flow of 25 mW m^{-2}. Data are from Sclater et al. (1980).

values correlate better with a constant mantle heat flow of 25 mW m^{-2} for ages between 500 Myr and 2.5 Gyr. This correlation is strong evidence that there is an additional heat input to the base of the continental lithosphere caused by mantle convection; this heat input is very close to the mean mantle heat flow $q_m = 28$ mW m^{-2} given in Section 4–5.

4–17 Plate Cooling Model of the Lithosphere

As discussed earlier, observations show that the half-space cooling model does not predict the time evolution of the continental lithosphere. The continental lithosphere does not continue to thicken with age but instead approaches an equilibrium, time-independent thermal structure. This result requires heating of the base of the continental lithosphere by mantle convection. The surface heat flow data from the ocean basins given in Figure 4–25 show that there is also basal heating of the oceanic lithosphere. To account for this basal heating, we introduce the plate cooling model.

Consider the instantaneous heating or cooling of a finite thickness plate. Since our application is to the lithosphere, we take the plate thickness to be y_{L0}, the thickness of the lithosphere at large times. The infinitely long plate fills the region $0 \leq y \leq y_{L0}$. The temperature in the plate is a solution of the one-dimensional unsteady heat conduction equation (4–68).

Initially at $t = 0$ the plate is at the temperature T_1; for $t > 0$, the surface of the plate $y = 0$ is maintained at the temperature T_0. The base of the plate $y = y_{L0}$

is maintained at the temperature T_1. These conditions can be written as

$$
\begin{aligned}
T &= T_1 \quad \text{at} \quad t = 0, \qquad 0 \leq y \leq y_{L0} \\
T &= T_0 \quad \text{at} \quad y = 0, \qquad t > 0 \\
T &= T_1 \quad \text{at} \quad y = y_{L0}, \qquad t > 0.
\end{aligned}
\tag{4–129}
$$

The solution of Equation (4–68) that satisfies the initial and boundary conditions given in Equation (4–129) can be obtained in the form of an infinite series. A detailed derivation of the solution has been given by Carslaw and Jaeger (1959, p. 100). The result can be written as

$$
T = T_0 + (T_1 - T_0)\left[\frac{y}{y_{L0}} + \frac{2}{\pi}\sum_{n=1}^{\infty}\frac{1}{n}\exp\left(-\frac{\kappa n^2 \pi^2 t}{y_{L0}^2}\right)\sin\left(\frac{n\pi y}{y_{L0}}\right)\right].
\tag{4–130}
$$

At large times, $t \gg y_{L0}^2/\kappa$, an equilibrium linear temperature profile is established:

$$
T = T_0 + (T_1 - T_0)\frac{y}{y_{L0}}.
\tag{4–131}
$$

At small times, $t \ll y_{L0}^2/\kappa$, the half-space cooling solution given in Section 4–16 is recovered. The deviations from the half-space cooling solution are well approximated if only the first two terms of the expansion given in Equation (4–130) are retained, with the result

$$
T = T_0 + (T_1 - T_0)\left[\frac{y}{y_{L0}} + \frac{2}{\pi}\exp\left(-\frac{\kappa \pi^2 t}{y_{L0}^2}\right)\sin\left(\frac{\pi y}{y_{L0}}\right) + \frac{1}{\pi}\exp\left(-\frac{4\kappa \pi^2 t}{y_{L0}^2}\right)\sin\left(\frac{2\pi y}{y_{L0}}\right)\right].
\tag{4–132}
$$

We can obtain the surface heat flow q_0 as a function of age t from Equations (4–1) and (4–130) as

$$
q_0 = \frac{k(T_1 - T_0)}{y_{L0}}\left[1 + 2\sum_{n=1}^{\infty}\exp\left(-\frac{\kappa n^2 \pi^2 t}{y_{L0}^2}\right)\right].
\tag{4–133}
$$

For large times, $t \gg y_{L0}^2/\kappa$, the equilibrium value of the surface heat flow is

$$
q_{0e} = \frac{k(T_1 - T_0)}{y_{L0}}.
\tag{4–134}
$$

We can approximate the deviations of the surface heat flow from the half-space cooling result given in Equation (4–127) by retaining the first two terms of the expansion in Equation (4–133) with the result

$$
q_0 = \frac{k(T_1 - T_0)}{y_{L0}} \left[1 + 2\exp\left(-\frac{\kappa \pi^2 t}{y_{L0}^2}\right) \right.
$$
$$
\left. + 2\exp\left(-\frac{4\kappa \pi^2 t}{y_{L0}^2}\right) \right]. \tag{4–135}
$$

For large times $t \to \infty$ the heat flow from Equation (4–135) approaches the equilibrium heat flow in Equation (4–134).

We next compare the predictions of the plate model with the heat flow compilation as a function of seafloor age given in Figure 4–25. Comparisons are made for two values of plate thickness, $y_{L0} = 95$ and 125 km, and for other parameter values as before. For $y_{L0} = 95$ km, the equilibrium $t \to \infty$ heat flow from Equation (4–134) is $q_{0e} = 45$ mW m^{-2}, and for $y_{L0} = 125$ km, we have $q_{0e} = 34$ mW m^{-2}. At ages of less than about 50 Ma, the half-space cooling model and the plate models give the same results. At these ages, the thickness of the thermal boundary layer is less than the thickness of the plate, so the presence of a finite plate thickness has no effect. At older ages, the specified plate thickness restricts the growth of the thermal boundary layer and the heat flows predicted by the plate models are somewhat greater than those predicted by the half-space cooling model. Further discussion of the agreement between theories and observations will be postponed to Section 4–23.

4–18 The Stefan Problem

A number of important geological problems involve the solidification of magmas. We assume that the magma has a well-defined melt temperature at which the phase change from liquid to solid occurs. Associated with this phase change is a latent heat of fusion L. This is the amount of heat that is liberated upon the solidification of 1 kg of magma. Heat conduction problems involving phase changes differ from problems we have already solved in two major ways. First, we have to determine as part of the solution where the phase change boundary, that is, the interface between solid and liquid, is located. The position of this boundary obviously

changes as solidification proceeds. Second, we have to account for the latent heat of fusion, which is liberated at the solid–liquid interface as solidification takes place; this additional heat must be conducted away from the phase change boundary.

The first problem we consider is that of a horizontal layer of magma that is solidifying from its upper surface downward as a result of being cooled from above. We assume that the upper surface is maintained at a constant temperature T_0. An example of this would be the solidification of a lava flow. Because of heat loss to the surface the solid layer grows thicker with time. A lava flow also solidifies at its base. However, if we assume that the magma is extruded at its melt temperature, then as long as there is still a liquid region, the solidification from the top and bottom can be treated independently. This also means that the overall flow thickness is unimportant in describing the solidification process as long as a molten region is present. In this section, we will consider the solidification from above; in the next section, we will treat the solidification from below. The solidification of a lava flow from above is essentially identical with the freezing of a lake. This is the problem for which Stefan (1891) first obtained the solution developed below.

The problem we solve is illustrated in Figure 4–29. The flow has solidified to the depth $y = y_m(t)$. We assume that molten material of uniform temperature T_m lies everywhere below the growing surface layer. The fact that the molten region does not extend infinitely far below the surface is of no consequence to the solution. We must solve the heat conduction equation (4–68) in the space $0 \le y \le y_m(t)$ subject to the conditions $T = T_0$ at $y = 0$, $T = T_m$ at $y = y_m(t)$, and $y_m = 0$ at $t = 0$. The position of the solidification boundary is

4–29 Growth of a solid layer at the surface of a cooling lava flow.

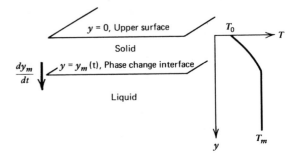

an a priori unknown function of time. As in the case of the sudden heating, or cooling, of a semi-infinite half-space, there is no length scale in this problem. For this reason, we once again introduce the dimensionless coordinate $\eta = y/2\sqrt{\kappa t}$ as in Equation (4–96); it is also convenient to introduce the dimensionless temperature $\theta = (T - T_0)/(T_m - T_0)$ as in Equation (4–93).

The dimensionless coordinate η is obtained by scaling the depth with the thermal diffusion length $\sqrt{\kappa t}$ because there is no other length scale in the problem. Similarly, the depth of the solidification interface y_m must also scale with the thermal diffusion length in such a way that $y_m/\sqrt{\kappa t}$ is a constant. In other words, the depth of the solidification boundary increases with time proportionately with the square root of time. We have used dimensional arguments to determine the functional form of the dependence of y_m on t, a nontrivial result. Because $\eta = y/2\sqrt{\kappa t}$ and y_m is proportional to $\sqrt{\kappa t}$, the solidification boundary corresponds to a constant value $\eta_m = y_m/2\sqrt{\kappa t}$ of the similarity coordinate η. We denote this constant value by $\eta_m = \lambda_1$. Thus we have

$$y_m = 2\lambda_1\sqrt{\kappa t}. \tag{4–136}$$

With our definitions of θ and η, the heat conduction equation for $\theta(\eta)$ is clearly identical to Equation (4–100), whose solution we already know to be proportional to $\mathrm{erf}(\eta)$. This form of solution automatically satisfies the condition $\theta = 0 (T = T_0)$ on $\eta = 0(y = 0)$. To satisfy the remaining condition that $\theta = 1(T = T_m)$ at $\eta = \eta_m(y = y_m) = \lambda_1$, we need simply choose the constant of proportionality appropriately. The solution is

$$\theta = \frac{\mathrm{erf}(\eta)}{\mathrm{erf}(\lambda_1)}. \tag{4–137}$$

Equation (4–137) determines the temperature in the solidified layer $0 \le y \le y_m$. In the molten region $y > y_m$, $T = T_m$ and $\theta = 1$.

The constant λ_1 is determined by requiring that the latent heat liberated at the solidification boundary be conducted vertically upward, away from the interface. The situation at the solidification boundary is illustrated in Figure 4–30. In time δt, the interface moves downward a distance $(dy_m/dt)\delta t$. In so doing, a mass per unit area $\rho(dy_m/dt)\delta t$ is solidified, thus releasing an amount of latent heat $\rho L(dy_m/dt)\delta t$ per unit area.

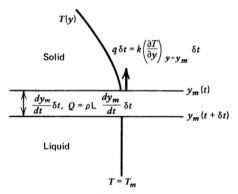

4–30 Latent heat released at the solidification boundary must be conducted upward through the solidified layer.

Conservation of energy requires that this heat release be conducted away from the boundary at precisely the rate at which it is liberated. The heat cannot be conducted downward because the magma is at a constant temperature; moreover, heat flows toward cooler temperatures that lie upward. Fourier's law gives the rate of upward heat conduction per unit time and per unit area at $y = y_m$ as $k(\partial T/\partial y)_{y=y_m}$. Multiplying this by δt and equating it to $\rho L(dy_m/dt)\delta t$ gives the equation for finding λ_1.

$$\rho L\frac{dy_m}{dt} = k\left(\frac{\partial T}{\partial y}\right)_{y=y_m}. \tag{4–138}$$

From Equation (4–136) the speed of the solidification boundary is

$$\frac{dy_m}{dt} = \frac{\lambda_1\sqrt{\kappa}}{\sqrt{t}}. \tag{4–139}$$

From Equation (4–137) the temperature gradient at $y = y_m$ is

$$\begin{aligned}\left(\frac{\partial T}{\partial y}\right)_{y=y_m} &= \left(\frac{d\theta}{d\eta}\right)_{\eta=\eta_m=\lambda_1}\left(\frac{\partial\eta}{\partial y}\right)(T_m - T_0)\\ &= \frac{(T_m - T_0)}{2\sqrt{\kappa T}}\frac{2}{\sqrt{\pi}}e^{-\lambda_1^2}\frac{1}{\mathrm{erf}\,\lambda_1}. \end{aligned} \tag{4–140}$$

Substituting Equations (4–139) and (4–140) into Equation (4–138), we get

$$\frac{L\sqrt{\pi}}{c(T_m - T_0)} = \frac{e^{-\lambda_1^2}}{\lambda_1\,\mathrm{erf}\,\lambda_1}, \tag{4–141}$$

a transcendental equation for determining λ_1. Given a numerical value for the left side of Equation (4–141),

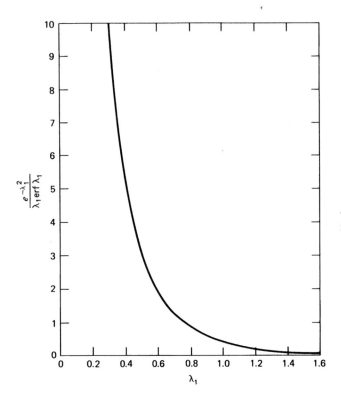

4–31 The right side of the transcendental equation for determining the growth of a solid layer at the surface of a cooling lava flow.

λ_1 can be found by iteratively calculating the right side of the equation until agreement is found. Alternatively, the right side of Equation (4–141) can be plotted as a function of λ_1, as in Figure 4–31, and the solution, for a particular value of the left side of the equation, can be found graphically.

This theory can be compared directly with observations. The thickness of the crusts on three lava lakes on the Hawaiian volcano Kilauea have been measured as functions of time. Eruptions produced lava lakes in the pit craters Kilauea Iki in 1959, Alae in 1963, and Makaopuhi in 1965. A photograph of the initial formation of the crust on the lava lake in the Alae pit crater is given in Figure 4–32. The thicknesses of the solidifying crusts on the three lava lakes are given as functions of time after the eruptions in Figure 4–33. For $L = 400$ kJ kg^{-1}, $c = 1$ kJ kg^{-1} K^{-1}, and $T_m - T_0 = 1050$ K, Equation (4–141) gives $\lambda_1 = 0.876$. With this value of λ_1 and $\kappa = 0.7$ mm^2 s^{-1}, we can determine the thickness of a solidifying crust as a function of time from Equation (4–136). The result plotted in Figure 4–33 shows quite good agreement between the observations and theory.

PROBLEM 4–37 A body of water at 0°C is subjected to a constant surface temperature of -10°C for 10 days. How thick is the surface layer of ice? Use $L = 320$ kJ kg^{-1}, $k = 2$ J m^{-1} s^{-1} K^{-1}, $c = 4$ kJ kg^{-1} K^{-1}, $\rho = 1000$ kg m^{-3}.

4–32 Photograph of the lava lake formed in the pit crater Alae during the 1963 eruption. A solid crust is just beginning to form on the magma (D. L. Peck 19, U.S. Geological Survey).

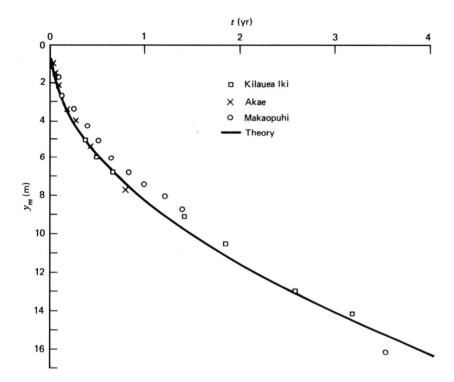

4-33 The thicknesses of the solidifying crusts on the lava lakes in the three pit craters Kilauea Iki (1959), Alae (1963), and Makaopuhi (1965) on the volcano Kilauea, Hawaii (Wright et al. 1976). The theoretical curve is from Equations (4–136) and (4–141).

PROBLEM 4-38 Scientists believe that early in its evolution, the Moon was covered by a magma ocean with a depth of 50 km. Assuming that the magma was at its melt temperature of 1500 K and that the surface of the Moon was maintained at 500 K, how long did it take for the magma ocean to solidify if it was cooled from the surface? Take $L = 320$ kJ kg^{-1}, $\kappa = 1$ mm^2 s^{-1}, and $c = 1$ kJ kg^{-1} K^{-1}.

PROBLEM 4-39 One of the estimates for the age of the Earth given by Lord Kelvin in the 1860s assumed that Earth was initially molten at a constant temperature T_m and that it subsequently cooled by conduction with a constant surface temperature T_0. The age of the Earth could then be determined from the present surface thermal gradient $(dT/dy)_0$. Reproduce Kelvin's result assuming $T_m - T_0 = 1700$ K, $c = 1$ kJ kg^{-1} K^{-1}, $L = 400$ kJ kg^{-1}, $\kappa = 1$ mm^2 s^{-1}, and $(dT/dy)_0 = 25$ K km^{-1}. In addition, determine the

thickness of the solidified lithosphere. Note: Since the solidified layer is thin compared with the Earth's radius, the curvature of the surface may be neglected.

PROBLEM 4-40 Consider the solidification near the upper surface of a lava flow. Compute the surface heat flux q_0 as a function of time. Integrate q_0 over time, and compare the result with the latent heat release up to that time, $\rho L y_m$.

PROBLEM 4-41 Generalize the solution for the solidification of the top of a lava lake to the situation where the lava is initially at a uniform temperature T_v greater than the solidification temperature T_m. Assume that the lava extends to great depth $y \to \infty$ and that $T \to T_v$ as $y \to \infty$ for all t. Also assume that $T = T_0$ at $y = 0$ for all t. Assume that the molten lava and the solidified layer near the surface have the same thermal properties.
HINT: You have to modify the energy balance condition at the solid–liquid interface to account for heat conduction in the liquid.

PROBLEM 4-42 The oceanic crust is believed to form from the solidification of a large magma chamber

4–34 (a) Photograph of a large sill on Finger Mountain, Victoria, Victoria Land, Antarctica (W. B. Hamilton 384, U.S. Geological Survey). (b) Photograph of a small dike offset along a joint (J. P. Lockwood 1, U.S. Geological Survey).

beneath the ridge crest. Use the Stefan solution to determine the width of the magma chamber at its base. Let $L = 400$ kJ kg^{-1}, $c = 1$ kJ kg^{-1} K^{-1}, $T_m - T_0 = 1300$ K, $u = 50$ mm yr^{-1}, $\kappa = 1$ mm^2 s^{-1}, and the thickness of the oceanic crust is 6 km.

PROBLEM 4–43 The mantle rocks of the asthenosphere from which the lithosphere forms are expected to contain a small amount of magma. If the mass fraction of magma is 0.05, determine the depth of the lithosphere–asthenosphere boundary for oceanic lithosphere with an age of 60 Ma. Assume $L = 400$ kJ kg^{-1}, $c = 1$ kJ kg^{-1} K^{-1}, $T_m = 1600$ K, $T_0 = 275$ K, and $\kappa = 1$ mm^2 s^{-1}.

4–19 Solidification of a Dike or Sill

A large fraction of the magma coming from the interior of the Earth does not reach the surface but instead solidifies as intrusive igneous bodies. Two of the simplest types of intrusive bodies are *sills* and *dikes*. A sill is a horizontal layer of solidified rock, and a dike is its vertical counterpart. These one-dimensional structures are illustrated in Figure 4–34. We will now consider the problem of the solidification of a dike or sill. The solidifying magma loses heat by conduction to the adjacent *country rock*.

Our model for dike or sill solidification is illustrated in Figure 4–35. The plane $y = 0$ defines the original magma–rock boundary. The dike or sill occupies the region $-2b < y < 0$. Initially at $t = 0$ the molten rock

in the dike is at its melt temperature T_m, and the wall rock is at the temperature T_0. At time $t = 0$, the dike begins solidifying at the interface $y = 0$. Figure 4–35 shows the temperature distribution initially at $t = 0$ and at a later time t_1. The liquid part of the dike $-2b < y < y_m(t)$ is still at temperature T_m, but the solidified part $y_m < y < 0$ has cooled below T_m. The surrounding rock near the dike has been warmed above T_0 by the release of the latent heat of fusion, but $T \to T_0$ far from the dike $y \to \infty$.

We assume that the physical properties of the country rock and solidified magma are the same. Therefore the temperature satisfies the one-dimensional, time-dependent heat conduction equation (4–68) in the region $y > y_m(t)$. The boundary conditions are that

4–35 Initial temperature distribution at $t = 0$ ($T = T_0$ for $y > 0$, $T = T_m$ for $-2b < y < 0$) and subsequent temperature distribution at $t = t_1$ when the solidification boundary is at $y = y_m(t)$.

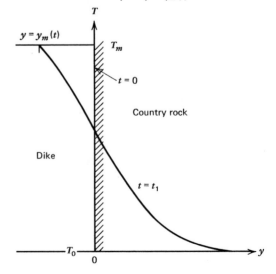

$T = T_m$ at $y = y_m(t)$ and $T \to T_0$ as $y \to \infty$; the initial condition is that $T = T_0$ for $y > 0$ and $y_m = 0$. Once again Equation (4–138) provides a balance between the heat conducted away from the solidification interface and the heat released by the solidification. We use the same method of solution as in the previous section and introduce the dimensionless variables $\eta = y/(2\sqrt{\kappa t})$ and $\theta = (T - T_0)/(T_m - T_0)$.

In this problem the position of the solidification boundary y_m is negative so that $\eta_m = y_m/(2\sqrt{\kappa t})$ is also negative. We denote this constant value by $\eta_m = -\lambda_2$. Thus we have

$$y_m = -2\lambda_2\sqrt{\kappa t}. \qquad (4\text{–}142)$$

The dimensionless temperature in the region $\eta > -\lambda_2$ satisfies Equation (4–100). We have previously shown that a solution of this equation is proportional to erfc η. Such a solution also satisfies the condition $\theta \to 0 (T \to T_0)$ as $\eta \to \infty (y \to \infty)$. In order to satisfy the condition that $\theta = 1$ $(T = T_m)$ at $\eta = \eta_m = -\lambda_2 (y = y_m)$, we need simply to choose the constant of proportionality appropriately. The solution is clearly

$$\theta = \frac{\text{erfc } \eta}{\text{erfc}(-\lambda_2)}. \qquad (4\text{–}143)$$

From the definition of erfc in Equation (4–112) and the property erf $(-x) = -$erf x we obtain

$$\text{erfc}(-\lambda_2) = 1 - \text{erf}(-\lambda_2) = 1 + \text{erf } \lambda_2. \qquad (4\text{–}144)$$

Substitution of Equation (4–144) into Equation (4–143) gives

$$\theta = \frac{\text{erfc } \eta}{1 + \text{erf } \lambda_2}. \qquad (4\text{–}145)$$

The temperature of the boundary between the country rock and the solidified magma $\eta = 0 (y = 0)$ is

$$\theta(0) = \frac{1}{1 + \text{erf } \lambda_2}. \qquad (4\text{–}146)$$

The temperature of this boundary is therefore constant while solidification is occurring.

In order to determine the constant λ_2, we must once again use the heat balance at the solidification boundary given in Equation (4–138). The speed of this boundary is obtained by differentiating Equation (4–142)

$$\frac{dy_m}{dt} = -\lambda_2 \left(\frac{\kappa}{t}\right)^{1/2}. \qquad (4\text{–}147)$$

The temperature gradient at $y = y_m$ is obtained by differentiating Equation (4–145).

$$\begin{aligned}
\left(\frac{\partial T}{\partial y}\right)_{y=y_m} &= \left(\frac{d\theta}{d\eta}\right)_{\eta=-\lambda_2} \left(\frac{\partial \eta}{\partial y}\right)(T_m - T_0) \\
&= \frac{-(T_m - T_0)}{(\pi \kappa t)^{1/2}} \frac{e^{-\lambda_2^2}}{(1 + \text{erf } \lambda_2)}.
\end{aligned} \qquad (4\text{–}148)$$

Substitution of Equations (4–147) and (4–148) into Equation (4–138) gives

$$\frac{L\sqrt{\pi}}{c(T_m - T_0)} = \frac{e^{-\lambda_2^2}}{\lambda_2(1 + \text{erf } \lambda_2)}. \qquad (4\text{–}149)$$

In terms of evaluating λ_2, this equation plays the same role that Equation (4–141) played in the Stefan problem. The right side of the equation is plotted as a function of λ_2 in Figure 4–36. Given a value for the left side of the equation, λ_2 may be determined graphically from the figure or more accurately by iterative numerical calculations.

The time t_s required to solidify a dike of width $2b$ can be obtained directly from Equation (4–142).

4–36 The right side of the transcendental Equation (4–149) for determining the motion of the solidification boundary.

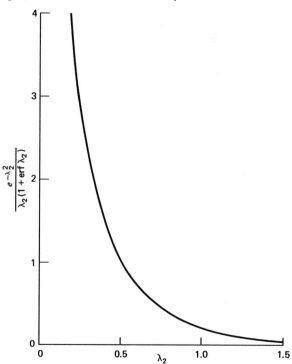

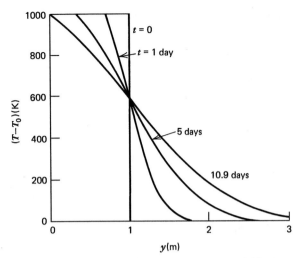

4-37 Temperature profiles at different times during dike solidification.

Solidification occurs symmetrically from the two sides of the dike so that

$$t_s = \frac{b^2}{4\kappa\lambda_2^2}. \tag{4-150}$$

At $t = t_s$ all the magma has solidified.

Let us again consider a numerical example. Taking $L = 320\,\text{kJ kg}^{-1}$, $T_m - T_0 = 1000\,\text{K}$, and $c = 1.2\,\text{kJ kg}^{-1}\,\text{K}^{-1}$, we find from Equation (4–149) (Figure 4–36) that $\lambda_2 = 0.73$. With this value of λ_2, $b = 1$ m, and $\kappa = 0.5\,\text{mm}^2\,\text{s}^{-1}$, we find from Equation (4–150) that the time required to solidify this intrusion is 10.9 days. The temperature at the boundary between the country rock and the solidified magma from Equation (4–139) is $T_0 + 590$ K. The temperature profiles at several times are given in Figure 4–37.

PROBLEM 4-44 Use the results of the sudden half-space heating problem, Equation (4–117), to estimate the time required for dike solidification by setting $Q = \rho L b$. How does this time compare with the 10.9 days computed in the example?

PROBLEM 4-45 Consider the following model for the cooling and solidification of an intrusive igneous body. Suppose that the region $y < 0$ is initially solid with constant temperature T_- and that the region $y > 0$ is initially liquid with constant temperature T_+. The igneous body cools and solidifies; a solid–liquid

interface at temperature T_m propagates into the region $y > 0$. The melting temperature T_m is less than T_+, but it is greater than T_-. Determine y_m, the position of the solidification boundary as a function of time t. Find T in the regions $y < 0$, $0 < y < y_m$, and $y > y_m$.

4-20 The Heat Conduction Equation in a Moving Medium: Thermal Effects of Erosion and Sedimentation

A number of important problems in geology involve moving boundaries. Examples include the solidification problems that we have just discussed; other examples involve erosion and sedimentation. One useful approach to the solution of moving boundary problems is to consider the boundary fixed and the material moving into the boundary. For this and other reasons it is worthwhile to develop the form of the equation of heat conduction for a moving medium. Let x, y be a fixed coordinate system, and assume that the medium moves in the positive x direction with velocity U. Let ξ, ζ be a coordinate system moving with the medium. The situation is sketched in Figure 4–38. The coordinates x, y and ξ, ζ are thus related by

$$x = \xi + Ut \qquad y = \zeta \frac{\partial^2 T}{\partial \zeta^2}. \tag{4-151}$$

The heat conduction equation for an observer moving with the medium is Equation (4–68) (generalized to two dimensions):

$$\left(\frac{\partial T}{\partial t}\right)_\xi = \kappa\left(\frac{\partial^2 T}{\partial \xi^2} + \frac{\partial^2 T}{\partial \zeta^2}\right). \tag{4-152}$$

To find the appropriate form of the heat conduction equation with respect to the fixed coordinate system, we need to relate partial derivatives with respect to ξ, ζ, t to partial derivatives with respect to x, y, t. From

4-38 Fixed (x, y) and moving (ξ, ζ) coordinate systems for the derivation of the heat conduction equation for a moving medium.

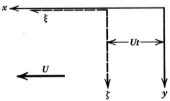

Equation (4–151) we have

$$\left(\frac{\partial T}{\partial t}\right)_{\xi} = \left(\frac{\partial T}{\partial t}\right)_{x} + \frac{\partial T}{\partial x}\frac{dx}{dt} = \left(\frac{\partial T}{\partial t}\right)_{x} + U\frac{\partial T}{\partial x},$$
(4–153)

$$\left(\frac{\partial T}{\partial \xi}\right)_{t} = \left(\frac{\partial T}{\partial x}\right)_{t} \qquad \left(\frac{\partial T}{\partial \zeta}\right)_{t} = \left(\frac{\partial T}{\partial y}\right)_{t}. \quad (4–154)$$

Thus Equation (4–152) can be rewritten as

$$\frac{\partial T}{\partial t} + U\frac{\partial T}{\partial x} = \kappa\left(\frac{\partial^2 T}{\partial x^2} + \frac{\partial^2 T}{\partial y^2}\right), \qquad (4–155)$$

where derivatives are understood to be taken with x or t held constant as appropriate. The term $U(\partial T/\partial x)$ is the advective derivative. An observer moving with the medium and measuring temperature on a recorder cannot distinguish between temperature variations resulting from motion through a spatially varying temperature field $U(\partial T/\partial x)$ and actual temporal variations in temperature $\partial T/\partial t$.

As an additional example of the use of Equation (4–155), consider the oceanic lithosphere cooling problem. With respect to an observer moving with a column of lithosphere, the relevant heat conduction equation is (4–152), a point of view we have already taken. Alternatively, one could take a larger view of the situation and consider a fixed observer viewing the whole spreading process and measuring x from the ridge. That observer sees a steady heat transfer problem described by Equation (4–155) with $\partial T/\partial t = 0$ (and vertical heat conduction only):

$$U\frac{\partial T}{\partial x} = \kappa\frac{\partial^2 T}{\partial y^2}. \qquad (4–156)$$

These alternative ways of approaching the problem are, of course, equivalent because the age of the seafloor is x/U.

PROBLEM 4–46 Assume that a half-space with a deep temperature T_∞ is being eroded at a constant velocity U. If the erosional surface is at a temperature T_0, determine the temperature as a function of the distance from the surface.

PROBLEM 4–47 Assume that a half-space $y > 0$ with a deep temperature T_∞ is being eroded such that $y_m = \alpha\sqrt{t}$, where y_m is the depth of the instantaneous surface measured from the location of the surface at

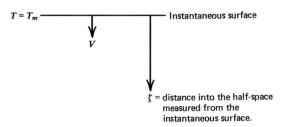

4-39 Model for the ablating meteorite problem.

$t = 0$. If the erosional surface is at a temperature T_0, determine $T(y, t)$ and the heat flow at the surface. HINT: Solve the problem in the y, t system and assume an artificial (unknown) temperature at $y = 0$.

PROBLEM 4–48 Suppose that upon entering the Earth's atmosphere, the surface of a meteorite has been heated to the melting point and the molten material is carried away by the flow. It is of interest to calculate the rate at which melting removes material from the meteorite. For this purpose, consider the following problem. The surface of a semi-infinite half-space moves downward into the half-space with constant velocity V, as indicated in Figure 4–39. The surface is always at the melting temperature T_m, and melted material above the instantaneous surface is removed from the problem. Assume that the surface of the half-space is melted by a constant heat flux q_m into the half-space from above the surface. Assume also that far from the melting surface the temperature is T_0; that is, $T \to T_0$ as $\zeta \to \infty$. Find the temperature distribution in the half-space as a function of time $T(\zeta, t)$, and determine V in terms of q_m and the thermodynamic properties of the rock. Account for the latent heat L required to melt the material.

4-21 One-Dimensional, Unsteady Heat Conduction in an Infinite Region

The problem solved in Section 4–19 provides the temperature distribution up until the time the dike or sill is completely solidified. To determine the subsequent thermal history, we must solve the problem of one-dimensional unsteady heat conduction in an infinite region with a specified initial temperature distribution.

If the temperature distribution at $t = 0$ is $\bar{T}(y)$, the temperature distribution at subsequent times is

$$T = \frac{1}{2\sqrt{\pi\kappa t}}\int_{-\infty}^{\infty}\bar{T}(y')e^{-(y-y')^2/4\kappa t}dy'. \qquad (4–157)$$

This result is known as *Laplace's solution*. For the dike or sill cooling problem, the temperature at the time of total solidification would be the initial temperature distribution $\bar{T}(y)$, and Equation (4–157) could then be used to determine the subsequent temperature distribution. To do this would require a numerical integration.

Instead, an approximate solution is possible if the temperature far from the dike is considered. The heat content of the dike per unit area of the dike–country rock interface is

$$Q = \rho[c(T_m - T_0) + L]2b \qquad (4\text{–}158)$$

For $|y| \gg b$, one can consider the dike to be a planar heat source located at $y = 0$ containing Q units of heat per unit area at $t = 0$. At later times this heat diffuses away from the origin, and we are interested in determining how the temperature evolves as a function of distance from the origin and time. The situation is sketched in Figure 4–40.

The temperature $T(y, t)$ must satisfy the one-dimensional, time-dependent heat conduction equation (4–68) subject to the conditions $T \to T_0$ as $|y| \to \infty$. An essential condition on the temperature distribution is that the heat content of all space must be the original heat content of the dike,

$$\rho c \int_{-\infty}^{\infty} (T - T_0)\, dy = 2\rho c \int_{0}^{\infty} (T - T_0)\, dy = Q,$$
$$(4\text{–}159)$$

for all t. In other words, the heat pulse supplied by the dike can spread out as it diffuses away from the origin, but no heat can be lost from the medium.

A nondimensional form of the solution must be possible. The only quantity with dimensions of length is the

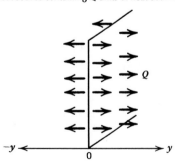

4–40 Planar heat source containing Q units of heat at $t = 0$.

thermal diffusion length $\sqrt{\kappa t}$ so that, once again, the solution depends on the similarity variable $\eta = y/2\sqrt{\kappa t}$. Because this problem has no imposed temperature drop, T must be made dimensionless with the specified initial heat content of the dike Q. A quantity with units of temperature obtained from Q is

$$\frac{Q}{2\rho c \sqrt{\kappa t}},$$

and the appropriate dimensionless temperature for this problem is

$$\theta \equiv \frac{T - T_0}{Q/(2\rho c \sqrt{\kappa t})}. \qquad (4\text{–}160)$$

θ as defined in this equation depends only on η. The integral constraint on temperature (4–159) can be written in terms of θ as

$$2\rho c \int_{0}^{\infty} \frac{Q}{2\rho c \sqrt{\kappa t}} \theta 2\sqrt{\kappa t}\, d\eta = Q$$

or

$$\int_{0}^{\infty} \theta\, d\eta = \frac{1}{2}. \qquad (4\text{–}161)$$

The heat conduction equation must be rewritten in terms of θ. From Equation (4–160) we have

$$\frac{\partial T}{\partial t} = \frac{Q}{2\rho c \sqrt{\kappa t}} \frac{d\theta}{d\eta}\left(\frac{-\eta}{2t}\right) + \frac{Q\theta}{2\rho c \sqrt{\kappa}}\left(-\frac{1}{2}\right)t^{-3/2}$$
$$= \frac{-Qt^{-3/2}}{4\rho c \sqrt{\kappa}}\left(\eta \frac{d\theta}{d\eta} + \theta\right). \qquad (4\text{–}162)$$

Also, from Equation (4–160) one obtains

$$\kappa \frac{\partial^2 T}{\partial y^2} = \frac{\kappa Q}{2\rho c \sqrt{\kappa t}} \frac{d^2\theta}{d\eta^2} \frac{1}{4\kappa t}. \qquad (4\text{–}163)$$

Upon equating (4–162) and (4–163), we find

$$-2\left(\eta \frac{d\theta}{d\eta} + \theta\right) = \frac{d^2\theta}{d\eta^2} \qquad (4\text{–}164)$$

or

$$-2\frac{d}{d\eta}(\eta\theta) = \frac{d^2\theta}{d\eta^2}, \qquad (4\text{–}165)$$

which can be integrated to give

$$-2\eta\theta = \frac{d\theta}{d\eta} + c_1. \qquad (4\text{–}166)$$

The constant c_1 must be zero because the temperature distribution must be symmetric about the plane $y = 0$.

This requires $d\theta/d\eta = 0$ at $\eta = 0$. Thus we have

$$-2\eta d\eta = \frac{d\theta}{\theta}, \tag{4–167}$$

which integrates to

$$\theta = c_2 e^{-\eta^2}. \tag{4–168}$$

From the integral constraint (4–161) we can find c_2 to be

$$\int_0^\infty c_2 e^{-\eta^2} d\eta = \frac{1}{2} = c_2 \frac{\sqrt{\pi}}{2}$$

or

$$c_2 = \frac{1}{\sqrt{\pi}}. \tag{4–169}$$

Finally, the temperature distribution is

$$T - T_0 = \frac{Q}{2\rho c \sqrt{\pi \kappa t}} e^{-y^2/4\kappa t}. \tag{4–170}$$

At distances that are large compared with the width of the initial temperature distribution, the time dependence of the temperature is independent of the initial temperature distribution and is proportional to the heat content of the region.

The temperature at any distance y as given by Equation (4–170) increases with time to a maximum value and then decreases. The time t_{max} when this maximum occurs can be obtained by setting the time derivative of Equation (4–170) equal to zero. The result is

$$t_{max} = \frac{y^2}{2\kappa}. \tag{4–171}$$

Except for a factor of 2, t_{max} is the thermal diffusion time corresponding to the distance y. Substitution of Equation (4–171) into Equation (4–170) gives the maximum temperature T_{max} as a function of y,

$$T_{max} = T_0 + \frac{Q}{\rho c y}\left(\frac{1}{2\pi e}\right)^{1/2}. \tag{4–172}$$

The maximum temperature is proportional to $1/y$.

Applying these results to the example given at the end of Section 4–19, we find $Q = 8.8 \times 10^9$ J m^{-2} with the parameter values given for that example and $\rho = 2900$ kg m^{-3}. For the temperature distribution given in Equation (4–170) to be valid, the time must be long compared with the solidification time of 10.9 days. The temperature profiles from Equation (4–170) at several times are given in Figure 4–41. The maximum temper-

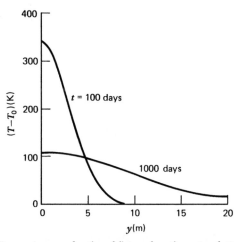

4–41 Temperatures as a function of distance from the center of a 2-m-wide intrusion at several times from Equation (4–170).

ature as a function of distance from the center of the dike, from Equation (4–172), is given in Figure 4–42. Calculations of this type can be used to determine the distance from an intrusion that low-temperature metamorphic reactions in the country rock can be expected.

4-22 Thermal Stresses

According to the laws of thermodynamics the equilibrium state of any material is determined by any two state variables. Examples of state variables include the

4–42 Maximum temperature as a function of distance from the center of a 2-m-wide intrusion from Equation (4–172).

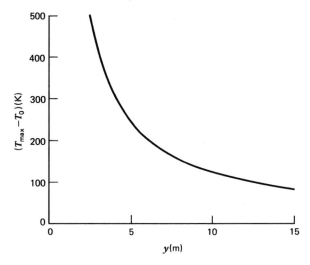

temperature T, pressure p, and density ρ. In thermodynamics it is often convenient to use the specific volume v (volume per unit mass) rather than the density; the two variables are related by

$$v = \frac{1}{\rho}. \tag{4–173}$$

As a state variable, the specific volume can be related to the pressure and temperature using the chain rule for partial differentiation

$$dv = \left(\frac{\partial v}{\partial T}\right)_p dT + \left(\frac{\partial v}{\partial p}\right)_T dp. \tag{4–174}$$

The subscript indicates the variable that is held constant during the differentiation; that is, $(\partial v/\partial T)_p$ is the partial derivative of volume with respect to temperature at constant pressure.

The two partial derivatives appearing in Equation (4–174) are related to well-known thermodynamic quantities. The *isothermal compressibility* β of a material is its fractional change in volume with pressure at constant temperature,

$$\beta = \frac{-1}{v}\left(\frac{\partial v}{\partial p}\right)_T, \tag{4–175}$$

and its *volumetric coefficient of thermal expansion* α_v is its fractional change in volume with temperature at constant pressure,

$$\alpha_v = \frac{1}{v}\left(\frac{\partial v}{\partial T}\right)_p. \tag{4–176}$$

The coefficients β and α_v are material properties that can be obtained from laboratory experiments. We previously saw in Equation (3–55) how β is related to the elastic properties of a material. Substitution of Equations (4–175) and (4–176) into Equation (4–174) yields

$$dv = -v\beta \, dp + v\alpha_v \, dT. \tag{4–177}$$

If a material is unconstrained, so that the pressure does not change ($dp = 0$) when the temperature and volume change, Equation (4–177) implies

$$dv = v\alpha_v \, dT \tag{4–178}$$

or

$$d\rho = -\rho\alpha_v \, dT. \tag{4–179}$$

If a material is confined, so that its volume cannot change ($dv = 0$), the changes in temperature and pressure are related by

$$dp = \frac{\alpha_v}{\beta} \, dT. \tag{4–180}$$

Typical values of α_v and β for rock are 3×10^{-5} K^{-1} and 10^{-11} Pa^{-1}, respectively. With these material properties and a temperature increase of 100 K, the increase in the confining pressure from Equation (4–180) is $\Delta p = 300$ MPa. The implication is that changes in temperature can lead to very large changes in pressure or stress.

When temperature changes occur, the laws of linear elasticity must be modified to include the thermally associated changes in volume. We have seen that a temperature change ΔT produces a volume change $\Delta v = v\alpha_v \Delta T$ in an unconstrained body. This change in volume is accompanied by the strains

$$\varepsilon_1 = \varepsilon_2 = \varepsilon_3 = \frac{-1}{3}\alpha_v \Delta T \tag{4–181}$$

if the body is isotropic. The minus sign on the right side of Equation (4–181) arises because of the sign convention that treats contraction strains as positive. The *linear coefficient of thermal expansion* α_l relates the thermally induced strains to the temperature change,

$$\varepsilon_1 = \varepsilon_2 = \varepsilon_3 = -\alpha_l \Delta T \quad \text{or} \quad \alpha_l = \frac{1}{3}\alpha_v, \tag{4–182}$$

so that the linear coefficient of thermal expansion is the change in the strain in the material per degree change in temperature.

The total strain in a body is the sum of the stress-associated strains and the temperature-associated strains. The stress-associated strains have been written in Equations (3–4) to (3–6). If to these we add the thermal strains of Equation (4–182), we obtain the total strain

$$\varepsilon_1 = \frac{1}{E}(\sigma_1 - \nu\sigma_2 - \nu\sigma_3) - \alpha_l \Delta T \tag{4–183}$$

$$\varepsilon_2 = \frac{1}{E}(-\nu\sigma_1 + \sigma_2 - \nu\sigma_3) - \alpha_l \Delta T \tag{4–184}$$

$$\varepsilon_3 = \frac{1}{E}(-\nu\sigma_1 - \nu\sigma_2 + \sigma_3) - \alpha_l \Delta T. \tag{4–185}$$

For a state of uniaxial stress we take $\sigma_1 = \sigma$ and $\sigma_2 = \sigma_3 = 0$. From Equations (4–183) to (4–185) we obtain

$$\varepsilon_1 = \frac{\sigma}{E} - \alpha_l \Delta T \tag{4-186}$$

$$\varepsilon_2 = \varepsilon_3 = \frac{-v}{E}\sigma - \alpha_l \Delta T. \tag{4-187}$$

For plane stress, $\sigma_3 = 0$, and the equations of thermal elasticity reduce to

$$\varepsilon_1 = \frac{1}{E}(\sigma_1 - v\sigma_2) - \alpha_l \Delta T \tag{4-188}$$

$$\varepsilon_2 = \frac{1}{E}(\sigma_2 - v\sigma_1) - \alpha_l \Delta T \tag{4-189}$$

$$\varepsilon_3 = -\frac{v}{E}(\sigma_1 + \sigma_2) - \alpha_l \Delta T. \tag{4-190}$$

For a state of isotropic stress $\sigma_1 = \sigma_2 = \sigma_3 = p$, $\varepsilon_1 = \varepsilon_2 = \varepsilon_3 = \Delta/3$, and by adding Equations (4–183) to (4–185) we find

$$\Delta = \frac{3}{E}(1 - 2v)p - 3\alpha_l \Delta T. \tag{4-191}$$

We previously identified the isothermal compressibility in Equation (3–55) as

$$\beta = \frac{3}{E}(1 - 2v). \tag{4-192}$$

This together with $\alpha_l = \alpha_v/3$ and $\Delta = -dv/v$ shows that Equation (4–191) is equivalent to Equation (4–177).

In Section 4–14, Equation (4–89), we obtained the temperature distribution in a semi-infinite half-space due to time periodic variation of the surface temperature. Assuming that this half-space is a uniform elastic medium, we can determine the resultant *thermal stresses*. Take the half-space to be confined in the horizontal directions so that $\varepsilon_1 = \varepsilon_2 = 0$ and to be unconstrained in the vertical direction so that $\sigma_3 = 0$. From Equations (4–188) and (4–189) we find

$$\sigma_1 = \sigma_2 = \frac{E\alpha_l \Delta T}{1 - v}. \tag{4-193}$$

The temperature ΔT is measured relative to the temperature at which the stress is zero. For the periodic heating of a semi-infinite half-space we assume that at $T = T_0$, the average temperature, the stress is zero. Therefore substitution of Equation (4–89) into Equation (4–193) gives

$$\sigma_1 = \sigma_2 = \frac{E\alpha_l \Delta T}{(1 - v)} \exp\left(-y\sqrt{\frac{\omega}{2\kappa}}\right)\cos\left(\omega t - y\sqrt{\frac{\omega}{2\kappa}}\right), \tag{4-194}$$

where ΔT is the actual amplitude of the periodic surface temperature variation about the average temperature. The maximum thermal stress is obtained by setting $y = t = 0$ in Equation (4–194),

$$\sigma_{\max} = \frac{E\alpha_l \Delta T}{1 - v}. \tag{4-195}$$

Let us take as typical values for rock $E = 60$ GPa, $v = 0.25$, and $\alpha_l = 10^{-5}$ K^{-1}. If $\Delta T = 100$ K, we find that $\sigma_{\max} = 80$ MPa.

In Section 3–4 we determined the elastic stresses resulting from sedimentation and erosion. It was shown that the addition or removal of overburden caused significant deviatoric stresses. However, when overburden is added or removed, the temperature at a given depth changes, and as a result thermal stresses are generated. Because the equations of thermal elasticity are linear, the thermal stresses can be added to those previously obtained.

We first consider near-surface rocks that have been buried to a depth h. If sufficient time has elapsed to reestablish the normal geothermal gradient β, the temperature increase of the surface rocks is βh. Again assuming no horizontal strain, the thermal stress from Equation (4–193) is

$$\sigma_1 = \sigma_2 = \frac{E\alpha_l \beta h}{(1 - v)}. \tag{4-196}$$

The elastic stress due to the addition of the overburden was given in Equation (3–24). Addition of the thermal stress to the elastic stress gives

$$\sigma_1 = \sigma_2 = \frac{h}{(1 - v)}(\rho g v + E\alpha_l \beta). \tag{4-197}$$

To determine the deviatoric stresses after sedimentation, we determine the pressure at depth h, noting that $\sigma_3 = \rho g h$; the result is

$$p = \frac{1}{3}(\sigma_1 + \sigma_2 + \sigma_3) = \frac{(1 + v)}{3(1 - v)}\rho g h + \frac{2}{3}\frac{E h \alpha_l \beta}{(1 - v)}. \tag{4-198}$$

The deviatoric stresses are obtained by subtracting this expression for p from Equation (4–197) and from $\sigma_3 = \rho g h$:

$$\sigma_1' = \sigma_2' = -\frac{(1 - 2v)}{3(1 - v)}\rho g h + \frac{E h \alpha_l \beta}{3(1 - v)} \tag{4-199}$$

$$\sigma_3' = \frac{2}{3}\frac{(1 - 2v)}{(1 - v)}\rho g h - \frac{2}{3}\frac{E h \alpha_l \beta}{(1 - v)}. \tag{4-200}$$

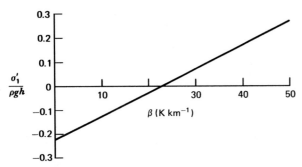

4–43 Differential stress resulting from the addition of h km of overburden to an initially unstressed surface.

The differential horizontal stresses due to the elastic effect are extensional; because of the thermal effect they are compressional. Figure 4–43 gives $\sigma_1'/\rho g h$ as a function of β for $E = 60$ GPa, $\nu = 0.25$, $\alpha_l = 10^{-5}$ K^{-1}, $g = 10$ m s^{-2}, and $\rho = 2700$ kg m^{-3}. The thermal effect is seen to be more important than the elastic effect for thermal gradients greater than 23 K km^{-1}. Because this is a typical thermal gradient in continental areas, the thermal and overburden stresses are likely to be comparable under most conditions of sedimentation.

Consider next what the surface stress is after h km of overburden have been eroded. As in Section 3–4 we assume that the initial stresses are lithostatic so that Equation (3–30) gives the nonthermal part of the surface stress. We also assume that a new thermal equilibrium has been established subsequent to the removal of surface material. After the erosion of h km the change in temperature of the surface rocks is $-\beta h$. Again assuming no horizontal strain, we find the surface thermal stress from Equation (4–193) to be

$$\sigma_1 = \sigma_2 = -\frac{E\alpha_l \beta h}{(1 - \nu)}. \tag{4–201}$$

Adding the surface thermal stress to the surface elastic stress due to the removal of overburden, Equation (3–30), we obtain

$$\sigma_1 = \sigma_2 = \frac{h}{(1 - \nu)}[(1 - 2\nu)\rho g - E\alpha_l \beta]. \tag{4–202}$$

As a consequence of erosion, the elastic effect causes surface compression, and the thermal effect causes surface extension. If $\sigma_1'/\rho g h$ given in Figure 4–43 is multiplied by -3, one obtains $\sigma_1/\rho g h$ due to erosion. Thus it is clear that surface thermal stress and surface stresses resulting directly from removal of surface material are

comparable for typical values of the geothermal gradient.

PROBLEM 4–49 Determine the surface stress after 10 km of erosion. Take $E = 60$ GPa, $\nu = 0.25$, $\alpha_l = 10^{-5}$ K^{-1}, $\rho = 2700$ kg m^{-3}, and $\beta = 20$ K km^{-1}.

PROBLEM 4–50 If $\alpha_v = 3 \times 10^{-5}$ K^{-1} and the temperature of the Earth increased by 100 K, what would the change in radius be?

There is an important distinction between *renewable* and *nonrenewable stresses*. Thermal and erosional stresses are permanently relieved by irreversible creep and are examples of nonrenewable stresses. Examples of renewable stresses include bending stresses in the lithosphere, the lithosphere stresses that drive plate tectonics, and the crustal stresses that support mountain ranges. These stresses are not relieved by a small amount of creep.

4–23 Ocean Floor Topography

We previously solved for the thermal structure of the lithosphere adjacent to ocean ridges in Equation (4–125). We also demonstrated in the previous section that the cooling of the oceanic lithosphere causes the density of lithospheric rock to increase. The relationship between density and temperature is given in Equation (4–179). In addition, we introduced the principle of *isostasy* in Section 2–2.

The principle of isostasy states that any vertical column of material has the same mass per unit area between the surface and some *depth of compensation*. This is equivalent to the assumption that the lithostatic pressure at some depth is the same over a large horizontal area. However, as shown in Figure 4–23, columns of mantle material at different ages do not contain the same mass per unit area. The older column contains more dense, cold lithosphere than the younger one; the extra weight of the older lithosphere causes it to subside. Mantle material below the lithosphere flows away to accommodate the subsidence, and the ocean fills in the hole created at the surface. Figure 4–44 shows the oceanic lithosphere with the overlying ocean increasing in depth with distance from the ridge. The two columns of ages t_1 and t_2 now have the same mass per unit area because the older column contains more water to offset the added weight of dense lithospheric rock.

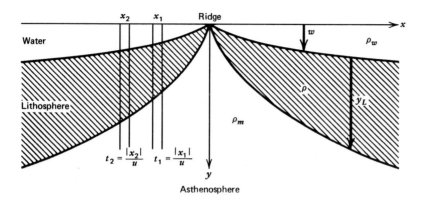

4–44 The principle of isostasy requires the ocean to deepen with age to offset the thermal contraction in the lithosphere.

The ability of the mantle rock beneath the lithosphere to behave as a fluid on geological time scales is the key to the isostatic adjustment of the oceanic lithosphere. By calculating the masses per unit area in vertical columns extending from the surface to the base of the lithosphere and requiring that these be the same for columns of all ages, we can derive a formula for the depth of the ocean floor w as a function of age t or distance from the ridge x.

The mass per unit area in a column of any age is

$$\int_0^{y_L} \rho \, dy + w\rho_w,$$

where y_L is the thickness of the lithosphere and ρ_w is the density of water. At the ridge crest, $\rho = \rho_m$ the deep mantle density, and the mass of a column of vertical height $w + y_L$ is $\rho_m(w + y_L)$. Isostasy requires that

$$\rho_m(w + y_L) = \int_0^{y_L} \rho \, dy + w\rho_w \qquad (4\text{–}203)$$

or

$$w(\rho_w - \rho_m) + \int_0^{y_L} (\rho - \rho_m) \, dy = 0. \qquad (4\text{–}204)$$

The first term in Equation (4–204) represents a negative mass because the water with density ρ_w is less dense than the mantle rock that it has replaced because of the subsidence of the seafloor a distance w. The second term in the equation represents a positive mass because thermal contraction in the cooling lithosphere causes the density ρ to be higher than the reference hot mantle rock density ρ_m. Introducing the volume coeffi-

cient of thermal expansion from Equation (4–179), we can write

$$\rho - \rho_m = \rho_m \alpha_v (T_1 - T). \qquad (4\text{–}205)$$

Upon substitution of the temperature profile from Equation (4–125) into Equation (4–205) and that result into Equation (4–204), we obtain

$$w(\rho_m - \rho_w) = \rho_m \alpha_v (T_1 - T_0)$$
$$\times \int_0^\infty \text{erfc}\left[\frac{y}{2}\left(\frac{u_0}{\kappa x}\right)^{1/2}\right] dy. \quad (4\text{–}206)$$

Because $\rho \to \rho_m$ and $T \to T_1$ at the base of the lithosphere, the limit on the integral has been changed from $y = y_L$ to $y = \infty$. We can rewrite Equation (4–206) by using the similarity variable $\eta = (y/2) \times (u_0/\kappa x)^{1/2}$ so that

$$w = \frac{2\rho_m \alpha_v (T_1 - T_0)}{(\rho_m - \rho_w)}\left(\frac{\kappa x}{u_0}\right)^{1/2} \int_0^\infty \text{erfc}(\eta) \, d\eta. \qquad (4\text{–}207)$$

The definite integral has the value

$$\int_0^\infty \text{erfc}(\eta) \, d\eta = \frac{1}{\sqrt{\pi}}, \qquad (4\text{–}208)$$

so that

$$w = \frac{2\rho_m \alpha_v (T_1 - T_0)}{(\rho_m - \rho_w)}\left(\frac{\kappa x}{\pi u_0}\right)^{1/2}. \qquad (4\text{–}209)$$

Equation (4–209) predicts that the depth of the ocean increases with the square root of the distance from the ridge or the square root of the age of the ocean floor. This theoretical result is compared with seafloor depths in Figure 4–45. The results shown are from Deep Sea Drilling Project (DSDP) and Ocean Drilling Project

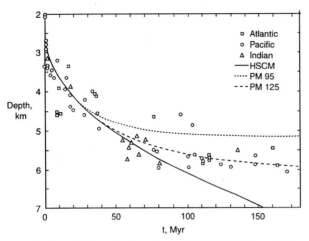

4–45 Seafloor depth as a function of age in the Atlantic, Pacific, and Indian Oceans. Data are from DSDP and ODP drill sites on normal ocean crust and depths have been corrected for sediment cover (Johnson and Carlson, 1992). Comparisons are made with the half-space cooling model (HSCM) from Equation (4–209) and the plate model from Equation (4–211) with $y_{L0} = 95$ km (PM 95) and $y_{L0} = 125$ km (PM 125).

(ODP) drill sites (Johnson and Carlson, 1992). Corrections have been made for sediment cover and results are given for the Atlantic, Pacific, and Indian Oceans. Predicted seafloor depths are included for the half-space cooling model, Equation (4–209), assuming $\rho_m = 3300$ kg m^{-3}, $\rho_w = 1000$ kg m^{-3}, $\kappa = 1$ mm^2 s^{-1}, $T_1 - T_0 = 1300$ K, and $\alpha_v = 3 \times 10^{-5}$K^{-1}. In addition, the ridge depth is 2.5 km. For seafloor ages greater than about 80 Myr, the predicted values are systematically deeper than the observed values. This divergence is taken as evidence of the basal heating of old oceanic lithosphere.

A model that approximates basal heating of the lithosphere is the plate cooling model introduced in Section 4–17. The temperature distribution in the oceanic lithosphere according to the plate cooling model was given in Equation (4–130). Substitution of this temperature distribution into Equation (4–205) and further substitution of the resulting density distribution into Equation (4–204) give

$$w = \frac{\rho_m \alpha_v (T_1 - T_0) y_{L0}}{2(\rho_m - \rho_w)} \left[1 - \frac{4}{\pi} \int_0^1 \sum_{n=1}^{\infty} \frac{1}{n} \exp \left(-\frac{\kappa n^2 \pi^2 t}{y_{L0}^2} \right) \sin(n\pi y') \, dy' \right]. \quad (4\text{--}210)$$

Evaluation of the integral in Equation (4–210) leads to

$$w = \frac{\rho_m \alpha_v (T_1 - T_0) y_{L0}}{(\rho_m - \rho_w)} \left[\frac{1}{2} - \frac{4}{\pi^2} \sum_{m=0}^{\infty} \frac{1}{(1+2m)^2} \exp \left(-\frac{\kappa (1+2m)^2 \pi^2 t}{y_{L0}^2} \right) \right]. \quad (4\text{--}211)$$

Only the terms with $n = 1, 3, 5, \ldots$ in Equation (4–210) contribute to the result in Equation (4–211) since the terms with $n = 2, 4, 6, \ldots$ integrate to zero. For large times, $t \gg y_{L0}^2/\kappa$, the equilibrium depth w_e is given by

$$w_e = \frac{\rho_m \alpha_v (T_1 - T_0) y_{L0}}{2(\rho_m - \rho_w)}. \quad (4\text{--}212)$$

This is the equilibrium depth of the old ocean basins beneath the ridge crests. This relation provides a constraint on the thickness of the plate in the plate cooling model. In comparing the predictions of this model with observations, we consider plate thicknesses $y_{L0} = 95$ and 125 km. With $\rho_m = 3300$ kg m^{-3}, $\rho_w = 1000$ kg m^{-3}, $T_1 - T_0 = 1300$ K, and $\alpha_v = 3 \times 10^{-5}$ K^{-1}, we find from Equation (4–212) that $w_e = 2.7$ km for $y_{L0} = 95$ km and $w_e = 3.5$ km for $y_{L0} = 125$ km. With the depth of ocean ridges equal to 2.5 km, the corresponding equilibrium depths of ocean basins are 5.2 km and 6 km, respectively.

We can approximate the deviations of bathymetry from the half-space cooling result given in Equation (4–209) by retaining the first term of the expansion given in Equation (4–211), with the result

$$w = \frac{\rho_m \alpha_v (T_1 - T_0) y_{L0}}{\rho_m - \rho_w} \left[\frac{1}{2} - \frac{4}{\pi^2} \exp \left(-\frac{\kappa \pi^2 t}{y_{L0}^2} \right) \right]. \quad (4\text{--}213)$$

The $n = 2$ term in the temperature distribution, Equation (4–130), does not contribute to the bathymetry. Seafloor subsidence predicted by the plate cooling model is compared with observations and the half-space cooling model in Figure 4–45 for plate thicknesses of $y_{L0} = 95$ km (PM 95) and $y_{L0} = 125$ km (PM 125). The results for a plate thickness $y_{L0} = 125$ km are in excellent agreement with the data. While a thickness of $y_{L0} = 95$ km is in good agreement with the heat flow data (as shown in Figure 4–25), a thickness of $y_{L0} = 125$ km is in good agreement with the subsidence data (Figure 4–45). Because there is generally less scatter in seafloor bathymetry than in heat flow, we prefer the value $y_{L0} = 125$ km although we recognize there is considerable uncertainty in the choice.

The plate model is clearly an idealization of the oceanic lithosphere. There is no well-defined "lower plate boundary" in the mantle. The flattening of the cooling curves can be attributed to the basal heating of the oceanic lithosphere. For $y_{L0} = 125$ km, the required basal heating from Equation (4–135) is $q_m = 34$ mW m^{-2}, which is quite close to our preferred value for the basal heating of the continental lithosphere, $q_m = 28$ mW m^{-2}. The difference between the basal heating flux $q_m = 34$ mW m^{-2} and the mean oceanic heat flux $q_o = 101$ mW m^{-2} we will refer to as the plate tectonic or subduction flux $q_s = 67$ mW m^{-2}.

From this value of the basal heat flux for the continental lithosphere and with the area of the continents including continental margins $A_c = 2.0 \times 10^8$ km^2, we find that the total basal heating of the continental lithosphere is $Q_{mc} = 0.56 \times 10^{13}$ W. From the value of the basal heat flux for the oceanic lithosphere and with the area of the oceans including marginal basins $A_o = 3.1 \times 10^8$ km^2, we find that the total basal heating of the oceanic lithosphere is $Q_{mo} = 1.05 \times 10^{13}$ W. Thus, the basal heating of the entire lithosphere is $Q_m = 1.61 \times 10^{13}$ W, which represents 36% of the total global heat flux $Q_g = 4.43 \times 10^{13}$ W. From the estimate of the fraction of the oceanic heat flow directly associated with subduction ($q_s = 67$ mW m^{-2}) we find that the total heat flux associated with subduction is $Q_s = 2.08 \times 10^{13}$ W. The total global heat flux can be divided into three contributions: (1) radiogenic heat production in the continental crust $Q_r = 0.74 \times 10^{13}$ W (16.7%), (2) basal heating of the lithosphere $Q_m = 1.61 \times 10^{13}$ W (36.3%), and (3) subduction of the oceanic lithosphere $Q_s = 2.08 \times 10^{13}$ W (47%).

These results show that basal heating of the lithosphere is quantitatively large. There are two competing hypotheses for this basal heating. The first is heat transfer from mantle plumes impinging on the base of the lithosphere, and the second is small-scale or secondary convection in the lower lithosphere and underlying asthenosphere. It is generally accepted that mantle plumes are a source of basal heating so that the only question is the magnitude of this heating. We will address this question in Chapter 6. Small-scale or secondary convection is associated with an instability in the lower lithosphere arising from the strong temperature dependence of the lithospheric viscosity. It is a form of delamination or foundering of the lithosphere and contrasts with the instability of the entire lithosphere that is manifest as lithospheric subduction at an ocean trench. The amount of heat transported by secondary convection near the base of the lithosphere is highly uncertain.

PROBLEM 4–51 Assume that the temperature in the subducting lithosphere is given by Equation (4–131). Show that the plate tectonic heat flux associated with subduction Q_s is given by

$$Q_s = \frac{1}{2}\rho_m c y_{L0} F_s,$$

where ρ_m is mantle density, c is the specific heat, and F_s is rate of seafloor subduction. Determine Q_s taking $\rho_m = 3300$ kg m^{-1}, $c = 1$ kJ kg^{-1} K^{-1}, $y_{L0} = 125$ km, and $F_s = 0.090$ m^2 s^{-1}.

PROBLEM 4–52 The ocean ridges are made up of a series of parallel segments connected by transform faults, as shown in Figure 1–13. Because of the difference of age there is a vertical offset on the fracture zones. Assuming the theory just derived is applicable, what is the vertical offset (a) at the ridge crest and (b) 100 km from the ridge crest in Figure 4–46 ($\rho_m = 3300$ kg m^{-3}, $\kappa = 1$ mm^2 s^{-1}, $\alpha_v = 3 \times 10^{-5}$ K^{-1}, $T_1 - T_0 = 1300$ K, $u = 50$ mm yr^{-1}).

PROBLEM 4–53 Because of its cooling, the seafloor subsides relative to a continent at a passive continental margin. Determine the velocity of subsidence if $\rho_m = 3300$ kg m^{-3}, $\kappa = 1$ mm^2 s^{-1}, $T_1 - T_0 = 1300$ K, $\alpha_v = 3 \times 10^{-5}$ K^{-1}, and the age is 20 Ma.

PROBLEM 4–54 The influence of a small amount of partial melt on the lithosphere–asthenosphere boundary has been considered in Problem 4–43. Determine the contribution of this small degree of partial melt

4–46 Diagram for Problem 4-52.

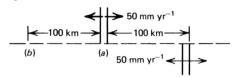

to the subsidence of the seafloor as a function of age. The density of the magma is ρ_l, its mass fraction is x, and the latent heat of fusion is L. If $x = 0.05$, $L = 400 \text{ kJ kg}^{-1}$, $c = 1 \text{ kJ kg}^{-1} \text{ K}^{-1}$, $T_1 - T_0 = 1350 \text{ K}$, $\kappa = 1 \text{ mm}^2 \text{ s}^{-1}$, $\rho_m = 3300 \text{ kg m}^{-3}$, $\rho_l = 2800 \text{ kg m}^{-3}$, and $\alpha_v = 3 \times 10^{-5} \text{ K}^{-1}$, determine the ratio of the subsidence due to solidification to the subsidence due to thermal contraction.

4-24 Changes in Sea Level

Changes in sea level are well documented from studies of sedimentation. On short time scales ($\sim 10^4$ to 10^5 years) sea level changes can be explained by variations in the volume of the polar ice sheets. However, on longer time scales ($\sim 10^7$ to 10^8 years) the magnitudes of sea level changes are too large to be understood in this way. If the polar ice sheets were completely melted, the water added to the oceans would increase sea level by about 80 m. Yet compared with its present level, the sea has been hundreds of meters higher during the last 550 Ma, as shown by the record in Figure 4–47 (the value at $t = 0$ of 80 m above present sea level accounts for the water in the polar ice sheets). Sea level in the Cretaceous (80 Ma) was 300 m higher than it is today, and water flooded about 40% of the present area of the continents. These large, long-term changes in sea level are attributed to changes in the average depth $\bar{w}$

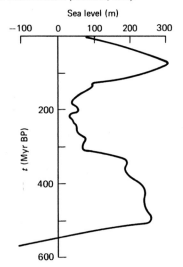

4–47 Height of the sea surface in the past relative to present sea level from studies of sedimentation (Vail et al., 1978).

of the seafloor below the level of the ridge crests. If this average depth decreases as a consequence of a decrease in the average age τ of subduction, the volume of water contained in the deep ocean basins decreases and the height of the sea above the ridge crests h, that is, sea level, increases. This is supported by magnetic anomaly studies that show that the Cretaceous was a time when there were more ridges and the ridges were spreading more rapidly than at present. Since 75 Ma, an extensive ridge system has been subducted beneath western North America. Thus, on average, the present seafloor is older and deeper than the seafloor at 80 Ma. The present deep ocean basins hold more water than the basins during the Cretaceous, and the sea surface today lies well below its level in that period.

Changes in the configuration of the seafloor cannot change the total volume of water in the oceans. If we neglect the changes that occur in the area of the oceans as sea level changes, a change $\delta \bar{w}$ in the mean depth of the ocean basins below the ridge crests produces an opposite change δh in the height of the sea above the ridge crests

$$\delta h = -\delta \bar{w}. \tag{4–214}$$

The mean depth of the ocean floor is

$$\bar{w} = \frac{1}{\tau} \int_0^\tau w \, dt. \tag{4–215}$$

By substituting for w from Equation (4–209), we get

$$\bar{w} = \frac{2\rho_m \alpha_v (T_1 - T_0)}{(\rho_m - \rho_w)} \left(\frac{\kappa}{\pi}\right)^{1/2} \left(\frac{1}{\tau} \int_0^\tau t^{1/2} \, dt\right)$$

$$= \frac{4}{3} \frac{\rho_m \alpha_v (T_1 - T_0)}{(\rho_m - \rho_w)} \left(\frac{\kappa \tau}{\pi}\right)^{1/2}. \tag{4–216}$$

The mean depth of the ocean basins is directly proportional to the square root of the mean age of subduction. Sea level changes are therefore related to changes in the average age at which subduction occurs by

$$\delta h = -\frac{4}{3} \frac{\rho_m \alpha_v (T_1 - T_0)}{(\rho_m - \rho_w)} \left(\frac{\kappa}{\pi}\right)^{1/2} \delta(\tau^{1/2}). \tag{4–217}$$

Equation (4–217) enables us to estimate the mean age of subduction during the Cretaceous. We take $\rho_m = 3300 \text{ kg m}^{-3}$, $\alpha_v = 3 \times 10^{-5} \text{ K}^{-1}$, $\kappa = 1 \text{ mm}^2 \text{ s}^{-1}$, $T_1 - T_0 = 1300 \text{ K}$, $\rho_w = 1000 \text{ kg m}^{-3}$, the present mean age of subduction equal to 120.8 Myr, and $\delta h = 220$ m

(80 m of the 300-m sea level rise is attributed to water presently locked up in polar ice). The average age at which seafloor subducted in the Cretaceous is found to be 100 Myr.

Sea level changes due to changes in the mean subduction age of the seafloor can be used to infer past variations in the mean oceanic heat flow $\bar{q}_0$. By combining Equations (4–128) and (4–217), we find

$$\delta\left(\frac{1}{\bar{q}_0}\right) = -\frac{3\pi}{8}\frac{(\rho_m - \rho_w)}{k\rho_m\alpha_v(T_1 - T_0)^2}\delta h. \qquad (4\text{–}218)$$

We will see that the changes that have occurred in mean oceanic heat flow $\delta\bar{q}_0$ are small compared with the present value $\bar{q}_{0\rho}$. The left side of Equation (4–218) can therefore be approximated by $-\delta\bar{q}_0/\bar{q}_{0\rho}^2$, and we can write the percentage variation in mean oceanic heat flow as

$$\frac{\delta\bar{q}_0}{\bar{q}_{0\rho}} = \frac{3\pi}{8}\frac{(\rho_m - \rho_w)\bar{q}_{0\rho}}{k\rho_m\alpha_v(T_1 - T_0)^2}\delta h. \qquad (4\text{–}219)$$

Higher sea levels in the past imply larger values of the mean oceanic heat flux. This is expected from the association of higher sea levels with a younger seafloor. With the previous parameter values, $k = 3.3$ W m^{-1} K^{-1}, and $\bar{q}_{0\rho} = 87$ mW m^{-2}, we find that a 26 m increase in sea level is associated with a 1% increase in the mean oceanic heat flux.

The fractional changes in average oceanic heat flow inferred from the sea level data of Figure 4–47 with Equation (4–219) and the above parameter values are shown in Figure 4–48. The figure also shows the increase in average oceanic heat flow that would be expected in the past if the heat lost through the oceans was proportional to the increased rate of heat production from the radioactive isotopes in the mantle – see Equation (4–8). The inferred fractional changes that have occurred in the average oceanic heat flux during the past 550 Ma are about 10%. These variations in oceanic heat flow are not attributable to larger radiogenic heat production rates in the past. Instead they are statistical variations associated with changes in the geometry and the mean spreading rate of the oceanic ridge system.

PROBLEM 4-55 What would be the decrease in sea level due to a 10% reduction in the area of the continents? Assume the depth of deep ocean basins to be 5 km.

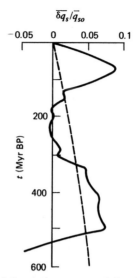

4–48 Fractional variations in the mean oceanic heat flow in the past 550 Ma inferred from the sea level data in Figure 4–47 and Equation (4–219). The dashed line is the expected increase in the mean oceanic heat flow due to the larger rate of radiogenic heat production in the past.

4-25 **Thermal and Subsidence History of Sedimentary Basins**

Subsidence of the Earth's surface often results in the formation of *sedimentary basins*. We can explain the subsidence history of many sedimentary basins by essentially the same model that we used to understand the cooling, thickening, and subsidence of the oceanic lithosphere. The model is illustrated in Figure 4–49. Consider a region of the Earth that is hot, either because of seafloor spreading or extensive volcanism. Initially ($t = 0$) there is no sediment, and the basement has a temperature T_1 and a density ρ_m. Surface cooling causes subsidence as the basement rocks cool and contract. We assume that sediments fill the basin caused by the subsidence; that is, the region $0 < y < y_{SB}$. This assumption requires an adequate supply of

4–49 Sedimentary basin model.

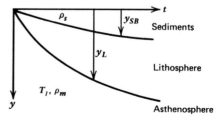

sediment to prevent the formation of a deep ocean basin.

As long as the thickness of the sediment y_{SB} is much smaller than the lithosphere thickness y_L, we can carry over the results of the cooling oceanic lithosphere calculation. Therefore the depth of the sedimentary basin is given by Equation (4–209) with ρ_s, the density of the sediments, replacing ρ_w, and t, the age of the basin, replacing x/u_0, the age of the oceanic lithosphere; the result is

$$y_{SB} = \frac{2\rho_m \alpha_m (T_1 - T_0)}{(\rho_m - \rho_s)} \left(\frac{\kappa_m t}{\pi} \right)^{1/2}, \qquad (4\text{–}220)$$

where the subscripts m on α and κ emphasize that these properties refer to the mantle rocks and not the sediment (α_m is α_v for the mantle rocks). The depth of the sedimentary basin is proportional to the square root of time. The subsidence of the basin is caused primarily by the cooling of the basement or lithospheric rocks. The cooling of the sedimentary rocks is a negligible effect when $y_{SB}/y_L \ll 1$.

It is of interest to determine the subsidence history of a sedimentary layer that was deposited at a time t_s after the initiation of subsidence. At the time t_s the basement lies at a depth given by setting $t = t_s$ in Equation (4–220). Assuming no compaction of the sediments, the layers deposited at time t_s will always be this distance above the basement. However, the depth of the basement at time t is given directly by Equation (4–220). Therefore the depth to the sediments deposited at time $t = t_s$ at a later time t, denoted by y_s, is given by the difference between the depth to basement at t and t_s; that is,

$$y_s = \frac{2\rho_m \alpha_m (T_1 - T_0)}{(\rho_m - \rho_s)} \left(\frac{\kappa_m}{\pi} \right)^{1/2} \left(t^{1/2} - t_s^{1/2} \right). \quad (4\text{–}221)$$

The depth to sedimentary layers deposited at various times is given in Figure 4–50. These curves were calculated assuming that $\rho_m = 3300 \, \text{kg m}^{-3}$, $\kappa_m = 1 \, \text{mm}^2$ s^{-1}, $T_1 - T_0 = 1300 \, \text{K}$, $\alpha_m = 3 \times 10^{-5} \, \text{K}^{-1}$, and $\rho_s = 2500 \, \text{kg m}^{-3}$.

Because the sedimentary layer is thin, the temperature–depth profile is essentially linear in the sediments. The sedimentary layer must transport the heat from the cooling basement rocks. Denoting this heat flux by q_0,

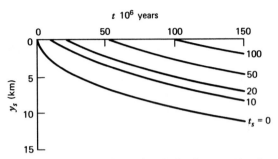

4–50 Depths to sedimentary layers deposited at times t_s as functions of time.

we have, from Equation (4–127),

$$q_0 = \frac{k_m (T_1 - T_0)}{\sqrt{\pi \kappa_m t}}. \qquad (4\text{–}222)$$

From Fourier's law of heat conduction, we know that in the sediments

$$q_0 = k_s \left(\frac{dT}{dy} \right)_s, \qquad (4\text{–}223)$$

where $(dT/dy)_s$ is the constant geothermal gradient in the sediments. By combining these last two equations, we get

$$\left(\frac{dT}{dy} \right)_s = \frac{k_m}{k_s} \frac{(T_1 - T_0)}{\sqrt{\pi \kappa_m t}}. \qquad (4\text{–}224)$$

Thus the temperature distribution in the sediments is

$$T_s = T_0 + \frac{k_m}{k_s} \frac{(T_1 - T_0)}{\sqrt{\pi \kappa_m t}} y. \qquad (4\text{–}225)$$

The temperature of a sedimentary layer deposited at time t_s at a subsequent time t is given by substituting Equation (4–221) into Equation (4–225)

$$T_{SL} = T_0 + \frac{2}{\pi} \frac{k_m}{k_s} \frac{\rho_m \alpha_m (T_1 - T_0)^2}{(\rho_m - \rho_s)} \left(1 - \sqrt{\frac{t_s}{t}} \right). \qquad (4\text{–}226)$$

The thermal history of a sedimentary layer can be used to determine whether organic material in the sediments has been converted to petroleum.

The Los Angeles basin is a relatively small sedimentary basin with a width of about 50 km and a length of about 75 km. The basin is a pull-apart structure associated with the San Andreas fault system. It is probably similar to the small spreading centers that offset transform faults in the Gulf of California. During the

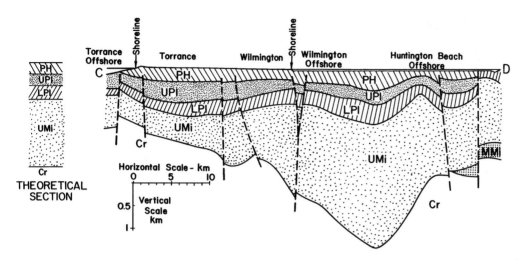

4–51 Cross section of the southwest block of the Los Angeles basin. (From *California Oil and Gas Fields*, Vol. 2, California Division of Oil and Gas, Report TR12, Sacramento, 1974). The sedimentary layers are Pleistocene–Holocene (PH), Upper Pliocene (UPl), Lower Pliocene (LPl), Upper Miocene (UMi), Middle Miocene (MMi), and Cretaceous or older basement (Cr). Also shown is the theoretical section from Equation (4–221).

initiation of the basin subsidence, volcanism was occurring. Volcanic rocks from drill holes in the basin have ages of 10 to 15 Ma. This volcanism was probably similar to the volcanism presently occurring in the Imperial Valley of southern California.

Since volcanism ceased at about 10 Ma, subsidence has continued. It is reasonable to assume that the volcanism thinned the lithosphere and that the subsequent subsidence is associated with the conductive cooling and thickening of the lithosphere. The structure of the basin is clearly complex, with considerable faulting. Although these faults add complexity to the basin, they are also likely to allow free vertical subsidence of the various fault-bounded blocks.

Let us apply our analysis of the thermal subsidence of sedimentary basins to the southwest block, which extends roughly from Santa Monica to Long Beach and is the site of several major oil fields. A cross section of this part of the basin is given in Figure 4–51. The depths of various sedimentary units in the Wilmington oil field are given as a function of their ages in Figure 4–52. The predicted depths of these sedimentary units are given by Equation (4–221). Taking $\rho_m = 3300$ kg m^{-3}, $\rho_s = 2500$ kg m^{-3}, $\alpha_m = 3 \times 10^{-5}$ K^{-1}, $T_1 - T_0 = 1200$ K, and $\kappa_m = 1$ mm^2 s^{-1}, we obtain the solid curve given

in Figure 4–52. The predicted theoretical section is also given in Figure 4–52. Reasonably good agreement is obtained, although considerable tectonic structure is clearly associated with the formation of the basin and subsequent motion on the San Andreas fault now located to the east.

The present thermal gradient in the basin is predicted by Equation (4–224). With $k_m = 3.3$ W m^{-1} K^{-1}, $k_s = 2$ W m^{-1} K^{-1}, and the other parameter values as before, we find $(dT/dy)_s = 59$ K km^{-1}. The measured surface thermal gradients in the Wilmington oil field are in the range 48 to 56 K km^{-1}. Again reasonably good agreement is obtained.

PROBLEM 4–56 Assume that the continental lithosphere satisfies the half-space cooling model. If a continental region has an age of 1.5×10^9 years, how much subsidence would have been expected to occur

4–52 The crosses are the depths to the boundaries between stratigraphic units in the Wilmington oil field at the ages of the boundaries. The solid line is the subsidence predicted by Equation (4–221).

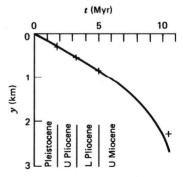

in the last 300 Ma? Take $\rho_m = 3300$ kg m^{-3}, $\kappa = 1$ mm^2 s^{-1}, $T_m - T_0 = 1300$ K, and $\alpha_v = 3 \times 10^{-5}$ K^{-1}. Assume that the subsiding lithosphere is being covered to sea level with sediments of density $\rho_s = 2500$ kg m^{-3}.

PROBLEM 4–57 If petroleum formation requires temperatures between 380 and 430 K, how deep would you drill in a sedimentary basin 20 Ma old? Assume $T_0 = 285$ K, $T_1 = 1600$ K, $\kappa_m = 1$ mm^2 s^{-1}, $k_s = 2$ W m^{-1} K^{-1}, and $k_m = 3.3$ W m^{-1} K^{-1}.

In Section 2–2 we introduced the crustal stretching model to explain the subsidence of a sedimentary basin. In this model the subsidence is caused by the thinning of the continental crust. The model was illustrated in Figure 2–4; a section of continental crust with an initial width w_0 stretched by a stretching factor α to a final width $w_b = \alpha w_0$ from Equation (2–6). In order to conserve the volume of the crust the initial thickness of the crust h_{cc} is reduced to $h_{cb} = h_{cc}/\alpha$ as given by Equation (2–8). The resulting depth of the sedimentary basin h_{sb} is given by Equation (2–10).

We now extend the crustal stretching model by assuming that the continental lithosphere within the sedimentary basin is also mechanically stretched and thinned by the same stretching factor α as the crust. We assume that the shape of the temperature profile in the lithosphere remains unchanged but that its thickness is reduced by the factor $1/\alpha$.

We assume that before stretching, the temperature distribution in the lithosphere is given by Equation (4–124). By introducing the thickness of the unstretched lithosphere y_{L0} from Equation (4–126), we can rewrite Equation (4–124) as

$$\frac{T_1 - T}{T_1 - T_0} = \text{erfc}(1.16y/y_{L0}). \tag{4–227}$$

In order to conserve the volume of the lithosphere we require

$$y_{Lb} = \frac{y_{L0}}{\alpha}, \tag{4–228}$$

where y_{Lb} is the thickness of the stretched lithosphere. The temperature distribution in the stretched and thinned lithosphere is given by

$$\frac{T_1 - T}{T_1 - T_0} = \text{erfc}(1.16y\alpha/y_{L0}). \tag{4–229}$$

Whereas the thinning of the crust produces subsidence,

the thinning of the lithosphere inhibits subsidence. The thinned continental lithosphere is hotter and less dense than the original lithosphere leading to a thermal uplift.

Application of the principle of isostasy to the base of the continental lithosphere gives

$$(\rho_{cc} - \rho_m)h_{cc} - \rho_m\alpha_v(T_1 - T_0)\int_0^\infty \text{erfc}\left(\frac{1.16y}{y_{L0}}\right) dy$$

$$= (\rho_s - \rho_m)h_{sb} + (\rho_{cc} - \rho_m)\frac{h_{cc}}{\alpha} - \rho_m\alpha_v(T_1 - T_0)$$

$$\times \int_0^\infty \text{erfc}\left(\frac{1.16\alpha y}{y_{L0}}\right) dy. \tag{4–230}$$

Evaluation of the integrals using Equation (4–208) gives the thickness of the sedimentary basin h_{sb} in terms of the stretching factor α as

$$h_{sb} = \left[\frac{(\rho_m - \rho_{cc})}{(\rho_m - \rho_s)}h_{cc} - \frac{1}{1.16\sqrt{\pi}}\frac{\rho_m\alpha_v(T_1 - T_0)y_{L0}}{(\rho_m - \rho_s)}\right]$$

$$\times\left(1 - \frac{1}{\alpha}\right). \tag{4–231}$$

The dependence of the basin thickness on α is the same as that given in Equation (2–10) for crustal thinning alone. Taking the same parameter values used in Section 2–2 ($h_{cc} = 35$ km, $\rho_m = 3300$ kg m^{-3}, $\rho_{cc} = 2800$ kg m^{-3}, and $\rho_s = 2500$ kg m^{-3}) along with $\alpha_v = 3 \times 10^{-5}$ K^{-1}, $T_1 - T_0 = 1300$ K, and $y_{L0} = 150$ km, we find that $h_{sb} = 10.1$ km in the limit $\alpha \to \infty$. The subsidence associated with crustal thinning alone would be 22 km, so the lithospheric thinning reduces the subsidence considerably.

In the analysis just given, we implicitly assumed that crust and lithosphere are stretched in a time interval that is short compared with the thermal time constant κ/y_L^2 of the thinned lithosphere. After the initial thinning of the crust and lithosphere, the lithosphere will thicken because of the loss of heat to the surface. This cooling and thickening of the lithosphere will lead to further thermal subsidence. With the assumption that the temperature profile in the thickening lithosphere as a function of time is given by Equation (4–124), the thickness of the sedimentary basin as a function of time is given by

$$h_{sb} = h_{cc}\left(\frac{\rho_m - \rho_{cc}}{\rho_m - \rho_s}\right)\left(1 - \frac{1}{\alpha}\right) - \frac{y_{L0}\rho_m\alpha_v(T_1 - T_0)}{1.16\sqrt{\pi}(\rho_m - \rho_s)}$$

$$\times\left[1 - \left(\frac{1}{\alpha^2} + \frac{2.32^2\kappa t}{y_{L0}^2}\right)^{1/2}\right]. \tag{4–232}$$

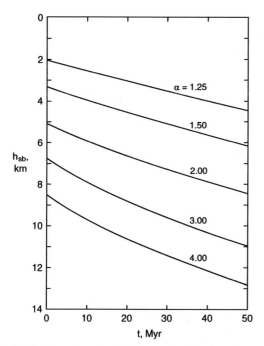

4-53 Depth of the sedimentary basin h_{sb} as a function of age t for several values of the stretching factor α.

The thickness of the sedimentary basin as a function of time is given in Figure 4–53 for several values of the stretching factor and for the same parameters as before with $\kappa = 1 \, \text{mm}^2 \, \text{s}^{-1}$. When the thickness of the thinned lithosphere increases to its initial value y_{L0}, the total subsidence will be that given by Equation (2–10).

PROBLEM 4–58 Assume that the continental crust and lithosphere have been stretched by a factor $\alpha = 2$. Taking $h_{cc} = 35 \, \text{km}$, $y_{L0} = 125 \, \text{km}$, $\rho_m = 3300 \, \text{kg m}^{-3}$, $\rho_{cc} = 2750 \, \text{kg m}^{-3}$, $\rho_s = 2550 \, \text{kg m}^{-3}$, $\alpha_v = 3 \times 10^{-5} \, \text{K}^{-1}$, and $T_1 - T_0 = 1300 \, \text{K}$, determine the depth of the sedimentary basin. What is the depth of the sedimentary basin when the thermal lithosphere has thickened to its original thickness?

PROBLEM 4–59 Assume that the continental crust and lithosphere have been stretched by a factor $\alpha = 4$. Taking $h_{cc} = 35 \, \text{km}$, $y_{L0} = 150 \, \text{km}$, $\rho_m = 3300 \, \text{kg m}^{-3}$, $\rho_{cc} = 2700 \, \text{kg m}^{-3}$, $\rho_s = 2450 \, \text{kg m}^{-3}$, $\alpha_v = 3 \times 10^{-5} \, \text{K}^{-1}$, and $T_1 - T_0 = 1250 \, \text{K}$, determine the depth of the sedimentary basin. What is the depth of the sedimentary basin when the thermal lithosphere has thickened to its original thickness?

PROBLEM 4–60 The compression model for a continental mountain belt considered in Problem 2–6 can be extended to include the compression of the lithosphere. Assuming that the temperature in the lithosphere after compression is given by

$$\frac{T_1 - T}{T_1 - T_0} = \text{erfc}\left(\frac{1.16 y}{\beta y_{L0}}\right) \tag{4–233}$$

show that the height of the mountain belt is given by

$$h = \left[\frac{(\rho_m - \rho_{cc})}{\rho_m} h_{cc} - \frac{\alpha_v (T_1 - T_0) y_{L0}}{1.16 \sqrt{\pi}}\right](\beta - 1). \tag{4–234}$$

Assuming $\beta = 2$, $h_{cc} = 35 \, \text{km}$, $\rho_m = 3300 \, \text{kg m}^{-3}$, $\rho_{cc} = 2800 \, \text{kg m}^{-3}$, $\alpha_v = 3 \times 10^{-5}$, $T_1 - T_0 = 1300 \, \text{K}$, and $y_{L0} = 150 \, \text{km}$, determine the height of the mountain belt and the thickness of the crustal root.

4-26 Heating or Cooling a Semi-Infinite Half-Space by a Constant Surface Heat Flux

So far we have been primarily concerned with heat conduction problems in which temperature boundary conditions are specified. In some geological applications it is appropriate to specify boundary conditions on the heat flux. If we take the partial derivative of the unsteady heat conduction equation (4–68) with respect to y and substitute Fourier's law (4–1), we obtain

$$\frac{\partial q}{\partial t} = \kappa \frac{\partial^2 q}{\partial y^2}. \tag{4–235}$$

The heat flux satisfies the same diffusion equation as does temperature.

We now consider the heating of a semi-infinite half-space by the constant addition of heat at its surface, $q = q_0$ at $y = 0$. Initially at $t = 0$ the temperature in the half-space is constant $T = T_0$, and there is no heat flow $q(0) = 0$. This problem is solved by Equation (4–235) with the boundary conditions

$$q = 0 \quad \text{at} \quad t = 0, \qquad y > 0$$
$$q = q_0 \quad \text{at} \quad y = 0, \qquad t > 0$$
$$q \to 0 \quad \text{as} \quad y \to \infty, \qquad t > 0. \tag{4–236}$$

This problem is identical with the sudden heating or cooling of a semi-infinite half-space. Equation (4–235) together with conditions (4–236) is equivalent to

Equation (4–94) and conditions (4–95) if we identify θ as q/q_0. The solution from Equation (4–112) is

$$q = q_0 \operatorname{erfc} \eta, \tag{4–237}$$

with η defined by Equation (4–96). In order to find the temperature, we substitute Fourier's law (4–1) into Equation (4–237) with the result

$$\frac{\partial T}{\partial y} = -\frac{q_0}{k} \operatorname{erfc} \eta = -\frac{q_0}{k} \operatorname{erfc}\left(\frac{y}{2\sqrt{\kappa t}}\right). \tag{4–238}$$

We can integrate Equation (4–238) using the boundary condition $T \to T_0$ as $y \to \infty$. We find

$$
\begin{aligned}
T &= T_0 - \frac{q_0}{k} \int_\infty^y \operatorname{erfc}\left(\frac{y'}{2\sqrt{\kappa t}}\right) dy' \\
&= T_0 + \frac{q_0}{k} \int_y^\infty \operatorname{erfc}\left(\frac{y'}{2\sqrt{\kappa t}}\right) dy' \\
&= T_0 + \frac{2q_0\sqrt{\kappa t}}{k} \int_\eta^\infty \operatorname{erfc} \eta' \, d\eta'. \tag{4–239}
\end{aligned}
$$

After an integration by parts we can express the temperature as

$$
\begin{aligned}
T &= T_0 + \frac{2q_0}{k}\sqrt{\kappa t}\left\{\frac{e^{-\eta^2}}{\sqrt{\pi}} - \eta \operatorname{erfc} \eta\right\} \\
&= T_0 + \frac{2q_0}{k}\left\{\sqrt{\frac{\kappa t}{\pi}}\, e^{-y^2/4\kappa t} - \frac{y}{2} \operatorname{erfc} \frac{y}{2\sqrt{\kappa t}}\right\}. \tag{4–240}
\end{aligned}
$$

The surface temperature T_s is obtained by setting $y = 0$ in Equation (4–240)

$$T_s = T_0 + \frac{2q_0}{k}\left(\frac{\kappa t}{\pi}\right)^{1/2}. \tag{4–241}$$

This formula gives the increase in the surface temperature due to the uniform addition of heat to a half-space.

PROBLEM 4–61 The heat loss from the Earth's surface q_s due to radiation is given by

$$q_s = \sigma T^4, \tag{4–242}$$

where $\sigma = 0.567 \times 10^{-7}$ W m^{-2} K^{-4} is the *Stefan-Boltzmann constant*, and T is the absolute temperature. Assuming that $T = 300$ K, $k = 2$ W m^{-1} K^{-1}, and $\kappa = 0.8$ mm^2 s^{-1}, use this heat loss to determine the cooling of the Earth's surface during 12 hr of night. (Assume q is constant, a reasonable approximation, and use the half-space cooling model in this section.)

4–27 Frictional Heating on Faults: Island Arc Volcanism and Melting on the Surface of the Descending Slab

As noted in Section 1–4, ocean trenches where subduction is occurring usually have parallel chains of active volcanoes overlying the descending lithosphere. Since the subduction process returns cold lithospheric rocks into the interior of the Earth, a subduction zone would be expected to have low temperatures and low surface heat flows. It is quite surprising, therefore, that extensive volcanism is associated with subduction zones.

One explanation for the high temperatures required for volcanism is frictional heating on the fault zone between the descending lithosphere and the overlying mantle. That this fault zone is the site of many large earthquakes is indicative of a large stress on the fault. When slip occurs in the presence of a large stress, significant frictional heating occurs. If the mean stress on the fault is τ and the mean velocity of the descending plate is u, the mean rate of heat production on the fault, per unit area of the fault, is

$$q = u\tau. \tag{4–243}$$

To assess the influence of fault heating on the descending lithosphere, let us consider the simplified geometry illustrated in Figure 4–54. The surface plate approaches the trench with a velocity u at an angle ϕ to the normal to the trench and descends into the mantle at an angle θ to the horizontal. The linear chain of active volcanoes lies at a distance d_v above the slip zone.

4–54 Geometry of the descending plate. (*a*) Side view. (*b*) Vertical view.

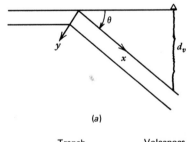

(a)

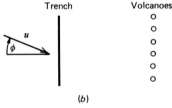

(b)

An x, y coordinate system is set up in the descending plate as shown.

The solution for constant heat addition to a uniform half-space can be used for this problem if several assumptions are made:

1. A substantial fraction of the heat produced on the fault zone is lost to the underlying descending lithosphere. This is a good approximation because the cold descending plate is the dominant heat sink.
2. The initial thermal structure of the lithosphere can be neglected. Because thermal conduction problems are linear in temperature, their solutions can be superimposed. The heat addition problem can be treated independently of the ambient conduction problem as long as the required boundary conditions are not violated.
3. Time t in the transient conduction problem is replaced by

$$ t = \frac{x}{u \cos \phi}. \tag{4–244} $$

Substitution of Equations (4–243) and (4–244) into Equation (4–241) gives the temperature on the slip zone T_{sz} as

$$ T_{sz} = T_0 + \frac{2\tau}{k} \left(\frac{u\kappa x}{\pi \cos \phi} \right)^{1/2}. \tag{4–245} $$

For surface volcanism to occur, the temperature on the slip zone beneath the volcanoes (with $d_v = x \sin \theta$) must equal the melt temperature of the rock T_m. From Equation (4–245) we find

$$ T_m = T_0 + \frac{2\tau}{k} \left(\frac{\kappa d_v u}{\pi \cos \phi \sin \theta} \right)^{1/2}. \tag{4–246} $$

As a typical example of a trench system we take $T_m - T_0 = 1200\,\text{K}, u = 100\,\text{mm yr}^{-1}, k = 4\,\text{W m}^{-1}\,\text{K}^{-1}, \kappa = 1\,\text{mm}^2\,\text{s}^{-1}, d_v = 125\,\text{km}, \theta = 45°,$ and $\phi = 0°$. From Equation (4–246) we find that the mean stress level required to produce the necessary heating is $\tau = 180\,\text{MPa}$. Although this is a high stress, it may be a reasonable value when relatively cool rocks are carried to depths where the lithostatic pressure is high. Stress levels on faults are considered in some detail in Chapter 8.

PROBLEM 4–62 Assume a constant sliding velocity u_f on a fault during an earthquake that results in a frictional heat production $u_f \tau$ (τ is the stress on the fault). If $u_f = 10\,\text{m s}^{-1}$, the total displacement $d = 4\,\text{m}, \tau = 10\,\text{MPa}, k = 4\,\text{W m}^{-1}\,\text{K}^{-1},$ and $\kappa = 1\,\text{mm}^2\,\text{s}^{-1}$, what is the temperature increase on the fault during the earthquake as predicted by Equation (4–245)?

PROBLEM 4–63 The amount of heat generated by friction on a fault during an earthquake is given by $Q = b\tau$, where b is the slip on the fault and τ is the mean stress on the fault. If $b = 3\,\text{m}$ and $\tau = 10\,\text{MPa}$, what is the maximum temperature increase 1 m from the fault due to friction on the fault ($\rho = 2700\,\text{kg m}^{-3}, c = 1\,\text{kJ kg}^{-1}\,\text{K}^{-1}$)?

4–28 Mantle Geotherms and Adiabats

The thermal structure of the upper mantle is dominated by the large temperature gradients in the lithosphere. The thermal structure of the oceanic lithosphere was determined in Sections 4–16 and 4–17; the temperature–depth relation is given in Equation (4–125) or Equation (4–130). A representative geotherm for the oceanic lithosphere was plotted in Figure 4–27 for $t = 60.4\,\text{Myr}, T_1 - T_0 = 1300\,\text{K},$ and $\kappa = 1\,\text{mm}^2\,\text{s}^{-1}$. The thermal structure of stable continental crust was determined in Section 4–8 and given in Figure 4–12.

Beneath the thermal boundary layer that defines the lithosphere, heat transport is primarily by convection. Details of this convection and the creep mechanisms responsible for the fluidlike behavior of hot, solid mantle rock are discussed in later chapters. For our purposes it is sufficient to know that in the interior of a vigorously convecting fluid the mean temperature increases with depth approximately along an *adiabat*. The *adiabatic temperature gradient* in the mantle is the rate of increase of temperature with depth as a result of compression of the rock by the weight of the overlying material. If an element of material is compressed and reduced in volume by increasing pressure, it will also be heated as a result of the work done by the pressure forces during the compression. If there is no transfer of heat into or out of the element during this process, the compression is said to be adiabatic, and the associated temperature rise is the adiabatic increase in temperature.

The change in density with pressure under adiabatic conditions is given by the adiabatic compressibility

$$ \beta_a = \frac{1}{\rho} \left(\frac{\partial \rho}{\partial p} \right)_s. \tag{4–247} $$

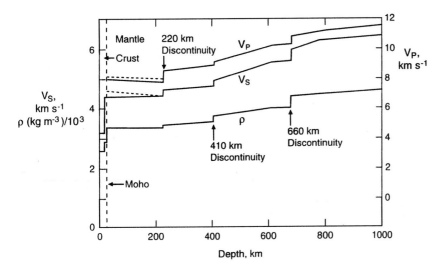

4–55 Seismic velocities V_p and V_s and the density ρ are given as a function of depth.

The subscript s means that the entropy s is constant. A reversible adiabatic process is a constant entropy or *isentropic* process. For a solid the adiabatic compressibility is somewhat smaller than the isothermal compressibility defined in Equation (4–175) because the temperature increases with pressure in an adiabatic process so there is some thermal expansion. If we assume that the adiabatic compressibility is a constant, we can integrate Equation (4–247) with the boundary condition $\rho = \rho_0$ at $p = 0$ to give

$$\rho = \rho_0 e^{\beta_a p}. \tag{4–248}$$

The increase in pressure with depth is given by

$$\frac{dp}{dy} = \rho g. \tag{4–249}$$

For the Earth's mantle we can reasonably assume that the gravitational acceleration g is a constant. By combining Equations (4–248) and (4–249) and integrating with g constant and the boundary condition $p = 0$ at $y = 0$, we obtain

$$p = \frac{-1}{\beta_a} \ln(1 - \rho_0 g \beta_a y) \tag{4–250}$$

$$\rho = \frac{\rho_0}{1 - \rho_0 g y \beta_a}. \tag{4–251}$$

These expressions for pressure and density as functions of depth are not completely satisfactory approximations to the actual pressure and density in the man-

tle. The dependence of the mantle density on depth is given in Figure 4–55. The values were deduced from the seismic velocities, which are also shown in Figure 4–55 and are tabulated in Section F of Appendix 2. The density discontinuity near a depth of 410 km is attributed to a solid–solid phase change of the mineral olivine, the dominant mineral in the mantle. Laboratory studies have shown that olivine transforms to a denser spinel structure at a pressure of 13.5 GPa and a temperature of about 1700 K. The density increase due to this phase change is $\Delta \rho = 200 - 300$ kg m^{-3}. Laboratory studies have also indicated that the density discontinuity near a depth of 660 km is caused by a transformation of the spinel structure to perovskite and magnesiowüstite. At a pressure of 23.1 GPa this transformation takes place at a temperature of about 1875 K. The density increase due to this transformation is $\Delta \rho \approx 400$ kg m^{-3}. These density discontinuities cannot be modeled using Equation (4–247). In addition, the adiabatic compressibility decreases with increasing pressure in the mantle from a near-surface value of 8.7×10^{-12} Pa^{-1} to a value of 1.6×10^{-12} Pa^{-1} at the core–mantle boundary. For these reasons Equation (4–251) is a relatively poor approximation for the Earth's mantle.

We now return to our discussion of the mantle geotherm. As already noted, the temperature gradient beneath the near-surface thermal boundary layer (the lithosphere) is very near the adiabatic gradient due to mantle convection. The adiabatic temperature gradient can be calculated from the thermodynamic relation between entropy per unit mass s, temperature, and pressure

$$ds = \frac{c_p}{T}dT - \frac{\alpha_v}{\rho}dp, \qquad (4-252)$$

where c_p is the specific heat at constant pressure and ds, dT, and dp are infinitesimal changes in entropy, temperature, and pressure. The entropy change in an adiabatic process is zero if the process is also reversible. Thus the rate of increase of temperature with pressure in an adiabatic, reversible process is obtained by putting $ds = 0$ in Equation (4–252), from which it follows that

$$\left(\frac{dT}{dp}\right)_s = \left(\frac{\alpha_v T}{\rho c_p}\right). \qquad (4-253)$$

We assume that Equation (4–253) is valid in the vigorously convecting compressible mantle in which heat conduction and other *irreversible processes* can be neglected.

If a material is strictly incompressible, pressure forces cannot change the volume of an element of the material. Accordingly, there can be no *adiabatic compressional heating* of an incompressible material; its adiabatic temperature gradient is zero. Rocks, however, are sufficiently compressible so that the large increases in pressure with depth in the mantle produce significant adiabatic increases of temperature with depth.

The adiabatic temperature gradient in the Earth $(dT/dy)_s$ can be found by multiplying $(dT/dp)_s$ from Equation (4–253) by dp/dy from Equation (4–249):

$$\left(\frac{dT}{dy}\right)_s = \frac{\alpha_v g T}{c_p}. \qquad (4-254)$$

For the near-surface values $\alpha_v = 3 \times 10^{-5}$ K^{-1}, $T = 1600$ K, $c_p = 1$ kJ kg^{-1} K^{-1}, and $g = 10$ m s^{-2}, Equation (4–235) yields $(dT/dy)_s = 0.5$ K km^{-1}. At greater depths the volume coefficient of thermal expansion is considerably smaller. To extend the temperature profile in the oceanic lithosphere given in Figure 4–27 to greater depths in the upper mantle, we assume that $(dT/dy)_s = 0.3$ K km^{-1}. Figure 4–56 shows the oceanic upper mantle geotherm to a depth of 400 km.

The upper mantle geotherm beneath the continents is not as well understood as the one beneath the oceans. One way to model the temperature distribution in the continental lithosphere would be to apply the same

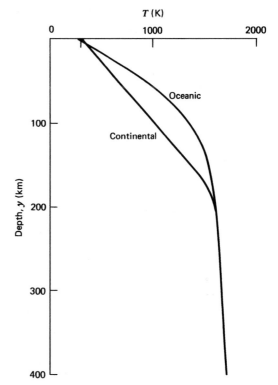

4–56 Representative oceanic and continental shallow upper mantle geotherms.

one-dimensional half-space cooling solution that we applied earlier to the oceanic lithosphere. However, as shown in Figure 4–28, the predicted mantle heat flows are considerably lower than the observed values. Also, if old continental lithosphere continued to cool, it would also continue to subside according to Equation (4–209) (see Problem 4–56). The result would be continental cratons overlain by a continuously thickening sedimentary cover. This condition has not been observed, so there must be a heat input into the base of the old continental lithosphere that retards further cooling and allows old continental lithosphere to tend toward a steady-state temperature profile. The input of heat to the base of the lithosphere is attributed either to mantle plumes impinging on the base of the lithosphere or to secondary convection in the lower lithosphere or to both, as discussed in Section 4–23.

The influence of near-surface radioactivity on continental surface heat flow has been considered in Section 4–8. From Equation (4–29) and the data given in Figure 4–11 we found that the heat flow beneath the near-surface layer of heat-producing elements is about 37 mW m^{-2}. We assume that heat production beneath the near-surface radioactive layer can be neglected and that the thermal structure of the continental lithosphere has reached a steady state. Therefore, it is appropriate to assume the heat flow through the continental lithosphere beneath the near-surface heat-producing layer, q_m, is constant. The resulting geotherm in the continental lithosphere is given in Figure 4–56 for $q_m = 37$ mW m^{-2} and $k = 3.35$ W m^{-1} K^{-1}. The thickness of the continental lithosphere is about 200 km.

Our discussion so far has centered on the thermal state of the shallow upper mantle; the geotherms in Figure 4–56 extend only to a depth of 400 km. If the entire mantle were homogeneous and strongly convecting, the adiabatic temperature gradient given by Equation (4–254) would be a good approximation of the slope of the temperature profile throughout the mantle. We have noted, however, that the distribution of density with depth has significant discontinuities near depths of 410 and 660 km (see Figure 4–55).

The density discontinuity at 410 km is associated with the transformation of olivine to a spinel structure.

The phase change from olivine to spinel is *exothermic* with a heat of reaction $L = 90$ kJ kg^{-1}. For adiabatic flow downward through the phase change, the temperature of the mantle rock increases by

$$\Delta T = \frac{L}{c_p}. \qquad (4\text{–}255)$$

The heat released by the phase change increases the temperature of the rock. For $c_p = 1$ kJ kg^{-1} K^{-1}, the increase in temperature is 90 K at a depth of 410 km. This increase in temperature with depth for adiabatic flow is shown in Figure 4–57, where the whole mantle geotherm is given. We will show that the 410-km phase change enhances mantle convection; that the associated density boundary does not block mantle convection is indicated by the descent of the subducted lithosphere through this depth.

The density discontinuity at a depth of 660 km is attributed to the transformation of the spinel structure to perovskite and magnesiowüstite. This transformation is endothermic with a heat of reaction $L = -70$ kJ kg^{-1}. The heat absorbed by this reaction cools the rock. From Equation (4–255) with $c_p = 1$ kJ kg^{-1} K^{-1}, the decrease

4–57 Mantle geotherms are given for whole-mantle convection "(Curve *a*) and layered mantle convection" (Curve *b*). The range of values for the mantle solidus and the minimum temperatures in a subducted slab are also given.

in temperature is 70 K at a depth of 660 km. We will show that the 660-km transformation is expected to retard flow through this boundary.

Deep-focus earthquakes provide conclusive evidence that there is active mantle convection to depths of 660 km. Since the lower mantle is expected to contain significant concentrations of radioactive isotopes, we expect that mantle convection will occur in the lower mantle in order to transport the resulting heat.

Three alternative models for mantle convection have been proposed:

1. Whole mantle convection. If significant amounts of subducted lithosphere can enter the lower mantle beneath 660 km, then there must be a complementary mantle upwelling. In this case the geotherm for the entire mantle is likely to be adiabatic. The expected geotherm is illustrated in Figure 4–57 as curve *a*. The primary arguments against whole mantle convection come from chemical geodynamic studies, which we will discuss in Chapter 10.
2. Layered mantle convection. Two separate convection systems are operating in the upper and lower mantle. This would be the case if the density discontinuity at a depth of 660 km completely blocks convection. An upper convective system associated with plate tectonics would be restricted to the upper 660 km of the mantle; a lower, separate system would operate between a depth of 660 km and the core–mantle boundary. In this case a thermal boundary layer would be expected to develop at a depth of 660 km similar to the lithosphere. However, it is very difficult to estimate the change in temperature associated with this boundary layer. An expected geotherm for layered mantle convection is given as curve *b* in Figure 4–57. Although deep-focus earthquakes do not occur at depths greater than 660 km, studies using mantle tomography indicate that at least some subducted slabs penetrate through this boundary. This is taken as convincing evidence that there is significant material transport between the upper and lower mantle.
3. Hybrid models. Hybrid models have been proposed that involve a strong time dependence and/or a barrier to convection within the lower mantle. If the 660-km seismic discontinuity acts as a partial barrier to mantle convection, then mantle "avalanches" may

be triggered that would lead to a strongly time-dependent mantle convection. Dense subducted lithosphere could "pile up" on the 660-km deep seismic discontinuity until a finite-amplitude instability resulted in a mantle "overturn" or avalanche. Episodic mantle overturns have been proposed as an explanation for apparent episodicities in the geological record. It has also been proposed that there is a compositional barrier to whole mantle convection within the lower mantle. Studies using seismic tomography have been used to argue in favor of such a barrier having considerable topography and time dependence.

A constraint on the temperature at the base of the mantle is the seismic evidence that the outer core is liquid. This evidence consists mainly of the inability of shear waves to propagate through the outer core. Measured velocities of seismic compressional waves in the outer core indicate that, although the outer core is primarily composed of iron, it must also contain significant concentrations of one or more other constituents, the most likely of which is sulfur. The melting temperature for the iron–sulfur eutectic mixture at the core–mantle boundary is estimated to be 3200 K. This is an approximate minimum value for the temperature at the core–mantle boundary. The adiabatic lower mantle geotherm in Figure 4–57 is in approximate agreement with this constraint.

Just as an upper mantle thermal boundary layer, the lithosphere, intervenes between the surface and the interior adiabatic state of the mantle, a lower mantle thermal boundary layer is expected to exist just above the core–mantle boundary. Seismic studies have confirmed the existence of this boundary layer, which is referred to as the D″-layer. The D″-layer has a complex structure with a thickness of 150 to 300 km. Laboratory studies indicate that the solidus temperature of a perovskite–magnesiowüstite assemblage at the core–mantle boundary would be about 4300 K. In addition to showing the two geotherms associated with whole mantle convection *a* and layered mantle convection *b*, Figure 4–57 gives the range of values for the mantle solidus as well as the temperature increases associated with the D″-layer. Although the required heat flux through the D″-layer can be estimated, the stability of the layer which would give its thickness is difficult to determine.

There may also be compositional stratification in this layer.

PROBLEM 4-64 How much heat is conducted along the adiabat of Figure 4–57 at depths of 1000 and 2000 km? At the core–mantle interface? Use $k = 4$ W m^{-1} K^{-1}, $\alpha_v = 1.5 \times 10^{-5}$ K^{-1}, $g = 10$ m s^{-2}, $c_p = 1$ kJ kg^{-1} K^{-1}.

PROBLEM 4-65 If the rate at which heat flows out of the core (J s^{-1}) is 10% of the rate at which heat is lost at the Earth's surface, how large is the mean temperature drop across the lower mantle thermal boundary layer in terms of the mean temperature drop across the upper mantle thermal boundary layer? Assume that the heat transport across a boundary layer can be calculated from Fourier's law of heat conduction in the simple form of Equation (4–3). Also assume that the upper and lower mantle boundary layers have the same thicknesses.

4-29 Thermal Structure of the Subducted Lithosphere

The subduction of the cold oceanic lithosphere into the deep mantle is a primary mechanism for the transport of heat from the interior of the Earth to its surface. Hot mantle rock comes to the surface at accretional plate boundaries (ocean ridges) and is cooled by heat loss to the seafloor. The result is a cold thermal "boundary layer," the oceanic lithosphere. The thermal structure of this boundary layer was determined in Sections 4–16 and 4–17. The cold subducted lithosphere is gradually heated and eventually becomes part of the convecting mantle. Upward convective heat transfer through the mantle involves the sinking of cold thermal anomalies (descending lithosphere at ocean trenches) and the rising of hot thermal anomalies (mantle plumes). The density differences associated with the lateral temperature variations provide the driving force for the mantle convective circulation. In this section we discuss the temperature distribution in the subducted oceanic lithosphere.

Isotherms in a lithosphere descending at an angle of 45° into the mantle are shown in Figure 4–58. Since the subducted lithosphere was formed on the seafloor, its initial thermal structure upon subduction is given by Equation (4–125). The dependence of temperature upon depth prior to subduction is the oceanic geotherm given in Figure 4–56. As the subducted lithosphere descends into the mantle, frictional heating occurs at its upper boundary. The effects of frictional heating were studied in Section 4–26. As discussed there, the temperature distribution due to frictional heating – Equation (4–240) – can be superimposed on the initial temperature distribution to give the isotherms in the slab. The result is shown in Figure 4–58.

4-58 Isotherms (°C) in a typical descending lithosphere. The 410-km phase change is elevated in the subducted lithosphere. The position of the slip zone is also shown.

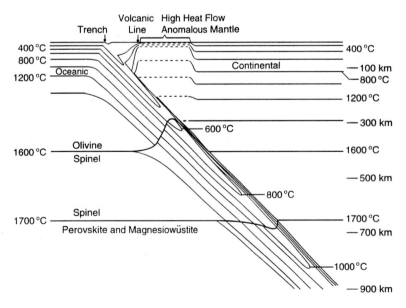

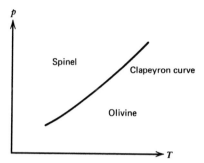

4–59 The Clapeyron or equilibrium curve separating two phases of the same material.

The low temperatures in the descending lithosphere cause it to have a higher density than the surrounding mantle. The higher density results in a body force driving the descending lithosphere downward. This body force is important in driving the plates. An additional downward body force on the descending slab is provided by the distortion of the olivine–spinel phase boundary in the slab, as shown in Figure 4–58.

The olivine–spinel phase boundary is elevated in the descending lithosphere as compared with its position in the surrounding mantle because the pressure at which the phase change occurs depends on temperature. Figure 4–59 is a sketch of the *Clapeyron curve*, which gives the pressures and temperatures at which two phases of the same material, such as olivine and spinel, are in equilibrium. Actually, the olivine–spinel transition is not *univariant*; it occurs over a range of temperatures and pressures. It is sufficient for our purposes here, however, to treat the phase change as occurring at a single temperature corresponding to a single pressure. The two phases can coexist at any point T, p lying on the Clapeyron curve.

The slope of the Clapeyron curve γ is defined by

$$\gamma \equiv \left(\frac{dp}{dT}\right)_{pc}. \qquad (4\text{--}256)$$

If we divide this equation by Equation (4–249), we obtain the change dy in the vertical location of the phase transition corresponding to a change in temperature dT

$$\left(\frac{dy}{dT}\right)_{pc} = \frac{\gamma}{\rho g}. \qquad (4\text{--}257)$$

For the olivine to spinel phase change, the slope of the Clapeyron curve is positive. Since dT is negative for the lower temperatures in the interior of the descending

lithosphere, dy is negative, and the olivine–spinel phase change occurs at a shallower depth (lower pressure) in the slab.

With $\gamma = 2$ MPa K^{-1}, $\rho = 3600$ kg m^{-3}, and $g = 10$ m s^{-2}, we find from Equation (4–257) that $(dy/dT)_{pc} = 0.055$ km K^{-1}. If we take the maximum temperature difference across the slab to be $\Delta T = 800$ K, we find that the elevation of the olivine–spinel phase boundary in the descending lithosphere is about 44 km. This elevation is illustrated in Figure 4–58. Since spinel is about 280 kg m^{-3} denser than olivine, the additional mass of the elevated spinel in the descending lithosphere provides a significant body force for driving the plates in addition to the downward body force provided by the thermal contraction of the lithosphere.

This approach can also be applied to the transition of spinel to perovskite and magnesiowüstite. In this case the slope of the Clapeyron curve is negative and the transition occurs at a deeper depth (higher pressure) in the slab. With $\gamma = -2.5$ MPa K^{-1}, $\rho = 3700$ kg m^{-3} and $g = 10$ m s^{-2}, we find from Equation (4–257) that $(dy/dT)_{pc} = -0.07$ km K^{-1}. If we take the maximum temperature across the slab to be $\Delta T = 750$ K, we find that the depression of this phase transition is 52 km. Since perovskite–magnesiowüstite is about 400 kg m^{-3} denser than spinel, the buoyancy of the depressed spinel provides a significant body force that inhibits convection through the 660-km boundary.

PROBLEM 4–66 Estimate the downward body force on the slab per unit length of trench due to the elevation of the olivine–spinel phase boundary in Figure 4–58. Assume $\rho(\text{spinel}) - \rho(\text{olivine}) = 300$ kg m^{-3}. Estimate the downward body force on the slab per unit length of trench due to thermal contraction by integrating over the temperature distribution in Figure 4–58. Assume $\alpha_v = 3 \times 10^{-5}$ K^{-1}, and consider the densification of the slab only to depths of 660 km.

4–30 Culling Model for the Erosion and Deposition of Sediments

The erosion and deposition of sediments are responsible for the formation and evolution of many landforms. A classic example is an alluvial fan caused by the deposition of sediments on a horizontal surface. Cross sections of alluvial fans often resemble the form of the complementary error function given in Figure 4–21.

This similarity suggests that sediment deposition may be modeled using the heat equation.

The use of the heat equation to model sediment erosion and deposition was first proposed by W. E. H. Culling (1960) and this approach is known as the Culling model. The basic hypothesis is that the down slope flux of sediments S is linearly proportional to the slope so that

$$S = -K \frac{\partial h}{\partial x}, \qquad (4\text{-}258)$$

where h is the elevation of topography above a base level, x is the horizontal distance, and K is a constant that is called a *transport coefficient*. The sediment flux S is the volume of sediment transported per unit time per unit width. In terms of the analogy with the heat equation (4–68), the flux equation (4–258) is directly analogous to Fourier's law given in Equation (4–1).

Consider an element of topography of width δx. The flux of sediment out of this element at $x + \delta x$ is $S(x + \delta x)$ and the flux of sediment into this element at x is $S(x)$. Using Equation (4–258) and the same expansion given in Equations (4–9) and (4–10) we have

$$S(x + \delta x) - S(x) = \delta x \frac{\partial S}{\partial x} = -\delta x K \frac{\partial^2 h}{\partial x^2}, \qquad (4\text{-}259)$$

where we have assumed K to be a constant. If there is a net flow of sediment into the element, there must be a change in elevation h given by

$$\delta x \frac{\partial h}{\partial t}.$$

Since a net flux of sediment out of the element leads to a decrease in elevation, we have

$$\frac{\partial h}{\partial t} = K \frac{\partial^2 h}{\partial x^2}, \qquad (4\text{-}260)$$

which is identical to the one-dimensional, time-dependent heat conduction equation (4–68).

Let us apply the Culling model to the progradation of a river delta into a quiet basin with a horizontal flow. Sediments are supplied to the delta by the river forming it. Sediments are deposited near the landward edge of the delta and are transported down the front of the delta by creep and shallow landslides. Our simple one-dimensional model is illustrated in Figure 4–60. The delta front is assumed to prograde forward at a constant

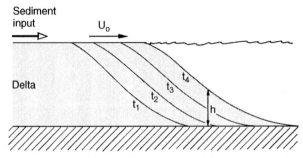

4–60 Illustration of the one-dimensional model for a prograding river delta. It is assumed that the delta progrades seaward at a constant velocity U_0; its position at successive times t_1 to t_4 is illustrated. The height of the prograding delta above the basin floor is h.

velocity U_0 and its position at successive times $t_1, t_2, t_3,$ and t_4 is shown.

We utilize the approach given in Section 4–19 to solve this problem. Let

$$\xi = x - U_0 t \qquad (4\text{-}261)$$

be a coordinate that is moving with the front of the delta. The shape of the delta is independent of time in this coordinate system and substitution of Equation (4–261) into Equation (4–260) gives

$$-U_0 \frac{dh}{d\xi} = K \frac{d^2 h}{d\xi^2}, \qquad (4\text{-}262)$$

with the boundary conditions $h = h_0$ at $\xi = 0$ and $h \to 0$ as $\xi \to \infty$ where h_0 is the height of the landward edge of the delta front. The solution of Equation (4–262) is

$$h = A \exp\left(-\frac{U_0 \xi}{K}\right) + B, \qquad (4\text{-}263)$$

where A and B are constants. When the boundary conditions are satisfied we obtain

$$h = h_0 \exp\left(-\frac{U_0 \xi}{K}\right). \qquad (4\text{-}264)$$

Substitution of Equation (4–261) into Equation (4–264) gives

$$h = h_0 \exp\left[-\frac{U_0}{K}(x - U_0 t)\right]. \qquad (4\text{-}265)$$

The height of the delta front above the floor decreases exponentially with distance from the shore. A plot of height versus distance from the shore is given in Figure 4–61.

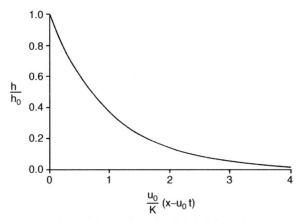

4–61 Dependence of the nondimensional height h/h_0 on the nondimensional distance from shore $U_0(x-U_0t)/K$ from Equation (4–265).

Comparisons with progradation data are obtained using the delta front slope. This slope at $\xi = 0$ is given by

$$\left(\frac{\partial h}{\partial x}\right)_{\xi=0} = -\frac{U_0 h}{K}. \tag{4–266}$$

Using this relation, we can obtain the transport coefficient from the progradation velocity U_0 and the morphology of the delta. As a specific example, consider the southwest pass segment of the Mississippi River delta. Longitudinal profiles of this delta front are shown in Figure 4–62. Taking $h = 107$ m, $U_0 = 76$ m yr^{-1}, and $(\partial h/\partial x)_{\xi=0} = -0.0096$, we find from Equation (4–266) that $K = 8.5 \times 10^5$ m^2 yr^{-1}.

PROBLEM 4–67 Consider a simplified one-dimensional model for the formation of an alluvial fan. Assume

4–62 Cross-sectional profiles of the Mississippi River delta (southwest passage) at various times showing its progradation (Fisk et al., 1954).

that there is a uniform flux of sediment S_0 over a vertical cliff, forming a one-dimensional, time-dependent alluvial fan. Assume that the Culling theory is applicable and use the methods of Section 4–25 to show that

$$h = \frac{2S_0}{K}\left\{\left(\frac{Kt}{\pi}\right)^{1/2}\exp\left(-\frac{x^2}{4Kt}\right)\right.$$
$$\left. - \frac{x}{2}\mathrm{erfc}\left(\frac{x}{2\left(Kt^{1/2}\right)}\right)\right\}. \tag{4–267}$$

Also show that the height of the alluvial fan at the cliff ($x = 0$) is given by

$$h_0 = 2S_0\left(\frac{t}{\pi K}\right)^{1/2} \tag{4–268}$$

and that the slope of the alluvial fan at the cliff is given by

$$\left(\frac{\partial h}{\partial x}\right)_{x=0} = -\frac{S_0}{K} \tag{4–269}$$

and

$$h_0 = -2\left(\frac{\partial h}{\partial x}\right)_{x=0}\left(\frac{Kt}{\pi}\right)^{1/2}. \tag{4–270}$$

For the alluvial fan beneath the San Gabriel Mountains in Pasadena, California, it is appropriate to take $h_0 = 400$ m, $(\partial h/\partial x)_{x=0} = -0.075$, and $t = 10^6$ years. What is the corresponding transport coefficient K?

PROBLEM 4–68 The Culling model can also be applied to the erosion and deposition of a fault scarp. Assume that a vertical fault scarp of height h_0 forms at $t = 0$ and $x = 0$ and subsequently erodes symmetrically. At $t = 0$, $h = h_0$ for $x < 0$ and $h = 0$ for $x > 0$. For $t > 0$, $h = h_0/2$ at $x = 0$, the region $x < 0$ erodes and deposition occurs in $x > 0$. Assume that both erosion and

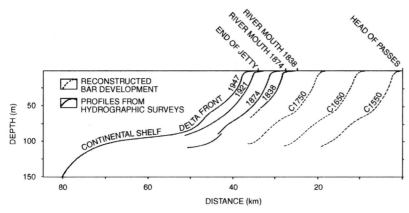

deposition are governed by Eq. (4–260) with K prescribed. Show that the height of the topography h is given by

$$h = \frac{h_0}{2}\text{erfc}\left(\frac{x}{2\sqrt{Kt}}\right). \tag{4–271}$$

Also show that slope at $x = 0$ is given by

$$\left(\frac{\partial h}{\partial x}\right)_{x=0} = \frac{-h_0}{2(\pi Kt)^{1/2}}. \tag{4–272}$$

An earthquake is known to have occurred 400 years ago; with $h_0 = 3$ m and $(\partial h/\partial x)_{x=0} = -0.5$, what is the value of the transport coefficient K?

The morphology of sedimentary landforms such as prograding river deltas, alluvial fans, eroding fault scarps, and eroding shorelines are often in good agreement with solutions of the heat equation. However, derived values of the transport coefficient K have considerable variability. This is not surprising because submarine sediment transport is very different from subaerial sediment transport. Also, both are very dependent on climate, weather, and rock type.

REFERENCES

Carslaw, H. S., and J. C. Jaeger (1959), *Conduction of Heat in Solids*, 2nd edition, Oxford University Press, Oxford, 510 p.

Culling, W. E. H. (1960), Analytical theory of erosion, *J. Geol.* **68**, 336–344.

Fisk, H. N., E. McFarlan, C.R. Kolb, and L. J. Wilbert (1954), Sedimentary framework of the modern Mississippi Delta, *J. Sedimen. Petrol.* **24**, 76–99.

Johnson, H. P., and R. L. Carlson (1992), Variation of sea floor depth with age: A test of models based on drilling results, *Geophys. Res. Lett.* **19**, 1971–1974.

Leeds, A. R., L. Knopoff, and E. G. Kausel (1974), Variations of upper mantle structure under the Pacific Ocean, *Science* **186**, 141–143.

Lister, C. R. B., J. G. Sclater, E. E. Davis, H. Villinger, and S. Nagihara (1990), Heat flow maintained in ocean basins of great age: Investigations in the north-equatorial west Pacific, *Geophys. J. Int.* **102**, 603–630.

Pollack, H. N., S. J. Hurter, and J. R. Johnson (1993), Heat flow from the Earth's interior: Analysis of the global data set, *Rev. Geophys.* **31**, 267–280.

Sclater, J. G., C. Jaupart, and D. Galson (1980), The heat flow through oceanic and continental crust and the heat loss of the Earth, *Rev. Geophys. Space Phys.* **18**, 269–311.

Stefan, J. (1891), Uber die Theorie der Eisbildung, insbesondere uber die Eisbildung im Polarmeere, *Ann. Physik Chem.* **42**, 269–286.

Vail, P. R., R. M. Mitchum, and S. Thompson (1978), Seismic stratigraphy and global changes of sea level, in *Seismic Stratigraphy: Applications to Hydrocarbon Exploration*, C. F. Payton, ed., American Association of Petroleum Geologists, Memoir 26, pp. 83–97.

Wright, T. L., D. L. Peck, and H. R. Shaw (1976), Kilauea lava lakes: Natural laboratories of study of cooling, crystallization and differentiation of basaltic magma, in *The Geophysics of the Pacific Ocean Basin and its Margin*, G. H. Sutton, M. H. Manghnani, and R. Moberly, eds., American Geophysical Union, Washington, D. C., pp. 375–390.

COLLATERAL READING

Burchfield, J. D., *Lord Kelvin and the Age of the Earth* (Science History Publications, New York, 1975), 260 pages.

A historical account of scientific attempts to determine the age of the Earth. The book focuses on Kelvin's influence and the debate between physicists and geologists between the mid-1800s and the early 1900s.

Carslaw, H. S., and J. C. Jaeger, *Conduction of Heat in Solids*, 2nd edition (Oxford University Press, Oxford, 1959), 510 pages.

A classic textbook on the mathematical theory of heat conduction in solids. It describes fundamental mathematical techniques for solving time-dependent heat conduction problems in a variety of geometries. The book contains an extensive compilation of solutions to boundary value problems often encountered in geological and geophysical applications.

FIVE

Gravity

5-1 Introduction

The force exerted on an element of mass at the surface of the Earth has two principal components. One is due to the gravitational attraction of the mass in the Earth, and the other is due to the rotation of the Earth. Gravity refers to the combined effects of both gravitation and rotation. If the Earth were a nonrotating spherically symmetric body, the gravitational acceleration on its surface would be constant. However, because of the Earth's rotation, topography, and internal lateral density variations, the *acceleration of gravity g* varies with location on the surface. The Earth's rotation leads mainly to a latitude dependence of the surface acceleration of gravity. Because rotation distorts the surface by producing an *equatorial bulge* and a *polar flattening*, gravity at the equator is about 5 parts in 1000 less than gravity at the poles. The Earth takes the shape of an *oblate spheroid*. The gravitational field of this spheroid is the reference gravitational field of the Earth. Topography and density inhomogeneities in the Earth lead to local variations in the surface gravity, which are referred to as *gravity anomalies*.

The mass of the rock associated with topography leads to surface gravity anomalies. However, as we discussed in Chapter 2, large topographic features have low-density crustal roots. Just as the excess mass of the topography produces a positive gravity anomaly, the low-density root produces a negative gravity anomaly. In the mid-1800s it was observed that the gravitational attraction of the Himalayan Mountains was considerably less than would be expected because of the positive mass of the topography. This was the first evidence that

the crust–mantle boundary is depressed under large mountain belts.

A dramatic example of the importance of crustal thickening is the absence of positive gravity anomalies over the continents. The positive mass anomaly associated with the elevation of the continents above the ocean floor is reduced or *compensated* by the negative mass anomaly associated with the thicker continental crust. We will show that compensation due to the hydrostatic equilibrium of thick crust leads in the first approximation to a zero value for the surface gravity anomaly. There are mechanisms for compensation other than the simple thickening of the crust. An example is the subsidence of the ocean floor due to the thickening of the thermal lithosphere, as discussed in Section 4–23.

Gravity anomalies that are correlated with topography can be used to study the flexure of the elastic lithosphere under loading. Short wavelength loads do not depress the lithosphere, but long wavelength loads result in flexure and a depression of the Moho. Gravity anomalies can also have important economic implications. Ore minerals are usually more dense than the country rock in which they are found. Therefore, economic mineral deposits are usually associated with positive gravity anomalies. Major petroleum occurrences are often found beneath salt domes. Since salt is less dense than other sedimentary rocks, salt domes are usually associated with negative gravity anomalies.

As we will see in the next chapter, mantle convection is driven by variations of density in the Earth's mantle. These variations produce gravity anomalies at the Earth's surface. Thus, measurements of gravity at the Earth's surface can provide important constraints on the flow patterns within the Earth's interior. However, it must be emphasized that the surface gravity does not provide a unique measure of the density distribution within the Earth's interior. Many different internal density distributions can give the same surface distributions of gravity anomalies. In other words, inversions of gravity data are non-unique.

5-2 Gravitational Acceleration External to the Rotationally Distorted Earth

The gravitational force exerted on a mass m' located at point P outside the Earth by a small element of mass dm in the Earth is given by *Newton's law of gravitation.*

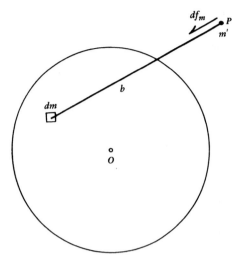

5–1 Force on a mass m' due to the gravitational attraction of an infinitesimal element of mass dm in the Earth.

As shown in Figure 5–1, the gravitational attraction df_m in the direction from P to dm is given by

$$df_m = \frac{Gm'\,dm}{b^2},\qquad(5\text{–}1)$$

where G is the *universal gravitational constant* $G = 6.673 \times 10^{-11}\ \mathrm{m^3\ kg^{-1}\ s^{-2}}$ and b is the distance between dm and the point P. The infinitesimal gravitational acceleration at P due to the attraction of dm is the force per unit mass exerted on m' in the direction of P:

$$dg_m = \frac{df_m}{m'}.\qquad(5\text{–}2)$$

By combining Equations (5–1) and (5–2) we obtain

$$dg_m = \frac{G\,dm}{b^2}.\qquad(5\text{–}3)$$

If the distribution of mass in the Earth were known exactly, the gravitational attraction of the Earth on a unit mass outside the Earth could be obtained by summing or integrating dg_m over the entire distribution. Suppose, for example, that the entire mass of the Earth M were concentrated at its center. The gravitational acceleration at a distance r from the center would then be directed radially inward and, according to Equation (5–3), it would be given by

$$g_m = \frac{GM}{r^2}.\qquad(5\text{–}4)$$

Following the generally accepted sign convention, we

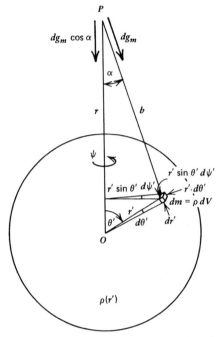

5–2 Geometry for the calculation of the gravitational acceleration at a point outside a spherically symmetric mass distribution.

take g_m to be positive, even though it is directed in the $-r$ direction.

We next determine the gravitational acceleration outside a spherical body with a density distribution that is a function of radius only, $\rho = \rho(r')$. The geometry is illustrated in Figure 5–2. It is clear from symmetry considerations that the gravitational acceleration g_m at a point P outside the mass distribution is directed radially inward and depends only on the distance r of point P from the center of the sphere. For convenience, we let the line from P to O be the polar axis of a spherical coordinate system r, θ, ψ. The gravitational acceleration at P due to an element of mass dm located in the sphere at r', θ', ψ' is directed along the line from P to dm and is given by Equation (5–3). The component of this gravitational acceleration along the line from P to O is

$$\frac{G\cos\alpha\,dm}{b^2}.$$

The net radially inward gravitational acceleration at P is found by integrating this expression over the entire mass distribution:

$$g_m = G\int\frac{\cos\alpha\,dm}{b^2}.\qquad(5\text{–}5)$$

The element of mass dm is the product of the volume element dV with the density $\rho(r')$ at the location of dV

$$dm = \rho(r')\,dV. \tag{5-6}$$

The element of volume can be expressed in spherical coordinates as

$$dV = r'^2 \sin\theta'\,d\theta'\,d\psi'\,dr'. \tag{5-7}$$

The integral over the spherical mass distribution in Equation (5–5) can thus be written

$$g_m = G \int_0^a \int_0^\pi \int_0^{2\pi} \frac{\rho(r')r'^2 \sin\theta' \cos\alpha\,d\psi'\,d\theta'\,dr'}{b^2}, \tag{5-8}$$

where a is the radius of the model Earth. The integral over ψ' is 2π, since the quantities in the integrand of Equation (5–8) are independent of ψ'. To carry out the integration over r' and θ', we need an expression for $\cos\alpha$. From the law of cosines we can write

$$\cos\alpha = \frac{b^2 + r^2 - r'^2}{2rb}. \tag{5-9}$$

Because the expression for $\cos\alpha$ involves b rather than θ', it is more convenient to rewrite Equation (5–8) so that the integration can be carried out over b rather than over θ'. The law of cosines can be used again to find an expression for $\cos\theta'$:

$$\cos\theta' = \frac{r'^2 + r^2 - b^2}{2rr'}. \tag{5-10}$$

By differentiating Equation (5–10) with r and r' held constant, we find

$$\sin\theta'd\theta' = \frac{b\,db}{rr'}. \tag{5-11}$$

Upon substitution of Equations (5–9) and (5–11) into Equation (5–8), we can write the integral expression for g_m as

$$g_m = \frac{\pi G}{r^2} \int_0^a r'\rho(r') \int_{r-r'}^{r+r'} \left\{ \frac{r^2 - r'^2}{b^2} + 1 \right\} db\,dr'. \tag{5-12}$$

The integration over b gives $4\,r'$ so that Equation (5–12) becomes

$$g_m = \frac{4\pi G}{r^2} \int_0^a dr' r'^2 \rho(r'). \tag{5-13}$$

Since the total mass of the model is given by

$$M = 4\pi \int_0^a dr' r'^2 \rho(r'), \tag{5-14}$$

the gravitational acceleration is

$$g_m = \frac{GM}{r^2}. \tag{5-15}$$

The gravitational acceleration of a spherically symmetric mass distribution, at a point outside the mass, is identical to the acceleration obtained by concentrating all the mass at the center of the distribution. Even though there are lateral density variations in the Earth and the Earth's shape is distorted by rotation, the direction of the gravitational acceleration at a point external to the Earth is very nearly radially inward toward the Earth's center of mass, and Equation (5–15) provides an excellent first approximation for g_m.

PROBLEM 5–1 For a point on the surface of the Moon determine the ratio of the acceleration of gravity due to the mass of the Earth to the acceleration of gravity due to the mass of the Moon.

The rotational distortion of the Earth's mass adds a small latitude-dependent term to the gravitational acceleration. This term depends on the excess mass in the rotational equatorial bulge of the Earth. The observed latitude dependence of g_m can thus be used to determine this excess mass. In addition, this effect must be removed from observed variations in surface gravity before the residual gravity anomalies can properly be attributed to density anomalies in the Earth's interior. The model we use to calculate the contribution of rotational distortion to gravitational acceleration is sketched in Figure 5–3. The Earth is assumed to be flattened at the poles and bulged at the equator because of its rotation with angular velocity ω. The mass distribution is assumed to be symmetrical about the rotation axis. Because of the departure from spherical symmetry due to rotation, the gravitational acceleration at a point P outside the Earth has both radial and tangential components. The radial component is the sum of GM/r^2 and the term g'_r due to rotational distortion of the mass distribution; the tangential component g'_t is entirely due to the rotationally induced departure from spherical symmetry. Following our previous sign convention both GM/r^2 and g'_r are positive if directed

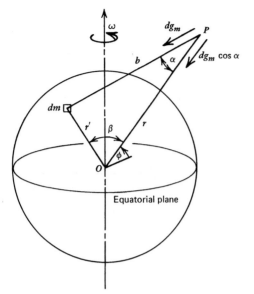

5-3 Geometry for calculating the contribution of rotational distortion to the gravitational acceleration.

inward. Since rotation modifies the otherwise spherically symmetric model Earth only slightly, g_r' and g_t' are small compared with GM/r^2.

The total gravitational acceleration is

$$\left\{ \left(\frac{GM}{r^2} + g_r'\right)^2 + g_t'^2 \right\}^{1/2}$$

$$= \left\{ \left(\frac{GM}{r^2}\right)^2 + 2\left(\frac{GM}{r^2}\right)g_r' + g_r'^2 + g_t'^2 \right\}^{1/2}. \quad (5\text{-}16)$$

It is appropriate to neglect the quadratic terms because the magnitudes of g_r' and g_t' are much less than GM/r^2. Therefore the gravitational acceleration is given by

$$\left\{ \left(\frac{GM}{r^2}\right)^2 + 2\left(\frac{GM}{r^2}\right)g_r' \right\}^{1/2}$$

$$= \left(\frac{GM}{r^2}\right)\left\{ 1 + \frac{2g_r'}{GM/r^2} \right\}^{1/2}$$

$$= \left(\frac{GM}{r^2}\right)\left\{ 1 + \frac{g_r'}{GM/r^2} \right\} = \frac{GM}{r^2} + g_r'. \quad (5\text{-}17)$$

Equation (5–17) shows that the tangential component of the gravitational acceleration is negligible; the net gravitational acceleration at a point P external to a rotationally distorted model Earth is essentially radially inward to the center of the mass distribution.

The radial gravitational acceleration for the rotationally distorted Earth model can be obtained by integrating Equation (5–5) over the entire mass distribution. We can rewrite this equation for g_m by substituting expression (5–9) for $\cos \alpha$ with the result

$$g_m = \frac{G}{2r^2} \int \left\{ \frac{r}{b} + \frac{r^3}{b^3}\left(1 - \frac{r'^2}{r^2}\right) \right\} dm. \quad (5\text{-}18)$$

The three distances appearing in the integral of Equation (5–18) r, r', and b are the sides of the triangle connecting O, P, and dm in Figure 5–3. It is helpful for carrying out the integration to eliminate b from the integrand in terms of r, r', and the angle β, which is opposite the side of length b in this triangle. From the law of cosines we can write

$$b^2 = r^2 + r'^2 - 2rr' \cos \beta, \quad (5\text{-}19)$$

which can be rearranged as

$$\frac{r}{b} = \left\{ 1 + \frac{r'^2}{r^2} - \frac{2r'}{r}\cos\beta \right\}^{-1/2}. \quad (5\text{-}20)$$

Upon substituting Equation (5–20) into Equation (5–18), we get

$$g_m = \frac{G}{2r^2} \int \left\{ 1 + \frac{r'^2}{r^2} - \frac{2r'}{r}\cos\beta \right\}^{-1/2}$$

$$\times \left\{ 1 + \left(1 - \frac{r'^2}{r^2}\right)\left(1 + \frac{r'^2}{r^2} - \frac{2r'}{r}\cos\beta\right)^{-1} \right\} dm.$$

$$(5\text{-}21)$$

An analytic evaluation of the integral in Equation (5–21) is not possible. The integration is complicated because both r' and β vary with the position of dm. However, the integration can be made tractable by approximating the integrand with a power series in r'/r and retaining terms only up to $(r'/r)^2$. For P outside the mass distribution, $r'/r < 1$. We will show that the expansion in powers of (r'/r) is equivalent to an expansion of the gravitational field in powers of a/r. This approximation yields an expression for g_m that is sufficiently accurate for our purposes. Using the formulas

$$(1+\varepsilon)^{-1/2} \approx 1 - \frac{\varepsilon}{2} + \frac{3\varepsilon^2}{8} + \cdots \quad (5\text{-}22)$$

$$(1+\varepsilon)^{-1} \approx 1 - \varepsilon + \varepsilon^2 + \cdots, \quad (5\text{-}23)$$

which are approximately valid for $\varepsilon < 1$, we find

$$g_m = \frac{G}{r^2} \int \left\{ 1 + \frac{2r'}{r} \cos \beta + \frac{3r'^2}{r^2} \left(1 - \frac{3}{2} \sin^2 \beta \right) \right\} dm. \tag{5-24}$$

The integrations in Equation (5–24) can be carried out in terms of well-known physical properties of a mass distribution. The first term is just the integral of dm over the entire mass. The result is simply M. The integral of $r' \cos \beta$ over the mass distribution is a first moment of the distribution. It is by definition zero if the origin of the coordinate system is the center of mass of the distribution. Thus Equation (5–24) becomes

$$g_m = \frac{GM}{r^2} + \frac{3G}{r^4} \int r'^2 \left(1 - \frac{3}{2} \sin^2 \beta \right) dm. \tag{5-25}$$

The first term on the right of Equation (5–25) is the gravitational acceleration of a spherically symmetric mass distribution. The second term is the modification due to rotationally induced *oblateness* of the body. If higher order terms in Equations (5–24) and (5–23) had been retained, the expansion given in Equation (5–25) would have been extended to include terms proportional to r^{-5} and higher powers of r^{-1}.

We will now express the integral appearing in Equation (5–25) in terms of the moments of inertia of an axisymmetric body. We take C to be the moment of inertia of the body about the rotational or z axis defined by $\theta = 0$. This moment of inertia is the integral over the entire mass distribution of dm times the square of the perpendicular distance from dm to the rotational axis. The square of this distance is $x'^2 + y'^2$ so that we can write C as

$$C \equiv \int (x'^2 + y'^2) \, dm = \int r'^2 \sin^2 \theta' \, dm \tag{5-26}$$

because

$$x' = r' \sin \theta' \cos \psi' \tag{5-27}$$

$$y' = r' \sin \theta' \sin \psi'. \tag{5-28}$$

The moment of inertia about the x axis, which is defined by $\theta = \pi/2$, $\psi = 0$, is

$$A \equiv \int (y'^2 + z'^2) \, dm$$

$$= \int r'^2 (\sin^2 \theta' \sin^2 \psi' + \cos^2 \theta') \, dm \tag{5-29}$$

because

$$z' = r' \cos \theta'. \tag{5-30}$$

Similarly, the moment of inertia about the y axis, which is defined by $\theta = \pi/2$, $\psi = \pi/2$, is

$$B \equiv \int (x'^2 + z'^2) \, dm$$

$$= \int r'^2 (\sin^2 \theta' \cos^2 \psi' + \cos^2 \theta') \, dm. \tag{5-31}$$

For a body that is axisymmetric about the rotation or z axis, $A = B$. The addition of Equations (5–26), (5–29), and (5–31) together with the assumption of axisymmetry gives

$$A + B + C = 2 \int r'^2 \, dm = 2A + C. \tag{5-32}$$

This equation expresses the integral of $r'^2 dm$ appearing in Equation (5–25) in terms of the moments of inertia of the body.

We will next derive an expression for the integral of $r'^2 \sin^2 \beta \, dm$. Because of the axial symmetry of the body there is no loss of generality in letting the line OP in Figure 5–3 lie in the xz plane. With the help of Equation (5–32) we rewrite the required integral as

$$\int r'^2 \sin^2 \beta \, dm = \int r'^2 (1 - \cos^2 \beta) \, dm$$

$$= A + \frac{1}{2} C - \int r'^2 \cos^2 \beta \, dm. \tag{5-33}$$

The quantity $r' \cos \beta$ is the projection of r' along OP. But this is also

$$r' \cos \beta = x' \cos \phi + z' \sin \phi, \tag{5-34}$$

where ϕ is the latitude or the angle between OP and the xy plane. Note that y' has no projection onto OP, since OP is in the xz plane. We use Equation (5–34) to rewrite the integral of $r'^2 \cos^2 \beta$ in the form

$$\int r'^2 \cos^2 \beta \, dm = \cos^2 \phi \int x'^2 \, dm$$

$$+ \sin^2 \phi \int z'^2 \, dm$$

$$+ 2 \cos \phi \sin \phi \int x' z' \, dm. \tag{5-35}$$

For an axisymmetric body,

$$\int x'^2 \, dm = \int y'^2 \, dm. \tag{5-36}$$

This result and Equation (5–26) give

$$\int x'^2\,dm = \frac{1}{2}\int (x'^2 + y'^2)\,dm = \frac{1}{2}C. \qquad (5\text{–}37)$$

The integral of $z'^2\,dm$ can be evaluated by using Equations (5–26) and (5–32)

$$\int z'^2\,dm = \int (x'^2 + y'^2 + z'^2)\,dm - \int (x'^2 + y'^2)\,dm$$

$$= \int r'^2\,dm - \int (x'^2 + y'^2)\,dm$$

$$= A - \frac{1}{2}C. \qquad (5\text{–}38)$$

With mass symmetry about the equatorial plane we have

$$\int x'z'\,dm = \int r'^2 \cos\theta' \sin\theta' \cos\psi'\,dm = 0. \quad (5\text{–}39)$$

Substitution of Equations (5–37) to (5–39) into Equation (5–35) yields

$$\int r'^2 \cos^2\beta\,dm = \frac{1}{2}C\cos^2\phi + \left(A - \frac{1}{2}C\right)\sin^2\phi. \qquad (5\text{–}40)$$

When Equations (5–33) and (5–40) are combined, we find, using $\sin^2\phi + \cos^2\phi = 1$, that

$$\int r'^2 \sin^2\beta\,dm = A\cos^2\phi + C\sin^2\phi. \qquad (5\text{–}41)$$

The gravitational acceleration is finally obtained by substituting Equations (5–32) and (5–41) into Equation (5–25):

$$g_m = \frac{GM}{r^2} - \frac{3G(C-A)}{2r^4}(3\sin^2\phi - 1). \qquad (5\text{–}42)$$

Equation (5–42) is a simplified form of *MacCullagh's formula* for an axisymmetric body. The moment of inertia about the rotational axis C is larger than the moment of inertia about an equatorial axis A because of the rotational flattening of the body. It is customary to write the difference in moments of inertia as a fraction J_2 of Ma^2, that is

$$C - A = J_2 Ma^2, \qquad (5\text{–}43)$$

where a is the Earth's equatorial radius. In terms of J_2, g_m is

$$g_m = \frac{GM}{r^2} - \frac{3GMa^2 J_2}{2r^4}(3\sin^2\phi - 1). \qquad (5\text{–}44)$$

The Earth's gravitational field can be accurately determined from the tracking of artificial satellites. The currently accepted values are:

$$a = 6378.137 \text{ km}$$

$$GM = 3.98600440 \times 10^{14} \text{ m}^3\text{s}^{-2}$$

$$J_2 = 1.0826265 \times 10^{-3}. \qquad (5\text{–}45)$$

Although a satellite is acted upon only by the Earth's gravitational acceleration, an object on the Earth's surface is also subjected to a centrifugal acceleration due to the Earth's rotation.

5-3 Centrifugal Acceleration and the Acceleration of Gravity

The force on a unit mass at the surface of the Earth due to the rotation of the Earth with angular velocity ω is the *centrifugal acceleration* g_ω. It points radially outward along a line perpendicular to the rotation axis and passing through P, as shown in Figure 5–4, and is given by

$$g_\omega = \omega^2 s, \qquad (5\text{–}46)$$

where s is the perpendicular distance from P to the rotation axis. If r is the radial distance from P to the center of the Earth and ϕ is the latitude of point P, then

$$s = r\cos\phi \qquad (5\text{–}47)$$

5-4 Centrifugal acceleration at a point on the Earth's surface.

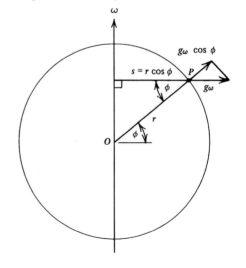

and

$$g_\omega = \omega^2 r \cos \phi. \tag{5–48}$$

The currently accepted value for the Earth's angular velocity is

$$\omega = 7.292115 \times 10^{-5} \text{ rad s}^{-1}.$$

PROBLEM 5–2 Determine the ratio of the centrifugal acceleration to the gravitational acceleration at the Earth's equator.

The gravitational and centrifugal accelerations of a mass at the Earth's surface combine to yield the acceleration of gravity g. Because $g_\omega \ll g_m$, it is appropriate to add the radial component of the centrifugal acceleration to g_m to obtain g; see Equations (5–16) and (5–17). As shown in Figure 5–4, the radial component of centrifugal acceleration points radially outward. In agreement with our sign convention that inward radial accelerations are positive, the radial component of the centrifugal acceleration is

$$g'_r = -g_\omega \cos \phi = -\omega^2 r \cos^2 \phi. \tag{5–49}$$

Therefore, the acceleration of gravity g is the sum of g_m in Equation (5–44) and g'_r:

$$g = \frac{GM}{r^2} - \frac{3GMa^2 J_2}{2r^4}(3 \sin^2 \phi - 1) - \omega^2 r \cos^2 \phi. \tag{5–50}$$

Equation (5–50) gives the radially inward acceleration of gravity for a point located on the surface of the model Earth at latitude ϕ and distance r from the center of mass.

5–4 The Gravitational Potential and the Geoid

By virtue of its position in a gravitational field, a mass m' has *gravitational potential energy*. The energy can be regarded as the negative of the work done on m' by the gravitational force of attraction in bringing m' from infinity to its position in the field. The *gravitational potential V* is the potential energy of m' divided by its mass. Because the gravitational field is *conservative*, the potential energy per unit mass V depends only on the position in the field and not on the path through which a mass is brought to the location. To calculate V for the rotationally distorted model Earth,

we can imagine bringing a unit mass from infinity to a distance r from the center of the model along a radial path. The negative of the work done on the unit mass by the gravitational field of the model is the integral of the product of the force per unit mass g_m in Equation (5–44) with the increment of distance dr (the acceleration of gravity and the increment dr are oppositely directed):

$$V = \int_\infty^r \left\{ \frac{GM}{r'^2} - \frac{3GMa^2 J_2}{2r'^4}(3 \sin^2 \phi - 1) \right\} dr' \tag{5–51}$$

or

$$V = -\frac{GM}{r} + \frac{GMa^2 J_2}{2r^3}(3 \sin^2 \phi - 1). \tag{5–52}$$

In evaluating V, we assume that the potential energy at an infinite distance from the Earth is zero. The gravitational potential adjacent to the Earth is negative; Earth acts as a potential well. The first term in Equation (5–52) is the gravitational potential of a point mass. It is also the gravitational potential outside any spherically symmetric mass distribution. The second term is the effect on the potential of the Earth model's rotationally induced oblateness. A *gravitational equipotential surface* is a surface on which V is a constant. Gravitational equipotentials are spheres for spherically symmetric mass distributions.

PROBLEM 5–3 (a) What is the gravitational potential energy of a 1-kg mass at the Earth's equator? (b) If this mass fell toward the Earth from a large distance where it had zero relative velocity, what would be the velocity at the Earth's surface? (c) If the available potential energy was converted into heat that uniformly heated the mass, what would be the temperature of the mass if its initial temperature $T_0 = 100$ K, $c = 1$ kJ kg^{-1} K^{-1}, $T_m = 1500$ K, and $L = 400$ kJ kg^{-1}?

A comparison of Equations (5–44) and (5–52) shows that V is the integral of the radial component of the gravitational acceleration g_m with respect to r. To obtain a *gravity potential U* which accounts for both gravitation and the rotation of the model Earth, we can take the integral with respect to r of the radial component of the acceleration of gravity g in Equation (5–50) with the result that

$$U = -\frac{GM}{r} + \frac{GMa^2 J_2}{2r^3}(3\sin^2\phi - 1)$$
$$- \frac{1}{2}\omega^2 r^2 \cos^2\phi. \qquad (5\text{–}53)$$

A *gravity equipotential* is a surface on which U is a constant. Within a few meters the sea surface defines an equipotential surface. Therefore, elevations above or below sea level are distances above or below a reference equipotential surface.

The reference equipotential surface that defines sea level is called the *geoid*. We will now obtain an expression for the geoid surface that is consistent with our second-order expansion of the gravity potential given in Equation (5–53). The value of the surface gravity potential at the equator is found by substituting $r = a$ and $\phi = 0$ in Equation (5–53) with the result

$$U_0 = -\frac{GM}{a}\left(1 + \frac{1}{2}J_2\right) - \frac{1}{2}a^2\omega^2. \qquad (5\text{–}54)$$

The value of the surface gravity potential at the poles must also be U_0 because we define the surface of the model Earth to be an equipotential surface. We substitute $r = c$ (the Earth's polar radius) and $\phi = \pm\pi/2$ into Equation (5–53) and obtain

$$U_0 = -\frac{GM}{c}\left[1 - J_2\left(\frac{a}{c}\right)^2\right]. \qquad (5\text{–}55)$$

The *flattening (ellipticity)* of this geoid is defined by

$$f \equiv \frac{a - c}{a}. \qquad (5\text{–}56)$$

The flattening is very slight; that is, $f \ll 1$. In order to relate the flattening f to J_2, we set Equations (5–54) and (5–55) equal and obtain

$$1 + \frac{1}{2}J_2 + \frac{1}{2}\frac{a^3\omega^2}{GM} = \frac{a}{c}\left[1 - J_2\left(\frac{a}{c}\right)^2\right]. \qquad (5\text{–}57)$$

Substituting $c = a(1 - f)$ and the neglecting quadratic and higher order terms in f and J_2, because $f \ll 1$ and $J_2 \ll 1$, we find that

$$f = \frac{3}{2}J_2 + \frac{1}{2}\frac{a^3\omega^2}{GM}. \qquad (5\text{–}58)$$

Taking $a^3\omega^2/GM = 3.46139 \times 10^{-3}$ and $J_2 = 1.0826265 \times 10^{-3}$ from Equation (5–45), we find from Equation (5–58) that $f = 3.3546 \times 10^{-3}$. Retention of

higher order terms in the theory gives the more accurate value

$$f = 3.35281068 \times 10^{-3} = \frac{1}{298.257222}. \qquad (5\text{–}59)$$

It should be emphasized that Equation (5–58) is valid only if the surface of the planetary body is an equipotential.

The shape of the model geoid is nearly that of a spherical surface; that is, if r_0 is the distance to the geoid,

$$r_0 \approx a(1 - \varepsilon), \qquad (5\text{–}60)$$

where $\varepsilon \ll 1$. By setting $U = U_0$ and $r = r_0$ in Equation (5–53), substituting Equation (5–54) for U_0 and Equation (5–60) for r_0, and neglecting quadratic and higher order terms in f, J_2, $a^3\omega^2/GM$, and ε, we obtain

$$\varepsilon = \left(\frac{3}{2}J_2 + \frac{1}{2}\frac{a^3\omega^2}{GM}\right)\sin^2\phi. \qquad (5\text{–}61)$$

The substitution of Equation (5–61) into Equation (5–60) gives the approximate model equation for the geoid as

$$r_0 = a\left\{1 - \left(\frac{3}{2}J_2 + \frac{1}{2}\frac{a^3\omega^2}{GM}\right)\sin^2\phi\right\} \qquad (5\text{–}62)$$

or

$$r_0 = a(1 - f\sin^2\phi). \qquad (5\text{–}63)$$

The nondimensional quantity $a^3\omega^2/GM$ is a measure of the relative importance of the centrifugal acceleration due to the rotation of the Earth compared with the gravitational attraction of the mass in the Earth. The rotational contribution is about 0.33% of the mass contribution.

In the preceding analysis we considered only terms linear in J_2 and $a^3\omega^2/GM$. In order to provide a reference geoid against which geoid anomalies are measured, it is necessary to include higher order terms. By convention, the *reference geoid* is a *spheroid (ellipsoid of revolution)* defined in terms of the equatorial and polar radii by

$$\frac{r_0^2 \cos^2\phi}{a^2} + \frac{r_0^2 \sin^2\phi}{c^2} = 1. \qquad (5\text{–}64)$$

The *eccentricity e* of the spheroid is given by

$$e \equiv \left(\frac{a^2 - c^2}{a^2}\right)^{1/2} = (2f - f^2)^{1/2}. \qquad (5\text{–}65)$$

It is the usual practice to express the reference geoid in terms of the equatorial radius and the flattening with the result

$$\frac{r_0^2 \cos^2 \phi}{a^2} + \frac{r_0^2 \sin^2 \phi}{a^2(1-f)^2} = 1 \qquad (5\text{-}66)$$

or

$$r_0 = a\left[1 + \frac{(2f - f^2)}{(1-f)^2} \sin^2 \phi\right]^{-1/2}. \qquad (5\text{-}67)$$

If Equation (5-67) is expanded in powers of f and if terms of quadratic and higher order in f are neglected, the result agrees with Equation (5-63). Equation (5-67) with $a = 6378.137$ km and $f = 1/298.257222$ defines the reference geoid.

The difference in elevation between the measured geoid and the reference geoid ΔN is referred to as a *geoid anomaly*. A map of geoid anomalies is given in Figure 5-5. The maximum geoid anomalies are around 100 m; this is about 0.5% of the 21-km difference between the equatorial and polar radii. Clearly, the measured geoid is very close to having the spheroidal shape of the reference geoid.

The major geoid anomalies shown in Figure 5-5 can be attributed to density inhomogeneities in the Earth. A comparison with the distribution of surface plates given in Figure 1-1 shows that some of the major anomalies can be directly associated with plate tectonic phenomena. Examples are the geoid highs over New Guinea and Chile–Peru; these are clearly associated with subduction. The excess mass of the dense subducted lithosphere causes an elevation of the geoid. The negative geoid anomaly over China may be associated with the continental collision between the Indian and Eurasian plates and the geoid low over the Hudson Bay in Canada may be associated with postglacial rebound (see Section 6-10). The largest geoid anomaly is the negative geoid anomaly off the southern tip of India, which has an amplitude of 100 m. No satisfactory explanation has been given for this geoid anomaly, which has no surface expression. A similar unexplained negative geoid anomaly lies off the west coast of North America.

The definition of geoid anomalies relative to the reference geoid is somewhat arbitrary. The reference geoid itself includes an averaging over density

5-5 Geoid height (EGM96) above reference ellipsoid WGS84 (Lemoine et al., 1998).

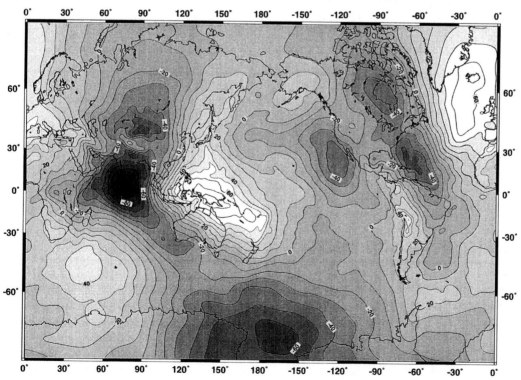

anomalies within the Earth. An alternative approach is to define geoid anomalies relative to a hydrostatic geoid. The Earth is assumed to have a layered structure in terms of density, but each layer is in hydrostatic equilibrium relative to the rotation of the Earth. The anomaly map is significantly different for the two approaches, but the major features remain unaffected.

One of the primary concerns in geodesy is to define topography and bathymetry. Both are measured relative to "sea level." Sea level is closely approximated by an equipotential surface corresponding to a constant value of U. As we have discussed, geoid anomalies relative to a reference spheroidal surface can be as large as 100 m. Thus, if we define sea level by a global spheroid we would be in error by this amount. Topography (and bathymetry) in any local area must be measured relative to a surface that approximates the local sea level (equipotential surface).

PROBLEM 5-4 Assume a large geoid anomaly with a horizontal scale of several thousand kilometers has a mantle origin and its location does not change. Because of continental drift the passive margin of a continent passes through the anomaly. Is there a significant change in sea level associated with the passage of the margin through the geoid anomaly? Explain your answer.

The anomaly in the potential of the gravity field measured on the reference geoid ΔU can be related directly to the geoid anomaly ΔN. The *potential anomaly* is defined by

$$\Delta U = U_{m0} - U_0, \qquad (5\text{--}68)$$

where U_{m0} is the measured potential at the location of the reference geoid and U_0 is the reference value of the potential defined by Equation (5–54). The potential on the measured geoid is U_0, as shown in Figure 5–6. It

can be seen from the figure that U_0, U_{m0}, and ΔN are related by

$$U_0 = U_{m0} + \left(\frac{\partial U}{\partial r}\right)_{r=r_0} \Delta N, \qquad (5\text{--}69)$$

because $\Delta N/a \ll 1$. Recall from the derivation of Equation (5–53) that we obtained the potential by integrating the acceleration of gravity. Therefore, the radial derivative of the potential in Equation (5–69) is the acceleration of gravity on the reference geoid. To the required accuracy we can write

$$\left(\frac{\partial U}{\partial r}\right)_{r=r_0} = g_0, \qquad (5\text{--}70)$$

where g_0 is the reference acceleration of gravity on the reference geoid. Just as the measured potential on the reference geoid differs from U_0, the measured acceleration of gravity on the reference geoid differs from g_0. However, for our purposes we can use g_0 in Equation (5–69) for $(\partial U/\partial r)_{r=r_0}$ because this term is multiplied by a small quantity ΔN. Substitution of Equations (5–69) and (5–70) into Equation (5–68) gives

$$\Delta U = -g_0 \Delta N. \qquad (5\text{--}71)$$

A local mass excess produces an outward warp of gravity equipotentials and therefore a positive ΔN and a negative ΔU. Note that the measured geoid essentially defines sea level. Deviations of sea level from the equipotential surface are due to lunar and solar tides, winds, and ocean currents. These effects are generally a few meters.

The reference acceleration of gravity on the reference geoid is found by substituting the expression for r_0 given by Equation (5–62) into Equation (5–50) and simplifying the result by neglecting quadratic and higher order terms in J_2 and $a^3\omega^2/GM$. One finds

$$g_0 = \frac{GM}{a^2}\left(1 + \frac{3}{2}J_2\cos^2\phi\right) + a\omega^2(\sin^2\phi - \cos^2\phi). \qquad (5\text{--}72)$$

To provide a standard *reference acceleration of gravity* against which gravity anomalies are measured, we must retain higher order terms in the equation for g_0. Gravity anomalies are the differences between measured values of g on the reference geoid and g_0. By international agreement in 1980 the reference gravity

5–6 Relationship of measured and reference geoids and geoid anomaly ΔN.

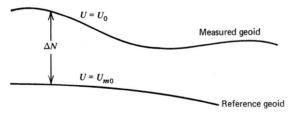

field was defined to be

$$
\begin{aligned}
g_0 = 9.7803267715(1 &+ 0.0052790414 \sin^2 \phi \\
&+ 0.0000232718 \sin^4 \phi \\
&+ 0.0000001262 \sin^6 \phi \\
&+ 0.0000000007 \sin^8 \phi),
\end{aligned}
\tag{5-73}
$$

with g_0 in m s^{-2}. This is known as the 1980 *Geodetic Reference System (GRS) (80) Formula.* The standard reference gravity field given by Equation (5–73) is of higher order in ϕ than is the consistent quadratic approximation used to specify both g_0 in Equation (5–72) and r_0 in Equation (5–67). The suitable SI unit for gravity anomalies is mm s^{-2}.

PROBLEM 5-5 Determine the values of the acceleration of gravity at the equator and the poles using GRS 80 and the quadratic approximation given in Equation (5–72).

PROBLEM 5-6 By neglecting quadratic and higher order terms, show that the gravity field on the reference geoid can be expressed in terms of the gravity field at the equator g_e according to

$$
g_0 = g_e \left[1 + \left(2 \frac{\omega^2 a^3}{GM} - \frac{3}{2} J_2 \right) \sin^2 \phi \right].
\tag{5-74}
$$

PROBLEM 5-7 What is the value of the acceleration of gravity at a distance b above the geoid at the equator $(b \ll a)$?

5-5 Moments of Inertia

MacCullagh's formula given in Equation (5–42) relates the gravitational acceleration of an oblate planetary body to its principal moments of inertia. Thus, we can use the formula, together with measurements of a planet's gravitational field by flyby or orbiting spacecraft, for example, to constrain the moments of inertia of a planet. Since the moments of inertia reflect a planet's overall shape and internal density distribution, we can use the values of the moments to learn about a planet's internal structure. For this purpose it is helpful to have expressions for the moments of inertia of some simple bodies such as spheres and spheroids.

The principal moments of inertia of a spherically symmetric body are all equal, $A = B = C$, because the mass distribution is the same about any axis passing through the center of the body. For simplicity, we will determine the moment of inertia about the polar axis defined by $\theta = 0$. For a spherical body of radius a, substitution of Equations (5–6) and (5–7) into Equation (5–26) gives

$$
C = \int_0^{2\pi} \int_0^{\pi} \int_0^a \rho(r') r'^4 \sin^3 \theta' \, dr' \, d\theta' \, d\psi'. \tag{5-75}
$$

Integration over the angles ψ' and θ' results in

$$
\int_0^{2\pi} d\psi' = 2\pi
$$

and

$$
\int_0^{\pi} \sin^3 \theta' \, d\theta' = \left[\frac{1}{3} \cos^3 \theta' - \cos \theta' \right]_0^{\pi} = \frac{4}{3},
$$

so that Equation (5–75) becomes

$$
C = \frac{8\pi}{3} \int_0^a \rho(r') r'^4 \, dr'. \tag{5-76}
$$

For a spherical body with a constant density ρ_0, the integration of Equation (5–76) gives

$$
C = \frac{8\pi}{15} \rho_0 a^5. \tag{5-77}
$$

Because the mass of the sphere is

$$
M = \frac{4}{3} \pi a^3 \rho_0, \tag{5-78}
$$

the moment of inertia is also given by

$$
C = \frac{2}{5} M a^2. \tag{5-79}
$$

The dimensionless polar moments of inertia of the Earth and Moon are listed in Table 5–1. The value $C/Ma^2 = 0.3307$ for the Earth is considerably less than the value 0.4 that Equation (5–79) gives for a constant-density spherical planet. This difference is clearly associated with the Earth's high-density core. The value $C/Ma^2 = 0.3935$ for the Moon is close to the value for a constant-density planet, but does not rule out a small (radius less than about 300 km) metallic core.

PROBLEM 5-8 Consider a spherical body of radius a with a core of radius r_c and constant density ρ_c surrounded by a mantle of constant density ρ_m. Show that the moment of inertia C and mass M are given

TABLE 5–1 **Values of the Dimensionless Polar Moment of Inertia, J_2, and the Polar Flattening for the Earth, Moon, Mars, and Venus**

	Earth	Moon	Mars	Venus
C/Ma^2	0.3307007	0.3935	0.366	0.33
$J_2 \equiv \dfrac{1}{Ma^2}\left(C - \dfrac{A+B}{2}\right)$	1.0826265×10^{-3}	2.037×10^{-4}	1.96045×10^{-3}	4.458×10^{-6}
$f \equiv \dfrac{2}{(a+b)}\left(\dfrac{a+b}{2} - c\right)$	$3.35281068 \times 10^{-3}$	1.247×10^{-3}	6.4763×10^{-3}	—

by

$$C = \frac{8\pi}{15}\left[\rho_c r_c^5 + \rho_m\left(a^5 - r_c^5\right)\right] \tag{5–80}$$

$$M = \frac{4\pi}{3}\left[\rho_c r_c^3 + \rho_m\left(a^3 - r_c^3\right)\right]. \tag{5–81}$$

Determine mean values for the densities of the Earth's mantle and core given $C = 8.04 \times 10^{37}$ kg m^2, $M = 5.97 \times 10^{24}$ kg, $a = 6378$ km, and $r_c = 3486$ km.

We will next determine the principal moments of inertia of a constant-density spheroid defined by

$$r_0 = \frac{ac}{\left(a^2\cos^2\theta + c^2\sin^2\theta\right)^{1/2}}. \tag{5–82}$$

This is a rearrangement of Equation (5–64) with the colatitude θ being used in place of the latitude ϕ. By substituting Equations (5–6) and (5–7) into Equations (5–26) and (5–29), we can write the polar and equatorial moments of inertia as

$$C = \rho \int_0^{2\pi} \int_0^{r_0} \int_0^{\pi} r'^4 \sin^3\theta' \, d\theta' \, dr' \, d\psi' \tag{5–83}$$

$$A = \rho \int_0^{2\pi} \int_0^{r_0} \int_0^{\pi} r'^4 \sin\theta'$$
$$\times \left(\sin^2\theta' \sin^2\psi' + \cos^2\theta'\right) d\theta' \, dr' \, d\psi', \tag{5–84}$$

where the upper limit on the integral over r' is given by Equation (5–82) and $B = A$ for this axisymmetric body. The integrations over ψ' and r' are straightforward and yield

$$C = \frac{2}{5}\pi\rho a^5 c^5 \int_0^{\pi} \frac{\sin^3\theta' \, d\theta'}{\left(a^2\cos^2\theta' + c^2\sin^2\theta'\right)^{5/2}} \tag{5–85}$$

$$A = \frac{1}{2}C + \frac{2}{5}\pi\rho a^5 c^5 \int_0^{\pi} \frac{\cos^2\theta' \sin\theta' \, d\theta'}{\left(a^2\cos^2\theta' + c^2\sin^2\theta'\right)^{5/2}}. \tag{5–86}$$

The integrals over θ' can be simplified by introducing the variable $x = \cos\theta'$ ($dx = -\sin\theta' \, d\theta'$, $\sin\theta' = (1 - x^2)^{1/2}$) with the result

$$C = \frac{2}{5}\pi\rho a^5 c^5 \int_{-1}^{1} \frac{(1 - x^2)\, dx}{\left[c^2 + (a^2 - c^2)x^2\right]^{5/2}} \tag{5–87}$$

$$A = \frac{1}{2}C + \frac{2}{5}\pi\rho a^5 c^5 \int_{-1}^{1} \frac{x^2 \, dx}{\left[c^2 + (a^2 - c^2)x^2\right]^{5/2}}. \tag{5–88}$$

From a comprehensive tabulation of integrals we find

$$\int_{-1}^{1} \frac{dx}{\left\{c^2 + (a^2 - c^2)x^2\right\}^{5/2}} = \frac{2}{3}\frac{(2a^2 + c^2)}{c^4 a^3} \tag{5–89}$$

$$\int_{-1}^{1} \frac{x^2 \, dx}{\left\{c^2 + (a^2 - c^2)x^2\right\}^{5/2}} = \frac{2}{3}\frac{1}{c^2 a^3}. \tag{5–90}$$

By substituting Equations (5–89) and (5–90) into Equations (5–87) and (5–88), we obtain

$$C = \frac{8}{15}\pi\rho a^4 c \tag{5–91}$$

$$A = \frac{4}{15}\pi\rho a^2 c(a^2 + c^2). \tag{5–92}$$

These expressions for the moments of inertia can be used to determine J_2 for the spheroid. The substitution of Equations (5–91) and (5–92) into the definition of J_2 given in Equation (5–43), together with the equation for the mass of a constant-density spheroid

$$M = \frac{4\pi}{3}\rho a^2 c, \tag{5–93}$$

yields

$$J_2 = \frac{1}{5}\left(1 - \frac{c^2}{a^2}\right). \tag{5–94}$$

Consistent with our previous assumption that $J_2 \ll 1$ and $(1 - c/a) \ll 1$ this reduces to

$$J_2 = \frac{2}{5}\left(1 - \frac{c}{a}\right) = \frac{2f}{5}. \quad (5\text{–}95)$$

Equation (5–95) relates J_2 to the flattening of a constant-density planetary body. The deviation of the near-surface layer from a spherical shape produces the difference in polar and equatorial moments of inertia in such a body. For a planet that does not have a constant density, the deviation from spherical symmetry of the density distribution at depth also contributes to the difference in moments of inertia.

If the planetary surface is also an equipotential surface, Equation (5–58) is valid. Substitution of Equation (5–95) into that relation gives

$$f = \frac{5}{4}\frac{a^3\omega^2}{GM} \quad (5\text{–}96)$$

or

$$J_2 = \frac{1}{2}\frac{a^3\omega^2}{GM}. \quad (5\text{–}97)$$

These are the values of the flattening and J_2 expected for a constant-density, rotating planetary body whose surface is a gravity equipotential.

Observed values of J_2 and f are given in Table 5–1. For the Earth $J_2/f = 0.3229$ compared with the value 0.4 given by Equation (5–95) for a constant-density body. The difference can be attributed to the variation of density with depth in the Earth and the deviations of the density distribution at depth from spherical symmetry.

For the Moon, where a constant-density theory would be expected to be valid, $J_2/f = 0.16$. However, both J_2 and f are quite small. The observed difference in mean equatorial and polar radii is $(a + b)/2 - c = 2$ km, which is small compared with variations in lunar topography. Therefore the observed flattening may be influenced by variations in crustal thickness. Because the Moon is tidally coupled to the Earth so that the same side of the Moon always faces the Earth, the rotation of the Moon is too small to explain the observed value of J_2. However, the present flattening may be a relic of a time when the Moon was rotating more rapidly. At that time the lunar lithosphere may have thickened enough so that the strength of the elastic lithosphere was sufficient to preserve the rotational flattening.

For Mars, $a^3\omega^2/GM = 4.59 \times 10^{-3}$ and $J_2 = 1.960 \times 10^{-3}$. From Equation (5–58) the predicted value for the dynamic flattening is 5.235×10^{-3}. This compares with the observed flattening of 6.4763×10^{-3}. Again the difference may be attributed to the preservation of a fossil flattening associated with a higher rotational velocity in the past. The ratio of J_2 to the observed flattening is 0.3027; this again is considerably less than the value of 0.4 for a constant-density planet from Equation (5–95).

PROBLEM 5–9 Assuming that the difference in moments of inertia $C - A$ is associated with a near-surface density ρ_m and the mass M is associated with a mean planetary density $\bar{\rho}$, show that

$$J_2 = \frac{2}{5}\frac{\rho_m}{\bar{\rho}}f. \quad (5\text{–}98)$$

Determine the value of ρ_m for the Earth by using the measured values of J_2, $\bar{\rho}$, and f. Discuss the value obtained.

PROBLEM 5–10 Assume that the constant-density theory for the moments of inertia of a planetary body is applicable to the Moon. Determine the rotational period of the Moon that gives the measured value of J_2.

PROBLEM 5–11 Take the observed values of the flattening and J_2 for Mars and determine the corresponding period of rotation. How does this compare with the present period of rotation?

5–6 Surface Gravity Anomalies

Mass anomalies on and in the Earth's crust are a primary source of surface gravity anomalies. Let us first consider the surface gravity anomalies caused by buried bodies of anomalous density. Examples include localized mineral deposits that usually have excess mass associated with them and igneous intrusions that often have an associated mass deficiency. The gravity anomaly due to a body of arbitrary shape and density distribution can be obtained by integrating Equation (5–3) over the body. However, it is generally impossible to carry out the necessary integrals except for the simplest shapes, and numerical methods are usually required.

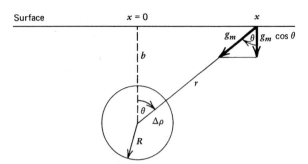

5-7 The gravitational attraction due to a sphere of anomalous density $\Delta\rho$ and radius R buried at a depth b beneath the surface.

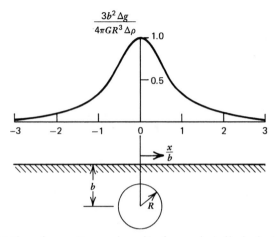

5-8 The surface gravity anomaly resulting from a spherical body of radius R whose center is at a depth b, as in Equation (5–102).

As a specific example of a buried body we consider a buried sphere of radius R with a uniform density anomaly $\Delta\rho$, as shown in Figure 5–7. It should be emphasized that the effective density in determining the surface gravity anomaly caused by a buried body is the density difference between the body and the surrounding rock. From Equation (5–15), the gravitational acceleration due to the spherical mass anomaly at a distance r from its center $(r > R)$ is

$$g_m = \frac{4\pi G R^3 \Delta\rho}{3r^2}. \tag{5-99}$$

This acceleration is directed toward the center of the sphere if $\Delta\rho$ is positive (see Figure 5–7). Because the gravitational acceleration due to the buried body is small compared with Earth's gravitational acceleration, the surface gravity anomaly Δg is just the vertical component of the surface gravitational acceleration of the body; see Equations (5–16) and (5–17). From Figure (5–7) we can write

$$\Delta g \equiv g_m \cos\theta, \tag{5-100}$$

where θ is indicated in the figure. Gravity anomalies are measured positive downward. For a point on the surface,

$$\cos\theta = \frac{b}{r} = \frac{b}{(x^2 + b^2)^{1/2}}, \tag{5-101}$$

where x is the horizontal distance between the surface point at which Δg is measured and the center of the sphere and b is the depth to the sphere's center. Substituting Equations (5–99) and (5–101) into Equation (5–100), we obtain

$$\Delta g = \frac{4\pi G R^3 \Delta\rho b}{3r^3} = \frac{4\pi G R^3 \Delta\rho}{3} \frac{b}{(x^2 + b^2)^{3/2}}. \tag{5-102}$$

The resulting gravity anomaly is plotted in Figure 5–8.

A specific example of a surface gravity anomaly caused by a density anomaly at depth is the gravity anomaly over a salt dome off the Gulf coast of the United States. A contour map of the surface gravity anomaly is given in Figure 5–9a. Measurements of the gravity on the cross section AA are given in Figure 5–9b. The measurements are compared with the theoretical gravity anomaly computed from Equation (5–102) taking $b = 6$ km and $4\pi G R^3 \Delta\rho/3b^2 = 0.1$ mm s^{-2}. Assuming that salt has a density of 2200 kg m^{-3} and that the mean density of the sediments is 2400 kg m^{-3}, we find that $R = 4.0$ km. This would appear to be a reasonable radius for an equivalent spherical salt dome.

PROBLEM 5–12 A gravity profile across the Pyramid No. 1 ore body near Pine Point, Northwest Territories, Canada, is shown in Figure 5–10. A reasonable fit with Equation (5–102) is obtained taking $b = 200$ m and $4\pi G R^3 \Delta\rho/3b^2 = 0.006$ mm s^{-2}. Assume that the gravity anomaly is caused by lead–zinc ore with a density of 3650 kg m^{-3} and that the country rock has a density of 2650 kg m^{-3}. Estimate the tonnage of lead–zinc ore, assuming a spherical body. The tonnage established by drilling in this ore body was 9.2 million tons.

PROBLEM 5–13 Show that the gravity anomaly of an infinitely long horizontal cylinder of radius R with

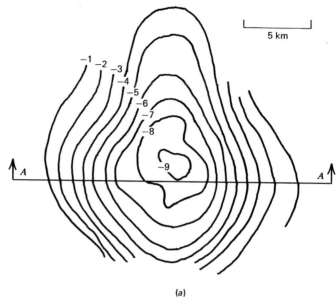

(a)

5-9 (a) Contour map (0.01 mm s^{-2} contours) of the surface gravity anomaly over a salt dome 125 miles southeast of Galveston, Texas, near the outer edge of the continental shelf (Nettleton, 1957). (b) Measurements of gravity on section AA from (a) compared with a theoretical fit based on Equation (5–102).

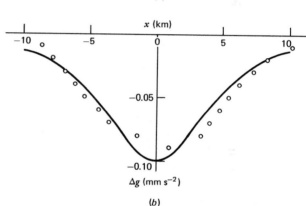

(b)

anomalous density $\Delta\rho$ buried at depth b beneath the surface is

$$\Delta g = \frac{2\pi G R^2 \Delta\rho b}{(x^2 + b^2)}, \qquad (5\text{–}103)$$

where x is the horizontal distance from the surface measurement point to the point on the surface directly over the cylinder axis. What is the maximum gravity anomaly caused by a long horizontal underground tunnel of circular cross section with a 10-m radius driven through rock of density 2800 kg m^{-3} if the axis of the tunnel lies 50 m below the surface?

PROBLEM 5–14 Calculate the gravity anomaly for a buried infinitely long horizontal line of excess mass γ per unit length by taking the limit of expression (5–103) as $R \rightarrow 0$ and $\Delta\rho \rightarrow \infty$ such that $\pi R^2 \Delta\rho \rightarrow \gamma$.

The result is

$$\Delta g = \frac{2G\gamma b}{x^2 + b^2}, \qquad (5\text{–}104)$$

where x is the horizontal distance from the surface observation point to the point directly above the line source. By integrating Equation (5–104), show that the gravity anomaly of a buried infinite slab of mass excess $\Delta\rho$ and thickness h is

$$\Delta g = 2\pi G h \Delta\rho. \qquad (5\text{–}105)$$

Note that the anomaly of the infinite slab depends only on its density excess and thickness but not on its depth of burial.

PROBLEM 5–15 Integrate Equation (5–104) to find the gravity anomaly, at $x = 0$, of the buried mass sheet

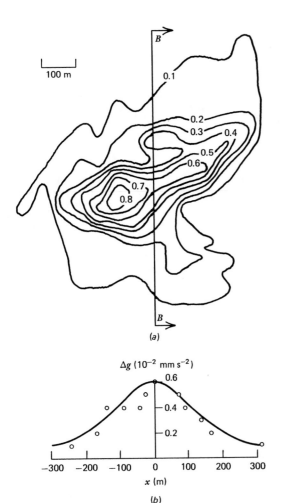

(a)

(b)

5-10 (a) Contour map (10^{-2} mm s^{-2} contours) of the surface gravity anomaly over the Pyramid No. 1 ore body (Seigel et al., 1968). (b) Gravity measurements on section BB from (a) compared with a theoretical fit based on Equation (5-102).

shown in Figure 5-11. The sheet extends infinitely far in the z direction and has an excess density σ per unit area. The surface gravity anomaly at $x = 0$ is given by

$$\Delta g = 2G\sigma\theta, \tag{5-106}$$

where θ is the angle defined in Figure 5-11.

5-7 Bouguer Gravity Formula

In the previous section we were concerned with surface gravity anomalies caused by buried bodies of anomalous density. Another important source of surface

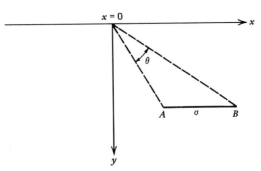

5-11 A buried sheet AB of excess mass σ per unit area.

gravity anomalies is the gravitational effect of the mass associated with topography. In general this effect can be determined by direct integration of Equation (5-3); however, such a procedure usually requires numerical calculations and is extremely tedious. Because almost all topography has a relatively shallow slope, we can derive an approximate expression for the gravitational effect of topography as well as other shallow density anomalies.

To determine the gravitational attraction of the topography immediately beneath an observer, we will consider an observer who is located a distance b above the upper surface of a cylindrical disk of radius R and thickness h, as illustrated in Figure 5-12. The observer is on the axis of the disk whose density ρ depends on the vertical coordinate y but not on the radial coordinate r, $\rho = \rho(y)$. By symmetry, the net gravitational attraction

5-12 Coordinate system used to determine the gravitational attraction of a circular disk at a point along its axis.

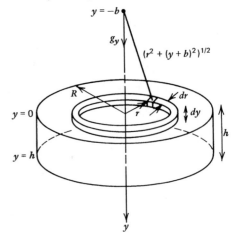

at the position of the observer due to a mass ring of vertical thickness dy and radial thickness dr is vertically downward along the axis of the cylinder. According to Equation (5–3) it is given by

$$dg_y = \frac{(2\pi r\,dr\,dy)(\rho)G}{[r^2 + (y + b)^2]}\left\{\frac{y + b}{[r^2 + (y + b)^2]^{1/2}}\right\}.$$
(5–107)

The various factors in Equation (5–107) are $2\pi r\,dr\,dy$, the volume of the ring; $r^2 + (y + b)^2$, the square of the distance between the observer and an element of the ring; and the quantity in braces, the cosine of the angle θ (see Figure 5–12) required to obtain the vertical component of the attraction of an element of the ring. Thus, the vertical component of the gravitational attraction of the entire disk on the axis of the disk at a distance b above its upper surface is given by

$$g_y = 2\pi G \int_0^h \int_0^R \frac{(b + y)r\rho(y)\,dr\,dy}{[r^2 + (b + y)^2]^{3/2}}.$$
(5–108)

We can readily integrate Equation (5–108) with respect to r to obtain

$$g_y = 2\pi G \int_0^h \rho(y)\left(1 - \frac{b + y}{[R^2 + (b + y)^2]^{1/2}}\right)dy.$$
(5–109)

An approximate result applicable to the situation of slowly varying topography and other shallow density anomalies is obtained by taking the limit $R \to \infty$ in Equation (5–109), such that

$$g_y = 2\pi G \int_0^h \rho(y)\,dy.$$
(5–110)

This is the *Bouguer gravity formula*. It relates the surface gravity anomaly at a point to the mass excess or deficiency beneath that point ($\int_0^h \rho\,dy$ is the mass per unit surface area of the circular disk.) The result is independent of the distance b at which the observer is above the anomalous mass. Equation (5–110) is a good approximation as long as the horizontal scale over which the density changes is large compared with both h and b.

The Bouguer gravity formula is particularly useful in obtaining the gravity anomaly due to topography. If topography has a height h and a density ρ_c, the resulting gravity anomaly from Equation (5–110) is

$$\Delta g = 2\pi \rho_c GH.$$
(5–111)

This result is identical to the formula we derived for the gravity anomaly of a buried infinite slab; see Equation (5–105). With $\rho_c = 2670$ kg m^{-3} the gravity anomaly for each kilometer of elevation is $\Delta g = 1.12$ mm s^{-2}.

In deriving the Bouguer gravity formula we have assumed a planar geometry. Using this formula as a topographic correction is a good approximation only if the wavelength of the topography is small compared with the radius of the Earth.

PROBLEM 5–16 A seamount with a density of 2900 kg m^{-3} rests on the seafloor at a depth of 5 km. What is the expected surface gravity anomaly if the seamount just reaches the sea surface? (Assume the width to height ratio of the seamount is large and that it does not deflect the seafloor on which it rests.)

PROBLEM 5–17 Integrate Equation (5–109) to show that the gravity anomaly due to a vertical cylinder of constant anomalous density $\Delta\rho$ on the axis of the cylinder a distance b above its upper surface is

$$\Delta g = 2\pi G\Delta\rho\{h + (b^2 + R^2)^{1/2}$$
$$- [(b + h)^2 + R^2]^{1/2}\}.$$
(5–112)

PROBLEM 5–18 A volcanic plug of diameter 10 km has a gravity anomaly of 0.3 mm s^{-2}. Estimate the depth of the plug assuming that it can be modeled by a vertical cylinder whose top is at the surface. Assume that the plug has density of 3000 kg m^{-3} and the rock it intrudes has a density of 2800 kg m^{-3}.

PROBLEM 5–19 The lunar gravity field has been determined by the tracking of orbiting spacecraft. Figure 5–13 is a contour map of the gravity anomalies on the near side of the Moon at an altitude of 100 km above the surface. The most noticeable features are the positive anomalies coincident with the circular mare basins. These are the *lunar mascons*. Determine the surface density of the anomalous mass associated with Mare Serenitatis centered at about 30°N, 17°E.

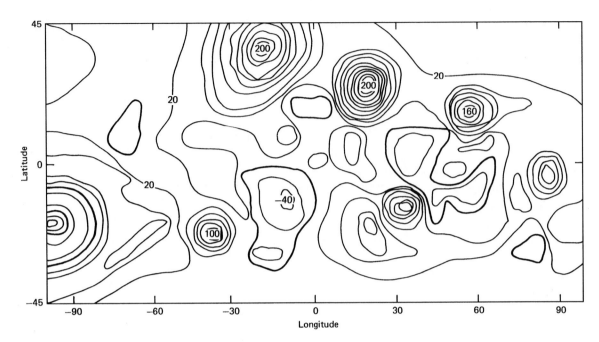

5–13 Lunar gravity anomalies at an altitude of 100 km (Sjogren, 1977). The values are in 10^{-2} mm s^{-2} and the contour interval is 0.2 mm s^{-2}.

5–8 Reductions of Gravity Data

Scientists measure gravity using a variety of gravimeters. Almost all these measuring devices are based on the simple principle that a spring is deflected as the gravitational acceleration acting on a mass attached to the spring varies. Gravimeters can easily measure variations in the gravity field of 1 part in 10^7 or 0.001 mm s^{-2}. When a surface gravity measurement is made, a series of corrections are applied in order to obtain the surface gravity anomaly. First the reference gravity field g_0 given by Equation (5–73) is subtracted out. This subtraction includes a *latitude correction*.

If the gravity measurement is carried out at an elevation h, a correction is also applied to account for the variation of gravity with elevation; this is known as the *elevation correction*. Using Equation (5–15), for example, we can relate the value of the gravitational acceleration at an elevation h above the reference geoid to the value g_0 on the reference geoid by

$$g = g_0 \frac{r_0^2}{(r_0 + h)^2} \approx g_0\left(1 - \frac{2h}{r_0}\right), \tag{5–113}$$

where r_0 is the radial position of the reference geoid given by Equation (5–67). Therefore the elevation correction Δg_h is

$$\Delta g_h = \frac{2h g_0}{r_0}. \tag{5–114}$$

The Δg_h is added to the measured gravity if the measurement is made at a point above the reference geoid. With $g_0 = 9.78$ m s^{-2} and $r_0 = 6378$ km the elevation correction at an elevation of 1 km is 3.07 mm s^{-2}. Often this correction is referred to as the *free-air correction*. When a gravity measurement has been corrected for latitude and elevation, the result is a *free-air gravity anomaly* Δg_{fa}.

At short wavelengths free-air gravity anomalies correlate strongly with local topography. To remove the gravitational attraction of the local topography, the Bouguer gravity formula is used. The *Bouguer gravity anomaly* Δg_B is given by

$$\Delta g_B = \Delta g_{fa} - 2\pi G \rho_c h. \tag{5–115}$$

This simple correction is effective in removing topographic influences if the correct crustal density is chosen and if the topography is not too steep. A typical value for the crustal density is $\rho_c = 2670$ kg m^{-3}. If steep topography is present near the measurement site, additional *terrain corrections* must be applied.

PROBLEM 5–20 The surface gravity at a measuring site is 9.803243 m s^{-2}. The site has a latitude 43°32′16″N and an elevation of 542.3 m. Obtain the free-air and Bouguer gravity anomalies.

5-9 Compensation

Although the Bouguer gravity formula is effective in removing the gravitational influence of local (short wavelength) topography, it is not effective in removing the influence of regional (long wavelength) topography. The reason for this has already been discussed in Section 3–14. A mountain or valley with a small horizontal scale, say 10 km, can be supported by the elastic lithosphere without deflection. Therefore the presence of the mountain or valley does not influence the density distribution at depth. However, the load due to a mountain range with a large horizontal scale, say 1000 km, deflects the lithosphere downward. Since the Moho is generally embedded in the lithosphere, it is also deflected downward. Because crustal rocks are lighter than mantle rocks, this results in a low-density "root" for the mountain ranges with a large horizontal scale. The mass associated with the topography of the mountains is *compensated* at depth by the low-density root.

Because the Bouguer gravity correction for topography does not account for this negative root, Bouguer gravity anomalies over mountain ranges are strongly negative. We have shown in Equation (3–115) that the negative mass of the mountain root cancels the positive mass of the mountain in the long-wavelength limit. The Bouguer gravity formula, Equation (5–110), relates the surface gravity anomaly to the net mass excess or deficiency beneath an observer. Because the condition of isostasy (hydrostatic equilibrium) gives no net mass difference, we expect that long-wavelength free-air gravity anomalies over mountain ranges are near zero. This is in fact the case.

An example of the free-air and Bouguer gravity anomalies associated with a mountain range is given in Figure 5–14. The free-air gravity anomaly is proportional to the short-wavelength topography, but it does not show any structure associated with the long-wavelength topography. The Bouguer correction removes the influence of the short-wavelength topography and smooths the profile. However, the Bouguer gravity anomaly is strongly negative, reflecting the negative density root of the long-wavelength topography.

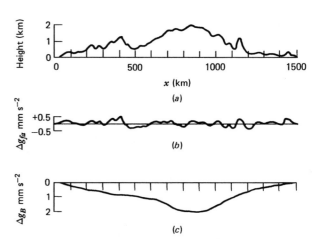

5–14 Free-air (*b*) and Bouguer (*c*) gravity anomalies associated with the topography given in (*a*).

We turn next to a quantitative consideration of the influence of lithospheric flexure on compensation and surface gravity anomalies. Before doing so, however, let us determine the surface gravity anomaly caused by a periodic density variation at depth. This will allow us to determine the surface gravity effect of a displacement of the Moho.

5-10 The Gravity Field of a Periodic Mass Distribution on a Surface

The Bouguer gravity formula derived in Section 5–7 gives the vertical component of the gravitational acceleration due to a layer of anomalous mass for an observer located immediately above the layer. This approximate formula is valid only if the observer's height above the mass layer is small relative to the layer thickness and the scale of any horizontal density variations in the layer. In this section we are interested in determining the gravitational acceleration due to a mass layer with horizontal density variations when the observer is at a height above the mass layer that is comparable to the scale of the lateral density variations in the layer. For this purpose we will consider the situation sketched in Figure 5–15. The mass is assumed to be concentrated in a layer of negligible thickness at $y = 0$. The mass per unit area of the layer σ is

$$\sigma = \lim_{h \to 0} \int_0^h \rho(y)\, dy. \qquad (5\text{–}116)$$

The layer extends to infinity in the positive and negative

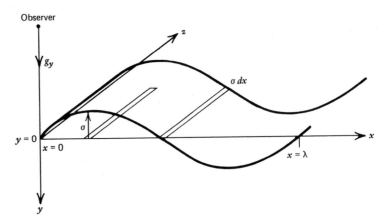

5–15 Gravitational acceleration due to a surface mass layer with horizontally varying density distribution.

z directions, and the surface density of mass is taken to vary periodically in the x direction,

$$\sigma = \sigma_0 \sin \frac{2\pi x}{\lambda}, \qquad (5\text{–}117)$$

where σ_0 is the amplitude of the surface density variation and λ is its wavelength. Just above the surface layer, at $y = 0-$, the vertical component of the gravitational acceleration of the layer is given by the Bouguer formula, Equation (5–110),

$$g_y = 2\pi G\sigma_0 \sin \frac{2\pi x}{\lambda} \qquad \text{at } y = 0 - . \qquad (5\text{–}118)$$

We are interested in determining how g_y depends on y.

The gravitational potential V associated with a mass distribution satisfies Laplace's equation outside the mass distribution. For the two-dimensional mass distribution considered here we can write

$$\frac{\partial^2 V}{\partial x^2} + \frac{\partial^2 V}{\partial y^2} = 0. \qquad (5\text{–}119)$$

The vertical component of the gravitational acceleration is related to the potential by

$$g_y = -\frac{\partial V}{\partial y}. \qquad (5\text{–}120)$$

Thus, if we take the derivative of Equation (5–119) with respect to y and substitute Equation (5–120), we obtain

$$\frac{\partial^2 g_y}{\partial x^2} + \frac{\partial^2 g_y}{\partial y^2} = 0. \qquad (5\text{–}121)$$

The vertical component of the gravitational acceleration also satisfies Laplace's equation. We can find g_y by solving the differential Equation (5–121) subject to the boundary condition imposed by Equation (5–118)

and the requirement that g_y goes to zero for observers infinitely far from the mass sheet,

$$g_y(y \to -\infty) = 0. \qquad (5\text{–}122)$$

We have already solved Laplace's equation for a half-space with periodic boundary conditions in Section 4–12. By direct analogy with Equation (4–58) we can write the solution to the problem of this section as

$$g_y = 2\pi G\sigma_0 \sin \frac{2\pi x}{\lambda} e^{2\pi y/\lambda}. \qquad (5\text{–}123)$$

The gravity anomaly decays exponentially with distance $(-y)$ from the mass layer. The length scale for the decay is $\lambda/2\pi$. Equation (5–123) provides a formula for the upward, or downward, continuation of gravity anomalies measured at a particular level above an anomalous mass distribution. Because any function of x can be Fourier-analyzed into periodic contributions of different wavelengths, any observed $g_y(x)$ can be continued to an arbitrary level by applying Equation (5–123) to the individual Fourier components of $g_y(x)$ and then reconstructing the new $g_y(x)$ by superposition of the modified components.

5–11 Compensation Due to Lithospheric Flexure

In Section 3–14 we considered the flexure of the lithosphere under periodic loading associated with the additional weight of topography. To simplify the analysis, we assumed periodic topography of the form

$$h = h_0 \sin \frac{2\pi x}{\lambda}. \qquad (5\text{–}124)$$

From Equations (3–110) and (3–111) we can write the deflection of the lithosphere w due to the loading of the topography as

$$w = \frac{h_0}{\left[\frac{\rho_m}{\rho_c} - 1 + \frac{D}{\rho_c g}\left(\frac{2\pi}{\lambda}\right)^4 \right]} \sin\frac{2\pi x}{\lambda}, \quad (5\text{–}125)$$

where ρ_c is the density of the crust, ρ_m is the density of the mantle, and D is the flexural rigidity of the lithosphere.

There are two contributions to the surface free-air gravity anomaly. The first is the contribution of the topography. From the Bouguer gravity formula, Equation (5–111), this is given by

$$\Delta g_t = 2\pi \rho_c G h_0 \sin\frac{2\pi x}{\lambda}. \quad (5\text{–}126)$$

The second is the contribution of the deflection of the Moho. The vertical deflection of the Moho is equal to the vertical deflection of the lithosphere because the Moho is assumed to be a compositional change embedded in the lithosphere. The anomalous surface mass density associated with the deflection of the Moho is

$$\sigma = (\rho_c - \rho_m)w$$
$$= \frac{-(\rho_m - \rho_c)h_0}{\left[\frac{\rho_m}{\rho_c} - 1 + \frac{D}{\rho_c g}\left(\frac{2\pi}{\lambda}\right)^4 \right]} \sin\frac{2\pi x}{\lambda}. \quad (5\text{–}127)$$

However, the Moho is buried at a mean depth b_m. Accordingly, the vertical component of the surface gravitational field due to the deflection of the Moho is obtained from Equation (5–123),

$$\Delta g_m = -\frac{2\pi G(\rho_m - \rho_c)h_0 e^{-2\pi b_m/\lambda}}{\left[\frac{\rho_m}{\rho_c} - 1 + \frac{D}{\rho_c g}\left(\frac{2\pi}{\lambda}\right)^4 \right]} \sin\frac{2\pi x}{\lambda}. \quad (5\text{–}128)$$

The surface free-air gravity anomaly is found by adding Equations (5–126) and (5–128) with the result

$$\Delta g_{fa} = \Delta g_t + \Delta g_m$$
$$= 2\pi \rho_c G \left[1 - \frac{e^{-2\pi b_m/\lambda}}{\left\{ 1 + \frac{D}{(\rho_m - \rho_c)g}\left(\frac{2\pi}{\lambda}\right)^4 \right\}} \right]$$
$$\times\, h_0 \sin\frac{2\pi x}{\lambda}. \quad (5\text{–}129)$$

From Equation (5–115) the surface Bouguer gravity anomaly is

$$\Delta g_B = \frac{-2\pi \rho_c G e^{-2\pi b_m/\lambda}}{\left[1 + \frac{D}{(\rho_m - \rho_c)g}\left(\frac{2\pi}{\lambda}\right)^4 \right]} h_0 \sin\frac{2\pi x}{\lambda}. \quad (5\text{–}130)$$

For short-wavelength topography,

$$\lambda \ll 2\pi \left[\frac{D}{(\rho_m - \rho_c)g} \right]^{1/4}, \quad (5\text{–}131)$$

the free-air gravity anomaly is

$$\Delta g_{fa} = 2\pi G \rho_c h_0 \sin\frac{2\pi x}{\lambda}, \quad (5\text{–}132)$$

and the Bouguer gravity anomaly is

$$\Delta g_B = 0. \quad (5\text{–}133)$$

The mass of the local topography is uncompensated, and the Bouguer gravity anomaly is zero.

For long-wavelength topography,

$$\lambda \gg 2\pi \left[\frac{D}{(\rho_m - \rho_c)g} \right]^{1/4} \quad (5\text{–}134)$$

and

$$\lambda \gg b_m, \quad (5\text{–}135)$$

the free-air gravity anomaly is

$$\Delta g_{fa.} = 0, \quad (5\text{–}136)$$

and the Bouguer gravity anomaly is

$$\Delta g_B = -2\pi \rho_c G h_0 \sin\frac{2\pi x}{\lambda}. \quad (5\text{–}137)$$

As shown in Equation (3–115), the surface topography is totally compensated.

The condition of isostasy, or total compensation, requires that the total mass in vertical columns be equal. This is the condition of hydrostatic equilibrium. In terms of the density distribution in the lithosphere this condition can be written as

$$\int_0^h \Delta\rho(y)dy = 0. \quad (5\text{–}138)$$

Because this is the integral that appears in the Bouguer formula, Equation (5–110), the free-air gravity anomaly associated with compensated topography is identically zero.

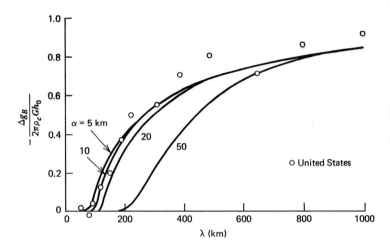

5-16 Correlation of Bouguer gravity anomalies with topography for the United States (Dorman and Lewis, 1972) compared with Equation (5–130).

The correlation of Bouguer gravity anomalies with topography is shown as a function of wavelength for the United States in Figure 5–16. It is seen that topography with a wavelength less than about 100 km is not compensated. Also included in Figure 5–16 is the predicted correlation of the Bouguer gravity anomaly with topography given by Equation (5–130). In making this comparison, we have taken $\rho_m = 3400$ kg m^{-3}, $\rho_c = 2700$ kg m^{-3}, $b_m = 30$ km, and $\alpha = [4D/(\rho_m - \rho_c)g]^{1/4} = 5$, 10, 20, and 50 km. Although there is considerable scatter in the data, reasonable agreement is obtained taking $\alpha \approx 20$ km or $D \approx 10^{21}$ N m. With $E = 60$ GPa and $\nu = 0.25$, this value of D implies, from Equation (3–72), that the thickness of the elastic lithosphere is about 6 km. A large fraction of the significant topography in the United States is in the West, where much of the area has high heat flow and active volcanism; therefore it is not too surprising that the derived thickness of the elastic lithosphere is small.

The ratio of the Bouguer gravity anomaly Δg_B to topography h as given in Figure 5–16 is often referred to as the *admittance*. This is a term used when the correlation between two quantities is obtained as a function of wavelength. It must be emphasized that any correlation between gravity and topography in ancient terrains must be considered critically. Erosion eliminates topography on a wide range of time scales. If a lithospheric plate is loaded by topography and that topography is subsequently eroded, then, according to the theory given here, the associated gravity anomaly is eliminated. However, if the thickness of the lithosphere changes between the time when topography is

created and when it is removed, isostatic displacements of the Moho can be preserved even though the associated topography is removed. The result is a *buried load* that causes a gravity anomaly. Several regions in the United States have significant gravity anomalies in areas of very flat topography. For example, the buried load believed to be responsible for the origin of the Michigan basin as discussed in Section 3–18 produces a strong surface gravity anomaly but no significant surface topography.

5-12 Isostatic Geoid Anomalies

In the previous section we showed that mass anomalies in the crust and upper mantle that extend over distances greater than a few hundred kilometers are completely compensated. We would like to learn how the mass deficiencies that balance the topographic mass excesses in isostatically compensated regions are distributed with depth. Although gravity anomalies can tell us that a region is isostatically compensated, they are not the best sources of information on the dependence of density on depth. This is because the net mass density σ defined in Equation (5–116) is identically zero in regions of isostatic compensation. The free-air gravity anomaly is approximately proportional to σ, according to the Bouguer formula (5–110), and thus $\Delta g_{fa} \approx 0$ for isostatically compensated topography. The nearly zero value of Δg_{fa} in an isostatically compensated region tells us only that $\int_0^h \Delta\rho\, dy = 0$; an infinite number of density distributions $\rho(y)$ satisfy this integral constraint.

In this section we show that geoid anomalies, or perturbations in the Earth's gravity equipotential surface, are nonzero in isostatically compensated regions and that they measure the *dipole moment of the density distribution*

$$\int_0^h \Delta\rho(y) y\, dy.$$

Thus they provide additional information on the distribution of density with depth and on the mechanisms of compensation in the lithosphere.

To calculate the geoid or gravity equipotential of topography and other shallow density anomalies, we proceed as in Section 5–7, where we derived the Bouguer gravity formula for such features. Recall that the starting point of that derivation was the determination of the gravitational acceleration of a cylindrical disk at a point on the axis of the disk a distance b above the top surface of the disk (see Figure 5–12). Let us begin by developing an expression for the gravitational potential of the disk at the same point of observation. We can find the gravitational potential of the cylindrical disk in Figure 5–12 by integrating the formula for the potential dU of a mass ring of radius r, cross section $dr\, dy$, and density $\rho(y)$ over the volume of the disk. From the discussion in Section (5–4) it is clear that dU is given by

$$dU = -\frac{G 2\pi r \Delta\rho(y)\, dr\, dy}{[r^2 + (y+b)^2]^{1/2}} \qquad (5\text{–}139)$$

because all the mass in the ring is at the same distance from the point on the axis at $y = -b$. Thus, the expression (5–139) for the potential anomaly ΔU on the axis of the disk a distance b above its upper surface is given by

$$\Delta U = -2\pi G \int_0^h \int_0^R \frac{r\, \Delta\rho(y)\, dr\, dy}{[r^2 + (b+y)^2]^{1/2}}. \qquad (5\text{–}140)$$

Integration with respect to r yields

$$\Delta U = -2\pi G \int_0^h \Delta\rho(y) \big\{ [R^2 + (b+y)^2]^{1/2} - (b+y) \big\}\, dy. \qquad (5\text{–}141)$$

We again assume that the density anomalies are slowly varying horizontally and take the limit $R \to \infty$. To do this, we expand the R-dependent term of the integrand of Equation (5–141) in powers of $1/R$ using the

binomial theorem

$$[R^2 + (b+y)^2]^{1/2} = R\left(1 + \frac{(b+y)^2}{R^2}\right)^{1/2}$$

$$\approx R\left(1 + \frac{1}{2}\left(\frac{b+y}{R}\right)^2 + \cdots\right). \qquad (5\text{–}142)$$

Equation (5–141) becomes

$$\Delta U = -2\pi G \left\{ R \int_0^h \Delta\rho(y)\, dy - \int_0^h (y+b)\Delta\rho(y)\, dy + \cdots \right\}. \qquad (5\text{–}143)$$

However, the condition of isostasy requires, from Equation (5–138), that the first integral in Equation (5–143) be zero. Therefore Equation (5–143) reduces to

$$\Delta U = 2\pi G \int_0^h y\Delta\rho(y)\, dy. \qquad (5\text{–}144)$$

The gravitational potential anomaly due to a shallow, long wavelength isostatic density distribution is proportional to the dipole moment of the density distribution beneath the point of measurement.

The anomaly in the geopotential has been related to the geoid anomaly in Equation (5–71). Substitution of Equation (5–144) into Equation (5–71) gives

$$\Delta N = -\frac{2\pi G}{g} \int_0^h y\Delta\rho(y)\, dy. \qquad (5\text{–}145)$$

Geoid height anomalies associated with long-wavelength isostatic density anomalies are directly proportional to the dipole moment of the density distribution. The dipole moment of $\Delta\rho(y)$ is nonzero, and the first moment of the density, that is, the net mass, is zero for isostatic density distributions. Thus observed geoid height anomalies are a direct measure of the lowest order nonzero moment of the density–depth profile. Geoid anomalies are directly measurable over oceanic areas because the surface of the ocean closely conforms to the geoid. Accurate *geoid height contour maps* over oceanic areas have been compiled by measuring sea surface heights with satellite altimeters and correcting for the small disturbing effects of currents, tides, etc. An example is given in Figure 5–17. A comparison with

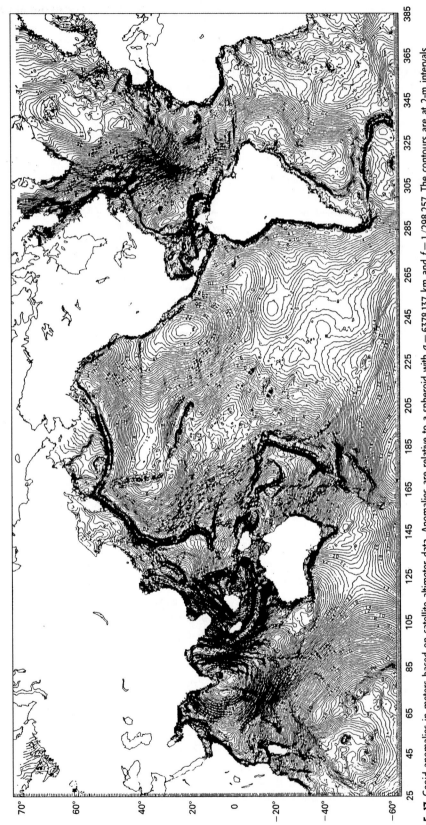

5-17 Geoid anomalies in meters based on satellite altimeter data. Anomalies are relative to a spheroid with $a = 6378.137$ km and $f = 1/298.257$. The contours are at 2-m intervals (Marsh et al., 1986).

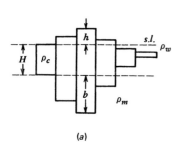

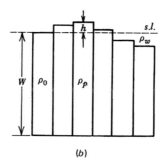

(a)   (b)

5–18 Airy (*a*) and Pratt (*b*) models of isostatic compensation.

the satellite-derived worldwide geoid anomaly map given in Figure 5–5 shows that measurements of sea surface topography give much more short-wavelength detail.

PROBLEM 5–21 Show that the result in Equation (5–145) is independent of the origin of the coordinate y. HINT: Use the isostatic condition given in Equation (5–138).

5–13 Compensation Models and Observed Geoid Anomalies

The density compensation required by isostasy can be accomplished in several ways. We discuss three simplified models for compensation; clearly, compensation in the lithosphere may be a complex combination of these models. We previously discussed in Section 2–2 compensation by the depression of the Moho as a consequence of topographic loading. This is known as *Airy compensation* and is illustrated in Figure 5–18*a*. The density of the crust ρ_c and the mantle ρ_m are assumed to be constant. The thickness of continental crust with zero elevation, with respect to sea level, is H; crust with an elevation h has a crustal root of thickness b. From the principle of hydrostatic equilibrium we find

$$b = \frac{\rho_c h}{\rho_m - \rho_c}. \tag{5–146}$$

If the height of the topography is negative and it is covered with water, then

$$b = \left(\frac{\rho_c - \rho_w}{\rho_m - \rho_c}\right)h. \tag{5–147}$$

Taking continental crust with zero elevation as the reference, the geoid anomaly associated with compen-

sated positive topography from Equation (5–145) is

$$\Delta N = -\frac{2\pi G}{g}\left\{\int_H^{H+b} y(\rho_c - \rho_m)\,dy + \int_{-h}^0 y\rho_c\,dy\right\}$$
$$= \frac{\pi G}{g}\rho_c\left\{2Hh + \frac{\rho_m}{(\rho_m - \rho_c)}h^2\right\}, \tag{5–148}$$

where Equation (5–146) has been used to substitute for the thickness of the root b. In writing Equation (5–148), we measured y positive downward from sea level. Because a crust of thickness H with zero elevation lying above a mantle of density ρ_m is the reference state, the anomalous density of positive topography is ρ_c, and the anomalous density of a crustal root is $\rho_c - \rho_m$. For topography below sea level (h negative) the geoid anomaly is given by

$$\Delta N = \frac{-2\pi G}{g}\left\{\int_0^{-h} dy\,y(\rho_w - \rho_c)\right.$$
$$\left. + \int_{H+b}^{H} dy\,y(\rho_m - \rho_c)\right\}$$
$$= \frac{\pi G}{g}(\rho_c - \rho_w)\left\{2Hh + \left(\frac{\rho_m - \rho_w}{\rho_m - \rho_c}\right)h^2\right\}. \tag{5–149}$$

The Airy geoid anomaly based on Equations (5–148) and (5–149) with $\rho_m = 3300$ kg m^{-3}, $\rho_c = 2800$ kg m^{-3}, and $H = 30$ km is given in Figure 5–19 as a function of elevation. A geoid anomaly of about 5 m is expected for each kilometer of elevated topography and a negative anomaly of less than 2 m is anticipated for each kilometer of topography below sea level.

The observed geoid anomaly across the Atlantic continental margin of North America at 40.5°N is shown in Figure 5–20*a*. This geoid anomaly was

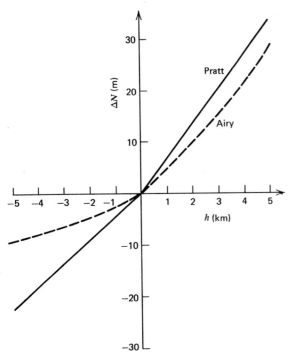

5-19 Geoid anomaly as a function of topographic elevation above and below sea level. For Pratt compensation $\rho_0 = 3100$ kg m^{-3} and $W = 100$ km. For Airy compensation $\rho_m = 3300$ kg m^{-3}, $\rho_c = 2800$ kg m^{-3}, and $H = 30$ km.

obtained using a radar altimeter from a satellite to determine the position of the sea surface. Let us assume that Airy isostasy is applicable across this passive continental margin and determine the predicted

5-20 (a) Observed geoid anomaly across the Atlantic continental margin of North America at 40.5°N compared with the predicted anomaly from Equation (5–149). (b) The distribution of density used in the calculation.

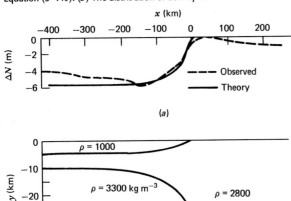

geoid anomaly from Equation (5–149). In making the comparison shown in Figure 5–20a, we assumed $\rho_c = 2800$ kg m^{-3}, $\rho_m = 3300$ kg m^{-3}, and $H = 30$ km. The assumed density distribution corresponding to the observed bathymetry is given in Figure 5–20b. Quite good agreement between observation and theory is obtained, even though the wavelength of the geoid anomaly is relatively small. This is evidence that passive continental margins are close to isostatic equilibrium.

PROBLEM 5-22 It is sometimes a better approximation to divide the continental crust into upper and lower crusts. If the lower crust has a constant thickness b_L and a density ρ_{cL} and the upper crust has a variable thickness with a density ρ_{cU}, determine the geoid anomaly associated with isostatically compensated positive topography.

PROBLEM 5-23 Consider the formation of a sedimentary basin on the seafloor. Suppose isostatic compensation is achieved by the displacement of mantle material of density ρ_m. Show that sediment thickness s is related to water depth d by

$$s = \frac{(\rho_m - \rho_w)}{(\rho_m - \rho_s)}(D - d), \qquad (5\text{–}150)$$

where D is the initial depth of the sediment-free ocean. What is the maximum possible thickness of the sediment if $\rho_s = 2500$ kg m^{-3}, $\rho_m = 3300$ kg m^{-3}, and $D = 5$ km?

An alternative model for isostatic compensation uses horizontal variations in density over a prescribed depth W. This is known as *Pratt compensation* and is illustrated in Figure 5–18b. The variable density ρ_p is related to the elevation above sea level by

$$\rho_p = \rho_0 \left(\frac{W}{W + h} \right), \qquad (5\text{–}151)$$

where ρ_0 is the reference density corresponding to zero elevation and W is referred to as the *depth of compensation*. For topography below sea level (h negative) the variable density is given by

$$\rho_p = \frac{\rho_0 W + \rho_w h}{W + h}. \qquad (5\text{–}152)$$

Again taking continental crust with zero elevation as the reference, the geoid anomaly associated with compensated positive topography is

$$\Delta N = \frac{-2\pi G}{g} \left\{ \int_{-h}^{0} \rho_p y \, dy + \int_{0}^{W} (\rho_p - \rho_0) y \, dy \right\}$$
$$= \frac{\pi G}{g} \rho_0 W h, \qquad (5\text{–}153)$$

where we have used Equation (5–151) to eliminate ρ_p. Similarly, the geoid anomaly of compensated negative topography is

$$\Delta N = \frac{-2\pi G}{g} \left\{ \int_{0}^{-h} (\rho_w - \rho_0) y \, dy \right.$$
$$\left. + \int_{-h}^{W} (\rho_p - \rho_0) y \, dy \right\}$$
$$= \frac{\pi G}{g} (\rho_0 - \rho_w) W h. \qquad (5\text{–}154)$$

The geoid anomaly is linearly dependent on the topography. With $\rho_0 = 3100$ kg m^{-3} and $W = 100$ km, the geoid–topography ratio (GTR) = 6.6 m km^{-1} for positive topography from Equation (5–153). Similarly, with $\rho_w = 1000$ kg m^{-3} and these same values of ρ_0 and W, the GTR = 4.5 m km^{-1} for negative topography from Equation (5–154). The geoid anomaly of the Pratt model is shown in Figure 5–19 as a function of topographic elevation.

Hotspot swells are areas of anomalously shallow topography associated with hotspot volcanics (see Section 1–6). Two examples are the Hawaiian swell in the Pacific Ocean and the Bermuda swell in the Atlantic Ocean. The dependence of the observed geoid anomaly across each of these swells as a function of anomalous bathymetry is given in Figure 5–21.

One hypothesis for the origin of this anomalously shallow topography is the thickening of the oceanic crust. Assuming a reference thickness of the oceanic crust to be $H = 6$ km with $\rho_c = 2900$ kg m^{-3} and $\rho_m = 3300$ kg m^{-3}, the predicted geoid anomaly from Equation (5–149) is given in Figure 5–21. Clearly the observed geoid anomalies are much larger than those predicted by crustal thickening.

In Figure 5–21 we also compare the observed geoid anomalies across the Hawaiian and Bermudas wells with the predicted geoid anomalies due to Pratt compensation. The Pratt geoid anomaly is obtained from Equation (5–154) with $\rho_0 = 3300$ kg m^{-3} and $W = 75$,

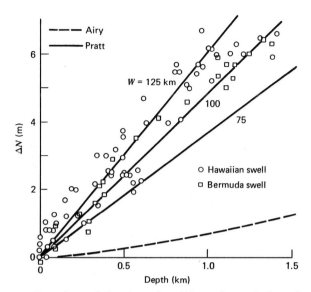

5–21 Dependence of the observed geoid anomalies on bathymetric anomalies across the Hawaiian swell (Crough, 1978) and across the Bermuda swell (Haxby and Turcotte, 1978) compared with the predicted dependence for crustal thickening (Airy compensation) and Pratt compensation with various depths of compensation.

100, and 125 km. Within the scatter of the data, good agreement is obtained for Pratt compensation with a depth of compensation of about 100 km. If we accept the Pratt model to be applicable, the conclusion is that the mantle rocks beneath the Hawaiian and Bermuda swells have anomalously low densities to depths of approximately 100 km.

A third type of isostatic compensation is *thermal isostasy*. This has been considered in Section 4–23 and is relevant to the oceanic lithosphere. The oceanic lithosphere is created from hot mantle rock (temperature T_1) at an ocean ridge. The lithosphere cools and thickens by heat transfer to the surface. As the oceanic lithosphere cools its density increases; as a result it subsides. We refer to this type of subsidence as thermal isostasy.

With the oceanic ridge crest taken as the reference density distribution, the geoid anomaly associated with the subsiding thermally compensated oceanic lithosphere can be written, using Equation (5–145), as

$$\Delta N = \frac{-2\pi G}{g} \left\{ \int_{-w}^{0} y(\rho_w - \rho_m) \, dy \right.$$
$$\left. + \int_{0}^{\infty} y(\rho - \rho_m) \, dy \right\}. \qquad (5\text{–}155)$$

The first term in Equation (5–155) can be integrated directly, and the second term can be rewritten by using Equation (4–205) relating density to temperature. The result is

$$\Delta N = \frac{-2\pi G}{g} \left\{ \frac{(\rho_m - \rho_w)w^2}{2} + \alpha \rho_m \int_0^\infty y(T_1 - T)\, dy \right\}. \tag{5–156}$$

By using Equation (4–209) for the ocean floor depth w and Equation (4–125) for the temperature distribution in the lithosphere, we can obtain the following simple formula for the geoid anomaly over a spreading ridge:

$$\Delta N = -\frac{2\pi G \rho_m \alpha (T_1 - T_0)\kappa}{g}$$
$$\times \left\{ 1 + \frac{2\rho_m \alpha (T_1 - T_0)}{\pi(\rho_m - \rho_w)} \right\} t. \tag{5–157}$$

This geoid anomaly is a linear function of the age of the seafloor. Taking $\rho_m = 3300$ kg m^{-3}, $\kappa = 1$ mm^2 s^{-1}, $T_1 - T_0 = 1200$ K, and $\alpha = 3 \times 10^{-5}$ K^{-1}, we find that the geoid anomaly decreases at the constant rate of 0.16 m Myr^{-1}. The geoid anomaly calculated from Equation (5–157) is compared with a measured geoid anomaly across the mid-Atlantic ridge in Figure 5–22. Clearly, good agreement is obtained.

PROBLEM 5–24 The mean geoid height over the continents is very nearly equal to the mean geoid height over the ocean basins. The positive geoid anomaly associated with the thicker continental crust is nearly cancelled by the negative geoid anomaly associated

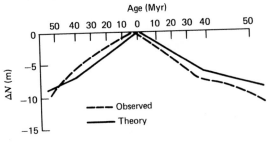

5–22 The observed geoid anomaly across the mid-Atlantic ridge at 44.5°N (referenced to the ridge crest) compared with the predicted anomaly from Equation (5–157).

with the thicker continental lithosphere. Assume that the two contributions are equal and determine the thickness of the continental lithosphere. Also assume that the temperature profiles in both the oceanic and continental crusts are given by Equation (4–124). Take $\rho_m = 3300$ kg m^{-3}, $\rho_c = 2800$ kg m^{-3}, $H = 35$ km, $y_{LO} = 100$ km, $\alpha = 3 \times 10^{-5}$ K^{-1}, and the depth of the ocean basins is 5.5 km. Neglect the contribution of the ocean crust.

PROBLEM 5–25 Determine the geoid offset across the fracture zone considered in Problem 4–52 (illustrated in Figure 4–46) assuming the applicability of the half-space cooling model. Use the parameter values given in Problem 4–52.

In Section 4–17 we introduced the plate cooling model as an alternative to the half-space cooling model. The geoid anomaly associated with the subsiding, thermally compensated oceanic lithosphere as predicted by the plate model is obtained by substituting Equation (4–130) for the temperature distribution into Equation (5–156). After carrying out the necessary integrals we obtain

$$\Delta N = -\frac{2\pi G}{g} \left\{ \frac{(\rho_1 - \rho_w)w^2}{2} + \alpha \rho_1 (T_1 - T_0) y_{L0}^2 \right.$$
$$\left. \times \left[\frac{1}{6} + \frac{2}{\pi^2} \sum_{n=1}^\infty \frac{(-1)^n}{n^2} \exp\left(-\frac{\kappa n^2 \pi^2 t}{y_{L0}^2} \right) \right] \right\}, \tag{5–158}$$

with w given by Equation (4–211). For large times, $t \gg y_{L0}^2/\kappa$, the equilibrium value of the geoid ΔN_e is given by

$$\Delta N_e = -\frac{2\pi G \alpha \rho_1 (T_1 - T_0) y_{L0}^2}{g} \left[\frac{1}{6} + \frac{\alpha \rho_1 (T_1 - T_0)}{8(\rho_1 - \rho_w)} \right]. \tag{5–159}$$

This is the predicted difference in the geoid between ocean ridges and ocean basins. For the parameter values used above and $y_{L0} = 95$ km, we find $\Delta N_e = -8.63$ m; with $y_{L0} = 125$ km, we find $\Delta N_e = -14.9$ m.

Again, we approximate the deviation of the geoid from the equilibrium value by retaining the first terms of the expansions given in Equations (4–211) and

(5–159) with the result

$$\Delta N = -\frac{2\pi G\rho_1\alpha(T_1 - T_0)y_{L0}^2}{g}\left\{\left[\frac{1}{6} + \frac{\alpha\rho_1(T_1 - T_0)}{8(\rho_1 - \rho_w)}\right]\right.$$
$$-\frac{2}{\pi^2}\left[1 + \frac{\rho_1\alpha(T_1 - T_0)}{(\rho_1 - \rho_w)}\right]\exp\left(-\frac{\kappa\pi^2 t}{y_{L0}^2}\right)$$
$$+\frac{8\rho_1\alpha(T_1 - T_0)}{\pi^4(\rho_1 - \rho_w)}\exp\left(-\frac{2\kappa\pi^2 t}{y_{L0}^2}\right)$$
$$\left.+\frac{1}{2\pi^2}\exp\left(-\frac{4\kappa\pi^2 t}{y_{L0}^2}\right)\right\}. \qquad (5\text{–}160)$$

5-14 Forces Required to Maintain Topography and the Geoid

In Section 2–2 we determined the horizontal force in the lithosphere required to maintain differences in topography by integrating the lithostatic pressure over the thickness of the lithosphere. This problem was illustrated in Figure 2–8. The resulting horizontal stress component was given in Equation (2–17). We will now show that this force difference is proportional to the difference in geoid height between the two points considered.

We consider a section of continental crust and lithosphere as illustrated in Figure 5–23 with a vertical distribution of density $\rho(y)$ to a depth of compensation h; reference lithosphere has a constant density ρ_m. Isostasy requires that

$$\int_0^h \rho(y)\,dy = b\rho_m. \qquad (5\text{–}161)$$

The horizontal force within the continental crust F_1 is obtained by integrating the lithostatic pressure over the depth of compensation with the result

$$F_1 = \int_0^h g\left[\int_0^y \rho(y')dy'\right]dy. \qquad (5\text{–}162)$$

5-23 Force balance on a section of continental crust and lithosphere.

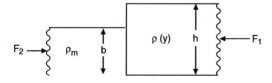

The horizontal force F_2 in the reference lithosphere is

$$F_2 = \int_0^b g\rho_m y\,dy = \frac{1}{2}g\rho_m b^2. \qquad (5\text{–}163)$$

The net horizontal force on the lithosphere F_R is

$$F_R = F_1 - F_2$$
$$= g\int_0^h\left[\int_0^y \rho(y')\,dy'\right]dy - \frac{1}{2}g\rho_m b^2. \qquad (5\text{–}164)$$

The integral in Equation (5–164) can be evaluated using the method of integration by parts, which, in general, gives

$$\int_a^b f(y)\frac{dg(y)}{dy}\,dy = f(b)g(b) - f(a)g(a)$$
$$-\int_a^b g(y)\frac{df(y)}{dy}\,dy. \qquad (5\text{–}165)$$

If we let

$$f(y) = \int_0^y \rho(y')\,dy', \quad g(y) - y,$$

then using Equation (5–165) to evaluate the integral in Equation (5–164) gives

$$\int_0^h\left[\int_0^h \rho(y')\,dy'\right]dy = h\int_0^h \rho(y)\,dy$$
$$-\int_0^h y\rho(y)\,dy. \qquad (5\text{–}166)$$

The isostasy condition, Equation (5–161), can be used to further simplify Equation (5–166) with the result

$$\int_0^h\left[\int_0^h \rho(y')\,dy'\right]dy = hb\rho_m - \int_0^h y\rho(y)\,dy. \qquad (5\text{–}167)$$

Upon substituting this result into Equation (5–164) we find

$$F_R = g\left[\rho_m\left(hb - \frac{1}{2}b^2\right) - \int_0^h y\rho(y)\,dy\right]. \qquad (5\text{–}168)$$

We next evaluate the difference in the gravitational potential between sections 1 and 2 using Equation (5–144) and obtain

$$\Delta U = U_1 - U_2 = 2\pi G \left[\int_0^{h-b} y\rho(y)\,dy \right.$$
$$+ \int_{h-b}^h y[\rho(y) - \rho_m]\,dy \bigg]$$
$$= 2\pi G \left[\int_0^h y\rho(y)\,dy - \rho_m \left(hb - \frac{1}{2}b^2 \right) \right].$$
$$(5\text{–}169)$$

A comparison of Equations (5–168) and (5–169) using Equation (5–71) gives

$$F_R = -\frac{g}{2\pi G}\Delta U = \frac{g^2}{2\pi G}\Delta N. \qquad (5\text{–}170)$$

Thus, the horizontal body force on the lithosphere is proportional to the surface geoid anomaly. Although this result was derived for a specific configuration, the result is generally valid under the same conditions that Equation (5–144) is valid.

For example, determine the ridge push force on the oceanic lithosphere assuming the validity of the plate cooling model. The difference in geoid between an ocean ridge and the adjacent ocean basin was given in Equation (5–159). Substitution of this into Equation (5–170) gives the ridge push force per unit ridge length:

$$F_{RP} = g\alpha\rho_m(T_1 - T_0)y_{L0}^2 \left[\frac{1}{6} + \frac{\alpha\rho_m(T_1 - T_0)}{8(\rho_m - \rho_w)} \right]. \qquad (5\text{–}171)$$

For the parameter values we have used previously and $y_{L0} = 125$ km we find that $F_{RP} = 3.41 \times 10^{12}$ N m^{-1}. If this force is distributed uniformly over a 100-km depth, the resulting compressional stress in the lithosphere is $\sigma_{xx} = 34.1$ MPa.

REFERENCES

Crough, S. T. (1978), Thermal origin of midplate hot-spot swells, *Geophys. J. Royal Astron. Soc.* **55**, 451–469.

Dorman, L. M., and B. T. R. Lewis (1972), Experimental isostasy. 3. Inversion of the isostatic Green's function and lateral density changes, *J. Geophys. Res.* **77**, 3068–3077.

Haxby, W. F., and D. L. Turcotte (1978), On isostatic geoid anomalies, *J. Geophys. Res.* **83**, 5473–5478.

Lemoine, F. G., S. C. Kenyon, J. K. Factor, R. G. Trimmer, N. K. Palvis, D. S. Chinn, C. M. Cox, S. M. Klosko, S. B. Luthcke, M. H. Torrence, Y. M. Wang, R. G. Williamson, E. C. Palvis, R. H. Rapp, and T. R. Olson (1998), The development of the joint NASA GSFC and the National Imagery and Mapping Agency (NIMA) Geopotential Model EGM96, NASA/TP-1998-206861.

Marsh, J. G., A. C. Brenner, B. D. Beckley, and T. V. Martin (1986), Global mean sea surface based on Seasat altimetry data, *J. Geophys. Res.* **91**, 3501–3506.

Nettleton, L. L. (1957), Gravity survey over a Gulf Coast continental shelf mound, *Geophysics* **22**, 630–642.

Seigel, H. O., H. L. Hill, and J. G. Baird (1968), Discovery case history of the Pyramid ore bodies Pine Point, Northwest Territories, Canada, *Geophysics* **33**, 645–656.

Sjogren, W. L. (1977), Lunar gravity determinations and their implications, *Phil. Trans. R. Soc. Lond. A* **285**, 219–226.

COLLATERAL READING

Bomford, G., *Geodesy*, 4th ed. (Oxford University Press, London, 1980), 561 pages.

See collateral reading list for Chapter 2.

Dobrin, M. B., *Introduction to Geophysical Prospecting*, 3rd ed. (McGraw-Hill Book Company, New York, 1976), 630 pages.

A textbook on the fundamental principles and techniques of geophysical prospecting. Principal emphasis is on gravity, magnetic, and seismic techniques. One chapter deals with electrical prospecting methods. The book is intended for advanced undergraduate and graduate students in geology, mining engineering, petroleum engineering, and mineral exploration.

Garland, G. D., *Introduction to Geophysics* (W. B. Saunders Company, Philadelphia, 1979), 494 pages.

A textbook on the physics of the solid Earth suitable for graduate and advanced undergraduate students. It extensively discusses seismology, gravity, magnetism, heat transfer, the Earth's thermal state, and geodynamics. There are appendixes with basic data, problems at the end of each chapter, and bibliographical citations to original research papers.

Grant, F. S., and G. F. West, *Interpretation Theory in Applied Geophysics* (McGraw-Hill Book Company, New York, 1965), 584 pages.

A basic textbook in applied geophysics presenting the mathematical and physical foundations for interpreting observational data in the areas of seismology, gravimetric and magnetometric surveying, and electrical and electromagnetic exploration. Six chapters on seismology introduce seismic methods, elastic waves in layered media, analysis of seismic records, seismic interpretation, and reflection and refraction of spherical waves. Six chapters on gravity and magnetic methods cover potential field theory, reduction and interpretation of gravity data, quantitative interpretations of gravity and magnetic anomalies, and rock magnetism. The final six chapters deal with electrical conduction and electromagnetic induction methods.

Heiskanen, W. A., and F. A. Vening Meinesz, *The Earth and Its Gravity Field* (McGraw-Hill Book Company, New York, 1958), 470 pages.

One of the first textbooks to discuss the Earth's gravity field and the information it provides about the structure and mechanical properties of the interior. Chapters cover the internal constitution of the Earth, gravity field, gravity potential, equilibrium figure, gravity measurements, reduction of gravity measurements, isostasy, gravity anomalies, physical geodesy, deviations from isostasy, and convection currents in the Earth.

Kaula, W. M., *An Introduction to Planetary Physics* (John Wiley & Sons, New York, 1968), 490 pages.

A basic textbook on planetary physics for graduate students. While several chapters deal with aspects of the Earth's interior, the emphasis is on all the terrestrial planets. In addition to the standard topics such as gravity, seismology, and magnetism, chapters also discuss the dynamics of the Earth–Moon system, the dynamics of the solar system, the geology of the Moon and Mars, remote sensing of the planets, meteorites, and planetary origins. Each chapter contains problems for the student.

Pick, M., J. Picha, and V. Vyskocil, *Theory of the Earth's Gravity Field* (Elsevier Scientific Publishing Company, Amsterdam, 1973), 538 pages.

A fundamental textbook on gravimetry for graduate students in solid Earth geophysics and geodesy. The coverage of the subject is extensive and includes potential theory, relative measurements of the acceleration of gravity, gravity anomalies and their interpretations, gravimetry and the Earth's internal structure, the geoid, the Earth's figure, tides, and astronomical aspects. A lengthy appendix describes the mathematical techniques employed in the book.

Stacey, F. D., *Physics of the Earth*, 3rd ed. (Brookfield Press, Brisbane, 1992), 513 pages.

A fundamental textbook on geophysics for graduate and advanced undergraduate students. Topics include the Earth as a part of the solar system, radioactivity and the age of the Earth, the Earth's rotation, gravity, tides, seismology, the Earth's internal heat, geomagnetism, paleomagnetism, and tectonics. There are tables of useful data, appendixes on special topics, and problems for the student.

Torge, W., *Geodesy*, 2nd ed. (Walter de Gruyter, Berlin, 1991), 253 pages.

A basic textbook on the theoretical aspects of geodesy. Subjects include the gravity field of the Earth, geodetic reference systems, methods of measurement, methods of evaluation, and geodetic networks.

SIX

Fluid Mechanics

6-1 Introduction

Any material that flows in response to an applied stress is a *fluid*. Although solids acquire a finite deformation or strain upon being stressed, fluids deform continuously under the action of applied forces. In solids, stresses are related to strains; in fluids, stresses are related to *rates of strain*. Strains in solids are a consequence of spatial variations or gradients in the displacements of elements from their equilibrium positions. Strain rates in fluids are a result of gradients in the velocities or rates of displacement of fluid elements. Velocity gradients are equivalent to strain rates, so stresses in fluids are related to velocity gradients. The equation connecting stresses with velocity gradients in a fluid is known as the rheological law for the fluid. The simplest fluid, and as a consequence the one most often studied, is the *Newtonian* or *linear fluid*, in which the rate of strain or velocity gradient is directly proportional to the applied stress; the constant of proportionality is known as the *viscosity*. We deal only with Newtonian viscous fluids throughout this chapter. Non-Newtonian fluid behavior is discussed in Chapter 7. *Fluid mechanics* is the science of fluid motion. It uses the basic principles of *mass*, *momentum*, and *energy conservation* together with the rheological or *constitutive law* for the fluid to describe how the fluid moves under an applied force.

Many problems involving fluid mechanics arise in geodynamics. Obvious examples involve flows of groundwater and magma. Groundwater flows through underground channels known as *aquifers*. If the aquifers are sufficiently deep and pass through rock sufficiently hot, hot springs may result. In areas of active

volcanism the groundwater may be heated above the *boiling point* and *geysers* result. In some cases steam or very hot water is trapped at depth; such deposits may serve as *reservoirs for geothermal power plants*. The circulation of seawater through the oceanic crust is similar in many respects to the flow of groundwater on land. Seawater can become so hot in passing through crustal rocks near an oceanic ridge that *submarine hot springs* develop.

Geochemical studies show that magmas flowing from surface volcanoes have in some cases originated at depths of 100 km or more. Studies of extinct volcanoes show that the magma flows through *volcanic conduits* at shallow depths. These conduits have the form of nearly circular pipes or two-dimensional channels. Mechanisms for the flow of magma at depths exceeding 5 km are a subject of considerable controversy. Alternative hypotheses involve propagating *fractures*, large bodies of ascending magma, and continuous conduits.

In terms of geodynamics, however, one of our principal interests is *mantle convection*. The fluid behavior of the mantle is responsible for *plate tectonics* and *continental drift*; it plays a dominant role in determining the thermal structure of the Earth. An understanding of *thermal convection* is essential to the understanding of fundamental geodynamic processes. When a fluid is heated from within or from below and cooled from above, thermal convection can occur. The hot fluid at depth is *gravitationally unstable* with respect to the cool fluid near the upper surface. *Buoyancy forces* drive the convective flow.

On many scales crustal rocks appear to have been folded. *Folding* can be attributed to the fluid behavior of these rocks. A fluid instability can also explain the formation of *salt domes* due to the *diapiric upwelling* of a buried layer of salt. The salt is gravitationally unstable because of its low density.

6-2 One-Dimensional Channel Flows

The movement of the plates over the surface of the Earth represents a flow of mantle rock from accreting plate boundaries to subduction zones. A complementary flow of mantle rock from subduction zones to accreting plate boundaries must occur at depth. One model for this counterflow assumes that it is confined to the asthenosphere immediately below the lithosphere. Interpretations of postglacial rebound data suggest the presence of a thin (about 100 km thick), low-viscosity region beneath the oceanic lithosphere. In addition, seismic studies show that there is a region beneath the lithosphere in which the seismic velocities are low and the seismic waves, particularly shear waves, are attenuated. This layer, the seismic low-velocity zone, has a thickness of about 200 km. Although the presence of a seismic low-velocity zone is not direct evidence of the existence of a low-viscosity region, the physical circumstances responsible for the reduction in seismic wave speeds and the attenuation of the waves (high temperature, small amounts of partial melting) also favor the formation of a low-viscosity region. Any flow in an asthenosphere would be approximately horizontal because of the large horizontal distances involved (the dimensions of lithospheric plates are thousands of kilometers) compared with the small vertical dimension of the region. Thus we consider the one-dimensional flow of a Newtonian viscous fluid in a channel between parallel plates as a model for asthenospheric flow.

Figure 6–1 is a sketch of a one-dimensional channel flow. The fluid moves with velocity u in the x direction in a channel of thickness h. The horizontal velocity varies only with the vertical coordinate; that is, $u = u(y)$, where y is the distance from the upper boundary ($y = 0$). The flow may occur as a result of either an applied horizontal *pressure gradient* $(p_0 - p_1)/l$ (l is the horizontal length of a section of the channel, p_1 is the pressure at the entrance to the section, and p_0 is the pressure at the section exit) or the prescribed motion of one of the walls (where we assume that the upper boundary $y = 0$ has the given speed $u = u_0$ and the lower boundary $y = h$ is motionless).

The flow may also be driven by a combination of a pressure gradient and a prescribed wall velocity. As a result of the *shear*, or gradient in the velocity profile,

a *shear stress* τ (force per unit area) is exerted on horizontal planes in the fluid and at the channel walls. For a Newtonian fluid with constant *viscosity* μ the shear stress at any location in the channel is given by

$$\tau = \mu \frac{du}{dy}. \qquad (6\text{–}1)$$

The shear stress defined in Equation (6–1) is the tangential stress on a surface whose outer normal points in the y direction. The viscosity of a Newtonian fluid is the constant of proportionality between shear stress and strain rate or velocity gradient. The more viscous the fluid, the larger the stress required to produce a given shear.

The viscosities of some common fluids are listed in Table 6–1. The SI unit of viscosity is the Pascal second (Pa s). The ratio μ/ρ (ρ is the density of the fluid) occurs frequently in fluid mechanics. It is known as the *kinematic viscosity* v of a fluid

$$v = \frac{\mu}{\rho}. \qquad (6\text{–}2)$$

The quantity μ is the dynamic viscosity. The SI unit of kinematic viscosity is square meter per second (m^2 s^{-1}). The kinematic viscosity is a diffusivity, similar to the thermal diffusivity κ. While κ describes how heat diffuses by molecular collisions, v describes how momentum diffuses. The ratio of v to κ is a dimensionless quantity known as the *Prandtl number*, Pr

$$\mathrm{Pr} \equiv \frac{v}{\kappa}. \qquad (6\text{–}3)$$

A fluid with a small Prandtl number diffuses heat more rapidly than it does momentum; the reverse is true for a fluid with a large value of Pr. Table 6–1 also lists the kinematic viscosities, thermal diffusivities, and Prandtl numbers of a variety of fluids.

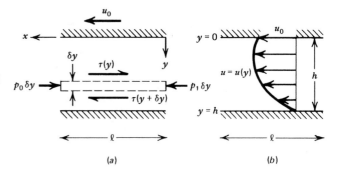

6–1 (a) The force balance on a layer of fluid in a channel with an applied pressure gradient. (b) A typical velocity profile.

TABLE 6–1 **Transport Properties of Some Common Fluids at** 15°C **and Atmospheric Pressure**

	Viscosity μ (Pa s)	Kinematic Viscosity ν (m² s⁻¹)	Thermal Diffusivity κ (m² s⁻¹)	Prandtl Number Pr
Air	1.78×10^{-5}	1.45×10^{-5}	2.02×10^{-5}	0.72
Water	1.14×10^{-3}	1.14×10^{-6}	1.40×10^{-7}	8.1
Mercury	1.58×10^{-3}	1.16×10^{-7}	4.2×10^{-6}	0.028
Ethyl alcohol	1.34×10^{-3}	1.70×10^{-6}	9.9×10^{-8}	17.2
Carbon tetrachloride	1.04×10^{-3}	6.5×10^{-7}	8.4×10^{-8}	7.7
Olive oil	0.099	1.08×10^{-4}	9.2×10^{-8}	1,170
Glycerine	2.33	1.85×10^{-3}	9.8×10^{-8}	18,880

The flow in the channel in Figure 6–1 is determined by the *equation of motion*. This is a mathematical statement of the *force balance* on a layer of fluid of thickness δy and horizontal length l (see Figure 6–1). The net *pressure force* on the element in the x direction is

$$(p_1 - p_0)\,\delta y.$$

This is the force per unit depth of the channel in the direction normal to the plane in Figure 6–1. Since the shear stress as well as the velocity is a function only of y, the shear force on the upper boundary of the layer in the x direction is

$$-\tau(y)l,$$

and the shear force on the lower boundary of the layer in the x direction is

$$\tau(y + \delta y)l = \left(\tau(y) + \frac{d\tau}{dy}\,\delta y\right)l. \tag{6–4}$$

The net force on the layer must be zero. This gives

$$(p_1 - p_0)\,\delta y + \left\{\tau(y) + \frac{d\tau}{dy}\,\delta y\right\}l - \tau(y)l = 0, \tag{6–5}$$

which in the limit $\delta y \to 0$ becomes

$$\frac{d\tau}{dy} = -\frac{(p_1 - p_0)}{l}. \tag{6–6}$$

The right side of Equation (6–6) is the horizontal pressure gradient in the channel

$$\frac{dp}{dx} = -\frac{(p_1 - p_0)}{l} \tag{6–7}$$

in terms of which the equation of motion can be written

$$\frac{d\tau}{dy} = \frac{dp}{dx}. \tag{6–8}$$

With $p_1 > p_0$, a pressure difference tending to move the fluid in the positive x direction, the pressure gradient dp/dx is negative. The pressure drop in a channel is often expressed in terms of a *hydraulic head* H given by

$$H \equiv \frac{(p_1 - p_0)}{\rho g}. \tag{6–9}$$

The hydraulic head is the height of fluid required to hydrostatically provide the applied pressure difference $p_1 - p_0$.

An equation for the velocity can be obtained by substituting the expression for τ from Equation (6–1) into Equation (6–8). We obtain

$$\mu \frac{d^2u}{dy^2} = \frac{dp}{dx}. \tag{6–10}$$

Integration of this equation gives

$$u = \frac{1}{2\mu}\frac{dp}{dx}y^2 + c_1 y + c_2. \tag{6–11}$$

To evaluate the constants, we must satisfy the boundary conditions that $u = 0$ at $y = h$ and $u = u_0$ at $y = 0$. These boundary conditions are known as *no-slip boundary conditions*. A viscous fluid in contact with a solid boundary must have the same velocity as the boundary. When these boundary conditions are satisfied,

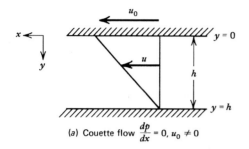

(a) Couette flow $\frac{dp}{dx} = 0, u_0 \neq 0$

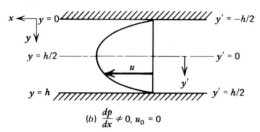

(b) $\frac{dp}{dx} \neq 0, u_0 = 0$

6–2 One-dimensional channel flows of a constant viscosity fluid.

Equation (6–11) becomes

$$u = \frac{1}{2\mu}\frac{dp}{dx}(y^2 - hy) - \frac{u_0 y}{h} + u_0. \qquad (6\text{–}12)$$

If the applied pressure gradient is zero, $p_1 = p_0$ or $dp/dx = 0$, the solution reduces to the linear velocity profile

$$u = u_0\left(1 - \frac{y}{h}\right). \qquad (6\text{–}13)$$

This simple flow, sketched in Figure 6–2a, is known as *Couette flow*. If the velocity of the upper plate is zero, $u_0 = 0$, the velocity profile is

$$u = \frac{1}{2\mu}\frac{dp}{dx}(y^2 - hy). \qquad (6\text{–}14)$$

When we rewrite this in terms of distance measured from the centerline of the channel y', where

$$y' = y - \frac{h}{2}, \qquad (6\text{–}15)$$

we find

$$u = \frac{1}{2\mu}\frac{dp}{dx}\left(y'^2 - \frac{h^2}{4}\right). \qquad (6\text{–}16)$$

The velocity profile is a parabola that is symmetric about the centerline of the channel, as shown in Figure 6–2b.

PROBLEM 6–1 Show that the mean velocity in the channel is given by

$$\bar{u} = -\frac{h^2}{12\mu}\frac{dp}{dx} + \frac{u_0}{2}. \qquad (6\text{–}17)$$

PROBLEM 6–2 Derive a general expression for the shear stress τ at any location y in the channel. What are the simplified forms of τ for Couette flow and for the case $u_0 = 0$?

PROBLEM 6–3 Find the point in the channel at which the velocity is a maximum.

PROBLEM 6–4 Consider the steady, unidirectional flow of a viscous fluid down the upper face of an inclined plane. Assume that the flow occurs in a layer of constant thickness h, as shown in Figure 6–3. Show that the velocity profile is given by

$$u = \frac{\rho g \sin\alpha}{2\mu}(h^2 - y^2), \qquad (6\text{–}18)$$

where y is the coordinate measured perpendicular to the inclined plane ($y = h$ is the surface of the plane), α is the inclination of the plane to the horizontal, and g is the acceleration of gravity. First show that

$$\frac{d\tau}{dy} = -\rho g \sin\alpha, \qquad (6\text{–}19)$$

and then apply the no-slip condition at $y = h$ and the *free-surface condition*, $\tau = 0$, at $y = 0$. What is the mean velocity in the layer? What is the thickness of a layer whose rate of flow down the incline (per unit width in the direction perpendicular to the plane in Figure 6–3) is Q?

6–3 Unidirectional flow of a constant thickness layer of viscous fluid down an inclined plane.

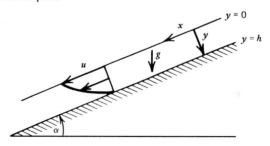

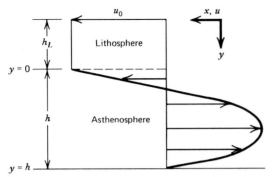

6–4 Velocity profile associated with the asthenospheric counterflow model.

6-3 Asthenospheric Counterflow

One model for the flow in the mantle associated with the movement of the surface plates is a *counterflow* immediately beneath the lithosphere, as shown in Figure 6–4. The lithosphere is assumed to be a rigid plate of thickness h_L moving with velocity u_0. Beneath the lithosphere is an asthenosphere of thickness h and uniform viscosity μ. At the base of the asthenosphere we assume that the mantle is stationary; that is, $u = 0$. The boundary conditions $u = 0$ at $y = h$ and $u = u_0$ at $y = 0$ were satisfied in writing Equation (6–12) so that this equation gives the flow in the asthenosphere. *Conservation of mass* requires that the flow of material in the $+x$ direction in the lithosphere must be balanced by a counterflow in the asthenosphere. For this model the net horizontal flow as illustrated in Figure 6–4 is zero. Quantitatively this can be written

$$u_0 h_L + \int_0^h u \, dy = 0, \qquad (6\text{–}20)$$

where the first term is the *flux of material* in the lithosphere and the second term is the flux of material in the asthenosphere (per unit distance perpendicular to the plane of the figure). By substituting Equation (6–12) into Equation (6–20) and integrating, we obtain

$$u_0 h_L - \frac{h^3}{12\mu} \frac{dp}{dx} + \frac{u_0 h}{2} = 0, \qquad (6\text{–}21)$$

where dp/dx is the horizontal pressure gradient in the asthenosphere. We can solve Equation (6–21) for the pressure gradient that satisfies the counterflow condition

$$\frac{dp}{dx} = \frac{12\mu u_0}{h^2} \left(\frac{h_L}{h} + \frac{1}{2} \right). \qquad (6\text{–}22)$$

Substitution of this result into Equation (6–12) gives the velocity profile in the asthenosphere,

$$u = u_0 \left\{ 1 - \frac{y}{h} + 6 \left(\frac{h_L}{h} + \frac{1}{2} \right) \left(\frac{y^2}{h^2} - \frac{y}{h} \right) \right\}. \qquad (6\text{–}23)$$

This velocity profile is illustrated in Figure 6–4; it is independent of the viscosity.

The shear stress on the base of the lithosphere τ_{LA} due to the counterflow in the asthenosphere can be evaluated directly using Equations (6–1) and (6–23). The result is

$$\tau_{LA} = -\frac{2\mu u_0}{h} \left(2 + 3\frac{h_L}{h} \right). \qquad (6\text{–}24)$$

The minus sign in Equation (6–24) indicates that the asthenosphere exerts a drag force on the base of the lithosphere tending to oppose its motion. For $\mu = 4 \times 10^{19}$ Pa s (a possible value for the viscosity of the asthenosphere), $h_L = 100$ km, $h = 200$ km, and $u_0 = 50$ mm yr^{-1}, we get 2.2 MPa for the magnitude of the shear stress on the base of the lithosphere from Equation (6–24).

The asthenospheric counterflow considered in this section requires that the pressure in the asthenosphere increase with x; that is, p must increase in the direction of *seafloor spreading*. This increase in pressure with distance from a ridge could only be provided by a *hydrostatic head* associated with topography; that is, the ocean floor would have to rise with distance from the ridge. The situation is sketched in Figure 6–5. The pressure in the asthenosphere a distance b beneath the ridge is given by the hydrostatic formula as

$$p = \rho_w g w + \rho g (w_r - w + b), \qquad (6\text{–}25)$$

where ρ_w is the density of seawater, w is the depth of the ocean a distance x from the ridge, ρ is the mantle

6–5 The asthenospheric counterflow model requires the seafloor to rise with distance from a ridge in order to supply the pressure required to drive the return flow toward the ridge in the asthenosphere.

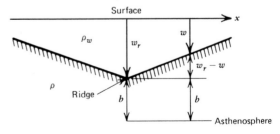

density, and w_r is the depth of the ocean at the ridge. By differentiating Equation (6–25) with respect to x, we can relate the slope of the seafloor to the horizontal pressure gradient in the asthenosphere:

$$\frac{dp}{dx} = -(\rho - \rho_w)g\frac{dw}{dx}. \quad (6\text{–}26)$$

A positive dp/dx requires a negative dw/dx or an ocean depth that decreases with x. By combining Equations (6–22) and (6–26), we can solve for the slope of the seafloor required by the asthenospheric counterflow model. We find

$$\frac{dw}{dx} = -\frac{12\mu u_0}{(\rho - \rho_w)gh^2}\left(\frac{h_L}{h} + \frac{1}{2}\right). \quad (6\text{–}27)$$

For $\rho_w = 1000$ kg m^{-3}, $\rho = 3300$ kg m^{-3}, $g = 10$ m s^{-2}, and the other parameter values given above, the slope of the seafloor is $dw/dx = -7.2 \times 10^{-4}$. Across the width of the Pacific Ocean, $x = 10{,}000$ km, this would give a decrease in depth of 7.2 km. However, no systematic decrease in ocean depth as one moves to the northwest in the Pacific has been observed.

The pressure gradient required to drive the asthenospheric counterflow would also result in a gravity anomaly. We can determine the value of the anomaly Δg using the Bouguer gravity formula, Equation (5–111), which combined with Equation (6–27) gives

$$\frac{d(\Delta g)}{dx} = \frac{24\pi G\mu u_0}{gh^2}\left(\frac{h_L}{h} + \frac{1}{2}\right). \quad (6\text{–}28)$$

For the preceding example we find that $d\Delta g/dx = 10^{-10}$ s^{-2}. Across the width of the Pacific this gives a gravity anomaly of 7.2 mm s^{-2}, which also has not been observed. We conclude, therefore, that the shallow counterflow model for mantle convection is not correct and that significant convective flows occur beneath the asthenosphere.

PROBLEM 6–5 For an asthenosphere with a viscosity $\mu = 4 \times 10^{19}$ Pa s and a thickness $h = 200$ km, what is the shear stress on the base of the lithosphere if there is no counterflow $(\partial p/\partial x = 0)$? Assume $u_0 = 50$ mm yr^{-1} and that the base of the asthenosphere has zero velocity.

PROBLEM 6–6 Assume that the base stress obtained in Problem 6–5 is acting on 6000 km of lithosphere with a thickness of 100 km. What tensional stress in the lithosphere ($h_L = 100$ km) must be applied at a trench to overcome this basal drag?

6–4 Pipe Flow

With subsequent applications to flows in aquifers and volcanic conduits in mind, we next consider viscous flow through a circular pipe. The pipe has a radius R and a length l, as illustrated in Figure 6–6. The flow is driven by the pressure difference $(p_1 - p_0)$ applied between the sections a distance l apart. We assume that the velocity of the fluid along the pipe u depends only on distance from the center of the pipe r. The form of the velocity profile $u(r)$ can be found by writing a force balance on a cylindrical *control volume* of radius r and length l, as shown in Figure 6–6. The net pressure force on the ends of the cylindrical control volume is $(p_1 - p_0)\pi r^2$; this is a force along the cylinder axis in the direction of flow. Since there can be no net force on the control volume if the flow is *steady*, this pressure force must be balanced by the shear force acting on the cylindrical surface of the control volume. The shear stress on the cylindrical surface $\tau(r)$ exerts a net *frictional force* $-2\pi rl\tau(r)$ on the control volume (τ is a negative quantity). The force balance equation is thus

$$\pi r^2(p_1 - p_0) = -2\pi rl\tau \quad (6\text{–}29)$$

or

$$\tau = \frac{r}{2}\frac{dp}{dx}, \quad (6\text{–}30)$$

where dp/dx is the pressure gradient along the pipe (Equation (6–7)).

6–6 Poiseuille flow through a circular pipe.

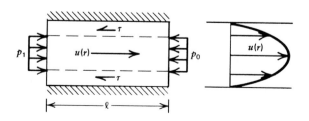

In the cylindrical geometry in Figure 6–6, the shear stress τ is directly proportional to the radial gradient of the velocity u

$$\tau = \mu \frac{du}{dr}. \tag{6-31}$$

As in Equation (6–1), the viscosity μ is the constant of proportionality. By substituting Equation (6–31) into Equation (6–30), we obtain an expression for the slope of the velocity profile,

$$\frac{du}{dr} = \frac{r}{2\mu} \frac{dp}{dx}, \tag{6-32}$$

which can be integrated to give

$$u = -\frac{1}{4\mu} \frac{dp}{dx} (R^2 - r^2). \tag{6-33}$$

We used the condition $u = 0$ at $r = R$ in obtaining Equation (6–33). The velocity profile in the pipe is a paraboloid of revolution; it is known as *Poiseuille flow*.

The maximum velocity in the pipe u_{max} occurs at $r = 0$. From Equation (6–33) it is given by

$$u_{max} = -\frac{R^2}{4\mu} \frac{dp}{dx}. \tag{6-34}$$

Because dp/dx is negative when $p_1 > p_0$, u_{max} is a positive quantity. The *volumetric flow rate Q* through the pipe is the total volume of fluid passing a cross section per unit time. The flow through an annulus of thickness dr and radius r occurs at the rate $2\pi r\, dr\, u(r)$; Q is the integral of this over a cross section

$$Q = \int_0^R 2\pi r u \, dr. \tag{6-35}$$

Upon substituting Equation (6–33) into Equation (6–35) and carrying out the integration, we get

$$Q = -\frac{\pi R^4}{8\mu} \frac{dp}{dx}. \tag{6-36}$$

If we divide Q by the cross-sectional area of the pipe πR^2, we obtain the mean velocity $\bar{u}$ in the pipe

$$\bar{u} = -\frac{R^2}{8\mu} \frac{dp}{dx}. \tag{6-37}$$

By comparing Equations (6–34) and (6–37), we see that

$$\bar{u} = \frac{1}{2} u_{max}. \tag{6-38}$$

The mean and maximum velocities in the pipe are directly proportional to the pressure gradient and inversely proportional to the viscosity. This result is valid as long as the flow is *laminar*.

It is often convenient in fluid mechanics to work in terms of *dimensionless variables*. The relation between the mean velocity in the pipe and the pressure gradient [Equation (6–37)] can be put into standard dimensionless form by introducing two quantities: a dimensionless pressure gradient or *friction factor f* and the *Reynolds number* Re. The friction factor is defined as

$$f \equiv \frac{-4R}{\rho \bar{u}^2} \frac{dp}{dx}, \tag{6-39}$$

and the Reynolds number is given by

$$\mathrm{Re} \equiv \frac{\rho \bar{u} D}{\mu}, \tag{6-40}$$

where $D = 2R$ is the pipe diameter. Using Equations (6–39) and (6–40), we can rewrite Equation (6–37) as

$$f = \frac{64}{\mathrm{Re}}. \tag{6-41}$$

The inverse dependence of the friction factor on the Reynolds number in laminar flow is shown in Figure 6–7.

At sufficiently high Reynolds numbers, observed pressure drops become considerably higher than those given by laminar theory. The flow in the pipe becomes unsteady with *random eddies*. This is known as *turbulent flow*. The qualitative difference between laminar

6–7 Dependence of the friction factor f on the Reynolds number Re for laminar flow, from Equation (6–41), and for turbulent flow, from Equation (6–42).

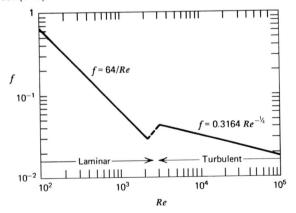

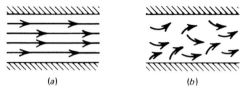

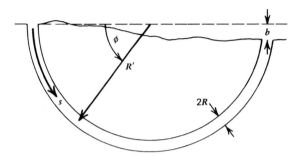

6–8 Illustration of the difference between (*a*) laminar and (*b*) turbulent flow. Laminar flow is steady, and the fluid flows parallel to the walls: lateral transport of momentum takes place on a molecular scale. Turbulent flow is unsteady and has many time-dependent eddies and swirls. These eddies are much more effective in the lateral transport of momentum than are molecular processes. Therefore, the friction factor (pressure drop) in turbulent flow is larger at a prescribed Reynolds number (flow velocity).

6–9 A semicircular aquifer with a circular cross section (a toroid). A hydrostatic head *b* is available to drive the flow.

and turbulent flow is illustrated in Figure 6–8. The principal advantage of the formulation of the problem in nondimensional form is that the *transition to turbulent flow* occurs at Re $\approx$ 2200 independent of the pipe radius, flow velocity, or type of fluid considered (viscosity and density). The mean velocity corresponding to the transition Reynolds number of 2200 is 22 mm s^{-1} for water with a viscosity of 10^{-3} Pa s flowing in a 0.1-m-diameter pipe. This illustrates that most flows of ordinary liquids and gases are in the turbulent regime.

No theoretical equivalent to the Newtonian relationship between shear stress and rate of strain as given in Equation (6–1) or Equation (6–31) exists for turbulent flow. It is found empirically that

$$f = 0.3164 \, \text{Re}^{-1/4} \qquad (6\text{–}42)$$

in the turbulent flow regime. This result is also shown in Figure 6–7 along with the transition from laminar to turbulent flow.

PROBLEM 6–7 Determine the Reynolds number for the asthenospheric flow considered in Problem 6–5. Base the Reynolds number on the thickness of the flowing layer and the mean velocity ($u_0 = 50$ mm yr^{-1} and $\rho = 3200$ kg m^{-3}). This problem illustrates that the viscosity of mantle rock is so high that the Reynolds number is generally small.

6–5 Artesian Aquifer Flows

Naturally occurring springs are usually due to the flow of groundwater from a high elevation to a low elevation. The flow takes place through an *aquifer* or *permeable formation*. Figure 6–9 shows an idealized model of an aquifer in the shape of a semicircle of radius R', a form suggested by the geometry of porous layers in folded sedimentary rock. The entrance of the aquifer lies a distance b above the exit, and its cross section is assumed to be circular with radius R. The hydrostatic pressure head available to drive flow through the aquifer is $\rho g b$, where ρ is the density of water. Since the overall length of the aquifer is $\pi R'(R' \gg b)$, the driving pressure gradient is

$$\frac{dp}{ds} = \frac{-\rho g b}{\pi R'}, \qquad (6\text{–}43)$$

where s is distance along the aquifer. The volumetric flow rate produced by this pressure gradient can be calculated from Equation (6–36) if the flow through the aquifer is laminar. The result of substituting Equation (6–43) into Equation (6–36), identifying dp/ds as dp/dx, is

$$Q = \frac{\rho g b R^4}{8 \mu R'}. \qquad (6\text{–}44)$$

If the flow is turbulent, we can determine Q by using the empirical relation (6–42) between f and Re. The first step is to recast Equation (6–42) into dimensional form using the definitions of f and Re. We find

$$\frac{-4R}{\rho \bar{u}^2} \frac{dp}{dx} = 0.3164 \left(\frac{\mu}{\rho \bar{u} 2R} \right)^{1/4}. \qquad (6\text{–}45)$$

The result of rearranging Equation (6–45) so as to determine $\bar{u}$ is

$$\bar{u} = \left(\frac{4 \times 2^{1/4}}{0.3164} \right)^{4/7} \left(-\frac{1}{\rho} \frac{dp}{dx} \right)^{4/7} R^{5/7} \left(\frac{\rho}{\mu} \right)^{1/7}. \qquad (6\text{–}46)$$

Because Q is $\pi R^2 \bar{u}$, we obtain the volumetric flow rate through the aquifer for turbulent flow by multiplying Equation (6–46) by πR^2 and substituting for

$(-1/\rho)\,(dp/dx)$ from Equation (6–43). One finds

$$Q = 7.686\left(\frac{gb}{R'}\right)^{4/7}\left(\frac{\rho}{\mu}\right)^{1/7}R^{19/7}. \qquad (6\text{–}47)$$

PROBLEM 6–8 A spring has a flow of 100 liters per minute. The entrance to the spring lies 2 km away from the outlet and 50 m above it. If the aquifer supplying the spring is modeled according to Figure 6–9, find its cross-sectional radius. What is the average velocity? Is the flow laminar or turbulent?

6–6 Flow Through Volcanic Pipes

Another example of naturally occurring pipe flow is the flow of magma through volcanic conduits of nearly circular cross section. The upward flow of magma is driven by the natural buoyancy of the lighter magma relative to the denser surrounding rock. At a depth h the *lithostatic pressure* in the rock is $\rho_s gh$, where ρ_s is the rock density. At the same depth the *hydrostatic pressure* in a stationary column of magma is $\rho_l gh$, where ρ_l is the magma density. Assuming that the lithostatic and hydrostatic pressures are equal in the pipe, the pressure gradient available to drive the magma up to the surface is $-(\rho_s - \rho_l)g$. The assumption of equal lithostatic and hydrostatic pressures in the pipe is equivalent to assuming that the walls of the pipe are free to deform as the magma is driven upward. The volumetric flow Q driven by this pressure gradient through a volcanic pipe of radius R is, from Equation (6–36),

$$Q = \frac{\pi}{8}\frac{(\rho_s - \rho_l)g R^4}{\mu}, \qquad (6\text{–}48)$$

if the flow is laminar. From Equation (6–46) and $Q = \pi R^2 \bar{u}$, the volumetric flow for turbulent conditions is

$$Q = 14.8\frac{R^{19/7}[(\rho_s - \rho_l)g]^{4/7}}{\rho_l^{3/7}\mu^{1/7}}. \qquad (6\text{–}49)$$

PROBLEM 6–9 Determine the rate at which magma flows up a two-dimensional channel of width d under the buoyant pressure gradient $-(\rho_s - \rho_l)g$. Assume laminar flow.

6–7 Conservation of Fluid in Two Dimensions

We now extend our studies of viscous fluid flow to two dimensions. We consider a general flow in the xy plane with the corresponding *velocity components* u and v.

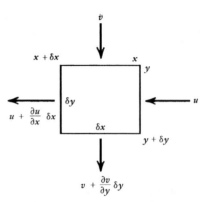

6–10 Flow across the surfaces of an infinitesimal rectangular element.

The spatial variations of these two velocity components are constrained by the need to conserve fluid. We consider a rectangular control volume with dimensions δx and δy, as illustrated in Figure 6–10. The flow rate per unit area in the x direction at x is u. The flow rate per unit area at $x + \delta x$ is

$$u(x + \delta x) = u + \frac{\partial u}{\partial x}\delta x. \qquad (6\text{–}50)$$

The net flow rate out of the region between x and $x + \delta x$ per unit area normal to the x direction is

$$u + \frac{\partial u}{\partial x}\delta x - u = \frac{\partial u}{\partial x}\delta x. \qquad (6\text{–}51)$$

Similarly, flow in the y direction (vertically downward) yields a net volume flow per unit area normal to the y direction out of the region between y and $y + \delta y$ given by

$$v + \frac{\partial v}{\partial y}\delta y - v = \frac{\partial v}{\partial y}\delta y. \qquad (6\text{–}52)$$

To find the net rate at which fluid flows out of the rectangular region shown in Figure 6–10, we must combine the flows in the two directions. The net outward flow rate in the x direction is $(\partial u/\partial x)\,\delta x$ times the area of the face across which the flow occurs, which is δy multiplied by a unit dimension in the direction normal to the diagram. The net outward flow rate in the x direction is thus $(\partial u/\partial x)\,\delta x\,\delta y$. Similarly the net outward flow rate in the y direction is $(\partial v/\partial y)\,\delta y\,\delta x$. The total net outward flow rate per unit area of the rectangle is

$$\frac{\partial u}{\partial x} + \frac{\partial v}{\partial y}.$$

If the flow is steady (time-independent), and there are

no density variations to consider, then there can be no net flow into or out of the rectangle. The *conservation of fluid* or *continuity equation* is

$$\frac{\partial u}{\partial x} + \frac{\partial v}{\partial y} = 0. \tag{6-53}$$

This is the form of the continuity equation appropriate to an *incompressible fluid*.

6–8 Elemental Force Balance in Two Dimensions

The forces acting on the control volume in Figure 6–10 must be in balance. Included in the force balance are the pressure forces, *viscous forces*, and *gravity force*. We neglect the *inertial force* associated with the *acceleration* of a fluid element. This is appropriate for the slow motion of very viscous or high Prandtl number fluids. The Earth's mantle behaves as a highly viscous fluid on geologic time scales. The viscosity of the mantle is about 10^{21} Pa s; its density and thermal diffusivity are about 4000 kg m^{-3} and 1 mm^2 s^{-1}. Thus the Prandtl number of the Earth's mantle is about 10^{23}. The balance of pressure, viscous, and gravity forces and the neglect of inertial forces are equivalent to the application of *Newton's second law of motion* to a fluid element with the neglect of its acceleration. It is also equivalent to a statement of momentum conservation.

The pressure forces acting on an infinitesimal rectangular element of fluid are illustrated in Figure 6–11. Because pressure is force per unit area, $p\,\delta y$ (times a unit length in the direction normal to the plane of the figure) is the force acting to the left on the face of the rectangle located at x, for example. Pressure forces act perpendicular to surfaces and are directed into the volume enclosed by the surface. The net pressure force on

6–11 Pressure forces acting on an infinitesimal rectangular fluid element.

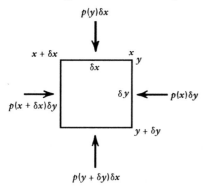

the element in the x direction per unit area of the fluid element is

$$\frac{p(x)\,\delta y - p(x+\delta x)\,\delta y}{\delta x\,\delta y} = -\frac{[p(x+\delta x) - p(x)]}{\delta x}, \tag{6-54}$$

which by virtue of a simple Taylor series expansion is

$$-\frac{\partial p}{\partial x}.$$

Thus, only if there is a pressure gradient in the x direction will there be any net pressure force on the fluid element in this direction. If there is no such pressure variation, the pressure forces on opposite sides of the element will simply cancel each other and there will be no net effect. Similarly, the net pressure force on the element in the y direction per unit area of the element is

$$-\frac{\partial p}{\partial y}.$$

The *gravitational body force* on a fluid element is its mass times the *acceleration of gravity*. The mass of the fluid element in Figure 6–11 is $\rho\,\delta x\,\delta y$ (times a unit length in the third dimension) and, accordingly, ρg is the force of gravity per unit area of the element (and per unit length in the third dimension). We assume that gravity acts in the positive y direction. Thus the net gravitational force per unit area of the element is in the y direction and is ρg.

Only the viscous forces acting on the element in Figure 6–11 remain to be discussed. These are shown in Figure 6–12. Viscous forces on the element act both parallel and perpendicular to the surfaces. The quantities τ_{xy} and τ_{yx} are *viscous shear stresses*, that is, viscous forces per unit area that act parallel to the surfaces of the element, and τ_{xx} and τ_{yy} are *viscous normal stresses*, that is, viscous forces per unit area that act perpendicular to the element's surfaces. The stresses are considered positive in the directions shown in the figure. The sign convention adopted here for the viscous stresses τ is standard in the fluid mechanics literature; it is opposite to the sign convention adopted in Chapters 2 and 3 for the stresses σ.

Clearly, if there is to be no net *torque* about the center of the fluid element, then

$$\tau_{xy} = \tau_{yx}. \tag{6-55}$$

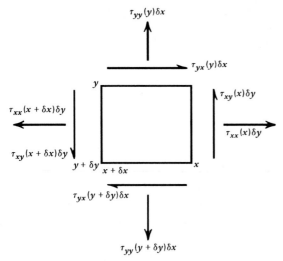

6-12 Viscous forces acting on an infinitesimal two-dimensional rectangular fluid element.

The net viscous force in the x direction per unit cross-sectional area of the element is

$$\frac{\tau_{xx}(x+\delta x)\,\delta y - \tau_{xx}(x)\,\delta y}{\delta x\,\delta y} + \frac{\tau_{yx}(y+\delta y)\,\delta x - \tau_{yx}(y)\,\delta x}{\delta x\,\delta y},$$

which, with a Taylor series expansion, simplifies to

$$\frac{\partial \tau_{xx}}{\partial x} + \frac{\partial \tau_{yx}}{\partial y}.$$

Similarly, the net viscous force in the y direction per unit cross-sectional area of the element is

$$\frac{\partial \tau_{yy}}{\partial y} + \frac{\partial \tau_{xy}}{\partial x}.$$

For an ideal Newtonian viscous fluid, the viscous stresses are linearly proportional to the velocity gradients. The generalization of Equation (6–1) to two dimensions yields

$$\tau_{xx} = 2\mu \frac{\partial u}{\partial x} \tag{6–56}$$

$$\tau_{yy} = 2\mu \frac{\partial v}{\partial y} \tag{6–57}$$

$$\tau_{yx} = \tau_{xy} = \mu\left(\frac{\partial u}{\partial y} + \frac{\partial v}{\partial x}\right), \tag{6–58}$$

where μ is again the dynamic viscosity. If $v = 0$ and $u = u(y)$, then τ_{xx} and τ_{yy} are zero and

$$\tau_{yx} = \tau_{xy} = \mu \frac{du}{dy}, \tag{6–59}$$

which is identical to Equation (6–1). The total normal stress is the sum of the pressure and the viscous stress; that is,

$$\sigma_{xx} = p - \tau_{xx} = p - 2\mu \frac{\partial u}{\partial x} \tag{6–60}$$

$$\sigma_{yy} = p - \tau_{yy} = p - 2\mu \frac{\partial v}{\partial y}. \tag{6–61}$$

The minus signs in front of τ_{xx} and τ_{yy} in these equations are the result of the opposite sign conventions adopted for σ and τ. The viscous stress is the only contribution to the shear stress.

When we use Equations (6–56) to (6–58) to rewrite the expressions already derived for the viscous forces on a small rectangular fluid element, we find that the viscous forces in the x and y directions per unit cross-sectional area of the element are, for constant viscosity,

$$2\mu \frac{\partial^2 u}{\partial x^2} + \mu\left(\frac{\partial^2 u}{\partial y^2} + \frac{\partial^2 v}{\partial x\,\partial y}\right)$$

and

$$2\mu \frac{\partial^2 v}{\partial y^2} + \mu\left(\frac{\partial^2 v}{\partial x^2} + \frac{\partial^2 u}{\partial y\,\partial x}\right),$$

respectively. Both these expressions can be further simplified by using the continuity equation. By differentiating Equation (6–53), we obtain

$$\frac{\partial^2 v}{\partial x\,\partial y} = -\frac{\partial^2 u}{\partial x^2} \tag{6–62}$$

$$\frac{\partial^2 u}{\partial y\,\partial x} = -\frac{\partial^2 v}{\partial y^2}. \tag{6–63}$$

Using Equations (6–62) and (6–63) for the mixed partial derivatives, we arrive at

$$\mu\left(\frac{\partial^2 u}{\partial x^2} + \frac{\partial^2 u}{\partial y^2}\right)$$

and

$$\mu\left(\frac{\partial^2 v}{\partial x^2} + \frac{\partial^2 v}{\partial y^2}\right)$$

as the expressions for the net viscous forces per unit cross-sectional area in the x and y directions, respectively.

We determine the force balance equations for an incompressible fluid with very large viscosity undergoing steady flow in two dimensions by adding the pressure, gravity, and viscous forces together and equating their sum to zero. For the x direction we obtain

$$0 = -\frac{\partial p}{\partial x} + \mu\left(\frac{\partial^2 u}{\partial x^2} + \frac{\partial^2 u}{\partial y^2}\right), \qquad (6\text{–}64)$$

and for the y direction the equation is

$$0 = -\frac{\partial p}{\partial y} + \rho g + \mu\left(\frac{\partial^2 v}{\partial x^2} + \frac{\partial^2 v}{\partial y^2}\right). \qquad (6\text{–}65)$$

Gravity acts only in the y direction, of course. To eliminate the hydrostatic pressure variation in Equation (6–65), we introduce

$$P = p - \rho g y. \qquad (6\text{–}66)$$

The pressure P is the pressure generated by fluid flow. Substitution of Equation (6–66) into Equations (6–64) and (6–65) yields

$$0 = -\frac{\partial P}{\partial x} + \mu\left(\frac{\partial^2 u}{\partial x^2} + \frac{\partial^2 u}{\partial y^2}\right) \qquad (6\text{–}67)$$

$$0 = -\frac{\partial P}{\partial y} + \mu\left(\frac{\partial^2 v}{\partial x^2} + \frac{\partial^2 v}{\partial y^2}\right). \qquad (6\text{–}68)$$

6–9 The Stream Function

We can satisfy the incompressible continuity equation in two dimensions if we introduce a *stream function* ψ defined such that

$$u = -\frac{\partial \psi}{\partial y} \qquad (6\text{–}69)$$

$$v = \frac{\partial \psi}{\partial x}. \qquad (6\text{–}70)$$

Substituting Equations (6–69) and (6–70) into Equation (6–53) yields

$$-\frac{\partial^2 \psi}{\partial x\,\partial y} + \frac{\partial^2 \psi}{\partial y\,\partial x} = 0 \qquad (6\text{–}71)$$

because the order of differentiation is interchangeable. Substituting Equations (6–69) and (6–70) into Equations (6–67) and (6–68) gives

$$0 = \frac{\partial P}{\partial x} + \mu\left(\frac{\partial^3 \psi}{\partial x^2\,\partial y} + \frac{\partial^3 \psi}{\partial y^3}\right) \qquad (6\text{–}72)$$

$$0 = -\frac{\partial P}{\partial y} + \mu\left(\frac{\partial^3 \psi}{\partial x^3} + \frac{\partial^3 \psi}{\partial y^2\,\partial x}\right). \qquad (6\text{–}73)$$

We can eliminate the pressure from these equations and obtain a single differential equation for ψ if we take

the partial derivative of Equation (6–72) with respect to y and the partial derivative of Equation (6–73) with respect to x and add. The result is

$$0 = \frac{\partial^4 \psi}{\partial x^4} + 2\frac{\partial^4 \psi}{\partial x^2\,\partial y^2} + \frac{\partial^4 y}{\partial y^4}. \qquad (6\text{–}74)$$

This is the *biharmonic equation*. In terms of the *Laplacian operator* ∇^2,

$$\nabla^2 = \frac{\partial^2}{\partial x^2} + \frac{\partial^2}{\partial y^2}, \qquad (6\text{–}75)$$

we can write the biharmonic equation for the stream function in the form

$$\nabla^4 \psi = 0. \qquad (6\text{–}76)$$

For two-dimensional flows of a very viscous fluid the stream function satisfies the biharmonic equation.

The stream function can be given a physical interpretation in terms of the volumetric flow rate between any two points in an incompressible, steady, two-dimensional flow. Consider two points A and B separated by an infinitesimal distance δs, as shown in Figure 6–13. The flow across AB can be calculated from the flows across AP and PB because conservation of mass requires zero net flow into or out of the infinitesimal triangle PAB. The volumetric flow rate across AP into the triangle per unit distance normal to the figure is $u\,\delta y$; similarly the flow rate across PB out of the triangle is $v\,\delta x$. The net flow rate out of PAB is thus $-u\,\delta y + v\,\delta x$; this must be equal to the volumetric flow rate (per unit distance in the third dimension) into PAB across AB. In terms of the stream function, $-u\,\delta y + v\,\delta x$ can be written

$$-u\,\delta y + v\,\delta x = \frac{\partial \psi}{\partial y}\,\delta y + \frac{\partial \psi}{\partial x}\,\delta x = d\psi. \qquad (6\text{–}77)$$

6–13 Volumetric flow rate between points A and B.

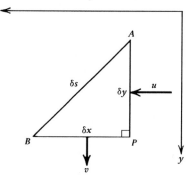

Thus, the small difference $d\psi$ is the volumetric flow rate between any two points separated by the infinitesimal distance δs. If the points are separated by an arbitrary distance, the integral of $d\psi$ between the points

$$\int_A^B d\psi = \psi_B - \psi_A \qquad (6\text{--}78)$$

gives the volumetric flow rate between the points; that is, the difference between the values of the stream function at any two points is the volumetric rate of flow across any line drawn between the points. The absolute value of the stream function is arbitrary; only the differences in ψ between points has physical significance.

PROBLEM 6–10 Determine the stream function for the general one-dimensional channel flow discussed in Section 6–2. Find ψ for the asthenospheric counterflow model in Section 6–3. Discuss the physical meaning of ψ in all these cases.

6-10 Postglacial Rebound

Important information on the fluid behavior of the Earth's mantle comes from studies of the dynamic response of the mantle to *loading* and *unloading* at the surface. Mountains depress the underlying crust–mantle boundary as discussed in Section 2–2. However, mountain building is so slow that dynamic effects can be neglected; that is, the mantle beneath a mountain is in essential hydrostatic equilibrium throughout the life cycle of the mountain. The growth and melting of ice sheets, on the other hand, occur sufficiently fast so that dynamic effects are important in the adjustment of the mantle to the changing surface load. The thick ice sheet that covers Greenland has depressed the surface several kilometers so that it is below sea level in places. The load of the ice sheet has forced mantle rock to flow laterally, allowing the Earth's surface beneath the ice to subside. During the last great ice age Scandinavia was covered with a thick ice sheet that caused considerable subsidence of the surface. When the ice sheet melted about 10,000 years ago the surface rebounded. The rate of rebound has been determined by dating elevated beaches. We will now show how these data can be used to determine the solid-state viscosity of the mantle. The process of subsidence and rebound under the loading and unloading of an ice sheet is illustrated in Figure 6–14.

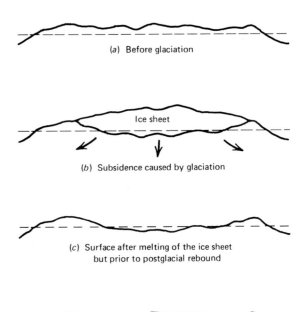

6-14 Subsidence due to glaciation and the subsequent postglacial rebound.

To determine the response of the Earth's mantle to the removal of an ice load, we consider the flow in a semi-infinite, viscous fluid half-space ($y > 0$) subjected to an initial periodic surface displacement. We assume the initial displacement of the surface is given by

$$w_m = w_{mo} \cos 2\pi x / \lambda, \qquad (6\text{--}79)$$

where λ is the wavelength and $w_m \ll \lambda$. The displacement of the surface w leads to a horizontal pressure gradient due to the hydrostatic load similar to that given in Equation (6–26). When the surface is displaced upward (negative w), the pressure is positive. This corresponds to a positive load, and fluid is driven away from this region as the displacement decreases. When the surface is displaced downward (positive w), the pressure is negative. This corresponds to the case when a load has been removed and fluid is driven into this region as the displacement decreases.

The return of the surface to an undeformed ($w = 0$) state is governed by the viscous flow in the half-space. The flow can be determined by solving the biharmonic equation for the stream function. Since the initial surface displacement is of the form $\cos 2\pi x/\lambda$, it is reasonable to anticipate that ψ must also vary periodically with x in a similar fashion. However, since ψ and w are not simply related, it is a priori uncertain whether ψ varies as $\cos 2\pi x/\lambda$, $\sin 2\pi x/\lambda$, or some combination thereof. It turns out that ψ is directly proportional to $\sin 2\pi x/\lambda$; we assume this at the outset to simplify the discussion. However, it would only require some additional algebra to carry through the solution assuming that ψ is an arbitrary combination of $\sin 2\pi x/\lambda$ and $\cos 2\pi x/\lambda$. Thus we apply the method of separation of variables and take

$$\psi = \sin \frac{2\pi x}{\lambda} Y(y), \qquad (6\text{–}80)$$

where $Y(y)$ is to be determined. By substituting this form of ψ into the biharmonic equation (6–74), we obtain

$$\frac{d^4 Y}{dy^4} - 2\left(\frac{2\pi}{\lambda}\right)^2 \frac{d^2 Y}{dy^2} + \left(\frac{2\pi}{\lambda}\right)^4 Y = 0. \qquad (6\text{–}81)$$

Solutions of the constant coefficient differential equation for Y are of the form

$$Y \propto \exp(my). \qquad (6\text{–}82)$$

If we substitute this function for Y in Equation (6–81), we find that m is a solution of

$$m^4 - 2\left(\frac{2\pi}{\lambda}\right)^2 m^2 + \left(\frac{2\pi}{\lambda}\right)^4 = \left[m^2 - \left(\frac{2\pi}{\lambda}\right)^2\right]^2 = 0$$

$$(6\text{–}83)$$

or

$$m = \pm \frac{2\pi}{\lambda}. \qquad (6\text{–}84)$$

These two values of m provide two possible solutions for Y

$$\exp\left(\frac{2\pi y}{\lambda}\right) \quad \text{and} \quad \exp\left(\frac{-2\pi y}{\lambda}\right).$$

Because the differential equation for Y is of fourth order, these two solutions are incomplete. Two additional solutions are required. It can be verified by direct substitution that

$$y \exp\left(\frac{2\pi y}{\lambda}\right) \quad \text{and} \quad y \exp\left(\frac{-2\pi y}{\lambda}\right)$$

also satisfy Equation (6–81). The general solution for Y is the sum of these four solutions; it can be written

$$\psi = \sin \frac{2\pi x}{\lambda}\left(Ae^{-2\pi y/\lambda}\right.$$
$$\left. + Bye^{-2\pi y/\lambda} + Ce^{2\pi y/\lambda} + Dye^{2\pi y/\lambda}\right), \qquad (6\text{–}85)$$

where the four arbitrary constants A, B, C, and D are determined by the appropriate boundary conditions.

We first require the solution to be finite as $y \to \infty$ so that $C = D = 0$. The formula for the stream function simplifies to

$$\psi = \sin \frac{2\pi x}{\lambda} e^{-2\pi y/\lambda}(A + By). \qquad (6\text{–}86)$$

The velocity components u and v can be obtained by differentiating ψ according to Equations (6–69) and (6–70). We find

$$u = \sin \frac{2\pi x}{\lambda} e^{-2\pi y/\lambda}\left\{\frac{2\pi}{\lambda}(A + By) - B\right\} \qquad (6\text{–}87)$$

and

$$v = \frac{2\pi}{\lambda} \cos \frac{2\pi x}{\lambda} e^{-2\pi y/\lambda}(A + By). \qquad (6\text{–}88)$$

Because the part of the mantle that behaves as a fluid is overlain with a rigid lithosphere, we force the horizontal component of the velocity to be zero at $y = w$; that is, we apply the no-slip condition at the upper boundary of the fluid half-space. However, because the vertical displacement of this boundary is small, $w \ll \lambda$, it is appropriate to apply this condition at $y = 0$. By setting $u = 0$ at $y = 0$ in Equation (6–87), we find that

$$B = \frac{2\pi A}{\lambda} \qquad (6\text{–}89)$$

and

$$\psi = A \sin \frac{2\pi x}{\lambda} e^{-2\pi y/\lambda}\left(1 + \frac{2\pi y}{\lambda}\right) \qquad (6\text{–}90)$$

$$u = A\left(\frac{2\pi}{\lambda}\right)^2 y e^{-2\pi y/\lambda} \sin \frac{2\pi x}{\lambda} \qquad (6\text{–}91)$$

$$v = A \frac{2\pi}{\lambda} \cos \frac{2\pi x}{\lambda} e^{-2\pi y/\lambda}\left(1 + \frac{2\pi y}{\lambda}\right). \qquad (6\text{–}92)$$

To evaluate the final constant A, we must equate the hydrostatic pressure head associated with the topography w to the normal stress at the upper boundary of the fluid half-space. The former quantity is $-\rho g w$, and the latter, from Equation (6–61), is $p - 2\mu(\partial v/\partial y)$. Because the surface displacement is small, it is appropriate to equate these stresses

$$-\rho g w = p - 2\mu \frac{\partial v}{\partial y} \qquad \text{at} \qquad y = 0. \qquad (6\text{–}93)$$

To apply condition (6–93), we must first calculate the pressure and the displacement at $y = 0$.

The pressure can be found by inserting expression (6–91) for u into the horizontal force balance (6–64). One obtains

$$\frac{\partial p}{\partial x} = -2\mu A \left(\frac{2\pi}{\lambda}\right)^3 \sin \frac{2\pi x}{\lambda}, \qquad (6\text{–}94)$$

at $y = 0$. This can be integrated with respect to x to give

$$p = 2\mu A \left(\frac{2\pi}{\lambda}\right)^2 \cos \frac{2\pi x}{\lambda}, \qquad (6\text{–}95)$$

at $y = 0$. We also need $(\partial v/\partial y)$ at $y = 0$ for Equation (6–93). This is easily found by differentiating Equation (6–92) with respect to y and then evaluating the result at $y = 0$. We get

$$\left(\frac{\partial v}{\partial y}\right)_{y=0} = 0. \qquad (6\text{–}96)$$

Condition (6–93) thus simplifies to

$$w_{y=0} = \frac{-2\mu A}{\rho g} \left(\frac{2\pi}{\lambda}\right)^2 \cos \frac{2\pi x}{\lambda}. \qquad (6\text{–}97)$$

The surface displacement w is related to the velocity field by the fact that the time derivative of w is just the vertical component of the surface velocity

$$\left(\frac{\partial w}{\partial t}\right)_{y=w} = v_{y=w}. \qquad (6\text{–}98)$$

Again, because the vertical displacement of the surface is small ($w \ll \lambda$), we can write

$$\left(\frac{\partial w}{\partial t}\right)_{y=0} = v_{y=0}. \qquad (6\text{–}99)$$

From Equation (6–92) we have

$$v_{y=0} = A \frac{2\pi}{\lambda} \cos \frac{2\pi x}{\lambda} \qquad (6\text{–}100)$$

so that

$$\left(\frac{\partial w}{\partial t}\right)_{y=0} = A \frac{2\pi}{\lambda} \cos \frac{2\pi x}{\lambda} \qquad (6\text{–}101)$$

By combining Equations (6–97) and (6–101), we find that w at $y = 0$ satisfies

$$\frac{\partial w}{\partial t} = -w \frac{\lambda \rho g}{4\pi \mu} = -w \frac{\lambda g}{4\pi \nu}. \qquad (6\text{–}102)$$

This can be integrated, with the initial condition $w = w_m$ at $t = 0$, to give

$$w = w_m \exp\left(\frac{-\lambda \rho g t}{4\pi \mu}\right) = w_m \exp\left(\frac{-\lambda g t}{4\pi \nu}\right). \qquad (6\text{–}103)$$

The surface displacement decreases exponentially with time as fluid flows from regions of elevated topography to regions of depressed topography. Equation (6–103) can be rewritten as

$$w = w_m e^{-t/\tau_r}, \qquad (6\text{–}104)$$

where τ_r, the characteristic time for the exponential relaxation of the initial displacement, is given by

$$\tau_r = \frac{4\pi \mu}{\rho g \lambda} = \frac{4\pi \nu}{g \lambda}. \qquad (6\text{–}105)$$

The viscosity of the mantle can be estimated from Equation (6–105) once the *relaxation time* for postglacial rebound has been determined.

PROBLEM 6–11 Show that the constant of integration A in the above postglacial rebound solution is given by

$$A = -\left(\frac{\lambda}{2\pi}\right)^2 \frac{\rho g w_{m0}}{2\mu} e^{-t/\tau_r}. \qquad (6\text{–}106)$$

Quantitative information on the rate of postglacial rebound can be obtained from *elevated beach terraces*. Wave action over a period of time erodes a beach to sea level. If sea level drops or if the land surface is elevated, a fossil beach terrace is created, as shown in Figure 6–15. The age of a fossil beach can be obtained by radioactive dating using carbon 14 in shells and driftwood. The elevations of a series of dated beach terraces at the mouth of the Angerman River in Sweden are given in Figure 6–16. The elevations of these beach terraces are attributed to the postglacial rebound of Scandinavia since the melting of the ice sheet. The elevations have been corrected for changes in sea level. The uplift of the beach terraces is compared with the exponential time dependence given in Equation (6–104). We assume that uplift began 10,000 years ago so that t is measured forward from that time to the present.

6-15 Elevated beach terraces on Östergransholm, Eastern Gotland, Sweden. The contempory uplift rate is about 2 mm yr^{-1}. (Photographer and copyright holder, Arne Philip, Visby, Sweden; courtesy IGCP Project Ecostratigraphy.)

We also assume that $w_{m0} = 300$ m with 30 m of uplift to occur in the future; that is, we take $w = 30$ m at $t = 10^4$ years, the present. The solid line in Figure 6–16 is obtained with $\tau_r = 4400$ years. Except for the earliest times, there is quite good agreement with the data.

6-16 Uplift of the mouth of the Angerman River, Sweden, as a function of time before the present compared with the exponential relaxation model, Equation (6–104), for $w_{m_0} = 300$ m less 30 m of uplift yet to occur, $\tau_r = 4400$ years, and an initiation of the uplift 10,000 years ago.

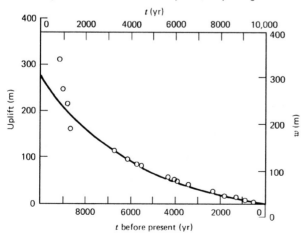

TABLE 6-2 **Distribution of Viscosity in the Mantle from Postglacial Rebound Studies**		
Region	**Depth (km)**	**Dynamic Viscosity (Pa s)**
Lithosphere	0–100	Elastic
Asthenosphere	100–175	4×10^{19}
	175–2848	10^{21}

This value of the relaxation time can be used to obtain a viscosity for the mantle using Equation (6–105). For the glaciation of Fennoscandia, a reasonable value for the wavelength is $\lambda = 3000$ km. Taking $\rho = 3300$ kg m^{-3} and $g = 10$ m s^{-2} along with $\tau_r = 4400$ years, we find that $\mu = 1.1 \times 10^{21}$ Pa s.

We have considered only the response to a spatially periodic surface displacement. Because the problem is linear, solutions can be superimposed in order to consider other distributions of surface displacement. However, more complete studies of postglacial rebound include the flexural rigidity of the elastic lithosphere and a depth-dependent mantle viscosity. If the ice sheets continue to melt during the period of rebound, the sea level will increase, and this must be taken into account. Available rebound data including changes in sea level are included on a worldwide basis. These studies require numerical solutions, and the results of one such effort are summarized in Table 6–2. We see that the mean mantle viscosity is in good agreement with the value we obtained using the approximate analytic solution.

PROBLEM 6-12 The ice sheet over Hudson Bay, Canada, had an estimated thickness of 2 km. At the present time there is a negative free-air gravity anomaly in this region of 0.3 mm s^{-2}.

a. Assuming that the ice (density of 1000 kg m^{-3}) was in isostatic equilibrium and displaced mantle rock with a density of 3300 kg m^{-3}, determine the depression of the land surface w_{m0}.

b. Assuming that the negative free-air gravity anomaly is due to incomplete rebound, determine w at the present time.

c. Applying the periodic analysis given above, determine the mantle viscosity. Assume that the ice sheet melted 10,000 years ago and that the appropriate wavelength for the Hudson Bay ice sheet was 5000 km.

d. Discuss the difference between the viscosity obtained in (c) and that obtained for Scandinavia.

6–11 Angle of Subduction

As discussed in Section 3–17, the oceanic lithosphere bends in a continuous manner as it is subducted at an ocean trench. The gravitational body force on the descending lithosphere is directed vertically downward. We might expect that under this body force the lithosphere would bend through 90° and descend vertically downward into the mantle. However, observations indicate that the oceanic lithosphere straightens out after subduction and descends at a finite angle of dip θ. (This was discussed in Section 1–4 and illustrated in Figure 1–9). The approximate dip angles associated with subduction at several ocean trenches are given in Table 6–3.

One explanation for why the lithosphere descends at an angle other than 90° is that pressure forces due to the induced flows in the mantle balance the gravitational body forces. This problem is illustrated in Figure 6–17. The pressure forces are due to the mantle flow induced by the motion of the descending lithosphere; they are flow pressures relative to the hydrostatic pressure. The dip of a subducting lithosphere is thus a consequence of the balance between the gravitational torque and the lifting pressure torque.

The pressure forces acting on a descending lithosphere can be calculated using the two-dimensional viscous corner flow model in Figure 6–18. The trench is located at $x = 0$. It is assumed that the surface $y = 0$,

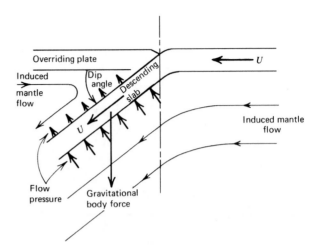

6–17 Forces acting on a descending lithosphere.

$x < 0$ moves with constant speed U toward the trench; the surface $y = 0$, $x > 0$ is stationary. The descending lithosphere is the line extending from the origin downward at the dip angle to the positive x axis; the velocity parallel to this line is U. Distance measured along this line is r. The line divides the viscous mantle into two corners: the arc corner and the oceanic corner. The motion of this line viscously drives a flow in the arc corner. The velocities of the dipping line and the surface induce a flow in the oceanic corner. We will solve for the motions in both corners and determine the flow pressures on the dipping line.

The stream functions for the corner flows in Figure 6–18 are solutions of the biharmonic Equation (6–74). For the corner flow geometry, we can write ψ in the form

$$\psi = (Ax + By) + (Cx + Dy)\arctan\frac{y}{x}, \qquad (6\text{–}107)$$

where A, B, C, and D are constants whose values are determined by boundary conditions. The problem in

TABLE 6–3 Approximate Dip Angles of Subduction at Several Island Arcs

Arc	Dip Angle
Central Chile	5°
Northern Chile	30°
Southern Chile	30°
Honshu	30°
Izu–Bonin	60°
Java	70°
New Hebrides	70°
Ryukyu	45°
West Indies	50°

6–18 Viscous corner flow model for calculating induced flow pressures on a descending lithosphere.

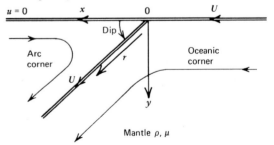

Figure 6–18 has two stream functions with distinct values of these constants because the arc and oceanic corners have different angles and different conditions on their bounding lines. It can be verified by direct substitution that Equation (6–107) is a solution of the biharmonic Equation (6–74). To do this, it is helpful to recall that

$$\frac{\partial}{\partial y} \arctan \left(\frac{y}{x} \right) = \frac{1}{(1 + y^2/x^2)} \frac{1}{x} = \frac{x}{x^2 + y^2} \tag{6-108}$$

and

$$\frac{\partial}{\partial x} \arctan \left(\frac{y}{x} \right) = \frac{1}{(1 + y^2/x^2)} \left(\frac{-y}{x^2} \right)$$
$$= \frac{-y}{x^2 + y^2}. \tag{6-109}$$

The velocity components corresponding to the stream function of Equation (6–107) follow from Equations (6–69) and (6–70):

$$u = -B - D \arctan \frac{y}{x} + (Cx + Dy) \left(\frac{-x}{x^2 + y^2} \right) \tag{6-110}$$

$$v = A + C \arctan \frac{y}{x} + (Cx + Dy) \left(\frac{-y}{x^2 + y^2} \right). \tag{6-111}$$

The pressure can be found by substituting Equation (6–110) into Equation (6–67) and integrating the resulting expression for $\partial P/\partial x$. Alternatively, Equations (6–68) and (6–111) can be used, in which case we integrate an expression for $\partial P/\partial y$, obtaining

$$P = \frac{-2\mu(Cx + Dy)}{(x^2 + y^2)}. \tag{6-112}$$

The pressure given by Equation (6–112) is the pressure relative to the hydrostatic pressure, that is, pressure associated with flow.

General expressions for the constants of integration are somewhat complicated; so we evaluate them for a particular value of the dip angle. As an example we choose a dip of $\pi/4$, representative of the Ryukyu arc. The boundary conditions for the arc corner are $u = v = 0$ on $y = 0$, $x > 0$, or

$$\arctan \frac{y}{x} = 0, \tag{6-113}$$

and $u = v = U\sqrt{2}/2$ on $y = x$, or

$$\arctan \frac{y}{x} = \frac{\pi}{4}. \tag{6-114}$$

Application of these conditions leads to the following expressions for the constants C and D in the arc corner:

$$C = \frac{-\pi U \sqrt{2}}{2(2 - \pi^2/4)} \tag{6-115}$$

$$D = \frac{-U\sqrt{2}(2 - \pi/2)}{(2 - \pi^2/4)}. \tag{6-116}$$

Thus the pressure in the arc corner is

$$P_{\text{arc corner}} = \frac{\mu U \sqrt{2} \{\pi x + (4 - \pi)y\}}{(2 - \pi^2/4)(x^2 + y^2)}. \tag{6-117}$$

If we evaluate this expression on $x = y$ and note that

$$x = y = \frac{r\sqrt{2}}{2} \tag{6-118}$$

on the dipping line, we find that the flow pressure on the top of the descending slab is

$$P = \frac{4\mu U}{(2 - \pi^2/4)r} = \frac{-8.558\mu U}{r}. \tag{6-119}$$

The negative value of the flow pressure on the top of the descending slab gives the effect of a *suction force* tending to lift the slab against the force of gravity. The pressure force varies as $1/r$ along the upper surface of the slab and therefore has a singularity in this idealized model as $r \to 0$. However, the *lifting torque* on the slab is the integral of the product rP over the upper surface of the slab. The lifting torque per unit distance along the top of the slab is a constant; the torque on the slab is thus proportional to its length.

The boundary conditions for the oceanic corner are $u = U$, $v = 0$ on $y = 0$, $x < 0$, or

$$\arctan \frac{y}{x} = \pi, \tag{6-120}$$

and $u = v = U\sqrt{2}/2$ on $y = x$, or

$$\arctan \frac{y}{x} = \frac{\pi}{4}. \tag{6-121}$$

By substituting Equations (6–110) and (6–111) into

Equations (6–120) and (6–121), we find

$$C = \frac{U}{(9\pi^2/4 - 2)}\left\{2 - \frac{\sqrt{2}}{(1 + 3\pi/2)}\left(\frac{3\pi}{2} + \frac{9\pi^2}{4}\right)\right\} \tag{6–122}$$

$$D = \frac{U}{(9\pi^2/4 - 2)}\left\{\sqrt{2}\left(2 + \frac{3\pi}{2}\right) - 2\left(1 + \frac{3\pi}{2}\right)\right\}. \tag{6–123}$$

The flow pressure in the oceanic corner is found by substituting these values of C and D into Equation (6–112). When the resulting expression is evaluated on the dipping line, we obtain

$$P = \frac{\mu U}{r}\left(\frac{3\pi\sqrt{2} - 4}{9\pi^2/4 - 2}\right) = \frac{0.462\mu U}{r} \tag{6–124}$$

for the flow pressure on the bottom of the descending slab. The positive value of P means that the induced pressure on the bottom of the slab also exerts a lifting torque on the slab. The torque per unit distance along the slab is a constant. The net lifting torque on the slab is the sum of the torques exerted by pressures on the top and bottom of the slab. A comparison of Equations (6–119) and (6–124) shows that the torque exerted by the suction pressure in the arc corner far outweighs the lifting effect of pressure on the bottom of the slab.

PROBLEM 6–13 Calculate the velocities in the arc and oceanic corners formed by a slab descending into the mantle with speed U and dip angle $\pi/4$.

PROBLEM 6–14 Derive expressions for the lifting torques on the top and bottom of a slab descending into the mantle with speed U at a dip angle of $60°$.

6–12 Diapirism

In the previous two sections we applied the equations of slow viscous flow to the mantle. We now turn to several problems involving the flow of crustal rocks. We first consider *diapirism*, or the buoyant upwelling of relatively light rock. As the lighter rock rises into the heavier overlying rock, a *diapir* of the lighter rock is formed. One example of diapirism is the formation of a *salt dome*, which occurs because salt is less dense than other typical sedimentary rocks. The process of salt dome formation is illustrated in Figure 6–19. Initially a layer of salt is deposited at the surface by evaporation of seawater (Figure 6–19a). Subsequent sedimentation buries this layer under other heavier sedimentary rocks such

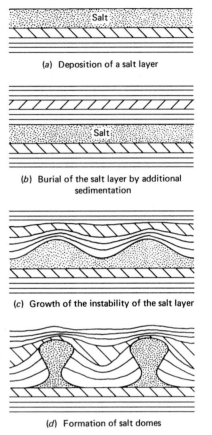

(a) Deposition of a salt layer

(b) Burial of the salt layer by additional sedimentation

(c) Growth of the instability of the salt layer

(d) Formation of salt domes

6–19 Diapiric formation of salt domes due to the gravitational instability of a light salt layer buried beneath heavier sedimentary rocks.

as shales and sandstones (Figure 6–19b). At shallow depths the strength of the salt layer is sufficient to prevent gravitational instability from inducing flow. As the depth of the salt layer increases with the further deposition of sediments, the temperature of the salt increases because of the geothermal gradient. Thermally activated creep processes then allow the salt to flow upward to be replaced by the heavier overlying sedimentary rocks. Eventually the upward flow of the salt creates a series of salt domes. Figure 6–20 is a photograph of salt domes in the Zagros Mountains of Iran.

Salt domes are important in the search for gas and oil. The deformation of the rocks above salt domes results in the formation of impermeable traps for the upward migrating oil and gas. Many oil and gas fields are found above salt domes.

There are other examples of diapirism in crustal rocks. In mountain belts high heat flow and volcanism

6-20 Satellite photograph of salt domes (dark circular areas) in the Zagros Mountains of Iran (NASA STS 047-151-035).

heat lower crustal rocks to sufficiently high temperatures so that they can freely flow by solid-state creep processes. If the heated rocks at depth are lighter than the overlying rocks, the deeper rocks will flow upward to form diapirs in a manner similar to the formation of salt domes. When the mountain belt is subsequently eroded, the diapirism is evident in the surface metamorphic rocks.

We apply the same type of analysis as was used in our study of postglacial rebound to investigate diapirism. The geometry of the problem is shown in Figure 6-21. A fluid layer with a thickness b and a density ρ_1 overlies a second fluid layer also of thickness b but with a density ρ_2. Both fluid layers have the same viscosity μ. The upper boundary of the top layer and the lower boundary of the bottom layer are rigid surfaces. Because we are interested in the case of instability, we take $\rho_1 > \rho_2$. The gravitational instability of heavy fluid

6-21 The Rayleigh–Taylor instability of a dense fluid overlying a lighter fluid.

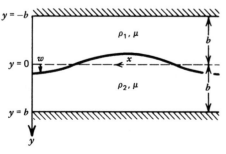

overlying light fluid is known as the *Rayleigh–Taylor instability*.

The undisturbed interface between the superposed fluid layers is taken to be at $y = 0$. Thus, $y = -b$ and $y = b$ are the upper and lower rigid boundaries, respectively. As a consequence of the gravitational instability, the interface between the fluids distorts and motions occur in the fluid layers. The displacement of the disturbed fluid interface is denoted by w. We assume that w is given by Equation (6–79). The stream function ψ_1 for the flow in the upper fluid layer has the form of Equation (6–85), which we rewrite here using hyperbolic functions instead of the exponentials

$$\psi_1 = \sin\frac{2\pi x}{\lambda}\left(A_1\cosh\frac{2\pi y}{\lambda} + B_1\sinh\frac{2\pi y}{\lambda}\right.$$
$$\left. + C_1 y\cosh\frac{2\pi y}{\lambda} + D_1 y\sinh\frac{2\pi y}{\lambda}\right). \quad (6\text{–}125)$$

Similarly, the stream function ψ_2 for the lower layer is

$$\psi_2 = \sin\frac{2\pi x}{\lambda}\left(A_2\cosh\frac{2\pi y}{\lambda} + B_2\sinh\frac{2\pi y}{\lambda}\right.$$
$$\left. + C_2 y\cosh\frac{2\pi y}{\lambda} + D_2 y\sinh\frac{2\pi y}{\lambda}\right). \quad (6\text{–}126)$$

The velocity components in the layers are found by differentiating these equations for ψ_1 and ψ_2 according to Equations (6–69) and (6–70):

$$u_1 = -\frac{2\pi}{\lambda}\sin\frac{2\pi x}{\lambda}\left\{\left(A_1 + C_1 y + \frac{\lambda D_1}{2\pi}\right)\sinh\frac{2\pi y}{\lambda}\right.$$
$$\left. + \left(B_1 + D_1 y + \frac{\lambda C_1}{2\pi}\right)\cosh\frac{2\pi y}{\lambda}\right\} \quad (6\text{–}127)$$

$$v_1 = \frac{2\pi}{\lambda}\cos\frac{2\pi x}{\lambda}\left\{(A_1 + C_1 y)\cosh\frac{2\pi y}{\lambda}\right.$$
$$\left. + (B_1 + D_1 y)\sinh\frac{2\pi y}{\lambda}\right\} \quad (6\text{–}128)$$

$$u_2 = -\frac{2\pi}{\lambda}\sin\frac{2\pi x}{\lambda}\left\{\left(A_2 + C_2 y + \frac{\lambda D_2}{2\pi}\right)\sinh\frac{2\pi y}{\lambda}\right.$$
$$\left. + \left(B_2 + D_2 y + \frac{\lambda C_2}{2\pi}\right)\cosh\frac{2\pi y}{\lambda}\right\} \quad (6\text{–}129)$$

$$v_2 = \frac{2\pi}{\lambda}\cos\frac{2\pi x}{\lambda}\left\{(A_2 + C_2 y)\cosh\frac{2\pi y}{\lambda}\right.$$
$$\left. + (B_2 + D_2 y)\sinh\frac{2\pi y}{\lambda}\right\}. \quad (6\text{–}130)$$

Among the boundary conditions we have for evaluating the constants of integration are the no-slip conditions on $y = \pm b$,

$$u_1 = v_1 = 0 \quad \text{on} \quad y = -b \quad (6\text{–}131)$$

$$u_2 = v_2 = 0 \quad \text{on} \quad y = b, \quad (6\text{–}132)$$

and continuity of u and v across the interface. For small displacements of the interface, $w \ll \lambda$, it is appropriate to require continuity of u and v at the undisturbed location of the interface, $y = 0$. Thus we require

$$u_1 = u_2 \quad \text{and} \quad v_1 = v_2 \quad \text{on} \quad y = 0. \quad (6\text{–}133)$$

By applying these boundary conditions to Equations (6–127) to (6–130), we obtain

$$B_1 + \frac{\lambda C_1}{2\pi} = B_2 + \frac{\lambda C_2}{2\pi} \quad (6\text{–}134)$$

$$A_1 = A_2 \quad (6\text{–}135)$$

$$\left(A_1 - bC_1 + \frac{\lambda D_1}{2\pi}\right)\tanh\frac{2\pi b}{\lambda} = B_1 - bD_1 + \frac{\lambda C_1}{2\pi} \quad (6\text{–}136)$$

$$(B_1 - bD_1)\tanh\frac{2\pi b}{\lambda} = A_1 - bC_1 \quad (6\text{–}137)$$

$$\left(A_2 + bC_2 + \frac{\lambda D_2}{2\pi}\right)\tanh\frac{2\pi b}{\lambda} = -B_2 - bD_2 - \frac{\lambda C_2}{2\pi} \quad (6\text{–}138)$$

$$(B_2 + bD_2)\tanh\frac{2\pi b}{\lambda} = -A_2 - bC_2. \quad (6\text{–}139)$$

Shear stress must also be continuous across the interface between the fluid layers. For $w \ll \lambda$ and for equal viscosities across the interface this condition can be written

$$\frac{\partial u_1}{\partial y} + \frac{\partial v_1}{\partial x} = \frac{\partial u_2}{\partial y} + \frac{\partial v_2}{\partial x} \quad \text{on} \quad y = 0, \quad (6\text{–}140)$$

where Equation (6–58) has been used for the shear stress. Since v is continuous at $y = 0$, so is $\partial v/\partial x$, and Equation (6–140) simplifies to

$$\frac{\partial u_1}{\partial y} = \frac{\partial u_2}{\partial y} \quad \text{on} \quad y = 0. \quad (6\text{–}141)$$

Equation (6–141) requires that

$$\left(A_1 + \frac{\lambda D_1}{2\pi}\right) + \frac{\lambda D_1}{2\pi} + D_1 = \left(A_2 + \frac{\lambda D_2}{2\pi}\right) + \frac{\lambda D_2}{2\pi} + D_2 \quad (6\text{–}142)$$

or, with $A_1 = A_2$,

$$D_1 = D_2. \quad (6\text{–}143)$$

By subtracting Equations (6–137) and (6–139) and combining the result with the difference between Equations (6–136) and (6–138), we obtain

$$0 = (C_1 + C_2)\left\{1 + \frac{2\pi b}{\lambda}\left(\tanh\frac{2\pi b}{\lambda} - \coth\frac{2\pi b}{\lambda}\right)\right\}. \quad (6\text{–}144)$$

Equation (6–144) can be satisfied for arbitrary $2\pi b/\lambda$ only if

$$C_1 = -C_2. \quad (6\text{–}145)$$

If we add Equations (6–137) and (6–139) and make use of Equation (6–145), we also deduce that

$$B_1 = -B_2. \quad (6\text{–}146)$$

By using Equations (6–145) and (6–146) to simplify Equation (6–134), we get

$$B_1 = -\frac{\lambda C_1}{2\pi}. \quad (6\text{–}147)$$

All the constants of integration can now be determined in terms of A_1 by solving Equations (6–136), (6–137), and (6–147). After some algebraic manipulation we find that the stream function in the upper layer is

$$\begin{aligned}
\psi_1 = {} & A_1 \sin\frac{2\pi x}{\lambda}\cosh\frac{2\pi y}{\lambda} \\
& + A_1\sin\frac{2\pi x}{\lambda}\left\{\frac{y}{b}\left(\frac{\lambda}{2\pi b}\right)\tanh\frac{2\pi b}{\lambda}\sinh\frac{2\pi y}{\lambda}\right. \\
& + \left(\frac{y}{b}\cosh\frac{2\pi y}{\lambda} - \frac{\lambda}{2\pi b}\sinh\frac{2\pi y}{\lambda}\right) \\
& \times \left.\left(\frac{\lambda}{2\pi b} + \frac{1}{\sinh(2\pi b/\lambda)\cosh(2\pi b/\lambda)}\right)\right\} \\
& \times \left\{\frac{1}{\sinh(2\pi b/\lambda)\cosh(2\pi b/\lambda)}\right. \\
& \left. - \left(\frac{\lambda}{2\pi b}\right)^2\tanh\frac{2\pi b}{\lambda}\right\}^{-1}. \quad (6\text{–}148)
\end{aligned}$$

The expression for ψ_2 is obtained by replacing y with $-y$ in Equation (6–148).

The solution for the stream function can be used to obtain an equation for the motion of the interface. The time rate of change of the interface displacement $\partial w/\partial t$ must be equal to the vertical component of the fluid velocity at the interface. If this condition were not satisfied, a void would be created between the fluid layers. Because the interface displacement is small, this condition can be written

$$\frac{\partial w}{\partial t} = v_{y=0}. \tag{6–149}$$

The vertical velocity v can be evaluated by differentiating Equation (6–148) with respect to x. If this is done, and the result evaluated at $y = 0$, we can rewrite Equation (6–149) as

$$\frac{\partial w}{\partial t} = \frac{2\pi A_1}{\lambda} \cos \frac{2\pi x}{\lambda}. \tag{6–150}$$

To eliminate the constant A_1 from the equation of motion of the interface, we need to incorporate an essential aspect of the physics of the problem into the analysis. This is the buoyancy force brought into play by the displacement of the interface. Figure 6–22 compares two columns of fluid, one with the interface in the undisturbed location and the other with the interface displaced downward. Because of the interface displacement, fluid of density ρ_1 replaces fluid of density ρ_2 between $y = 0$ and $y = w$. The additional weight of this fluid $(\rho_1 - \rho_2)gw$ is felt as a normal stress or pressure on the disturbed interface. It must be balanced by the net normal stress on the interface due to flow pressure and normal viscous stress. It is sufficient to determine these stresses on $y = 0$ because of the small interface displacement. According to Equation (6–57) the normal viscous stress on $y = 0$ is $2\mu(\partial v/\partial y)_{y=0}$. By differentiating Equation (6–148) with respect to x and y and evaluating the result on $y = 0$, we see that this quantity is zero. Thus the buoyancy force per unit area due to the displacement of the interface is balanced solely by the net flow pressure exerted on the interface. This

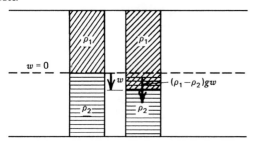

6–22 The buoyancy force associated with the displacement of the interface.

condition can be written

$$(\rho_1 - \rho_2)gw = (P_2 - P_1)_{y=0}. \tag{6–151}$$

Equation (6–151) provides a second relationship between w and the flow field that allows us to relate w to A_1 and thereby convert Equation (6–150) into an equation for w.

The flow pressure on $y = 0$ in the upper layer can be found by substituting Equation (6–148) into Equation (6–72) with the result

$$(P_1)_{y=0} = \frac{2\mu A_1}{b}\left(\frac{2\pi}{\lambda}\right)\left(\frac{\lambda}{2\pi b} + \frac{1}{\sinh\frac{2\pi b}{\lambda}\cosh\frac{2\pi b}{\lambda}}\right)$$
$$\times \left(\frac{1}{\sinh\frac{2\pi b}{\lambda}\cosh\frac{2\pi b}{\lambda}} - \left(\frac{\lambda}{2\pi b}\right)^2\right.$$
$$\left.\times \tanh\frac{2\pi b}{\lambda}\right)^{-1} \times \cos\frac{2\pi x}{\lambda}. \tag{6–152}$$

By carrying through the same procedure using ψ_2, we find

$$(P_2)_{y=0} = -(P_1)_{y=0}. \tag{6–153}$$

Equation (6–151) becomes

$$(\rho_1 - \rho_2)gw = -2(P_1)_{y=0}. \tag{6–154}$$

Equation (6–154) shows that with a heavy fluid above a light one $(\rho_1 > \rho_2)$, a downward displacement of the interface $(w > 0)$ causes a negative pressure in the upper fluid layer. This tends to produce a further downward displacement of the interface leading to instability of the configuration. Upon substituting Equation (6–152) into Equation (6–154), we get

$$(\rho_1 - \rho_2)gw = -\frac{4\mu A_1}{b}\left(\frac{2\pi}{\lambda}\right)\cos\frac{2\pi x}{\lambda}$$
$$\times \left(\frac{\lambda}{2\pi b} + \frac{1}{\sinh(2\pi b/\lambda)\cosh(2\pi b/\lambda)}\right)$$
$$\times \left(\frac{1}{\sinh(2\pi b/\lambda)\cosh(2\pi b/\lambda)}\right.$$
$$\left.- \left(\frac{\lambda}{2\pi b}\right)^2\tanh\frac{2\pi b}{\lambda}\right)^{-1}. \tag{6–155}$$

By solving this equation for A_1 and substituting the resulting expression into Equation (6–150), we finally

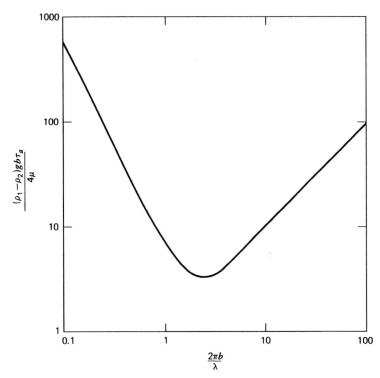

6-23 Dimensionless growth time of a disturbance as a function of dimensionless wave number for the Rayleigh–Taylor instability.

arrive at

$$
\frac{\partial w}{\partial t} = \frac{(\rho_1 - \rho_2)gb}{4\mu}
$$

$$
\times \frac{\left(\left(\frac{\lambda}{2\pi b}\right)^2 \tanh \frac{2\pi b}{\lambda} - \frac{1}{\sinh \frac{2\pi b}{\lambda} \cosh \frac{2\pi b}{\lambda}}\right)}{\left(\frac{\lambda}{2\pi b} + \frac{1}{\sinh \frac{2\pi b}{\lambda} \cosh \frac{2\pi b}{\lambda}}\right)} w.
$$

$$(6\text{–}156)$$

The solution of this equation is

$$
w = w_0 e^{t/\tau_a},
\tag{6–157}
$$

with

$$
\tau_a = \frac{4\mu}{(\rho_1 - \rho_2)gb}
$$

$$
\times \frac{\left(\frac{\lambda}{2\pi b} + \frac{1}{\sinh \frac{2\pi b}{\lambda} \cosh \frac{2\pi b}{\lambda}}\right)}{\left(\left(\frac{\lambda}{2\pi b}\right)^2 \tanh \frac{2\pi b}{\lambda} - \frac{1}{\sinh \frac{2\pi b}{\lambda} \cosh \frac{2\pi b}{\lambda}}\right)}.
$$

$$(6\text{–}158)$$

The quantity τ_a is the growth time (for $\rho_1 > \rho_2$) of a disturbance. Its value depends on the wavelength λ of the interface distortion. Figure 6–23 is a plot of the dimen-

sionless growth time $(\rho_1 - \rho_2)gb\tau_a/4\mu$ as a function of the dimensionless disturbance wave number $2\pi b/\lambda$. If heavy fluid lies on top ($\rho_1 > \rho_2$), the interface is always unstable; that is, $\tau_a > 0$. If light fluid lies on top ($\rho_1 < \rho_2$), τ_a is negative for all λ and the interface is stable. It can be shown from Equation (6–158) that for large wavelengths,

$$
\tau_a \rightarrow \frac{24\mu}{(\rho_1 - \rho_2)gb} \left(\frac{\lambda}{2\pi b}\right)^2.
\tag{6–159}
$$

For very small wavelengths,

$$
\tau_a \rightarrow \frac{4\mu}{(\rho_1 - \rho_2)gb} \left(\frac{2\pi b}{\lambda}\right).
\tag{6–160}
$$

These asymptotic behaviors of τ_a can be seen in Figure 6–23.

When the heavy fluid lies on top and the configuration is unstable, the disturbance with the shortest time constant grows and dominates the instability. The wavelength that gives the smallest value for τ_a is

$$
\lambda = 2.568b.
\tag{6–161}
$$

The rate of growth of this dominant disturbance is obtained by substituting Equation (6–161) into Equation

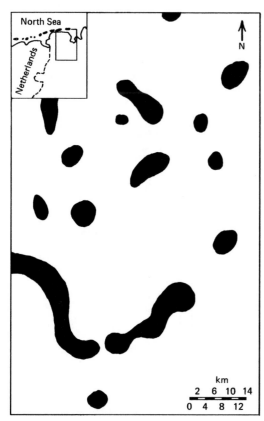

North Sea

Netherlands

N

km
2   6 10 14
0   4   8 12

6-24 Distribution of salt domes in northern Germany.

(6–158) with the result

$$\tau_a = \frac{13.04\mu}{(\rho_1 - \rho_2)gb}. \tag{6-162}$$

The instability takes longer to grow, the more viscous the fluids and the smaller the density difference. Although we have considered only the stability problem for small displacements, we expect that the wavelength of the most rapidly growing small disturbance closely corresponds to the spacing between fully developed diapirs. A map view showing the distribution of salt domes in the north of Germany is given in Figure 6–24. The depth to the salt layer is about 5 km, and the spacing of the salt domes is about 10 to 15 km, in good agreement with Equation (6–161).

PROBLEM 6–15 A layer of salt at a depth of 3 km with a density of 2150 kg m^{-3} lies beneath sediments with a density of 2600 kg m^{-3}. The salt layer is known to have doubled the amplitude of its instability in

100,000 years. Estimate the equivalent viscosity of the system.

PROBLEM 6–16 Suppose that the 660-km density discontinuity in the mantle corresponds to a compositional change with lighter rocks lying above more dense ones. Estimate the minimum decay time for a disturbance to this boundary. Assume $\rho = 4000$ kg m^{-3}, $\Delta\rho = 100$ kg m^{-3}, and $\mu = 10^{21}$ Pa s.

PROBLEM 6–17 Discuss how you would modify the analysis of the previous section to account for a viscosity difference between the two fluid layers.

6-13 Folding

Folding of crustal rock occurs on all scales. On the largest scale, folding results in a series of parallel mountain ranges. This was illustrated in Figure 1–41. On this large scale a fold that is concave upward is referred to as a *synclinorium*. Rocks folded in this manner are usually sedimentary rocks, and younger rocks are ordinarily found in the flexure formed of older rocks. Also on this large scale a fold whose flanks diverge downward is referred to as an *anticlinorium*. Erosion of large-scale folds often results in a *valley and ridge topography* such as that found in Pennsylvania and West Virginia (see Figure 1–42). In this case the valleys are the result of the erosion of shales, whereas the ridges are composed of more resistant sandstones.

Folds are found in both sedimentary and metamorphic rocks on scales ranging down to a few centimeters. Folding occurs under a wide variety of conditions, but it is often associated with *compressional tectonics*. Two important questions concerning folding are why does the rock deform and why does it deform in such a manner as to produce folds. It is perhaps surprising that at relatively low temperatures sedimentary rocks flow to produce folds rather than fracture. Although the rheology of folded sedimentary rocks is not fully understood, *pressure solution creep* is thought to play an important role. Sedimentary rocks are often saturated with water. The *solubility* of minerals such as quartz in the water is a function of pressure as well as temperature. When differential stresses are applied to the rock, the minerals dissolve in regions of high stress and are deposited in regions of low stress. The result is a deformation of the rock. Pressure solution creep of sedimentary rocks can

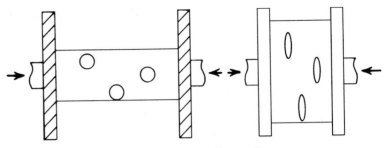

(a) Compression of a uniform medium

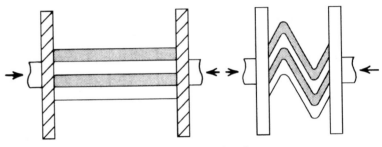

(b) Folding of a layered medium

result in a linear relationship between stress and rate of strain and, therefore, a Newtonian fluid behavior. A viscosity can be determined. A more detailed discussion of pressure solution creep is given in the next chapter.

Folded sedimentary or metamorphic rocks usually have a preexisting layered structure. There usually is considerable variation in the material properties of adjacent layers when folding occurs. If a uniform medium is subjected to compression, it will be uniformly squeezed, as illustrated in Figure 6–25a. However, if the medium is composed of a series of weak and strong layers, folding will occur, as shown in Figure 6–25b. The strong layers are referred to as being *competent*; an example is a limestone. The weak layers are referred to as being *incompetent*; an example is a sandstone. It should be noted that strength and resistance to erosion are not necessarily related.

One approach to the quantitative study of folding is to consider an *elastic* (competent) *layer* of thickness h embedded between two semi-infinite Newtonian viscous fluids (incompetent). An end load $\bar{P}$ on the elastic layer may cause it to buckle; however, its deformation will be restricted by the confining fluids. This problem is illustrated in Figure 6–26.

6–25 (a) Compression of a uniform medium. This type of compression can often be identified in rocks by the flattening of spherical inclusions. (b) Folding of a layered medium composed of strong (competent) and weak (incompetent) members under compression.

We have already considered the deformation of a thin elastic plate under end loading in Section 3–11. The applicable differential equation is Equation (3–74). The vertical component of the normal stress due to flow in the fluids above and below the plates can be used to determine the force per unit area $q(x)$ on the plate. The fluids occupy semi-infinite half-spaces. We assume that the deformation of the plate is given by

$$w = w_m \cos(2\pi x/\lambda)e^{t/\tau_a}. \qquad (6\text{--}163)$$

6–26 An elastic plate of thickness h is embedded between two viscous fluids with viscosity μ. An end load $\bar{P}$ is applied to the elastic plate until it buckles (folds).

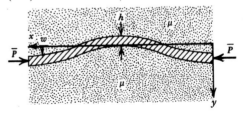

Because the plate forms the boundaries of the fluid half-spaces, these boundaries have sinusoidally varying shapes. This situation is identical with the one we encountered in our study of postglacial rebound in Section 6–10. We can use the results of that section to advantage here.

By symmetry, the solutions above and below the plate are identical. We consider the solution below the plate and measure y positive downward from the base of the plate, as illustrated in Figure 6–26. The appropriate solution of the biharmonic equation is Equation (6–85). The condition that the velocities be finite as $y \to \infty$ again requires $C = D = 0$. The rigidity of the elastic plate requires that $u = 0$ on the plate, and because we again assume $w \ll \lambda$, this boundary condition can be applied at $y = 0$. Therefore Equation (6–90) is applicable in the fluid below the plate. From Equation (6–95), the pressure P_b on the base of the plate (evaluated at $y = 0$) is given by

$$P_b = 2 A \mu \left(\frac{2\pi}{\lambda} \right)^2 \cos \frac{2\pi x}{\lambda}. \tag{6–164}$$

This can be rewritten in terms of w by using Equation (6–101)

$$P_b = 2\mu \left(\frac{2\pi}{\lambda} \right) \frac{\partial w}{\partial t}. \tag{6–165}$$

The pressure P_T acting downward on the top of the plate is related to the pressure P_b acting upward on the base of the plate by

$$P_T(x) = -P_b(x). \tag{6–166}$$

This is a consequence of the symmetry of the flows above and below the plate; we found an identical result – Equation (6–153) – in the previous section. There is no normal viscous stress on the plate because $\partial v / \partial y$ vanishes on $y = 0$ according to Equation (6–96). Thus the net normal stress on the plate is

$$q = P_T - P_b = -2 P_b. \tag{6–167}$$

By substituting Equation (6–165) into this equation, we obtain

$$q(x, t) = -4\mu \left(\frac{2\pi}{\lambda} \right) \frac{\partial w(x, t)}{\partial t}. \tag{6–168}$$

With the force per unit area acting on the elastic plate now determined, we can write the equation for the deflection of the plate – Equation (3–74) – as

$$D \frac{\partial^4 w}{\partial x^4} + \bar{P} \frac{\partial^2 w}{\partial x^2} = -4\mu \left(\frac{2\pi}{\lambda} \right) \frac{\partial w}{\partial t}, \tag{6–169}$$

where D, you recall, is the flexural rigidity of the plate; see Equation (3–72). Upon substituting Equation (6–163) into (6–169), we find

$$\tau_a = \frac{-4\mu}{\frac{2\pi}{\lambda} \left[D \left(\frac{2\pi}{\lambda} \right)^2 - \bar{P} \right]}. \tag{6–170}$$

The wavelength corresponding to the smallest value of τ_a is obtained by setting the derivative of τ_a with respect to λ equal to zero; the result is

$$\lambda = 2\pi \left(\frac{3D}{\bar{P}} \right)^{1/2}. \tag{6–171}$$

This is the wavelength of the most rapidly growing disturbance. Upon substituting Equation (3–72) for D into (6–171) and writing

$$\bar{P} = \sigma h, \tag{6–172}$$

where σ is the stress in the elastic layer associated with the end load, we get

$$\lambda = \pi h \left\{ \frac{E}{\sigma (1 - \nu^2)} \right\}^{1/2}. \tag{6–173}$$

It is expected that when folds develop in an elastic layer of rock surrounded by rock exhibiting fluid behavior, the initial wavelength of the folds has the dependence on the thickness of the elastic layer and the applied stress given by Equation (6–173).

The observed dependence of fold wavelength on the thickness of the dominant member of a fold is given in Figure 6–27 for a wide variety of folds. Excellent agreement with Equation (6–173) is obtained for $\sigma (1 - \nu^2)/E = 10^{-2}$. For $E = 50$ GPa and $\nu = 0.25$ for sedimentary rocks, this gives $\sigma = 530$ MPa. Although this is a high stress, it is likely to be about the same as the compressional strength of many sedimentary rocks when they are buried to a depth of 2 to 5 km.

As the amplitude of a fold increases, its wavelength decreases somewhat, and the bending stress in the elastic member exceeds the *yield strength* of the rock. The elastic member then either fractures or plastically yields

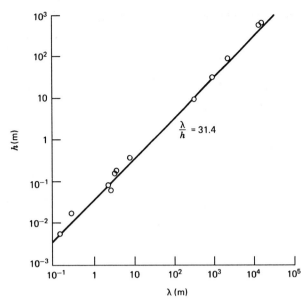

6-27 Dependence of the thickness of the dominant member in a fold on the wavelength of the fold compared with Equation (6–173). Data points are from Currie et al. (1962).

(a)

(b)

6-28 (a) Angular or chevron folds near Copiapo in Atacama Province, northern Chile. The folded resistant layers of silty limestone alternate with less resistant but more competent layers of sandstone (K. Sagerstrom 563, U.S. Geological Survey.) (b) Rounded fold in amphibolite near Salmon River, Idaho (W. B. Hamilton 377, U.S. Geological Survey.)

at the points of maximum bending moment that are at $x = \pm\frac{1}{2}n\lambda$, $n = 0, 1, 2, \ldots$. If *plastic bending* occurs, an *angular or chevron fold* would be expected, as illustrated in Figure 6–28a. Folds with nearly straight limbs of this type are often observed. A further analysis of this type of plastic bending is given in the next chapter.

Although many observed folds exhibit a plastic deformation in the dominant or competent member of the fold, there are many other cases in which a rounded structure is observed. For a rounded fold such as that illustrated in Figure 6–28b the dominant member has probably also been deformed in a fluidlike manner. An alternative approach to the theory of folding is to assume that the competent layer is a Newtonian fluid with a viscosity μ_1. It is embedded between two semi-infinite fluids with a viscosity μ_0, and $\mu_1 \gg \mu_0$. This mechanism, first proposed by Maurice Biot, is often referred to as the *Biot theory of folding*.

To analyze the viscous folding problem, it is necessary to develop the theory for the bending of a free or isolated plate of viscosity μ. We follow the derivation for the bending of a free elastic plate given in Section 3–9. Recall that the key aspect of that derivation was the determination of the bending moment M given by

Equation (3–61):

$$M = \int_{-h/2}^{h/2} \sigma_{xx} y \, dy. \tag{6–174}$$

The longitudinal stress σ_{xx} in a viscous plate is given by Equation (6–60). For a free plate, σ_{yy} must vanish on its surfaces, and if the plate is thin, we can take $\sigma_{yy} = 0$ throughout the plate, as in the elastic plate derivation. From Equation (6–61) with $\sigma_{yy} = 0$ we obtain

$$p = 2\mu \frac{\partial v}{\partial y}. \tag{6–175}$$

The incompressible continuity Equation (6–53) gives $\partial v / \partial y = -\partial u / \partial x$, and we can rewrite Equation (6–175) as

$$p = -2\mu \frac{\partial u}{\partial x}. \tag{6–176}$$

By substituting Equation (6–176) into Equation (6–60) in order to eliminate the pressure, we obtain

$$\sigma_{xx} = -4\mu \frac{\partial u}{\partial x}. \tag{6–177}$$

This is the relationship between the bending stress σ_{xx} and the rate of longitudinal strain $\partial u / \partial x$ for a thin viscous plate. It is analogous to Equation (3–64), which relates the fiber stress σ_{xx} to the strain ε_{xx} in a thin elastic plate.

Equation (6–174) for the bending moment in the viscous plate becomes

$$M = -4\mu \int_{-h/2}^{h/2} \frac{\partial u}{\partial x} y \, dy. \tag{6–178}$$

By direct analogy with Equation (3–70) the rate of strain $\partial u / \partial x$ is given by

$$\frac{\partial u}{\partial x} = y \frac{\partial^3 w}{\partial x^2 \partial t}. \tag{6–179}$$

The sign of this equation is opposite to that of Equation (3–70), since the rate of strain $\partial u / \partial x$ and the strain rate $\dot{\varepsilon}_{xx}$ have opposite signs. If we substitute Equation (6–179) into (6–178) and carry out the integration, we get

$$M = -\frac{\mu h^3}{3} \frac{\partial^3 w}{\partial x^2 \partial t}. \tag{6–180}$$

Upon substituting the second derivative with respect to x of (6–180) into (3–60), we obtain the general equation

for the bending of a thin viscous plate,

$$\frac{\mu h^3}{3} \frac{\partial^5 w}{\partial x^4 \partial t} = q - \bar{P} \frac{\partial^2 w}{\partial x^2}. \tag{6–181}$$

Solutions of this equation give the vertical displacement w of a viscous plate as a function of time.

As a specific example, consider a free viscous plate of length L embedded at one end with a concentrated load V_a applied at its other end, as in Figure 3–17. Since $\bar{P} = q = 0$, Equation (6–181) reduces to

$$\frac{\mu h^3}{3} \frac{\partial^5 w}{\partial x^4 \partial t} = 0. \tag{6–182}$$

Integrating twice with respect to x yields

$$\frac{\mu h^3}{3} \frac{\partial^3 w}{\partial x^2 \partial t} = -M = f_1(t)x + f_2(t), \tag{6–183}$$

where $f_1(t)$ and $f_2(t)$ are constants of integration that can depend on time. Because the overall torque balance given in Equation (3–78),

$$M = V_a(x - L), \tag{6–184}$$

must also be applicable to the viscous plate, we can identify f_1 and f_2 as

$$f_1 = -V_a \qquad f_2 = V_a L. \tag{6–185}$$

Equation (6–183) thus takes the form

$$\frac{\mu h^3}{3} \frac{\partial^3 w}{\partial x^2 \partial t} = -V_a x + V_a L. \tag{6–186}$$

We integrate this equation twice more with respect to x and satisfy the boundary conditions for an embedded plate, $w = \partial w / \partial x = 0$ at $x = 0$, to get

$$\frac{\mu h^3}{3} \frac{\partial w}{\partial t} = \frac{V_a x^2}{2} \left(L - \frac{x}{3} \right). \tag{6–187}$$

A final integration with respect to time and application of the initial condition $w = 0$ at $t = 0$ gives

$$w = \frac{3}{2} \frac{V_a x^2}{\mu h^3} \left(L - \frac{x}{3} \right) t. \tag{6–188}$$

A comparison of Equations (6–188) and (3–83) shows that the deflection of the viscous plate has the same spatial dependence as the deflection of the elastic plate. This is a general correspondence between the behavior of viscous and elastic plates. However, although the deflection of the elastic plate is time-independent, the deflection of the viscous plate increases linearly with time.

We return now to the viscous folding problem by considering the buckling of a viscous plate contained between two semi-infinite viscous fluids. If the approximation $\sigma_{yy} = 0$ that we made in our derivation of the bending moment of a free viscous plate is to be applicable to this situation, the plate viscosity μ_1 must be much larger than the viscosity μ_0 of the surrounding half-spaces. In this case, Equation (6–181) governs the time-dependent displacement of the plate, which we can take to be of the form (6–163). The responses of the semi-infinite fluids to the deformation of the viscous plate are identical with their responses to the bending of an elastic plate. Therefore the force per unit area on the viscous plate is given by Equation (6–168). Upon substituting Equation (6–168) into (6–181), we obtain

$$\frac{\mu_1 h^3}{3} \frac{\partial^5 w}{\partial x^4 \partial t} = -\frac{8\pi \mu_0}{\lambda} \frac{\partial w}{\partial t} - \bar{P}\frac{\partial^2 w}{\partial x^2}. \tag{6–189}$$

With w given by Equation (6–163) we must have

$$\tau_a = \frac{1}{\bar{P}}\left[\frac{2\lambda}{\pi}\mu_0 + \frac{4\pi^2}{3\lambda^2}\mu_1 h^3\right]. \tag{6–190}$$

The wavelength corresponding to the smallest value of τ_a is obtained by setting the derivative of τ_a with respect to λ equal to zero; the result is

$$\lambda = 2\pi h \left(\frac{1}{6}\frac{\mu_1}{\mu_0}\right)^{1/3}. \tag{6–191}$$

This is the wavelength of the most rapidly growing mode. A comparison of this result with the observed dependence of the wavelength of the competent layer on its thickness given in Figure 6–27 shows good agreement for $\mu_1/\mu_0 = 750$.

PROBLEM 6–18 In the examples of folding just considered we assumed that the competent rock adhered to the incompetent rock. If the layers are free to slip, show that the wavelength of the most rapidly growing disturbance in an elastic layer of rock contained between two semi-infinite viscous fluids is given by

$$\lambda = \pi h[E/\sigma(1 - v^2)]^{1/2}. \tag{6–192}$$

The *free slip condition* is equivalent to a zero shear stress condition at the boundaries of the elastic layer.

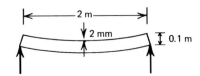

6–29 Sketch for Problem 6–20.

PROBLEM 6–19 In the folding examples, we assumed that the competent rock adhered to the incompetent rock. If the layers are free to slip, show that the wavelength of the most rapidly growing disturbance in a fluid layer of viscosity μ_1 contained between two semi-infinite fluids of viscosity $\mu_0, \mu_1 \gg \mu_0$, is given by

$$\lambda = 2\pi h(\mu_1/6\mu_0)^{1/3}. \tag{6–193}$$

The free slip condition is equivalent to a zero shear stress condition at the boundaries between the layers.

PROBLEM 6–20 A marble plate 0.1 m thick and 2 m long is simply supported at its ends, as shown in Figure 6–29. In 150 years the center has sagged 2 mm. Assuming that the plate behaves as a Newtonian fluid, determine the viscosity of the marble ($\rho = 2300$ kg m^{-3}).

PROBLEM 6–21 A marble plate 0.2 m thick and 5 m long is embedded at one end. In 200 years the free end has sagged 5 mm, as shown in Figure 6–30. Assuming the plate behaves as a Newtonian fluid, determine the viscosity of the marble ($\rho = 2300$ kg m^{-3}).

6–14 Stokes Flow

A solid body will rise or fall through a fluid if its density is different from the density of the fluid. If the body is less dense, the buoyancy force will cause it to rise; if the body is more dense, it will fall. If the fluid is very viscous, the Reynolds number Re based on the size of the body, the velocity at which the body moves through the fluid, and the viscosity of the fluid will be small. In the limit Re $\ll 1$ inertia forces can be

6–30 Sketch for Problem 6–21.

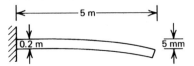

neglected, and Equations (6–53), (6–67), and (6–68) are applicable.

If the body has a spherical shape, a relatively simple solution can be obtained in the limit of a very viscous fluid. The resulting flow is known as *Stokes flow*. This problem has a number of geological applications. One is to obtain an estimate for the velocity of magmas as they rise through the lithosphere. Many basaltic lava flows contain *xenoliths*, chunks of solid rock that were entrained in the magma as it rose through the lithosphere. The solid xenoliths are carried with the magma when it is erupted on the Earth's surface. Because the viscosity of the magma and the density difference between the xenolith and magma can be estimated, the size of the largest observed xenolith can be used to estimate the magma ascent velocity.

Stokes solution can also be used to estimate the size of a mantle heterogeneity that can be entrained in mantle flows. One model for the ascent of magma in the mantle is that magma "bubbles" rise under the buoyancy force. Stokes solution can be used to estimate the rate of magma ascent as a function of the size of the magma bubble.

Let us derive an expression for the velocity of ascent or descent U of a spherical body in a constant-viscosity fluid with a different density. We first calculate the net force or drag exerted by the fluid on the sphere and then equate this force to the buoyancy force responsible for the sphere's motion. For the purpose of calculating the drag on the sphere due to its steady motion through the "fluid" we can consider the sphere to be fixed and have the fluid move past the sphere. We will not discuss the transient period during which the sphere accelerates to its final steady or terminal velocity.

The sphere of radius a is centered at the origin of a spherical coordinate system (r, θ, ϕ), as illustrated in Figure 6–31. The fluid approaches the sphere at $z = \infty$ with velocity $-U$ in the z direction. The viscosity of the fluid is μ. The flow is clearly axisymmetric about the z axis. Thus, neither the velocity nor the pressure p of the fluid depends on the azimuthal angle ϕ. In addition there is no azimuthal component of fluid motion; that is, the only nonzero components of fluid velocity are the radial velocity u_r and the meridional velocity u_θ, as shown in Figure 6–31. The continuity equation and the equations of motion for the slow, steady, axisymmetric flow of a viscous incompressible fluid are, in spherical

polar coordinates with $u_\phi = 0$,

$$0 = \frac{1}{r^2}\frac{\partial}{\partial r}(r^2 u_r) + \frac{1}{r \sin\theta}\frac{\partial}{\partial\theta}(\sin\theta u_\theta) \tag{6–194}$$

$$0 = -\frac{\partial p}{\partial r} + \mu\left\{\frac{1}{r^2}\frac{\partial}{\partial r}\left(r^2\frac{\partial u_r}{\partial r}\right)\right.$$
$$+ \frac{1}{r^2\sin\theta}\frac{\partial}{\partial\theta}\left(\sin\theta\frac{\partial u_r}{\partial\theta}\right) - \frac{2u_r}{r^2}$$
$$\left. - \frac{2}{r^2\sin\theta}\frac{\partial}{\partial\theta}(u_\theta\sin\theta)\right\} \tag{6–195}$$

$$0 = -\frac{1}{r}\frac{\partial p}{\partial\theta} + \mu\left\{\frac{1}{r^2}\frac{\partial}{\partial r}\left(r^2\frac{\partial u_\theta}{\partial r}\right)\right.$$
$$+ \frac{1}{r^2\sin\theta}\frac{\partial}{\partial\theta}\left(\sin\theta\frac{\partial u_\theta}{\partial\theta}\right)$$
$$\left. + \frac{2}{r^2}\frac{\partial u_r}{\partial\theta} - \frac{u_\theta}{r^2\sin^2\theta}\right\}. \tag{6–196}$$

These are the axisymmetric equivalents of Equations (6–53), (6–67), and (6–68). We must obtain a solution subject to the condition that the fluid velocity approaches the uniform velocity $-U$ in the z direction as $r \to \infty$. The radial and meridional components of the uniform velocity are $-U\cos\theta$ and $U\sin\theta$, respectively. Therefore we can write

$$u_r \to -U\cos\theta \quad \text{and} \quad u_\theta \to U\sin\theta \quad \text{as} \quad r \to \infty. \tag{6–197}$$

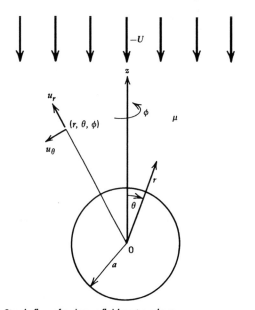

6–31 Steady flow of a viscous fluid past a sphere.

We must also satisfy the no-slip velocity boundary condition on $r = a$

$$u_r = u_\theta = 0 \quad \text{on} \quad r = a. \quad (6\text{–}198)$$

The nature of the boundary conditions suggests that we try a solution of the form

$$u_r = f(r)\cos\theta \quad \text{and} \quad u_\theta = g(r)\sin\theta. \quad (6\text{–}199)$$

If we substitute Equations (6–199) into (6–194) to (6–196), we obtain

$$g = \frac{-1}{2r}\frac{d}{dr}(r^2 f) \quad (6\text{–}200)$$

$$0 = -\frac{\partial p}{\partial r} + \frac{\mu\cos\theta}{r^2}\left\{\frac{d}{dr}\left(r^2\frac{df}{dr}\right) - 4(f+g)\right\} \quad (6\text{–}201)$$

$$0 = -\frac{\partial p}{\partial \theta} + \frac{\mu\sin\theta}{r}\left\{\frac{d}{dr}\left(r^2\frac{dg}{dr}\right) - 2(f+g)\right\}. \quad (6\text{–}202)$$

We can eliminate the pressure by differentiating Equation (6–201) with respect to θ and subtracting the derivative of Equation (6–202) with respect to r to obtain

$$0 = \frac{1}{r^2}\frac{d}{dr}\left(r^2\frac{df}{dr}\right) - \frac{4(f+g)}{r^2}$$
$$+ \frac{d}{dr}\left\{\frac{1}{r}\frac{d}{dr}\left(r^2\frac{dg}{dr}\right) - \frac{2(f+g)}{r}\right\}. \quad (6\text{–}203)$$

The solutions of Equations (6–200) and (6–203) for the functions f and g can be found as simple powers of r. Thus we let

$$f = cr^n, \quad (6\text{–}204)$$

where c is a constant. Equation (6–200) gives

$$g = \frac{-c(n+2)}{2}r^n. \quad (6\text{–}205)$$

By substituting Equations (6–204) and (6–205) into (6–203), we find that n must satisfy

$$n(n+3)(n-2)(n+1) = 0 \quad (6\text{–}206)$$

or

$$n = 0, -3, 2, -1. \quad (6\text{–}207)$$

The functions f and g are thus linear combinations of r^0, r^{-3}, r^2, and r^{-1}

$$f = c_1 + \frac{c_2}{r^3} + \frac{c_3}{r} + c_4 r^2 \quad (6\text{–}208)$$

$$g = -c_1 + \frac{c_2}{2r^3} - \frac{c_3}{2r} - 2c_4 r^2, \quad (6\text{–}209)$$

where c_1, c_2, c_3, and c_4 are constants. The velocity components u_r and u_θ are given by

$$u_r = \left(c_1 + \frac{c_2}{r^3} + \frac{c_3}{r} + c_4 r^2\right)\cos\theta \quad (6\text{–}210)$$

$$u_\theta = \left(-c_1 + \frac{c_2}{2r^3} - \frac{c_3}{2r} - 2c_4 r^2\right)\sin\theta. \quad (6\text{–}211)$$

Since u_r and u_θ must satisfy conditions (6–197) as $r \to \infty$, it is clear that

$$c_1 = -U \quad \text{and} \quad c_4 = 0. \quad (6\text{–}212)$$

The no-slip condition on $r = a$, Equation (6–198), requires

$$c_2 = \frac{-a^3 U}{2} \quad c_3 = \frac{3aU}{2}. \quad (6\text{–}213)$$

The final expressions for u_r and u_θ are

$$u_r = U\left(-1 - \frac{a^3}{2r^3} + \frac{3a}{2r}\right)\cos\theta \quad (6\text{–}214)$$

$$u_\theta = U\left(1 - \frac{a^3}{4r^3} - \frac{3a}{4r}\right)\sin\theta. \quad (6\text{–}215)$$

The pressure associated with this flow can be found by substituting Equations (6–214) and (6–215) into (6–196) and integrating with respect to θ

$$p = \frac{3\mu a U}{2r^2}\cos\theta. \quad (6\text{–}216)$$

Both pressure forces and viscous forces act on the surface of the sphere. By symmetry, the net force on the sphere must be in the negative z direction. This net force is the drag D on the sphere. We first calculate the contribution of the pressure forces to the drag. The pressure force on the sphere acts in the negative radial direction. The component of this force in the negative z direction is, per unit area of the surface,

$$p\cos\theta = \frac{3\mu U}{2a}\cos^2\theta. \quad (6\text{–}217)$$

The pressure drag D_p is obtained by integrating the product of this force per unit area with the surface area element $2\pi a^2 \sin\theta \, d\theta$ over the entire surface of the sphere

$$D_p = 3\pi\mu a U \int_0^\pi \sin\theta \cos^2\theta \, d\theta = 2\pi\mu a U. \tag{6-218}$$

We next compute the net viscous drag D_v on the sphere. The viscous stresses acting on an area element of the sphere's surface are the radial viscous stress τ_{rr}

$$(\tau_{rr})_{r=a} = 2\mu \left(\frac{\partial u_r}{\partial r}\right)_{r=a} \tag{6-219}$$

and the tangential stress $\tau_{r\theta}$

$$(\tau_{r\theta})_{r=a} = \mu \left\{ r\frac{\partial}{\partial r}\left(\frac{u_\theta}{r}\right) + \frac{1}{r}\frac{\partial u_r}{\partial \theta} \right\}_{r=a}. \tag{6-220}$$

By substituting Equations (6–214) and (6–215) into these expressions, we find that the viscous stresses on the surface of the sphere are

$$(\tau_{rr})_{r=a} = 0 \tag{6-221}$$

$$(\tau_{r\theta})_{r=a} = \frac{3\mu U \sin\theta}{2a}. \tag{6-222}$$

The nonzero tangential stress $\tau_{r\theta}$ is a force per unit area in the θ direction. The component of this force per unit area in the negative z direction is

$$\tau_{r\theta}\sin\theta = \frac{3\mu U \sin^2\theta}{2a}. \tag{6-223}$$

The viscous drag D_v is found by integrating the product of this quantity with the surface area element $2\pi a^2 \sin\theta \, d\theta$ over the entire surface of the sphere

$$D_v = 3\pi\mu a U \int_0^\pi \sin^3\theta \, d\theta = 4\pi\mu a U. \tag{6-224}$$

The total drag on the sphere is the sum of the pressure drag and the viscous drag

$$D = D_p + D_v = 6\pi\mu a U. \tag{6-225}$$

This is the well-known Stokes formula for the drag on a sphere moving with a small constant velocity through a viscous incompressible fluid. Stokes resistance law is often written in dimensionless form by normalizing the drag with the product of the pressure $\frac{1}{2}\rho_f U^2$ (ρ_f is the density of the fluid) and the cross-sectional area of

the sphere πa^2. The dimensionless *drag coefficient* c_D is thus

$$c_D \equiv \frac{D}{\frac{1}{2}\rho_f U^2 \pi a^2} = \frac{12}{(\rho_f U a)/\mu} = \frac{24}{\text{Re}}, \tag{6-226}$$

where the Reynolds number is given by

$$\text{Re} = \frac{\rho_f U (2a)}{\mu}. \tag{6-227}$$

The Stokes drag formula can be used to determine the velocity of a sphere rising buoyantly through a fluid by equating the drag to the gravitational driving force. If the density of the sphere ρ_s is less than the density of the fluid ρ_f, the net upward buoyancy force according to Archimedes principle is

$$F = (\rho_f - \rho_s)g\left(\frac{4}{3}\pi a^3\right). \tag{6-228}$$

We set this equal to the drag on the sphere $6\pi\mu a U$ and solve for the upward velocity U to obtain

$$U = \frac{2(\rho_f - \rho_s)ga^2}{9\mu}. \tag{6-229}$$

It should be emphasized that this result is valid only if the Reynolds number is less than 1.

For larger values of the Reynolds number the flow of a fluid about a sphere becomes quite complex. Vortices are generated, and the flow becomes unsteady. The measured dependence of the drag coefficient for a sphere on Reynolds number is given in Figure 6–32. This dependence applies to any type of fluid as long as it is incompressible. Figure 6–32 also shows the result for Stokes flow from Equation (6–226). We see that Stokes flow is a valid approximation for Re < 1. The sharp drop in the drag coefficient at Re = 3×10^5 is associated with the transition to turbulent flow. The dependence of c_D on Re for a sphere given in Figure 6–32 is similar to the dependence of f on Re for pipe flow given in Figure 6–7. In terms of the drag coefficient, the upward velocity of a sphere from Equations (6–226), (6–227), and (6–229) is given by

$$U = \left[\frac{8}{3}\frac{ag(\rho_f - \rho_s)}{c_D \rho_f}\right]^{1/2}. \tag{6-230}$$

The drag coefficient can be obtained from the value of the Reynolds number and Figure 6–32.

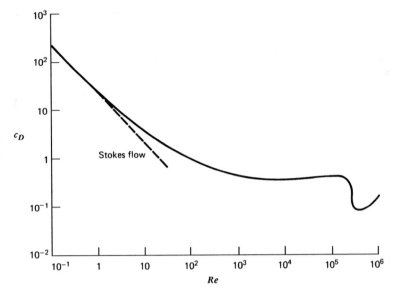

6–32 Dependence of the drag coefficient C_D for a sphere on Reynolds number. The solid line is the measured dependence, and the dashed line is the result from Stokes formula, Equation (6–226).

We can now obtain an estimate for the velocity of magma ascent through the lithosphere. Refractory peridotite xenoliths with a maximum dimension of about 0.3 m have been found in the basaltic lavas erupted in 1801 at Hualalai Volcano, Hawaii. These solid xenoliths were entrained in the lava as it flowed through the lithosphere. An upper limit on the size of the xenoliths that can be entrained is obtained by setting the relative velocity U equal to the flow velocity of the magma. A reasonable estimate for the viscosity of the basaltic magma is 10 Pa s. Also assuming $\rho_s - \rho_m = 600$ kg m^{-3} and $a = 0.15$ m, we find from Equation (6–229) that $U = 3$ m s^{-1} (10.8 km per hr). The corresponding value of the Reynolds number from Equation (6–227) with $\rho_f = 2700$ kg m^{-3} is 243. Therefore the Stokes formula is only approximately valid. Using Equation (6–230) and the empirical correlation given in Figure 6–32, we find $U = 0.87$ m s^{-1} and Re = 70. This is still quite a high velocity. It indicates that magma can penetrate a 100-km-thick lithosphere in about 32 hr.

It is also of interest to determine whether a body with a different density will be entrained in mantle convection. Taking a typical mantle velocity to be 10 mm yr^{-1}, $\Delta\rho = 100$ kg m^{-3}, $\mu = 10^{21}$ Pa s, and $g = 10$ m s^{-2}, we find from Equation (6–229) that spherical bodies with radii less than 38 km will be entrained in mantle flows. The conclusion is that sizable inhomogeneous bodies can be carried with the mantle rocks during mantle convection.

One model for magma migration is that sizable magma bodies move through the mantle because of the differential buoyancy of the liquid. The velocity of a spherical bubble of low-viscosity fluid moving through a high-viscosity fluid because of buoyancy is given by

$$U = \frac{a^2 g(\rho_f - \rho_b)}{3\mu_f}, \qquad (6\text{–}231)$$

where ρ_b is the density of the fluid in the bubble, ρ_f is the density of the surrounding fluid, and μ_f is the viscosity of the ambient fluid. See Problem 6–23 for an explanation of the difference between Equations (6–231) and (6–229). Taking $a = 0.5$ km, $\rho_f - \rho_b = 600$ kg m^{-3}, and $\mu = 10^{21}$ Pa s, we find that $U = 0.016$ mm yr^{-1}. Even for a relatively large magma body the migration velocity is about 13 orders of magnitude smaller than that deduced from the entrainment of xenoliths.

Another order of magnitude calculation also shows that this calculated velocity is unreasonably small. At a velocity of 0.016 mm yr^{-1} it would take the magma body about 10 Gyr to migrate 100 km. This is clearly an unreasonable length of time. It is also necessary that a magma body reach the Earth's surface without solidification if surface volcanism is to occur. An order of magnitude estimate of the time it takes to solidify a magma body of a minimum dimension a is the thermal time constant a^2/κ. If we take a time of 32 hr derived above from xenolith entrainment and $\kappa = 1$ mm^2 s^{-1}, we estimate the minimum dimension of a magma body that penetrates the lithosphere to be 10 cm.

If magma does not penetrate the lithosphere by diapirism, an alternative mechanism must be found. One possibility that has been proposed is hydrofracturing. Liquid under pressure can fracture rock. It has been suggested that the pressure caused by the differential buoyancy of magma can result in the propagation of a fracture through the lithosphere along which the magma migrates.

PROBLEM 6–22 The Stokes drag D on a sphere can only depend on the velocity of the sphere U, its radius a, and the viscosity μ and density ρ of the fluid. Show by dimensional analysis that

$$\frac{D}{\rho U^2 a^2} = f\left(\frac{\rho U a}{\mu}\right), \tag{6–232}$$

where f is an arbitrary function. Because the equations of slow viscous flow are linear, D can only be directly proportional to U. Use this fact together with Equation (6–232) to conclude that

$$D \propto \mu U a. \tag{6–233}$$

PROBLEM 6–23 Consider a spherical bubble of a low-viscosity fluid with density ρ_b rising or falling through a much more viscous fluid with density ρ_f and viscosity μ_f because of a buoyancy force. For this problem the appropriate boundary conditions at the surface of the sphere, $r = a$, are $u_r = 0$ and $\tau_{r\theta} = 0$.

Using Equations (6–210), (6–211), and (6–220) show that

$$u_r = U\left(-1 + \frac{a}{r}\right)\cos\theta \tag{6–234}$$

$$u_\theta = U\left(1 - \frac{1}{2}\frac{a}{r}\right)\sin\theta. \tag{6–235}$$

By integrating Equation (6–196), show that on $r = a$,

$$p = \frac{\mu_f U}{a}\cos\theta. \tag{6–236}$$

The drag force is obtained by carrying out the integral

$$D = 2\pi a^2 \int_0^\pi \left(p - 2\mu_f \frac{\partial u_r}{\partial r}\right)_{r=a}\cos\theta\sin\theta\,d\theta. \tag{6–237}$$

Show that

$$D = 4\pi\mu_f a U, \tag{6–238}$$

and demonstrate that the terminal velocity of the bubble in the fluid is

$$U = \frac{a^2 g(\rho_f - \rho_b)}{3\mu_f}. \tag{6–239}$$

6-15 Plume Heads and Tails

A simple steady-state model for the ascent of a plume head through the mantle is given in Figure 6–33. The plume head is modeled as a spherical diapir whose velocity is given by the Stokes flow solution. The mantle rock in the plume head is hotter, less dense, and less viscous than the surrounding mantle rock. We utilize the solution to Problem 6–23 and write the terminal velocity U of the ascending spherical diapir from Equation (6–239) as

$$U = \frac{a^2 g(\rho_m - \rho_p)}{3\mu_m}, \tag{6–240}$$

where a is the radius of the diapir, ρ_p is the density of the hot plume rock, ρ_m is the density of the surrounding rock, and μ_m is the viscosity of the surrounding mantle rock. We take T_p to be the mean temperature of the plume rock and T_1 to be the temperature of the surrounding mantle rock. From Equation (4–172) we write

$$\rho_p - \rho_m = -\rho_m \alpha_v (T_p - T_1). \tag{6–241}$$

6-33 Illustration of the plume model.

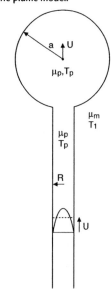

Substitution of Equation (6–241) into (6–240) gives

$$U = \frac{a^2 g \rho_m \alpha_v (T_p - T_1)}{3 \mu_m}, \qquad (6\text{–}242)$$

as the ascent velocity of the plume head.

The plume tail is modeled as a cylindrical pipe and the buoyancy driven volume flux Q_p of plume rock is given by Equation (6–48)

$$Q_p = \frac{\pi}{8} \frac{(\rho_m - \rho_p) g R^4}{\mu_p}. \qquad (6\text{–}243)$$

where R is the radius of the plume tail and μ_p is the viscosity of the plume rock. A measure of the strength of a plume is the buoyancy flux B, which is defined by

$$B = Q_p(\rho_m - \rho_p). \qquad (6\text{–}244)$$

A combination of Equations (6–241), (6–243), and (6–244) gives

$$B = \frac{\pi}{8} \frac{g R^4 \rho_m^2 (T_p - T_1)^2 \alpha_v^2}{\mu_p}. \qquad (6\text{–}245)$$

The total heat flux in a plume Q_H is related to the volume flux by

$$Q_H = \rho_m c_p (T_p - T_1) Q_p, \qquad (6\text{–}246)$$

where c_p is the specific heat at constant pressure. A combination of Equations (6–241), (6–244), and (6–246) gives

$$Q_H = \frac{c_p B}{\alpha_v}. \qquad (6\text{–}247)$$

This relation can be used to convert a plume buoyancy flux to a heat flux.

In our steady-state model the plume head neither gains nor loses fluid; this requires that the mean flow velocity in the plume tail equals the ascent velocity of the plume head U. Thus we have

$$Q_p = \pi R^2 U. \qquad (6\text{–}248)$$

Once the plume flux B has been specified along with the other parameters, the radius of the plume tail R can be determined from Equation (6–245), the heat flux in the plume from Equation (6–246), the ascent velocity of the plume head U from Equation (6–248), and the radius of the plume head from Equation (6–242).

As we pointed out in Section 1–6, hotspots that are attributed to mantle plumes are associated with topographic swells. The Hawaiian swell was illustrated in Figure 1–20. The buoyancy flux associated with a mantle plume can be determined from the rate of hotspot swell formation. We hypothesize that the excess mass associated with the swell is compensated by the mass deficit of the hot (light) plume rock impinging on the base of the lithosphere. Thus the buoyancy flux B associated with the plume is given by

$$B = (\rho_m - \rho_w) A_s u_p, \qquad (6\text{–}249)$$

where ρ_m is the mantle density, ρ_w is the water density (assuming the swell is covered by water), A_s is the cross-sectional area of the swell in a vertical cross section perpendicular to the plume track, and u_p is the plate speed relative to a "fixed" hotspot reference frame.

As a specific example consider the Hawaiian hotspot. From Figure 1–19 we have $u_p = 90$ mm yr^{-1}, from Figure 1–20 we have $A_s = 1.13$ km^2, and taking $\rho_m - \rho_w = 2300$ kg m^{-3} we find $B = 7.4 \times 10^3$ kg s^{-1}. Taking $c_p = 1.25$ kJ kg^{-1} K^{-1} and $\alpha_v = 3 \times 10^{-5}$ K^{-1}, the plume heat flux from Equation (6–247) $Q_H = 3 \times 10^{11}$ W; this represents slightly less than 1% of the total surface heat flux. The radius of the Hawaiian plume R can be obtained from Equation (6–245). Taking $B = 7.4 \times 10^3$ kg s^{-1}, $\mu_p = 10^{19}$ Pa s, $T_p - T_1 = 200$ K, $\alpha_v = 3 \times 10^{-5}$ K^{-1}, $\rho_m = 3300$ kg m^{-3}, and $g = 9.8$ m s^{-2}, we find that the plume radius $R = 84$ km. This is relatively small and explains why plumes are very difficult to observe seismically.

From Equations (6–241) and (6–244) and the parameter values given above, the volume flux in the Hawaiian plume $Q_p = 12$ km^3 yr^{-1}. It is of interest to compare this volume flux with the volume flux of basalt Q_v required to create the Hawaiian Islands and seamount chain. It is estimated that $Q_v = 0.1$ km^3 yr^{-1}, thus it was necessary to melt only about 1% of the plume flux to generate the hotspot volcanics at Hawaii. From Equation (6–248) we find that the mean ascent velocity in the plume $U = 0.54$ m yr^{-1}.

The buoyancy fluxes for forty-three mantle plumes are given in Table 6–4. The total buoyancy flux for these plumes $B = 58.5 \times 10^3$ kg s^{-1}. Taking $c_p = 1.25$ kJ kg^{-1} K^{-1} and $\alpha_v = 3 \times 10^{-5}$ K^{-1}, the total plume heat flux

TABLE 6-4 Values of the Buoyancy Flux Associated with Forty-Three Hotspot Swells

Hotspot	B, Buoyancy Flux (10^3 kg s^{-1})
Afar, Ethiopia	1.2
Ascenscion	0.9
Australia, East	0.9
Azores	1.1
Baja, California	0.3
Bermuda	1.3
Bouvet	0.4
Bowie Seamount	0.6
Canary Islands	1.0
Cape Verde	1.0
Caroline Islands	1.6
Crozet	0.5
Darfur	0.4
Discovery Seamount	0.4
East African	0.6
Easter Island	3.3
Ethiopia	1.0
Fernando	0.7
Galapagos Islands	1.0
Great Meteor Seamount	0.4
Hawaii	7.4
Hoggar Mountains, Algeria	0.6
Iceland	1.4
Juan de Fuca/Cobb Seamount	0.3
Juan Fernandez	1.6
Kerguelen	0.4
Louisville	2.0
MacDonald Seamount	3.6
Marquesas Islands	4.0
Martin	0.6
Meteor	0.4
Pitcairn Islands	2.5
Réunion	1.4
Samoa	1.6
San Felix	2.0
St. Helena	0.4
Tahiti	4.6
Tasman, Central	0.9
Tasman, East	0.9
Tibesti, Chad	0.3
Tristan de Cunha	1.1
Vema Seamount	0.4
Yellowstone	1.5
Total	**58.5**

from Equation (6–247) $Q_H = 0.244 \times 10^{13}$ W. This represents 5.5% of the total global heat flow $Q = 4.43 \times 10^{13}$ W. In Section 4–23 we estimated that the basal heating of the oceanic and continental lithosphere $Q_m = 1.58 \times 10^{13}$ W. Thus our derived plume heat flux is only 15% of the total heat flux associated with the basal heating of the lithosphere. This missing heat flux can be attributed either to plumes that impinge on the base of the lithosphere but are too small to have a surface expression or to secondary mantle convection involving the lower part of the lithosphere.

The relationship between the Réunion hotspot and the flood basalt province of the Deccan Traps was illustrated in Figure 1–22. This basalt province is associated with the plume head that initiated the plume tail responsible for the plume track that now terminates in the Réunion hotspot. We now estimate the quantitative aspects of the Réunion mantle plume and plume head.

From Table 6–4, the present buoyancy flux of the Réunion plume $B = 1.4 \times 10^3$ kg s^{-1}. With the same parameter values used before we find from Equation (6–243) that the radius of the plume conduit $R = 55$ km; from Equations (6–241) and (6–244) we find that the volume flux $Q_p = 2.2$ km^3 yr^{-1}; and from Equation (6–248) we find that the mean ascent velocity in the plume $U = 0.23$ m yr^{-1}. We make the assumption that the strength of the Réunion plume has remained constant for the last 60 Myr that it has been active. Taking $\mu_m = 10^{21}$ Pa s, we find from Equation (6–242) that the radius of the plume head $a = 336$ km. The corresponding volume of the plume head $V_{PH} = 1.2 \times 10^8$ km^3. The volume of basalts in the Deccan Traps $V_B \approx 1.5 \times 10^6$ km^3. Thus it was necessary to melt about one percent of the plume head to form the flood basalts of the Deccan Traps. This is the same melt fraction that we previously obtained for the volcanics of the Hawaiian hotspot. Assuming that the volume flux of the Réunion plume $Q_p = 2.2$ km^3 yr^{-1} and has remained constant over the 60 Myr lifetime of the plume, the total volume flux through the plume tail has been 1.3×10^8 km^3. This is essentially equal to the volume of the plume head. For the ascent velocity of the plume head U equal to 0.23 m yr^{-1}, it would take about 12 Myr for the plume head to ascend from the core–mantle boundary to the Earth's surface.

PROBLEM 6–24 Determine the radius of the plume conduit, the volume flux, the heat flux, the mean ascent velocity, and the plume head volume for the Azores plume. Assume that $T_p - T_1 = 200$ K, $\alpha_v = 3 \times 10^{-5}$ K^{-1}, $\mu_p = 10^{19}$ Pa s, $\rho_m = 3300$ kg m^{-3}, $\mu_m = 10^{21}$ Pa s, and $c_p = 1.25$ kJ kg^{-1} K^{-1}.

PROBLEM 6–25 Determine the radius of the plume conduit, the volume flux, the heat flux, the mean ascent velocity, and the plume head volume for the Tahiti plume. Assume that $T_p - T_1 = 200$ K, $\alpha_v = 3 \times 10^{-5}$ K^{-1}, $\mu_p = 10^{19}$ Pa s, $\rho_m = 3300$ kg m^{-3}, $\mu_m = 10^{21}$ Pa s, and $c_p = 1.25$ kJ kg^{-1} K^{-1}.

6–16 Pipe Flow with Heat Addition

We now turn to problems involving both fluid flow and heat transfer. As our first example we will treat the flow in a pipe with heat addition or heat loss, a situation relevant to the heating of water in an aquifer. We consider the heat balance on a thin cylindrical shell of fluid in the pipe. The thickness of the shell is δr, and its length is δx, as illustrated in Figure 6–34. The heat conducted out of the cylindrical surface at $r + \delta r$ per unit time is

$$2\pi(r + \delta r)\,\delta x q_r(r + \delta r),$$

where $q_r(r + \delta r)$ is the radial *heat flux* at $r + \delta r$. The heat conducted into the shell across its inner cylindrical surface is

$$2\pi r \delta x q_r(r)$$

per unit time. Because δr is small, we can expand $q_r(r + \delta r)$ as

$$q_r(r + \delta r) = q_r(r) + \frac{\partial q_r}{\partial r}\delta r + \cdots.$$

By neglecting higher powers of δr, we can write the net rate at which heat is conducted into the cylindrical shell

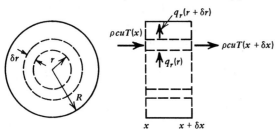

6-34 Heat balance on a small cylindrical shell in a circular pipe.

through its inner and outer surfaces as

$$2\pi\delta x[rq_r(r) - (r + \delta r)q_r(r + \delta r)]$$
$$= -2\pi\delta x\left(r\frac{\partial q_r}{\partial r} + q_r\right)\delta r. \tag{6–250}$$

In cylindrical coordinates, the radial heat flux q_r is related to the radial temperature gradient $\partial T/\partial r$ by Fourier's law of heat conduction (see Problem 4–21),

$$q_r = -k\frac{\partial T}{\partial r}, \tag{6–251}$$

where k is the thermal conductivity of the fluid. Expression (6–250) for the net effect of radial heat conduction can thus be rewritten in terms of the temperature as

$$2\pi\delta x\,\delta r k\left(r\frac{\partial^2 T}{\partial r^2} + \frac{\partial T}{\partial r}\right).$$

The amount of heat convected out of the shell at $x + \delta x$ by the velocity $u(r)$ per unit time is given by

$$2\pi r\,\delta r u\rho c T(x + \delta x),$$

and the amount of heat convected into the shell at x per unit time is given by

$$2\pi r\,\delta r u\rho c T(x).$$

By using the first two terms of a Taylor series expansion for $T(x + \delta x)$, we find that the net rate at which fluid carries heat out of the shell is

$$2\pi r\,\delta r u\rho c[T(x + \delta x) - T(x)] = 2\pi r\,\delta r u\rho c\frac{\partial T}{\partial x}\delta x. \tag{6–252}$$

If the flow is steady so that the temperature of the fluid does not change with time and if axial heat conduction is unimportant compared with advection of heat by the flow, the net effects of radial heat conduction and axial heat advection must balance. Therefore we can equate the right side of Equation (6–252) with the modified form of the right side of Equation (6–250) to obtain

$$\rho uc\frac{\partial T}{\partial x} = k\left(\frac{\partial^2 T}{\partial r^2} + \frac{1}{r}\frac{\partial T}{\partial r}\right). \tag{6–253}$$

By equating axial heat advection to radial heat conduction, we also tacitly assumed that *viscous dissipation* or *frictional heating* in the fluid is negligible.

We can determine the temperature distribution in the pipe using Equation (6–253) for the laminar flow considered in Section 6–4. The velocity as a function of radius can be expressed in terms of the mean velocity $\bar{u}$ by combining Equations (6–33) and (6–37) to give

$$u = 2\bar{u}\left[1 - \left(\frac{r}{R}\right)^2\right]. \tag{6–254}$$

We consider the case in which the wall temperature of the pipe T_w is changing linearly along its length; that is,

$$T_w = C_1 x + C_2, \tag{6–255}$$

where C_1 and C_2 are constants. Accordingly, we assume that the temperature of the fluid is given by

$$T = C_1 x + C_2 + \theta(r) = T_w + \theta(r). \tag{6–256}$$

(In this situation, the net contribution of axial heat conduction to the heat balance of a small cylindrical shell vanishes identically.) Thus θ is the difference between the fluid temperature and the wall temperature. Substitution of Equations (6–254) and (6–256) into (6–253) yields

$$2\rho c\bar{u}\left[1 - \left(\frac{r}{R}\right)^2\right]C_1 = k\left(\frac{d^2\theta}{dr^2} + \frac{1}{r}\frac{d\theta}{dr}\right). \tag{6–257}$$

The boundary conditions are

$$T = T_w \quad \text{at} \quad r = R \tag{6–258}$$

and

$$q_r = 0 \quad \text{at} \quad r = 0. \tag{6–259}$$

The latter condition is required because there is no line source or sink of heat along the axis of the pipe. Condition (6–258) is satisfied if

$$\theta_{r=R} = 0, \tag{6–260}$$

and Condition (6–259), with the aid of Fourier's law (6–251), becomes

$$\left(\frac{d\theta}{dr}\right)_{r=0} = 0. \tag{6–261}$$

The solution of Equation (6–257) that satisfies these boundary conditions is

$$\theta = -\frac{\rho c\bar{u}C_1 R^2}{8k}\left\{3 - 4\frac{r^2}{R^2} + \frac{r^4}{R^4}\right\}. \tag{6–262}$$

The heat flux to the wall q_w can be found by substituting Equation (6–262) into Fourier's law (6–251) and evaluating the result at $r = R$. One finds

$$q_w = -\tfrac{1}{2}\rho c\bar{u}RC_1. \tag{6–263}$$

The heat flux is thus a constant, independent of x. If C_1 is positive, the wall temperature increases in the direction of flow, and heat flows through the wall of the pipe into the fluid. If C_1 is negative, the wall temperature decreases in the direction of flow, and heat flows out of the fluid into the wall of the pipe. The heat flux to the wall can be expressed in a convenient way by introducing a *heat transfer coefficient h* between the wall heat flux and the excess fluid temperature according to

$$q_w = h(\bar{T} - T_w) = h\bar{\theta}, \tag{6–264}$$

where the overbar represents an average over the cross section of the pipe. The average is weighted by the flow per unit area, that is, the velocity through an annular area at radius r. Thus the flow-weighted average excess fluid temperature is

$$\bar{\theta} = \frac{2\pi \int_0^R \theta u r\, dr}{\pi R^2 \bar{u}} = \frac{-11\rho c\bar{u}C_1 R^2}{48k}. \tag{6–265}$$

By combining Equations (6–263) to (6–265), we find that the heat transfer coefficient for laminar flow in a circular pipe is

$$h = \frac{48k}{11D}, \tag{6–266}$$

where $D = 2R$ is the pipe diameter. Equation (6–266) is valid only for Reynolds numbers less than about 2200. At higher values of the Reynolds number the flow is turbulent.

The fluid mechanics literature commonly introduces a dimensionless measure of the heat transfer coefficient known as the *Nusselt number* Nu. For pipe flow with heat addition,

$$\text{Nu} \equiv \frac{hD}{k} = \frac{48}{11} = 4.36. \tag{6–267}$$

The Nusselt number measures the efficiency of the heat transfer process. If the temperature difference $\bar{T} - T_w$ were established across a stationary layer of fluid of thickness D and thermal conductivity k, the conductive heat flux q_c would be

$$q_c = \frac{k(\bar{T} - T_w)}{D} = \frac{q_w k}{Dh}. \tag{6-268}$$

Thus the Nusselt number can be written

$$\text{Nu} = \frac{q_w}{q_c}. \tag{6-269}$$

Therefore, heat transfer with fluid flow through the pipe is 4.36 times more efficient than conductive heat transport through an equivalent stationary fluid layer across which the same temperature difference is applied.

PROBLEM 6-26 Consider unidirectional flow driven by a constant horizontal pressure gradient through a channel with stationary plane parallel walls, as discussed in Section 6-2. Determine the temperature distribution in the channel, the wall heat flux, the heat transfer coefficient, and the Nusselt number by assuming, as in the pipe flow problem above, that the temperature of both walls and the fluid varies linearly with distance x along the channel. You will need the form of the temperature equation in two dimensions that balances horizontal heat advection against vertical heat conduction, as given in Equation (4-156).

6-17 Aquifer Model for Hot Springs

We can use the results of the previous section to study the heating of water flowing through an aquifer surrounded by hot rocks. We again consider the semicircular aquifer with circular cross section illustrated in Figure 6-9. If we balance the heat convected along the aquifer against the heat lost or gained by conduction to the walls, we can write

$$\pi R^2 \rho c \bar{u} \frac{d\bar{T}}{ds} = 2\pi Rh(T_w - \bar{T}), \tag{6-270}$$

where s is the distance measured along the aquifer from the entrance, $\bar{u}$ is the mean velocity in the aquifer, $\bar{T}$ is the flow-averaged temperature of the aquifer fluid, and T_w is the temperature of the aquifer wall rock. We assume laminar flow so that the heat transfer coefficient h is given by Equation (6-266). The coordinate s can be related to the angle ϕ (see Figure 6-9) by

$$s = R'\phi. \tag{6-271}$$

We assume that the wall temperature of the aquifer can be related to the local geothermal gradient β by

$$T_w = R'\beta \sin \phi + T_0, \tag{6-272}$$

where T_0 is the surface temperature and β is constant. Equation (6-272) assumes that the flow in the aquifer does not affect the temperature of the adjacent rock. Substitution of Equations (6-266), (6-271), and (6-272) into (6-270) yields

$$\frac{R^2 \rho c \bar{u}}{R'} \frac{d\bar{T}}{d\phi} = \frac{48}{11} k(R'\beta \sin \phi + T_0 - \bar{T}). \tag{6-273}$$

This equation can be simplified through the introduction of the *Péclet number* Pe defined by

$$\text{Pe} = \frac{\rho c \bar{u} R}{k}. \tag{6-274}$$

The Péclet number is a dimensionless measure of the mean velocity of the flow through the aquifer. It is related to the dimensionless parameters Re and Pr already introduced. Since the thermal diffusivity κ is $k/\rho c$, Pe can be written as

$$\text{Pe} = \frac{\bar{u} R}{\kappa}. \tag{6-275}$$

Using the definition of the Reynolds number Re in Equation (6-40) and the Prandtl number Pr in Equation (6-3) we can further rewrite Equation (6-275) as

$$\text{Pe} = \frac{1}{2} \frac{\bar{u} 2R}{\nu} \frac{\nu}{\kappa} = \frac{1}{2} \text{Re Pr}. \tag{6-276}$$

The simplification of Equation (6-273) is also facilitated by the introduction of a dimensionless temperature θ defined by

$$\theta = \frac{\bar{T} - T_0}{\beta R'}. \tag{6-277}$$

With Equations (6-274) and (6-277) we can put (6-273) into the form

$$\frac{11}{48} \frac{R}{R'} \text{Pe} \frac{d\theta}{d\phi} + \theta = \sin \phi. \tag{6-278}$$

This is a linear first-order differential equation that can be integrated using an *integrating factor*. With the boundary condition that the water entering the aquifer is at the surface temperature, $\bar{T} = T_0$ or $\theta = 0$ at $\phi = 0$,

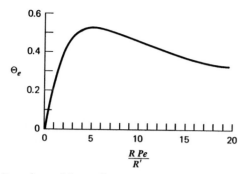

6–35 Dependence of the nondimensional temperature at the exit of the hot spring on the nondimensional flow rate through the aquifer.

the solution can be written

$$
\theta = \left[\frac{48R'}{11\,RPe} \sin\phi - \cos\phi + \exp\left(-\frac{48}{11} \frac{R'}{RPe} \phi \right) \right]
$$
$$
\times \left(\frac{48}{11} \frac{R'}{RPe} \right) \left[1 + \left(\frac{48R'}{11\,RPe} \right)^2 \right]^{-1}. \quad (6\text{–}279)
$$

The nondimensional temperature θ_e at the exit of the aquifer, $\phi = \pi$, is given by

$$
\theta_e = \frac{\left[\exp\left(-\frac{48}{11} \frac{R'\pi}{RPe} \right) + 1 \right] \frac{48}{11} \frac{R'}{RPe}}{1 + \left(\frac{48R'}{11\,RPe} \right)^2}. \quad (6\text{–}280)
$$

The nondimensional exit temperature is plotted as a function of RPe/R' in Figure 6–35. It is seen that the exit temperature of the *hot spring* is a maximum for $RPe/R' = 5$. Thus, for given values of all parameters other than $\bar{u}$, there is a particular flow rate through the aquifer that maximizes the exit temperature of the water. The maximum exit temperature is about one-half the maximum wall temperature at the base of the aquifer because $\theta_e = 1/2$ corresponds to $\bar{T}_e = T_0 + \frac{1}{2}\beta R'$, and T_w at $\phi = \pi/2$ is $T_0 + \beta R'(T_0 \ll \beta R')$.

To better understand why there is a maximum exit temperature, we will show the mean temperature of the water in the aquifer as a function of position in Figure 6–36 for three flow rates. The dimensionless wall or rock temperature,

$$
\theta_w = \frac{T_w - T_0}{\beta R'}, \quad (6\text{–}281)
$$

is also given in the figure. For a low flow rate, $RPe/R' = 1$, for example, the water temperature fol-

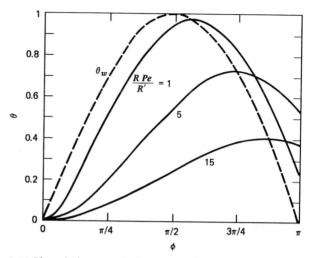

6–36 Dimensionless mean water temperature in the aquifer as a function of position for three nondimensional flow rates. The dashed line is the dimensionless aquifer wall temperature.

lows the wall temperature because of the large heat transfer, and the exit temperature is low. For very slow flow, $RPe/R' \to 0$, the water temperature equals the wall temperature $\theta = \theta_w = \sin\phi$, the exit temperature equals the entrance temperature, and there is no hot spring. For a high flow rate, $RPe/R' = 15$, for example, there is very little heat transfer, and the water does not heat up. In the limit $RPe/R' \to \infty$ the water temperature everywhere in the aquifer equals the entrance temperature, and there is no hot spring. The case of maximum exit temperature, $RPe/R' = 5$ and $\theta_e = 0.52$ is also shown in Figure 6–36.

Although the analysis given here has been greatly simplified, the results are applicable to the more general problem in which the temperature distribution in the rock through which the aquifer passes must also be determined. This requires a solution of Laplace's equation. Also, the transition to turbulence must be considered. The more complete solutions require numerical methods. However, the results show that the maximum temperature to expect from a hot spring is about one-half the temperature obtained by extrapolating the regional geothermal gradient to the base of the aquifer, similar to the result obtained here.

PROBLEM 6–27 Verify by direct substitution that Equation (6–279) is the solution of (6–278).

PROBLEM 6–28 The results of this section were based on the assumption of a laminar heat transfer coefficient for the aquifer flow. Because this requires Re < 2200, what limitation is placed on the Péclet number?

6–18 Thermal Convection

As discussed in Section 1–14, plate tectonics is a consequence of thermal convection in the mantle driven largely by radiogenic heat sources and the cooling of the Earth. When a fluid is heated, its density generally decreases because of thermal expansion. A fluid layer that is heated from below or from within and cooled from above has dense cool fluid near the upper boundary and hot light fluid at depth. This situation is gravitationally unstable, and the cool fluid tends to sink and the hot fluid rises. This is thermal convection. The phenomenon is illustrated in Figure 1–61.

Appropriate forms of the continuity, force balance, and temperature equations for two-dimensional flow are required for a quantitative study of thermal convection. Density variations caused by thermal expansion lead to the buoyancy forces that drive thermal convection. Thus it is essential to account for density variations in the gravitational body force term of the conservation of momentum or force balance equation. In all other respects, however, the density variations are sufficiently small so that they can be neglected. This is known as the *Boussinesq approximation*. It allows us to use the incompressible conservation of fluid equation (6–53). The force balance equations (6–64) and (6–65) are also applicable. However, to account for the buoyancy forces, we must allow for small density variations in the vertical force balance, Equation (6–65), by letting

$$\rho = \rho_0 + \rho', \tag{6–282}$$

where ρ_0 is a reference density and $\rho' \ll \rho_0$. Equation (6–65) can then be written

$$0 = -\frac{\partial p}{\partial y} + \rho_0 g + \rho' g + \mu \left(\frac{\partial^2 v}{\partial x^2} + \frac{\partial^2 v}{\partial y^2} \right). \tag{6–283}$$

We can eliminate the hydrostatic pressure corresponding to the reference density by introducing

$$P = p - \rho_0 g y \tag{6–284}$$

as in Equation (6–66). The horizontal and vertical equations of motion, Equations (6–64) and (6–283), become

$$0 = -\frac{\partial P}{\partial x} + \mu \left(\frac{\partial^2 u}{\partial x^2} + \frac{\partial^2 u}{\partial y^2} \right) \tag{6–285}$$

$$0 = -\frac{\partial P}{\partial y} + \rho' g + \mu \left(\frac{\partial^2 v}{\partial x^2} + \frac{\partial^2 v}{\partial y^2} \right). \tag{6–286}$$

Density variations caused by temperature changes are given by Equation (4–179)

$$\rho' = -\rho_0 \alpha_v (T - T_0), \tag{6–287}$$

where α_v is the volumetric coefficient of thermal expansion and T_0 is the reference temperature corresponding to the reference density ρ_0. Substitution of Equation (6–287) into Equation (6–286) gives

$$0 = -\frac{\partial P}{\partial y} + \mu \left(\frac{\partial^2 v}{\partial x^2} + \frac{\partial^2 v}{\partial y^2} \right) - g \rho_0 \alpha_v (T - T_0). \tag{6–288}$$

The last term in this equation is the buoyancy force per unit volume. The gravitational buoyancy term depends on temperature. Thus the velocity field cannot be determined without simultaneously solving for the temperature field. Therefore we require the heat equation that governs the variation of temperature.

The energy balance must account for heat transport by both conduction and convection. Consider the small two-dimensional element shown in Figure 6–37. Since the thermal energy content of the fluid is $\rho c T$ per unit volume, an amount of heat $\rho c T u \, \delta y$ is transported across the right side of the element by the velocity

6–37 Heat transport across the surfaces of an infinitesimal rectangular element by convection.

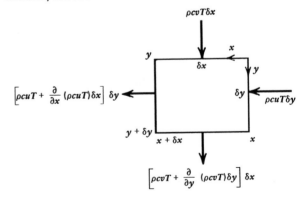

component u in the x direction. This is an energy flow per unit time and per unit depth or distance in the dimension perpendicular to the figure. If $\rho c T u$ is the energy flux at x, then $\rho c T u + \partial/\partial x (\rho c T u) \delta x$ is the energy flow rate per unit area at $x + \delta x$. The net energy advected out of the elemental volume per unit time and per unit depth due to flow in the x direction is thus

$$\left[\left\{\rho c T u + \frac{\partial}{\partial x}(\rho c T u)\,\delta x\right\} - \rho c T u\right]\delta y$$

$$= \frac{\partial}{\partial x}(\rho c T u)\,\delta x\,\delta y. \tag{6-289}$$

The same analysis applied in the y direction gives

$$\left[\left(\rho c T v + \frac{\partial}{\partial y}\{\rho c T v\}\,\delta y\right) - \rho c T v\right]\delta x$$

$$= \frac{\partial}{\partial y}(\rho c T v)\,\delta x\,\delta y \tag{6-290}$$

for the net rate at which heat is advected out of the element by flow in the y direction per unit depth. Thus, the net rate of heat advection out of the element by flow in both directions is

$$\left[\frac{\partial}{\partial x}(\rho c T u) + \frac{\partial}{\partial y}(\rho c T v)\right]\delta x\,\delta y$$

per unit depth. We have already derived the expression for the net rate at which heat is conducted out of the element, per unit depth, in Equation (4–49); it is

$$-k\left(\frac{\partial^2 T}{\partial x^2} + \frac{\partial^2 T}{\partial y^2}\right)\delta x\,\delta y.$$

Conservation of energy states that the combined transport of energy out of the elemental volume by conduction and convection must be balanced by the change in the energy content of the element. The thermal energy of the fluid is $\rho c T$ per unit volume. Thus, this quantity changes at the rate

$$\frac{\partial}{\partial t}(\rho c T)\,\delta x\,\delta y$$

per unit depth of fluid. By combining the effects of conduction, convection, and *thermal inertia*, we obtain

$$\frac{\partial}{\partial t}(\rho c T) - k\left(\frac{\partial^2 T}{\partial x^2} + \frac{\partial^2 T}{\partial y^2}\right)$$

$$+ \frac{\partial}{\partial x}(\rho c u T) + \frac{\partial}{\partial y}(\rho c v T) = 0. \tag{6-291}$$

By treating ρ and c as constants and noting that

$$\frac{\partial}{\partial x}(uT) + \frac{\partial}{\partial y}(vT) = u\frac{\partial T}{\partial x} + v\frac{\partial T}{\partial y} + T\left(\frac{\partial u}{\partial x} + \frac{\partial v}{\partial y}\right)$$

$$= u\frac{\partial T}{\partial x} + v\frac{\partial T}{\partial y} \tag{6-292}$$

(the last step following as a consequence of the continuity equation) and $\kappa = k/\rho c$, we finally arrive at the heat equation for two-dimensional flows

$$\frac{\partial T}{\partial t} + u\frac{\partial T}{\partial x} + v\frac{\partial T}{\partial y} = \kappa\left(\frac{\partial^2 T}{\partial x^2} + \frac{\partial^2 T}{\partial y^2}\right). \tag{6-293}$$

In deriving Equation (6–293), we have neglected some factors that contribute to a general energy balance but are negligible in our present application. These include frictional heating in the fluid associated with the resistance to flow and compressional heating associated with the work done by pressure forces in moving the fluid. We have already derived and used simplified forms of this equation in Section 4–20.

6-19 Linear Stability Analysis for the Onset of Thermal Convection in a Layer of Fluid Heated from Below

The layer of fluid illustrated in Figure 6–38 is heated from below; that is, its upper surface $y = -b/2$ is maintained at the relatively cold reference temperature T_0 and its lower boundary $y = b/2$ is kept at the relatively hot temperature $T_1 (T_1 > T_0)$. We assume that there are no heat sources in the fluid. Buoyancy forces tend to drive convection in the fluid layer. Fluid near the heated lower boundary becomes hotter and lighter than the overlying fluid and tends to rise. Similarly, fluid near the colder, upper boundary is denser than the fluid

6-38 Two-dimensional cellular convection in a fluid layer heated from below.

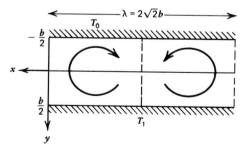

below and tends to sink. However, the motion does not take place for small temperature differences across the layer because the fluid's viscous resistance to flow must be overcome. We use the equations of the preceding section to determine the conditions required for convection to occur, such as the minimum temperature difference.

In the absence of convection, that is, for $T_1 - T_0$ sufficiently small, the fluid is stationary ($u = v = 0$), and we can assume that a steady ($\partial/\partial t = 0$) conductive state with $\partial/\partial x = 0$ exists. The energy equation (6–293) then simplifies to

$$\frac{d^2 T_c}{dy^2} = 0, \tag{6–294}$$

where the subscript c indicates that this is the conduction solution. The solution of Equation (6–294) that satisfies the boundary conditions $T = T_0$ at $y = -b/2$ and $T = T_1$ at $y = +b/2$ is the linear temperature profile

$$T_c = \frac{T_1 + T_0}{2} + \frac{(T_1 - T_0)}{b} y. \tag{6–295}$$

If one imagines gradually increasing the temperature difference across the layer ($T_1 - T_0$), the stationary conductive state will persist until $T_1 - T_0$ reaches a critical value at which even the slightest further increase in temperature difference will cause the layer to become unstable and convection to occur. Thus, at the *onset of convection* the fluid temperature is nearly the conduction temperature profile and the temperature difference T',

$$T' \equiv T - T_c = T - \frac{(T_1 + T_0)}{2} - \frac{(T_1 - T_0)}{b} y, \tag{6–296}$$

is arbitrarily small. The convective velocities u', v' are similarly infinitesimal when motion first takes place.

The form of the energy equation that pertains to the onset of convection can be written in terms of T' by solving Equation (6–296) for T and substituting into Equation (6–293). One gets

$$\frac{\partial T'}{\partial t} + u' \frac{\partial T'}{\partial x} + v' \frac{\partial T'}{\partial y} + \frac{v'(T_1 - T_0)}{b}$$
$$= \kappa \left(\frac{\partial^2 T'}{\partial x^2} + \frac{\partial^2 T'}{\partial y^2} \right). \tag{6–297}$$

Because T', u', v' are small quantities, the *nonlinear terms* $u' \partial T'/\partial x$ and $v' \partial T'/\partial y$ on the left side of Equation (6–297) are much smaller than the remaining linear terms in the equation. Thus they can be neglected and Equation (6–297) can be written as

$$\frac{\partial T'}{\partial t} + \frac{v'}{b}(T_1 - T_0) = \kappa \left(\frac{\partial^2 T'}{\partial x^2} + \frac{\partial^2 T'}{\partial y^2} \right). \tag{6–298}$$

The neglect of the nonlinear terms, the terms involving products of the small quantities u', v', and T', is a standard mathematical approach to problems of stability. Our analysis for the conditions in the fluid layer at the onset of convection is known as a *linearized stability analysis*. It is a valid approach for the study of the onset of convection when the motions and the thermal disturbance are infinitesimal.

To summarize, the equations for the small perturbations of temperature T', velocity u', v', and pressure P' when the fluid layer becomes unstable are

$$\frac{\partial u'}{\partial x} + \frac{\partial v'}{\partial y} = 0 \tag{6–299}$$

$$0 = -\frac{\partial P'}{\partial x} + \mu \left(\frac{\partial^2 u'}{\partial x^2} + \frac{\partial^2 u'}{\partial y^2} \right) \tag{6–300}$$

$$0 = -\frac{\partial P'}{\partial y} - \rho_0 \alpha_v g T' + \mu \left(\frac{\partial^2 v'}{\partial x^2} + \frac{\partial^2 v'}{\partial y^2} \right) \tag{6–301}$$

$$\frac{\partial T'}{\partial t} + \frac{v'}{b}(T_1 - T_0) = \kappa \left(\frac{\partial^2 T'}{\partial x^2} + \frac{\partial^2 T'}{\partial y^2} \right). \tag{6–302}$$

From the second term on the right side of the vertical force balance equation (6–301), it is seen that we have taken the buoyancy force at any point in the layer to depend only on the departure of the fluid temperature from the basic conduction temperature at the point. The conduction temperature profile of the stationary state is the reference temperature profile against which buoyancy forces are determined.

Equations (6–299) to (6–302) are solved subject to the following boundary conditions. We assume that the surfaces $y = \pm b/2$ are isothermal and that no flow occurs across them; that is,

$$T' = v' = 0 \quad \text{on} \quad y = \pm \frac{b}{2} \tag{6–303}$$

If the boundaries of the layer are solid surfaces, then

$$u' = 0 \quad \text{on} \quad y = \pm \frac{b}{2}. \quad (6\text{–}304)$$

This is the no-slip condition requiring that there be no relative motion between a viscous fluid and a bounding solid surface at the solid–fluid interface. If the surfaces $y = \pm b/2$ are free surfaces, that is, if there is nothing at $y = \pm b/2$ to exert a shear stress on the fluid, u' need not vanish on the boundaries. Instead, the shear stress τ'_{yx} must be zero on $y = \pm b/2$. From Equation (6–58) this requires

$$\frac{\partial u'}{\partial y} + \frac{\partial v'}{\partial x} = 0 \quad \text{on} \quad y = \pm \frac{b}{2}. \quad (6\text{–}305)$$

Conditions (6–305) can be simplified even further because $v' = 0$ on $y = \pm b/2$ for any x and consequently $\partial v'/\partial x \equiv 0$ on $y = \pm b/2$. The free surface boundary conditions are therefore

$$\frac{\partial u'}{\partial y} = 0 \quad \text{on} \quad y = \pm \frac{b}{2}. \quad (6\text{–}306)$$

A simple analytic solution can be obtained for the linearized stability problem if the free surface conditions (6–306) are adopted.

We once again introduce the stream function defined in Equations (6–69) and (6–70). Thus the conservation Equation (6–299) is automatically satisfied, and Equations (6–300) to (6–302) can be written

$$0 = -\frac{\partial P'}{\partial x} - \mu \left(\frac{\partial^3 \psi'}{\partial x^2 \, \partial y} + \frac{\partial^3 \psi'}{\partial y^3} \right) \quad (6\text{–}307)$$

$$0 = -\frac{\partial P'}{\partial y} - \rho_0 g \alpha_v T' + \mu \left(\frac{\partial^3 \psi'}{\partial x^3} + \frac{\partial^3 \psi'}{\partial y^2 \, \partial x} \right) \quad (6\text{–}308)$$

$$\frac{\partial T'}{\partial t} + \frac{1}{b}(T_1 - T_0) \frac{\partial \psi'}{\partial x} = \kappa \left(\frac{\partial^2 T'}{\partial x^2} + \frac{\partial^2 T'}{\partial y^2} \right). \quad (6\text{–}309)$$

Eliminating the pressure from (6–307) and (6–308) yields

$$0 = \mu \left(\frac{\partial^4 \psi'}{\partial x^4} + 2 \frac{\partial^4 \psi'}{\partial x^2 \, \partial y^2} + \frac{\partial^4 \psi'}{\partial y^4} \right) - \rho_0 g \alpha_v \frac{\partial T'}{\partial x}. \quad (6\text{–}310)$$

The problem has now been reduced to the solution of two simultaneous partial differential equations (6–309) and (6–310) for the two variables ψ' and T'.

Because these equations are linear equations with constant coefficients, we can solve them by the method of separation of variables. The boundary conditions (6–303) and (6–306) are automatically satisfied by solutions of the form

$$\psi' = \psi'_0 \cos \frac{(\pi y)}{b} \sin \left(\frac{2\pi x}{\lambda} \right) e^{\alpha' t} \quad (6\text{–}311)$$

$$T' = T'_0 \cos \frac{(\pi y)}{b} \cos \left(\frac{2\pi x}{\lambda} \right) e^{\alpha' t}. \quad (6\text{–}312)$$

The velocity and temperature perturbations described by these equations are horizontally periodic disturbances with wavelength λ and maximum amplitudes ψ'_0 and T'_0. The value of α' determines whether or not the disturbances will grow in time. For α' positive, the disturbances will amplify, and the heated layer is convectively unstable. For α' negative, the disturbances will decay in time, and the layer is stable against convection. We can determine α' by substituting Equations (6–311) and (6–312) into Equations (6–309) and (6–310). We find

$$\left(\alpha' + \frac{\kappa \pi^2}{b^2} + \frac{\kappa 4\pi^2}{\lambda^2} \right) T'_0 = -\frac{(T_1 - T_0)2\pi}{\lambda b} \psi'_0 \quad (6\text{–}313)$$

$$\mu \left(\frac{4\pi^2}{\lambda^2} + \frac{\pi^2}{b^2} \right)^2 \psi'_0 = -\frac{2\pi}{\lambda} \rho_0 g \alpha_v T'_0. \quad (6\text{–}314)$$

The disturbance amplitudes ψ'_0 and T'_0 can be eliminated from these equations by division, yielding an equation that can be solved for α'. The *growth rate α'* is found to be

$$\alpha' = \frac{\kappa}{b^2} \left\{ \left(\frac{\rho_0 g \alpha_v b^3 (T_1 - T_0)}{\mu \kappa} \right) \left(\frac{\frac{4\pi^2 b^2}{\lambda^2}}{\left(\frac{4\pi^2 b^2}{\lambda^2} + \pi^2 \right)^2} \right) \right. $$
$$\left. - \left(\pi^2 + \frac{4\pi^2 b^2}{\lambda^2} \right) \right\}. \quad (6\text{–}315)$$

The dimensionless growth rate $\alpha' b^2 / \kappa$ is seen to depend on only two quantities, $2\pi b/\lambda$, a dimensionless wave number, and a dimensionless combination of parameters known as the *Rayleigh number* Ra

$$\text{Ra} = \frac{\rho_0 g \alpha_v (T_1 - T_0) b^3}{\mu \kappa}. \quad (6\text{–}316)$$

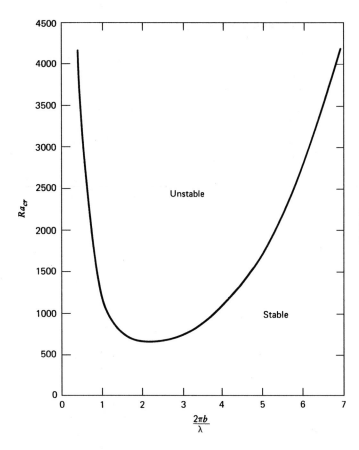

6–39 Critical Rayleigh number Ra_{cr} for the onset of convection in a layer heated from below with stress-free boundaries as a function of dimensionless wave number $2\pi b/\lambda$.

In terms of the Rayleigh number we can write Equation (6–315) as

$$\frac{\alpha' b^2}{\kappa} = \frac{\mathrm{Ra}\frac{4\pi^2 b^2}{\lambda^2} - \left(\pi^2 + \frac{4\pi^2 b^2}{\lambda^2}\right)^3}{\left(\pi^2 + \frac{4\pi^2 b^2}{\lambda^2}\right)^2}. \qquad (6\text{–}317)$$

The growth rate is positive and there is instability if

$$\mathrm{Ra} > \frac{\left(\pi^2 + \frac{4\pi^2 b^2}{\lambda^2}\right)^3}{\frac{4\pi^2 b^2}{\lambda^2}}. \qquad (6\text{–}318)$$

The growth rate is negative and there is stability if Ra is less than the right side of Equation (6–318). Convection just sets in when $\alpha' = 0$, which occurs when

$$\mathrm{Ra} \equiv \mathrm{Ra_{cr}} = \frac{\left(\pi^2 + \frac{4\pi^2 b^2}{\lambda^2}\right)^3}{\frac{4\pi^2 b^2}{\lambda^2}} \qquad (6\text{–}319)$$

The *critical value of the Rayleigh number* Ra_{cr} marks the onset of convection. If Ra < Ra_{cr}, disturbances will decay with time; if Ra > Ra_{cr}, perturbations will grow exponentially with time.

According to Equation (6–319), the critical Rayleigh number is a function of the wavelength of the disturbance. Figure 6–39 shows how Ra_{cr} depends on $2\pi b/\lambda$. If the Rayleigh number and disturbance wavelength are such that the point lies above the curve, the perturbation of wavelength λ is unstable; if the point lies below the curve, convection cannot occur with disturbances of wavelength λ. For example, if Ra = 2000, all disturbances with $0.8 \lesssim 2\pi b/\lambda \lesssim 5.4$ are convectively unstable. However, convection cannot occur for $2\pi b/\lambda \lesssim 0.8$ and $2\pi b/\lambda \gtrsim 5.4$. Figure 6–39 shows that there is a minimum value of Ra_{cr}. If Ra lies below the minimum value, all disturbances decay, the layer is stable, and convection cannot occur.

The value of $2\pi b/\lambda$ at which Ra_{cr} is a minimum can be obtained by setting the derivative of the right side of Equation (6–319) with respect to $2\pi b/\lambda$ equal to zero.

One obtains

$$\frac{\partial \mathrm{Ra}_{cr}}{\partial \left(\frac{2\pi b}{\lambda}\right)} = \left[\frac{4\pi^2 b^2}{\lambda^2} 3 \left(\pi^2 + \frac{4\pi^2 b^2}{\lambda^2} \right)^2 2 \left(\frac{2\pi b}{\lambda} \right) \right.$$
$$\left. - \left(\pi^2 + \frac{4\pi^2 b^2}{\lambda^2} \right)^3 2 \left(\frac{2\pi b}{\lambda} \right) \right]$$
$$\times \left(\frac{4\pi^2 b^2}{\lambda^2} \right)^{-2} = 0 \qquad (6\text{-}320)$$

or

$$\frac{2\pi b}{\lambda} = \frac{\pi}{\sqrt{2}}. \qquad (6\text{-}321)$$

The value of the wavelength corresponding to the smallest value of the critical Rayleigh number is

$$\lambda = 2\sqrt{2}\, b. \qquad (6\text{-}322)$$

Substitution of this value for the wavelength back into Equation (6–319) gives the minimum critical Rayleigh number

$$\min(\mathrm{Ra}_{cr}) = \frac{27\pi^4}{4} = 657.5. \qquad (6\text{-}323)$$

The requirement that Ra exceed Ra_{cr} for convection to occur can be restated in a number of more physical ways. One can think of the temperature difference across the layer as having to exceed a certain minimum value or the viscosity of the fluid as having to lie below a critical value before convection sets in. If Ra is increased from 0, for example, by increasing $T_1 - T_0$, other quantities remaining fixed, convection sets in when Ra reaches 657.5 (for heating from below with stress-free boundaries), and the aspect ratio of each convection cell is $\sqrt{2}$, as shown in Figure 6–38. The minimum value of Ra_{cr} and the disturbance wavelength for which Ra_{cr} takes the minimum value must be determined numerically for no-slip velocity boundary conditions. For that case, $\min \mathrm{Ra}_{cr} = 1707.8$ and $\lambda = 2.016b$.

The linear stability analysis for the onset of convection can also be carried out for a fluid layer heated uniformly from within and cooled from above. The lower boundary is assumed to be insulating; that is, no heat flows across the boundary. Once again the fluid near the upper boundary is cooler and more dense than the fluid beneath. Therefore buoyancy forces can drive fluid motion provided they are strong enough to overcome the viscous resistance. This type of instability is directly applicable to the Earth's mantle because the interior of the Earth is heated by the decay of the radioactive elements and the near-surface rocks are cooled by heat conduction to the surface. These near-surface rocks are cooler and more dense than the hot mantle rocks at depth. The appropriate Rayleigh number for a fluid layer heated from within is

$$\mathrm{Ra}_H = \frac{\alpha_v \rho_0^2 g H b^5}{k \mu \kappa}, \qquad (6\text{-}324)$$

where H is the rate of internal heat generation per unit mass. For no-slip velocity boundary conditions, the minimum critical Rayleigh number is 2772, and the associated value of $2\pi b/\lambda$ is 2.63; for free-slip conditions, $\min \mathrm{Ra}_{cr} = 867.8$, and the associated value of $2\pi b/\lambda$ is 1.79.

We can estimate the value of this Rayleigh number for the mantle of the Earth. Based on the postglacial rebound studies, we take $\mu = 10^{21}$ Pa s. For the rock properties we take $k = 4$ W m^{-1} K^{-1}, $\kappa = 1$ mm^2 s^{-1}, and $\alpha_v = 3 \times 10^{-5}$ K^{-1}. We assume $g = 10$ m s^{-2} and an average density $\rho_0 = 4000$ kg m^{-3}. Based on our discussion of the distribution of heat sources in the mantle (see Chapter 4) we take $H = 9 \times 10^{-12}$ W kg^{-1}. If convection is restricted to the upper mantle, it is reasonable to take $b = 700$ km. We find that $\mathrm{Ra}_H = 2 \times 10^6$. If we apply the same values to the entire mantle and take $b = 2880$ km, we find that $\mathrm{Ra}_H = 2 \times 10^9$. In either case the calculated value for the Rayleigh number is much greater than the minimum critical value. It was essentially this calculation that led Arthur Holmes to propose in 1931 that thermal convection in the mantle was responsible for driving continental drift.

PROBLEM 6–29 Estimate the values of the Rayleigh numbers for the mantles of Mercury, Venus, Mars, and the Moon. Assume heat is generated internally at the same rate it is produced in the Earth. Use the same values for μ, k, κ, and α_v as used above for the Earth's mantle. Obtain appropriate values of ρ_0, g, and b from the discussion in Chapter 1.

PROBLEM 6–30 Calculate the exact minimum and maximum values of the wavelength for disturbances that are convectively unstable at Ra = 2000. Consider a fluid layer heated from below with free-slip boundary conditions.

PROBLEM 6–31 Formulate the linear stability problem for the onset of convection in a layer of fluid heated from within. Assume that the boundaries are stress-free. Take the upper boundary to be isothermal and the lower boundary to be insulating. Carry the formulation to the point where the solution to the problem depends only on the integration of a single ordinary differential equation for the stream function subject to appropriate boundary conditions.

6-20 A Transient Boundary-Layer Theory for Finite-Amplitude Thermal Convection

The linear stability theory given in the previous section determines whether thermal convection occurs. However, it is not useful in determining the structure of the convection when the Rayleigh number exceeds the critical value. Because it is linear, the stability analysis cannot predict the magnitude of *finite-amplitude convective flows*. To do this, it is necessary to solve the full nonlinear equations, which, in general, can only be done numerically. In the following, however, we present approximate solutions that are valid when the Rayleigh number is large and convection is vigorous.

For large values of the Rayleigh number, a convecting fluid layer of thickness b heated from below is largely isothermal. By symmetry, the isothermal core temperature T_c of the bulk of the fluid is given by

$$T_c = T_0 + \tfrac{1}{2}(T_1 - T_0), \qquad (6\text{–}325)$$

where the upper boundary is maintained at the temperature T_0 and the lower boundary at the temperature T_1. The thermal gradient between the cold upper boundary at temperature T_0 and the core at temperature T_c occurs across a thin *thermal boundary layer* adjacent to the upper boundary. The thermal gradient between core at temperature T_c and the hot lower boundary at temperature T_1 occurs across a thin thermal boundary layer adjacent to the lower boundary.

We first consider a *boundary-layer stability* approach to the thickening thermal boundary layers. We assume that initially the isothermal core fluid is in contact with the boundaries of the fluid layer. Subsequently the fluid adjacent to the hot lower boundary is heated forming a hot thermal boundary layer and the fluid adjacent to the cold upper boundary layer is cooled forming a cold thermal boundary layer. The boundary layers thicken until they become gravitationally unstable and separate from the boundaries. The hot lower boundary layer ascends into the isothermal core heating it, and the cold upper boundary layer descends into the isothermal core cooling it. The end of the boundary-layer growth is determined by a stability analysis of the boundary layers similar to the stability analysis of a fluid layer carried out in the previous section. The process is then assumed to repeat.

The transient growth of the two thermal boundary layers can be described by the one-dimensional heat conduction analysis of Section 4–15. Initially, at $t = 0$, the core fluid with temperature T_c is assumed to be in contact with the boundaries. Subsequently, conductive heat losses to the surface result in the development of thin thermal boundary layers. Because of symmetry we consider only the growth of the cold thermal boundary layer adjacent to the upper boundary. The results can be directly applied to the hot thermal boundary layer adjacent to the lower boundary. The temperature distribution in the cold thermal boundary layer as a function of time from Equation (4–113) is

$$\frac{T_c - T}{T_c - T_0} = \operatorname{erfc}\left(\frac{y}{2\sqrt{\kappa t}}\right). \qquad (6\text{–}326)$$

A similar expression can be written for the thickening hot boundary layer on the lower boundary. The thickness of the thermal boundary layer from Equation (4–115) is

$$y_T = 2.32(\kappa t)^{1/2}. \qquad (6\text{–}327)$$

The thickness increases with the square root of time since the boundary layer was established. The basic assumption in this approach is that a linear stability analysis can be applied to the boundary layers. We assume that the boundary layers thicken until the stability condition is satisfied, at which time they break away from the boundary surfaces to be replaced by isothermal core fluid and the process repeats. The breakaway condition is assumed to be given by the stability analysis for a fluid layer developed in Section 6–19. The applicable Rayleigh number, given by an expression similar to Equation (6–316), is based on the boundary-layer thickness, i.e., b in Equation (6–316) is replaced by y_T

from Equation (6–327). Also the relevant temperature difference is

$$T_c - T_0 = \tfrac{1}{2}(T_1 - T_0), \qquad (6\text{–}328)$$

from Equation (6–325). The critical value of the Rayleigh number $\mathrm{Ra}_{y_T,\mathrm{cr}}$ gives a critical value for the boundary layer thickness $y_{T,\mathrm{cr}}$

$$\mathrm{Ra}_{y_T,\mathrm{cr}} = \frac{\rho_0 \alpha_v g (T_1 - T_0) y_{T,\mathrm{cr}}^3}{2\mu\kappa}. \qquad (6\text{–}329)$$

The critical value of this Rayleigh number for free-surface boundary conditions is $\mathrm{Ra}_{y_T,\mathrm{cr}} = 657.5$.

From Equations (6–327) and (6–329), the time at which boundary layer breakaway occurs t_c is given by

$$t_c = \frac{1}{5.38\kappa}\left[\frac{2\mu\kappa\,\mathrm{Ra}_{y_T,\mathrm{cr}}}{\rho_0\alpha_v g(T_1 - T_0)}\right]^{2/3}. \qquad (6\text{–}330)$$

The mean heat flow q across the fluid layer during the time t_c from Equation (4–117) is

$$\bar{q} = \frac{2k(T_c - T_0)}{(\pi\kappa t_c)^{1/2}} = \frac{k(T_1 - T_0)}{(\pi\kappa t_c)^{1/2}} \qquad (6\text{–}331)$$

The combination of Equations (6–330) and (6–331) gives

$$\bar{q} = 1.31k(T_1 - T_0)\left[\frac{\rho_0\alpha_v g(T_1 - T_0)}{2\mu\kappa\,\mathrm{Ra}_{y_T,\mathrm{cr}}}\right]^{1/3}. \qquad (6\text{–}332)$$

The Nusselt number Nu is defined in Equation (6–267) as the ratio of the convective heat flow across the layer $\bar{q}$ to the heat flow q_c that conduction would transport,

$$\mathrm{Nu} \equiv \frac{\bar{q}}{q_c}, \qquad (6\text{–}333)$$

and

$$q_c = \frac{k(T_1 - T_0)}{b}. \qquad (6\text{–}334)$$

Upon substituting Equations (6–332) and (6–334) into Equation (6–333), we obtain

$$\mathrm{Nu} = 1.04\left(\frac{\mathrm{Ra}}{\mathrm{Ra}_{y_T,\mathrm{cr}}}\right)^{1/3}. \qquad (6\text{–}335)$$

where Ra is the Rayleigh number based on the full layer thickness b and the overall temperature difference $(T_1 - T_0)$ as defined in Equation (6–316). We find that the Nusselt number is proportional to the Rayleigh

number to the one-third power. With $\mathrm{Ra}_{y_T,\mathrm{cr}} = 657.5$ we find

$$\mathrm{Nu} = 0.120\mathrm{Ra}^{1/3}. \qquad (6\text{–}336)$$

Although this is an approximate solution, the dependence of the Nusselt number on the Rayleigh number is generally valid for vigorous thermal convection in a fluid layer heated from below with free-surface boundary conditions.

It is of interest to apply this boundary-layer stability analysis directly to the problem of thermal convection in the upper mantle. The depth of deep earthquakes associated with the descending lithosphere at ocean trenches (about 660 km) provides a minimum thickness for the convecting part of the mantle. We assume that mantle convection is restricted to the upper 700 km of the mantle and evaluate the Rayleigh number in Equation (6–316), with $b = 700$ km, $\rho_0 = 3700\,\mathrm{kg\,m^{-3}}$, $g = 10\,\mathrm{m\,s^{-2}}$, $\alpha_v = 3\times10^{-5}\,\mathrm{K^{-1}}$, $T_1 - T_0 = 1500$ K, $\kappa = 1\,\mathrm{mm^2\,s^{-1}}$, and $\mu = 10^{21}$ Pa s; we obtain $\mathrm{Ra} = 5.7\times10^5$. The mean surface heat flux $\bar{q}$ is given by

$$\bar{q} = \frac{k(T_1 - T_0)}{b}\mathrm{Nu} = \frac{0.120k(T_1 - T_0)\,\mathrm{Ra}^{1/3}}{b}. \qquad (6\text{–}337)$$

With the same parameter values and $k = 4\,\mathrm{W\,m^{-1}\,K^{-1}}$, we find $\bar{q} = 85\,\mathrm{mW\,m^{-2}}$. This is remarkably close to the the Earth's mean surface heat flow of $87\,\mathrm{mW\,m^{-2}}$ given in Section 4–4. However, such excellent agreement must be considered fortuitous.

Equation (6–330) for the time at which boundary-layer breakaway occurs can be rewritten using $\mathrm{Ra}_{y_T,\mathrm{cr}} = 657.5$ as

$$t_c = \frac{22.3b^2}{\kappa\,\mathrm{Ra}^{2/3}}. \qquad (6\text{–}338)$$

Substituting the values given above for upper mantle convection we find $t_c = 50.5$ Myr. This is about one-half the mean age of subduction given in Figure 4–26.

The boundary-layer stability approach can also be applied to a fluid layer that is heated from within and cooled from above. In this case there is only a single thermal boundary layer on the upper boundary of the fluid layer. The mean heat flow out of the upper boundary $\bar{q}$ is related to the heat generation per unit mass in the layer H by

$$\bar{q} = \rho_0 H b. \qquad (6\text{–}339)$$

Applying Equations (6–630) and (6–331) to the upper boundary layer only, we have

$$\bar{q} = 2.62k(T_1 - T_0)\left[\frac{\rho_0\alpha_v g(T_1 - T_0)}{\mu\kappa\,\mathrm{Ra}_{y_T,\mathrm{cr}}}\right]^{1/3}, \quad (6\text{–}340)$$

where T_1 is now the temperature of both the lower boundary and the isothermal core. By combining Equations (6–339) and (6–340), we can solve for the temperature of the isothermal core with the result

$$T_1 - T_0 = \left(\frac{\rho_0 Hb}{2.62k}\right)^{3/4}\left(\frac{\mu\kappa\,\mathrm{Ra}_{y_T,\mathrm{cr}}}{\rho_0\alpha_v g}\right)^{1/4}, \quad (6\text{–}341)$$

where $T_1 - T_0$ is also the temperature rise across the fluid layer.

The efficiency with which convection cools the fluid layer can be assessed by comparing the temperature rise across the internally heated layer given by Equation (6–341) with that which would be obtained if all the internally generated heat were removed only by conduction ($T_{1_c} - T_0$). The dimensionless temperature ratio,

$$\theta = \frac{T_1 - T_0}{T_{1_c} - T_0}, \quad (6\text{–}342)$$

is thus a measure of convective efficiency for the internally heated fluid layer. The smaller θ is, the more efficient convection is in removing the heat produced in the fluid. Without convection, the temperature rise across the layer would be (see Section 4–6)

$$T_{1_c} - T_0 = \frac{\rho_0 Hb^2}{2k}. \quad (6\text{–}343)$$

By substituting Equations (6–341) and (6–343) into (6–342) we find that

$$\theta = 0.97\left(\frac{\mathrm{Ra}_{y_T,\mathrm{cr}}}{\mathrm{Ra}_H}\right)^{1/4} \quad (6\text{–}344)$$

where Ra_H is the Rayleigh number defined for a fluid layer heated from within in Equation (6–324). The nondimensional temperature difference between the isothermal core and the upper boundary decreases as convection becomes more vigorous with increasing Rayleigh number. Taking $\mathrm{Ra}_{y_T,\mathrm{cr}} = 657.5$ we find

$$\theta = 4.91\mathrm{Ra}_H^{-1/4}. \quad (6\text{–}345)$$

Again, the dependence of the dimensionless tempera-

ture on Rayleigh number is generally valid for a vigorously convecting fluid layer heated from within and cooled from above with free surface boundary conditions, although the constant of proportionality is model dependent. The dimensional temperature of the isothermal core is given by

$$T_1 - T_0 = \frac{phb^2}{2k}\theta = 2.45\frac{pHb^2}{k}\mathrm{Ra}_H^{-1/4}. \quad (6\text{–}346)$$

The boundary layer stability analysis for thermal convection in a uniformly heated fluid layer cooled from above can be applied to thermal convection that occurs throughout the whole mantle. For this case we take $b = 2880$ km. Based on the discussion in Chapter 4 we assume $H = 9 \times 10^{-12}$ W kg^{-1}. We also take $\alpha_v = 3 \times 10^{-5}$ K^{-1}, $\rho_0 = 4700$ kg m^{-3}, $g = 10$ m s^{-2}, $k = 4$ W m^{-1} K^{-1}, $\kappa = 1$ mm^2 s^{-1}, and $\mu = 10^{21}$ Pa s and obtain $\mathrm{Ra}_H = 3 \times 10^9$ from Equation (6–324). From Equation (6–346) and these parameter values we have $T_1 - T_0 = 918$ K. This is about a factor of 2 too low.

The boundary-layer stability results just discussed give episodic bursts of convection. This is clearly quite different than the steady-state subduction that occurs on the Earth. However, as discussed in Section 1–20, episodic subduction has been proposed to explain the global resurfacing that occurred on Venus about 500 Ma ago.

6-21 A Steady-State Boundary-Layer Theory for Finite-Amplitude Thermal Convection

We will now develop a thermal boundary-layer analysis of vigorous steady convection in a fluid layer heated from below. We will limit our considerations to very large Prandtl numbers so that the inertia terms in the momentum equations can be neglected. The boundary layer structure and coordinate system are illustrated in Figure 6–40. The flow is divided into cellular *two-dimensional rolls* of width $\lambda/2$; alternate rolls rotate in opposite directions. The entire flow field is highly viscous. On the cold upper boundary a thin thermal boundary layer forms. When the two cold boundary layers from adjacent cells meet, they separate from the boundary and form a cold descending *thermal plume*. Similarly, a hot thermal boundary layer forms on the lower boundary of the cell. When two hot boundary layers meet from adjacent cells, they form a hot ascending

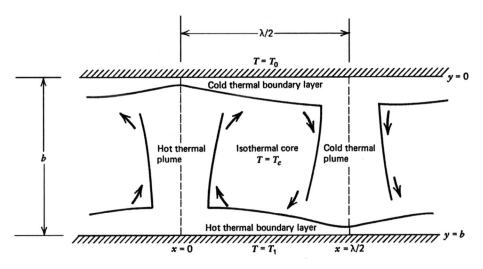

6–40 Boundary-layer structure of two-dimensional thermal convection cells in a fluid layer heated from below.

plume. The buoyancy forces in the ascending and descending plumes drive the flow. The core of each cell is a nearly isothermal viscous rotational flow. As pointed out in our discussion of the boundary-layer stability analysis, symmetry requires that the temperature T_c in the nearly isothermal core must be the mean of the two boundary temperatures and Equation (6–325) applies. We now carry out a quantitative calculation for the velocities in the fluid layer and the amount of heat transported by the motions.

Although an exact solution for the steady-state boundary-layer model requires numerical methods, we can obtain an analytic solution by making a number of approximations. The results will then be compared with more rigorous solutions. We first obtain the structure of the cold thermal boundary layer adjacent to the upper boundary of the fluid layer. For this calculation we let $y = 0$ be the upper boundary and measure y positive downward. We also let x be the horizontal coordinate and let $x = 0$ be at the center of the ascending plume (Figure 6–40). To obtain an analytic solution, we assume that the horizontal fluid velocity at the upper boundary is a constant u_0. The horizontal fluid velocity is actually zero at $x = 0$ and $\lambda/2$ and increases to a maximum near $x = \lambda/4$. The constant velocity u_0 is an average of the actual horizontal velocity on the upper boundary. We have already solved for the thermal structure of this boundary layer in Section 4–16. From

Equation (4–124) the temperature distribution in the cold thermal boundary layer is

$$\frac{T_c - T}{T_c - T_0} = \mathrm{erfc}\left[\frac{y}{2}\left(\frac{u_0}{\kappa x}\right)^{1/2}\right]. \qquad (6\text{–}347)$$

As discussed previously in Section 4–16, a direct association can be made between the cold thermal boundary layer of a thermal convection cell and the thickening oceanic lithosphere. By integrating the surface heat flux (4–127) across the width of the cell, that is, from $x = 0$ to $x = \lambda/2$, we obtain the total rate rate of heat flow Q out of the top of the cell per unit distance along the axis of the roll,

$$Q = 2k(T_c - T_0)\left(\frac{u_0\lambda}{2\pi\kappa}\right)^{1/2}. \qquad (6\text{–}348)$$

At the boundary between two cells the cold thermal boundary layers from two adjacent cells turn through $90°$ to form a cold, symmetrical descending thermal plume. This process is directly analogous to the subduction of the oceanic lithosphere at an ocean trench (although lithospheric subduction is not a symmetrical process). Because very little heat conduction can occur during this transition from a thermal boundary layer to a thermal plume, the distribution of temperature in the newly formed plume is the same as in the boundary layer. As in the case of the horizontal velocity in the cold boundary layer, we assume that the vertical (descending) velocity in the cold plume is a constant v_0. However, the velocity v_0 may differ from the velocity u_0. Since the convected heat in the plume just after its formation must equal the convected heat just before its formation, the thickness of the plume relative to the boundary layer must be in the ratio u_0/v_0. Therefore the temperature distribution in the plume just as it is formed is given by

$$\frac{T_c - T}{T_c - T_o} = \text{erfc}\left[\frac{v_0}{2u_0}\left(\frac{\lambda}{2} - x\right)\left(\frac{2u_0}{\kappa\lambda}\right)^{1/2}\right], \quad (6\text{–}349)$$

with a similar expression for the other half of the symmetrical plume formed from the adjacent cell.

As the plume descends along the boundary between two adjacent cells, its temperature distribution can be obtained by using the temperature given in Equation (6–349) as the initial temperature distribution in Laplace's solution of the heat conduction equation – see Equation (4–157). Laplace's solution for a time-dependent problem can be applied to the descending plume by identifying t as y/v_0. This is analogous to our use of a time-dependent solution of the heat conduction equation for the structure of the cold surface thermal boundary layer (see Section 4–16). In that problem we identified t as x/u_0.

The temperature distribution in the descending plume can be used to calculate the total downward gravitational body force on the plume due to its negative buoyancy relative to the isothermal core. The downward buoyancy force per unit volume on an element of the plume is

$$\rho_0 g \alpha_v (T_c - T).$$

Thus

$$f_b = -\rho_0 g \alpha_v \int_{\lambda/2}^{-\infty} (T_c - T)\, dx \quad (6\text{–}350)$$

is the downward buoyancy body force per unit depth and per unit distance along the roll axis on one-half of the cold plume. It is appropriate to replace the integral across the finite width of the plume with the infinite integral, since $T \rightarrow T_c$ at the edge of the plume; see, for example, Equation (4–206) for a similar change of limits. The total downward buoyancy body force F_b on the descending plume is obtained by integrating f_b along the vertical extent of the plume from $y = 0$ to $y = b$; that is,

$$F_b = \int_0^b f_b\, dy, \quad (6\text{–}351)$$

where F_b is a force per unit length of the plume along the roll axis.

The integral in Equation (6–350) is proportional to the heat content of a slice of the plume of thickness dy. Since no heat is added to the descending plume along its length, this heat content is a constant. This also follows from the fact that the integral of the temperature distribution given by Laplace's solution is a constant independent of t (see Section 4–21). Therefore, the buoyancy body force on the plume per unit depth f_b is independent of y and

$$F_b = f_b b. \quad (6\text{–}352)$$

Because f_b is a constant, we can evaluate it anywhere along the plume, that is, at any depth, by carrying out the integration in Equation (6–350). This is most conveniently done just after the plume forms, where Equation (6–349) provides an expression for the temperature in the plume. Substitution of Equation (6–349) into (6–350) with $x' = \lambda/2 - x$ yields

$$\begin{aligned}
f_b &= \rho_0 g \alpha_v (T_c - T_0) \int_0^\infty \text{erfc}\left\{\frac{v_0 x'}{2u_0}\left(\frac{2u_0}{\kappa\lambda}\right)^{1/2}\right\} dx' \\
&= 2\rho_0 g \alpha_v (T_c - T_0)\frac{u_0}{v_0}\left(\frac{\kappa\lambda}{2u_0}\right)^{1/2}\int_0^\infty \text{erfc}\, z\, dz \\
&= 2\rho_0 g \alpha_v (T_c - T_0)\frac{u_0}{v_0}\left(\frac{\kappa\lambda}{2\pi u_0}\right)^{1/2}. \quad (6\text{–}353)
\end{aligned}$$

Thus the total downward gravitational body force F_b on one-half of the symmetrical plume is

$$F_b = f_b b = 2\rho_0 g \alpha_v b (T_c - T_0)\frac{u_0}{v_0}\left(\frac{\kappa\lambda}{2\pi u_0}\right)^{1/2}. \quad (6\text{–}354)$$

So far we have considered only the cold thermal boundary layer and plume. However, the problem is entirely symmetrical, and the structures of the hot thermal boundary layer and plume are identical with their cold counterparts when $T_c - T_0$ is replaced by $T_c - T_1$. The total upward body force on the ascending hot plume is equal to the downward body force on the cold descending plume and is given by Equation (6–354).

Determination of the viscous flow in the isothermal core requires a solution of the biharmonic equation. However, an analytic solution cannot be obtained for

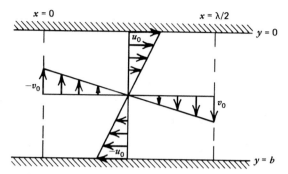

6-41 Linear velocity profiles used to model the core flow in a convection cell. The areas under the triangles are equal to conserve fluid.

the boundary conditions of this problem. Therefore we approximate the core flow with the linear velocity profiles shown in Figure 6–41; that is, we take

$$u = u_0 \left(1 - 2\frac{y}{b}\right) \qquad (6\text{–}355)$$

$$v = -v_0 \left(1 - 4\frac{x}{\lambda}\right). \qquad (6\text{–}356)$$

To conserve fluid, we require

$$\frac{v_0 \lambda}{2} = u_0 b. \qquad (6\text{–}357)$$

This balance is also illustrated in Figure 6–41. It must be emphasized that the assumed velocity profiles do not satisfy the required boundary conditions on the velocity components. For example, the condition $u = 0$ at $x = 0, \lambda/2$ is not satisfied. However, the assumed profiles are reasonable approximations to the actual flow near the center of the cell.

The shear stress on the vertical boundaries of the core flow is given by Equation (6–58) as

$$\tau_{cv} = \mu \frac{\partial v}{\partial x} = \mu \frac{4v_0}{\lambda}, \qquad (6\text{–}358)$$

and the shear stress on the horizontal boundaries is given by

$$\tau_{ch} = \mu \frac{\partial u}{\partial(-y)} = \mu \frac{2u_0}{b}. \qquad (6\text{–}359)$$

The derivative with respect to $-y$ occurs in Equation (6–359), since the derivative must be in the direction of the outward normal to the surface. For the horizontal area at the top of the cell this is the negative y direction. The rate at which work is done on each vertical bound-

ary by the shear stress is $b\tau_{cv}v_0$ per unit distance parallel to the roll axis. The rate of doing work is the product of force and velocity; see also Equation (4–243). The rate at which work is done on each horizontal boundary is $(\lambda/2)\tau_{ch}u_0$. The rate at which the buoyancy force does work on each of the plumes is $F_b v_0$. The rate at which work is done on the plumes by the gravitational body forces must equal the rate at which work is done on the boundaries by the viscous forces; this gives

$$2F_b v_0 = 2b\tau_{cv}v_0 + \lambda\tau_{ch}u_0. \qquad (6\text{–}360)$$

Substitution of Equations (6–354), (6–358), and (6–359) into Equation (6–360) yields

$$\rho_0 g \alpha_v u_0 (T_c - T_0) \left(\frac{\kappa\lambda}{2\pi u_0}\right)^{1/2} = \frac{2v_0^2 \mu}{\lambda} + \frac{u_0^2 \mu \lambda}{2b^2}. \qquad (6\text{–}361)$$

After eliminating the core temperature using Equation (6–346) and the vertical velocity using Equation (6–357), we solve for the horizontal velocity and obtain

$$u_0 = \frac{\kappa}{b} \frac{\left(\frac{\lambda}{2b}\right)^{7/3}}{\left(1 + \frac{\lambda^4}{16b^4}\right)^{2/3}} \left(\frac{\text{Ra}}{2\sqrt{\pi}}\right)^{2/3}, \qquad (6\text{–}362)$$

where the Rayleigh number Ra appropriate to a fluid layer heated from below has been defined in Equation (6–316).

Having determined the mean velocity along the upper boundary of the cell, we can now find the total rate of heat flow through the cell Q from Equation (6–348)

$$Q = \frac{k(T_1 - T_0)}{2^{1/3}\pi^{2/3}} \frac{\left(\frac{\lambda}{2b}\right)^{5/3}}{\left(1 + \frac{\lambda^4}{16b^4}\right)^{1/3}} \text{Ra}^{1/3}. \qquad (6\text{–}363)$$

The Nusselt number Nu is defined as the ratio of the heat flow rate with convection Q to the heat flow rate by conduction Q_c in the absence of convection [see also Equation (6–333)]

$$\text{Nu} = \frac{Q}{Q_c}, \qquad (6\text{–}364)$$

where

$$Q_c = \frac{k(T_1 - T_0)}{b} \frac{\lambda}{2}. \qquad (6\text{–}365)$$

Upon substituting Equations (6–363) and (6–365) into Equation (6–364), we obtain

$$\text{Nu} = \frac{1}{2^{1/3}\pi^{2/3}} \frac{\left(\frac{\lambda}{2b}\right)^{2/3}}{\left(1 + \frac{\lambda^4}{16b^4}\right)^{1/3}} \text{Ra}^{1/3}. \tag{6–366}$$

The *aspect ratio* of the cells, that is, the ratio of the horizontal width to the vertical thickness, $\lambda/2b$, remains unspecified. According to linear stability theory, the aspect ratio of the most rapidly growing disturbance is $\lambda/(2b) = \sqrt{2}$; see Equation (6–322). However, for finite-amplitude convection we determine the aspect ratio for which the Nusselt number is a maximum. This is the aspect ratio of the cells that is most effective in transporting heat across the fluid layer at a fixed value of the Rayleigh number. We therefore require

$$\frac{\partial \text{Nu}}{\partial(\lambda/2b)} = 0 \tag{6–367}$$

and find that

$$\frac{\lambda}{2b} = 1. \tag{6–368}$$

For this value of the aspect ratio the horizontal velocity is

$$u_0 = 0.271\frac{\kappa}{b}\text{Ra}^{2/3} \tag{6–369}$$

and the Nusselt number is

$$\text{Nu} = 0.294\text{Ra}^{1/3}. \tag{6–370}$$

It is of interest to compare this result with that obtained using the transient boundary-layer theory given in Equation (6–336). Both approximate solutions give the same power law dependence of the Nusselt number on the Rayleigh number but the numerical constants differ by about a factor of two, 0.120 versus 0.294. Numerical calculations show that the value of this constant should be 0.225.

It is appropriate to apply the steady-state, boundary-layer analysis of thermal convection in a fluid layer heated from below to the problem of thermal convection in the upper mantle. As before we take $b = 700$ km, $\rho_0 = 3700$ kg m^{-3}, $g = 10$ m s^{-2}, $\alpha_v = 3 \times 10^{-5}$ K^{-1}, $T_1 - T_0 = 1500$ K, $\kappa = 1$ mm^2 s^{-1}, and $\mu = 10^{21}$ Pa s and from Equation (6–316) again find that Ra $= 5.7 \times 10^5$.

The mean surface heat flux $\bar{q}$ is given by

$$\begin{aligned} \bar{q} &= \frac{2Q}{\lambda} = \frac{2Q_c}{\lambda}\text{Nu} = \frac{k(T_1 - T_0)}{b}\text{Nu} \\ &= \frac{k(T_1 - T_0)}{b}(0.294)\text{Ra}^{1/3}. \end{aligned} \tag{6–371}$$

[Compare with Equation (6–337)]. With the above parameter values and $k = 4$ W m^{-1} K^{-1}, $\bar{q}$ is 200 mW m^{-2}. This is about 2.3 times larger than the observed mean heat flow of 87 mW m^{-2}. From Equation (6–369) the mean horizontal velocity u_0 is 84 mm yr^{-1}. This is about twice the mean surface velocity associated with plate tectonics.

The steady-state boundary-layer theory can also be applied to a fluid layer that is heated from within and cooled from above. The flow is again divided into counterrotating, two-dimensional cells with dimensions b and $\lambda/2$. A cold thermal boundary layer forms on the upper boundary of each cell. When the two cold boundary layers from adjacent cells meet, they separate from the boundary to form a cold descending thermal plume. However, for the fluid layer heated from within there is no heat flux across the lower boundary. Therefore no hot thermal boundary layer develops on the lower boundary, and there are no hot ascending plumes between cells. This flow is illustrated in Figure 6–42. In the boundary-layer approximation, we can assume that all fluid that is not in the cold thermal boundary layers and plumes has the same temperature T_1. The temperature T_1 is not known a priori and must be determined as part of the solution to the convection problem.

The temperature distribution in the upper cold thermal boundary layer is given by Equation (6–347), and the total rate at which heat flows out of the top of each cell Q is given by Equation (6–348). In the layer there

6–42 The boundary-layer structure of two-dimensional thermal convection cells in a fluid layer heated from within and cooled from above.

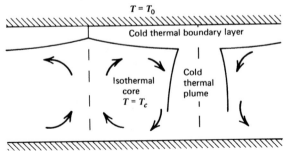

is a uniform heat production H per unit mass. Thus, the total heat production in a cell is $\rho_0 H b \lambda / 2$. Because we assume a steady state, Q must equal the rate of heat generation in the cell

$$\frac{\rho_0 H b \lambda}{2} = 2k(T_1 - T_0) \left(\frac{u_0 \lambda}{2 \pi \kappa} \right)^{1/2}. \quad (6\text{--}372)$$

The calculation of the total buoyancy force on the descending cold plume is also the same as in the previous problem; F_b is given by Equation (6–354). This problem, however, has only the single plume. The rate of doing work on the boundaries is the same as in the previous problem. However, the energy input comes only from the single plume. By equating the rate of energy input to a cell to the rate of doing work on the boundaries, we find

$$\rho_0 g \alpha_v (T_1 - T_0) u_0 \left(\frac{\kappa \lambda}{2 \pi u_0} \right)^{1/2} = \frac{4 v_0^2 \mu}{\lambda} + \frac{u_0^2 \mu \lambda}{b^2}. \quad (6\text{--}373)$$

Substitution of Equation (6–357) to eliminate v_0 and Equation (6–372) to eliminate $T_1 - T_0$ yields

$$u_0 = \frac{\kappa}{2b} \frac{\left(\frac{\lambda}{2b} \right)^2}{\left(1 + \frac{\lambda^4}{16b^4} \right)^{1/2}} Ra_H^{1/2}, \quad (6\text{--}374)$$

where the Rayleigh number for a fluid layer heated from within has been defined in Equation (6–324).

We can solve for the temperature of the core T_1 by substituting Equation (6–374) into Equation (6–372):

$$T_1 - T_0 = \left(\frac{\pi}{2} \right)^{1/2} \frac{\rho_0 H b^2}{k} \frac{\left(1 + \frac{\lambda^4}{16b^4} \right)^{1/4}}{\left(\frac{\lambda}{2b} \right)^{1/2}} Ra_H^{-1/4}, \quad (6\text{--}375)$$

where $T_1 - T_0$ is also the temperature rise across the fluid layer. To assess the efficiency with which convection cools the fluid layer we again introduce the dimensionless temperature ratio θ defined in Equation (6–342). This is the ratio of the temperature difference across the layer with convection to the temperature difference without convection, the latter was given in Equation (6–343). By substituting Equation (6–375) into Equation (6–342), we find

$$\theta = (2\pi)^{1/2} \frac{\left(1 + \frac{\lambda^4}{16b^4} \right)^{1/4}}{\left(\frac{\lambda}{2b} \right)^{1/2}} Ra_H^{-1/4}. \quad (6\text{--}376)$$

The dimensionless temperature ratio θ is a function of the cell aspect ratio $\lambda / 2b$. The cell aspect ratio that minimizes θ is found by setting

$$\frac{\partial \theta}{\partial (\lambda / 2b)} = 0. \quad (6\text{--}377)$$

This gives

$$\frac{\lambda}{2b} = 1, \quad (6\text{--}378)$$

the same value that was obtained for the layer heated from below. With an aspect ratio of unity, the horizontal velocity and dimensionless temperature ratio are

$$u_0 = 0.354 \frac{\kappa}{b} Ra_H^{1/2} \quad (6\text{--}379)$$

and

$$\theta = 2.98 Ra_H^{-1/4}. \quad (6\text{--}380)$$

The nondimensional temperature difference between the isothermal core and the upper boundary decreases as convection becomes more vigorous with increasing Rayleigh number. We compare this result with that obtained using the transient boundary-layer analysis given in Equation (3–344). Again both approximate solutions give the same power law dependence of the nondimensional temperature difference on the Rayleigh number but the numerical constants again differ by about a factor of two, 4.91 versus 2.98.

It is again of interest to apply the steady-state boundary-layer analysis of thermal convection in a uniformly heated fluid layer cooled from above to the problem of thermal convection in the whole mantle. We again take $b = 2880$ km, $H = 9 \times 10^{-12}$ W kg^{-1}, $\rho_0 = 4700$ kg m^{-3}, $g = 10$ m s^{-2}, $k = 4$ W m^{-1} K^{-1}, $\kappa = 1$ mm^2 s^{-1}, and $\mu = 10^{21}$ Pa s and from Equation (6–324) find that $Ra_H = 3 \times 10^9$. Equations (6–341), (6–342), and (6–380) give $T_1 - T_0 = 550$ K. This is about a factor of 4 low. From Equation (6–379) we find that $u_0 = 210$ mm yr^{-1}, which is about a factor of 4 too high.

The parameterizations of Nusselt number and nondimensional temperature obtained above have been for free-slip surface boundary conditions. Most laboratory experiments are carried out with no-slip surface boundary conditions because walls are required to confine the fluid. For a fluid layer heated from below with no-slip wall boundary conditions it is found experimentally that

$$\text{Nu} = 0.131\text{Ra}^{0.3}, \qquad (6\text{--}381)$$

and for a fluid layer heated from within and cooled from above,

$$\theta = \frac{5.95}{\text{Ra}_H^{0.23}}. \qquad (6\text{--}382)$$

These results are similar to those obtained above.

PROBLEM 6–32 An excellent approximation to the Nusselt number–Rayleigh number relation for a fluid layer heated from below as in Equation (6–381) can be derived from the θ–Ra relation in Equation (6–382) for an internally heated fluid layer as follows. Write Nu as

$$\text{Nu} = \frac{\bar{q}}{k(T_1 - T_0)/b}. \qquad (6\text{--}383)$$

Identify $\bar{q}$, the heat flux through the upper boundary with $\rho_0 Hb$, the total rate of heat generation in the internally heated layer. Take $T_1 - T_0$ equal to twice the temperature rise across the internally heated convecting layer. Rewrite Equation (6–383) as

$$\text{Nu} = \frac{\rho_0 Hb^2}{2k(T_c - T_0)}. \qquad (6\text{--}384)$$

Eliminate H from Equation (6–384) by using Equations (6–341), (6–342), and (6–382). Introduce the Rayleigh number for heated-from-below convection based on the temperature difference $T_c - T_0$. You should obtain

$$\text{Nu} = 0.12\text{Ra}^{0.3}, \qquad (6\text{--}385)$$

an almost identical result with Equation (6–381).

PROBLEM 6–33 Consider convection in a fluid layer heated from below. The mean surface heat flux $\bar{q}$ is transferred through the cold thermal boundary layer by conduction. Therefore we can write

$$\bar{q} = \frac{k(T_c - T_0)}{\delta}, \qquad (6\text{--}386)$$

where δ is a characteristic thermal boundary layer thickness. Show that

$$\frac{\delta}{b} = 1.7\text{Ra}^{-1/3}. \qquad (6\text{--}387)$$

Calculate δ for an upper mantle convection cell given the parameter values used in the discussion in this section.

PROBLEM 6–34 In what ways are surface plates and descending slabs different from the thermal boundary layers and descending plumes of two-dimensional convection cells in layers of ordinary viscous fluids heated from below or from within?

PROBLEM 6–35 Suppose that convection extends through the entire mantle and that 10% of the mean surface heat flow originates in the core. If the surface thermal boundary layer and the boundary layer at the core–mantle interface have equal thicknesses, how does the temperature rise across the lower mantle boundary layer compare with the temperature increase across the surface thermal boundary layer?

PROBLEM 6–36 Apply the two-dimensional boundary-layer model for heated-from-below convection to the entire mantle. Calculate the mean surface heat flux, the mean horizontal velocity, and the mean surface thermal boundary-layer thickness. Assume $T_1 - T_0 = 3000$ K, $b = 2880$ km, $k = 4$ W m^{-1} K^{-1}, $\kappa = 1$ mm^2 s^{-1}, $\alpha_v = 3 \times 10^{-5}$ K^{-1}, $g = 10$ m s^{-2}, and $\rho_0 = 4000$ kg m^{-3}.

6-22 The Forces that Drive Plate Tectonics

In Section 6–21 we saw that thermal convection in a fluid layer heated from within has many similarities to mantle convection. The thermal boundary layer adjacent to the cooled upper surface can be directly associated with the oceanic lithosphere. The separation of the boundary layer to form a cold descending plume is associated with the subduction of the lithosphere at an ocean trench. Just as the gravitational body force on the cold plume drives the convective flow, the gravitational body force on the descending lithosphere at a trench is most likely important in driving plate tectonics.

The gravitational body force F_{b1} on the descending lithosphere due to its temperature deficit relative to the adjacent mantle can be evaluated with Equation (6–354). Because of the rigidity of the lithosphere, $u_0 = v_0$. This also follows from Equation (6–357) for an aspect ratio $\lambda/2b = 1$. The equation for F_{b1} is thus

$$F_{b1} = 2\rho_0 g\alpha_v b(T_c - T_0)\left(\frac{\kappa\lambda}{2\pi u_0}\right)^{1/2}. \qquad (6\text{--}388)$$

In using this expression, we have neglected the heating of the descending lithosphere by friction, as discussed in Section 4–27. A principal uncertainty in evaluating the gravitational body force is the depth of the convection cell b; this is equivalent to the length of the descending lithosphere beneath trenches. Based on the distribution of earthquakes that extend to a depth of about 700 km, we take $b = 700$ km. Also taking $\rho_0 = 3300$ kg m^{-3}, $g = 10$ m s^{-2}, $\alpha_v = 3 \times 10^{-5}$ K^{-1}, $T_c - T_0 = 1200$ K, $\kappa = 1$ mm^2 s^{-1}, $u_0 = 50$ mm yr^{-1}, and $\lambda = 4000$ km, we obtain $F_{b1} = 3.3 \times 10^{13}$ N m^{-1}. This is a force per unit length parallel to the trench.

Another force on the descending lithosphere is due to the elevation of the olivine–spinel phase change (see Section 4–29). The position of the phase change boundary in the descending lithosphere is sketched in Figure 6–43. The phase change occurs at a depth in the surrounding mantle where the temperature is T_{os}. Because the descending lithosphere is colder than the mantle, the phase change occurs at lower pressure or shallower depth in the slab. Because the temperature of the descending lithosphere T_s at the depth where the mantle phase change occurs depends on position $T_s = T_s(x')$, the phase change boundary elevation h_{os} also depends on position $h_{os}(x')$. The downward gravitational body force on the descending lithosphere due to the phase boundary elevation F_{b2} is thus

$$F_{b2} = g \Delta \rho_{os} \int_{x'=0}^{x'=x'_s} h_{os}(x')\, dx', \qquad (6\text{–}389)$$

where $\Delta \rho_{os}$ is the positive density difference between

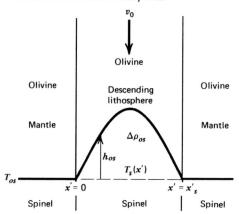

6–43 Elevation of the olivine–spinel phase change in the descending lithosphere contributes to the downward body force.

the phases. The elevation of the phase boundary is given by

$$h_{os} = \frac{\gamma(T_{os} - T_s)}{\rho_0 g}, \qquad (6\text{–}390)$$

where γ is *slope of the Clapeyron curve* (dp/dT). Substitution of Equation (6–390) into (6–389) yields

$$F_{b2} = \frac{\gamma \Delta \rho_{os}}{\rho_0} \int_{x'=0}^{x'=x'_s} (T_{os} - T_s)\, dx'. \qquad (6\text{–}391)$$

The integral in Equation (6–391) is the integrated temperature deficit in the descending lithosphere at the depth in the mantle where the olivine–spinel phase change occurs. This can be evaluated using the boundary-layer model discussed previously. In calculating f_b in Equation (6–353), we found that the integrated temperature deficit per unit depth of the descending plume is a constant. Its value, from Equation (6–353), with $u_0 = v_0$, is $f_b / \rho_0 g \alpha_v$ or

$$2(T_c - T_0)\left(\frac{\kappa \lambda}{2\pi u_0}\right)^{1/2}.$$

Using this for the value of the integral in Equation (6–391), we find

$$F_{b2} = \frac{2(T_c - T_0)\gamma \Delta \rho_{os}}{\rho_0}\left(\frac{\kappa \lambda}{2\pi u_0}\right)^{1/2}. \qquad (6\text{–}392)$$

With $\Delta \rho_{os} = 270$ kg m^{-3}, $\gamma = 4$ MPa K^{-1}, and the other parameter values given above, we obtain $F_{b2} = 1.6 \times 10^{13}$ N m^{-1}. The body force due to elevation of the olivine–spinel phase change is about half the body force due to thermal contraction. The total body force on the descending lithosphere is $F_b = 4.9 \times 10^{13}$ N m^{-1}. This force is often referred to as *trench pull*. If the force is transmitted to the surface plate as a tensional stress in an elastic lithosphere with a thickness of 50 km, the required tensional stress is 1 GPa, clearly a very high stress.

A force is also exerted on the surface plates at ocean ridges. The elevation of the ridges establishes a *pressure head* that drives the flow horizontally away from the center of the ascending plume. This *ridge push* can also be thought of as *gravitational sliding*. A component of the gravitational field causes the surface plate to slide downward along the slope between the ridge crest and the deep ocean basin.

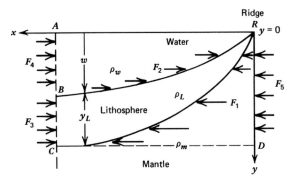

6–44 Horizontal forces acting on a section of the ocean, lithosphere, and mantle at an ocean ridge.

The force exerted on a surface plate due to the elevation of an ocean ridge can be evaluated from the force balance in Figure 6–44. We consider the horizontal forces on a section of the ocean, lithosphere, and underlying mantle, as shown in the figure. All pressure forces are referenced to the ridge crest ($y = 0$). The water layer above the ridge crest has a constant depth and exerts no net horizontal force. The integrated horizontal force on the base of the lithosphere F_1 can be determined from the equilibrium of section RCD of the mantle. The net horizontal pressure force on RD, F_5, must equal F_1. The force F_5 is easily obtained by integrating the lithostatic pressure beneath the ridge crest:

$$F_5 = F_1 = \int_0^{w+y_L} \rho_m g y \, dy, \tag{6–393}$$

where ρ_m is the mantle density. This can be rewritten as

$$F_1 = g \int_0^w \rho_m y \, dy + g \int_0^{y_L} \rho_m (w + \bar{y}) \, d\bar{y}, \tag{6–394}$$

where

$$\bar{y} = y - w. \tag{6–395}$$

The integrated pressure force on the upper surface of the lithosphere F_2 is equal to F_4, the net pressure force on AB, because the section of water RAB must be in equilibrium. Thus we can integrate the hydrostatic pressure in the water to obtain

$$F_2 = F_4 = \int_0^w \rho_w g y \, dy, \tag{6–396}$$

where ρ_w is the water density. The horizontal force F_3

acting on the section of lithosphere BC is the integral of the pressure in the lithosphere P_L

$$F_3 = \int_0^{y_L} P_L \, d\bar{y}, \tag{6–397}$$

where

$$P_L = \rho_w g w + \int_0^{\bar{y}} \rho_L g \, d\bar{y}' \tag{6–398}$$

and ρ_L is the density in the lithosphere. Substituting Equation (6–398) into Equation (6–397) gives

$$F_3 = \int_0^{y_L} \left\{ \rho_w g w + \int_0^{\bar{y}} \rho_L g \, d\bar{y}' \right\} d\bar{y}. \tag{6–399}$$

The net horizontal force on the lithosphere adjacent to an ocean ridge F_R is obtained by combining Equations (6–394), (6–396), and (6–399)

$$F_R = F_1 - F_2 - F_3 = g \int_0^w (\rho_m - \rho_w) y \, dy$$
$$+ g \int_0^{y_L} \left\{ (\rho_m - \rho_w) w + \rho_m \bar{y} - \int_0^{\bar{y}} \rho_L \, d\bar{y}' \right\} d\bar{y}. \tag{6–400}$$

We substitute the isostatic relation from Equation (4–204) and the identity

$$\rho_m \bar{y} = \int_0^{\bar{y}} \rho_m \, d\bar{y}' \tag{6–401}$$

to give

$$F_R = g(\rho_m - \rho_w)\frac{w^2}{2} + g \int_0^{y_L} \left\{ \int_0^{\infty} (\rho_L - \rho_m) \, d\bar{y}' \right.$$
$$\left. - \int_0^{\bar{y}} (\rho_L - \rho_m) \, d\bar{y}' \right\} d\bar{y}'$$
$$= g(\rho_m - \rho_w)\frac{w^2}{2}$$
$$+ g \int_0^{\infty} \left\{ \int_{\bar{y}}^{\infty} (\rho_L - \rho_m) \, d\bar{y}' \right\} d\bar{y}'. \tag{6–402}$$

As in Equation (4–206) the limit y_L has been replaced by ∞ because the integrals are convergent. Substitution of Equations (4–205) and (4–124) and

$$z = \frac{1}{2}\bar{y}\left(\frac{u_0}{\kappa x}\right)^{1/2} \tag{6–403}$$

yields

$$F_R = g(\rho_m - \rho_w)\frac{w^2}{2} + g\rho_m\alpha_v(T_1 - T_0)$$
$$\times \frac{4\kappa x}{u_0} \int_0^\infty \left(\int_z^\infty \mathrm{erfc}z' \, dz' \right) dz, \qquad (6\text{–}404)$$

where T_1 is the mantle temperature. The repeated integral of the complementary error function has the value $1/4$. By substituting for w from Equation (4–209), we finally arrive at

$$F_R = g\rho_m\alpha_v(T_1 - T_0)\left[1 + \frac{2}{\pi}\frac{\rho_m\alpha_v(T_1 - T_0)}{(\rho_m - \rho_w)} \right]\frac{\kappa x}{u_0}$$
$$= g\rho_m\alpha_v(T_1 - T_0)\left[1 + \frac{2}{\pi}\frac{\rho_m\alpha_v(T_1 - T_0)}{(\rho_m - \rho_w)} \right]\kappa t, \qquad (6\text{–}405)$$

where t is the age of the seafloor.

The horizontal forces required to maintain topography were derived in Section 5–14 and the resulting ridge push force for the plate cooling model was given in Equation (5–171). The geoid anomaly ΔN associated with the half-space cooling model was given in Equation (5–157). Substitution of Equation (5–157) into Equation (6–405) gives

$$F_R = \frac{g^2 \Delta N}{2\pi G}. \qquad (6\text{–}406)$$

This result was previously given in Equation (5–170).

From Equation (6–405) the force due to the elevation of the ocean ridge is proportional to the age of the lithosphere. Taking $g = 10$ m s^{-2}, $\rho_m = 3300$ kg m^{-3}, $\rho_w = 1000$ kg m^{-3}, $\kappa = 1$ mm^2 s^{-1}, $T_1 - T_0 = 1200$ K, and $\alpha_v = 3 \times 10^{-5}$ K^{-1}, we find that the total ridge push on 100 Myr old oceanic lithosphere is 3.9×10^{12} N m^{-1}. This is a force per unit length parallel to the ridge. This ridge push force is in quite good agreement with the value obtained in Section 5–14 for the plate cooling model with $y_{L0} = 125$ km. The force in the equilibrated ocean basin was found to be 3.41×10^{12} N m^{-1}.

Ridge push is thus an order of magnitude smaller than trench pull. However, trench pull may be mostly offset by large resistive forces encountered by the descending lithosphere as it penetrates the mantle. The net force at the trench is probably comparable to ridge push.

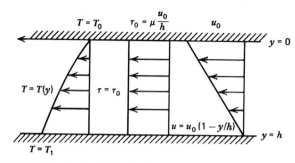

6–45 Frictional heating in Couette flow.

6–23 Heating by Viscous Dissipation

Throughout the discussion in this chapter we have neglected the effects of *viscous dissipation* or frictional heating. In this section we will calculate the temperature rise and the heat flux produced by viscous dissipation in a simple Couette flow (see Section 6–2) between plane parallel walls, as shown in Figure 6–45.

The velocity profile

$$u = u_0 \left(1 - \frac{y}{h} \right) \qquad (6\text{–}407)$$

is produced by the constant shear stress,

$$\tau = \mu \frac{du}{d(-y)} = \frac{\mu u_0}{h} = \tau_0, \qquad (6\text{–}408)$$

applied at the upper surface ($y = 0$) of the channel. The derivative with respect to $-y$ occurs in Equation (6–408) because the outer normal to a horizontal area at the upper boundary of the fluid points in the negative y direction.

The rate at which shear forces do work on the entire fluid layer, per unit horizontal area, is given by the product of the shear stress τ_0 and the velocity of the upper boundary u_0, that is,

$$\frac{\mu u_0^2}{h}.$$

If we average this over the entire fluid layer, we get

$$\frac{\mu u_0^2}{h^2},$$

the rate of shear heating per unit volume. This heating rate per unit volume is constant because the shear stress is constant and the velocity profile is linear. For example, we could have calculated the rate at which shear forces do work on the fluid in the lower half of the channel, $\mu u_0^2/2h$, upon averaging this over the appropriate volume of fluid, unit horizontal area $\times h/2$, we still get $\mu u_0^2/h^2$ for the volumetric heating rate due to viscous dissipation. This volumetric heating rate can be identified with the internal volumetric heat production rate ρH in Equation (4–12) to obtain the equation for the temperature distribution in the channel

$$k\frac{d^2 T}{dy^2} = -\frac{\mu u_0^2}{h^2}. \qquad (6\text{–}409)$$

A straightforward integration of this equation with the boundary conditions $T = T_0$ at $y = 0$ and $T = T_1$ at $y = h$ gives

$$T = T_0 + \frac{y}{h}\left\{T_1 - T_0 + \frac{\mu u_0^2}{2k}\right\} - \frac{\mu u_0^2}{2k}\frac{y^2}{h^2}. \qquad (6\text{–}410)$$

This can be written in the convenient dimensionless form

$$\theta = \frac{T - T_0}{T_1 - T_0} = \frac{y}{h}\left\{1 + \frac{\mu u_0^2/2k}{T_1 - T_0}\right\} - \frac{y^2}{h^2}\left(\frac{\mu u_0^2/2k}{T_1 - T_0}\right). \qquad (6\text{–}411)$$

The temperature distribution in the channel is governed by the single dimensionless parameter

$$\frac{\mu u_0^2/2k}{(T_1 - T_0)}.$$

This can be written as $1/2$ times the product of the Prandtl number Pr and a dimensionless parameter known as the *Eckert number*,

$$E \equiv \frac{u_0^2}{c_p(T_1 - T_0)}, \qquad (6\text{–}412)$$

where c_p is the specific heat at constant pressure. Thus we can write

$$\frac{1}{2}\text{PrE} = \frac{\mu u_0^2/2k}{(T_1 - T_0)} \qquad (6\text{–}413)$$

and

$$\theta = \frac{y}{h}\left(1 + \frac{\text{PrE}}{2}\right) - \frac{y^2}{h^2}\left(\frac{\text{PrE}}{2}\right). \qquad (6\text{–}414)$$

The dimensionless temperature θ is plotted in Figure 6–46 for several values of Pr E. The conduction profile in the absence of frictional heating is the straight line for Pr E = 0. The temperatures in excess of this linear pro-

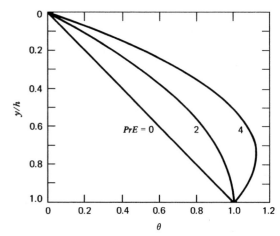

6–46 Dimensionless temperature distribution in a frictionally heated Couette flow.

file are a consequence of viscous dissipation. The slopes of the temperature profiles indicate that so much heat is generated by friction when Pr E = 4 that heat flows out of the channel at both boundaries. Normally, with $T_1 > T_0$, heat flows into the channel at the lower boundary. The excess temperature due to frictional heating θ_e is obtained by subtracting the linear profile from Equation (6–414)

$$\theta_e = \left(\frac{\text{PrE}}{2}\right)\left(\frac{y}{h}\right)\left(1 - \frac{y}{h}\right). \qquad (6\text{–}415)$$

The maximum excess temperature is found by differentiating θ_e with respect to y/h and setting the result to zero. The maximum θ_e occurs at $y/h = 1/2$ and

$$\theta_e^{\max} = \frac{\text{PrE}}{8}. \qquad (6\text{–}416)$$

The upward heat flux at the upper boundary q_0 is found by using Fourier's law – Equation (4–1) – and Equation (6–410)

$$q_0 = k\left(\frac{dT}{dy}\right)_{y=0} = \frac{k}{h}\left\{T_1 - T_0 + \frac{\mu u_0^2}{2k}\right\}. \qquad (6\text{–}417)$$

The excess upward heat flux q_e at $y = 0$ is clearly

$$q_e = \frac{\mu u_0^2}{2h}. \qquad (6\text{–}418)$$

If all the frictionally generated heat flowed out of the upper boundary, q_e would equal $\mu u_0^2/h$; half the shear heating in the channel flows out through the lower boundary. The ratio of the excess heat flowing through the upper boundary q_e to the heat flux without viscous dissipation $q_c = k(T_1 - T_0)/h$ is

$$\frac{q_e}{q_c} = \frac{1}{2}\mathrm{Pr}\,E. \qquad (6\text{–}419)$$

We can use the results of this section to quantify the effects of frictional heating in an asthenospheric shear flow, for example. With $\mu = 4 \times 10^{19}$ Pa s, $u_0 = 50$ mm yr^{-1}, $k = 4$ W m^{-1} K^{-1}, and $T_1 - T_0 = 300$ K, we find $\mathrm{Pr}\,E/2 = 0.04$. Thus the maximum additional temperature rise due to shear heating would be 1% of the temperature rise across the asthenosphere or about 3 K in this example; see Equation (6–416). The excess heat flux to the surface would be 4% of the heat conducted across the asthenosphere in the absence of dissipation. These results show that frictional heating effects in mantle shear flows are generally small. However, they can be important, especially if the viscosity is larger than assumed in our numerical example. We discuss frictional heating again in Chapter 8 in connection with thermally activated creep on a fault zone.

PROBLEM 6–37 Show that half of the frictionally generated heat flows out of the lower boundary of the channel in the Couette flow example in this section.

PROBLEM 6–38 Consider frictional heating in a Couette flow with an isothermal upper boundary and an insulated lower boundary. Determine the temperature profile in the channel and the excess upward heat flow at the upper boundary due to the shear heating. What is the temperature of the lower boundary as a consequence of the frictional heating? Compare the temperature rise across this channel with the maximum temperature rise in a channel with equal wall temperatures.

6–24 Mantle Recycling and Mixing

The plate tectonic cycle is an inherent component of mantle convection. The surface plates are the lithosphere and the lithosphere is the upper thermal boundary layer of mantle convection cells. The oceanic lithosphere is created at mid-ocean ridges and is recycled back into the mantle at subduction zones.

Mid-ocean ridges migrate over the surface of the Earth in response to the kinematic constraints of plate tectonics. Mantle rock ascends passively beneath an ocean ridge in response to seafloor spreading and be-

comes partially molten due to the decrease in pressure on ascent. The magma percolates through the residual solid and then solidifies to form the oceanic crust, with an average thickness of ~6 km. The result is a two-layer structure for the rigid oceanic lithosphere. The upper part of the lithosphere is the solidified magma of the oceanic crust and the lower part is the complementary residual solid in the upper mantle. The residual solid also has a vertical stratification. The uppermost mantle rock is highly depleted in the low-melting-temperature basaltic component, and it grades into undepleted mantle over a depth range of ~50 km.

Isotopic and trace element studies of mid-ocean ridge basalts (MORB) show that they are remarkably uniform and systematically depleted in incompatible trace elements with respect to bulk Earth values. This indicates that, on average, the mantle source of MORB is a well-stirred depleted chemical reservoir on the scale at which it is sampled by mid-ocean ridge processes. However, heterogeneities do persist in this reservoir, as indicated by variations in MORB. Large-scale heterogeneities are evident in variations between average Atlantic Ocean MORB and average Indian Ocean MORB. In addition, small-scale heterogeneities are evident in deviations from average MORB. These heterogeneities are most evident when the mantle is sampled locally, as it is at young Pacific Ocean seamounts.

The depleted mantle source of MORB is complementary to the enriched continental crust. Incompatible elements are partitioned into the continents by the volcanic processes responsible for the formation of the continents; this occurs primarily at island arcs. When the oceanic lithosphere is subducted, the oceanic crust is partially melted; the resulting enriched magmas ascend to the surface and form island-arc volcanoes leaving a more strongly depleted oceanic lithosphere. The complementary nature of the continental crust and the MORB source reservoir requires that this depleted oceanic lithosphere, from which the continental crust has been extracted, be mixed into the MORB source region.

Atomic diffusion plays a role in the homogenization of the mantle only on scales of a meter or less because the solid-state diffusion coefficient is so small. Values of the relevant diffusion coefficients are estimated to be in the range $D = 10^{-18} - 10^{-20}$ m^2 s^{-1}. Over the age of the Earth, 4.5×10^9 yr, the corresponding range of diffusion lengths is 0.3–0.03 m. We conclude that the subducted lithosphere is mixed back into the mantle by convection, but that diffusive mixing is significant only on small scales. This process of convective homogenization is known as kinematic mixing and has been extensively studied in polymer science. The mantle is composed of a matrix of discrete, elongated layers of subducted oceanic lithosphere. Each layer has its own isotopic, chemical, and age identity. The older the layer the more it will have been elongated by mantle flows; on average, the older layers will be thinner. The mantle thus has the appearance of a marble cake. The marble cake comprises the enriched oceanic crust, which has been partially depleted by subduction zone volcanism, and the complementary, highly depleted upper mantle.

Approximately the upper 60 km of the lithosphere is processed by the plate tectonic cycle. We first ask the question: What fraction of the mantle has been processed by the plate tectonic cycle since the Earth was formed? We consider the two limiting cases of layered mantle convection (above a depth of 660 km) and whole mantle convection.

To simplify the analysis we assume that the rate $\dot{M}$ at which mass is processed into a layered structure at ocean ridges is constant, and that the subducted rock is uniformly distributed throughout the mantle (upper mantle). We define M_p to be the primordial unprocessed mass in the mantle reservoir. The rate of loss of this primordial mass by processing at ocean ridges is given by

$$\frac{dM_p}{dt} = -\frac{M_p}{M_m}\dot{M}, \qquad (6\text{–}420)$$

where M_m is the mass of the mantle participating in the plate tectonic convective cycle – the whole mantle for whole mantle convection and the upper mantle for layered mantle convection. The ratio $M_p(t)/M_m$ is the fraction of the mantle reservoir that has not been processed at an ocean ridge. Upon integration with the initial condition $M_p = M_m$ at $t = 0$ we obtain

$$M_p = M_m e^{-t/\tau_p}, \qquad (6\text{–}421)$$

where

$$\tau_p = \frac{M_m}{\dot{M}} \qquad (6\text{–}422)$$

is the characteristic time for processing the mantle in the plate tectonic cycle. The processing rate $\dot{M}$ is given by

$$\dot{M} = \rho_m h_p \frac{dS}{dt}, \qquad (6\text{–}423)$$

where ρ_m is the mantle density, h_p is the thickness of the layered oceanic lithosphere structure, and dS/dt is the rate at which new surface plate area is created (or subducted). Taking $dS/dt = 0.0815$ m^2 s^{-1} (see Figure 4–26), $h_p = 60$ km, and $\rho_m = 3300$ kg m^{-3}, we obtain $\dot{M} = 1.61 \times 10^7$ kg s^{-1}. For layered mantle convection ($M_m = 1.05 \times 10^{24}$ kg) the characteristic time for processing the mantle from Equation (6–422) is $\tau_p = 2$ Gyr; for whole mantle convection $\tau_p = 8$ Gyr. The fraction of primordial unprocessed mantle M_p/M_m obtained from Equation (6–421) is given as a function of time t in Figure 6–47 for both layered and whole mantle convection. For layered mantle convection 10.5% of the upper mantle is unprocessed at the present time while for whole mantle convection 57% is unprocessed.

This analysis was carried out assuming a constant rate of recycling. As shown in Section 4–5 the rate of radioactive heat generation in the Earth H was higher in the past. To extract this heat from the Earth's interior, the rate of plate tectonics was probably also higher in the past. The time dependence of the radioactive heat generation as given in Figure 4–4 can be approximated by the relation

$$H = H_0 e^{\lambda(t_e - t)}, \qquad (6\text{–}424)$$

6–47 Fraction of the mantle reservoir that has not been processed by the plate tectonic cycle M_p/M_m during a period of time t. (a) Whole mantle convection. (b) Layered mantle convection. The dashed lines are for a constant processing rate and the solid lines are a rate that decreases exponentially with time.

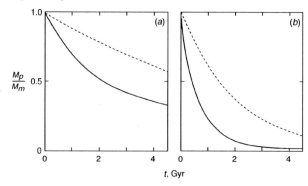

where H_0 is the present rate of heat production, t_e is the present value of the time t, and λ is the average decay constant for the mixture of radioactive isotopes in the mantle. From the results given in Figure 4–4 we take $\lambda = 2.77 \times 10^{-10}$ yr^{-1}. Assuming that the rate $\dot{M}$ at which mass is processed into a layered structure at ocean ridges is proportional to the rate of heat generation given in Equation (6–424) we write

$$\dot{M} = \dot{M}_0 e^{\lambda(t_e - t)}, \tag{6–425}$$

where $\dot{M}_0$ is the present rate of processing. Substitution of Equation (6–425) into Equation (6–420) gives

$$\frac{dM_p}{dt} = -\frac{M_p}{M_m} \dot{M}_0 e^{\lambda(t_e - t)}. \tag{6–426}$$

Integration of Equation (6–426) with the initial condition $M_p = M_m$ at $t = 0$ gives

$$M_p = M_m \exp\left[\frac{-1}{\tau_{p0}\lambda}\left(e^{\lambda t_e} - e^{\lambda(t_e - t)}\right)\right], \tag{6–427}$$

where

$$\tau_{p0} = \frac{M_m}{\dot{M}_0} \tag{6–428}$$

is again the characteristic time for processing the mantle in the plate tectonic cycle. For layered mantle convection we again have $\tau_{p0} = 2$ Gyr and for whole mantle convection we have $\tau_{p0} = 8$ Gyr. With $\lambda = 2.77 \times 10^{-10}$ yr^{-1}, the fraction of primordial unprocessed mantle M_p/M_m obtained from Equation (6–427) is given as a function of time t in Figure 6–47 for both layered and whole mantle convection. For layered mantle convection 1% of the upper mantle is unprocessed at the present time while for whole mantle convection 33% is unprocessed. The time-dependent processing is more efficient, as expected. However, in all cases substantial fractions of the mantle reservoir have been processed by the plate tectonic cycle.

The layered oceanic lithosphere is subducted back into the mantle at oceanic trenches. The cold subducted lithosphere is heated by conduction from the surrounding mantle on a time scale of ≈ 50 Myr. The heated and softened subducted lithosphere is then entrained in the mantle convective flows and is subjected to the fluid deformation. With the assumption that the subducted layered lithosphere behaves passively, it is subject to kinematic mixing.

We next quantify the rate of kinematic mixing in the mantle. We consider the problem of layer stretching. As stated before, we hypothesize that the subducted oceanic crust becomes entrained in the convecting mantle and is deformed by the strains associated with thermal convection. Kinematic mixing can occur by both shear strains and normal strains. We first consider the thinning of a passive layer in a uniform shear flow. Initially we take the one-dimensional channel flow (Couette flow, see Figure 6–2a) of width h as illustrated in Figure 6–48. The passive layer has an initial width δ_0 and is assumed to be vertical with a length L. The linear velocity profile from Equation (6–13) is

$$u = \dot{\varepsilon}(h - y), \tag{6–429}$$

where the strain rate $\dot{\varepsilon} = u_0/h$. At a subsequent time t the top of the layer has moved a distance $u_0 t$ while the bottom boundary remains in place. The total length of the strip is now

$$L = h[1 + (\dot{\varepsilon}t)^2]^{1/2}. \tag{6–430}$$

However, to conserve the mass of material in the strip we require

$$h\delta_0 = L\delta. \tag{6–431}$$

6–48 Illustration of the kinematic stretching of a passive layer of material in a uniform shear flow. (a) Initially at $t = 0$ the layer is vertical and has a thickness δ_0 and a length h. (b) At a subsequent time t the layer has been stretched in the horizontal flow to reduced thickness δ and an increased length L.

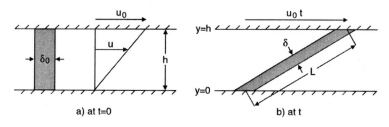

a) at t=0   b) at t

Substitution of Equation (6–430) into (6–431) gives

$$\frac{\delta}{\delta_0} = \frac{1}{[1 + (\dot{\varepsilon}t)^2]^{1/2}}. \tag{6–432}$$

And for large strains, $\dot{\varepsilon}t \gg 1$, this becomes

$$\frac{\delta}{\delta_0} = (\dot{\varepsilon}t)^{-1}. \tag{6–433}$$

Using Equation (6–433) we can determine how long it takes to thin the subducted oceanic crust ($\delta_0 = 6$ km) to a thickness $\delta = 10$ mm. For whole mantle convection we take $\dot{\varepsilon} = 50$ mm yr^{-1}/2886 km $= 5.5 \times 10^{-16}$ s^{-1} and find that $t = 3.5 \times 10^4$ Gyr. For layered mantle convection we take $\dot{\varepsilon} = 50$ mm yr^{-1}/660 km $= 2.4 \times 10^{-15}$ s^{-1} and find that $t = 7.9 \times 10^3$ Gyr. Clearly this type of mixing is very inefficient.

In the relatively complex flows associated with mantle convection, normal strains may also be important for mixing. An idealized flow that illustrates normal strain is the two-dimensional stagnation point flow illustrated in Figure 6–49. In this flow

$$u = \dot{\varepsilon}x \tag{6–434}$$

$$v = -\dot{\varepsilon}y, \tag{6–435}$$

where u is the x-component of velocity and v is the y-component of velocity. The strain rate $\dot{\varepsilon}$ is independent of time. In the upper half-space ($y > 0$) there is a uniform downward flow and in the lower half-space ($y < 0$) there is a uniform upward flow. These vertical flows converge on $y = 0$. There is a complementary divergent horizontal flow. In the right half-space

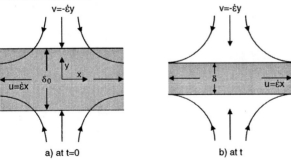

6–49 Illustration of the kinematic stretching of a passive layer of material in a uniform stagnation point flow. The converging vertical flow is given by Equation (6–435) and the diverging horizontal flow is given by Equation (6–434). (*a*) Initially at $t = 0$ the passive horizontal layer has a thickness δ_0. (*b*) At a subsequent time t the layer has been stretched and its thickness has been reduced to δ.

($x > 0$) there is a uniform divergent flow to the right. In the left half-space there is a uniform divergent flow to the left. This steady stagnation flow satisfies the governing continuity Equation (6–53) and force balance equations (6–67) and (6–68).

PROBLEM 6–39 Obtain the stream function corresponding to the two-dimensional stagnation point flow given in Equations (6–434) and (6–435). Show that this stream function satisfies the biharmonic Equation (6–74).

We again consider the thinning of a passive layer by the stagnation point flow. This passive layer initially occupies the region $-\delta_0/2 \le y \le \delta_0/2$. The deformation of this layer is uniform in x direction and the change of the layer thickness δ with time is given by

$$\frac{1}{2}\frac{d\delta}{dt} = v(\delta) = -\dot{\varepsilon}\delta. \tag{6–436}$$

Integration with the initial condition $\delta = \delta_0$ at $t = 0$ gives

$$\delta = \delta_0 e^{-2\dot{\varepsilon}t}. \tag{6–437}$$

The stagnation point flow stretches and thins the passive layer. With normal strains the passive layer thins exponentially with time. Normal strains are much more effective in layer thinning than shear strains.

Based on Equation (6–437), we can determine how long it takes for normal strains to thin the subducted oceanic crust ($\delta_0 = 6$ km) to a thickness of 10 mm. We again take $\dot{\varepsilon} = 5.5 \times 10^{-16}$ s^{-1} for whole mantle convection and find $t = 380$ Myr. For layered mantle convection we take $\dot{\varepsilon} = 2.4 \times 10^{-15}$ s^{-1} and find $t = 88$ Myr. Thus, normal strains can thin the oceanic lithosphere to thicknesses that can be homogenized by diffusion in reasonable lengths of geological time ($\approx 10^8$ yr).

The first question that arises in discussing the geological implications of the mixing hypothesis is whether there is direct observational evidence of an imperfectly mixed mantle. Allègre and Turcotte (1986) argued that the "marble cake structure" associated with imperfect mixing can be seen in high-temperature peridotites (also called orogenic lherzolite massifs), which represent samples of the Earth's mantle. Typical locations include Beni Bousera in Morocco, Rhonda in Spain,

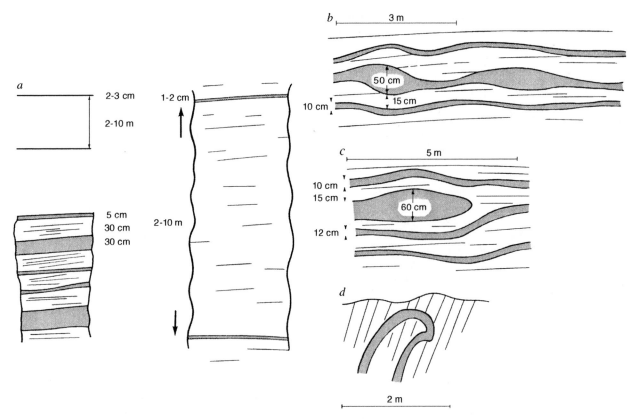

6-50 Occurrences of pyroxenite layers in the Beni Bousera high-temperature peridotite, Morocco. Grey, pyroxenite; white, lherzolite with foliation. (*a*) Occurrences in an outcrop with no folding; (*b*)–(*d*), occurrences with folding and boudinage.

and Lherz in France. These rocks consist primarily of depleted lherzolite. Embedded in this matrix are bands of pyroxenite comprising a few percent of the massif. Trace element studies of these bands indicate that they were originally basaltic in composition.

These characteristics led to the postulate that the bands are former samples of oceanic crust that have been subducted and deformed by convective shear before emplacement into their current locations. The bands range in thickness from a few meters to a few centimeters and some have been extensively folded. Essentially, no stripes are found with thicknesses of 1 cm or less, probably because stripes of this width have been destroyed by diffusive processes. According to this hypothesis, they have undergone 5 to 6 orders of magnitude of thinning from an initial thickness of 6 km. Figure 6–50 illustrates some examples.

Further evidence for the marble cake structure of the mantle comes from eclogitic xenoliths associated with basaltic volcanism and kimberlites. In some kimberlites, diamonds are found in the eclogite nodules. These "eclogitic" diamonds have been found to have carbon isotope ratios characteristic of sediments. A number of authors have suggested that subducted carbonates are one source of diamonds. Graphitized diamonds have been found in the pyroxenite bands of the Beni Bousera high-temperature peridotite in Morocco. These observations are completely consistent with the hypothesis that sediments are entrained in the subducted oceanic crust. During convective mixing in the deep interior some carbonate sediments are transformed to diamonds as the oceanic crust in which they are embedded is stretched and thinned.

REFERENCES

Allègre, C. J., and D. L. Turcotte (1986), Implications of a two-component marble-cake mantle, *Nature* **323**, 123–127.

Currie, J. B., H. W. Patnode, and R. P. Trump (1962), Development of folds in sedimentary strata, *Geol. Soc. Am. Bull.* **73**, 655–674.

COLLATERAL READING

Acheson, D. J., *Elementary Fluid Dynamics* (Oxford University Press, Oxford, 1990), 397 pages.

This is an intermediate level textbook that covers a broad range of topics in fluid dynamics. Topics of particular interest include the Navier–Stokes equations, very viscous flows, and fluid instabilities.

Batchelor, G. K., *An Introduction to Fluid Dynamics* (Cambridge University Press, London, 1967), 615 pages.

A modern classic on the fundamentals of fluid mechanics written for the student with a good foundation in applied mathematics and a familiarity with vector analysis and tensor notation. The topics covered include the physical properties of fluids, kinematics of flow fields, equations governing the motions of fluids, flows of a uniform incompressible viscous fluid, flows at large Reynolds numbers (boundary-layer theory), and the theory of irrotational flows. Discussions of low Reynolds number flows are particularly useful for geological applications. Many sections contain challenging exercises for the student.

Cathles, L. M., *The Viscosity of the Earth's Mantle* (Princeton University Press, Princeton, New Jersey, 1975), 386 pages.

A research monograph describing how the viscosity of the mantle has been inferred from the geological data on uplift and subsidence of the Earth's surface in response to the melting of the last great ice sheets that covered Canada, Fennoscandia, and Siberia and the addition of this water to the oceans. The book presents the basic theory necessary for modeling the Earth's isostatic adjustment to these Pleistocene load redistributions as that of a self-gravitating, viscoelastic sphere. The geological data are reviewed and applied to the models to infer a mantle with a nearly uniform viscosity of 10^{21} Pa s.

Chandrasekhar, S., *Hydrodynamic and Hydromagnetic Stability* (Oxford University Press, London, 1961), 652 pages.

A classic treatise on the stability of fluids subjected to adverse distributions of temperature and density, angular velocity, angular momentum, shear, gravity, and capillarity. The onset of thermal convection in fluid layers heated from below is treated with particular completeness and rigor. The required level of mathematical expertise is high. The student should be familiar with vector analysis, tensor notation, and partial differential equations.

Faber, T. E., *Fluid Dynamics for Physicists* (Cambridge University Press, Cambridge, 1995), 440 pages.

This is a comprehensive treatment of the fundamentals of fluid dynamics at a moderately advanced level. It covers a wide variety of topics with many applications.

Furbish, D. J., *Fluid Physics in Geology* (Oxford University Press, New York, 1997), 476 pages.

This is a relatively advanced level textbook that treats a wide variety of problems in fluid mechanics with geological applications. Both fluid flows and flows in porous media receive attention.

Johnson, A. M., *Physical Process in Geology* (Freeman, Cooper, San Francisco, 1970), 577 pages.

A basic textbook on mechanics and its applications to geological problems. Emphasis is placed on the development of flexure theory from the fundamental principles governing the behavior of elastic materials. Theories of plate and beam bending are applied to the geological problem of folding. Stresses and failure criteria are discussed in the context of understanding patterns of faults, joints, and dikes. A final chapter deals with the rheology and flow of ice, lava, and debris. There are exercises for the student and the mathematical level is not demanding given the nature of the subject.

Johnson, A. M., *Styles of Folding: Mechanics and Mechanisms of Folding of Natural Elastic Materials* (Elsevier, Amsterdam, 1977), 406 pages.

A specialized book detailing the research of the author and his colleagues on folding. Many of the chapters are individual research papers previously published in various journals. Chapter 1 reviews the literature on folding. Chapter 2 discusses the folding of bedded sandstones in Arches National Monument, Utah, and attempts to understand these folds using Biot's theory for buckling viscous layers. The next several chapters discuss more sophisticated attempts to model the observed characteristics of folds including their kink-like and chevron-like appearances.

Lamb, H., *Hydrodynamics*, 6th edition (Cambridge University Press, Cambridge, 1932), 738 pages.

A classic treatise on the fundamentals of hydrodynamics, the dynamics of inviscid fluids. The major subjects include the equations of motion, irrotational flows motions of a liquid in two and three dimensions, motions of solids through a liquid, vortex motions, tidal waves, surface waves, waves of expansion, effects of viscosity, and rotating liquid masses. The level of mathematical sophistication is high.

Langlois, W. E., *Slow Viscous Flow* (Macmillan Company, New York, 1964), 229 pages.

A book on low Reynolds number fluid dynamics for the applied mathematician. Familiarity with differential equations and multiple integrals is assumed. The first chapter introduces cartesian tensors that are used extensively throughout the text. Chapter 2 discusses the basic equations and boundary conditions of viscous flow theory. Remaining chapters treat exact solutions of the equations, pipe flow, flow past a sphere, plane flow, rotary flow, and lubrication theory.

Ramberg, H., *Gravity, Deformation and the Earth's Crust* (Academic Press, London, 1967), 214 pages.

A textbook on laboratory and theoretical modeling of geologic structures resulting from tectonic deformation of the crust. The role of gravity in tectonics is emphasized. Among the problems discussed are dome structures, buckling of horizontal layers, buckling of the crust, spreading and folding, buoyant rise of magma, subsiding bodies, and isostatic adjustment. An easily readable, relatively nonmathematical discussion of the dynamics of gravity tectonics.

Schlichting, H., *Boundary Layer Theory*, 6th edition (McGraw-Hill, New York, 1968), 747 pages.

A classic textbook on boundary-layer phenomena written principally for engineers. The first part of the book discusses the fundamentals of viscous fluid flow and introduces the concept of a boundary layer. Exact solutions of the Navier–Stokes equations of motion are presented for flows in pipes and channels and other geometries. A chapter is devoted to very slow motions and introduces the hydrodynamic theory of lubrication and Hele–Shaw flow, both topics having applications in geophysics. A number of chapters discuss boundary-layer theory in great detail, including thermal boundary layers in forced and natural flows. Other sections deal with turbulence and turbulent boundary layers. A reasonable degree of mathematical sophistication is required.

Tritton, D. J., *Physical Fluid Dynamics*, 2nd edition (Clarendon Press, Oxford, 1988), 519 pages.

This is an outstanding introductory textbook in fluid mechanics. The author is a geophysicist, therefore many of the topics covered are directly relevant to geodynamics. Pipe and channel flows, flows past circular cylinders, and free convective flows are treated. The basic equations are introduced with concepts of similarity. Viscous flows, stratified flows, flows in rotating fluids, turbulence, double diffusive convection, and dynamical chaos are considered in depth.

White, F. M., *Viscous Fluid Flow* (McGraw-Hill, New York, 1974), 725 pages.

A textbook for senior undergraduate or first-year graduate students in engineering dealing with the analysis of viscous flows. A knowledge of basic fluid mechanics, vector notation, and differential equations is assumed. The first two chapters cover the properties of fluids and the basic equations governing their behavior. Chapters 3 and 4 discuss methods of analysis of laminar flows. Chapter 5 treats the stability of laminar flows and their transition to turbulence. Chapter 6 deals with incompressible turbulent flows, while the final chapter covers compressible laminar and turbulent boundary layers. Problems are given at the end of each chapter.

SEVEN

Rock Rheology

7-1 Introduction

At atmospheric pressure and room temperature most rocks are brittle; that is, they behave nearly elastically until they fail by fracture. Cracks or fractures in rock along which there has been little or no relative displacement are known as *joints*. They occur on all scales in both sedimentary and igneous rocks. Joints are commonly found in sets defining parallel or intersecting patterns of failure related to local stress orientations. The breakdown of surface rocks by erosion and weathering is often controlled by systems of joints along which the rocks are particularly weak and susceptible to disintegration and removal. These processes in turn enhance the visibility of the jointing. Igneous rocks often develop joints as a result of the thermal stresses associated with cooling and contraction. Columnar jointing in basaltic lava flows (Figure 7–1) and parallel jointing in granitic rocks (Figure 7–2) are examples.

Faults are fractures along which there has been relative displacement. Faults also occur on all scales; examples of faults have already been given in Figures 1–58 and 4–34*b* and another example is given in Figure 7–3. The mechanical aspects of faulting are discussed in the next chapter.

Although fracture is important in shallow crustal rock at low temperatures and pressures, there are many circumstances in which rock behaves as a ductile material. In determining the transition from brittle to ductile behavior, pressure, temperature, and strain rate are important. If the confining pressure of rock is near the brittle strength of the rock, a transition from brittle to ductile behavior will occur. To model this behavior of crustal and mantle rocks, it is often appropriate to use an idealized *elastic–perfectly plastic rheology*. An elastic–perfectly plastic material exhibits a linear elastic behavior until a yield stress is reached. The material can then be deformed plastically an unlimited amount at this stress.

At temperatures that are a significant fraction of the melt temperature the atoms and dislocations in a crystalline solid become sufficiently mobile to result in creep when the solid is subjected to deviatoric stresses. At very low stresses diffusion processes dominate, and the crystalline solid behaves as a Newtonian fluid with a viscosity that depends exponentially on pressure and the inverse absolute temperature. At higher stresses the motion of dislocations becomes the dominant creep process resulting in a non-Newtonian or nonlinear fluid behavior that also has an exponential pressure and inverse absolute temperature dependence. Mantle convection and continental drift are attributed to these thermally activated creep processes.

The exponential dependence of the rheology on the inverse absolute temperature is particularly important in understanding the role of mantle convection in transporting heat. The temperature dependence of the rheology acts as a thermostat to regulate the mantle temperature. Any tendency of the mean mantle temperature to increase is offset by an associated reduction in mantle viscosity, an increase in convective vigor, and a more efficient outward transport of heat. Similarly, a decrease in mantle temperature tends to increase mantle viscosity, reduce convective flow velocities, and decrease the rate of heat transfer. As a result of the sensitive feedback between mean mantle temperature and rheology, relatively small changes in temperature can produce large changes in heat flux, and the temperature is consequently buffered at a nearly constant value.

Creep processes are also important in the lower lithosphere, where they can relax elastic stresses. Such behavior can be modeled with a rheological law that combines linear elasticity and linear (Newtonian) or nonlinear viscosity. A material that behaves both elastically and viscously is known as a *viscoelastic medium*. Viscoelastic relaxation can be used to determine the thickness of the elastic upper part of the lithosphere.

Folding is evidence that crustal rocks also exhibit ductile behavior under stress. Examples of folding have been given in Figure 6–28. *Pressure solution creep* is a mechanism that can account for the ductility of crustal

7–1 Columnar jointing in a basalt due to thermal contraction in the Devil's Postpile National Monument, California (University of Colorado, Boulder).

7–2 Ordovician diabase sill with cooling joints, Rodeo, San Juan Province, Argentina (photograph courtesy of Richard Allmendinger).

rocks at relatively low temperatures and pressures. The process involves the dissolution of minerals in regions of high pressure and their precipitation in regions of low pressure. As a result creep of the rock occurs.

7–2 Elasticity

At low stress levels and relatively high strain rates, rock behaves elastically. The linear relations between components of stress and strain associated with elastic deformation have been given in Equations (3–1) to (3–3) or (3–4) to (3–6). The elastic behavior of a crystalline solid arises from the interatomic forces maintaining each atom in its lattice position. These forces resist any attempt to move the atoms farther apart or closer together. If the crystalline lattice is compressed, the interatomic forces resist the compression; if the crys-

talline lattice is placed under tension, the interatomic forces resist the expansion. This situation is illustrated in Figure 7–4, which shows the energy of a crystal lattice U as a function of the separation of the atoms b. When the atoms are infinitely far apart, the energy of the lattice is zero. Long-range attractive forces – for example, the electrostatic Coulomb forces between ions of opposite sign – act to reduce the separation between atoms; they cause the energy of the lattice to decrease with decreasing lattice spacing. The repulsive forces that come into play at short range prevent the atoms from approaching too closely; they cause the lattice energy to increase with decreasing lattice spacing. The total energy of the lattice is the sum of these two contributions. It exhibits a minimum value U_0 when the lattice spacing has its equilibrium value b_0. There are no forces on the atoms of the lattice when they occupy their equilibrium positions. The energy $-U_0$ is known as the *binding or cohesive energy* of the lattice. It is the energy required to break up the lattice and disperse its atoms to infinity.

The compressibility β of a solid provides a direct measure of these interatomic forces because it gives

7-3 Offsets of trees in an orchard caused by the $m = 7.5$ Guatemala City earthquake February 4, 1976. This earthquake resulted in the deaths of 23,000 people (U.S. Geological Survey).

the pressure required to change the volume, or lattice spacing, of the solid. In the case of ionic solids such as sodium chloride (NaCl) we can derive a simple formula connecting lattice spacing, lattice energy, and compressibility because of the relatively simple lattice structures of such solids and the known character of the attractive energy resulting from ionic bonding. Figure 7–5 shows the configuration of the NaCl lattice. Every ion is surrounded by six nearest neighbors of opposite sign; this is the *coordination number* for the NaCl lattice. The nearest neighbor distance between Na^+ and Cl^- ions is b. The atomic volume, that is, the volume per atom, is b^3. Since there are two atoms per

molecule in NaCl, the molecular volume V, or volume per molecule, is

$$V = 2b^3. \tag{7–1}$$

It is standard practice to characterize a lattice by its molecular or molar properties.

To derive the formula connecting β, b_0, and U_0 for an ionic solid, we recall the definition of compressibility provided in Equation (3–50). Compressibility is the ratio of the fractional change in volume dV/V caused by a change in pressure dp to the pressure change, that is,

$$\beta = \frac{1}{V} \frac{dV}{dp}. \tag{7–2}$$

The reciprocal of the compressibility is the bulk modulus K

$$K \equiv \frac{1}{\beta} = V \frac{dp}{dV}. \tag{7–3}$$

Recall that we consider a decrease in volume to be a positive quantity. Thus with $dp > 0$, dV is positive and β is also a positive quantity. The work done by

7-4 Lattice energy U as a function of lattice spacing b.

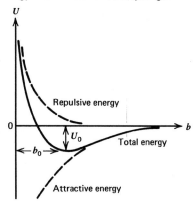

7-5 Lattice structure of NaCl.

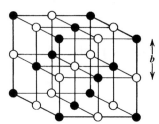

the pressure forces in compressing an elastic solid an amount dV is pdV; this work is stored as elastic strain energy dU in the crystal lattice, that is,

$$pdV = dU \qquad (7\text{-}4)$$

or

$$p = \frac{dU}{dV}. \qquad (7\text{-}5)$$

Upon substituting this expression for the pressure into Equation (7–3), we obtain

$$\frac{1}{\beta} = V\frac{d^2U}{dV^2}, \qquad (7\text{-}6)$$

where U is the lattice energy per molecule just as V is the molecular volume. By taking the derivative of Equation (7–1) with respect to the spacing b, we obtain

$$dV = 6b^2 db \qquad (7\text{-}7)$$

and

$$\frac{1}{\beta} = \frac{1}{18b}\frac{d^2U}{db^2}. \qquad (7\text{-}8)$$

The compressibility β_0 is the value of β at $p = 0$ when the equilibrium lattice spacing is b_0. Thus we have

$$\frac{1}{\beta_0} = \frac{1}{18b_0}\left(\frac{d^2U}{db^2}\right)_{b=b_0}. \qquad (7\text{-}9)$$

The equilibrium bulk modulus is directly proportional to the second derivative of the lattice energy per molecule with respect to the lattice spacing.

We need a model of the lattice potential energy in order to calculate d^2U/db^2. For an ionic lattice such as NaCl we can write

$$U = \frac{-z^2e^2A}{4\pi\varepsilon_0 b} + Be^{-b/s}, \qquad (7\text{-}10)$$

where the first term is the attractive *Coulomb energy* of the lattice, the second term is an approximate representation of the short-range repulsive potential, z is the number of electron charges e on each ion, ε_0 is the permittivity of free space, and A, B, and s are constants. The charge on an electron e is 1.602×10^{-19} coulomb and $\varepsilon_0 = 8.85 \times 10^{-12}$ farad m^{-1}. The constant A, known as the *Madelung constant*, depends only on lattice structure and can be calculated for any ionic

lattice. For NaCl, A is 1.7476 and $z = 1$. The calculation of A is straightforward. The Coulomb energy of an Na$^+$ ion and a Cl$^-$ ion separated by a distance b is $-e^2/4\pi\varepsilon_0 b$. The Coulomb energy of any ion pair is given by a similar formula employing the appropriate separation distance and using a plus sign for ions of the same sign. Since the regular geometrical structure of the lattice allows any separation distance to be calculated in terms of b, the Coulomb energy of any ion pair can be written as a positive or negative multiple of $-e^2/4\pi\varepsilon_0 b$. Thus, the total energy obtained by summing the Coulomb potential energies between any one ion and all other ions in the lattice can be written in the form $-e^2 A/4\pi\varepsilon_0 b$.

The second derivative of Equation (7–10) with respect to lattice spacing b gives

$$\left(\frac{d^2U}{db^2}\right)_{b=b_0} = \frac{-2z^2e^2A}{4\pi\varepsilon_0}\frac{1}{b_0^3} + \frac{B}{s^2}e^{-b_0/s}. \qquad (7\text{-}11)$$

The constants B and s can be eliminated from this expression by noting that dU/db is zero at the equilibrium spacing $b = b_0$

$$\left(\frac{dU}{db}\right)_{b=b_0} = 0 = \frac{z^2e^2A}{4\pi\varepsilon_0}\frac{1}{b_0^2} - \frac{B}{s}e^{-b_0/s} \qquad (7\text{-}12)$$

and $U = U_0$, the lattice potential energy per molecule, at $b = b_0$

$$U_0 = \frac{-z^2e^2A}{4\pi\varepsilon_0 b_0} + Be^{-b_0/s}. \qquad (7\text{-}13)$$

Equation (7–13) yields

$$Be^{-b_0/s} = U_0 + \frac{z^2e^2A}{4\pi\varepsilon_0 b_0}, \qquad (7\text{-}14)$$

which, together with Equation (7–12), gives

$$s = \frac{4\pi\varepsilon_0 b_0^2}{z^2e^2A}\left(U_0 + \frac{z^2e^2A}{4\pi\varepsilon_0 b_0}\right). \qquad (7\text{-}15)$$

By combining Equations (7–9), (7–11), (7–14), and (7–15), we obtain

$$\beta_0 = -18b_0^4\left(\frac{4\pi\varepsilon_0}{z^2e^2A}\right)\frac{\left(U_0 + \frac{z^2e^2A}{4\pi\varepsilon_0 b_0}\right)}{\left(2U_0 + \frac{z^2e^2A}{4\pi\varepsilon_0 b_0}\right)}, \qquad (7\text{-}16)$$

which, upon rearrangement, gives

$$-U_0 = \frac{9V_0 + \beta_0 \left(\frac{z^2 e^2 A}{4\pi \varepsilon_0 b_0}\right)}{2\beta_0 + 9V_0 \left(\frac{4\pi \varepsilon_0 b_0}{z^2 e^2 A}\right)}, \qquad (7\text{–}17)$$

where, from Equation (7–1), V_0 is the equilibrium molecular volume $2b_0^3$.

Equation (7–17) provides a means of calculating the binding energy per molecule in an ionic solid from basic information on lattice structure and measurements of density and compressibility. Because the binding energy is the energy required to disperse the lattice, we can compare the prediction of this equation with independent thermodynamic determinations of $-U_0$. For NaCl we have already noted that $A = 1.7476$ and $z = 1$. In addition, $\beta_0 = 4.26 \times 10^{-11}$ Pa^{-1}. Since the density of NaCl is 2163 kg m^{-3} and its molecular weight is 58.448, its molar volume is 2.702×10^{-5} m^3. (A mole of a substance has a mass equal to its molecular weight times 10^{-3} kg.) Because a mole of NaCl contains Avogadro's number ($N_0 = 6.023 \times 10^{23}$) of molecules, its molecular volume is $V_0 = 4.486 \times 10^{-29}$ m^3. With V_0 given by $2b_0^3$ we find $b_0 = 2.82 \times 10^{-10}$ m. When these values of β_0, V_0, b_0, z, and A are substituted into Equation (7–17), we obtain $-U_0 = 1.262 \times 10^{-18}$ J molecule^{-1} = 7.6×10^5 J mole^{-1} = 760 kJ mole^{-1}. Thermodynamic data give 773 kJ mole^{-1} for the binding energy of NaCl.

There is good agreement between the calculated and measured values of binding energy for ionic crystals because the forces between the ions in the lattice are short-range repulsive forces and Coulomb forces. However, other forces not accounted for in this theory are important in the lattices of oxides and silicates. These forces include van der Waals forces, covalent bonds, and dipole and higher order multipole forces. The forces associated with covalent bonding are generally the most important of the nonionic forces. Theoretical estimates of lattice binding energy must include the energy in these bonds.

PROBLEM 7–1 Compute the binding energy of CsCl. Use $\beta_0 = 5.95 \times 10^{-11}$ Pa^{-1}, $\rho_0 = 3988$ kg m^{-3}, and $A = 1.7627$. The molecular weight of CsCl is 168.36, and thermodynamic data give $-U_0 = 660$ kJ mole^{-1}.

TABLE 7–1 **Elastic and Structural Data for Computing the Lattice Binding Energies of Several Oxides**

Crystal	V (10^{-30} m^3)	A	β (10^{-12} Pa^{-1})
BeO	13.77	1.886	4.636
MgO	18.67	1.748	6.254
CaO	27.83	1.748	9.532
ZnO	23.74	1.905	7.199

PROBLEM 7–2 Calculate the binding energies of the oxides BeO, MgO, CaO, and ZnO using the data in Table 7–1. Account only for the Coulomb and repulsive energies.

PROBLEM 7–3 A theoretical estimate of the strength of a crystalline solid is its binding energy per unit volume. Evaluate the strength of forsterite if its binding energy is 10^3 kJ mole^{-1} and its mean atomic volume is 6.26×10^{-6} m^3 mole^{-1}. The presence of grain boundaries and dislocations weakens a crystalline solid considerably below its theoretical strength.

The atoms of a crystalline solid do not, of course, occupy fixed lattice positions. The lattice sites are the mean locations of the atoms, which oscillate about their equilibrium positions. The energy in these oscillatory motions is the internal thermal energy of the solid. The temperature of a crystalline solid is a measure of its internal energy and the vigor of the oscillations of its atoms. The oscillations of the atoms in a crystalline solid are a consequence of the interatomic forces tending to bind the atoms to their lattice sites. In their equilibrium positions the atoms experience no net force. However, the displacement of an atom from its lattice site results in a force on the atom tending to restore it to its equilibrium position. To a first approximation, that is, for small excursions of an atom from its equilibrium location, the restoring force is linearly proportional to the displacement of the atom and results in simple harmonic motion of the atom about its equilibrium position. As for the thermal energy of a lattice, consider the atoms as being interconnected by elastic springs with linear restoring forces. The lattice thermal energy resides in the kinetic energy of the oscillating atoms and the elastic strain energy or potential energy of the springs.

The equation of simple harmonic motion of an atom about its equilibrium position is, in one dimension,

$$m\frac{d^2x}{dt^2} + \bar{k}x = 0, \tag{7-18}$$

where m is the mass of the atom, x is its displacement from the equilibrium position ($x = 0$), and $\bar{k}$ is the "spring constant," a measure of the strength of the interatomic restoring forces. The force on the atom is $-\bar{k}x$; it always points in the direction of the equilibrium position $x = 0$. A solution for the position and velocity ($v = dx/dt$) of the atom is

$$x = x_0 \sin\left(\frac{2\pi t}{\tau}\right) \tag{7-19}$$

$$v = v_0 \cos\left(\frac{2\pi t}{\tau}\right), \tag{7-20}$$

where the period τ of the oscillation is given by

$$\tau = 2\pi\left(\frac{m}{\bar{k}}\right)^{1/2}, \tag{7-21}$$

and the maximum velocity v_0 and maximum displacement x_0 are related according to

$$v_0 = \frac{2\pi}{\tau}x_0 = \left(\frac{\bar{k}}{m}\right)^{1/2}x_0. \tag{7-22}$$

The period of the oscillation τ is known as the *Einstein period*; its reciprocal is the *Einstein frequency*.

The instantaneous kinetic energy of the atom is $\frac{1}{2}mv^2$. The kinetic energy varies between 0, when the atom is at the farthest point of its excursion, and $\frac{1}{2}mv_0^2$ when the atom is passing through its equilibrium position. The mean kinetic energy of the atom during a single oscillation period is

$$\frac{1}{2}m\overline{v^2} = \frac{m}{2\tau}\int_0^\tau v^2\,dt = \frac{mv_0^2}{2\tau}\int_0^\tau \cos^2\left(\frac{2\pi t}{\tau}\right)dt$$
$$= \frac{mv_0^2}{4} = \frac{\bar{k}x_0^2}{4}. \tag{7-23}$$

The mean kinetic energy is one-half the maximum kinetic energy. To obtain the potential energy of the atom, we need only integrate the product of the force acting on the atom $-\bar{k}x$ with the infinitesimal displacement dx from the equilibrium position to any location x. The

instantaneous potential energy of the atom is

$$-\int_0^x (-\bar{k}x)\,dx = \frac{1}{2}\bar{k}x^2, \tag{7-24}$$

where the minus sign in front of the integral accounts for the fact that the force and the displacement are in opposite directions. The potential energy varies between 0, when the atom is passing through its equilibrium position, and $\frac{1}{2}\bar{k}x_0^2$, when the atom is farthest away from the origin. The mean potential energy of the atom during an oscillatory cycle is

$$\frac{1}{2}\bar{k}\,\overline{x^2} = \frac{\bar{k}}{2\tau}\int_0^\tau x^2\,dt$$
$$= \frac{\bar{k}x_0^2}{2\tau}\int_0^\tau \sin^2\left(\frac{2\pi t}{\tau}\right)dt = \frac{\bar{k}x_0^2}{4}. \tag{7-25}$$

The mean potential energy is one-half the maximum potential energy. A comparison of Equations (7-23) and (7-25) shows that the mean kinetic and potential energies of the one-dimensional harmonic oscillator are equal. The total instantaneous energy of the atom is

$$\frac{1}{2}mv^2 + \frac{1}{2}\bar{k}x^2 = \frac{1}{2}mv_0^2\cos^2\left(\frac{2\pi t}{\tau}\right)$$
$$+ \frac{1}{2}\bar{k}x_0^2\sin^2\left(\frac{2\pi t}{\tau}\right)$$
$$= \frac{1}{2}\bar{k}x_0^2\left[\cos^2\left(\frac{2\pi t}{\tau}\right) + \sin^2\left(\frac{2\pi t}{\tau}\right)\right]$$
$$= \frac{1}{2}\bar{k}x_0^2 = \frac{1}{2}mv_0^2. \tag{7-26}$$

Equation (7-26) is a statement of conservation of energy. At any point in its oscillation, the energy of the atom is divided between kinetic and potential energies, but the sum of these energies is always a constant.

In an actual lattice, an individual atom can oscillate in any of three basic orthogonal directions. Kinetic and potential energies can be associated with vibrations in each of the three directions. There are thus six contributions to the total energy of an atom; the atom is said to have six degrees of freedom. Not all the atoms in the lattice oscillate with the same amplitude. There is a spectrum of vibrational amplitudes or energies, as we will discuss in more detail in the next section. However, according to the principle of *equipartition of energy*, when an average is taken over all the atoms of a lattice, the energies associated with the six degrees of freedom are all equal and each is given by $\frac{1}{2}kT$, where k is Boltzmann's constant ($k = 1.3806 \times 10^{-23}$ J K^{-1}) and T is the absolute temperature. A proof of the validity of the equipartition principle requires the solution of the Schrödinger equation for the quantum mechanical behavior of atoms.

The internal energy of the solid per unit mass e is therefore given by

$$e = \frac{6\left(\frac{1}{2}kT\right)}{m} = \frac{3kT}{m} = \frac{3(kN_0)T}{(mN_0)} = \frac{3RT}{M_a}, \quad (7\text{--}27)$$

where $R = N_0 k$ is the universal gas constant ($R = 8.314510$ J mole^{-1} K^{-1}) and $M_a = mN_0$ is the mean mass of Avogadro's number of atoms in the crystal, that is, the mean atomic weight. The specific heat of the solid c is defined to be the change in internal energy with temperature

$$c = \frac{de}{dT}. \quad (7\text{--}28)$$

From Equations (7–27) and (7–28) the specific heat c of a crystalline solid is

$$c = 3\frac{R}{M_a}. \quad (7\text{--}29)$$

This is the *law of Dulong and Petit*; it is a good approximation for all crystalline solids at moderate temperatures.

As an example, consider forsterite, Mg_2SiO_4, whose molecular weight is 140.73. The mass of a mole of forsterite is 140.73×10^{-3} kg. Because a mole of forsterite contains $7N_0$ atoms, the mean weight of N_0 atoms is 20.1×10^{-3} kg. Thus M_a equals 20.1×10^{-3} kg mole^{-1}, and c, according to Equation (7–29), is 1.24×10^3 J kg^{-1} K^{-1}. The measured value of c at standard conditions of temperature and pressure is 840 J kg^{-1} K^{-1}.

PROBLEM 7–4 According to the law of Dulong and Petit the specific heats of solids should differ only because of differences in M_a. Calculate M_a and c for $MgSiO_3$ and MgO. The measured values of c at standard conditions of temperature and pressure are 815 J kg^{-1} K^{-1} for $MgSiO_3$ and 924 J kg^{-1} K^{-1} for MgO.

PROBLEM 7–5 Obtain an order of magnitude estimate for the spring constant $\bar{k}$ associated with the interatomic forces in a silicate crystal such as forsterite by assuming $\bar{k} \sim Eb$, where E is Young's modulus and b is the average interatomic spacing. Young's modulus for forsterite is 1.5×10^{11} Pa. Obtain a value for b by assuming b^3 is the mean atomic volume. The density of forsterite is 3200 kg m^{-3}. Estimate the maximum amplitude of vibration of an atom in a forsterite crystal at a temperature of 300 K. How does it compare with the mean interatomic spacing? What is the Einstein frequency at this temperature? The spring constant may also be estimated from the compressibility of forsterite using $\bar{k} \sim 3b/\beta$, where the factor of 3 arises from the relation between fractional volume changes and fractional changes in length. How does this estimate of $\bar{k}$ compare with the previous one? The compressibility of forsterite is 0.8×10^{-11} Pa^{-1}.

If the atoms of a crystalline solid were perfect harmonic oscillators, the amplitudes of their vibrations would increase with increasing temperature, but the mean distances between the atoms would remain constant; that is, there would be no change in volume with temperature. The thermal expansion of a crystalline solid is thus a direct consequence of the anharmonicity in the vibrations of its atoms. The anharmonicity of the thermal motions is, in turn, a result of the asymmetry of the lattice potential energy about its minimum value. A qualitative understanding of the phenomenon of thermal expansion can be obtained from Figure 7–6 which shows the potential energy of a lattice U as a function of the spacing b of its atoms, as in Figure 7–4.

7–6 The asymmetry of the lattice potential energy about its minimum results in the expansion and contraction of solids with changes in temperature.

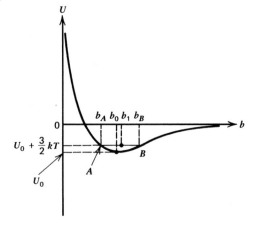

The minimum in the potential energy curve gives the equilibrium lattice spacing b_0 at zero temperature. At any nonzero value of temperature, the thermal energy $3kT$ is partitioned equally between the kinetic and potential energies of the atoms. Thus the potential energy of the lattice exceeds the zero temperature lattice potential energy U_0 by $3kT/2$. Accordingly, the lattice spacing can lie between b_A and b_B, as shown in Figure 7–6, where b_A and b_B are the values of the lattice spacing at the intersections of the lattice potential energy curve with the energy $U_0 + \frac{3}{2}kT$. The mean lattice spacing b_1 is

$$b_1 = \tfrac{1}{2}(b_A + b_B). \tag{7–30}$$

Clearly, because the potential energy curve is skewed about its minimum, b_1 exceeds b_0; that is, the lattice expands with increasing temperature. If the lattice potential energy curve were symmetric about its minimum, b_1 would equal b_0 and there would be no change in mean lattice spacing, or volume of the crystalline solid, with temperature.

One can derive a simple expression for the mean lattice spacing in terms of the skewness of the lattice potential energy function. Sufficiently near the minimum in the U versus b curve, the lattice energy can be written as a power series

$$U = U_0 + a\delta^2 - f\delta^3, \tag{7–31}$$

where

$$\delta = b - b_0. \tag{7–32}$$

A positive value of f ensures that the energy rises more steeply to the left of the minimum than it does to the right of this point. The intersections of the potential energy function with the energy $U_0 + \frac{3}{2}kT$ are obtained by substituting this value of the energy in Equation (7–31) with the result

$$\tfrac{3}{2}kT = a\delta^2 - f\delta^3. \tag{7–33}$$

The solutions of Equation (7–33), correct to first order in f (the asymmetry in the potential function is assumed to be small), are

$$\delta_A = \left(\frac{3kT}{2a}\right)^{1/2} + f\left(\frac{3kT}{4a^2}\right) \tag{7–34}$$

$$\delta_B = -\left(\frac{3kT}{2a}\right)^{1/2} + f\left(\frac{3kT}{4a^2}\right). \tag{7–35}$$

The mean lattice spacing is thus

$$\begin{aligned} b_1 &= \frac{1}{2}(b_A + b_B) = \frac{1}{2}(b_A - b_0 + b_B - b_0) + b_0 \\ &= \frac{1}{2}(\delta_A + \delta_B) + b_0 = \frac{3fkT}{4a^2} + b_0. \end{aligned} \tag{7–36}$$

The lattice expands with increasing temperature if f is a nonzero positive quantity, as is the case for the potential in Figure 7–6. The linear coefficient of thermal expansion,

$$\alpha_l = \frac{1}{b_0}\left(\frac{\partial b_1}{\partial T}\right), \tag{7–37}$$

is one-third of the volumetric thermal expansivity α_v – see Equation (4–175) – so that

$$\alpha_v = \frac{3}{b_0}\frac{\partial b_1}{\partial T}. \tag{7–38}$$

By substituting Equation (7–36) into Equation (7–38), we obtain

$$\alpha_v = \frac{9kf}{4a^2 b_0}. \tag{7–39}$$

The thermal expansion coefficient of a crystalline solid is directly proportional to the skewness of the lattice potential energy function about its minimum; to this order of approximation it is also independent of temperature.

The coefficients a and f in the expansion of U about its equilibrium value U_0 can be determined for an ionic solid from the exact expression for the lattice potential energy – Equation (7–10). A Taylor series expansion of the right side of Equation (7–10) in powers of $(b - b_0)$ yields, with the help of Equation (7–13),

$$\begin{aligned} U = U_0 &+ (b - b_0)^2\left\{-\frac{z^2 e^2 A}{4\pi\varepsilon_0 b_0^3} + \frac{Be^{-b_0/s}}{2s^2}\right\} \\ &+ (b - b_0)^3\left\{\frac{z^2 e^2 A}{4\pi\varepsilon_0 b_0^4} - \frac{Be^{-b_0/s}}{6s^3}\right\}. \end{aligned} \tag{7–40}$$

A comparison of Equations (7–31), (7–32), and (7–40) enables us to identify a and f as

$$a = -\frac{z^2 e^2 A}{4\pi\varepsilon_0 b_0^3} + \frac{Be^{-b_0/s}}{2s^2} \tag{7–41}$$

$$f = -\frac{z^2 e^2 A}{4\pi\varepsilon_0 b_0^4} + \frac{Be^{-b_0/s}}{6s^3}. \tag{7–42}$$

By substituting for $Be^{-b_0/s}$ and s from Equations (7–14) and (7–15), and by using (7–16) for β_0 and (7–17) for U_0, we find

$$a = \frac{9b_0}{\beta_0} \tag{7–43}$$

$$f = \frac{z^2 e^2 A}{24\pi \varepsilon_0 b_0^4 \beta_0^2}\left[\left\{2\beta_0 + 9V_0\left(\frac{4\pi\varepsilon_0 b_0}{z^2 e^2 A}\right)\right\}^2 - 6\beta_0^2\right]. \tag{7–44}$$

Upon substituting these expressions for a and f into Equation (7–39), we can write the equation for the thermal expansivity of an ionic crystalline solid as

$$\alpha_v = \frac{z^2 e^2 kA}{216\pi \varepsilon_0 b_0 V_0^2}\left[\left\{2\beta_0 + 9V_0\left(\frac{4\pi\varepsilon_0 b_0}{z^2 e^2 A}\right)\right\}^2 - 6\beta_0^2\right]. \tag{7–45}$$

For NaCl we have seen that $z = 1$, $A = 1.7476$, $b_0 = 2.82 \times 10^{-10}$ m, $V_0 = 4.486 \times 10^{-29}$ m^3, and $\beta_0 = 4.26 \times 10^{-11}$ Pa^{-1}. When these values are used in Equation (7–45), we obtain $\alpha_v = 2.26 \times 10^{-5}$ K^{-1}. This is an estimate of the zero temperature value of thermal expansivity. The measured value of the coefficient of thermal expansion of NaCl at 300 K is about 5 times larger; at 100 K the measured value of α_v is only 3 times larger.

PROBLEM 7–6 Calculate the thermal expansivity of CsCl from the data contained in Problem 7–1.

The elastic behavior of a crystalline solid is limited to relatively low temperatures T and pressures p. At higher values of T and p other microscopic physical processes occur that allow the solid to creep. We now proceed to discuss these other deformation mechanisms.

7–3 Diffusion Creep

At very low stress levels, creep deformation in rocks occurs predominantly by diffusion. *Diffusion creep* results from the diffusion of atoms through the interiors of crystal grains when the grains are subjected to stress. As a result of the diffusion, the grains deform leading to rock strain. We will derive an expression for the dependence of strain rate on stress using a simple model for a crystalline solid. Our result will show that diffusion creep leads to a Newtonian fluid behavior. A viscosity for the solid also will be derived. Diffusion can also occur along grain boundaries. In some cases this process dominates over diffusion through the interiors of grains. When grain boundary diffusion controls crystal deformation, the process is known as *Coble creep*.

In the previous section we described how the mean kinetic energy associated with the thermal motions of atoms in a crystal lattice is $\frac{3}{2}kT$ per atom, corresponding to $\frac{1}{2}kT$ for vibrations in each of the three orthogonal directions. Individual atoms have kinetic energies that are distributed about this mean; some atoms have higher kinetic energies than the mean, and some have lower energies. The number of atoms per unit volume dn_ϕ with kinetic energies between ϕ and $\phi + d\phi$ is given by

$$dn_\phi = \frac{2\pi n\phi^{1/2}}{(\pi kT)^{3/2}} \exp\left(\frac{-\phi}{kT}\right) d\phi, \tag{7–46}$$

where n is the total number of atoms per unit volume. This is the *Maxwell–Boltzmann distribution* of energy; it has a wide application to all forms of matter.

PROBLEM 7–7 Verify that the total number of atoms per unit volume in the Maxwell–Boltzmann distribution is n by integrating Equation (7–46) over all energies between 0 and ∞.

PROBLEM 7–8 The average kinetic energy $\bar{\phi}$ of an atom in a crystalline solid is given by

$$\bar{\phi} = \frac{1}{n}\int_0^\infty \phi \, dn_\phi. \tag{7–47}$$

Verify that the Maxwell–Boltzmann distribution gives $\bar{\phi} = \frac{3}{2}kT$ by carrying out the integration.

Each atom in the crystalline solid is bound to its lattice site by the interatomic forces discussed in the previous section. An atom is effectively in a potential well; if its kinetic energy exceeds the potential energy barrier ϕ_0 confining it to its lattice position, it is able to escape the site and move elsewhere in the lattice. In the context of the one-dimensional simple harmonic oscillator model, an atom whose vibrational amplitude x_0 matches the lattice spacing b must certainly be considered to have escaped from its site in the lattice. The barrier potential energy ϕ_0 can thus be equated with the maximum potential energy $\frac{1}{2}\bar{k}b^2$ of such an atom

$$\phi_0 = \frac{1}{2}\bar{k}b^2. \tag{7–48}$$

The spring constant of the interatomic forces is a measure of the barrier potential energy. The Einstein period of an atom with sufficient energy to escape its lattice site can be related to ϕ_0 by solving Equation (7–48) for $\bar{k}$ and substituting into (7–21). The result is

$$\tau = 2\pi b \left(\frac{m}{2\phi_0} \right)^{1/2}. \tag{7–49}$$

The number of atoms per unit volume in a crystal that have kinetic energies greater than ϕ_0, n_{ϕ_0}, can be found by integrating Equation (7–46) from $\phi = \phi_0$ to $\phi = \infty$

$$n_{\phi_0} = \int_{\phi_0}^{\infty} dn_\phi = \frac{2\pi n}{(\pi kT)^{3/2}} \int_{\phi_0}^{\infty} \phi^{1/2} \exp\left(\frac{-\phi}{kT} \right) d\phi. \tag{7–50}$$

By introducing $s = (\phi/kT)^{1/2}$, we can rewrite this equation as

$$n_{\phi_0} = \frac{4n}{\pi^{1/2}} \int_{(\frac{\phi_0}{kT})^{1/2}}^{\infty} s^2 e^{-s^2} ds, \tag{7–51}$$

which can be integrated by parts to give

$$\begin{aligned} n_{\phi_0} &= \frac{2n}{\pi^{1/2}} \left\{ (-se^{-s^2})_{(\frac{\phi_0}{kT})^{1/2}}^{\infty} + \int_{(\frac{\phi_0}{kT})^{1/2}}^{\infty} ds\, e^{-s^2} \right\} \\ &= n \left\{ 2 \left(\frac{\phi_0}{\pi kT} \right)^{1/2} e^{-\phi_0/kT} + \mathrm{erfc}\left(\frac{\phi_0}{kT} \right)^{1/2} \right\}. \end{aligned} \tag{7–52}$$

The ratio n_{ϕ_0}/n is the fraction of the atoms in a crystalline solid that have sufficient energy to jump the potential barrier between lattice sites. Because most lattice sites are occupied, the potential barrier ϕ_0 must greatly exceed the average kinetic energy of the atoms $\frac{3}{2}kT$. Thus, $\phi_0/kT \gg 1$, and we can simplify the expression for n_{ϕ_0}/n by using the asymptotic formula for the complementary error function,

$$\mathrm{erfc}\left(\frac{\phi_0}{kT} \right)^{1/2} \approx \left(\frac{kT}{\pi \phi_0} \right)^{1/2} e^{-\phi_0/kT}, \tag{7–53}$$

which is valid when the argument of erfc is large compared with unity. It is clear from Equation (7–53) that the second term in (7–52) is much smaller than the first and can be neglected when $\phi_0/kT \gg 1$. The simplified formula for n_{ϕ_0} is therefore

$$n_{\phi_0} = 2n \left(\frac{\phi_0}{\pi kT} \right)^{1/2} e^{-\phi_0/kT}. \tag{7–54}$$

The fraction of high energy atoms is a very sensitive function of temperature through its exponential dependence on the inverse absolute temperature.

At any finite temperature, some atoms have enough energy to escape from their lattice sites; therefore, some lattice positions will be empty. These open lattice sites are known as *vacancies*. A vacancy is one form of a *point defect* in a crystal. Other types of point defects are interstitial atoms that do not fit into the regular lattice structure and impurity atoms. Vacancies play an essential role in diffusion processes. Let us make the reasonable assumption that the number of vacancies per unit volume n_v in a crystal lattice is equal to the number of atoms that have sufficient energy to overcome the potential barrier. From Equation (7–54) we can therefore write

$$n_v = n_{\phi_0} = 2n \left(\frac{\phi_0}{\pi kT} \right)^{1/2} e^{-\phi_0/kT}. \tag{7–55}$$

If the lattice site adjacent to an atom is unoccupied, the atom can jump into the site when its energy exceeds the potential energy barrier ϕ_0. The site originally occupied by the atom becomes a vacancy. In effect, the vacancy has jumped into the adjacent lattice site. We are interested in the rate at which vacancies migrate or diffuse through a lattice because we shall see that the flow of vacancies from one region of a crystal to another can deform the crystal and lead to creep. The frequency $\Gamma_{v,+x}$ with which a given vacancy jumps into the adjacent lattice site in the $+x$ direction is the product of the probability that the atom in that site has the requisite barrier energy, n_{ϕ_0}/n, and the frequency with which the atom moves toward the vacancy, $1/6\tau$ (in an Einstein period τ the atom can move in any one of six directions):

$$\Gamma_{v,+x} = \frac{n_{\phi_0}}{6\tau n}. \tag{7–56}$$

By substituting Equation (7–55) for n_{ϕ_0} and (7–49) for τ into (7–56), we obtain

$$\Gamma_{v,+x} = \frac{\phi_0}{6\pi b} \left(\frac{2}{\pi mkT} \right)^{1/2} e^{-\phi_0/kT}. \tag{7–57}$$

The presence of a vacancy actually reduces the potential energy barrier against an adjacent atom jumping into the vacant site. However, this effect has not been taken into account in this approximate analysis.

Let us assume that a small gradient of vacancy density exists in the crystal in the x direction. The number of vacancies per unit volume is $n_v(x)$. The number of vacancies on the plane of atoms at x per unit area is $n_v(x)b$. The number of vacancies on the adjacent plane of atoms at $x + b$ per unit area is $n_v(x + b)b$. The net rate of flow of vacancies from the plane of atoms at x to the plane of atoms at $x + b$ is, per unit area,

$$J_{v,x} = \Gamma_{v,+x}n_v(x)b - \Gamma_{v,-x}n_v(x+b)b. \quad (7\text{-}58)$$

The derivation of $\Gamma_{v,+x}$ leading to Equation (7–57) could just as easily have been applied to the jump frequency of a vacancy into an adjacent lattice site in the $-x$ direction, $\Gamma_{v,-x}$. Thus $\Gamma_{v,+x} = \Gamma_{v,-x}$ and Equation (7–58) can be written

$$J_{v,x} = -b^2\Gamma_{v,+x}\left\{\frac{n_v(x+b) - n_v(x)}{b}\right\}. \quad (7\text{-}59)$$

If the spacing b between the planes of atoms is small compared with the scale of variation of n_v, it is a good approximation to write

$$\frac{n_v(x+b) - n_v(x)}{b} = \frac{dn_v}{dx}. \quad (7\text{-}60)$$

The expression for the vacancy flux can therefore be written

$$J_{v,x} = -b^2\Gamma_{v,+x}\frac{dn_v}{dx}. \quad (7\text{-}61)$$

The flux of vacancies by diffusion is directly proportional to the gradient of the vacancy number density. The minus sign in Equation (7–61) means that vacancies diffuse from regions of high vacancy concentration toward regions of low vacancy density (the vacancy density gradient points in the direction of increasing vacancy concentration). The magnitude of the constant of proportionality in Equation (7–61) is the *diffusion coefficient* D_v for vacancies

$$D_v = b^2\Gamma_{v,+x}. \quad (7\text{-}62)$$

In terms of the diffusion coefficient, the vacancy flux can be written

$$J_{v,x} = -D_v\frac{dn_v}{dx}, \quad (7\text{-}63)$$

a form known as *Fick's first law of diffusion*. By substituting Equation (7–57) into (7–62), we find that the vacancy diffusion coefficient is given by

$$D_v = \frac{\phi_0 b}{6\pi}\left(\frac{2}{\pi mkT}\right)^{1/2}e^{-\phi_0/kT}. \quad (7\text{-}64)$$

So far we have discussed only the diffusion of vacancies in a crystal lattice. If a lattice predominantly made up of atoms of species B contains a small nonuniform concentration of atoms of species A, the existence of vacancies will allow the atoms of species A to migrate or diffuse through the lattice, a process illustrated in Figure 7–7. Initially all atoms of minor species A are to the left of the dashed line (Figure 7–7a). Vacancies are denoted by blank spaces. After a time that is of the order of the product of the Einstein period and the ratio n_v/n, a significant fraction of the vacancies will have

7–7 Diffusion of a minor species in a crystalline solid. Initially (a) all atoms of minor species A are to the left of the dashed line. Vacancies are denoted by unfilled spaces. After a time that is of the order of an Einstein period multiplied by n_v/n, a significant fraction of the atoms will have randomly jumped into adjacent vacancies. This is illustrated in (b) with arrows denoting the directions in which atoms have jumped in order to create the vacancies. The further migration of atoms into vacant lattice sites is illustrated in (c) and (d). Species A will eventually become randomly mixed.

been filled by atoms that have randomly jumped from adjacent lattice sites. This is illustrated qualitatively in Figures 7–7b–d. Arrows denote the directions in which atoms have jumped. As a result of the random migration of atoms, species A becomes more evenly distributed in the region. In Figure 7–7d there are as many atoms of species A to the right of the dashed line as there are to the left. Therefore, an initially ordered distribution of minor species A (all atoms of species A to the left of the dashed line) has become disordered (atoms of species A randomly distributed throughout the region). The entropy of the system has increased, and the process cannot be reversed. Diffusion is an irreversible phenomenon.

To quantitatively describe the diffusion of a nonuniformly distributed minor species A, we need to determine the frequency $\Gamma_{A,+x}$ with which a given atom of species A jumps into a vacant lattice site in the $+x$ direction. The frequency is the product of three factors: the probability that the atom has the requisite energy, $n_{A,\phi_0}/n_A$ (n_A is the number density of atoms of species A and n_{A,ϕ_0} is the number of atoms of species A per unit volume with energy in excess of ϕ_0), the probability that the adjacent lattice site is empty n_v/n, and the frequency with which the atom moves toward the adjacent lattice site in the $+x$ direction, $1/6\tau$,

$$\Gamma_{A,+x} = \left(\frac{n_{A,\phi_0}}{n_A}\right)\left(\frac{n_v}{n}\right)\left(\frac{1}{6\tau}\right). \tag{7–65}$$

The fraction of atoms of any species with energy in excess of the barrier energy ϕ_0 depends only on temperature (see Equation (7–55)). Accordingly, we can equate the number density ratios

$$\frac{n_{A,\phi_0}}{n_A} = \frac{n_{\phi_0}}{n} \tag{7–66}$$

and write $\Gamma_{A,+x}$ as

$$\Gamma_{A,+x} = \frac{n_v}{n}\left(\frac{n_{\phi_0}}{6\tau n}\right). \tag{7–67}$$

Upon comparing Equations (7–56) and (7–67), we see that the frequency factors $\Gamma_{A,+x}$ and $\Gamma_{v,+x}$ are related by

$$\Gamma_{A,+x} = \frac{n_v}{n}\Gamma_{v,+x}. \tag{7–68}$$

The frequency with which a minor species atom jumps

in the $+x$ direction is smaller, by the factor n_v/n, than the frequency with which a vacancy jumps. The reason for this is that a minor species atom jumps in the $+x$ direction only when there is a vacancy in the adjacent position, while a vacancy makes the jump with an atom in the adjacent position. The likelihood that a vacancy is in the adjacent site is much smaller than the chance that an atom occupies the position.

We now assume that a concentration gradient of minor species A exists in a crystal in the x direction. The number of atoms of species A on the plane of atoms at x per unit area is $n_A(x)b$. The number of atoms of species A on the adjacent plane of atoms at $x + b$ per unit area is $n_A(x + b)b$. The net rate of flow of atoms of species A from the plane of atoms at x to the plane of atoms at $x + b$ is, per unit area,

$$J_{A,x} = \Gamma_{A,+x}n_A(x)b - \Gamma_{A,-x}n_A(x+b)b$$
$$= -b\Gamma_{A,+x}\{n_A(x+b) - n_A(x)\}, \tag{7–69}$$

where we have used the fact that $\Gamma_{A,+x} = \Gamma_{A,-x}$ just as $\Gamma_{v,+x} = \Gamma_{v,-x}$. By using Equation (7–68), we can rewrite the equation for the flux of minor species atoms as

$$J_{A,x} = -\frac{b^2 n_v}{n}\Gamma_{v,+x}\left\{\frac{n_A(x+b) - n_A(x)}{b}\right\}. \tag{7–70}$$

If the spacing b between planes of atoms is small compared with the scale of variation of n_A, it is a good approximation to write

$$\frac{n_A(x+b) - n_A(x)}{b} = \frac{dn_A}{dx}. \tag{7–71}$$

Upon substituting Equation (7–71) into (7–70), we obtain

$$J_{A,x} = -\frac{b^2 n_v}{n}\Gamma_{v,+x}\frac{dn_A}{dx}. \tag{7–72}$$

Minor species atoms diffuse down their concentration gradient in accordance with Fick's first law. The diffusion coefficient for the atoms D is given by

$$D = \frac{b^2 n_v}{n}\Gamma_{v,+x} = \frac{n_v}{n}D_v. \tag{7–73}$$

In terms of D, the flux of minor species atoms is

$$J_{A,x} = -D\frac{dn_A}{dx}. \tag{7–74}$$

TABLE 7–2 **Properties of Several Elements Including Diffusion Coefficient Parameters D_0 and E_a for the Given Radioactive Isotope**

	Aluminum	Copper	Magnesium	Silicon
Atomic mass	26.98	63.55	24.30	28.09
Density, kg m^{-3}	2700	8960	1740	2330
Melt temperature, K	933	1356	922	1683
Specific heat, J kg^{-1} K^{-1}	900	385	1017	703
Isothermal compressibility, Pa^{-1}	1.38×10^{-11}	0.73×10^{-11}	3.0×10^{-11}	1.0×10^{-11}
Volume coefficient of thermal expansion, K^{-1}	7.5×10^{-5}	5×10^{-5}	7.5×10^{-5}	0.9×10^{-5}
Diffusing radioactive isotope	Al^{27}	Cu^{67}	Mg^{28}	Si^{31}
Frequency factor, D_0, m^2 s^{-1}	1.7×10^{-4}	7.8×10^{-5}	1.5×10^{-4}	1.8×10^{-1}
Activation energy, E_a, kJ mole^{-1}	142	211	136	460

By combining Equations (7–55), (7–64), and (7–73), we put the diffusion coefficient for atoms in the form

$$D = \frac{\phi_0 b}{3\pi^2 kT} \left(\frac{2\phi_0}{m} \right)^{1/2} e^{-2\phi_0/kT}. \qquad (7–75)$$

An alternative expression for D employs the gas constant R, the mean atomic weight M_a, and the barrier energy per mole $E_0 = N_0\phi_0$,

$$D = \frac{E_0 b}{3\pi^2 RT} \left(\frac{2E_0}{M_a} \right)^{1/2} e^{-2E_0/RT}. \qquad (7–76)$$

In general, the diffusion coefficient is a function of both temperature T and pressure p, a dependence often expressed in the form

$$D = D_0 \exp\left(-\frac{E_a + pV_a}{RT} \right), \qquad (7–77)$$

where E_a is the activation energy per mole, V_a is the activation volume per mole, and D_0 is the frequency factor. The term pV_a takes account of the effect of pressure in reducing the number of vacancies and increasing the potential energy barrier between lattice sites. Note that Equation (7–77) neglects the temperature dependence of the frequency factor. We can also account for both the temperature and pressure dependences of the diffusion coefficient by the equation

$$D = D_0 e^{-aT_m/T}, \qquad (7–78)$$

where T_m is the melt temperature of the crystalline solid. The ratio T/T_m is referred to as the *homologous temperature*. The pressure dependence of the diffusion coefficient is accounted for through the pressure dependence of the melt temperature.

The parameters in Equation (7–77), E_a, V_a, and D_0, or the ones in (7–78), a, T_m, and D_0, are usually empirically determined. Diffusion coefficients for many crystalline solids can be obtained by using radioactive isotopes as tracers. The diffusion of the radioactive isotope through the crystal can be monitored and the diffusion coefficient thereby determined. Diffusion coefficient parameters for several elements are given in Table 7–2.

PROBLEM 7–9 Consider the one-dimensional diffusion of radioactive tracer atoms initially absent from a crystalline solid but deposited uniformly at time $t = 0$ on the surface $x = 0$ of the semi-infinite solid. The number of radioactive atoms deposited at $t = 0$ is C per unit surface area. Show that the concentration of radioactive atoms n (number per unit volume) in the solid must satisfy the diffusion equation

$$\frac{\partial n}{\partial t} = D \frac{\partial^2 n}{\partial x^2}. \qquad (7–79)$$

Equation (7–79) can be obtained by first deriving the equation of conservation of tracer atoms

$$\frac{\partial n}{\partial t} = -\frac{\partial J}{\partial x}, \qquad (7–80)$$

where we assume that tracer atoms diffuse in the x direction only. The actual decay of the tracer atoms has been ignored in formulating the mass balance. Solve Equation (7–79) subject to the initial and boundary conditions

$$n(x, t = 0) = 0 \qquad (7–81)$$

$$\int_0^\infty n(x, t) \, dx = C. \qquad (7–82)$$

TABLE 7–3 **Model Data for Determining the Diffusion Coefficient from a Radioactive Tracer Experiment**

Counts per Second	Section Thickness (10^{-6} m)
5020	104
3980	110
2505	101
1395	98
570	96

Show that $n(x, t)$ is given by

$$n(x, t) = \frac{C}{(\pi Dt)^{1/2}} \exp\left(\frac{-x^2}{4Dt}\right). \qquad (7\text{–}83)$$

We solved a similar heat diffusion problem in Section 4–21. Determine the diffusion coefficient for the laboratory data summarized in Table 7–3. The data were obtained by depositing a thin layer of tracer atoms on an Al surface, annealing for 30 hours, and then sectioning.

We now consider how the diffusion of atoms in a crystal in the presence of differential stress can result in creep. Diffusion creep is illustrated in Figure 7–8. A crystal is initially a cube of dimension h. The crystal is subjected to a compressional stress σ in the

x direction and an equal tensional stress $-\sigma$ in the y direction. These stresses cause atoms to diffuse from the crystal faces A and C to the crystal faces B and D. Alternatively we can consider the equivalent process of vacancy diffusion in the opposite direction. When a layer of atoms has been removed from faces A and C and added to faces B and D, the strain in the x direction is $\varepsilon_{xx} = 2b/h$, and the strain in the y direction is $\varepsilon_{yy} = -2b/h$. It should also be noted that atoms diffuse away from faces A and C toward faces E and F (on which no stresses are applied, $\sigma_{zz} = 0$) and diffuse away from faces E and F toward faces B and D. There will be no net loss or gain of atoms on faces E and F so that $\varepsilon_{zz} = 0$.

Because of the application of a compressional stress on face A (and face C) the number density of vacancies is decreased from n_v to n_{vA}. Similarly, the application of a tensional stress on face B (and face D) increases the number density of vacancies from n_v to n_{vB}. The number density of vacancies on faces E and F remains n_v because no stress is applied. The difference in vacancy densities on the faces of the cube results in a flux of vacancies from faces B and D to faces A and C and a corresponding flux of atoms from faces A and C to faces B and D.

The fluxes of vacancies (per unit area and time) from faces A and C to faces B and D are, from Equation (7–63),

$$J_{v,AB} = J_{v,AD} = J_{v,CB} = J_{v,CD} = \frac{D_v \sqrt{2}}{h}(n_{vB} - n_{vA}). \qquad (7\text{–}84)$$

In writing Equations (7–84), we have used $n_{vD} = n_{vB}$ and $n_{vC} = n_{vA}$ and the fact that $h/\sqrt{2}$ is the mean distance between adjacent faces of the cube. Similarly, the fluxes of vacancies from faces A and C to faces E and F and from faces E and F to faces B and D are

$$J_{v,AE} = J_{v,AF} = J_{v,CE} = J_{v,CF} = \frac{D_v \sqrt{2}}{h}(n_v - n_{vA}) \qquad (7\text{–}85)$$

$$J_{v,EB} = J_{v,ED} = J_{v,FB} = J_{v,FD} = \frac{D_v \sqrt{2}}{h}(n_{vB} - n_v). \qquad (7\text{–}86)$$

The area over which each flux occurs is $h^2/\sqrt{2}$, and the strain associated with the transfer of each vacancy

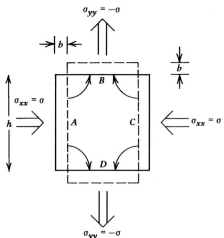

7-8 Diffusion of atoms in a cubic crystal of dimension h due to the application of a horizontal compressive stress and a vertical tensional stress. The shape of the crystal after the diffusion of a surface layer of atoms is illustrated by the dashed lines. The six faces of the cube are denoted by A, B, C, D, E, and F.

is $-b^3/h^3$. The rates of strain in the x and y directions are therefore given by

$$\dot{\varepsilon}_{yy} = -\frac{b^3}{h^3}\frac{h^2}{\sqrt{2}}(J_{v,AB} + J_{v,AD} + J_{v,CB} + J_{v,CD}$$
$$+ J_{v,EB} + J_{v,ED} + J_{v,FB} + J_{v,FD}) \qquad (7\text{–}87)$$

$$\dot{\varepsilon}_{xx} = \frac{b^3}{h^3}\frac{h^2}{\sqrt{2}}(J_{v,AB} + J_{v,AD} + J_{v,CB} + J_{v,CD}$$
$$+ J_{v,AE} + J_{v,AF} + J_{v,CE} + J_{v,CF}). \qquad (7\text{–}88)$$

The fluxes can be eliminated from these strain rate expressions by substitution of Equations (7–84) to (7–86). If in addition we use the fact that $n = 1/b^3$, we obtain

$$\dot{\varepsilon}_{yy} = -\frac{4D_v}{nh^2}(2n_{vB} - n_{vA} - n_v) \qquad (7\text{–}89)$$

$$\dot{\varepsilon}_{xx} = -\frac{4D_v}{nh^2}(n_v + n_{vB} - 2n_{vA}). \qquad (7\text{–}90)$$

We must now specify how the vacancy densities n_{vA} and n_{vB} that drive diffusion depend on the applied stresses. The isotropic vacancy density in the absence of an applied stress n_v is given by Equation (7–55). By direct analogy with the way pressure affects the diffusion constant (see Equation (7–77)) the stresses at the crystal boundaries modify the vacancy density according to

$$n_{vA} = 2n\left(\frac{E_0}{\pi RT}\right)^{1/2}\exp\left\{-\frac{(E_0 + \sigma V_a)}{RT}\right\} \qquad (7\text{–}91)$$

$$n_{vB} = 2n\left(\frac{E_0}{\pi RT}\right)^{1/2}\exp\left\{-\frac{(E_0 - \sigma V_a)}{RT}\right\}. \qquad (7\text{–}92)$$

If the applied stress is sufficiently small, then $\sigma V_a/RT \ll 1$, and we can write

$$\frac{n_{vA}}{n_v} = e^{-\sigma V_a/RT} \approx 1 - \frac{\sigma V_a}{RT} \qquad (7\text{–}93)$$

$$\frac{n_{vB}}{n_v} = e^{\sigma V_a/RT} \approx 1 + \frac{\sigma V_a}{RT}. \qquad (7\text{–}94)$$

By substituting Equations (7–73), (7–93), and (7–94) into Equations (7–89) and (7–90), we obtain

$$\dot{\varepsilon}_{xx} = -\dot{\varepsilon}_{yy} = \frac{12V_a D}{RTh^2}\sigma. \qquad (7\text{–}95)$$

Diffusion creep, also referred to as *Herring–Nabarro creep*, results in a linear relationship between strain

rate and stress. A Newtonian fluid exhibits a linear relationship between stress and velocity gradient; from Equation (6–56) we have

$$\tau_{xx} = 2\mu\frac{\partial u}{\partial x}. \qquad (7\text{–}96)$$

In our creep analysis we have $\sigma = \sigma_{xx} = \tau_{xx}$ and using Equation (2–83) we write

$$\frac{\partial u}{\partial x} = \frac{\partial}{\partial x}\left(\frac{\partial w_x}{\partial t}\right) = \frac{\partial}{\partial t}\left(\frac{\partial w_x}{\partial x}\right) = \frac{\partial \varepsilon_{xx}}{\partial t} = \dot{\varepsilon}_{xx}. \qquad (7\text{–}97)$$

Substitution of Equation (7–97) into Equation (7–96) gives

$$\sigma = 2\mu\dot{\varepsilon}_{xx}. \qquad (7\text{–}98)$$

From Equations (7–95) and (7–96) the viscosity of the crystalline solid is given by

$$\mu = \frac{RTh^2}{24V_a D}. \qquad (7\text{–}99)$$

By substituting for the diffusion coefficient from Equation (7–77), we can rewrite the formula for the viscosity associated with diffusion creep as

$$\mu = \frac{RTh^2}{24V_a D_0}\exp\left(\frac{E_a + pV_a}{RT}\right), \qquad (7\text{–}100)$$

or using Equation (7–78)

$$\mu = \frac{RTh^2}{24V_a D_0}\exp\left(\frac{aT_m}{T}\right). \qquad (7\text{–}101)$$

The Newtonian viscosity of diffusion creep is strongly temperature and pressure dependent. The temperature dependence of the preexponential factor in Equations (7–100) and (7–101) is virtually insignificant when compared with the highly sensitive dependence of the exponential of the inverse absolute temperature. Because of the dominance of the exponential factor, μ decreases markedly with an increase in temperature. The diffusion creep viscosity increases with pressure, as can be seen directly in Equation (7–100). Because T_m increases with pressure, the dependence of μ on p is also inherent in Equation (7–101).

So far in our discussion of diffusion creep we have assumed that the atoms diffuse through the interiors of mineral grains. However, diffusion occurs along grain boundaries as well. If the diffusion coefficient for grain boundary diffusion of vacancies D_{vb} is much larger than the coefficient for diffusion of vacancies through the interiors of grains, grain boundary diffusion may be the dominant manner in which creep occurs. For the diffusion of vacancies along grain boundaries the fluxes of vacancies from faces A and C to faces B and D (see Figure 7–8) are, from Equation (7–63),

$$J_{v,AB} = J_{v,AD} = J_{v,CB} = J_{v,CD} = \frac{D_{vb}}{h}(n_{vB} - n_{vA}), \tag{7–102}$$

where h is the mean distance of diffusion along the boundary. The specification of constant values of n_{vB}, n_{vA}, etc., on the grain boundaries is only an approximation because diffusion occurs along these same boundaries. The fluxes of vacancies from faces A and C to faces E and F and from faces E and F to faces B and D are, according to Equation (7–63),

$$J_{v,AE} = J_{v,AF} = J_{v,CE} = J_{v,CF} = \frac{D_{vb}}{h}(n_v - n_{vA}) \tag{7–103}$$

$$J_{v,EB} = J_{v,ED} = J_{v,FB} = J_{v,FD} = \frac{D_{vb}}{h}(n_{vB} - n_v). \tag{7–104}$$

The area over which each flux occurs is $h\delta$, where δ is the width of the grain boundary. As before, the strain associated with the transfer of each vacancy is $-b^3/h^3$. Therefore, we can write the net strain rates as we did previously in Equations (7–87) and (7–88) in the form

$$\dot{\varepsilon}_{yy} = -\frac{b^3}{h^3}h\delta(J_{v,AB} + J_{v,AD} + J_{v,CB} + J_{v,CD}$$
$$+ J_{v,EB} + J_{v,ED} + J_{v,FB} + J_{v,FD}) \tag{7–105}$$

$$\dot{\varepsilon}_{xx} = \frac{b^3}{h^3}h\delta(J_{v,AB} + J_{v,AD} + J_{v,CB} + J_{v,CD}$$
$$+ J_{v,AE} + J_{v,AF} + J_{v,CE} + J_{v,CF}). \tag{7–106}$$

Upon substitution of Equations (7–102) to (7–104) into (7–105) and (7–106), we obtain expressions for the strain rates that are analogous to those of Equations (7–89) and (7–90)

$$\dot{\varepsilon}_{yy} = -\frac{4\delta D_{vb}}{h^3 n}(2n_{vB} - n_{vA} - n_v) \tag{7–107}$$

$$\dot{\varepsilon}_{xx} = \frac{4\delta D_{vb}}{h^3 n}(n_v + n_{vB} - 2n_{vA}). \tag{7–108}$$

We can further simplify these equations, as we did previously, by substituting the formulas for n_{vA}/n_v and n_{vB}/n_v from Equations (7–93) and (7–94). In addition, we can introduce a diffusion coefficient for grain boundary diffusion of atoms D_b, by analogy with Equation (7–73), as

$$D_b = \frac{n_v}{n}D_{vb}. \tag{7–109}$$

The strain rates can then be written

$$\dot{\varepsilon}_{xx} = -\dot{\varepsilon}_{yy} = \frac{12V_a\delta D_b}{RTh^3}\sigma. \tag{7–110}$$

Grain boundary creep, or Coble creep, also yields a linear relationship between rate of strain and stress. The associated viscosity, from Equation (7–98), is

$$\mu = \frac{RTh^3}{24V_a\delta D_b}. \tag{7–111}$$

The grain boundary diffusion coefficient is also of the form given in Equation (7–77), that is,

$$D_b = D_{b0}\exp\left(-\frac{E_a + pV_a}{RT}\right), \tag{7–112}$$

so that the viscosity of grain boundary creep can be written

$$\mu = \frac{RTh^3}{24V_a\delta D_{b0}}\exp\left(\frac{E_a + pV_a}{RT}\right). \tag{7–113}$$

The dependence of viscosity on temperature and pressure when diffusion takes place along grain boundaries is the same as when diffusion occurs through the interiors of grains. The magnitude of the ratio $\delta D_b/hD$ determines whether grain boundary or intragranular diffusion dominates.

7-4 Dislocation Creep

In the previous section we saw how the migration of vacancies in crystalline solids leads to creep deformation. In this section we explain how the migration of dislocations also results in subsolidus creep. *Dislocations* are imperfections in the crystalline lattice structure. Although dislocations can be found in many complex forms, they can all be obtained by the superposition of two basic types. These are the edge and screw dislocations.

An edge dislocation in a cubic lattice is illustrated in Figure 7–9. A plane of atoms is present in part of

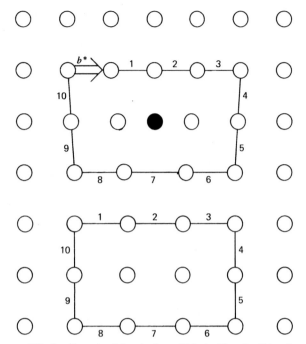

7–9 Side view of an edge dislocation in a cubic lattice. The edge dislocation is the line of atoms at the end of the additional plane of atoms in the upper part of the crystal. The edge dislocation, denoted by the solid circle, is perpendicular to the plane of the figure. Two Burgers circuits are also shown. The one in the lower part of the figure does not include the dislocation and is closed. The Burgers circuit in the upper part of the figure includes the edge dislocation. The ten steps in the two circuits are numbered, and the upper circuit does not close. The vector defining the lack of closure **b*** is the Burgers vector for this dislocation.

the crystal. The end of the plane, indicated by the solid circle in Figure 7–9, defines the line of atoms perpendicular to the figure that is the *edge dislocation*. The dislocation deforms the crystal lattice and produces stresses in it as a result. The adjacent planes of atoms are in compression above the dislocation and in tension below it.

Dislocations are defined in terms of the *Burgers vector*, which is a measure of the relative atomic motion (slip) that occurs when a dislocation line passes through a lattice. The surface that traces the motion of a dislocation line is the *glide surface*, and such surfaces are usually planar. Specification of the Burgers vector and the line direction fully defines a dislocation. The Burgers vector is determined by a Burgers circuit. A Burgers circuit that does not include a dislocation is illustrated in the lower part of Figure 7–9. The ten steps on the circuit are numbered, and the circuit closes. A Burgers circuit that includes the edge dislocation is shown in

the upper part of Figure 7–9. Again the ten steps on the circuit are numbered, but the circuit does not close. The vector defining the lack of closure is the Burgers vector **b***. For the simple cubic lattice with a single edge dislocation $|\mathbf{b}^*| = b$. The Burgers vector of an edge dislocation lies in the plane of the Burgers circuit.

The *screw dislocation* is an out-of-the-plane deformation of the crystal, as indicated in Figure 7–10. The appropriate Burgers circuit is also shown. After the circuit that includes the screw dislocation is completed, a displacement perpendicular to the plane of the circuit is required to close the circuit. Thus the Burgers vector is perpendicular to the plane of the Burgers circuit. If the Burgers circuit were continued, it would form a second circuit lying mainly in the layer of atoms behind the one illustrated before ending on a third layer. The further continuation of the Burgers circuit would constitute a spiral motion from one layer to another. Because this spiral motion resembles the threads on a screw, this out-of-the-plane crystal deformation is called a screw dislocation if the dislocation line is parallel to the Burgers vector. If the dislocation line is perpendicular to the Burgers vector, it is an edge dislocation. Most dislocations have both edge and screw components.

7–10 A screw dislocation in a cubic lattice constitutes a deformation that is out of the plane of atoms illustrated. The two atoms denoted by solid circles are essentially part of a second plane. The Burgers circuit indicated by the numbered steps naturally moves into this second plane. Therefore in order to close the circuit the Burgers vector **b*** must be perpendicular to the plane of atoms shown.

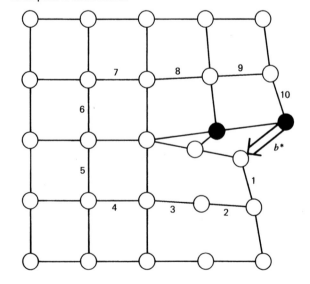

The two principal ways in which dislocations can contribute to creep are through *dislocation climb* and *dislocation slip*. We first consider dislocation climb, as illustrated in Figure 7–11 for a pair of edge dislocations. The process of dislocation climb for an edge dislocation refers to either a lengthening or a shortening of the extra plane of atoms defining the dislocation. The dislocation line moves by the addition of atoms. This is not a mass-conserving motion because it requires the diffusion of atoms from elsewhere in the lattice. For a crystal under horizontal compression and vertical tension, the edge dislocation defined by the additional vertical plane of atoms tends to shorten and the edge dislocation defined by the extra horizontal plane of atoms tends to lengthen. As a result, compressional strain occurs in the horizontal direction, and extensional strain occurs in the vertical direction. Figure 7–11 shows how the diffusion of the atoms b, c, and d from the extra vertical plane to the extra horizontal plane results in the deformation. The dislocation defined by the extra vertical plane of atoms climbs as a result of the process. The diffusion of atoms between dislocations is caused by the migration of vacancies as discussed in the previous section. Thus the analysis developed for diffusion creep can also be applied to the lengthening and shortening of crystals by dislocation climb.

Dislocation slip is illustrated in Figure 7–12 for an edge dislocation. The process involves the transfer of an edge dislocation to an adjacent plane of atoms as the result of a shear stress applied to the crystal. This motion conserves mass so that diffusion of atoms through the lattice is not required. Shear strain occurs as the dislocation sweeps across the lattice. Dislocation slip is a thermally activated process driven by a shear stress acting on the dislocation. Because diffusion through the lattice structure is not required, dislocation slip is a more rapid process than dislocation climb.

There are many alternative quantitative formulations for dislocation creep. All types of dislocation creep are thermally activated processes, at relatively low stress levels, so the rheology is exponentially dependent on the inverse absolute temperature and the pressure. Again the Maxwell–Boltzmann distribution gives the number of atoms that have sufficient energy to overcome the interatomic bonds restricting the motion of a dislocation. Different formulations yield differ-

ent power-law relations between strain rate and stress; however, all are non-Newtonian. As an example of one approach we assume that dislocations act as sources and sinks of vacancies just as grain boundaries do in diffusion creep. Equation (7–95) can then be used to relate the strain rate $\dot{\varepsilon}_{xx}$ or $\dot{\varepsilon}_{yy}$ to the stress σ if the grain size h is replaced by the mean spacing between dislocations h_d

$$\dot{\varepsilon}_{xx} = -\dot{\varepsilon}_{yy} = \frac{12 V_a D}{RT h_d^2} \sigma. \tag{7–114}$$

The mean spacing between dislocations is related to the volume density of dislocations n_d by

$$h_d = n_d^{-1/3}. \tag{7–115}$$

The dislocation density, or alternatively h_d, depends on stress and many other factors. Under a wide range of conditions,

$$h_d = \frac{b^* G}{\sigma}, \tag{7–116}$$

where b^* is the magnitude of the Burgers vector for the dislocations and G is the shear modulus. Upon substituting Equation (7–116) into Equation (7–114), we obtain

$$\dot{\varepsilon}_{xx} = -\dot{\varepsilon}_{yy} = \frac{12 V_a D}{RT b^{*2} G^2} \sigma^3 \tag{7–117}$$

as the relation between strain rate and stress for dislocation creep. Although dislocation creep gives a non-Newtonian fluid behavior, an effective viscosity μ_{eff} can still be defined (using Equation (7–98)) as the ratio of stress to twice the strain rate

$$\mu_{\text{eff}} = \frac{\sigma}{2\dot{\varepsilon}_{xx}} = \frac{RT b^{*2} G^2}{24 V_a D} \frac{1}{\sigma^2}. \tag{7–118}$$

By substituting Equation (7–77) for the diffusion coefficient into (7–118), we can rewrite μ_{eff} as

$$\mu_{\text{eff}} = \frac{RT b^{*2} G^2}{24 V_a D_0} \frac{1}{\sigma^2} \exp\left(\frac{E_a + p V_a}{RT}\right). \tag{7–119}$$

The effective viscosity of dislocation creep is inversely proportional to the square of the stress; it is also proportional to the exponential of the inverse absolute temperature and the pressure. The stress dependence of

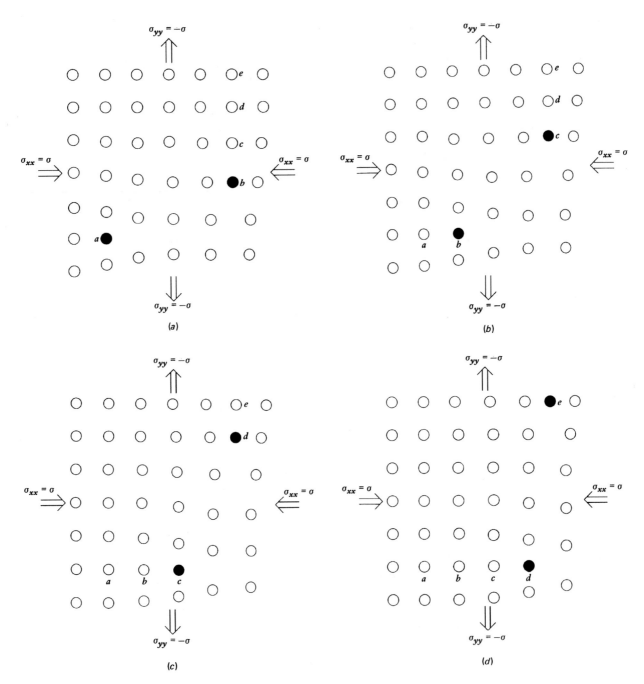

7–11 The process of dislocation climb. Because of the diffusion of the atoms b, c, and d from the extra vertical plane of atoms to the additional horizontal plane of atoms, the crystal is shortened in the x direction and lengthened in the y direction. This is the result of the tensional stress in the y direction and the compressional stress in the x direction that drives the diffusion of atoms between the two dislocations.

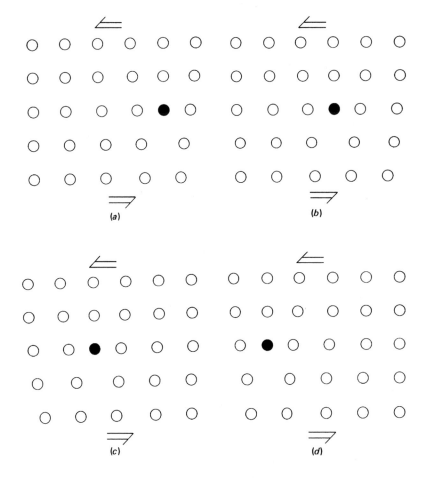

7-12 The process of dislocation slip for an edge dislocation involves the transfer of the dislocation to adjacent planes of atoms as the result of an applied shear stress.

μ_{eff} for dislocation creep facilitates deformation in regions of high stress.

7-5 Shear Flows of Fluids with Temperature- and Stress-Dependent Rheologies

We have seen in the previous sections that the viscosities of both diffusion creep and dislocation creep are directly proportional to the exponential of the inverse absolute temperature. Thus, the viscosity of the mantle has this strong temperature dependence no matter which of these mechanisms controls the subsolidus deformation of the mantle. In addition, if dislocation creep is the dominant mechanism, and we conclude this is likely to be the case in the next section, the effective viscosity of the mantle will be stress dependent as well. It is important then to consider how the strong temperature dependence and possible stress dependence of mantle viscosity influence convection and shear flow in the mantle. To do this, let us first consider the effects of temperature- and stress-dependent rheologies on some of the simple flows we discussed in Chapter 6. We will begin with an example of the channel flow of a fluid with stress-dependent viscosity. We will then devote the rest of the section to the more important effects of strongly temperature-dependent viscosity.

The rheological law given in Equation (7–117) is a particular example of non-Newtonian deformation known as power-law creep in which strain rate is proportional to a power n of the stress. The manner in which such a non-Newtonian rheology influences fluid motion can be readily illustrated by the simple example of the flow of a power-law fluid in a channel. We consider a channel of thickness h with stationary walls at $y = \pm h/2$. The flow is driven by a pressure difference $p_1 - p_0$ over the channel length L. The shear stress τ in the fluid satisfies Equation (6–6)

$$\frac{d\tau}{dy} = \frac{-(p_1 - p_0)}{L}. \tag{7–120}$$

For a power-law fluid, the shear stress and velocity gradient, or strain rate, are related by

$$\frac{du}{dy} = C_1 \tau^n, \tag{7–121}$$

with $n = 1, 3, 5, 7, \ldots$ and C_1 a positive constant. Because the shear du/dy can be positive or negative in a flow, n cannot be an even integer. Upon solving Equation (7–121) for τ and substituting into Equation (7–120), we get

$$\frac{1}{C_1^{1/n}} \frac{d}{dy} \left\{ \left(\frac{du}{dy} \right)^{1/n} \right\} = -\frac{(p_1 - p_0)}{L}. \tag{7–122}$$

A single integration with the symmetry condition $du/dy = 0$ at $y = 0$ yields

$$\frac{du}{dy} = -C_1 \left\{ \frac{p_1 - p_0}{L} \right\}^n y^n. \tag{7–123}$$

A second integration with the boundary condition $u = 0$ at $y = \pm h/2$ gives

$$u = \frac{C_1}{(n+1)} \left\{ \frac{p_1 - p_0}{L} \right\}^n \left\{ \left(\frac{h}{2} \right)^{n+1} - y^{n+1} \right\}. \tag{7–124}$$

The mean velocity in the channel is

$$\bar{u} = \frac{2}{h} \int_0^{h/2} u \, dy = \frac{C_1}{(n+2)} \left\{ \frac{p_1 - p_0}{L} \right\}^n \left(\frac{h}{2} \right)^{n+1} \tag{7–125}$$

and the ratio of the velocity to the mean velocity is

$$\frac{u}{\bar{u}} = \left(\frac{n+2}{n+1} \right) \left\{ 1 - \left(\frac{2y}{h} \right)^{n+1} \right\}. \tag{7–126}$$

Velocity profiles for $n = 1$ (Newtonian), 3, and 5 are given in Figure 7–13. We see that for increasing values of n, the gradients of the velocity become large near the walls where the shear stress is a maximum. A nearly rigid core flow develops where the shear stress is low.

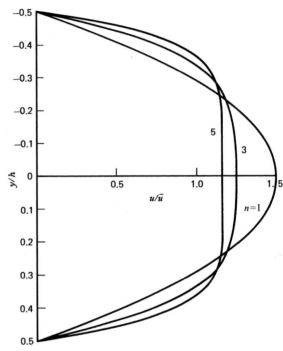

7-13 Velocity profiles in a channel for power-law fluid rheologies with $n = 1$ (Newtonian), 3, and 5.

The effective viscosity of the power-law fluid is proportional to τ^{1-n}. For large n, the viscosity is high where τ is small and low where τ is large. Because τ is small in the center of the channel, the fluid is highly viscous there. Near the walls where τ is high, μ_{eff} is low, and the velocity gradients are large. The plug-flow appearance of the velocity profiles for large n is a consequence of the stress dependence of the effective viscosity.

PROBLEM 7–10 Show that the effective viscosity μ_{eff} for the channel flow of a power-law fluid is given by

$$\mu_{\text{eff}} \equiv \frac{\tau}{du/dy} = \left(\frac{p_1 - p_0}{L} \right) \frac{h^2}{4(n+2)\bar{u}} \left(\frac{2y}{h} \right)^{1-n} \tag{7–127}$$

or

$$\frac{\mu_{\text{eff}}}{\mu_{\text{eff,wall}}} = \left(\frac{2y}{h} \right)^{1-n}, \tag{7–128}$$

where $\mu_{\text{eff,wall}}$ is the value of μ_{eff} at $y = \pm h/2$. Plot $\mu_{\text{eff}}/\mu_{\text{eff,wall}}$ as a function of y/h for $n = 1, 3$, and 5.

Because of its relevance to shear flow in the asthenosphere, we next consider the influence of a viscosity with an exponential dependence on the inverse absolute temperature on the Couette flow in Section 6–2. Recall that Couette flow takes place in an infinitely long channel whose upper boundary ($y = 0$) moves with velocity u_1 relative to its stationary lower boundary ($y = h$). There is no applied pressure gradient along the channel. We assume that the upper wall is maintained at temperature T_0 while the lower wall is kept at temperature $T_1 (T_1 > T_0)$. If account is taken of heating by viscous dissipation in the shear flow, the temperature dependence of the viscosity couples the temperature $T(y)$ and velocity profiles in the channel. Both quantities $T(y)$ and $u(y)$ must be determined simultaneously because one depends on the other. The velocity depends on T through the dependence of μ on T and T depends on u because frictional heating depends on the shear in the velocity profile. For simplicity, we will first treat a situation in which shear heating is negligible. This allows us to write the solution for the temperature in the channel as the simple linear profile

$$T = T_0 + (T_1 - T_0)\frac{y}{h} \qquad (7\text{–}129)$$

independent of $u(y)$.

We assume that the viscosity has the form given in Equation (7–100) and that the channel is thin enough so that the effect of pressure on velocity is unimportant. We also consider the temperature dependence of the preexponential factor as negligible compared with the temperature dependence of the exponential factor and write

$$\mu = Ce^{E_a/RT}, \qquad (7\text{–}130)$$

where C is a constant. The shear stress τ, which is a constant in the absence of a horizontal pressure gradient, is given by

$$\tau = \mu\frac{du}{dy} = Ce^{E_a/RT}\frac{du}{dy}. \qquad (7\text{–}131)$$

Upon substituting for T from Equation (7–129), we obtain an equation for du/dy by rearranging Equation (7–131)

$$\frac{du}{dy} = \frac{\tau}{C}\exp\left\{\frac{-E_a/R}{T_0 + (T_1 - T_0)\frac{y}{h}}\right\}. \qquad (7\text{–}132)$$

The solution for the velocity profile is found as the integral of Equation (7–132).

The integration can be carried out analytically if we assume that the temperature difference $T_1 - T_0$ is small compared with T_0. The argument of the exponential factor in Equation (7–132) can then be approximated as

$$\frac{-(E_a/RT_0)}{\left\{1 + \frac{(T_1 - T_0)}{T_0}\frac{y}{h}\right\}} \approx -\frac{E_a}{RT_0}\left\{1 + \frac{(T_1 - T_0)}{T_0}\frac{y}{h}\right\}, \qquad (7\text{–}133)$$

so that Equation (7–132) becomes

$$\frac{du}{dy} = \frac{\tau}{C}\exp\left\{\frac{-E_a}{RT_0}\right\}\exp\left\{\frac{E_a(T_1 - T_0)}{RT_0^2}\frac{y}{h}\right\}. \qquad (7\text{–}134)$$

Upon integrating this equation with the boundary condition $u = 0$ on $y = h$, we find

$$u = \frac{\tau h RT_0^2}{CE_a(T_1 - T_0)}\exp\left\{\frac{-2E_a T_0 + E_a T_1}{RT_0^2}\right\}$$
$$\times\left[\exp\left\{\frac{E_a(T_1 - T_0)}{RT_0^2}\left(\frac{y}{h} - 1\right)\right\} - 1\right]. \qquad (7\text{–}135)$$

By further requiring that $u = u_1$ at $y = 0$, we can rewrite the velocity profile in the somewhat simpler form

$$\frac{u}{u_1} = \frac{\exp\left\{\frac{-E_a(T_1 - T_0)}{RT_0^2}\left(1 - \frac{y}{h}\right)\right\} - 1}{\exp\left\{\frac{-E_a(T_1 - T_0)}{RT_0^2}\right\} - 1}. \qquad (7\text{–}136)$$

The shear stress and the velocity of the upper boundary are related by

$$u_1 = \frac{\tau h RT_0^2}{CE_a(T_1 - T_0)}\exp\left(\frac{-E_a}{RT_0}\right)$$
$$\times\left[1 - \exp\left\{\frac{E_a(T_1 - T_0)}{RT_0^2}\right\}\right]. \qquad (7\text{–}137)$$

Velocity profiles u/u_1 versus y/h are shown in Figure 7–14 for $(T_1 - T_0)/T_0 = 0.5$ and $E_a/RT_0 = 0$, 10, 20, and 30. We will see that the larger values of E_a/RT_0 are representative of the upper mantle; T_0 can be thought of as the temperature at the base of the rigid lithosphere while T_1 is the temperature at the base of the asthenosphere. The reasonable values $T_0 = 800°C$ and $T_1 = 1300°C$ give $(T_1 - T_0)/T_0 = 500/1073 \approx 0.5$.

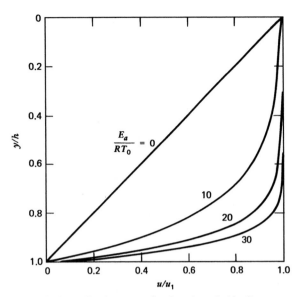

7–14 Velocity profiles for Couette flow in a channel with a linear temperature profile. The fluid's viscosity is proportional to the exponential of the inverse absolute temperature. The dimensionless temperature difference across the channel $(T_1 - T_0)/T_0$ is 0.5. E_a/RT_0 is the dimensionless activation energy parameter. The top wall $(y = 0)$ is cold $(T = T_0)$, and the bottom wall $(y = h)$ is hot $(T = T_1)$.

When the fluid viscosity is independent of temperature $(E_a/RT_0 = 0)$, the velocity profile is linear. As the viscosity becomes increasingly temperature dependent (larger values of E_a/RT_0), the shear in the velocity profile is confined to progressively narrower regions near the lower boundary where the fluid is hottest and the viscosity is the smallest. For the large values of E_a/RT_0 typical of the mantle, most of the fluid in the channel moves at the velocity of the upper boundary while the reduction in velocity occurs in a relatively hot low viscosity layer adjacent to the lower boundary. The upper part of the channel behaves as a nearly rigid extension of the overlying lithosphere, that is, it is really part of the lithosphere, while the lower part of the channel is a high shear, low viscosity asthenosphere.

The viscosity variation in the channel can be determined from Equation (7–129) and (7–130). The constant C can be eliminated by referencing the viscosity to its value at the upper boundary μ_0. From Equation (7–130) μ_0 is given by

$$\mu_0 = C e^{E_a/RT_0} \tag{7–138}$$

and C is

$$C = \mu_0 e^{-E_a/RT_0}. \tag{7–139}$$

The viscosity can thus be written

$$
\begin{aligned}
\mu &= \mu_0 \exp\left\{ \frac{E_a}{RT_0}\left(\frac{T_0}{T} - 1 \right) \right\} \\
&= \mu_0 \exp\left\{ \frac{E_a}{RT_0}\left(\left[1 + \left(\frac{T_1 - T_0}{T_0} \right) \frac{y}{h} \right]^{-1} - 1 \right) \right\}.
\end{aligned}
\tag{7–140}
$$

The viscosity profiles μ/μ_0 versus y/h are shown in Figure 7–15 for $(T_1 - T_0)/T_0 = 0.5$ and $E_a/RT_0 = 0, 10, 20$, and 30. The large reduction in viscosity in the hot lower portion of the channel that occurs for the higher values of E_a/RT_0 is apparent.

PROBLEM 7–11 Determine the shear stress in the channel. Assume $E_a/RT_0 = 20$, $(T_1 - T_0)/T_0 = 0.5$, $u_1 = 50$ mm yr^{-1}, $h = 100$ km, and $\mu_0 = 10^{24}$ Pa s.

PROBLEM 7–12 Consider an ice sheet of thickness h lying on bedrock with slope α, as shown in Figure 7–16. The ice will creep slowly downhill under the force of its own weight. Determine the velocity profile $u(y)$ in the ice. The viscosity of ice has the temperature dependence given in Equation (7–130). Assume that the temperature profile in the ice is linear with the surface temperature T_0 (at $y = 0$) and the bedrock–ice interface temperature T_1 (at $y = h$). Assume that there is no melting at the base of the ice sheet so that the no-slip condition applies; that is, $u = 0$ at $y = h$, and utilize the approximation given in Equation (7–133).

Frictional heating can have dramatic consequences on the shear flow of a fluid with a strongly temperature-dependent viscosity. A simple channel flow model suffices to demonstrate the effects. We again consider a situation in which flow is driven by a shear stress τ applied at the upper moving boundary $(y = 0)$ of a channel whose lower surface $(y = h)$ is fixed. With no pressure gradient along the channel, τ is a constant, independent of y, as before. In the present example we insulate the lower boundary of the channel so that all the excess heat in the fluid is generated internally solely by viscous dissipation. The heat generated by friction in the flow escapes through the upper boundary of the channel whose temperature is maintained at T_0. Equation (6–409) governing the temperature in a frictionally heated shear flow is valid even when viscosity is

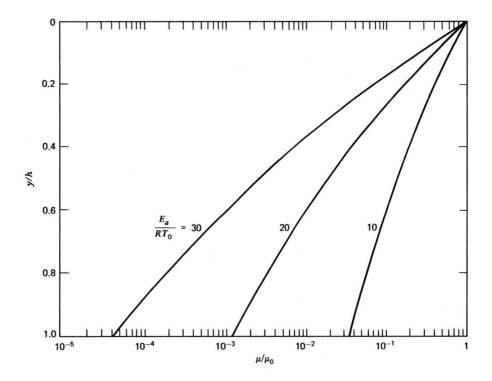

temperature dependent. The temperature in the channel is thus the solution of

$$k\frac{d^2 T}{dy^2} + \frac{\tau^2}{\mu} = 0, \tag{7–141}$$

together with the boundary conditions

$$T = T_0 \quad \text{on} \quad y = 0 \tag{7–142}$$

$$\frac{dT}{dy} = 0 \quad \text{on} \quad y = h. \tag{7–143}$$

With μ given by Equation (7–130), we can rewrite the temperature equation as

$$\frac{d^2 T}{dy^2} + \frac{\tau^2}{kC}e^{-E_a/RT} = 0. \tag{7–144}$$

Equations (7–142) to (7–144) define a nonlinear boundary value problem for temperature because of the dependence of the frictional heating term on $\exp(-E_a/RT)$. We can solve for $T(y)$ analytically if we only consider situations wherein frictional heating produces small temperature increases. Thus we set

$$T = T_0 + T', \tag{7–145}$$

where $T' \ll T_0$. The inverse of the temperature, which appears in the exponent of the shear heating term, is

7–15 Viscosity profiles for Couette flow with temperature-dependent viscosity.

approximately given by

$$T^{-1} = (T_0 + T')^{-1} = T_0^{-1}\left(1 + \frac{T'}{T_0}\right)^{-1}$$

$$\approx T_0^{-1}\left(1 - \frac{T'}{T_0}\right). \tag{7–146}$$

Upon substituting Equations (7–145) and (7–146) into (7–144), we get

$$\frac{d^2 T'}{dy^2} + \frac{2}{kC}e^{-(E_a/RT_0)}e^{(E_a T'/RT_0^2)} = 0. \tag{7–147}$$

7–16 An ice sheet of thickness h on bedrock sloping at angle α. The ice will creep downhill under its own weight.

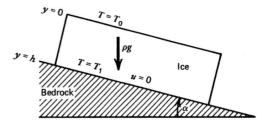

While the temperature rise due to frictional heating is small ($T'/T_0 \ll 1$), the associated decrease in viscosity, $\exp(-E_a T'/RT_0^2)$, may be quite large because $(E_a/RT_0)(T'/T_0)$ can be large. Consider $T'/T_0 = 0.5$ and $E_a/RT_0 = 30$, for example. The boundary conditions on the temperature rise T' are

$$T' = 0 \quad \text{on} \quad y = 0 \tag{7-148}$$

$$\frac{dT'}{dy} = 0 \quad \text{on} \quad y = h. \tag{7-149}$$

The solution for T' can be facilitated by introducing the dimensionless variables

$$\theta = \frac{E_a T'}{RT_0^2} \qquad \bar{y} = \frac{y}{h}. \tag{7-150}$$

The equation for the dimensionless temperature rise θ and its associated boundary conditions are

$$\frac{d^2\theta}{d\bar{y}^2} + \frac{\tau^2 h^2 E_a}{kCRT_0^2} e^{-(E_a/RT_0)} e^{\theta} = 0, \tag{7-151}$$

$$\theta = 0 \quad \text{on} \quad \bar{y} = 0 \tag{7-152}$$

$$\frac{d\theta}{d\bar{y}} = 0 \quad \text{on} \quad \bar{y} = 1. \tag{7-153}$$

The dimensionless coefficient of e^{θ} in Equation (7–151) is known as the *Brinkman number* Br

$$\text{Br} = \frac{\tau^2 h^2 E_a e^{-(E_a/RT_0)}}{kCRT_0^2}. \tag{7-154}$$

In terms of the Brinkman number, the differential equation for the dimensionless temperature increase is simply written as

$$\frac{d^2\theta}{d\bar{y}^2} + \text{Br}\, e^{\theta} = 0. \tag{7-155}$$

The entire temperature distribution in the frictionally heated shear flow is uniquely determined by the single dimensionless parameter Br.

The Brinkman number has a simple physical interpretation. The heat generated per unit horizontal area of the channel at the temperature T_0 is

$$\frac{\tau^2 h}{C e^{E_a/RT_0}}.$$

The conductive heat flux produced by a temperature rise RT_0^2/E_a across the channel is

$$\frac{k(RT_0^2/E_a)}{h},$$

where the temperature increase RT_0^2/E_a is just what is required to reduce the viscosity by the factor e. The Brinkman number, also sometimes known as the *Gruntfest number*, is the ratio of these two heat flows. Qualitatively, Br measures the ability of the fluid to conduct away the frictionally generated heat.

To solve for θ, we make the substitution

$$\phi = e^{\theta}. \tag{7-156}$$

By differentiating this expression, we find

$$\frac{d\phi}{d\bar{y}} = e^{\theta}\frac{d\theta}{d\bar{y}} = \phi\frac{d\theta}{d\bar{y}} \tag{7-157}$$

or

$$\frac{d\theta}{d\bar{y}} = \frac{1}{\phi}\frac{d\phi}{d\bar{y}}. \tag{7-158}$$

Thus we can rewrite Equation (7–155) as

$$\frac{d}{d\bar{y}}\left(\frac{1}{\phi}\frac{d\phi}{d\bar{y}}\right) = -\text{Br}\phi. \tag{7-159}$$

A rearrangement of this equation yields

$$\left(\frac{1}{\phi}\frac{d\phi}{d\bar{y}}\right)\frac{d}{d\bar{y}}\left(\frac{1}{\phi}\frac{d\phi}{d\bar{y}}\right) = -\text{Br}\frac{d\phi}{d\bar{y}}, \tag{7-160}$$

which can be integrated to give

$$\left(\frac{1}{\phi}\frac{d\phi}{d\bar{y}}\right)^2 = -2\text{Br}\phi + c_1, \tag{7-161}$$

where c_1 is a constant of integration. Because $d\theta/d\bar{y}$ is zero on $\bar{y} = 1$ from Equation (7–153), so is $d\phi/d\bar{y}$, and the constant c_1 must be

$$c_1 = 2\text{Br}\phi(1), \tag{7-162}$$

where $\phi(1)$ is ϕ at $\bar{y} = 1$. The equation for ϕ is thus

$$\frac{1}{\phi}\frac{d\phi}{d\bar{y}} = \{2\text{Br}(\phi(1) - \phi)\}^{1/2}. \tag{7-163}$$

The argument of the square root is always positive; that is, $\phi(1) \geq \phi$. This is because $\theta(1) \geq \theta$, a result that follows from the boundary conditions on θ and Equation (7–155). The curvature of θ versus $\bar{y}$ is always negative according to Equation (7–155). Thus θ increases from

0 on $\bar{y} = 0$ to a maximum value $\theta(1)$ on $\bar{y} = 1$ where $d\theta/d\bar{y} = 0$.

Equation (7–163) can be integrated by writing it in the form

$$dy = \frac{d\phi}{\phi\{2\mathrm{Br}[\phi(1) - \phi]\}^{1/2}}. \qquad (7\text{--}164)$$

Integration of this equation yields

$$\bar{y} = \frac{1}{[2\mathrm{Br}\phi(1)]^{1/2}}$$
$$\times \ln\left[c_2\left\{\frac{\{2\mathrm{Br}[\phi(1) - \phi]\}^{1/2} - [2\mathrm{Br}\phi(1)]^{1/2}}{\{2\mathrm{Br}[\phi(1) - \phi]\}^{1/2} + [2\mathrm{Br}\phi(1)]^{1/2}}\right\}\right] \qquad (7\text{--}165)$$

with c_2 a constant. From boundary condition (7–152) and Equation (7–156), we must have $\phi = 1$ on $\bar{y} = 0$. Thus the argument of the log in Equation (7–165) must be 1 when $\phi = 1$. This determines the constant c_2 and leads to the expression

$$\bar{y} = [2\mathrm{Br}\phi(1)]^{-1/2}$$
$$\times \ln\left[\left\{\frac{\{2\mathrm{Br}[\phi(1) - \phi]\}^{1/2} - [2\mathrm{Br}\phi(1)]^{1/2}}{\{2\mathrm{Br}[\phi(1) - \phi]\}^{1/2} + [2\mathrm{Br}\phi(1)]^{1/2}}\right\}\right.$$
$$\left.\times\left\{\frac{\{2\mathrm{Br}[\phi(1) - 1]\}^{1/2} + [2\mathrm{Br}\phi(1)]^{1/2}}{\{2\mathrm{Br}[\phi(1) - 1]\}^{1/2} - [2\mathrm{Br}\phi(1)]^{1/2}}\right\}\right]. \qquad (7\text{--}166)$$

An equation for $\phi(1)$, the exponential of the maximum temperature rise in the channel, can be found by setting $\bar{y} = 1$ in this equation. The result is

$$[2\mathrm{Br}\phi(1)]^{1/2}$$
$$= \ln\left[\frac{\{2\mathrm{Br}[\phi(1) - 1]\}^{1/2} + [2\mathrm{Br}\phi(1)]^{1/2}}{[2\mathrm{Br}\phi(1)]^{1/2} - \{2\mathrm{Br}[\phi(1) - 1]\}^{1/2}}\right] \qquad (7\text{--}167)$$

This can be rearranged in the form

$$\phi(1) = \frac{1}{1 - \left\{\frac{e^{\sqrt{2\mathrm{Br}\phi(1)}} - 1}{e^{\sqrt{2\mathrm{Br}\phi(1)}} + 1}\right\}^2} \qquad (7\text{--}168)$$

or

$$[\phi(1)]^{1/2} = \cosh\left(\frac{\mathrm{Br}\phi(1)}{2}\right)^{1/2}. \qquad (7\text{--}169)$$

Equation (7–169) is a transcendental equation for $\phi(1)$ as a function of the Brinkman number. The maximum temperature increase in the channel $\theta(1)$ is simply $\ln\phi(1)$. The most straightforward way to calculate $\phi(1)$

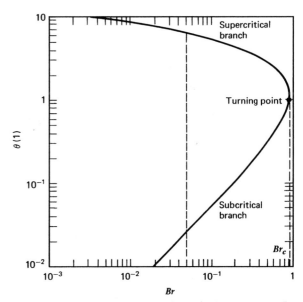

7-17 Maximum temperature $\theta(1)$ in a frictionally heated Couette flow with temperature-dependent viscosity and an adiabatic lower boundary as a function of the Brinkman number Br. There are two solutions, one on the subcritical branch and the other on the supercritical branch, for every value of Br between 0 and $\mathrm{Br}_c = 0.88$.

is to rewrite Equation (7–169) as

$$\mathrm{Br} = \frac{2\left(\frac{\mathrm{Br}\phi(1)}{2}\right)}{\left\{\cosh\left(\frac{\mathrm{Br}\phi(1)}{2}\right)^{1/2}\right\}^2}. \qquad (7\text{--}170)$$

Br can be calculated directly from Equation (7–170) for any given value of $[\mathrm{Br}\phi(1)/2]^{1/2}$. Thus, one assumes a value of the latter parameter, calculates Br, and then determines $\phi(1)$. The maximum temperature $\theta(1)$ calculated in this way is shown in Figure 7–17 as a function of Br. For Br = 0, there is no motion in the channel and $\theta(1) = 0$. As Br increases from zero, by increasing the shear stress applied to the upper wall of the channel, for example, $\theta(1)$ increases. In fact, by using Taylor series expansions of cosh and ln, one finds from Equation (7–169) that $\theta(1)$ increases as

$$\theta(1) \approx \tfrac{1}{2}\mathrm{Br} \qquad (7\text{--}171)$$

for Br $\ll 1$. As Br or the applied shear stress increases, the fluid adjacent to the upper wall of the channel moves faster, resulting in a larger shear and an increase in frictional heating. However, as the upper wall of the channel is pulled increasingly fast, a point is reached where an increase in the applied shear stress is no longer required. In fact, larger velocities and higher temperatures can be achieved by reducing the applied stress or Br. Thus, there are two distinct types of shear flow in the channel. For the class of flows referred to as subcritical, $\theta(1)$ increases with increasing Br. For supercritical flows, $\theta(1)$ increases with decreasing Br. The turning point in Figure 7–17, where Br equals its maximum value Br_c, separates the states on the lower subcritical branch of the $\theta(1)$ versus Br curve from the solutions along the upper supercritical branch.

Figure 7–17 shows that there are multiple solutions to the channel shear flow with frictional heating and temperature-dependent viscosity. For any value of Br between 0 and $Br_c = 0.88$ two possible flows exist, one on the subcritical branch and one on the supercritical branch. For the same value of applied shear stress there are two flows, one having considerably higher temperatures and velocities than the other. However, for a given velocity of the upper wall, or a prescribed temperature at the lower wall, there is a unique flow. Figure 7–17 also indicates that there are no steady channel flows for too large an applied stress, that is, Br > Br_c. Actually, a more exact analysis shows that there is a second turning point at high temperature and a third branch along which $\theta(1)$ again increases with Br. There is therefore a channel flow for any value of Br, but the temperatures along the third or hot branch are so large as to be physically unrealizable, that is, the material in the channel would melt long before reaching the high temperatures of the hot branch. Thus, application of too large a stress to the upper wall, that is, a stress giving Br > Br_c, leads to an instability phenomenon known as *thermal runaway*. Shear heating produces such large temperatures when Br exceeds Br_c that melting occurs. There have been suggestions that thermal runaway might occur under certain situations in the mantle, for example, in the asthenosphere or in the slip zone at the top of a descending slab. However, such possibilities must be regarded as speculative at the present time.

The existence of subcritical and supercritical solutions to shear flow with viscous dissipation and temperature-dependent viscosity can be understood as follows. On the subcritical branch there is relatively little frictional heating and only small increases in tem-

perature. Therefore, when the applied shear stress is increased, a subcritical flow responds as the flow of a constant viscosity fluid would – the shear du/dy and the amount of viscous dissipation both increase. Because of the increased heating, the temperature of the fluid rises and its viscosity drops. This reduces the increase in shear stress somewhat ($\tau = \mu\, du/dy$), but the decrease in viscosity is sufficiently small so that the increase in the shear du/dy dominates. On the supercritical branch large temperature increases occur due to frictional heating. Viscous dissipation and temperature decrease with an increase in shear stress because the viscosity variation with temperature is the controlling factor and μ must increase for τ to increase. This is only possible with a reduction in temperature.

PROBLEM 7–13 Compute the stress that gives a Brinkman number equal to the value at the turning point of Figure 7–17. Assume $E_a/RT_0 = 20$, $h = 100$ km, $\mu_0 = 10^{24}$ Pa s, $T_0 = 1073$ K, and $k = 4$ Wm^{-1} K^{-1}. What is the temperature of the lower boundary for this value of the stress?

PROBLEM 7–14 Calculate the heat flux through the upper boundary for the channel flow with a moving isothermal upper wall and a stationary insulated lower wall. Construct a plot of the heat flux as a function of the Brinkman number. The heat flux through the upper wall is due entirely to heat generated frictionally in the channel. This heat ultimately derives from the work done in moving the upper boundary. Use this principle to derive a formula for the velocity of the upper wall. Construct a plot of the velocity of the upper boundary as a function of Br. Derive expressions for the heat flux through the upper boundary, the velocity of the upper boundary, and the temperature of the lower boundary at the turning point. Calculate numerical values for these quantities for the parameters given in the previous problem.

7–6 Mantle Rheology

In Sections 7–3 and 7–4 we discussed two fundamental mechanisms for the subsolidus deformation of rocks – diffusion creep and dislocation creep. We do not know which of these mechanisms governs flow in the mantle; although, as the following discussion illustrates, we can make some informed guesses. If diffusion creep pertains, the mantle behaves as a Newtonian fluid. If dislocation creep applies, the mantle is a power-law fluid with n approximately equal to 3. While it is important to understand the rheology of the mantle, it is not crucial that we be able to distinguish between a rheological law with $n = 1$ and one with $n = 3$ to obtain a qualitatively correct picture of flow in the mantle. To be sure, if the mantle viscosity were the stress-dependent effective viscosity of dislocation creep, mantle motions would be quantitatively different from those of a Newtonian fluid. However, the temperature and pressure dependences of the viscosity, which are identical for diffusion creep and dislocation creep, are much more important in controlling mantle motions than is any possible stress dependence of mantle viscosity.

One source of information on the rheology of the mantle is the study of postglacial rebound data. As discussed in Section 6–10, these data have been interpreted in terms of a Newtonian fluid mantle with a viscosity of about 10^{21} Pa s. Although the mathematical analysis would be more complex, the rebound data could alternatively be interpreted in terms of a non-Newtonian fluid mantle with a power-law rheology. The inferred viscosity of the mantle would then be the stress-dependent effective viscosity in Equation (7–119). To properly interpret postglacial rebound data, it must be realized that the mantle flow associated with the rebound phenomenon is superimposed on the circulation associated with mantle convection. This superposition of strain rates and stresses is illustrated in Figure 7–18 for both linear and cubic rheologies. The mantle convection strain rate and stress are denoted by $\dot{\varepsilon}_m$ and σ_m, and the rebound strain rate and stress are

$\dot{\varepsilon}_r$ and σ_r. The total strain rate and stress $\dot{\varepsilon}$ and σ are

$$\dot{\varepsilon} = \dot{\varepsilon}_r + \dot{\varepsilon}_m \tag{7–172}$$

$$\sigma = \sigma_r + \sigma_m. \tag{7–173}$$

For the linear rheology

$$\sigma = \mu\dot{\varepsilon}, \tag{7–174}$$

so that

$$\sigma_r + \sigma_m = \mu(\dot{\varepsilon}_r + \dot{\varepsilon}_m). \tag{7–175}$$

But the strain rate and stress of mantle convection are separately related by

$$\sigma_m = \mu\dot{\varepsilon}_m. \tag{7–176}$$

By subtracting Equation (7–176) from (7–175), we find that

$$\sigma_r = \mu\dot{\varepsilon}_r \tag{7–177}$$

or

$$\mu = \frac{\sigma_r}{\dot{\varepsilon}_r}. \tag{7–178}$$

In the case of a Newtonian mantle, the viscosity inferred from the ratio of rebound stress to rebound strain rate is the actual mantle viscosity.

For the power-law rheology with $n = 3$, total strain rate and total stress are related by

$$\dot{\varepsilon} = C\sigma^3 \tag{7–179}$$

or

$$\dot{\varepsilon}_m + \dot{\varepsilon}_r = C(\sigma_m + \sigma_r)^3$$
$$= C\sigma_m^3\left(1 + \frac{\sigma_r}{\sigma_m}\right)^3. \tag{7–180}$$

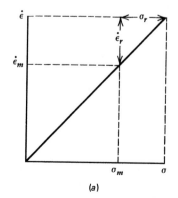
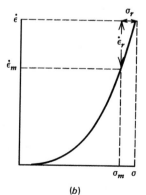

7-18 The strain rate and stress associated with postglacial rebound $\dot{\varepsilon}_r$ and σ_r, superimposed on the strain rate and stress associated with mantle convection $\dot{\varepsilon}_m$ and σ_m for (a) a linear rheology and (b) a cubic rheology.

(a)

(b)

If σ_r is small compared with σ_m, we can approximate the right side of Equation (7–180) as

$$\dot{\varepsilon}_m + \dot{\varepsilon}_r \approx C\sigma_m^3\left(1 + 3\frac{\sigma_r}{\sigma_m}\right) = C\sigma_m^3 + 3C\sigma_m^2\sigma_r. \tag{7–181}$$

The strain rate and stress of mantle convection also satisfy the rheological law

$$\dot{\varepsilon}_m = C\sigma_m^3. \tag{7–182}$$

Upon subtracting Equation (7–182) from (7–181), we obtain

$$\dot{\varepsilon}_r = \left(3C\sigma_m^2\right)\sigma_r. \tag{7–183}$$

The effective viscosity for mantle convection is

$$\mu_{\text{eff},m} = \frac{\sigma_m}{\dot{\varepsilon}_m} = \frac{1}{C\sigma_m^2}. \tag{7–184}$$

The effective viscosity corresponding to the rebound strain rate and stress is

$$\mu_{\text{eff},r} = \frac{\sigma_r}{\dot{\varepsilon}_r} = \frac{1}{3C\sigma_m^2} = \frac{1}{3}\mu_{\text{eff},m}. \tag{7–185}$$

In the non-Newtonian case, the strain rate and stress of postglacial rebound define an effective viscosity that is one-third of the effective viscosity associated with mantle convection.

A typical strain rate for mantle convection is obtained by dividing a velocity of 50 mm yr^{-1} by a depth of 700 km; one gets $\dot{\varepsilon}_m = 2.3 \times 10^{-15}$ s^{-1}. The product of this strain rate with the Newtonian viscosity of 10^{21} Pa s inferred from postglacial rebound data gives a mantle convection stress of 2.3 MPa. If the mantle is non-Newtonian and 10^{21} Pa s is the value of $\mu_{\text{eff},r}$, the effective viscosity of mantle convection is 3 times larger, and the mantle convection stress is 6.9 MPa. Considering the many uncertainties involved in deducing the viscosity and stress level in the mantle, a factor of 3 uncertainty associated with the rheological law is not too serious. Studies of postglacial rebound give important information on the rheology of the mantle, but it is doubtful that they can discriminate between a linear and a third-power rheology.

Another important source of information on mantle rheology is laboratory studies of creep. Since olivine is the primary mineral in the mantle, studies of the high-temperature creep of olivine are particularly relevant. The measured dependence of strain rate $\dot{\varepsilon}_{xx}$ or $-\dot{\varepsilon}_{yy}$ on

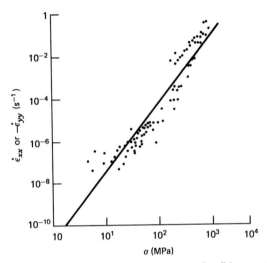

7–19 Observed dependence of strain rate on stress for olivine at a temperature of 1400°C. (Ashby and Verrall, 1978).

stress in dry olivine at a temperature of 1400°C is given in Figure 7–19. The relation

$$\dot{\varepsilon}_{xx} = -\dot{\varepsilon}_{yy} = C_1\sigma^3 e^{-E_a/RT} \tag{7–186}$$

with $C_1 = 4.2 \times 10^5$ MPa^{-3} s^{-1} and $E_a = 523$ kJ mol^{-1} is also shown. The data are well represented by this cubic power-law rheology. (The experimental data in Figure 7–19 were fit by an equation that neglects the weak temperature dependence of the preexponential constant. In addition, the data do not account for the effect of pressure on creep.) The agreement of the data with the theoretical relation for dislocation creep is taken as evidence that dislocation creep is the dominant deformation mechanism over the range of strain rates considered in the laboratory. It should be emphasized, however, that the smallest strain rate studied in the laboratory, approximately 10^{-8} s^{-1}, is some 7 orders of magnitude larger than mantle strain rates. Nevertheless, the theoretical basis for Equation (7–186) is reasonably sound so that its extrapolation to mantle strain rates should be justified.

Other geologic materials have been found to creep nonlinearly when deformed at high temperature in the laboratory. Table 7–4 lists the values of the rheological parameters in the relation

$$\dot{\varepsilon}_{xx} = -\dot{\varepsilon}_{yy} = C_1\sigma^n e^{-E_a/RT} \tag{7–187}$$

for ice, halite, and several crustal rocks including

TABLE 7–4 Rheological Parameter Values of Geologic Materials

Material	$C_1(MPa^{-n} S^{-1})$	n	$E_a(kJ\ mol^{-1})$
Ice	8.8×10^5	3	60.7
Halite	9.5×10^{-1}	5.5	98.3
Dry quartzite	6.7×10^{-12}	6.5	268
Wet quartzite	4.4×10^{-2}	2.6	230
Limestone	4.0×10^3	2.1	210
Maryland diabase	5.2×10^2	3	356

quartzite under wet and dry conditions, limestone, and Maryland diabase.

PROBLEM 7–15 Calculate the stresses required to deform olivine and the crustal rocks listed in Table 7–4 at the constant rate $\dot{\varepsilon}_{xx} = 10^{-15}\ s^{-1}$ for a series of temperatures between 700 and 1300°C, and construct a plot of σ vs. T. Compute the effective viscosities of these materials and plot μ_{eff} vs T. Assume that Equation (7–184), with parameter values determined by laboratory creep tests, is applicable at the very low strain rate of $10^{-15}\ s^{-1}$.

Experiments and theory indicate that a general form of the relationship between strain $\dot{\varepsilon}$ and deviatoric stress σ valid for both diffusion and dislocation creep is given by

$$\dot{\varepsilon}_{xx} = -\dot{\varepsilon}_{yy} = A\left(\frac{\sigma}{G}\right)^n \left(\frac{b}{h}\right)^m \exp\left(-\frac{E_a + pV_a}{RT}\right),$$

(7–188)

where A is the preexponential factor, G is the shear modulus, h is the grain size, and b is the lattice spacing. Presently preferred values for diffusion creep are $n = 1$ and $m = 2.5$ and for dislocation creep $n = 3.5$ and $m = 0$. This relation is in quite good agreement with our derived theoretical relations, Equation (7–95) for diffusion creep and Equation (7–117) for dislocation creep, with the diffusion coefficient given by Equation (7–77).

The parameter values for diffusion creep and dislocation creep in a dry upper mantle are given in Table 7–5. An important question is whether diffusion creep or dislocation creep is the applicable deformation mechanism in the upper mantle. The transition between diffusion creep and dislocation creep occurs when, for a given stress, the strain rates given by the two mechanisms are equal. In general, for a given stress, the deformation mechanism with the larger strain rate prevails. We can delineate the regimes of applicability of rival deformation mechanism, by using a deformation map, which gives stress as a function of temperature for several values of strain rate. A deformation map for a dry upper mantle with $p = 0$, based on Equation (7–188) and the parameter values in Table 7–5, is given in Figure 7–20. The diffusion creep values are based on a grain size $h = 3$ mm; this is a typical value for mantle rocks found in diatremes and in ophiolites. Dislocation creep is the applicable deformation mechanism for high stress levels and high temperatures, while diffusion creep is dominant for low stress levels and low temperatures. Uncertainties in flow law parameters lead to uncertainties of about an order of magnitude in deformation maps. Typical values of $\dot{\varepsilon}$ and T for mantle convection are $\dot{\varepsilon} = 10^{-15}\ s^{-1}$ and $T = 1600$ K; these values locate the solid circle in Figure 7–20 and correspond

TABLE 7–5 Parameter Values for Diffusion Creep and Dislocation Creep in a Dry Upper Mantle (Karato and Wu, 1993)*

Quantity	Diffusion Creep	Dislocation Creep
Preexponential factor A, s^{-1}	8.7×10^{15}	3.5×10^{22}
Stress exponent n	1	3.5
Grain size exponent m	2.5	0
Activation energy E_a, kJ mol^{-1}	300	540
Activation volume V_a, m^3 mol^{-1}	6×10^{-6}	2×10^{-5}

* Other relevant parameter values are $G = 80$ GPa, $b = 0.5$ nm, and $R = 8.3144$ J K^{-1} mol^{-1}.

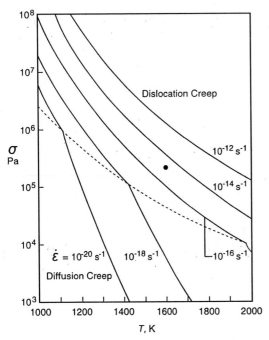

7-20 Deformation map for a dry upper mantle. The deviatoric stress σ is given as a function of temperature T for several strain rates $\dot{\varepsilon}$. The dashed line separates the dislocation creep regime from the diffusion creep regime. The solid circle represents a typical condition for mantle convection.

to $\sigma = 0.5$ MPa. This point clearly falls in the dislocation creep field of the upper mantle deformation map.

The generalized strain rate–deviatoric stress relation, Equation (7–188), can be used to generate a generalized viscosity relation valid for both diffusion and dislocation creep:

$$\mu = C \left(\frac{\sigma}{G} \right)^{1-n} \exp \left\{ \frac{E_a + pV_a}{RT} \right\}, \qquad (7\text{–}189)$$

where

$$C = \frac{1}{2} \left(\frac{G}{A} \right) \left(\frac{h}{b} \right)^m \qquad (7\text{–}190)$$

and the coefficient C depends upon both the rigidity G and the grain size h. For diffusion creep with $n = 1$, grain size $h = 3$ mm, and parameter values from Table 7–5 we have

$$\mu = C_1 \exp \left\{ \frac{E_a + pV_a}{RT} \right\}, \qquad (7\text{–}191)$$

with $C_1 = 4.05 \times 10^{11}$ Pa s. For dislocation creep with $n = 3.5$ and parameter values from Table 7–5 we have

$$\mu = C_2 \left(\frac{G}{\sigma} \right)^{2.5} \exp \left\{ \frac{E_a + pV_a}{RT} \right\}, \qquad (7\text{–}192)$$

with $C_2 = 1.14 \times 10^{-12}$ Pa s.

The viscosities from Equations (7–191) and (7–192) are given in Figure 7–21 as a function of temperature for shallow upper mantle conditions ($p = 0$). The result for diffusion creep (solid line) is independent of stress and results for dislocation creep (dashed lines) are given for $\sigma = 10^5$ and 10^6 Pa. For $\sigma = 10^6$ Pa deformation is due to dislocation creep for the entire range of temperatures considered. For $\sigma = 10^5$ Pa deformation is due to dislocation creep for $T > 1415$ K and to diffusion creep for $T < 1415$ K. For $\sigma = 10^4$ Pa deformation is due to diffusion creep for the entire range of temperatures considered. Typical upper mantle viscosity and temperature values are $\mu = 3 \times 10^{20}$ Pa s and

7-21 Dependence of the viscosity of a dry upper mantle on temperature is given for several stress levels. The solid line is for diffusion creep; the viscosity is not dependent on stress level. The dashed lines are for dislocation creep illustrating the dependence on the stress level. The solid circle represents a typical condition for mantle convection.

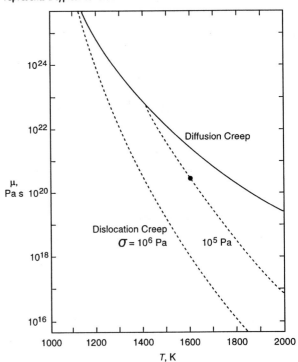

$T = 1600$ K; this condition (filled circle in Figure 7–21) lies in the dislocation creep field with $\sigma = 10^5$ Pa.

We have seen in Section 7–3 that the temperature and pressure dependences of the diffusion coefficient and therefore the strain rate could be written in the form

$$e^{-aT_m/T}$$

instead of

$$e^{-[(E_a + pV_a)/RT]}.$$

By equating the exponents of these expressions, we obtain

$$aT_m = \frac{E_a + pV_a}{R}. \tag{7–193}$$

Because $p = \rho g y$, we can rewrite this as

$$aRT_m = E_a + \rho g V_a y. \tag{7–194}$$

Upon differentiating with respect to depth, treating E_a and V_a as constants, we get

$$V_a = \frac{aR}{\rho g}\frac{dT_m}{dy}. \tag{7–195}$$

We can substitute Equation (7–195) into (7–193) and solve for a to obtain

$$a = \frac{E_a}{R\left(T_m - y\frac{dT_m}{dy}\right)}. \tag{7–196}$$

If we approximate the melting temperature by the linear profile

$$T_m = T_{m0} + y\frac{dT_m}{dy}, \tag{7–197}$$

where T_{m0} is the melting temperature at the surface, the parameter a is approximately

$$a \approx \frac{E_a}{RT_{m0}}. \tag{7–198}$$

For olivine, E_a is 523 kJ mol^{-1} and T_{m0} is 2140 K; the parameter a is thus 29.4. A reasonable value for the increase of the olivine melt temperature with depth in the upper mantle is 2 K km^{-1}. With $\rho = 3500$ kg m^{-3} we find that V_a in the upper mantle is 1.2×10^{-5} m^3 mol^{-1} from Equation (7–195). This is in good agreement with the empirically determined value of 1.34×10^{-5} m^3 mol^{-1} for olivine.

PROBLEM 7–16 Calculate mantle viscosity as a function of depth assuming $\mu \propto \exp\{(E_a + pV_a)/RT\}$. Use $E_a = 523$ kJ mol^{-1} and $V_a = 1.34 \times 10^{-5}$ m^3 mol^{-1}. Normalize the viscosity to the value 10^{21} Pa s at a depth of 150 km. Assume that a single rheological law applies over the entire depth of the mantle and that all rheological parameters and the mantle stress are constant with depth. Assume also that there are no viscosity changes across phase transitions. Use the models for T and p as functions of depth in the mantle developed in Section 4–28. Are your calculated values of μ consistent with the conclusion from postglacial rebound studies that viscosity does not increase substantially across the mantle? If not, which of the assumptions would you change in order to obtain a viscosity that is more nearly constant with depth?

7-7 Rheological Effects on Mantle Convection

In Chapter 6 we studied thermal convection in a Newtonian fluid with constant fluid properties. We developed a boundary-layer theory for convection at high Rayleigh number and showed that the boundary layer that grew adjacent to the upper cold boundary of the fluid was similar in structure to the oceanic lithosphere. Also, velocities obtained from the boundary-layer analysis were in reasonably good agreement with the velocities of the surface plates.

In this chapter we have shown that the mantle is likely to be a non-Newtonian fluid with an effective viscosity dependent on the exponential of the inverse absolute temperature and the pressure. It is important to consider how this rheology affects mantle convection. The studies of the preceding sections have indicated that the power-law rheology with $n \approx 3$ is likely to have a relatively minor influence, while the strong temperature dependence is certain to have important effects. For example, the temperature dependence of the rheology allows the lithosphere to develop rigidity as a consequence of the cold temperatures in the surface thermal boundary layer. However, the rigidity of the lithosphere has been incorporated into our analysis of its thermal evolution in Section 4–16, and this same thermal development has been applied to the fluid boundary layer in Section 6–19. Thus, this effect of temperature-dependent rheology does not directly modify the age dependences of such measurable

TABLE 7–6 **Approximate Aspect Ratios of Mantle Convection Cells**

Plate	Upper Mantle Convection	Whole Mantle Convection
Pacific	14	3.3
North American	11	2.6
South American	11	2.6
Indian	8	2.1
Nazca	6	1.6

quantities as oceanic heat flow, ocean floor topography, and the oceanic geoid.

One way in which the rigidity of the lithosphere has an important influence on mantle convection is by inhibiting subduction. A cold fluid boundary layer will separate from the upper boundary and sink more readily than an elastic plate. Thus the aspect ratios of mantle convection cells are generally larger than those of cells in a constant viscosity fluid. In Chapter 6 we found that the aspect ratios of convection cells in constant viscosity fluids were near unity. Table 7–6 lists approximate values of the aspect ratios of convection cells associated with the major tectonic plates for convection restricted to the upper mantle and for whole mantle convection. Because the gravitational instability of the lithosphere is inhibited by its rigidity, the aspect ratios of mantle convection cells are larger than unity.

If there are thermal boundary layers elsewhere in the mantle, for example, at the mantle–core interface, the temperature dependence of μ would produce strong viscosity variations across such layers. Because a core–mantle boundary layer would be hotter than the overlying mantle, the viscosity in such a boundary layer would be significantly reduced. Plumes represent another situation in which localized temperature contrasts could cause large associated variations in mantle viscosity. The lowered viscosity in a hot narrow mantle plume would facilitate the upwelling of plume material in the surrounding more viscous mantle. Mantle plumes could originate by a Rayleigh–Taylor or gravitational instability of hot, light, and relatively inviscid material in a mantle–core boundary layer.

It is possible that the mantle is divided into upper and lower convection systems. Such a division could be caused by a change in mantle composition. A compositional boundary would act as a barrier to thermal convection so that separate upper and lower mantle convection systems would be expected. A thermal boundary layer would develop between the systems, and the lower mantle would be expected to have a significantly higher temperature than the upper mantle. Associated with the higher temperature would be a lower viscosity. However, the postglacial rebound data, which suggest that the mantle has a nearly uniform viscosity, argue against separate upper and lower mantle convection systems.

The nearly uniform viscosity of the mantle can be understood in terms of its strong temperature and pressure dependences. The increase of temperature with depth in the mantle tends to decrease mantle viscosity with depth. However, the increase of pressure with depth tends to increase mantle viscosity with depth. These competing effects cancel each other, thereby producing a mantle with nearly constant viscosity.

The viscosity of the mantle can also be understood by considering the relation between the mantle geotherm and its solidus, as indicated by Equation (7–101). Figure 7–22 is a sketch of the geotherm, the solidus,

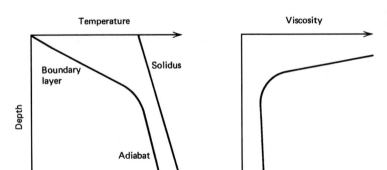

7–22 The closest approach of the geotherm to the solidus in the upper mantle leads to a weak viscosity minimum.

and the viscosity profile. The rapid increase in temperature across the surface thermal boundary layer brings the geotherm closer to the solidus as depth increases. The large associated decrease of viscosity with depth corresponds to the decrease in viscosity between the highly viscous, essentially rigid lithosphere and the underlying mantle. With a further increase in depth, T rises along an adiabat that increases slightly less steeply with depth than does the solidus. The ratio T_m/T therefore has a weak minimum in the upper mantle, and according to Equation (7–101) so does the viscosity. The region of the weak viscosity minimum may be associated with the asthenosphere, a zone that may decouple the lithosphere from the underlying mantle by a large shear in the mantle velocity. The velocity shear would be a direct consequence of the reduction of viscosity in a region of constant shear stress.

7–8 Mantle Convection and the Cooling of the Earth

The amount of heat escaping through the Earth's surface at the present time is due to the heat now being generated by the decay of radioactive isotopes in the Earth and to the cooling of the Earth. The decrease in the temperature of the Earth's interior with time is a consequence of the decay of its radiogenic heat sources. For example, 3 billion years ago the rate of heat production was about twice as great as it is today (see Section 4–5). As the heat generation decreases, the temperature of the convecting mantle also decreases. The strong temperature dependence of the mantle's rheology causes its viscosity to increase, and it convects less vigorously to transport the reduced amount of internally generated heat. In this section we develop a simple model of mantle cooling by the convection of a fluid with strongly temperature-dependent viscosity that allows us to estimate the rate of decrease of mantle temperature with time and the relative contributions of cooling and heat production to the present surface heat flow.

An upper limit to the rate at which the Earth's temperature is decreasing with time can be obtained by attributing the entire surface heat flow to the cooling of the Earth. The heat lost through the Earth's surface is the product of its surface area $4\pi a^2$ where a is the Earth's radius, with its mean surface heat flux $\bar{q}$. The thermal energy of the Earth is the product of its mass $\frac{4}{3}\pi a^3 \bar{\rho}$, where $\bar{\rho}$ is the Earth's mean density, with its mean specific heat $\bar{c}$ and its mean temperature $\bar{T}$. Upon equating the rate of heat flow through the surface to the rate of decrease of the Earth's thermal energy, we obtain

$$4\pi a^2 \bar{q} = -\frac{4}{3}\pi a^3 \bar{\rho}\,\bar{c}\frac{d\bar{T}}{dt} \tag{7–199}$$

or

$$\frac{d\bar{T}}{dt} = -\frac{3\bar{q}}{a\bar{\rho}\,\bar{c}}. \tag{7–200}$$

With $\bar{q} = 87\,\text{mW m}^{-2}$, $\bar{\rho} = 5520\,\text{kg m}^{-3}$, $\bar{c} = 920\,\text{J kg}^{-1}$ K^{-1}, and $a = 6371$ km, we find from Equation (7–200) that $d\bar{T}/dt = -254$ K Gyr^{-1}. The actual rate of cooling of the Earth must be smaller than this because a significant fraction of the surface heat loss comes from radioactive heat generation.

In order to determine the actual rate of mantle cooling, it is necessary to relate the temperature of the convecting mantle T_1 to its volumetric rate of heat production ρH. The temperature of an internally heated convecting layer has been discussed in Section 6–21. We apply the considerations of that section to the mantle and combine Equations (6–324), (6–342), (6–343), and (6–380) to obtain

$$\frac{2k(T_1 - T_0)}{\rho Hb^2} = 2.98\left(\frac{k\kappa\mu}{\alpha\rho^2 g H b^5}\right)^{1/4}, \tag{7–201}$$

where T_0 is the surface temperature and b is the depth of the mantle. This is the equation connecting mantle temperature and heat production. However, it is not as simple as it appears at first glance because the mantle viscosity μ is a strong function of T_1. From Equation (7–130) we can write

$$\mu = \mu_r \exp\left(\frac{E_a}{RT_1}\right), \tag{7–202}$$

where μ_r is a constant of proportionality. By substituting Equation (7–202) into (7–201), we get

$$\frac{2k(T_1 - T_0)}{\rho Hb^2} = 2.98\left(\frac{k\kappa\mu_r}{\alpha\rho^2 g H b^5}\right)^{1/4} \exp\left(\frac{E_a}{4RT_1}\right). \tag{7–203}$$

The radioactive heat sources in the mantle decay exponentially with time according to

$$H = H_0 e^{-\lambda(t-t_0)}, \tag{7–204}$$

where H_0 is the present rate of heat production, t_0 is the present value of the time t, and λ is the average decay constant for the mixture of radioactive isotopes in the mantle, see Equation (4–8). As H decreases, T_1 must also decrease to maintain the balance of heat production and convective heat transport expressed by Equation (7–203). Upon substituting Equation (7–204) into (7–203), we obtain

$$\frac{2k(T_1 - T_0)}{\rho H_0 b^2} = 2.98\left(\frac{k\kappa\mu_r}{\alpha\rho^2 g H_0 b^5}\right)^{1/4}$$
$$\times \exp\left(\frac{E_a}{4RT_1}\right)\exp\left(-\frac{3\lambda(t - t_0)}{4}\right). \tag{7–205}$$

This equation gives the dependence of the mantle temperature on time. It can be simplified considerably by noting that the present mantle temperature T_{10} must satisfy the equation at $t = t_0$

$$\frac{2k(T_{10} - T_0)}{\rho H_0 b^2} = 2.98\left(\frac{k\kappa\mu_r}{\alpha\rho^2 g H_0 b^5}\right)^{1/4}\exp\left(\frac{E_a}{4RT_{10}}\right). \tag{7–206}$$

The division of Equation (7–205) by Equation (7–206) yields

$$\frac{T_1 - T_0}{T_{10} - T_0} = \exp\left\{\frac{E_a}{4R}\left(\frac{1}{T_1} - \frac{1}{T_{10}}\right)\right\}$$
$$\times \exp\left\{-\frac{3\lambda}{4}(t - t_0)\right\}. \tag{7–207}$$

We can find the mantle cooling rate by differentiating Equation (7–207) with respect to time:

$$\frac{dT_1}{dt}\left(\frac{1}{T_{10} - T_0}\right) = \exp\left\{\frac{E_a}{4R}\left(\frac{1}{T_1} - \frac{1}{T_{10}}\right)\right\}$$
$$\times\left(-\frac{E_a}{4RT_1^2}\right)\frac{dT_1}{dt}\exp\left\{-\frac{3\lambda}{4}(t - t_0)\right\}$$
$$+\exp\left\{\frac{E_a}{4R}\left(\frac{1}{T_1} - \frac{1}{T_{10}}\right)\right\}\left(-\frac{3\lambda}{4}\right)$$
$$\times\exp\left\{-\frac{3\lambda}{4}(t - t_0)\right\}. \tag{7–208}$$

We can rewrite this equation by using Equation (7–207) to substitute for the exponential factors:

$$\frac{dT_1}{dt}\left(\frac{1}{T_{10} - T_0}\right) = \left(\frac{T_1 - T_0}{T_{10} - T_0}\right)\left(\frac{-E_a}{4RT_1^2}\right)\frac{dT_1}{dt}$$
$$-\frac{3\lambda}{4}\left(\frac{T_1 - T_0}{T_{10} - T_0}\right). \tag{7–209}$$

A further rearrangement yields

$$\frac{dT_1}{dt}\left\{1 + \frac{E_a}{4RT_1^2}(T_1 - T_0)\right\} = -\frac{3\lambda}{4}(T_1 - T_0). \tag{7–210}$$

The second term in the brackets on the left side of Equation (7–210) is much larger than unity because the term is approximately $E_a/4RT_1(T_0 \ll T_1)$, a quantity we have estimated to be about 10. The approximate mantle cooling rate is therefore

$$\frac{dT_1}{dt} = -3\lambda\left(\frac{RT_1^2}{E_a}\right). \tag{7–211}$$

The present cooling rate is

$$\frac{dT_1}{dt}(t = t_0) = -3\lambda\left(\frac{RT_{10}^2}{E_a}\right). \tag{7–212}$$

The rate at which the mantle is cooling is independent of its thickness, its present rate of heat generation, and the reference viscosity.

A numerical estimate of the mantle cooling rate based on Equation (7–212) depends on only three reasonably well-known mantle parameters: $-\lambda$, T_{10}, and E_a/RT_{10}. From the discussion of Section 7–6 we take $E_a/RT_{10} = 30$. The considerations in Section 4–28 give a mean mantle temperature $T_{10} = 2250$ K, and from Section 4–5 we obtain $\lambda = 2.77 \times 10^{-10}$ yr^{-1}. The cooling rate is found to be 62 K Gyr^{-1}. From Equation (7–200), this cooling rate contributes 21 mW m^{-2} to the mean surface heat flow of 87 mW m^{-2}. Thus the cooling of the Earth is responsible for about 25% of the Earth's heat loss, while 75% is attributable to radiogenic heating. There is little room for uncertainty in this conclusion. The mantle cools at a relatively slow rate because its temperature is buffered by the strong temperature dependence of its viscosity. As the rate of heat production in the mantle decreases, less vigorous convection is required to transport the heat to the Earth's surface. As a result the required Rayleigh number is less. However, the Rayleigh number is inversely proportional to the mantle viscosity, and this viscosity is an exponential function of the inverse absolute temperature. Therefore only a relatively small decrease in temperature suffices to produce the required increase in viscosity, decrease in Rayleigh number, and decrease in convective heat flux.

PROBLEM 7-17 The way in which subsolidus convection with temperature-dependent viscosity regulates the Earth's thermal history can be quantitatively assessed using the following simple model. Assume that the Earth can be characterized by the mean temperature $\bar{T}$ and that Equation (7–200) gives the rate of cooling. Let the model Earth begin its thermal evolution at time $t = 0$ with a high temperature $\bar{T}(0)$ and cool thereafter. Disregard the heating due to the decay of radioactive isotopes and assume that the Earth cools by vigorous subsolidus convection. Show that the mean surface heat flow $\bar{q}$ can be related to the mean temperature by

$$\bar{q} = 0.74k \left(\frac{\rho g \alpha_v}{\mu \kappa} \right)^{1/3} (\bar{T} - T_0)^{4/3}. \qquad (7\text{–}213)$$

Use Equations (6–316) and (6–337) and assume that the total temperature drop driving convection is twice the difference between the mean temperature $\bar{T}$ and the surface temperature T_0.

Following Equation (7–100), assume that the viscosity is given by

$$\mu = C\bar{T} \exp \left(\frac{E_a}{R\bar{T}} \right) \qquad (7\text{–}214)$$

and write the cooling formula as

$$\frac{d\bar{T}}{dt} = -\frac{2.2\kappa}{a} \left(\frac{\rho g \alpha_v}{C\kappa} \right)^{1/3} \bar{T} \exp \left(-\frac{E_a}{3R\bar{T}} \right). \qquad (7\text{–}215)$$

Equation (7–215) was obtained assuming $(\bar{T} - T_0)^{4/3} \approx \bar{T}^{4/3}$, a valid simplification since $T_0 \ll \bar{T}$. Integrate the cooling formula and show that

$$\text{Ei} \left(\frac{E_a}{3R\bar{T}} \right) - \text{Ei} \left(\frac{E_a}{3R\bar{T}(0)} \right) = \frac{2.2\kappa}{a} \left(\frac{\rho g \alpha_v}{C\kappa} \right)^{1/3}, \quad t \qquad (7\text{–}216)$$

where Ei is the exponential integral. Calculate and plot $\bar{T}/\bar{T}(0)$ versus t for representative values of the parameters in Equation (7–216). Discuss the role of the temperature dependence of the viscosity in the cooling history.

Note: The exponential integral Ei is distinct from the exponential integral E_1 defined in Problem 4–35 and listed in Table 8–4. Ei(x) is $-f_{-x}^{\infty}(e^{-t}/t)\,dt$, where f indicates that the path of integration excludes the origin and does not cross the negative real axis. In addition, x should be positive. Values of Ei(x) are given in Table 7–7.

TABLE 7-7 Values of the Exponential Integral
$\text{Ei}(x) = -f_{-x}^{\infty}(e^{-t}/t)\,dt.$

x	Ei(x)	xe^{-x}Ei(x)
0	$-\infty$	
0.01	−4.01793	
0.02	−3.31471	
0.03	−2.89912	
0.04	−2.60126	
0.05	−2.36788	
0.10	−1.62281	
0.20	−0.82176	
0.30	−0.30267	
0.40	0.10477	
0.50	0.45422	
0.60	0.76988	
0.70	1.06491	
0.80	1.34740	
0.90	1.62281	
1.0	1.89512	
1.2	2.44209	
1.4	3.00721	
1.6	3.60532	
1.8	4.24987	
2.0	4.95423	
2.5	7.07377	
3.0	9.93383	
3.5	13.92533	
4.0	19.63087	
4.5	27.93370	
5.0	40.18524	
6	85.98976	
7		1.22241
8		1.18185
9		1.15276
10		1.13147

7-9 Crustal Rheology

Near-surface rocks exhibit not only brittle behavior resulting in joints and faults, but also fluidlike deformation, as evidenced by the occurrence of folds at all spatial scales. Folding can be attributed to either plastic deformation or fluid behavior; there is observational evidence of both. Plastic deformation is discussed in Section 7–11. In this section we are concerned with how relatively cool crustal rocks can behave as a fluid.

The textures of many folded rocks indicate that the deformation that led to the folding was the result of diffusive mass transfer. However, studies of metamorphic reactions in the rocks show that the temperature at the time of folding was only a small fraction of the solidus temperature. Therefore the deformation could not have been the result of the thermally activated diffusion of atoms discussed in the previous section. Instead, it is inferred that the rate of diffusive mass transport was enhanced by the presence of an intergranular fluid film through a process known as *pressure solution* in which material is forced into solution in regions of high pressure or stress and is precipitated in regions of low pressure or stress. Pressure solution creep is similar to Coble creep in that they both involve mass transport along intergranular boundaries.

An example of deformation due to pressure solution is the compaction of sediments. Consider the collection of quartz sand grains shown in Figure 7–23a. The pore spaces between the sand grains are assumed to be filled with water. As long as the sand grains are more dense than water, the excess mass of the grains must be supported on the contacts between the grains and the pressure on the contacts exceeds the pressure in the water. The actual pressure at the contacts depends on their area and the elastic response of the grains. Because the pressure on the contacts is higher, quartz tends to dissolve on the contacts and be deposited on the free

7-23 Compaction and deformation of sand grains by pressure solution. (*a*) Initially undeformed grains with nearly point contacts. (*b*) Deformed grains with widened contacts due to minerals entering solution. The thick lines represent grain growth on free surfaces caused by mineral precipitation.

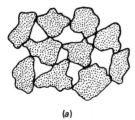

(a)

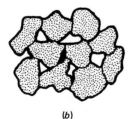

(b)

surfaces of the grains where the pressure is lower. In effect, silica diffuses through the intergranular film of water from the contacts where it dissolves to the free surfaces where it precipitates. This process of solution and precipitation leads to the structure in Figure 7–23b, which is well documented in sandstones.

Pressure solution is believed to play an important role in the continuum deformation of a wide variety of crustal rocks. Its occurrence has been verified in many folded crustal rocks. As long as water is present along grain boundaries, it can act as a solvent for the minerals constituting the grains. The dissolved minerals then diffuse along the grain boundaries from regions of high stress where the solubility is high to regions of low stress where the solubility is low. The diffusion of the dissolved minerals results in creep.

Pressure solution creep can be dealt with quantitatively in a manner analogous to the way in which grain boundary diffusion creep was treated in Section 7–3. We again consider a cubic crystal with an initial dimension h subjected to a compressive stress σ in the x direction and a tensional stress $-\sigma$ in the y direction, as shown in Figure 7–8. We assume that the crystal is completely surrounded by a water film in which the minerals of the crystal can dissolve. In the absence of an applied deviatoric stress there is an equilibrium concentration of minerals or solute C_{s0} in the water film; C_{s0} (kg of solute per kg of solution) depends on both pressure and temperature. Because the mass fraction of solute is a small quantity, the equilibrium number density of solute n_{s0} can be calculated from

$$n_{s0} = \frac{\rho_e C_{s0}}{M_s}, \qquad (7\text{–}217)$$

where ρ_e is the density of the solvent (water) and M_s is the molecular mass of the solute. In the presence of an applied deviatoric stress σ, the solute concentration C_s is

$$C_s = C_{s0} + C_s', \qquad (7\text{–}218)$$

where the stress dependence of C_s' is given by the empirical relation

$$C_s' = \frac{\sigma}{\sigma_s(T)}. \qquad (7\text{–}219)$$

The function $\sigma_s(T)$ has the approximate value of 300 MPa for the solubility of silica in water at 500°C.

Equation (7–219) shows that solubility increases under compression, that is, minerals dissolve, and decreases under tension, that is, minerals precipitate. The number density of solute n_s under the action of an applied stress is

$$n_s = n_{s0} + n_s' = \frac{\rho_e}{M_s}(C_{s0} + C_s')$$
$$= n_{s0} + \frac{\rho_e \sigma}{M_s \sigma_s}. \tag{7-220}$$

The solute number densities at the faces of the crystal in Figure 7–8 are therefore

$$n_A = n_C = n_{s0} + \frac{\rho_e \sigma}{M_s \sigma_s} \tag{7-221}$$

$$n_B = n_D = n_{s0} - \frac{\rho_e \sigma}{M_s \sigma_s} \tag{7-222}$$

$$n_E = n_F = n_{s0}. \tag{7-223}$$

The fluxes of solute molecules through the fluid film surrounding the crystal from faces A and C to faces B and D are, by analogy with Equation (7–102),

$$J_{AB} = J_{AD} = J_{CB} = J_{CD}$$
$$= \frac{D_s}{h}(n_A - n_B)$$
$$= \frac{D_s}{h}\left\{ \left(n_{s0} + \frac{\rho_e \sigma}{M_s \sigma_s}\right) - \left(n_{s0} - \frac{\rho_e \sigma}{M_s \sigma_s}\right)\right\}$$
$$= \frac{D_s}{h}\frac{2\rho_e \sigma}{M_s \sigma_s}, \tag{7-224}$$

where D_s is the diffusion coefficient for the solute in the solvent. Similarly, the fluxes of solute molecules from faces A and C to faces E and F, and from faces E and F to faces B and D, are, by analogy with Equations (7–103) and (7–104),

$$J_{AE} = J_{AF} = J_{CE} = J_{CF} = \frac{D_s}{h}(n_A - n_E)$$
$$= \frac{D_s}{h}\left\{ n_{s0} + \frac{\rho_e \sigma}{M_s \sigma_s} - n_{s0}\right\}$$
$$= \frac{D_s}{h}\frac{\rho_e \sigma}{M_s \sigma_s} \tag{7-225}$$

$$J_{EB} = J_{ED} = J_{FB} = J_{FD} = \frac{D_s}{h}(n_E - n_B)$$
$$= \frac{D_s}{h}\left\{ n_{s0} - \left(n_{s0} - \frac{\rho_e \sigma}{M_s \sigma_s}\right)\right\}$$
$$= \frac{D_s}{h}\frac{\rho_e \sigma}{M_s \sigma_s}. \tag{7-226}$$

The area over which each flux occurs is $h\delta$, where δ is the width of the grain boundary. The strain associated with the transfer of each atom is $b^3/h^3 = 1/n_s h^3$, where n_s is the number density of the solute. Therefore, by analogy with Equations (7–105) and (7–106), we can write the net strain rates as

$$\dot{\varepsilon}_{yy} = \frac{-h\delta}{n_s h^3}(J_{AB} + J_{AD} + J_{CB} + J_{CD} + J_{EB}$$
$$+ J_{ED} + J_{FB} + J_{FD}) \tag{7-227}$$
$$\dot{\varepsilon}_{xx} = \frac{h\delta}{n_s h^3}(J_{AB} + J_{AD} + J_{CB} + J_{CD} + J_{AE}$$
$$+ J_{AF} + J_{CE} + J_{CF}). \tag{7-228}$$

By substituting Equations (7–224) to (7–226) into (7–227) and (7–228), we obtain

$$\dot{\varepsilon}_{xx} = -\dot{\varepsilon}_{yy} = \frac{12\delta\rho_e D_s}{h^3 \rho_s \sigma_s}\sigma, \tag{7-229}$$

where ρ_s is the solute density $M_s n_s$.

Strain rate is linearly proportional to applied stress in pressure solution creep. Thus the deformation is equivalent to that of a Newtonian fluid with a viscosity

$$\mu_{ps} = \frac{h^3 \rho_s \sigma_s}{24\delta\rho_e D_s}. \tag{7-230}$$

For the pressure solution creep of quartz at 500°C we take $h = 2 \times 10^{-3}$ m, $D_s = 2.4 \times 10^{-8}$ m^2 s^{-1}, $\delta = 10^{-9}$ m, $\rho_s = 2700$ kg m^{-3}, $\rho_e = 1000$ kg m^{-3}, and $\sigma_s = 300$ MPa. The calculated value of the viscosity is $\mu_{ps} = 1.1 \times 10^{16}$ Pa s, a very low viscosity for crustal rocks at a temperature of 500°C. It should be emphasized that there are many uncertainties in the application of Equation (7–230). The value of the diffusion coefficient we used was determined in bulk experiments and its relevance to grain boundary films is in question. The value we used for the width of the grain boundary is only an estimate. Nevertheless, pressure solution creep is clearly an important deformation mechanism for crustal rocks and it can explain viscous folding of rocks at quite low temperatures.

7-10 Viscoelasticity

Seismic shear waves propagate through the Earth's mantle with relatively little attenuation. Therefore we conclude that the mantle is an elastic solid. However, we have shown conclusively that the crystalline solid behaves as a viscous fluid on geological time scales as short as the 10^4 yr characteristic of postglacial rebound. The mantle behaves as an elastic solid on time scales of the order of 1 to 10^4 seconds but behaves as a viscous fluid on time scales of 10^{11} to 10^{17} seconds.

A material that behaves elastically on short time scales and viscously on long time scales is known as a viscoelastic material. The Maxwell model of a viscoelastic material consists of a material in which the rate of strain $\dot{\varepsilon}$ is the superposition of a linear elastic strain rate $\dot{\varepsilon}_e$ produced by the rate of change of stress $\dot{\sigma}$ and a linear viscous strain rate $\dot{\varepsilon}_f$ produced by the stress σ. Consider such a medium subjected to the uniaxial stress σ. The elastic strain of the material as given by Equation (3–14) is

$$\varepsilon_e = \frac{\sigma}{E}, \tag{7–231}$$

where, it will be recalled, E is Young's modulus. The rate of strain of a Newtonian viscous fluid subjected to a deviatoric normal stress σ is given by Equation (6–60) as

$$\frac{d\varepsilon_f}{dt} = -\frac{\partial u}{\partial x} = \frac{\sigma}{2\mu}. \tag{7–232}$$

The minus sign connecting $\dot{\varepsilon}_f$ and $\partial u/\partial x$ arises from our sign convention that treats compressive strains as positive. The total strain ε is the sum of the elastic and fluid strains

$$\varepsilon = \varepsilon_e + \varepsilon_f. \tag{7–233}$$

The total rate of strain is therefore the sum of $\dot{\varepsilon}_f$ from Equation (7–232) and the time derivative $\dot{\varepsilon}_e$ of (7–231)

$$\frac{d\varepsilon}{dt} = \frac{1}{2\mu}\sigma + \frac{1}{E}\frac{d\sigma}{dt}. \tag{7–234}$$

This is the fundamental rheological law relating strain rate, stress, and rate of change of stress for a Maxwell viscoelastic material.

Let us consider what will happen if we suddenly apply a strain ε_0 to this viscoelastic medium at $t = 0$ and maintain the strain constant for $t > 0$. During the very rapid application of strain the time derivative terms in Equation (7–234) dominate and the material behaves elastically. Therefore, the initial stress σ_0 at $t = 0$ is

$$\sigma_0 = E\varepsilon_0. \tag{7–235}$$

Subsequently, there is no change in the strain, $d\varepsilon/dt = 0$, and Equation (7–234) reduces to

$$0 = \frac{1}{2\mu}\sigma + \frac{1}{E}\frac{d\sigma}{dt} \tag{7–236}$$

or

$$\frac{d\sigma}{\sigma} = -\frac{E}{2\mu}dt. \tag{7–237}$$

This equation can be integrated with the initial condition $\sigma = \sigma_0$ at $t = 0$ to give

$$\sigma = \sigma_0 \exp\left(-\frac{Et}{2\mu}\right). \tag{7–238}$$

The stress relaxes to $1/e$ of its original value in a time

$$\tau_{ve} = \frac{2\mu}{E}. \tag{7–239}$$

This is known as the *viscoelastic relaxation time*. The relaxation time for the asthenosphere can be estimated by taking $\mu = 4 \times 10^{19}$ Pa s and $E = 70$ GPa with the result $\tau_{ve} = 36$ years. As expected, this time is intermediate between the periods of seismic waves and the times associated with postglacial rebound.

An example of an instantaneous application of strain is the coseismic displacement on a fault. This displacement occurs in a few seconds, and a change in the regional state of stress occurs. If the temperature of the rock is sufficiently high, this stress is relaxed by solid-state creep processes.

PROBLEM 7–18 Determine the response of a Maxwell viscoelastic material to the sudden application of a stress σ_0 at time $t = 0$ assuming that the stress is maintained constant for $t > 0$. What is the initial value of the strain ε_0? Describe what will happen if the stress is removed at time $t = t_1 > 0$.

PROBLEM 7–19 Another model of viscoelastic behavior is the Kelvin model, in which the stress σ in the medium for a given strain ε and strain rate $\dot{\varepsilon}$ is the superposition of linear elastic and linear viscous stresses, σ_e and σ_f. Show that the rheological law for the Kelvin viscoelastic material is

$$\sigma = \varepsilon E + 2\mu\frac{d\varepsilon}{dt}. \tag{7–240}$$

Show also that the response of the Kelvin viscoelastic

material to the sudden application of a stress σ_0 at time $t = 0$ is

$$\varepsilon = \frac{\sigma_0}{E}\left(1 - e^{-t/\tau_{ve}}\right). \tag{7–241}$$

Assume that $\sigma = \sigma_0$ for $t > 0$. While stresses decay exponentially with time in a Maxwell material subjected to constant strain, strain relaxes in the same way in a Kelvin material subjected to constant stress.

PROBLEM 7–20 Generalize the rheological law, Equation (7–234), for a Maxwell viscoelastic material to a three-dimensional state of stress and strain by appropriately combining the linear elastic equations (3–4) to (3–6) with the linear viscous equations (6–60) and (6–61) and the obvious extension of the viscous equations to the third dimension. Use the idea that strain components add and stress components are identical to show that

$$\dot{\varepsilon}_1 = \frac{\dot{\sigma}_1}{E} - \frac{\nu}{E}(\dot{\sigma}_2 + \dot{\sigma}_3) + \frac{1}{2\mu}(\sigma_1 - p) \tag{7–242}$$

$$\dot{\varepsilon}_2 = \frac{\dot{\sigma}_2}{E} - \frac{\nu}{E}(\dot{\sigma}_1 + \dot{\sigma}_3) + \frac{1}{2\mu}(\sigma_2 - p) \tag{7–243}$$

$$\dot{\varepsilon}_3 = \frac{\dot{\sigma}_3}{E} - \frac{\nu}{E}(\dot{\sigma}_1 + \dot{\sigma}_2) + \frac{1}{2\mu}(\sigma_3 - p) \tag{7–244}$$

where

$$p = \tfrac{1}{3}(\sigma_1 + \sigma_2 + \sigma_3). \tag{7–245}$$

Determine the stresses and strains in a Maxwell viscoelastic medium in a state of uniaxial strain $\varepsilon_2 \neq 0$, $\varepsilon_1 = \varepsilon_3 \equiv 0$. Assume that a stress $\sigma_2 = \sigma_0$ is suddenly applied at $t = 0$ and that $\sigma_2 = \sigma_0$ for $t > 0$. Assume also that there is no preferred horizontal direction, that is, take $\sigma_1 = \sigma_3$. Prove that

$$\sigma_1 = \sigma_3 = \sigma_0\left\{1 + \frac{(2\nu - 1)}{(1 - \nu)}\exp\left(\frac{-Et}{6\mu(1 - \nu)}\right)\right\} \tag{7–246}$$

$$\varepsilon_2 = \frac{\sigma_0}{E}(1 - 2\nu)\left\{3 + \frac{2(2\nu - 1)}{(1 - \nu)}\exp\left(\frac{-Et}{6\mu(1 - \nu)}\right)\right\}. \tag{7–247}$$

Discuss the behavior of the Maxwell material in the limits $t \rightarrow 0$ and $t \rightarrow \infty$.

A simple viscoelastic model can be used to determine if the elastic stresses in the lithosphere are relaxed by subsolidus creep. We assume that relaxation of litho-

spheric stresses occurs by dislocation creep, and, accordingly, we modify the rheological law for the Maxwell solid by using the stress-dependent effective viscosity μ_{eff} for dislocation creep. From the rate of strain-stress relation for dislocation creep, Equation (7–186), we can write the effective viscosity as

$$\mu_{\text{eff}} = \frac{1}{2C_1\sigma^2}e^{E_a/RT}. \tag{7–248}$$

Upon substituting Equation (7–248) into (7–234), we obtain the viscoelastic relation

$$\frac{d\varepsilon}{dt} = C_1\sigma^3 e^{-E_a/RT} + \frac{1}{E}\frac{d\sigma}{dt}. \tag{7–249}$$

We again consider the case in which a constant strain is applied instantaneously at $t = 0$ with the resultant initial stress σ_0. Since the strain is constant, Equation (7–249) reduces to

$$0 = C_1\sigma^3 e^{-E_a/RT} + \frac{1}{E}\frac{d\sigma}{dt} \tag{7–250}$$

or

$$\frac{d\sigma}{\sigma^3} = -EC_1 e^{-E_a/RT}\,dt, \tag{7–251}$$

which can readily be integrated to yield

$$\sigma = \left\{\frac{1}{\sigma_0^2} + 2EC_1 t e^{-E_a/RT}\right\}^{-1/2}. \tag{7–252}$$

The time τ_r for the stress σ_0 to relax to one-half of its original value is

$$\tau_r = \frac{3}{2EC_1\sigma_0^2}e^{E_a/RT}. \tag{7–253}$$

If we base μ_{eff} on the initial stress σ_0,

$$\mu_{\text{eff},0} \equiv \frac{1}{2C_1\sigma_0^2}e^{E_a/RT}, \tag{7–254}$$

then the stress relaxation time can be written

$$\tau_r = \frac{3\mu_{\text{eff},0}}{E}, \tag{7–255}$$

which is closely analogous to Equation (7–239) for the viscoelastic relaxation time of a Maxwell material.

According to Equation (7–253), the stress relaxation time is a strong function of temperature, the rheological parameters, and the initial stress. Figure 7–24 illustrates these dependences for two sets of rheological parameters: dry olivine for which $C_1 = 4.2 \times 10^5$ MPa^{-3} s^{-1}, $E_a = 523$ kJ mol^{-1}, and wet olivine for which

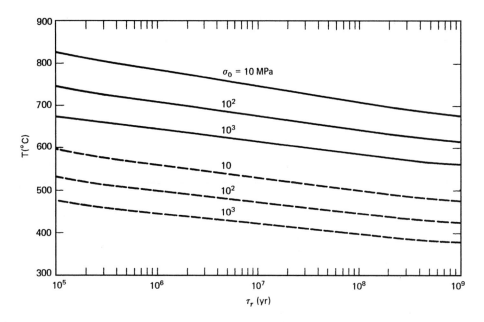

7–24 The temperature at which an initial stress relaxes to one-half of its original value as a function of time for several values of the initial stress. The solid curves are for a dry olivine rheology, and the dashed ones are for a wet olivine.

$C_1 = 5.5 \times 10^4\,\mathrm{MPa}^{-3}\,\mathrm{s}^{-1}$, $E_a = 398\,\mathrm{kJ\,mol}^{-1}$. In addition we assumed $E = 7 \times 10^4$ MPa. For relevant geological times (say 10^7 years) and stress levels (say 100 MPa), significant stress relaxation does not occur at temperatures less than about 675°C for dry olivine or 475°C for wet olivine. Thus, if the dry olivine rheology controls creep, the base of the elastic lithosphere is defined by the 675°C isotherm; if the wet olivine rheology pertains, the appropriate isotherm is 475°C. The thickness or base of the elastic oceanic lithosphere y_{EL} can therefore be determined as a function of its age t by using Equation (4–125) for the thermal structure. By choosing $T_0 = 0°$C, and $T_1 = 1300°$C, we obtain

$$y_{EL} = 2(\kappa t)^{1/2}\,\mathrm{erf}^{-1}\left(\tfrac{675}{1300}\right) = 1.0(\kappa t)^{1/2} \qquad (7\text{--}256)$$

for dry olivine and

$$y_{EL} = 2(\kappa t)^{1/2}\,\mathrm{erf}^{-1}\left(\tfrac{475}{1300}\right) = 0.68(\kappa t)^{1/2} \qquad (7\text{--}257)$$

for wet olivine. At depths in the thermal lithosphere greater than y_{EL}, elastic stresses are relieved by solid-state creep processes on geologically significant time scales. However, the rock still has sufficient rigidity so that the strain is small compared to unity and the lower thermal lithosphere is able to maintain its integrity.

The predicted thicknesses of the oceanic elastic lithosphere for the two rheologies are compared with observations in Figure 7–25. The data were obtained from studies of lithospheric flexure at ocean trenches and

7–25 Thicknesses of the oceanic lithosphere from flexure studies at ocean trenches, islands, and ridges as a function of age of the oceanic lithosphere at the time of loading (Calmant et al., 1990). The squares are data for the Atlantic Ocean, diamonds for the Indian Ocean, and triangles for the Pacific Ocean. The solid line curve defines the base of the elastic lithosphere for the dry olivine rheology, and the dashed curve gives the base for the wet olivine rheology.

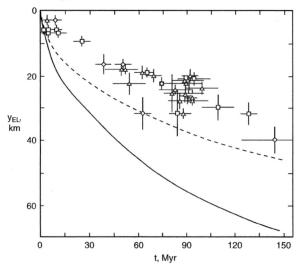

under the loads of islands, and from correlations of ocean ridge topography with gravity. Although there is considerable scatter, the observed thickness of the oceanic elastic lithosphere does appear to increase with its age. The predicted elastic lithosphere thicknesses are larger than the observed thicknesses, especially for the very old lithosphere. The model thickness for a wet olivine rheology fits the data much better than does that for a dry olivine rheology. Elastic stresses in the lithosphere are apparently relaxed at even lower temperatures than predicted by our particular wet olivine rheological formula.

7-11 Elastic–Perfectly Plastic Behavior

At low confining pressures rock behaves as a brittle material; that is, it fractures when a large stress is applied. However, when the confining pressure approaches a rock's brittle strength, a transition from brittle or elastic behavior to plastic behavior occurs, as shown in Figure 7–26. The elastic–plastic transformation takes place when the stress exceeds a critical value known as the yield stress σ_0. In the plastic regime the material yields and deforms irreversibly; upon loading, the stress–strain history follows path AB in Figure 7–27; upon unloading, path BC is followed. The unloading history follows a path essentially parallel to the initial elastic stress–strain line and results in an unrecoverable amount of strain associated with the plastic yielding. In general, the deformation of a material exhibiting an elastic–plastic transition depends on its entire loading history. Temperature also has a strong influence on elastic–plastic deformation. In particular,

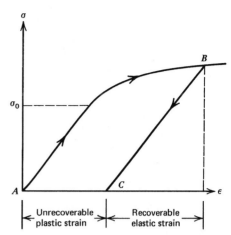

7–27 Stress–strain history for loading and unloading of an elastic–plastic material.

the yield stress usually decreases with increasing temperature. Most studies of elastic–plastic behavior generally assume that the stress–strain curves are independent of the rate of application of the load. Thus, the stress–strain relation is taken to be time independent.

An idealized representation of the behavior just described is the *elastic–perfectly plastic* rheology, in which the material behaves elastically at stresses less than the yield stress σ_0 and deforms without limit at the yield stress, as shown in Figure 7–28. On loading, the material follows the stress–strain path ABC. Along AB the linear elastic relation $\sigma = E\varepsilon$ applies. On BC $\sigma = \sigma_0$, and ε can be arbitrarily large. Upon unloading, the

7–28 The stress–strain relation for an elastic–perfectly plastic material.

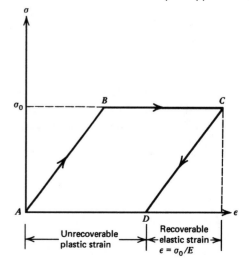

7–26 Deformation of a solid exhibiting an elastic–plastic transformation.

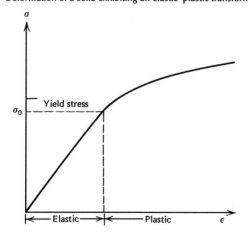

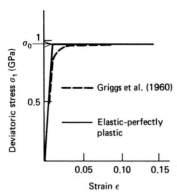

7–29 The elastic–perfectly plastic rheology is compared with the experimentally observed stress–strain behavior of dunite at a confining pressure of 500 MPa and a temperature of 800°C.

material behaves elastically in a manner unaffected by the plastic flow; that is, it follows path CD, which is parallel to AB. When the applied stress is reduced to zero, the elastic strain $\varepsilon = \sigma_0/E$ is recovered, but the plastic strain AD remains.

The elastic–perfectly plastic rheology is compared in Figure 7–29 with laboratory studies of the deformation of dunite at a confining pressure of 500 MPa and a temperature of 800°C. Dunite is a representative mantle rock, and its rheological behavior is in good agreement with the elastic–perfectly plastic model. A typical confining pressure required for the brittle–plastic transformation in rock is 500 MPa. This corresponds to a depth of 17 km in rock of average density 3000 kg m^{-3}. At depths greater than this, plastic yielding is expected at large deviatoric stress levels.

In the case of uniaxial loading the yield condition for plastic deformation is $\sigma = \sigma_0$. In the case of three-dimensional stress, however, the yield condition is more complicated. There are two criteria that are in general use. The *Tresca* or *maximum shear stress criterion* states that a solid yields when the maximum shear stress reaches a critical value σ^*. We noted in Section 2–4 that the maximum shear stress in a three-dimensional state of stress is one-half the difference between the minimum and maximum principal stresses, $\frac{1}{2}(\sigma_1 - \sigma_3)$. Thus the Tresca yield condition is

$$\sigma^* = \tfrac{1}{2}(\sigma_1 - \sigma_3). \tag{7–258}$$

Because this must reduce to the yield condition for uniaxial stress when $\sigma_2 = \sigma_3 = 0$, for example, we

can write

$$\sigma^* = \tfrac{1}{2}\sigma_1 = \tfrac{1}{2}\sigma_0. \tag{7–259}$$

The final form of the Tresca condition is therefore

$$\sigma_0 = \sigma_1 - \sigma_3. \tag{7–260}$$

The *von Mises criterion* asserts that plastic yielding occurs when

$$\sigma^{*2} = (\sigma_1 - \sigma_2)^2 + (\sigma_1 - \sigma_3)^2 + (\sigma_2 - \sigma_3)^2. \tag{7–261}$$

The right side of Equation (7–261) is a stress invariant; that is, its value is unchanged by the replacement of $\sigma_1, \sigma_2,$ and σ_3 with the values $\sigma_{xx}, \sigma_{yy},$ and σ_{zz} appropriate to any coordinate system. This criterion must also reduce to the condition $\sigma_0 = \sigma_1$ when $\sigma_2 = \sigma_3 = 0$ so that

$$\sigma^{*2} = 2\sigma_1^2 = 2\sigma_0^2. \tag{7–262}$$

Thus the von Mises criterion is

$$2\sigma_0^2 = (\sigma_1 - \sigma_2)^2 + (\sigma_1 - \sigma_3)^2 + (\sigma_2 - \sigma_3)^2. \tag{7–263}$$

PROBLEM 7–21 Determine the shear stress at which an elastic–perfectly plastic material yields in pure shear using (*a*) the Tresca criterion and (*b*) the von Mises criterion.

PROBLEM 7–22 Consider an elastic–perfectly plastic material loaded in plain strain (see Section 3–6) with $\varepsilon_3 = 0$ and $\sigma_2 = 0$. Use the von Mises criterion to determine the value of σ_1 at which yielding occurs, and determine the associated value of ε_1.

PROBLEM 7–23 Consider the state of stress $\sigma_{xx} = \sigma_{yy} = \sigma_{zz} = \sigma$ and $\sigma_{xy} = \sigma_{yx} = \tau, \sigma_{xz} = \sigma_{zx} = \sigma_{yz} = \sigma_{zy} = 0$. Determine the yield conditions on the basis of the Tresca and von Mises criteria. How does hydrostatic loading affect plastic yielding?

We now apply the elastic–perfectly plastic rheology to the bending of a plate. The purely elastic bending of a plate has been discussed in Chapter 3. The stress distribution in an elastic plate is given by Equations (3–64) and (3–70)

$$\sigma_{xx} = -\frac{Ey}{1 - v^2}\frac{d^2 w}{dx^2}. \tag{7–264}$$

The stress increases linearly with distance y from the center of the plate. The plate bends elastically until the stresses at the surface of the plate, $y = \pm h/2$, become sufficiently large for plastic yielding to occur. We denote the value of σ_{xx} at which yielding first takes place by $\sigma_{xx,0}$. The value of $\sigma_{xx,0}$ can be determined in terms of the yield stress σ_0 from the Tresca or von Mises criteria. The principal stresses in the plate are $\sigma_{xx}, \sigma_{yy},$ and σ_{zz}. We recall that $\sigma_{yy} = 0$ and $\varepsilon_{zz} = 0$ in the two-dimensional bending of a plate. These conditions give $\sigma_{zz} = \nu\sigma_{xx}$. Thus the three principal stresses in the plate, arranged in the order $\sigma_1 \geq \sigma_2 \geq \sigma_3$, are

$$\sigma_1 = \sigma_{xx} \qquad \sigma_2 = \sigma_{zz} = \nu\sigma_{xx} \qquad \sigma_3 = \sigma_{yy} = 0.$$
$$(7\text{–}265)$$

By substituting Equation (7–265) into the Tresca criterion (7–260), we get

$$\sigma_{xx,0} = \sigma_0. \qquad (7\text{–}266)$$

For the von Mises criterion we find

$$\sigma_{xx,0} = \frac{\sigma_0}{(1 - \nu + \nu^2)^{1/2}}. \qquad (7\text{–}267)$$

For $\nu = 0.25$, the von Mises criterion gives a value of $\sigma_{xx,0}$ that is only 8% larger than the value obtained from the Tresca criterion.

The plate curvature corresponding to the onset of plasticity is given by Equation (7–264)

$$\frac{d^2w}{dx^2} = -\frac{2\sigma_{xx,0}(1 - \nu^2)}{Eh}. \qquad (7\text{–}268)$$

The corresponding value of the bending moment which follows from Equation (3–71) is

$$M_0 = \frac{\sigma_{xx,0}h^2}{6}. \qquad (7\text{–}269)$$

If the bending moment in the plate exceeds M_0, the elastic solution is no longer valid because plastic deformation occurs.

Let us consider the state of stress in the plate when $M > M_0$. We assume as we did for purely elastic bending that transverse sections of the plate remain plane. Therefore the strain is still a linear function of the distance y from the center of the plate, and Equation (3–70), which is a purely geometrical result, remains valid. The interior part of the plate where strains are small remains elastic, but the outer parts deform plastically. In the elastic part of the plate, $|y| < y_0$, the fiber stress is proportional to the longitudinal strain and the strain is proportional to y, so that stress is also propor-

tional to y. In the plastic part of the plate, $|y| > y_0$, the stress has the constant value $\sigma_{xx,0}$. The stress distribution σ_{xx} is thus

$$\sigma_{xx} = \sigma_{xx,0} \qquad\qquad y_0 \leq y \leq \frac{h}{2}$$
$$= \sigma_{xx,0}\left(\frac{y}{y_0}\right) \qquad -y_0 \leq y \leq y_0 \qquad (7\text{–}270)$$
$$= -\sigma_{xx,0} \qquad\qquad -\frac{h}{2} \leq y \leq -y_0$$

The bending moment for the partially plastic plate is obtained by substituting Equation (7–270) into Equation (3–61) and integrating with the result

$$M = 2\left\{\int_0^{y_0} \sigma_{xx,0}\left(\frac{y}{y_0}\right)y\,dy + \int_{y_0}^{h/2} \sigma_{xx,0}y\,dy\right\}$$
$$= \frac{\sigma_{xx,0}h^2}{4}\left(1 - \frac{4y_0^2}{3h^2}\right). \qquad (7\text{–}271)$$

When $y_0 = 0$, the plate is entirely plastic. The maximum or critical bending moment M_c corresponding to this case is

$$M_c = \frac{\sigma_{xx,0}h^2}{4}. \qquad (7\text{–}272)$$

This is the maximum bending moment that the plate can transmit. The bending moment at the onset of plasticity M_0 corresponding to $y_0 = h/2$ is related to the maximum bending moment by

$$M_0 = \tfrac{2}{3}M_c. \qquad (7\text{–}273)$$

The bending moment in the plate can be increased 50% beyond the elastic limit before the maximum bending moment is reached. Stress distributions for various bending moments are given in Figure 7–30.

The curvature of the plate is related to the half-width of the elastic core y_0 and the yield stress $\sigma_{xx,0}$ by Equation (7–264), which gives

$$\frac{d^2w}{dx^2} = -\frac{\sigma_{xx,0}(1 - \nu^2)}{Ey_0}. \qquad (7\text{–}274)$$

Upon eliminating y_0 from Equations (7–271) and (7–274), we get

$$\frac{d^2w}{dx^2} = -\frac{\sigma_{xx,0}(1 - \nu^2)}{E\left(\frac{3}{4}h^2 - \frac{3M}{\sigma_{xx,0}}\right)^{1/2}}, \qquad (7\text{–}275)$$

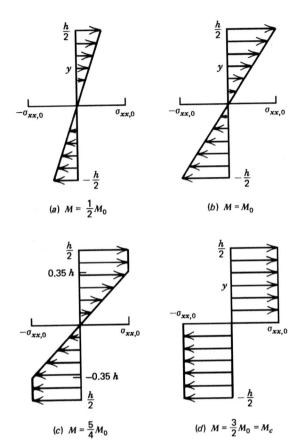

7-30 Stress profiles across a bending plate for various values of the moment. (a) Elastic bending. (b) Onset of plasticity, $\sigma_{xx} = \sigma_{xx,0}$ at $y = h/2$. (c) Partially plastic bending, $y_0 = 0.354h$. (d) Totally plastic bending, $y_0 = 0$.

a result that relates the curvature of the plate to the bending moment. This is the elastic–perfectly plastic equivalent of Equation (3–71). At the onset of plasticity

$$\left(\frac{d^2w}{dx^2}\right)_0 = -\frac{2\sigma_{xx,0}(1-\nu^2)}{Eh} = -\frac{\sigma_{xx,0}h^2}{6D}, \quad (7\text{–}276)$$

where the flexural rigidity D has been defined in Equation (3–72). Using this result, we can put Equation (7–275) in the convenient form

$$\frac{d^2w}{dx^2} \bigg/ \left(\frac{d^2w}{dx^2}\right)_0 = \left(3 - \frac{2M}{M_0}\right)^{-1/2}. \quad (7\text{–}277)$$

This dependence of the plate curvature on bending moment is given in Figure 7–31. The curvature approaches infinity as the bending moment approaches the critical

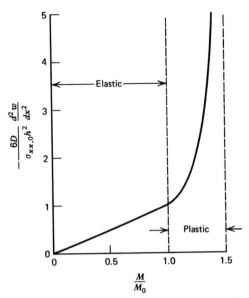

7-31 Dependence of the nondimensional plate curvature on the ratio of the bending moment to the bending moment at the onset of plasticity.

value $M_c = 1.5M_0$. This rapid increase in curvature is referred to as a *plastic hinge*.

Figure 3–35 shows that some ocean trench profiles are in good agreement with the elastic theory for the bending of plates. However, some are not, as illustrated by the profile across the Tonga trench given in Figure 7–32. This profile is compared with the predicted elastic profile given by Equation (3–159) taking $x_b = 60$ km and $w_b = 0.2$ km. The observed profile has a much larger curvature in the trench than the one predicted by elastic theory. If we attribute this additional curvature to plastic hinging, the analysis given above predicts that the excess curvature will develop where the bending moment is a maximum. This occurs at $x = 2x_0 - x_b$ according to Equation (3–160). Therefore we predict that a plastic hinge would develop at $x = -60$ km. This prediction agrees with the observations.

PROBLEM 7-24 Consider a long circular cylinder of elastic–perfectly plastic material that is subjected to a torque T at its outer surface $r = a$. The state of stress in the cylinder can be characterized by an azimuthal shear stress τ. Determine the torque for which an elastic core of radius c remains. Assume that the yield stress in shear is σ_0. In the elastic region the shear stress is proportional to the distance from the axis

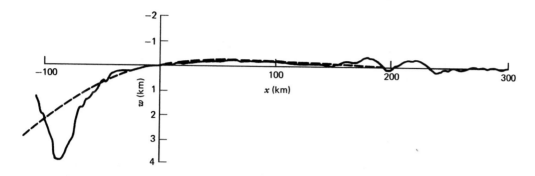

7-32 Observed profile across the Tonga trench compared with an elastic plate profile assuming $x_b = 60$ km and $w_b = 0.2$ km.

of the cylinder r. What is the torque for the onset of plastic yielding? What is the maximum torque that can be sustained by the cylinder?

REFERENCES

Ashby, M. F., and R. A. Verrall (1978), Micromechanisms of flow and fracture and their relevance to the rheology of the upper mantle, *Phil. Trans. Roy. Soc. London* **288A**, 59–95.

Calmant, S., J. Francheteau, and A. Cazenave (1990), Elastic layer thickening with age of the oceanic lithosphere: A tool for prediction of the age of volcanoes or oceanic crust, *Geophys. J. Int.* **100**, 59–67.

Griggs, D. T., F. J. Turner, and H. C. Heard (1960), Deformation of rocks at 500°C to 800°C, *Geol. Soc. Am. Memoir* **79**, 39–104.

Karato, S. I., and P. Wu (1993), Rheology of the upper mantle: A synthesis, *Science* **260**, 771–778.

COLLATERAL READING

Hill, R., *The Mathematical Theory of Plasticity* (Oxford University Press, London, 1950), 355 pages.

A fundamental textbook for engineers and applied mathematicians on the theory of plasticity. The student should be familiar with elasticity theory, cartesian tensors, and hyperbolic differential equations. The book presents the foundations of plasticity theory, solutions of elastic–plastic problems including bending and torsion of beams and bars, plane strain and slip-line theory, steady two-dimensional problems including sheet-drawing and sheet-extrusion, nonsteady two-dimensional problems such as indentation and hardness tests, problems with axial symmetry, and plastic anisotropy.

Hull, D., and D. J. Bacon, *Introduction to Dislocations*, 3rd edition (Pergamon Press, Oxford, 1984), 255 pages.

An account of the geometry, properties, and behavior of dislocations in crystals for advanced undergraduate students in metallurgy, engineering, and physics. The first part of the book describes the geometry, movement, and elastic properties of dislocations together with a discussion of the methods of observing and studying them. This is followed by a description of the more detailed features of dislocations in specific structures: face-centered cubic, hexagonal close-packed, body-centered cubic, ionic, layer and super-lattice structures. Other topics include jogs and the intersection of dislocations, origin and multiplication of dislocations, dislocation arrays and crystal boundaries, strength of annealed crystals, and strengthening by dislocations.

Nadai, A., *Theory of Flow and Fracture of Solids, Volume 2* (McGraw-Hill, New York, 1963), 705 pages.

This classic work on the deformation of solids emphasizes problems in which temperature and the time rate of permanent deformation play an important role. Part 1, on the principles of mechanical work, contains three chapters dealing with the theory of thermoelasticity, mechanical work associated with small finite strain, and extremum principles of work. The next six chapters, which comprise part 2, discuss elasticity and viscosity, plane strain and plane stress, axially symmetric stress distributions, and the bending of plates and viscoelastic beams. Part 3 contains five chapters on subsidence and postglacial uplift, thermal stresses and strains, residual stress, and flow of a generally viscous substance through a cylindrical tube. The next two parts discuss the theory of perfectly loose and of generally plastic substances and the creep of metals at elevated temperatures. The final part is a selection of problems in geomechanics.

Nicolas, A., and J. P. Poirier, *Crystalline Plasticity and Solid State Flow in Metamorphic Rocks* (John Wiley and Sons, London, 1976), 444 pages.

A textbook on the mechanics of deformation of minerals and rocks. Major chapter headings include structural analysis in metamorphic rocks, elements of solid mechanics and physical metallurgy, flow and annealing processes in crystals, plastic deformation of rock-forming minerals, development of textures and preferred orientations by plastic flow and recrystallization, strain analysis of field structures, interpretation of structures, textures, and preferred orientations in peridotites, large-scale flow in peridotites, and upper mantle geodynamics.

Poirier, J. P., *Introduction to the Physics of the Earth's Interior* (Cambridge University Press, Cambridge, 1991), 264 pages.

This is an advanced level treatise on the physics of the Earth's interior with an emphasis on mineral physics. Subjects covered include the thermodynamics of solids, elastic moduli, lattice vibrations, equations of state, melting, and transport properties.

Poirier, J. P., *Creep of Crystals* (Cambridge University Press, Cambridge, 1985), 260 pages.

This is an excellent and broad treatment of solid-state creep processes. Diffusion and dislocation mechanisms are considered along with the role of water, dynamic recrystallization, superplasticity, and transformation plasticity.

Prager, W., and P. G. Hodge, Jr., *Theory of Perfectly Plastic Solids* (John Wiley and Sons, New York, 1951), 264 pages.

An introduction to the mathematical theory of the behavior of perfectly plastic solids written for senior undergraduate or graduate students in engineering and applied mathematics. Following an introduction to the basic concepts of plasticity, the book treats trusses and beams, including the flexure of elastic–plastic beams, torsion of cylindrical and prismatic bars, problems involving plane strain, and extremum principles. Problems are given at the end of each chapter.

Ranalli, G., *Rheology of the Earth*, 2nd edition (Chapman and Hall, London, 1995), 413 pages.

A reference work that provides a comprehensive coverage of the deformation and flow of Earth materials from both the continuum mechanics and the microphysical viewpoints. The fundamental principles of elasticity, viscous flow, and plasticity are covered. The atomic bases for deformation are introduced including vacancies, diffusion, dislocations, and recrystallization. The concept of deformation maps is introduced.

Weertman, J., and J. R. Weertman, *Elementary Dislocation Theory* (Macmillan Company, New York, 1964), 213 pages.

A book on dislocation theory for advanced undergraduate students. Basic calculus and vector analysis are required. A large number of problems are given. The main subjects are description of a dislocation, the stress field around a dislocation, forces on a dislocation, dislocation reactions in crystals, dislocation multiplication, twinning, Peierls force, image forces, and interactions with point defects.

Wert, C. A., and R. M. Thomson, *Physics of Solids*, 2nd edition (McGraw-Hill, New York, 1970), 522 pages.

A book for advanced undergraduate engineering students on basic solid-state physics. The treatment is quantitative, but the mathematical analysis is on an intermediate level. The first part of the book discusses the crystal structure of solids and the properties that depend on structure. Topics include the geometry of perfect crystals, imperfections in crystals, heat capacity, vacancies and interstitials, point defects, diffusion, phase diagrams, and dislocations. The second part deals with the electronic structure of solids and covers metals, ionic crystals, covalent crystals, molecular crystals, and the physical properties of semiconductors. The remaining chapters are applications of electronic structure to the electrical and magnetic properties of metals, semiconductors, and ionic crystals. The subjects covered are transport properties and specific heat of metals, semiconductor devices, electromagnetic and dielectric behavior of solids, diamagnetism, paramagnetism, and ferromagnetism. Problems are given at the end of each chapter.

EIGHT

Faulting

8-1 Introduction

At low temperatures and pressures rock is a brittle material that will fail by fracture if the stresses become sufficiently large. Fractures are widely observed in surface rocks of all types. When a lateral displacement takes place on a fracture, the break is referred to as a fault. Surface faults occur on all scales. On the smallest scale the offset on a clean fracture may be only millimeters. On the largest scale the surface expression of a major fault is a broad zone of broken up rock known as a fault gouge; the width may be a kilometer or more, and the lateral displacement may be hundreds of kilometers.

Earthquakes are associated with displacements on many faults. Faults lock, and a displacement occurs when the stress across the fault builds up to a sufficient level to cause rupture of the fault. This is known as *stick–slip* behavior. When a fault sticks, elastic energy accumulates in the rocks around the fault because of displacements at a distance. When the stress on the fault reaches a critical value, the fault slips and an earthquake occurs. The elastic energy stored in the adjacent rock is partially dissipated as heat by friction on the fault and is partially radiated away as seismic energy. This is known as *elastic rebound*. Fault displacements associated with the largest earthquakes are of the order of 30 m.

The relative motions of the rigid plates are often accommodated on major faults. At ocean trenches, the oceanic lithosphere is being subducted beneath an adjacent oceanic or continental lithosphere along a dipping fault plane. The convergence of the two lithospheres results in thrust faulting and the occurrence of most of the world's great earthquakes. These earthquakes occur regularly in order to accommodate the continuous subduction process. Because the surface expression of these faults is at the base of an ocean trench, they are difficult to study in detail.

At accretional plate margins, extensive normal faulting occurs on the flanks of the ocean ridges. The lithosphere at ocean ridges is thin and weak, so the resulting earthquakes are small. Segments of the ocean ridge system are connected by transform faults. Strike–slip faulting occurs on these faults. The San Andreas fault is a major strike–slip fault that is a plate boundary in the continental lithosphere. This fault has good surface exposure and has been extensively studied; it is discussed in detail in Section 8–8. Extensive faulting of all types occurs in zones of continental collision. The great earthquakes that take place throughout China are associated with the broad zone of deformation resulting from the collision between the Indian and Eurasian plates. Further to the west the extension of this zone of continental collision causes extensive seismicity in Turkey, as discussed in Section 8–9.

8-2 Classification of Faults

We previously discussed the classification of faults in Section 1–13. Here we provide quantitative definitions of the different types of faults in terms of the relative magnitudes of the principal stresses. Because voids cannot open up deep in the Earth, displacements on faults occur parallel to the fault surface. For simplicity we assume that the fault surface is planar; in fact, faulting often occurs on curved surfaces or on a series of surfaces that are offset from one another.

We will first consider thrust faulting, which occurs when the oceanic lithosphere is thrust under the adjacent continental (or oceanic) lithosphere at an ocean trench. Thrust faulting also plays an important role in the compression of the lithosphere during continental collisions. Idealized thrust faults are illustrated in Figure 8–1. Compressional stresses cause displacement along a fault plane dipping at an angle β to the horizontal. As a result of the faulting, horizontal compressional strain occurs. Thrust faults can form in either of the two conjugate geometries shown in Figure 8–1a and b. The elevated block is known as the *hanging wall*, and the depressed block is called the *foot wall*. The upward movement of the hanging wall is also referred to as *reverse faulting*.

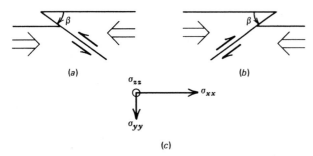

8–1 Thrust faulting. Two conjugate thrust faults with dip angles β are shown in (a) and (b). The principal stresses illustrated in (c) are all compressional with magnitudes $\sigma_{xx} > \sigma_{zz} > \sigma_{yy}$.

We assume that the stresses in the x, y, and z directions are the principal stresses. The vertical component of stress σ_{yy} is the overburden or lithostatic pressure

$$\sigma_{yy} = \rho g y. \tag{8–1}$$

The vertical deviatoric stress $\Delta\sigma_{yy}$ is zero. To produce the thrust faults in Figure 8–1, a compressional deviatoric stress applied in the x direction $\Delta\sigma_{xx}$ is required,

$$\Delta\sigma_{xx} > 0. \tag{8–2}$$

The horizontal compressional stress,

$$\sigma_{xx} = \rho g y + \Delta\sigma_{xx}, \tag{8–3}$$

therefore exceeds the vertical lithostatic stress

$$\sigma_{xx} > \sigma_{yy}. \tag{8–4}$$

For the fault geometry shown in Figure 8–1 it is appropriate to assume that there is no strain in the z direction. In this situation of plane strain we can use Equation (3–44) to relate the deviatoric stress component $\Delta\sigma_{zz}$ to $\Delta\sigma_{xx}$

$$\Delta\sigma_{zz} = \nu\Delta\sigma_{xx}. \tag{8–5}$$

The deviatoric stress in the z direction is also compressional, but its magnitude is a factor of ν less than the deviatoric applied stress. Therefore the horizontal compressional stress,

$$\sigma_{zz} = \rho g y + \Delta\sigma_{zz} = \rho g y + \nu\Delta\sigma_{xx}, \tag{8–6}$$

exceeds the vertical stress σ_{yy}, but it is smaller than the horizontal stress σ_{xx}. Thrust faults satisfy the condition

$$\sigma_{xx} > \sigma_{zz} > \sigma_{yy}. \tag{8–7}$$

The vertical stress is the least compressive stress.

Just as thrust faulting accommodates horizontal compressional strain, normal faulting accommodates horizontal extensional strain. Normal faulting occurs on the flanks of ocean ridges where new lithosphere is being created. Normal faulting also occurs in continental rift valleys where the lithosphere is being stretched. Applied tensional stresses can produce normal faults in either of the two conjugate geometries shown in Figure 8–2. The displacements on the fault planes dipping at an angle β to the horizontal lead to horizontal extensional strain. Normal faulting is associated with a state of stess in which the vertical component of stress is the lithostatic pressure $\sigma_{yy} = \rho g y$ and the applied deviatoric horizontal stress $\Delta\sigma_{xx}$ is tensional

$$\Delta\sigma_{xx} < 0. \tag{8–8}$$

The horizontal stress,

$$\sigma_{xx} = \rho g y + \Delta\sigma_{xx}, \tag{8–9}$$

is therefore smaller than the vertical stress,

$$\sigma_{yy} > \sigma_{xx}. \tag{8–10}$$

The plane strain assumption is again appropriate to the situation in Figure 8–2, and Equation (8–5) is applicable. Consequently, the deviatoric stress in the z direction $\Delta\sigma_{zz}$ is also tensional, but its magnitude is a factor of ν smaller than the deviatoric applied stress. The total stress,

$$\sigma_{zz} = \rho g y + \nu\Delta\sigma_{xx}, \tag{8–11}$$

is smaller than σ_{yy} but larger than σ_{xx}. Normal faults

8–2 Normal faulting. Two conjugate normal faults with angle of dip β are shown in (a) and (b). The principal stresses illustrated in (c) have magnitudes related by $\sigma_{yy} > \sigma_{zz} > \sigma_{xx}$.

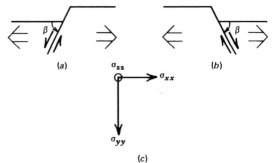

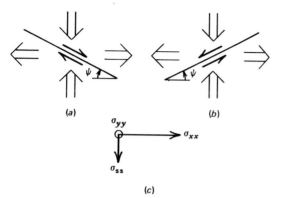

(a)   (b)

(c)

8–3 Strike-slip faulting. Two conjugate strike-slip faults inclined at an angle ψ to the direction of the principal stress σ_{xx} are shown in (a) and (b). The principal stresses illustrated in (c) are related by $\sigma_{zz} > \sigma_{yy} > \sigma_{xx}$.

satisfy the condition

$$\sigma_{yy} > \sigma_{zz} > \sigma_{xx}, \tag{8–12}$$

where the vertical stress is the maximum compressive stress. Both thrust faults and normal faults are also known as dip–slip faults because the displacement along the fault takes place on a dipping plane.

A strike–slip fault is a fault along which the displacement is strictly horizontal. Thus there is no strain in the y direction. The situation is one of plane strain with the nonzero strain components confined to the horizontal plane. Vertical views of two conjugate strike–slip faults are shown in Figure 8–3. The fault planes make an angle ψ with respect to the direction of the principal stress σ_{xx}. The fault illustrated in Figure 8–3a is right lateral and the one in Figure 8–3b is left lateral.

The state of stress in strike–slip faulting consists of a vertical lithostatic stress $\sigma_{yy} = \rho g y$ and horizontal deviatoric principal stresses that are compressional in one direction and tensional in the other. The case shown in Figure 8–3 has

$$\Delta\sigma_{xx} < 0 \qquad \Delta\sigma_{zz} > 0. \tag{8–13}$$

One can also have

$$\Delta\sigma_{xx} > 0 \qquad \Delta\sigma_{zz} < 0. \tag{8–14}$$

One horizontal stress will thus be larger than σ_{yy} while the other will be smaller. For the situation given by Equation (8–13) we have

$$\sigma_{zz} > \sigma_{yy} > \sigma_{xx}, \tag{8–15}$$

while Equation (8–14) gives

$$\sigma_{xx} > \sigma_{yy} > \sigma_{zz}. \tag{8–16}$$

For strike–slip faulting, the vertical stress is always the intermediate stress. A special case of strike–slip faulting occurs when

$$|\Delta\sigma_{xx}| = |\Delta\sigma_{zz}| = \tau_0. \tag{8–17}$$

This is the situation of pure shear discussed in Section 3–7. The stress τ_0 is the shear stress applied across the fault. In pure shear the angle ψ is $45°$.

The displacement on an actual fault is almost always a combination of strike–slip and dip–slip motion. However, one type of motion usually dominates.

8–3 Friction on Faults

Displacements on faults accommodate a substantial fraction of the strain occurring in the upper crust. These displacements sometimes occur in a continuous manner at tectonic velocities of tens of millimeters per year. This type of displacement is referred to as *fault creep*. However, it is much more common for the displacements on faults to occur during earthquakes. Between earthquakes the fault remains locked. This is known as stick–slip behavior.

A simple model for the stick–slip behavior of a fault is illustrated in Figure 8–4. We assume that the behavior of the fault is uniform with depth and neglect the forces at the bases of the adjacent plates. Figure 8–4a shows the situation after a major earthquake when the fault locks. The stress across the fault is τ_{fd}, the frictional stress that is operative on the fault at the end of faulting. A uniform relative velocity u_0 is applied at a distance b from the fault, and the shear strain increases with time according to $\varepsilon(t) = u_0 t / (4b)$ – see Equation (2–102) – for example, as shown in Figure 8–4b. The shear stress on the fault as a function of time t since the last displacement on the fault is therefore

$$\tau = \tau_{fd} + \frac{Gu_0 t}{2b}, \tag{8–18}$$

where G is the shear modulus (see Equation (3–49)). The locked fault can transmit any shear stress less than the static frictional stress τ_{fs}. When this stress is reached, slip occurs. Therefore, the time $t = t^*$ when the next

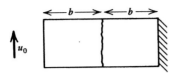

(a) After a major earthquake the fault sticks

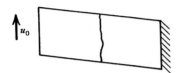

(b) Just prior to the next major earthquake

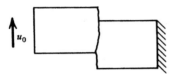

(c) After this major earthquake the fault locks
and the cycle repeats

8–4 Stick–slip behavior and elastic rebound on a fault.

displacement occurs on the fault is

$$t^* = \frac{2b}{Gu_0}(\tau_{fs} - \tau_{fd}). \tag{8–19}$$

The slip on the fault generates an earthquake. The displacement on the fault during the earthquake occurs in a few seconds so that the edges of the plates can be assumed to be stationary during this time. The accumulated shear strain $\varepsilon = u_0 t^*/4b$ is recovered by the plates in a process known as *elastic rebound*. The resulting displacement on the fault Δw is $2\varepsilon(2b)$ – see Equation (2–94) – or

$$\Delta w = 2\varepsilon(2b) = 4b\left(\frac{u_0 t^*}{4b}\right) = \frac{2b}{G}(\tau_{fs} - \tau_{fd}). \tag{8–20}$$

The quantity $\tau_{fs} - \tau_{fd}$ is the stress drop on the fault during the earthquake. After the earthquake, the fault locks and the cycle repeats, as shown in Figure 8–4c.

The displacement on a fault during an earthquake can be measured from the surface rupture. A typical value for a large earthquake is 5 m. It is difficult to determine the stress drop during an earthquake. Estimates of stress drops during large earthquakes range from $\tau_{fs} - \tau_{fd} = 1$ to 100 MPa. Taking G for crustal

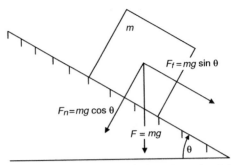

8–5 A block of mass m on an inclined surface. The angle θ is increased until the block slips. The component of the gravitational force mg normal to the surface is $mg\cos\theta$, the component parallel to the surface is $mg\sin\theta$.

rocks equal to 30 GPa, we find from Equation (8–20) that the distance b lies in the range 75 m to 7.5 km.

The static frictional stress is the stress on the fault when earthquake rupture initiates on the fault. During rupture, slip is occurring on the fault and the shear stress on the fault is the dynamic frictional stress. Stick–slip behavior occurs as long as the static frictional stress τ_{fs} is greater than the dynamic frictional stress τ_{fd}, $\tau_{fs} > \tau_{fd}$.

Extensive laboratory studies have been carried out to determine when slip will initiate on a contact surface. A simple example is a block of mass m sitting on an inclined surface as illustrated in Figure 8–5 (see also Problem 2–10). The angle θ is increased until the block begins to slip. The normal stress that the block exerts on the surface is

$$\sigma_n = \frac{mg\cos\theta}{A} \tag{8–21}$$

where A is the contact area of the block. The frictional shear stress on the surface required to keep the block from slipping is

$$\tau_f = \frac{mg\sin\theta}{A}. \tag{8–22}$$

Slip will occur when $\tau_f = \tau_{fs}$, the static frictional stress. Under a wide variety of conditions it is found experimentally that

$$\tau_{fs} = f_s\sigma_n, \tag{8–23}$$

where f_s is the *coefficient of static friction*. This relation is known as *Amonton's law*. The greater the normal stress, the harder it is to initiate sliding. The coefficient

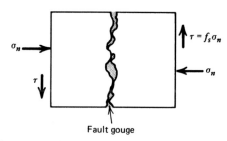

8–6 A shear stress with magnitude $|\tau| = f_s \sigma_n$ must be applied across a fault to initiate sliding when there is a normal compressive stress σ_n acting on the fault.

of friction depends weakly on the types of material in contact but is independent of the normal stress. The concept of friction was discussed briefly in Section 2–2.

PROBLEM 8-1 Assuming Amonton's law to be applicable with $f_s = 0.8$, determine the angle θ at which the block illustrated in Figure 8–5 will begin to slip.

Amonton's law is generally valid when two rough surfaces are in contact or when a granulated material such as a fault gouge is present between the surfaces. The law can be applied to a nearly planar fracture or fault, as sketched in Figure 8–6. A shear stress τ_{fs} given by Equation (8–23) must be applied parallel to the fault to cause sliding when the two sides of the fault are pressed together by the normal stress σ_n.

Laboratory data on the maximum shear stress to initiate sliding are given in Figure 8–7 for a wide variety of rocks including granites, gabbros, limestones, and sandstones at normal stresses up to 100 MPa. Although there is considerable scatter, good agreement is obtained for max $f_s = 0.85$.

The presence of water throughout much of the Earth's crust affects the frictional behavior of a fault. The pressure of water on a fault is referred to as the pore pressure p_w. The effective normal stress acting on a wet fault is the actual normal stress less the pore pressure. Therefore on a wet fault Amonton's law can be written

$$|\tau| = f_s(\sigma_n - p_w). \tag{8-24}$$

If the water is freely connected with the surface and there are no flow losses, the water pressure is the hydrostatic pressure $p_w = \rho_w g y$. Because the density of water ρ_w is considerably less than the density of rocks, the hydrostatic pressure is generally 35 to 50% of the

overburden or lithostatic pressure. In some cases, however, water is trapped, and the pore pressure can nearly equal or even exceed the overburden pressure. In these cases the shear stress resisting motion on a fault is low.

It is clear from Equation (8–24) that the injection of fluid can reduce the frictional resistance to an earthquake on a fault. There are many instances in which fluids pumped at high pressures into waste disposal wells have induced small earthquakes. When pumping ended, the earthquakes ceased. It has been suggested that large earthquakes could be prevented by the injection of fluids on major faults. The fluid pressure would reduce the maximum shear stress required for an earthquake. Displacement on the fault would be accommodated through a large number of small earthquakes rather than a few large earthquakes. The consequences of inducing a large earthquake have so far prevented a large-scale test of this suggestion.

8-4 Anderson Theory of Faulting

We now combine the results of the previous two sections and determine the angle of dip β of normal and thrust faults in terms of the coefficient of static friction f_s. As in Section 8–2 we assume that the horizontal stresses σ_{xx} and σ_{zz} and the vertical stress σ_{yy} are principal stresses and that the fault is a plane dipping at an angle β (see Figures 8–1 and 8–2). We again assume that the vertical stress σ_{yy} is the lithostatic pressure and that σ_{xx} is the sum of the lithostatic pressure and a tectonic deviatoric stress $\Delta\sigma_{xx}$,

$$\sigma_{yy} = \rho g y \tag{8-25}$$

$$\sigma_{xx} = \rho g y + \Delta\sigma_{xx}. \tag{8-26}$$

For thrust faulting $\Delta\sigma_{xx}$ is positive, and for normal faulting it is negative.

To apply Amonton's law, it is necessary to relate σ_{xx} and σ_{yy} to σ_n and τ. The geometry is illustrated in Figure 8–8. By comparing this figure with Figure 2–11, we see that σ_n and τ can be found from Equations (2–53) and (2–54) by equating σ_1 and σ_2 in those equations with σ_{xx} and σ_{yy}; the result is

$$\sigma_n = \tfrac{1}{2}(\sigma_{xx} + \sigma_{yy}) + \tfrac{1}{2}(\sigma_{xx} - \sigma_{yy})\cos 2\theta \tag{8-27}$$

$$\tau = -\tfrac{1}{2}(\sigma_{xx} - \sigma_{yy})\sin 2\theta, \tag{8-28}$$

where θ is the angle of the fault with respect to the

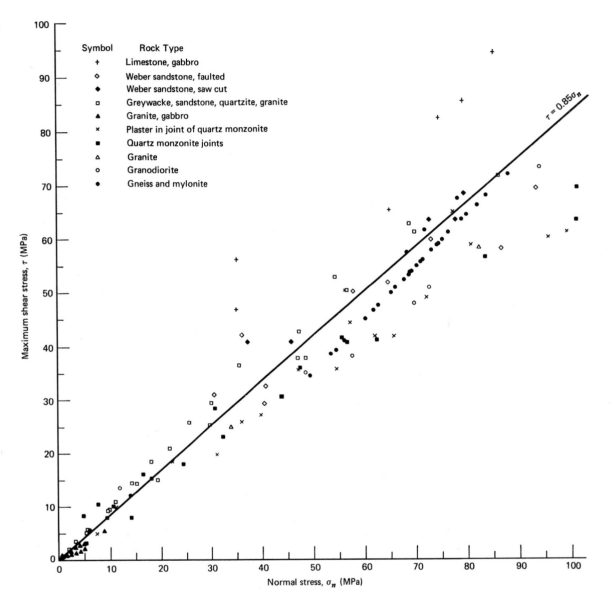

8–7 Maximum shear stress to initiate sliding as a function of normal stress for a variety of rock types. The linear fit defines a maximum coefficient of static friction max f_s equal to 0.85. Data from Byerlee (1977).

vertical, $\theta = \pi/2 - \beta$. Upon substituting Equations (8–25) and (8–26) into (8–27) and (8–28), we find that the normal and tangential stresses on the fault are

$$\sigma_n = \rho g y + \frac{\Delta\sigma_{xx}}{2}(1 + \cos 2\theta) \qquad (8\text{–}29)$$

$$\tau = -\frac{\Delta\sigma_{xx}}{2}\sin 2\theta. \qquad (8\text{–}30)$$

If we include the effect of pore pressure on the fault, these stresses are related by Amonton's law in the form of Equation (8–24). Substitution of Equations (8–29) and (8–30) into Equation (8–24) yields

$$\pm\frac{\Delta\sigma_{xx}}{2}\sin 2\theta = f_s\left\{\rho g y - p_w + \frac{\Delta\sigma_{xx}}{2}(1 + \cos 2\theta)\right\}, \qquad (8\text{–}31)$$

where the upper sign applies to thrust faults ($\Delta\sigma_{xx} > 0$) and the lower sign to normal faults ($\Delta\sigma_{xx} < 0$). Rearrangement of Equation (8–31) gives an expression for

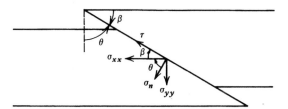

8-8 Principal stresses and normal and tangential stresses on a dip–slip fault.

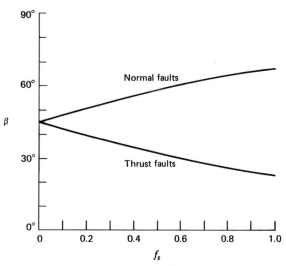

8-9 Dependence of the angle of dip β on the coefficient of friction f_s for normal and thrust faults.

the tectonic stress $\Delta\sigma_{xx}$ in terms of the angle of the fault with the vertical

$$\Delta\sigma_{xx} = \frac{2f_s(\rho gy - p_w)}{\pm\sin 2\theta - f_s(1 + \cos 2\theta)}. \qquad (8\text{–}32)$$

Continental crustal rocks contain many preexisting joints and faults. We hypothesize that under a tectonic stress these preexisting zones of weakness will be reactivated to form a dip–slip fault at an angle requiring the minimum value of the tectonic stress. In other words, thrust faulting and normal faulting will occur at angles that minimize $|\Delta\sigma_{xx}|$. The angle θ that gives the minimum value of $|\Delta\sigma_{xx}|$ in Equation (8–32) is determined by setting $d\Delta\sigma_{xx}/d\theta = 0$ with the result

$$\tan 2\theta = \mp\frac{1}{f_s}. \qquad (8\text{–}33)$$

This can be rewritten in terms of β as

$$\tan 2\beta = \pm\frac{1}{f_s}. \qquad (8\text{–}34)$$

The upper signs in these equations apply to thrust faults and the lower signs to normal faults. Figure 8–9 shows how the dip angles of normal and thrust faults depend on the coefficient of friction. Thrust faults dip less steeply than normal faults.

The tectonic stresses corresponding to these angles of dip are obtained by substituting Equation (8–33) into Equation (8–32)

$$\Delta\sigma_{xx} = \frac{\pm 2f_s(\rho gy - p_w)}{\left(1 + f_s^2\right)^{1/2} \mp f_s}. \qquad (8\text{–}35)$$

Again, the upper signs are used for thrust faults and the lower signs for normal faults. Figure 8–10 shows how the tectonic stress computed from Equation (8–35) varies with the coefficient of friction on normal and thrust faults for the case $p_w = \rho_w gy$, $\rho = 2700$ kg m^{-3},

$\rho_w = 1000$ kg m^{-3}, $g = 10$ m s^{-2}, and $y = 5$ km. Thrust faulting requires somewhat larger stresses, in absolute magnitude, than does normal faulting. Based on laboratory measurements, a typical value for the coefficient of friction would be $f_s = 0.85$ (see Figure 8–7). From Equation (8–34) the corresponding angle of dip for a thrust fault is $\beta = 24.8°$. At a depth of 5 km the deviatoric stress from Figure 8–10 is $\Delta\sigma_{xx} = 305$ MPa. The

8-10 Dependence of the deviatoric stress on the coefficient of static friction for thrust and normal faults with $p_w = \rho_w gy$, $\rho = 2,700$ kg m^{-3}, $\rho_w = 1,000$ kg m^{-3}, $g = 10$ m s^{-2}, and $y = 5$ km.

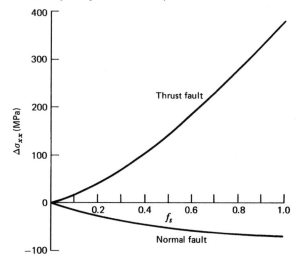

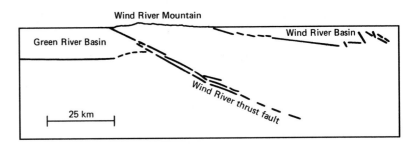

8–11 Deep structure of the Wind River thrust fault, Wyoming, is obtained by deep crustal seismic reflection profiling (Brewer et al., 1980).

angle of dip of a normal fault is $\beta = 65.2°$, and the tectonic stress is $\Delta\sigma_{xx} = -65$ MPa. The derivation of the angle of dip of dip–slip faults given above was developed by E. M. Anderson and is often referred to as the Anderson theory of faulting.

As an example of a major thrust fault, let us consider the Wind River thrust fault in Wyoming. The structure of this fault, determined using seismic reflection profiling, is illustrated in Figure 8–11; it is essentially the same as the structure shown in Figure 8–1. The elevated block is responsible for the uplift of the Wind River Mountains. This is a range of mountains 220 km long and 70 km wide; the highest peak has an elevation of 4267 m. The mountains are cored by Precambrian rock with an age of 2.7 Ga; they are representative of the deep crust. Clearly there has been considerable uplift and erosion. The depressed block to the west of the Wind River thrust fault is responsible for the formation of the Green River basin. This is a sedimentary basin with large petroleum reserves. The basin has a maximum depth of about 12 km. The Wind River Mountains are one of a series of Laramide ranges in Wyoming with an age of 50 to 70 Ma. They are clearly the result of crustal compression with the compressional strain being accommodated on a series of major thrust faults that are responsible for the formation of the mountain ranges and the adjacent sedimentary basins. On the Wind River thrust fault there has been at least 21 km of crustal shortening and 13 km of vertical uplift.

The seismic profiling illustrated in Figure 8–11 clearly traces the fault to a depth of 24 km and possibly as deep as 36 km. Over this depth range the angle of dip is nearly constant with an average value of 35°. Assuming that the Anderson theory of faulting is applicable, the coefficient of friction corresponding to this angle of dip is $f_s = 0.36$ from Equation (8–34). Although this value for the coefficient of friction is considerably less than that obtained in laboratory experiments, this lower value is probably applicable to the large-scale deformations associated with a major thrust fault. A substantial pore pressure could help explain the reduced value of f_s. With $f_s = 0.36$, the angle of dip of a normal fault is 55°. Typical angles of dip for the surface exposures of normal faults are 55 to 65°.

PROBLEM 8–2 Using the Anderson theory of faulting for the strike–slip fault illustrated in Figure 8–3 show that

$$\cot 2\psi = \pm f_s. \tag{8–36}$$

Here the upper sign applies if $\Delta\sigma_{xx} > 0$ and the lower sign applies if $\Delta\sigma_{xx} < 0$. Prove that this result is independent of the ratio $\Delta\sigma_{zz}/\Delta\sigma_{xx}$.

PROBLEM 8–3 A number of criteria have been proposed to relate the brittle fracture of rock to the state of stress. One of these is the *Coulomb–Navier criterion*, which states that failure occurs on a plane when the shear stress τ attains the value

$$|\tau| = S + \mu\sigma_n, \tag{8–37}$$

where S is the inherent shear strength of the rock and μ is the coefficient of internal friction. Consider a two-dimensional state of stress with principal stresses σ_1 and σ_2 and show that $|\tau| - \mu\sigma_n$ has a maximum value for a plane whose normal makes an angle θ to the larger principal stress given by

$$\tan 2\theta = \frac{-1}{\mu}. \tag{8–38}$$

Show also that the quantity $|\tau| - \mu\sigma_n$ for this plane is

$$|\tau| - \mu\sigma_n = \tfrac{1}{2}(\sigma_1 - \sigma_2)(1 + \mu^2)^{1/2} - \tfrac{1}{2}(\sigma_1 + \sigma_2)\mu. \tag{8–39}$$

According to the Coulomb–Navier criterion, failure will occur if this quantity equals S; that is, the failure criterion takes the form

$$\sigma_1\{(1+\mu^2)^{1/2} - \mu\} - \sigma_2\{(1+\mu^2)^{1/2} + \mu\} = 2S. \tag{8-40}$$

What is the compressive strength of the rock in terms of μ and S? From Equation (8–38) it is seen that θ must exceed $45°$, so that the direction of shear fracture makes an acute angle with σ_1. The Coulomb–Navier criterion is found to be reasonably valid for igneous rocks under compression.

8-5 Strength Envelope

The Anderson theory of faulting can also be used to find a strength envelope for the lithosphere. For example, let us consider the oceanic lithosphere. We assume that the failure stress in the upper part of the lithosphere is given by the Anderson theory of faulting from Equation (8–35). We evaluate this equation by taking the fluid pressure to be hydrostatic so that $p_w = \rho_w g y$, $\rho = 3300\ \mathrm{kg\ m^{-3}}$, $\rho_w = 1000\ \mathrm{kg\ m^{-3}}$, and the coefficient of friction $f_s = 0.6$, and show by the solid lines in Figure 8–12 the resulting tectonic stresses $\Delta\sigma_{xx}$ as a function of depth for compressional and tensional failures.

We further assume that the failure stress in the lower part of the oceanic lithosphere is given by the solid-state creep law for the mantle from Equation (7–192). To determine a stress, we must specify the temperature, pressure, and strain rate. We assume a linear thermal gradient $dT/dy = 25\ \mathrm{K\ km^{-1}}$ and neglect the pressure dependence at the shallow depths considered. The resulting stress envelopes due to solid-state creep are given as the dashed lines in Figure 8–12 for two strain rates, $\dot{\varepsilon} = 10^{-12}\ \mathrm{s^{-1}}$ and $\dot{\varepsilon} = 10^{-14}\ \mathrm{s^{-1}}$.

In defining the strength envelopes for the oceanic lithosphere given in Figure 8–12, we assume that the lower of the frictional stress or the creep stress determines the strength. We see that the maximum strength is at a depth of about 28 km for the conditions considered. The maximum compressional stress $\Delta\sigma_{xx}$ is about 1300 MPa and the maximum tensional strength $\Delta\sigma_{xx}$ is about -450 MPa. The total compressional force F_c that can be transmitted by the oceanic lithosphere is the area under the curve in Figure 8–12 and is approximately $2.2 \times 10^{13}\ \mathrm{N\ m^{-1}}$. Similarly, the total tensional force

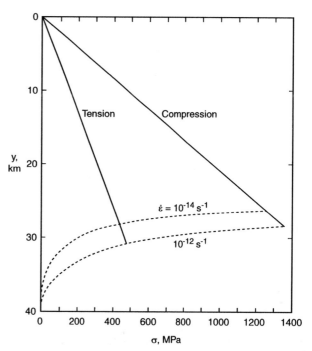

8–12 Strength envelope for the oceanic lithosphere. The solid lines are the strength of the lithosphere for the Anderson theory of faulting from Equation (8–35). The dashed lines are the stresses associated with solid-state creep in the lithosphere from Equation (7–192).

F_T that can be transmitted by the oceanic lithosphere is about $8 \times 10^{12}\ \mathrm{N\ m^{-1}}$. In Section 6–21 we estimated that the "ridge push" force $F_R = 3.9 \times 10^{12}\ \mathrm{N\ m^{-1}}$ and that the maximum "trench pull" force $F_B = 4.9 \times 10^{13}\ \mathrm{N\ m^{-1}}$. However, as noted in that section, the trench pull force is largely balanced by resistive forces encountered by the descending lithosphere so that the net force in the oceanic lithosphere is probably comparable to the ridge-push force. The strength of the oceanic lithosphere, as we have determined it, is sufficient to transmit this force through old oceanic lithosphere.

8-6 Thrust Sheets and Gravity Sliding

Displacements on thrust faults are an important mechanism for accommodating compressional strain in the continental crust. Another mechanism is the emplacement of long thin thrust sheets as illustrated in Figure 8–13. The continental crust is split into two parts, the upper brittle part of the crust A is overthrust over the adjacent upper brittle crust B. The lower part of the continental crust C is compressed plastically to about

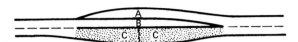

8–13 Compressional strain of continental crust can be accommodated by the emplacement of a thrust sheet A over the adjacent brittle upper crust B. The stippled region of the lower crust C is deformed plastically to about twice its original thickness.

twice its original thickness. The overthrust brittle crust constitutes a *thrust sheet*.

The mechanics of thrust sheet emplacement can be studied with the idealized wedge-shaped model shown in Figure 8–14. We assume that isostasy applies to the thrust sheet. The angles γ and β are therefore related by

$$\beta = \left(\frac{\rho_c}{\rho_m - \rho_c}\right)\gamma, \tag{8–41}$$

where we have also assumed that the angles are small so that $\tan\beta \approx \beta$ and $\tan\gamma \approx \gamma$. A horizontal static force balance on a section of the thrust sheet of length l leads to the conclusion that the net horizontal force on the base of the thrust sheet must equal the sum of the lithostatic pressure and the tectonic stress $\Delta\sigma_{xx}$ integrated over the thickness of the sheet at $x = l$ (see Figure 8–14). The latter quantity, denoted by F_1, is

$$F_1 = \int_{-\gamma l}^{\beta l} (\gamma l + y)\rho_c g\, dy + \int_{-\gamma l}^{\beta l} \Delta\sigma_{xx}\, dy$$
$$= \frac{\rho_c g}{2}(\gamma + \beta)^2 l^2 + \Delta\sigma_{xx}(\gamma + \beta)l. \tag{8–42}$$

Substitution of Equation (8–41) into (8–42) gives

$$F_1 = \frac{\rho_c g}{2}\left(\frac{\rho_m}{\rho_m - \rho_c}\right)^2 \gamma^2 l^2 + \Delta\sigma_{xx}\left(\frac{\rho_m}{\rho_m - \rho_c}\right)\gamma l. \tag{8–43}$$

To calculate the horizontal force acting on the base of the thrust sheet, we need to determine the normal and shear stresses on the basal fault. The lithostatic

8–14 A wedge-shaped model of a thrust sheet.

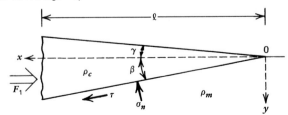

stress on the basal plane at a horizontal distance x from the apex of the wedge is $\rho_c g(\gamma + \beta)x$. Since the angles γ and β are small, σ_n on the basal plane is approximately equal to the lithostatic pressure

$$\sigma_n = \rho_c g(\gamma + \beta)x = \frac{\rho_c \rho_m}{(\rho_m - \rho_c)}\gamma g x. \tag{8–44}$$

The normal stresses on the basal plane exert a force whose horizontal component is

$$\int_0^l \sigma_n \beta\, dx = \frac{\rho_c^2 \rho_m g \gamma^2 l^2}{2(\rho_m - \rho_c)^2}, \tag{8–45}$$

where we have used the small angle approximation $\sin\beta \approx \beta$. Amonton's law, Equation (8–23), gives the shear stress acting on the basal fault during emplacement of the thrust sheet

$$\tau = \frac{f_s \rho_c \rho_m}{(\rho_m - \rho_c)}\gamma g x. \tag{8–46}$$

The shear stresses on the basal plane exert a force whose horizontal component is

$$\int_0^l \tau\, dx = \frac{f_s \rho_c \rho_m \gamma g l^2}{2(\rho_m - \rho_c)}, \tag{8–47}$$

where the small angle approximation $\cos\beta \approx 1$ has been used. The force balance on the thrust sheet is obtained by equating expression (8–43) to the sum of Equations (8–45) and (8–47) with the result

$$\Delta\sigma_{xx} = \frac{lg\rho_c(f_s - \gamma)}{2}. \tag{8–48}$$

If the friction coefficient f_s is greater than the slope γ of the thrust sheet, a compressive tectonic stress is required to emplace it. On the other hand, if the coefficient of friction is less than the slope γ of the thrust sheet, the gravitational body force on the base of the thrust sheet suffices for emplacement. This is *gravitational sliding*.

The Appalachian Mountains in the southeastern United States appear to be cored by a major thrust sheet some 250 km or more in width. The Appalachians are the remnants of a major mountain belt that resulted from a continental collision when the proto-Atlantic Ocean (Iapetus) closed. This ocean was created during the late Precambrian by the rifting of a supercontinent. In the early Cambrian a trench system developed off what is now the east coast of the United States. Remnants of the resulting island arc are seen in

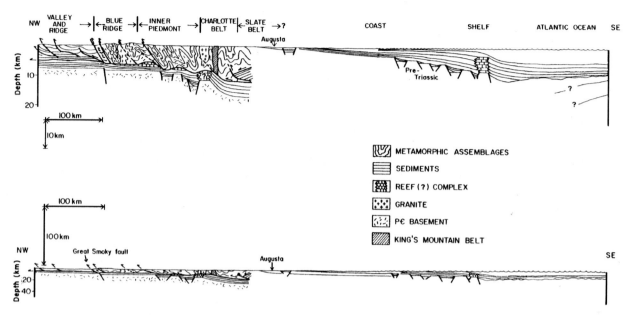

8–15 Cross section of the Appalachian Mountains of the southeastern United States showing the emplacement of an allochthonous thrust sheet from the southeast (Cook et al., 1979).

the Carolina slate belt (see Figure 8–15). The closure of the proto-Atlantic was completed in the Carboniferous and Permian (at about 250 to 300 Ma). The suture between proto-Africa and proto-North America is believed to lie east of the Carolina slate belt.

Apparently during the continental collision, crystalline rocks of proto-Africa and the island arcs were overthrust to the northwest over proto-North America as a major thrust sheet. The structure, as obtained from deep crustal seismic reflection profiling, is illustrated in Figure 8–15. Because the thrust sheet has been displaced a considerable distance, it is also referred to as an *allochthonous sheet*. The Valley and Ridge province to the northwest is composed of sediments that were pushed ahead of the sheet. The crystalline rocks of the Blue Ridge province are separated from the sedimentary rocks of the Valley and Ridge province by the Great Smoky thrust fault. This fault, which dips relatively steeply near the surface but flattens rapidly at depth, is the basal fault of the thrust sheet. The Brevard fault that separates the crystalline rocks of the Inner Piedmont province from the crystalline rocks of the Blue Ridge province is a thrust fault in the thrust sheet.

If the thrust sheet has a width $l = 250$ km and a maximum thickness of 15 km, a wedge model of the sheet has

$\gamma + \beta = 3.43°$. From Equation (8–41) with $\rho_c = 2600$ kg m^{-3} and $\rho_m = 3300$ kg m^{-3}, we find that $\gamma = 0.73°$ and $\beta = 2.70°$. The tectonic stress required to emplace the thrust sheet is obtained from Equation (8–48). The dependence of this tectonic stress on the coefficient of friction is given in Figure 8–16. We see that gravitational sliding will occur if $f_s < 0.0127$; reasonable stress levels require very low values for the coefficient of friction. Seismic studies indicate that much of the thrust sheet in the southern Appalachians is underlain by a thin layer of sediments. These sediments apparently provide a zone of weakness and a low coefficient of friction. A thrust sheet that is emplaced over a zone of weakness is also known as a *décollement*.

8–16 Dependence of the tectonic stress required for emplacement of the Appalachian thrust sheet on the coefficient of friction.

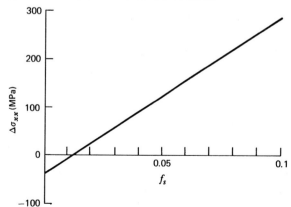

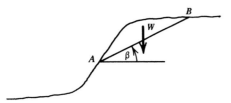

8-17 A rock slope with a potential slide surface *AB* making an angle β to the horizontal.

PROBLEM 8-4 Consider the stability of the rock slope sketched in Figure 8–17 against sliding along the plane *AB*. Assume Amonton's law is applicable and show that the condition for slope stability is

$$\tan \beta < f_s. \tag{8–49}$$

8-7 Earthquakes

We have previously discussed the earthquake cycle in terms of stick–slip behavior and elastic rebound. This behavior was illustrated in Figure 8–4. Due to tectonic motion, shear stress builds up on a locked fault until the failure stress is reached. At this time an earthquake occurs and the accumulated stress is relieved.

During an earthquake, the displacement on the fault takes place in a few seconds. The earthquake propagates along the fault at a velocity near the speed of sound in the rock (a few kilometers per second). The displacement on the fault generates seismic waves that propagate through the surrounding rock. A significant fraction of the stored elastic energy goes into the seismic waves; the remainder is dissipated as heat by friction on the fault.

To illustrate the earthquake cycle, we will consider the simple slider-block model illustrated in Figure 8–18. A block of rock of mass *m* rests on a surface. The contact area *A* represents the fault that will rupture to produce an earthquake. The mass is pressed against the surface by a normal force F_n. We take the normal stress on the fault to be the lithostatic pressure so that

$$F_n = \rho g h A, \tag{8–50}$$

where *h* is the mean depth of the fault being considered. The mass of the block *m* is given by

$$m = \rho A^{3/2}. \tag{8–51}$$

This is a cube with a linear dimension $A^{1/2}$.

We assume that the block is being pulled along the surface by a constant-velocity driver plate, the constant

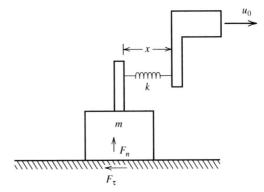

8-18 Slider-block model for fault behavior. The driver plate extends the spring at a constant velocity u_0 until the spring force kx equals the static friction force F_τ. At this time slip occurs and the cycle repeats.

velocity is u_0. The driver plate is attached to the block by a spring with a spring constant *k*. This spring force must be balanced by a resisting shear force on the surface. This surface shear force F_τ is given by

$$F_\tau = kx, \tag{8–52}$$

where *x* is the extension of the spring. In our model the spring represents the accumulation of elastic strain in the rock adjacent to the fault. The block in our model interacts with the surface through friction, which prevents the block from sliding (or a model earthquake from occurring) until a critical pulling force $F_{\tau s}$ is reached. The block sticks and the force in the spring increases until it equals the frictional resistance to sliding on the surface, and then slip occurs. The extension of the spring is analogous to the elastic strain in the rock adjacent to a fault. The slip is analogous to an earthquake on a fault. This is *stick–slip* behavior. The stored elastic strain in the spring is relieved; this is *elastic rebound*, as previously discussed in Section 8–3. When the block "sticks" the extension of the spring *x* is related to the constant velocity u_0 of the driver plate by $x = u_0 t$, where *t* is the time since the last slip event. From Equation (8–18) the shear stress on a fault is given by

$$\tau = \frac{Gx}{2b}, \tag{8–53}$$

where *b* is the distance from the fault where the uniform velocity is applied. For a fault it is appropriate to assume that this distance is approximately equal to the linear dimension of the fault $A^{1/2}$. Taking $b = A^{1/2}$ and using

Equation (8–53) we can write an expression for the shear force on the fault:

$$F_\tau = \tau A = \frac{G x A^{1/2}}{2}. \tag{8-54}$$

Thus from Equations (8–52) and (8–54) we have an expression for our model spring constant:

$$k = \frac{G A^{1/2}}{2}. \tag{8-55}$$

Assuming that the slip condition is given by Amonton's law, Equation (8–23), the static condition for the onset of sliding is

$$k x_s = f_s F_n. \tag{8-56}$$

This is the extension of the spring x_s required to initiate slip. Substitution of Equations (8–50) and (8–55) into Equation (8–56) shows that the accumulated displacement on the fault is given by

$$x_s = \frac{2 f_s \rho g h A^{1/2}}{G}. \tag{8-57}$$

This is the displacement that would occur on the fault if the shear stress on the fault was reduced to zero.

Once the block in Figure 8–18 starts to slip, there is still a frictional force resisting the motion. The simplest representation of this resistance is through a dynamic coefficient of friction f_d defined by

$$F_\tau = f_d F_n. \tag{8-58}$$

For stick–slip behavior to occur, the static coefficient of friction must be greater than the dynamic coefficient of friction, $f_s > f_d$. When $f_s < f_d$, stable sliding occurs and the block is pulled along the surface at the velocity u_0. Once sliding begins the equation of motion for the block is

$$m \frac{d^2 x}{dt^2} + k x = f_d F_n. \tag{8-59}$$

Sliding is analogous to an earthquake and it relieves the accumulated strain in the spring in analogy to elastic rebound. Substitution of Equations (8–51), (8–55), and (8–50) into Equation (8–59) gives

$$\frac{2\rho A}{G} \frac{d^2 x}{dt^2} + x = \frac{2 f_d \rho g h A^{1/2}}{G}. \tag{8-60}$$

In writing Equation (8–60) we assumed that the loading velocity of the driver plate u_0 is so slow that it can be

neglected during the sliding of the block. This is reasonable because an earthquake lasts only a few tens of seconds, whereas the interval between earthquakes on a fault is typically hundreds of years or more.

To study the motion of the block, we introduce the following nondimensional variables

$$\phi = \frac{f_s}{f_d} \qquad T = t \left(\frac{G}{2\rho A} \right)^{1/2}$$

$$X = \frac{x G}{2 f_s \rho g h A^{1/2}} \qquad U = \frac{u}{f_s g h} \left(\frac{G}{2\rho} \right)^{1/2}, \tag{8-61}$$

where $u = dx/dt$. In terms of these variables, the condition for the initiation of slip given by Equation (8–57) becomes

$$X = 1 \tag{8-62}$$

and the equation of motion (8–60) becomes

$$\frac{d^2 X}{dT^2} + X = \frac{1}{\phi}. \tag{8-63}$$

We assume that slip starts at $T = 0$ with $X = 1$ as given by Equation (8–62). It is also appropriate to assume that the initial slip velocity is zero so that $U = dX/dT = 0$ at $T = 0$. The solution of Equation (8–63) that satisfies these conditions is

$$X = \frac{1}{\phi} + \left(1 - \frac{1}{\phi} \right) \cos T \tag{8-64}$$

and the slip velocity is given by

$$U = \frac{dX}{dT} = -\left(1 - \frac{1}{\phi} \right) \sin T. \tag{8-65}$$

Sliding ends at $T = T_s = \pi$ when dX/dT is again zero. When the velocity is zero the friction jumps to its static value, preventing further slip. The position of the block at the end of slip is $X = (2/\phi) - 1$ so that the total nondimensional displacement of the block during the slip event is

$$\Delta X = \left(\frac{2}{\phi} - 1 \right) - 1 = 2 \left(\frac{1}{\phi} - 1 \right). \tag{8-66}$$

If $\phi = f_s / f_d$ is only slightly larger than 1, then ΔX is small and only a fraction of the stress (strain) is lost in the slip event. If $f_d \to 0$, we have $\phi \to \infty$ and $\Delta X = -2$. Because the dynamic friction is small, energy is conserved and the energy associated with the extension of the spring $X = 1$ is converted to energy associated

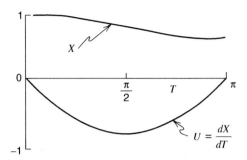

8-19 Dependence of the nondimensional slider-block position X and velocity dX/dT on time T during a slip event as given by Equations (8–64) and (8–65) for $\phi = 1.25$.

with the compression of the spring $X = -1$. This type of overshoot is rarely seen in actual earthquakes.

PROBLEM 8–5 In the slider-block model, what value of ϕ gives zero residual extension of the spring $X = 0$?

From Equation (8–65) the maximum nondimensional slip velocity U_{max} occurs at $T = \pi/2$ and is given by

$$U_{max} = \left(\frac{dX}{dT}\right)_{max} = -\left(1 - \frac{1}{\phi}\right). \qquad (8\text{–}67)$$

The dependences of X and U on T during slip are given in Figure 8–19 for $\phi = 1.25$. For this case $\Delta X = -0.4$ and $U_{max} = -0.2$.

After the slip event is completed, the spring again extends due to the velocity u_0 of the driver plate. The nondimensional velocity of extension U_0 from Equation (8–61) is

$$U_0 = \frac{u_0}{f_s g}\left(\frac{k}{m}\right)^{1/2}. \qquad (8\text{–}68)$$

The nondimensional time before the next slip event ΔT_e obtained from Equations (8–66) and (8–68) is

$$\Delta T_e = \frac{\Delta X_e}{U_0} = \frac{2}{U_0}\left(1 - \frac{1}{\phi}\right). \qquad (8\text{–}69)$$

At the end of this time another slip event occurs and the cycle repeats. Slip events occur periodically.

PROBLEM 8–6 The spring force on the slider block in Figure 8–18 at the time of slip initiation is $f_s F_n$. What is the spring force on the slider block at the end of slip?

PROBLEM 8–7 (a) Show that the work done by the driver plate during a stress accumulation phase is

$$W = \frac{2m^2 g^2 f_d}{k}(f_s - f_d).$$

(b) Show that this is also the work required during this time if there is stable sliding, i.e., if the block moves at the velocity u_0 with $F = f_d f_n$.

We will now use these results to approximate the actual behavior of a fault during an earthquake cycle. We first relate the displacement on a fault during an earthquake Δw to the displacement of the slider block using Equations (8–61) and (8–66) with the result

$$\Delta w = \frac{2 f_s \rho g h A^{1/2}}{G}\Delta X = \frac{4 f_s \rho g h A^{1/2}}{G}\left(1 - \frac{f_d}{f_s}\right). \qquad (8\text{–}70)$$

The maximum slip velocity on the fault u_{max} from Equations (8–61) and (8–67) is

$$u_{max} = f_s g h\left(\frac{2\rho}{G}\right)^{1/2} U_{max} = f_s g h\left(\frac{2\rho}{G}\right)^{1/2}\left(1 - \frac{f_d}{f_s}\right). \qquad (8\text{–}71)$$

The stress on the fault just prior to the earthquake from Equations (8–23) and (8–50) is $\tau_{fs} = f_s \rho g h$. The stress on the slider block is proportional to the nondimensional extension of the spring X. Prior to a slip event $X = 1$ and the change in X, ΔX, is given by Equation (8–66). We conclude that the fraction of the initial stress on a fault that is lost during an earthquake is equal to ΔX and the stress drop on the fault due to the earthquake $\Delta \tau$ is given by

$$\Delta \tau = \tau_{fs}\Delta X = 2 f_s \rho g h\left(1 - \frac{f_d}{f_s}\right), \qquad (8\text{–}72)$$

where $\Delta \tau$ is defined to be a positive quantity.

As an example, let us take the mean depth of the fault to be $h = 5$ km, $\rho = 2500$ kg m^{-3}, and $g = 9.8$ m s^{-2} and find from Equation (8–50) that the mean normal stress on the fault σ_n is 124 MPa. To specify the static frictional stress on the fault we must choose a coefficient of static friction and take $f_s = 0.05$. From Equation (8–23) we find that the static frictional stress on the fault τ_{fs} is 6.2 MPa. This is the stress on the fault just prior to rupture.

Our choice of such a low value for the coefficient of static friction requires an explanation. As discussed in Section 8–3, a typical value of f_s from laboratory experiments is 0.85. However, as we will show, such a high value for f_s requires large stress drops in earthquakes that are not consistent with observations. One explanation for this low value of f_s is a high pore pressure of water on faults. As seen in Equation (8–24), a high pore pressure p_w can greatly reduce the coefficient of static friction on a fault.

We must also choose a value for the ratio of static to dynamic friction ϕ. Again, values for the dynamic coefficient of friction are controversial. Slip velocities on faults are high and they cannot be simulated in the laboratory. We assume $\phi = 1.1$ and find from Equation (8–72) that the stress drop in the earthquake $\Delta\tau$ is 1.23 MPa, independent of the area of rupture A. We will show that observations confirm this. The residual stress on the fault after an earthquake is $\tau_{fd} = \tau_{fs} - \Delta\tau = 5$ MPa.

We will now consider an earthquake of a specified size. Let the rupture area $A = 100$ km^2 so that the characteristic linear dimension $A^{1/2}$ is 10 km. For the parameter values given before and $G = 3 \times 10^{10}$ Pa we find from Equation (8–70) that the displacement on the fault during an earthquake Δw is 0.82 m. From Equation (8–71) we find that the maximum slip velocity on the fault during rupture $u_{\max}$ is 0.1 m s^{-1}. This is a typical slip velocity on a fault during an earthquake.

The displacements of the rocks adjacent to a fault generate seismic waves during an earthquake. These waves are of two types: body waves that propagate through the interior of the Earth and surface waves that propagate along the surface. There are two types of body waves: P or compressional waves and S or shear waves. Displacements in compressional waves are in the direction of propagation and displacements in shear waves are perpendicular to the direction of propagation. There are also two types of surface waves: Love and Rayleigh waves. Surface waves are similar to wind-driven waves on an ocean.

The ground motions caused by seismic waves are the primary cause of damage when an earthquake occurs. Because the displacements in surface waves are restricted to a thin surface layer, they are primarily responsible for earthquake damage. Prior to the development of the seismograph in about 1900 the only measure of the strength of an earthquake was the felt

TABLE 8-1 **Typical Felt Intensity Scale**	
I	Not felt.
II	Felt by a few people.
III	Hanging objects sway.
IV	Windows and doors rattle.
V	Sleepers waken.
VI	Windows and glassware broken.
VII	Difficult to stand.
VIII	Branches broken from trees.
IX	Cracks in ground – general panic.
X	Large landslides – most masonry structures destroyed.
XI	Nearly total destruction.

intensity. An earthquake was quantified in terms of the effects it had on people, buildings, and the environment in the immediate vicinity of the rupture. A typical felt intensity scale (the modified Mercalli) is given in Table 8–1. The maximum felt intensities are usually found near the fault where the earthquake occurs; felt intensities decrease with distance from the fault. The felt intensity is influenced by the type and depth of soil; it is generally higher on deep, loose soils. Because the application of the felt intensity scale is subjective, care must be taken in its use.

The development of the seismograph made it possible to introduce a quantitative measure of earthquake size. The *magnitude m* of an earthquake is obtained from the measured surface displacements at seismic stations. The magnitude scale was originally defined in terms of the amplitude of ground motions at a specified distance from an earthquake. Typically, the *surface wave magnitude m_s* is based on the motions generated by surface waves (Love and Rayleigh waves) with a 20-s period, and the *body wave magnitude m_b* is based on the motions generated by body waves (P and S waves) having periods of 6.8 s. The magnitude scale is a popular measure of earthquake strength because of its logarithmic basis, which allows essentially all earthquakes to be classified on a scale of 0–10. Unfortunately, the two magnitude scales, m_s and m_b, are not equal for a given earthquake. Also, magnitudes are sensitive to the paths that seismic waves traverse and to conditions in the source region and near the seismic station. In addition, different types of earthquakes generate different surface displacements. Thus, published magnitudes for a given earthquake can differ by 10% or more.

An alternative approach to the quantification of the size of an earthquake is the *seismic moment M* defined by

$$M = GA\Delta w. \tag{8–73}$$

This seismic moment is the product of the shear modulus G of the rock in which the fault is embedded, the fault rupture area A, and the mean displacement across the fault Δw during the earthquake. The seismic moment is a measure of the strain released during the earthquake and is determined from seismograms recorded at a large number of seismic stations.

Although the accepted measure of the size of an earthquake is the seismic moment, the magnitude scale has received such wide acceptance that a *moment magnitude m* has been empirically defined in terms of the seismic moment M using the relation

$$\log M = 1.5m + 9.1. \tag{8–74}$$

This definition is consistent with the definition of the surface wave magnitude but not with the definition of the body wave magnitude. It is standard practice to use long-period (50–200 s) body and/or surface waves to directly determine the seismic moment M, and Equation (8–74) is used to obtain a moment magnitude. In the remainder of our discussion of earthquakes we will consider only the moment magnitude.

The moment magnitude can be related to the total energy in the seismic waves generated by the earthquake E_s through the empirical relation

$$\log E_s = 1.5m + 4.8, \tag{8–75}$$

where E_s is in joules. Table 8–2 relates the moment

magnitude of an earthquake to the felt intensity and the felt distance. Again, this is an approximate relation since the felt intensity will depend on the depth of the earthquake as well as its magnitude. Also, the felt intensity of an earthquake of a given magnitude varies from one location to another.

In terms of the simple slider-block model, the seismic moment for a model earthquake is obtained by substituting the displacement Δw from Equation (8–70) into the definition of the seismic moment given in Equation (8–73) with the result

$$M = 4 f_s \rho g h A^{3/2} \left(1 - \frac{f_d}{f_s}\right). \tag{8–76}$$

This result predicts that the seismic moment is proportional to the rupture area A raised to the 3/2 power. The seismic moments of a large number of earthquakes are given in Figure 8–20 as a function of the square

8-20 Dependence of the seismic moment M on the square root of the rupture area $A^{1/2}$ for a large number of earthquakes (Hanks, 1977). The solid line is the prediction given by Equation (8–76) taking $f_s = 0.05$, $\rho = 2500 \text{ kg m}^{-3}$, $g = 9.8 \text{ m s}^{-2}$, $h = 5 \text{ km}$, and $f_s/f_d = \phi = 1.1$.

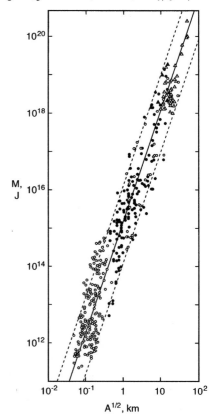

TABLE 8–2 **Typical Maximum Felt Intensities and Felt Distances for Earthquakes of Prescribed Magnitude (*m*)**

Magnitude (*m*)	Maximum Felt Intensity	Felt Distance (km)
2	I–II	0
3	III–IV	15
4	V	80
5	VI–VII	150
6	VIII	220
7	IX–X	400
8	XI	600

root of the rupture area. This figure also shows the prediction of Equation (8–76) for $f_s = 0.05$, $\rho = 2500$ kg m^{-3}, $g = 9.8$ m s^{-2}, $h = 5$ km, and $f_s/f_d = \phi = 1.1$. The corresponding stress drop from Equation (8–72) is 1.23 MPa. The prediction of the slider block model is in good agreement with the observed moments and rupture areas of earthquakes.

For the earthquake considered previously ($A^{1/2} = 10$ km), the moment M from Equation (8–76) is 2.23×10^{18} J. From Equation (8–74), the corresponding moment magnitude m is 6.17. And from Equation (8–75) the radiated elastic wave energy $E_s = 1.33 \times 10^{14}$ J.

We can estimate the elastic energy released in an earthquake using the slider-block model. From Equations (8–55) and (8–57), the energy stored in the spring prior to the earthquake is

$$E_s = \frac{1}{2}kx_s^2 = \frac{f_s^2 \rho^2 g^2 h^2 A^{3/2}}{G}. \tag{8–77}$$

From Equation (8–70) the energy in the spring after the earthquake is

$$E_r = \frac{1}{2}k(x_s - \Delta w)^2 = \frac{f_s^2 \rho^2 g^2 h^2 A^{3/2}}{G}\left(\frac{2f_d}{f_s} - 1\right)^2. \tag{8–78}$$

The energy released in the model earthquake is

$$\Delta E = E_s - E_r = \frac{4 f_s f_d \rho^2 g^2 h^2 A^{3/2}}{G}\left(1 - \frac{f_d}{f_s}\right). \tag{8–79}$$

For the earthquake with $A^{1/2} = 10$ km, we find $\Delta E = 4.13 \times 10^{14}$ J. This is higher than the seismic radiated energy given before, as expected. In our simple slider-block model the elastic energy lost in a slip event goes into frictional heating on the slipping surface. In a real earthquake a fraction of the lost energy is radiated in seismic waves and the remainder is converted to heat by friction on the fault.

PROBLEM 8–8 Compute the wave energy released in a magnitude 8.5 earthquake and compare it with the amount of heat lost through the surface of the Earth in an entire year.

PROBLEM 8–9 Data on the frequency with which earthquakes of a given magnitude occur can be sum-marized by the formula

$$\frac{dN}{dm} = 5.25 \times 10^7 \exp(-2.07m), \tag{8–80}$$

where dN is the number of earthquakes per year with magnitude between m and $m + dm$. Use Equations (8–75) and (8–80) to show that the rate of energy release $\dot{E}_{12}$ by earthquakes with magnitudes lying between m_1 and m_2 is

$$\dot{E}_{12}(\text{J yr}^{-1}) = 7.35 \times 10^{12}\{e^{1.25m_2} - e^{1.25m_1}\}. \tag{8–81}$$

If the largest earthquakes to occur have magnitude 8.5, compute the yearly release of wave energy. How does this compare with the energy released by a single magnitude 8.5 earthquake? What do you conclude about the relative contributions of large numbers of small earthquakes and small numbers of large earthquakes to the yearly release of seismic energy?

8–8 San Andreas Fault

Let us now turn our attention to two major strike–slip faults. First consider the San Andreas fault, which stretches almost the entire length of the state of California, as shown in Figure 8–21. Along much of its length the motion on the fault is primarily strike–slip. The San Andreas is recognized as a major boundary between the Pacific and North American plates. If the fault is to accommodate the relative motion between these two plates by strike–slip motion, it must lie on a small circle about the pole of rotation that defines the relative motion between the two plates. Two small circles drawn about the pole of rotation given in Table 1–6 are compared with the trend of the fault in Figure 8–21. The small circles are in excellent agreement with the trend of the fault along much of its length. This is strong evidence that a large fraction of the relative motion between the Pacific and North American plates occurs on the San Andreas fault. Studies of the relative motion between the two plates give a relative velocity of 46 mm yr^{-1} (see Section 1–8).

The San Andreas fault appears to be divided into four distinct sections that exhibit quite different be-havior. Some sections of the fault exhibit little seismic activity; the fault appears to be locked, and strain is being accumulated. The accumulated strain on these sections is relieved in great earthquakes. On other sec-tions of the fault, small earthquakes and aseismic

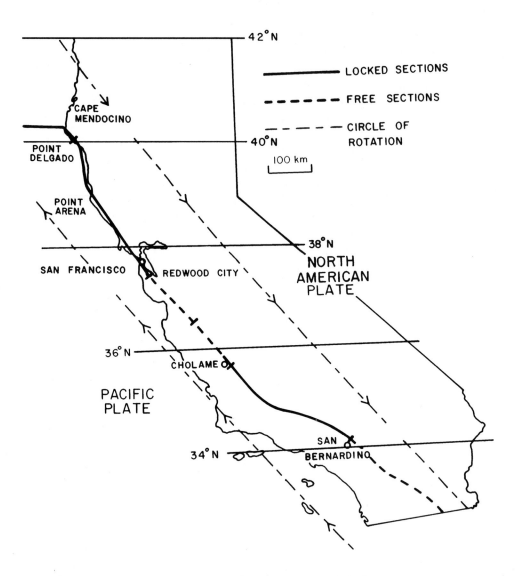

8-21 Surface trace of the locked and free sections of the San Andreas fault. Also shown are two small circles drawn about the pole of rotation for the motion of the Pacific plate relative to the North American plate.

creep relieve at least a fraction of the accumulating strain.

The northern terminus of the fault is near Cape Mendocino at the triple junction of the Pacific, North American, and Juan de Fuca plates. A northern locked section extends about 300 km from Cape Mendocino to near Redwood City. A fault break with surface displacements of about 4 m was reported along the entire length of this section in the 1906 San Francisco earthquake. There has been no reported fault creep or earthquakes on this section since 1906.

A central free section extends some 300 km from Redwood City to near Cholame. Fault creep and small earthquakes regularly occur on this section of the fault.

Surface displacements during the 1906 earthquake extended some 100 km into this section, but the magnitude of the surface displacement was considerably less than in the northern locked section. The rate of fault creep on this section of the fault is 20 to 30 mm yr^{-1}. This is less than the 46 mm yr^{-1} relative plate velocity given in Section 1–8.

A southern locked section of the fault extends some 350 km from near Cholame to near San Bernardino. There were reports of fault breaks along the entire length of this section during the 1857 southern California earthquake. Since then there have been no reports of earthquakes or fault creep on this section. The curvature of the fault north of San Bernardino probably plays an important role in the behavior of the fault in this section. The convergence of the plates at this bend in the fault requires a thrusting component, and the result is the uplift associated with the Transverse Ranges, a series of mountain belts in this area.

South of San Bernardino, the San Andreas fault splays off into a series of faults. Small and moderate earthquakes and fault creep occur on a number of these faults, and it appears that this section of the fault is not locked.

It is consistent with our present knowledge of the San Andreas fault to postulate that displacements on the locked sections of the fault occur during great earthquakes. Earthquakes on the two locked zones occur at different times, possibly alternately, and the central free section is a transition zone between the two locked zones.

The documented history of great earthquakes on the San Andreas fault is relatively short; the San Francisco earthquake on April 18, 1906, was certainly a great earthquake, with an estimated surface wave magnitude m_s of 8.25. Based on the surface offsets given in Figure 2–27, we take the mean displacement on the fault in this earthquake to be $\Delta w = 4$ m along a 200-km length of the fault. For a rupture depth of 15 km, the rupture area A is 3000 km^2. With $G = 3 \times 10^{10}$ Pa, the seismic moment M of this earthquake from Equation (8–73) is 3.6×10^{20} J. From Equation (8–74), the corresponding moment magnitude m is 7.6. Because very few seismic records are available to estimate the surface wave magnitude, the moment magnitude is the preferred magnitude value for this earthquake.

If the full relative plate velocity of $u_0 = 46$ mm yr^{-1} was accommodated on the San Andreas fault, the 4 m of displacement during the 1906 earthquake would represent ≈ 90 years of accumulated strain. However, tectonic displacements associated with the Pacific–North American plate boundary occur throughout the western United States. In particular, other important faults in California accommodate a fraction of the relative motion between the plates. Based on geodetic observations, it is estimated that about 35 mm yr^{-1} or 60%

of the total motion is accommodated directly on the San Andreas fault. With this rate of strain accumulation, the 1906 San Francisco earthquake represents 114 years of accumulated strain. There is recorded evidence of a large earthquake in northern California in 1838, but it has not been documented that this earthquake accommodated major displacements on the northern locked section of the fault.

A great earthquake occurred on the San Andreas fault north of Los Angeles on January 9, 1857. It is generally accepted that this was a great earthquake on the southern locked section, although there were no direct observations of surface displacements. Studies of apparent stream offsets attributed to this earthquake indicate a mean displacement Δw of 7 m and a magnitude $m_s \approx 8.3$. A second great historic earthquake occurred on this section on December 8, 1812. Sieh et al. (1989) dated the displaced layers of sediments in riverbeds adjacent to this southern locked section and concluded that great earthquakes had occurred on this section in the years 1480 ± 15, 1346 ± 17, and 1100 ± 65. The mean interval between these five earthquakes is 190 years; the longest interval was 332 years and the shortest was 45. In 2000, the interval since the last great earthquake is 143 years and the accumulated strain, assuming a relative velocity of 35 mm yr^{-1}, would result in a slip of 5 m if relieved by a great earthquake. Although great earthquakes occur on the southern locked section of the San Andreas fault fairly regularly, they are certainly not periodic. This behavior can be attributed to the complex interactions between the San Andreas fault and the many other major faults in the region that have large earthquakes.

The Parkfield section of the San Andreas fault just north of the southern locked section has a particularly interesting history of seismicity. During the past 150 years, earthquakes with $m \approx 6$ have occurred with remarkable regularity. Events occurred in 1857, 1881, 1901, 1922, 1934, and 1966. The mean interval between these earthquakes is 22 years and the shortest is 12 years and the longest is 32 years. The prediction that an $m \approx 6$ earthquake would occur on this section prior to 1992 with a 95% probability led to a massive instrumentation of the fault; the predicted earthquake had not yet occurred in 2001. Again, we have an example of earthquakes on a major fault occurring fairly regularly, but certainly not periodically.

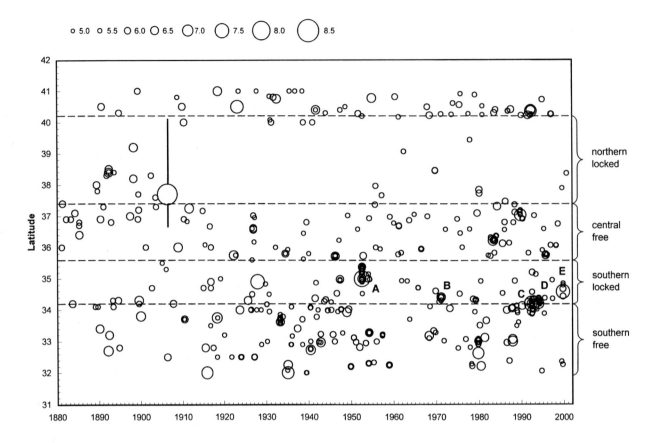

8–22 Earthquakes within 100 km of the San Andreas fault with magnitudes greater than $m = 5$ that occurred from 1880 to 2000 are given as a function of latitude. The solid line shows the surface rupture of the 1906 earthquake. The letters locate the Kern County (A), San Fernando (B), Landers (C), Northridge (D), and Hector Mine (E) earthquakes. The free and locked sections of the fault are located by the dashed lines.

The magnitudes of earthquakes associated with the San Andreas fault system are given in Figure 8–22 as a function of latitude for the period from 1880 to 2000. Earthquakes with a magnitude of 5 or larger that occurred within 100 km of the fault are included. The extent of the 1906 fault break and the division of the fault into sections are indicated. The reporting bias evident in the data can be attributed to the large increase in the number of seismographs since 1950, which has greatly improved the recording of earthquakes. Despite the variations in the quality of the data some clear trends are discernible. The most obvious is the reduction in the number of small and intermediate size earthquakes adjacent to the northern locked section following the 1906 earthquake. The first intermediate size earthquake to occur was the October 17, 1989, Loma Prieta earthquake ($m = 7.1$). In direct contrast, a number of intermediate size earthquakes occurred in this region prior to the 1906 earthquake. They include one in the Santa Cruz mountains in 1865 ($m \approx 6.5$), Hayward in 1868 ($m \approx 6.8$), Vacaville in 1892 ($m \approx 6.8$), and Mare Island

in 1898 ($m \approx 6.5$). There appears to have been a systematic activation of seismic activity prior to the great 1906 earthquake. The reduction in seismic activity following the 1906 earthquke is strong evidence that the stress level in the elastic lithosphere adjacent to the northern locked section of the San Andreas fault was significantly reduced by the 1906 earthquake.

Let us now turn our attention to the seismic activity adjacent to the southern locked section of the San Andreas fault as illustrated in Figure 8–22. No great earthquake has occurred on this section since 1857. A significant number of intermediate size earthquakes have occurred adjacent to this section since 1952. These include the July 21, 1952, Kern County earthquake ($m = 7.5$), the February 9, 1971, San Fernando earthquake ($m = 6.7$), the June 28, 1992, Landers earthquake ($m = 7.3$), the January 17, 1994, Northridge earthquake ($m = 6.7$), and the October 16, 1999, Hector Mine earthquake ($m = 7.1$). Indicated by letters in Figure 8–22, these events may be indicative of a systematic seismic activation preceding the next great earthquake on the southern locked section of the San Andreas fault.

8-9 North Anatolian Fault

Another example of a major strike–slip fault is the North Anatolian fault in Turkey. This fault is the site of a remarkable series of major earthquakes that began in 1939; the earthquakes swept along almost the entire 1000-km length of the fault from east to west. The tectonic setting of the North Anatolian fault is considerably more complex than that of the San Andreas fault. Turkey forms part of the broad collisional zone that extends from southern Europe through Asia to India. The tectonics of Turkey can be at least partially understood if a major part of it is assumed to consist of a small plate, the Anatolian plate, as illustrated in Figure 8–23. The collision between the Arabian plate and the Eurasian plate is wedging the Anatolian plate toward the west. The North Anatolian fault forms the northern boundary of the plate and the East Anatolian fault the southeastern boundary. The westerly movement of the plate causes right-lateral

TABLE 8-3 Characteristics of a Series of Earthquakes with Magnitudes Greater than 7 along the North Anàtolian Fault

Date	m	Length of Break (km)	Offset (m)
December 26, 1939	7.8	360	7.5
December 20, 1942	7.1	50	1.7
November 26, 1943	7.3	260	4.5
February 1, 1944	7.3	180	3.5
May 26, 1957	7.0	40	1.65
July 22, 1967	7.1	80	2.60
August 17, 1999	7.4	100	2.5

strike–slip displacements on the North Anatolian fault and left-lateral strike–slip displacements on the East Anatolian fault. Subduction of the African plate occurs at a series of arcuate trenches along the southern boundary of the Anatolian plate. The foundering of the African plate may result in tensional stresses in the Anatolian plate landward of the trenches. These stresses may also contribute to the westward movement of the Anatolian plate. To the west and east of the Anatolian plate the tectonics of the broad collisional zone become even more complex. To the east of the Anatolian plate the convergence between the Arabian and the Eurasian plates is accommodated in a broad zone of compression. The western part of the Anatolian plate merges into an extensive zone of extensional tectonics in western Turkey and Greece. Because of the complexity, plate tectonics provides relatively poor constraints on the relative velocity across the North Anatolian fault. This relative velocity is estimated to be about 15 mm yr^{-1}.

The magnitude of earthquakes associated with the North Anatolian fault system are given in Figure 8–24 as a function of longitude for the period 1880 to 2000. Also included are the surface breaks associated with a series of large earthquakes that have occurred since 1939. The dates, magnitudes, lengths of the surface breaks, and offsets for these earthquakes are summarized in Table 8–3.

The first and largest of these earthquakes was the Erzincan earthquake of December 26, 1939, with a magnitude of 7.8. The surface break extended from near the junction of the North and East Anatolian

8–23 The complex tectonic situation in Turkey. The Anatolian plate is bounded on the north by the North Anatolian fault and on the east by the East Anatolian fault.

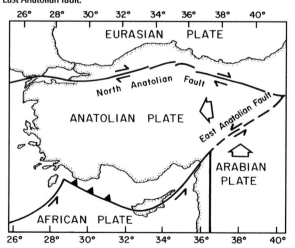

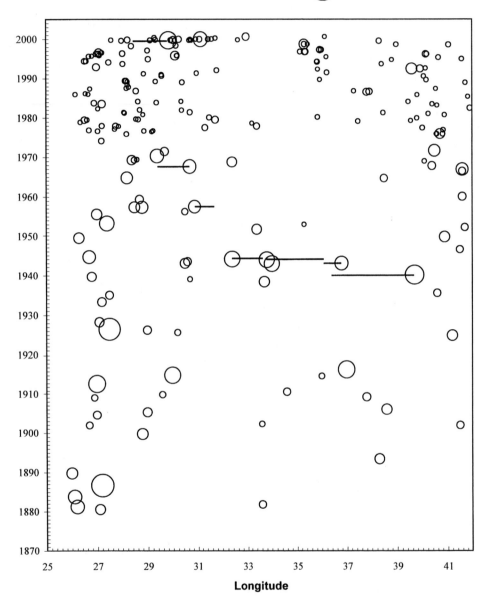

Anatolian Fault Catalog 1880 - 2000

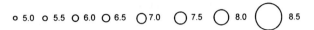

8-24 Earthquakes on and adjacent to the North Anatolian fault with magnitudes greater than $m = 5$ that occurred from 1880 to 2000 are given as a function of longitude. The solid lines are the surface ruptures of the sequence of large earthquakes that occurred on the North Anatolian fault between 1939 and 1999.

faults some 350 km to the west. This event was followed on December 20, 1942, by an earthquake on a more northerly branch of the fault near the western end of the 1939 surface break. Another large earthquake occurred on the northerly branch on November 26, 1943, extending the surface break 260 km to the west. This was followed on February 1, 1944, by another large earthquake that extended the surface break another 190 km to the west. Two large earthquakes occurred in 1957 and 1967 near the western termination of the 1944 surface break. The propagating sequence was extended to the west by the August 17, 1999, Izmit earthquake with a surface break of 100 km. This earthquake killed close to 20,000 people. The propagation of seismic activity along the fault can be attributed to the transfer of stress from a section of the fault on which an earthquake has occurred to an adjacent section. This behavior is quite different from that on the San Andreas fault.

Fault creep occurs on the North Anatolian fault at Ismetpasa, which is located midway along the surface rupture of the February 1, 1944, magnitude 7.3 earthquake. Between 1970 and 1980 the rate of fault creep was close to $10 \, \text{mm yr}^{-1}$. This observation of fault creep is in contrast to the San Andreas fault system, where fault creep apparently occurs on sections of the fault where great earthquakes do not occur. The fault creep on the North Anatolian fault and on the San Andreas fault are the only well-documented examples of aseismic fault creep at this time.

8-10 Some Elastic Solutions for Strike–Slip Faulting

In this section we will consider two mathematical solutions of the equations of elasticity that are applicable to displacements during an earthquake on a strike–slip fault. The models are idealized, but they illustrate some important aspects of the problem. In the first example we determine the strain field caused by a displacement on a fault. We consider the half-space $y > 0$ in which there is initially a uniform shear stress $\sigma_{xz,0}$ as shown in Figure 8–25a. A two-dimensional crack that extends from the surface to a depth $y = a$ is then introduced at $x = 0$. The introduction of the crack causes the displacements shown in Figure 8–25b. The resulting strain field models the strain due to an earthquake on a strike–slip fault. We assume that the only nonzero component of displacement is w_z in the z direction. After the introduction of the crack there are two nonzero components of shear stress: σ_{xz} (and σ_{zx}) and σ_{yz} (and σ_{zy}). To simplify the analysis, we neglect the frictional stress on the fault after the earthquake has occurred and assume $\sigma_{xz} = 0$ on $x = 0$, $0 \leq y \leq a$. The displacement and stress fields in this two-dimensional problem are independent of z.

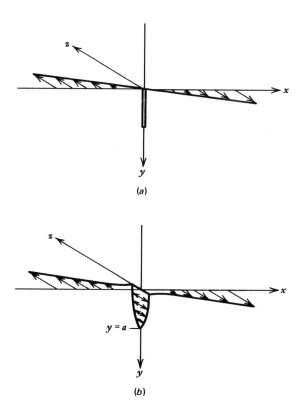

8-25 The displacement field due to faulting on a model strike–slip fault. The surface is at $y = 0$ and the fault is at $x = 0$, $0 \leq y \leq a$. Displacements are in the z direction. (a) Initially the half-space $y > 0$ is in a state of uniform shear stress $\sigma_{xz,0}$. (b) After the introduction of the crack the shear stress on the crack is zero ($\sigma_{xz} = 0$) and there is a displacement across the crack.

The determination of the stresses, strains, and displacements in the half-space $y > 0$ requires that we derive the differential equation expressing the equilibrium of forces on a small element of material. Figure 8–26 shows a small rectangular parallelepiped in the half-space with those nonzero stress components that exert forces in the z direction. Because the net force in the z direction on this small volume element must be zero, we can write

$$\{\sigma_{xz}(x + \delta x, y) - \sigma_{xz}(x, y)\} \, \delta y \, \delta z$$
$$+ \{\sigma_{yz}(x, y + \delta y) - \sigma_{yz}(x, y)\} \, \delta x \, \delta z = 0. \quad (8\text{–}82)$$

Upon expanding $\sigma_{xz}(x + \delta x, y)$ and $\sigma_{yz}(x, y + \delta y)$ in a Taylor series, performing the indicated subtractions and dividing by $\delta x \delta y \delta z$, we obtain the equation of equilibrium in the form

$$\frac{\partial \sigma_{xz}}{\partial x} + \frac{\partial \sigma_{yz}}{\partial y} = 0. \quad (8\text{–}83)$$

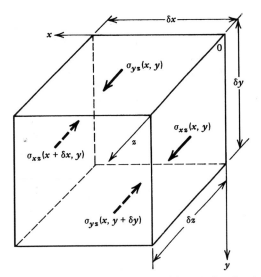

8–26 The nonzero stress components that exert forces in the z direction on a small rectangular parallelepiped in the half-space of the model strike–slip faulting problem of Figure 8–25.

Both σ_{xz} and σ_{yz} are related to the derivatives of the displacement w_z. From Equations (2–102), and (3–49), we obtain

$$\sigma_{xz} = G\frac{\partial w_z}{\partial x} \qquad \sigma_{yz} = G\frac{\partial w_z}{\partial y}. \qquad (8\text{–}84)$$

Substitution of Equation (8–84) into (8–83) yields

$$\frac{\partial^2 w_z}{\partial x^2} + \frac{\partial^2 w_z}{\partial y^2} = 0. \qquad (8\text{–}85)$$

The displacement w_z satisfies Laplace's equation.

The solution for the displacement prior to the introduction of the crack is simply

$$w_z = \frac{\sigma_{xz,0}}{G}x. \qquad (8\text{–}86)$$

After the introduction of the crack the solution to Laplace's equation must satisfy the boundary conditions

$$\frac{\partial w_z}{\partial y} = 0 \qquad \text{at } y = 0 \qquad (8\text{–}87)$$

$$\frac{\partial w_z}{\partial x} = 0 \qquad \text{at } x = 0, \quad 0 < y < a \qquad (8\text{–}88)$$

$$w_z = 0 \qquad \text{at } x = 0, \quad a < y \qquad (8\text{–}89)$$

$$w_z \to \frac{\sigma_{xz,0}}{G}x \qquad \text{as } x \to \infty. \qquad (8\text{–}90)$$

The first condition follows from Equation (8–84) and

the fact that the shear stress σ_{yz} must be zero at the surface $y = 0$. The second condition comes from Equation (8–84) together with the vanishing of the shear stress σ_{xz} on the crack. The third condition is a consequence of the overall geometry and the fact that the crack does not extend below $y = a$. Finally, the fourth condition is a requirement that the displacement far from the crack equal the initial displacement.

The solution of Laplace's equation that satisfies conditions (8–87) to (8–90) can only be obtained by using the mathematical theory of complex variables. We introduce the complex variable Z according to

$$Z = y + ix, \qquad (8\text{–}91)$$

where i is $\sqrt{-1}$. The real and imaginary parts of any function of Z are solutions of Laplace's equation. To solve the model strike–slip faulting problem, we must devise a function of Z whose real or imaginary part satisfies Equations (8–87) to (8–90). The function that does this is

$$w_z = \frac{\sigma_{xz,0}}{G}\text{Im}\{(Z^2 - a^2)^{1/2}\}, \qquad (8\text{–}92)$$

where Im denotes the imaginary part of the complex function. We can immediately verify that conditions (8–89) and (8–90) are satisfied. On the plane $x = 0$, $Z = y$ and W_z becomes

$$w_z = \frac{\sigma_{xz,0}}{G}\text{Im}\{(y^2 - a^2)^{1/2}\}. \qquad (8\text{–}93)$$

This is clearly zero for $y > a$ because $(y^2 - a^2)^{1/2}$ is a real number.

The surface displacement is obtained by setting $Z = ix$ in Equation (8–92) with the result

$$\begin{aligned} w_z &= \pm\frac{\sigma_{xz,0}}{G}\text{Im}\{(-x^2 - a^2)^{1/2}\} \\ &= \pm\frac{\sigma_{xz,0}}{G}\text{Im}\{i(x^2 + a^2)^{1/2}\} \\ &= \pm\frac{\sigma_{xz,0}}{G}\text{Re}\{(x^2 + a^2)^{1/2}\} \\ &= \pm\frac{\sigma_{xz,0}}{G}(x^2 + a^2)^{1/2}. \end{aligned} \qquad (8\text{–}94)$$

The plus sign is for $x > 0$, the minus sign is for $x < 0$, and Re denotes the real part of a complex function. In the limit $x \to \infty$ this reduces to

$$w_z \to \frac{\sigma_{xz,0}}{G}x. \qquad (8\text{–}95)$$

Thus conditions (8–89) and (8–90) are indeed satisfied by Equation (8–92). To check the other two conditions, we need to differentiate w_z with respect to x and y. The derivatives of the complex quantity $(Z^2 - a^2)^{1/2}$ are obtained by using the chain rule as follows

$$\frac{\partial}{\partial x}(Z^2 - a^2)^{1/2} = \frac{\partial Z}{\partial x}\frac{d}{dZ}(Z^2 - a^2)^{1/2}$$
$$= \frac{iZ}{(Z^2 - a^2)^{1/2}} \tag{8–96}$$

$$\frac{\partial}{\partial y}(Z^2 - a^2)^{1/2} = \frac{\partial Z}{\partial y}\frac{d}{dZ}(Z^2 - a^2)^{1/2}$$
$$= \frac{Z}{(Z^2 - a^2)^{1/2}}. \tag{8–97}$$

From Equations (8–92), (8–96), and (8–97) we obtain

$$\frac{\partial w_z}{\partial y} = \frac{\sigma_{xz,0}}{G}\,\text{Im}\left\{\frac{Z}{(Z^2 - a^2)^{1/2}}\right\} \tag{8–98}$$

$$\frac{\partial w_z}{\partial x} = \frac{\sigma_{xz,0}}{G}\,\text{Re}\left\{\frac{Z}{(Z^2 - a^2)^{1/2}}\right\}. \tag{8–99}$$

On $y = 0$ we have $Z = ix$ and Equation (8–98) becomes

$$\frac{\partial w_z}{\partial y} = \frac{\sigma_{xz,0}}{G}\,\text{Im}\left\{\frac{ix}{(-x^2 - a^2)^{1/2}}\right\}$$
$$= \frac{\sigma_{xz,0}}{G}\,\text{Im}\left\{\frac{ix}{i(x^2 + a^2)^{1/2}}\right\}$$
$$= \frac{\sigma_{xz,0}}{G}\,\text{Im}\left\{\frac{x}{(x^2 + a^2)^{1/2}}\right\} = 0. \tag{8–100}$$

Therefore condition (8–87) is satisfied. On $x = 0$, $Z = y$ and Equation (8–99) becomes

$$\frac{\partial w_z}{\partial x} = \frac{\sigma_{xz,0}}{G}\,\text{Re}\left\{\frac{y}{(y^2 - a^2)^{1/2}}\right\}. \tag{8–101}$$

This is clearly zero when $0 < y < a$ because $(y^2 - a^2)^{1/2}$ is an imaginary number. Thus Condition (8–88) is satisfied.

The shear stresses in the half-space after the introduction of the crack are obtained by combining Equations (8–84), (8–98), and (8–99)

$$\sigma_{xz} = \sigma_{xz,0}\,\text{Re}\left\{\frac{Z}{(Z^2 - a^2)^{1/2}}\right\} \tag{8–102}$$

$$\sigma_{yz} = \sigma_{xz,0}\,\text{Im}\left\{\frac{Z}{(Z^2 - a^2)^{1/2}}\right\}. \tag{8–103}$$

On the plane $x = 0$ the stresses are

$$\sigma_{xz} = 0 \qquad\qquad 0 \le y < a$$
$$= \frac{\sigma_{xz,0}\,y}{(y^2 - a^2)^{1/2}} \qquad y > a \tag{8–104}$$

$$\sigma_{yz} = \frac{-\sigma_{xz,0}\,y}{(a^2 - y^2)^{1/2}} \qquad 0 \le y < a$$
$$= 0 \qquad\qquad y > a. \tag{8–105}$$

As $y \to \infty$, $\sigma_{yz} \to 0$ and $\sigma_{xz} \to \sigma_{xz,0}$ on the plane of the crack. Both components of shear stress are infinite at the crack tip $x = 0$, $y = a$. This stress singularity is typical of crack problems. On the surface $y = 0$, $Z = ix$, and the stresses according to Equations (8–102) and (8–103) are

$$\sigma_{xz} = \frac{\pm\sigma_{xz,0}\,x}{(x^2 + a^2)^{1/2}} \tag{8–106}$$

$$\sigma_{yz} = 0. \tag{8–107}$$

As $x \to \pm\infty$ we again find $\sigma_{xz} \to \sigma_{xz,0}$.

We can use the solution obtained above to determine the surface displacement caused by a displacement on a strike–slip fault. Prior to the introduction of the crack the surface displacement is given by Equation (8–86). After the crack is introduced the surface displacement is given by Equation (8–94). The surface displacement caused by the introduction of the crack Δw_z is the difference between Equations (8–94) and (8–86)

$$\Delta w_z = \pm\frac{\sigma_{xz,0}}{G}\left[(x^2 + a^2)^{1/2} - |x|\right]. \tag{8–108}$$

Thus, the displacement across the fault (crack) is

$$\Delta w_{z0} = \frac{2a\sigma_{xz,0}}{G} \tag{8–109}$$

and the coseismic displacement as a function of the distance x from the fault can be written

$$\Delta w_z = \pm\frac{\Delta w_{z0}}{2}\left[\left(1 + \frac{x^2}{a^2}\right)^{1/2} - \frac{|x|}{a}\right]. \tag{8–110}$$

The prediction of Equation (8–110) for the dependence of coseismic surface displacement on distance from the fault is compared with data from the 1906 San Francisco earthquake in Figure 8–27. Measurements of surface displacement at several distances from the

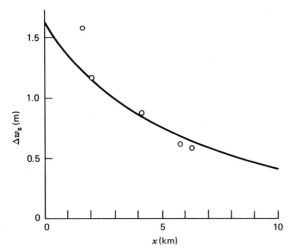

8-27 Surface displacements during the 1906 San Francisco earthquake as a function of distance from the San Andreas fault compared with the predicted displacements from Equation (8–110) for $\Delta w_{z0} = 3.2$ m and $a = 6$ km.

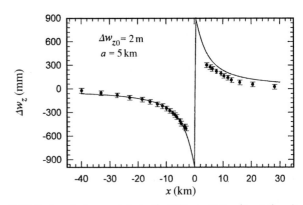

8-28 Surface displacements during the 1992 $m = 7.3$ Landers earthquake as a function of the distance x from the fault. Positive distances are to the northeast and negative distances to the southwest. The data points are from the synthetic aperture radar interferometry image in Figure 2–38 (Massonnet et al., 1993). The solid line gives the predicted displacements from Equation (8–110) with $\Delta w_{z0} = 2$ m and $a = 5$ km.

San Andreas fault are shown by the circles, while the solid curve is a plot of Equation (8–110) with $\Delta w_{z0} = 3.2$ m and $a = 6$ km. For $G = 30$ GPa, these values of Δw_{z0} and a correspond to a shear stress $\sigma_{xz,0} = 8$ MPa. Because earthquakes occur on the San Andreas fault to depths of 12 to 15 km, this value of a is probably too small by about a factor of 2. This disagreement can be attributed in part to the assumption that $\sigma_{xz} = 0$ for the entire depth of the fault after an earthquake. As discussed before, a nonzero residual stress is expected because of friction on the fault. The stress drop during the earthquake, though quite low, is probably reasonably accurate.

Coseismic horizontal displacements associated with the $m = 7.3$ Landers (California) earthquake of June 28, 1992, obtained from synthetic aperture radar interferometry (INSAR) are given in Figure 8–28 as a function of distance from the rupture. The INSAR pattern used to obtain these displacements was given in Figure 2–38. The surface displacements during this earthquake were primarily right-lateral strike–slip. The observed displacement data in Figure 8–28 are compared with the predicted displacements from Equation (8–110) for $\Delta w_{z0} = 2$ m and $a = 5$ km. For $G = 30$ GPa, these values of Δw_{z0} and a correspond to a shear stress $\sigma_{xz,0} = 6$ MPa from Equation (8–109).

These examples illustrate how surface displacements can be obtained when faulting occurs on a specified fault. The coseismic surface displacements associated with an earthquake are often measured by geodetic methods. If the fault plane is known from the location of earthquakes, then various distributions of displacement on the fault can be considered until the surface displacements are matched. The distributed displacements on the fault are known as *dislocations*. The application of this approach requires extensive numerical calculations. For an earthquake involving only local displacements, this procedure may be reasonably successful in determining the distribution of strain and the change in stress associated with the earthquake. However, for very large plate boundary earthquakes such as the 1906 San Francisco earthquake, displacements are expected to occur throughout the lithosphere.

Let us now consider a two-dimensional model for the cyclic accumulation and release of stress and strain on a strike–slip fault located at $x = 0$ and accommodating horizontal motion in the z direction between two lithospheric plates of thickness b, as illustrated in Figure 8–29. The initial situation shown in Figure 8–29a corresponds to a time after a major earthquake when the shear stress and shear strain associated with the interaction between the adjacent plates have been reduced to zero by both the main shock and the relaxation effects of aftershocks. The plates are subject to a

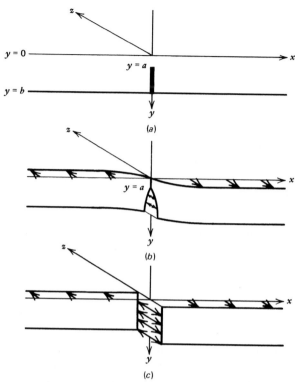

8-29 The displacement field due to the stick–slip behavior of two lithospheric plates of thickness b sliding past each other. (*a*) Initially there is no stress but the fault at $x = 0$ is locked to a depth a. (*b*) The plates are displaced but no displacement occurs on the locked part of the fault. (*c*) The locked part of the fault fails and the plates are uniformly displaced with respect to each other and the stress is zero. The cycle repeats.

uniform relative motion parallel to the fault. We assume that the fault is locked to a depth a and that no stress is transmitted across the fault at greater depths. The base of each plate is also assumed to be stress-free. The shear stress and shear strain accumulate as a result of the relative motion and maximize in the vicinity of the fault, as illustrated in Figure 8–29*b*. When the failure stress is reached on the locked fault, an earthquake occurs. The earthquake relieves the stress in the immediate vicinity of the locked fault and transfers stress to the lower plastic portion of the fault. The stresses on this deeper zone are relieved by plastic flow, a process that may be responsible for the decaying sequence of aftershocks that follows a major earthquake. In our model we assume that the accumulated stresses are totally relieved following the aftershock sequence, as illustrated in Figure 8–29*c*. Except for the finite displacement on

the plate boundary the state of zero shear stress and zero shear strain is identical with that illustrated in Figure 8–29*a*. The cyclic process is then repeated.

As in the previous example there is only one nonzero component of displacement w_z and two nonzero components of shear stress σ_{xz} (and σ_{zx}) and σ_{yz} (and σ_{zy}). The displacement w_z satisfies Laplace's equation (8–85). Initially the displacement and shear stresses are zero; after strain has accumulated the required boundary conditions are

$$\frac{\partial w_z}{\partial y} = 0 \quad \text{at } y = 0, \quad b \tag{8–111}$$

$$w_z = 0 \quad \text{at } x = 0, \quad 0 < y < a \tag{8–112}$$

$$\frac{\partial w_z}{\partial x} = 0 \quad \text{at } x = 0, \quad a < y < b. \tag{8–113}$$

The first condition is the vanishing of the shear stress σ_{yz} at the surface and at the base of the plate. Equation (8–112) requires the displacement to be zero on the locked portion of the fault. The third condition is the vanishing of the shear stress σ_{xz} on the plastic part of the fault. The solution to this problem in terms of the complex variable $Z = y + ix$ is

$$w_z = A \operatorname{Re} \left\{ \ln \left[\frac{\sin \frac{\pi Z}{2b} + \left(\sin^2 \frac{\pi Z}{2b} - \sin^2 \frac{\pi a}{2b} \right)^{1/2}}{\sin \frac{\pi a}{2b}} \right] \right\}, \tag{8–114}$$

where A is a constant of integration. The complex components of $\sin Z$, $\cos Z$, and $\ln Z$ are given by

$$\sin Z = \sin y \cosh x + i \cos y \sinh x \tag{8–115}$$

$$\cos Z = \cos y \cosh x - i \sin y \sinh x \tag{8–116}$$

$$\ln Z = \frac{1}{2} \ln(y^2 + x^2) + i \tan^{-1} \frac{x}{y}. \tag{8–117}$$

On the surface, $y = 0$ and $Z = ix$, the displacement is

$$w_z = A \ln \left\{ \frac{\sinh \frac{\pi x}{2b} + \left(\sinh^2 \frac{\pi x}{2b} + \sin^2 \frac{\pi a}{2b} \right)^{1/2}}{\sin \frac{\pi a}{2b}} \right\}. \tag{8–118}$$

At large distances from the fault, $\pi x/(2b) \gg 1$, the

surface displacement becomes

$$w_z \approx A \ln \left\{ \frac{2 \sinh \frac{\pi x}{2b}}{\sin \frac{\pi a}{2b}} \right\} \approx A \left\{ \ln e^{\pi x/2b} - \ln \sin \frac{\pi a}{2b} \right\}$$

$$\approx A \frac{\pi x}{2b}. \tag{8-119}$$

The two shear stress components are found by differentiating Equation (8–114) with the result

$$\sigma_{xz} = G \frac{\partial w_z}{\partial x}$$

$$= -\frac{\pi G A}{2b} \operatorname{Im} \left\{ \frac{\cos \frac{\pi Z}{2b}}{\left(\sin^2 \frac{\pi Z}{2b} - \sin^2 \frac{\pi a}{2b} \right)^{1/2}} \right\} \tag{8-120}$$

$$\sigma_{yz} = G \frac{\partial w_z}{\partial y} = \frac{\pi G A}{2b} \operatorname{Re} \left\{ \frac{\cos \frac{\pi Z}{2b}}{\left(\sin^2 \frac{\pi Z}{2b} - \sin^2 \frac{\pi a}{2b} \right)^{1/2}} \right\}. \tag{8-121}$$

On the surface, $y = 0$, $Z = ix$, the shear stress is

$$\sigma_{xz} = \frac{\pi G A}{2b} \frac{\cosh \frac{\pi x}{2b}}{\left(\sinh^2 \frac{\pi x}{2b} + \sin^2 \frac{\pi a}{2b} \right)^{1/2}}, \tag{8-122}$$

and σ_{yz} is zero on $y = 0$ as required by condition (8–111). At large distances from the fault, $\pi x/(2b) \gg 1$, the shear stress becomes

$$\sigma_{xz} \approx \frac{\pi G A}{2b} \operatorname{ctnh} \frac{\pi x}{2b} \approx \frac{\pi G A}{2b} \equiv \sigma_{xz,0}. \tag{8-123}$$

Far from the fault the surface shear stress is a constant. The constant of integration A is related to the shear stress applied across the fault at large distances from the fault. At the fault, $x = 0$, the surface shear stress is given by

$$\sigma_{xz} = \frac{\pi G A}{2b \sin(\pi a/2b)} = \frac{\sigma_{xz,0}}{\sin(\pi a/2b)}. \tag{8-124}$$

For small values of a/b the stress is strongly concentrated near the fault.

It is of interest to compare the results of this analysis with the measured surface velocities in the vicinity of the San Andreas fault. Observed velocity vectors for geodetic stations in southern California were given in Figure 2–37. These velocities were obtained using global positioning system (GPS) and very long baseline interferometry (VLBI) observations. The observed surface velocities as a function of distance from the San Andreas fault are given in Figure 8–30. The velocity on the fault trace is taken to be zero and the distances x measured perpendicular to the fault are positive to the northeast and negative to the southwest. Velocities on the Pacific plate are positive (to the northwest) and velocities on the North American plate are negative (to the southeast).

The theoretical formula for the surface velocity parallel to a strike–slip fault as a function of the distance from the fault $u_z(x)$ is obtained by taking the time derivative of surface displacement given in

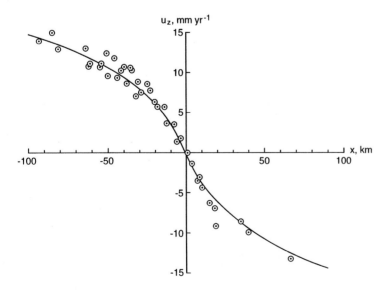

8-30 Surface velocity u_z as a function of distance x from the San Andreas fault in southern California. The data points are a compilation of velocity measurements using GPS, VLBI, and surface triangulation (after Shen et al., 1996). The solid curve is from Equation (8-128) taking the thickness of the elastic lithosphere $b = 75$ km, the depth of the locked portion of the fault $a = 10$ km, and requiring $u_z = \pm 23$ mm yr^{-1} at $x = \pm 135$ km.

Equation (8–118). The result is

$$u_z = \frac{\partial w_z}{\partial t} = \frac{dA}{dt}$$

$$\times \ln \left\{ \frac{\sinh \frac{\pi x}{2b} + \left(\sinh^2 \frac{\pi x}{2b} + \sin^2 \frac{\pi a}{2b}\right)^{1/2}}{\sin \frac{\pi a}{2b}} \right\}.$$

$$(8\text{–}125)$$

At large distances from the fault, Equation (8–119) indicates that

$$u_z \approx \frac{dA}{dt} \frac{\pi x}{2b}. \qquad (8\text{–}126)$$

The model parameter dA/dt can be related to the relative velocity u_r across the San Andreas fault by requiring the displacement rate $\partial w_z/\partial t$ to equal $u_r/2$ at a distance x_r from the fault. If we assume that x_r is sufficiently large so that Equation (8–126) is valid, we can write

$$\frac{u_r}{2} = \left(\frac{\partial w_z}{\partial t}\right)_{x=x_r} = \frac{\pi x_r}{2b} \frac{dA}{dt}. \qquad (8\text{–}127)$$

Substitution of Equation (8–127) into Equation (8–125) gives

$$u_z = \frac{b u_r}{\pi x_r} \ln \left\{ \frac{\sinh \frac{\pi x}{2b} + \left(\sinh^2 \frac{\pi x}{2b} + \sin^2 \frac{\pi a}{2b}\right)^{1/2}}{\sin \frac{\pi a}{2b}} \right\}.$$

$$(8\text{–}128)$$

To compare this result with the data from the San Andreas fault, let the thickness of the elastic lithosphere $b = 75$ km and the depth of the locked portion of the fault $a = 10$ km. We assume that the relative velocity across the San Andreas fault determined from plate tectonic studies, 46 mm yr^{-1}, is applied at a distance $x_r = 135$ km from the fault. The theoretical rate of strain accumulation using these parameters and Equation (8–128) is compared with the observations in Figure 8–30. We see that reasonable agreement between theory and experiment is obtained. The cyclic strain accumulation and release is restricted to the immediate vicinity of the fault. We previously reached this same conclusion in Section 8–6 using a more approximate analysis. In the next section we suggest that this restriction is due to the interaction of the elastic lithosphere with the viscous asthenosphere.

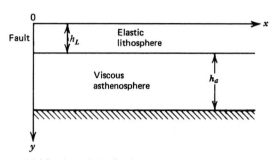

8–31 Model for determining the damping of cyclic strains on a fault by the interaction of an elastic lithosphere of thickness h_L with a Newtonian viscous asthenosphere of thickness h_a. The fault is at $x = 0$ and strikes in the z direction.

8–11 Stress Diffusion

The interaction of the viscous asthenosphere with the overlying elastic lithosphere causes the cyclic strains on a major strike–slip fault to be damped with distance from the fault. The effect can be demonstrated by the simple model illustrated in Figure 8–31. An elastic lithosphere of thickness h_L overlies a Newtonian viscous asthenosphere of thickness h_a. The fault, located at $x = 0$ strikes in the z direction. The cyclic behavior of the fault is modeled by a periodic displacement applied to the elastic lithosphere at $x = 0$. To simplify the analysis, we neglect the vertical variation of the displacement w_{zL} in the lithosphere. We also assume that there are no variations along the strike of the fault, that is, $\partial/\partial z = 0$. Thus we take $w_{zL} = w_{zL}(x)$.

The partial differential equation governing the behavior of the lithosphere can be derived by carrying out a force balance on a section of the lithosphere of width dx and unit length in the z direction, as illustrated in Figure 8–32. The displacement of the lithosphere parallel to the strike of the fault gives rise to the elastic shear stresses σ_{xz} on the vertical boundaries of the element and the viscous shear stress τ_{yz} on its base. The balance of forces on the element in the z direction gives

$$\tau_{yz}\, dx + \{\sigma_{xz}(x+dx) - \sigma_{xz}(x)\} h_L = 0 \qquad (8\text{–}129)$$

or

$$\tau_{yz} = -h_L \frac{\partial \sigma_{xz}}{\partial x}. \qquad (8\text{–}130)$$

Because the elastic shear stress is related to the strike displacement by

$$\sigma_{xz} = G \frac{\partial w_{zL}}{\partial x}, \qquad (8\text{–}131)$$

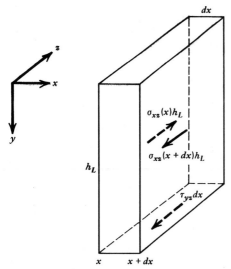

8-32 The force balance on a section of the lithosphere of width dx and unit length parallel to the strike of the fault.

where G is the shear modulus of the lithosphere (see Equations [2–102] and [3–49]), we can rewrite the force balance equation as

$$\tau_{yz} = -h_L G \frac{\partial^2 w_{zL}}{\partial x^2}. \tag{8–132}$$

To relate the viscous shear stress in the asthenosphere to w_{zL}, we assume that the velocity at the base of the asthenosphere is zero and that the viscous flow in the asthenosphere can be approximated by the linear Couette flow considered in Section 6–2. From Equations (6–1) and (6–13),

$$\tau_{yz} = \frac{-\mu}{h_a} \frac{\partial w_{zL}}{\partial t}, \tag{8–133}$$

where μ is the viscosity of the asthenosphere. Substitution of this expression into Equation (8–132) yields

$$\frac{\partial w_{zL}}{\partial t} = \frac{h_a h_L G}{\mu} \frac{\partial^2 w_{zL}}{\partial x^2}. \tag{8–134}$$

Both the displacement of the lithosphere and unsteady one-dimensional heat conduction are governed by the one-dimensional diffusion equation (see Section 4–13). The effective diffusivity for the displacement of the lithosphere is equal to $h_a h_L G / \mu$.

The spatial damping of cyclic displacements on the fault by the lithosphere–asthenosphere interaction can be demonstrated by solving Equation (8–134) for an applied periodic displacement at $x = 0$ of the form

$$w_{zL} = w_{zL,0} \cos \omega t, \tag{8–135}$$

where ω is the circular frequency of the applied displacement. The solution to the analogous heat conduction problem has previously been obtained in Section 4–14. By appropriate modifications of Equation (4–89), we find that the displacement of the lithosphere is given by

$$w_{zL} = w_{zL,0} \exp \left\{ -x \left(\frac{\omega \mu}{2 h_a h_L G} \right)^{1/2} \right\}$$
$$\times \cos \left\{ \omega t - x \left(\frac{\omega \mu}{2 h_a h_L G} \right)^{1/2} \right\}. \tag{8–136}$$

The amplitude of the displacement of the lithosphere decreases exponentially with distance from the fault; the displacement is damped to $1/e$ of its value at the fault in a distance d_s given by

$$d_s = \left(\frac{2 h_a h_L G}{\omega \mu} \right)^{1/2}. \tag{8–137}$$

To estimate d_s, we take $h_a = 100$ km, $h_L = 30$ km, $G = 30$ GPa, and $\mu = 4 \times 10^{19}$ Pa s. The frequency of the applied displacement at the fault is related to the period τ by $\omega = 2\pi/\tau$. As a typical time interval between great earthquakes we take $\tau = 150$ years; this gives $\omega = 1.33 \times 10^{-9}$ s^{-1}. From Equation (8–137) we find $d_s = 58$ km, which agrees with values we estimated earlier in the chapter. Although this solution is approximate and only periodic variations at the fault have been considered, more exact numerical calculations verify that fault displacements are damped a few hundred kilometers from the fault by the interaction between the elastic lithosphere and the viscous asthenosphere. A major earthquake on one part of a plate boundary is therefore unlikely to change the stress significantly on faults on other parts of the plate boundary. Thus great earthquakes are unlikely to trigger other great earthquakes at large distances.

8-12 Thermally Activated Creep on Faults

On near-surface fault zones the concept of a coefficient of friction is likely to be applicable. However, many faults extend deep into the lithosphere, where they are likely to behave plastically. In this section we consider

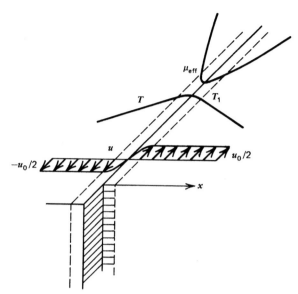

8-33 One-dimensional model for the structure of a fault zone on which there is steady-state creep.

the steady-state creep on deep fault zones. We previously suggested in Section 4–27 that frictional heating may be important on such fault zones. Therefore our analysis includes heating due to viscous dissipation.

Figure 8–33 illustrates the one-dimensional model we will use to determine the structure of the shear zone formed between two half-spaces moving in opposite directions parallel to the interface plane $x = 0$ with total relative velocity u_0. The center of the shear zone is the plane $x = 0$, and $|x|$ measures the distance normal to the fault. Sufficiently far from the fault the velocity u approaches $\pm u_0/2$. Frictional heating produces a temperature anomaly in the shear zone; the temperature T far from the fault plane must decrease linearly with distance from the fault in order to conduct away the heat generated by friction in the shear zone. The temperature will have a maximum value T_1 at the center of the shear zone. The symmetry of the model requires that T be symmetric and u be antisymmetric about $x = 0$; T and u depend only on x in this steady one-dimensional problem. We assume that dislocation creep with an effective viscosity μ_{eff} given by Equation (7–119) governs the deformation in the shear zone. The viscosity in the shear zone is substantially lower than it is far from the fault because of the strong temperature dependence of μ_{eff} and the temperature anomaly due to frictional

heating in the shear zone. The reduced viscosity in the shear zone facilitates the decoupling of the oppositely moving half-spaces and concentrates the shear into a relatively narrow region around the fault.

The equations governing the temperature and velocity in frictionally heated shear flows with temperature-dependent viscosity have already been discussed in Section 7–5. From Equation (7–120) we know that the shear stress τ in the shear zone is a constant if there is no pressure gradient along the fault plane. We group all the preexponential quantities in the viscosity Equation (7–119) except for the shear stress and the temperature into a constant C and write the equation for μ_{eff} as

$$\mu_{\text{eff}} = \frac{CT}{\tau^2} e^{E_a/RT}. \tag{8–138}$$

Since τ is a constant, the effective viscosity varies only with T and primarily with the exponential of the inverse absolute temperature.

The differential equation for the temperature is Equation (7–141); with $\tau = \mu_{\text{eff}} \, du/dx$ this equation takes the form

$$k\frac{d^2 T}{dx^2} + \tau\frac{du}{dx} = 0. \tag{8–139}$$

Because the shear stress is constant, Equation (8–139) can be integrated to give

$$k\frac{dT}{dx} + \tau u = 0. \tag{8–140}$$

The constant of integration is zero because $u = 0$ at $x = 0$ and symmetry requires $dT/dx = 0$ at $x = 0$. Substitution of $\tau = \mu_{\text{eff}} du/dx$ in Equation (8–140) yields

$$u \, du = \frac{-k \, dT}{\mu_{\text{eff}}}, \tag{8–141}$$

which can be further modified using Equation (8–138) for μ_{eff} to

$$d\left(\frac{u^2}{2}\right) = \frac{-k\tau^2}{CT} e^{-E_a/RT} \, dT. \tag{8–142}$$

Equation (8–142) can be integrated analytically using the exponential integral function E_1 first introduced in Problem 4–35 to obtain

$$u^2 = \frac{2k\tau^2}{C}\left\{E_1\left(\frac{E_a}{RT_1}\right) - E_1\left(\frac{E_a}{RT}\right)\right\}, \tag{8–143}$$

where the constant of integration has been evaluated by letting $u = 0$ and $T = T_1$ at the center of the shear

TABLE 8-4 Values of the Exponential Integral
$E_1(x) = \int_x^\infty (e^{-t}/t)\, dt$

x	$E_1(x)$	$xe^x E_1(x)$
0	∞	
0.01	4.03793	
0.02	3.35471	
0.03	2.95912	
0.04	2.68126	
0.05	2.46790	
0.10	1.82292	
0.20	1.22265	
0.30	0.90568	
0.40	0.70238	
0.50	0.55977	
0.60	0.45438	
0.70	0.37377	
0.80	0.31060	
0.90	0.26018	
1.0	0.21938	
1.2	0.15841	
1.4	0.11622	
1.6	0.08631	
1.8	0.06471	
2.0	0.04890	0.72266
2.5	0.02491	0.75881
3.0	0.01305	0.78625
3.5	0.00697	0.80787
4.0	0.00378	0.82538
4.5	0.00207	0.83989
5.0	0.00115	0.85211
6	3.6008×10^{-4}	0.87161
7	1.1548×10^{-4}	0.88649
8	3.7666×10^{-5}	0.89824
9	1.2447×10^{-5}	0.90776
10	4.1570×10^{-6}	0.91563
20	9.8355×10^{-11}	0.95437

zone. The exponential integral function is tabulated in Table 8–4 and plotted in Figure 8–34.

In the limit $x \to \infty$, $u \to u_0/2$, T decreases, and $E_1(E_a/RT) \to E_1(\infty) = 0$ (Table 8–4). Thus Equation (8–143) gives

$$u_0^2 = \frac{8k\tau^2}{C} E_1\left(\frac{E_a}{RT_1}\right), \qquad (8\text{–}144)$$

a relation that can be used to calculate the maximum temperature at the center of the shear zone as a function of the total relative velocity across the shear zone, the shear stress, and the rheological properties. In

Chapter 7 we noted that $E_a/RT_1 \gg 1$. An approximate formula for the exponential integral of a large quantity is

$$E_1\left(\frac{E_a}{RT_1}\right) \approx \frac{RT_1}{E_a} e^{-E_a/RT_1} \qquad (8\text{–}145)$$

(see Table 8–4). Therefore, Equation (8–144) can be approximated by

$$u_0^2 \approx \frac{8k\tau^2 RT_1}{C E_a} e^{-E_a/RT_1}. \qquad (8\text{–}146)$$

Figure 8–35 shows how the maximum temperature at the center of a mantle shear zone depends on the relative velocity across the zone for several values of shear stress. The curves were drawn using Equation (8–146) with $k = 4.2$ W m^{-1} K^{-1}, $R = 8.314$ J K^{-1} mol^{-1}, $C = 10^9$ Pa^3 s K^{-1}, and $E_a = 523$ kJ mol^{-1}. Temperatures of about 1300 K are required if dislocation creep is to accommodate relative velocities of tens of millimeters per year across mantle shear zones under applied shear stresses of about 1 MPa.

An equation for the temperature distribution in the shear zone can be obtained by substituting Equation (8–143) into Equation (8–140):

$$\frac{dT}{dx} = -\tau^2 \left(\frac{2}{kC}\right)^{1/2} \left\{ E_1\left(\frac{E_a}{RT_1}\right) - E_1\left(\frac{E_a}{RT}\right) \right\}^{1/2}. \qquad (8\text{–}147)$$

This equation is subject to the condition $T = T_1$ at $x = 0$. It can be integrated numerically or analytically if two approximations are made. First, we assume that the arguments of the exponential integral functions are large compared with 1. We can then use the approximation contained in Equation (8–145) and write

$$\frac{dT}{dx} = -\tau^2 \left(\frac{2}{kC}\right)^{1/2}$$
$$\times \left\{ \frac{RT_1}{E_a} e^{-E_a/RT_1} - \frac{RT}{E_a} e^{-E_a/RT} \right\}^{1/2}. \qquad (8\text{–}148)$$

If we then write $\theta = T_1 - T$ and assume $\theta \ll T_1$, we can approximate Equation (8–148) by

$$\frac{d\theta}{dx} = \frac{\tau}{k} \left\{ \frac{2kRT_1\tau^2 e^{-E_a/RT_1}}{C E_a} \right\}^{1/2} \left\{ 1 - e^{-E_a\theta/RT_1^2} \right\}^{1/2}. \qquad (8\text{–}149)$$

From Equation (8–146) we recognize that the first quantity in brackets on the right side of this expression

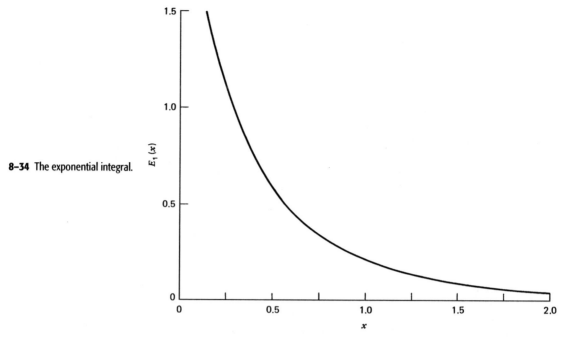

8-34 The exponential integral.

is $u_0^2/4$; thus we can simplify Equation (8–149) to

$$\frac{d\theta}{dx} = \frac{\tau u_0}{2k}\left\{1 - e^{-E_a\theta/RT_1^2}\right\}^{1/2}.$$ (8–150)

This equation can be integrated analytically by making

8-35 Dependence of the temperature at the center of a shear zone on the relative velocity across the shear zone for several values of the applied shear stress.

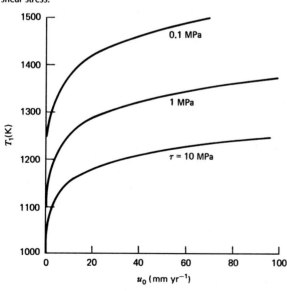

the substitution

$$s = e^{-E_a\theta/RT_1^2}$$ (8–151)

and we find

$$\frac{ds}{s(1-s)^{1/2}} = \frac{-\tau u_0 E_a}{2kRT_1^2}\,dx.$$ (8–152)

The integral of Equation (8–152) yields

$$T = T_1 + \frac{RT_1^2}{E_a}\ln\left\{\text{sech}^2\left(\frac{\tau u_0 E_a x}{4kRT_1^2}\right)\right\},$$ (8–153)

where the constant of integration has been chosen to satisfy $T = T_1$, $\theta = 0$, $s = 1$ on $x = 0$. By substituting Equation (8–153) into Equation (8–140), we obtain the velocity distribution in the shear zone as

$$u = \frac{u_0}{2}\tanh\left(\frac{\tau u_0 E_a x}{4kRT_1^2}\right).$$ (8–154)

In the limit $x \to \infty$,

$$\text{sech}\left(\frac{\tau u_0 E_a x}{4kRT_1^2}\right) \to 2e^{-\tau u_0 E_a x/4kRT_1^2}$$

and

$$\frac{dT}{dx} \to \frac{-\tau u_0}{2k}.$$

Also, as $x \to \infty$,

$$\tanh\left(\frac{\tau u_0 E_a x}{4kRT_1^2}\right) \to 1$$

and

$$u \to \frac{u_0}{2}.$$

PROBLEM 8–10 Define the half-width of the shear zone x_s as the value of x at which $u = 0.9(u_0/2)$.

a. Obtain an expression for the half-width from Equation (8–154).

b. What is the half-width of the shear zone if $u_0 = 100$ mm yr^{-1} and $\tau = 10$ MPa? Use the values of C, E_a, and k given previously.

REFERENCES

Brewer, J. A., S. B. Smithson, J. E. Oliver, S. Kaufman, and L. D. Brown (1980), The Laramide orogeny: Evidence from COCORP deep crustal seismic profiles in the Wind River mountains, Wyoming, *Tectonophysics* **62**, 165–189.

Byerlee, J. (1977), Friction of rocks, in *Experimental Studies of Rock Friction with Application to Earthquake Prediction*, J. F. Evernden, ed., pp. 55–77 (U.S. Geological Survey, Menlo Park, California).

Cook, F. A., D. S. Albaugh, L. D. Brown, S. Kaufman, J. E. Oliver, and R. D. Hatcher (1979), Thin-skinned tectonics in the crystalline southern Appalachians; COCORP seismic reflection profiling of the Blue Ridge and Inner Piedmont, *Geology* **7**, 563–567.

Hanks, T. C. (1977), Earthquake stress drops, ambient tectonic stresses and stresses that drive plate motions, *Pure Appl. Geophys.* **115**, 441–458.

Massonnet, D., M. Rossi, C. Carmona, F. Adragna, G. Peltzer, K. Feigi, and T. Rabauts (1993), The displacement field of the Landers earthquake mapped by radar interferometry, *Nature* **364**, 138–142.

Shen, Z. K., D. D. Jackson, and B. X. Ge (1996), Crustal deformation across and beyond the Los Angeles basin from geodetic measurements, *J. Geophys. Res.* **101**, 27, 957–27, 980.

Sieh, K., M. Stuiver, and D. Brillinger (1989), A more precise chronology of earthquakes produced by the San Andreas fault in southern California, *J. Geophys. Res.* **94**, 603–623.

COLLATERAL READING

Bullen, K. E., *An Introduction to the Theory of Seismology*, 3rd edition (Cambridge University Press, London, 1963), 381 pages.

A classic introductory textbook on seismology. The first eight chapters provide the essential background theory on the mechanics of deformable media and the transmission of seismic waves subject to various boundary conditions. This is followed by three chapters dealing with the gathering and treatment of instrumental data on earthquakes. The remaining parts of the book apply the theory and observational results to a variety of problems including the structure of the Earth's interior, long-period oscillations of the Earth, earthquake occurrence, nuclear explosions, and extraterrestrial seismology.

Hobbs, B. E., W. D. Means, and P. F. Williams, *An Outline of Structural Geology* (John Wiley and Sons, New York, 1976), 571 pages.

An introductory textbook on structural geology for undergraduate students covering the mechanical aspects of rock deformation, stress, strain, the response of rocks to stress, the microfabric of deformed rocks, crystal defects, microscopic mechanisms of deformation, undeformed rock structures, folding, features of folded rocks, foliations, lineations, brittle deformation, joints, faults, field methods, and structural associations of the Earth. A few of the chapters have problems with answers given in an appendix.

Jaeger, C., *Rock Mechanics and Engineering* (Cambridge University Press, London, 1979), 523 pages.

A textbook about the physical and mechanical properties of rocks and the engineering applications of rock mechanics. Part 1 discusses the development of rock mechanics and the geologists' input to this science. Part 2 deals with rock properties, in situ residual stresses, strains and failure of rock masses, theoretical approaches to determinations of stress–strain fields, and effects of interstitial water. Part 3 covers applications to rock slopes and rock slides, galleries, tunnels, mines, underground excavations, and dams. Part 4 describes case histories of accidents, dam disasters, and rock slides.

Lay, T., and T. C. Wallace, *Modern Global Seismology* (Academic Press, San Diego, 1995), 517 pages.

This is a comprehensive treatment of seismology and earthquakes at a moderately advanced level. Topics include body waves, surface waves, free oscillations, seismic sources, Earth structure, and earthquake mechanics.

Ramsay, J. G., *Folding and Fracturing of Rocks* (McGraw-Hill, New York, 1967), 568 pages.

A book outlining the basic theories of stress, strain, the properties of rocks, and rock deformation in geology. Chapter 1 introduces the methods used to analyze the orientation of structures. Chapters 2 to 6 develop the analysis of stress and strain from first principles. The various types of folds in rocks are described in Chapter 7 with an analysis of their formation and a discussion of their general tectonic environment. The final three chapters treat the deformation of linear structures, folding of obliquely inclined surfaces, and structural complexities associated with the superposition of two systems of folds.

Reid, H. F., The mechanics of the earthquake, in *The California Earthquake of April 18, 1906, Report of the State Earthquake Investigation Commission*, Volume 2 (Carnegie Institution of Washington, Washington, D.C., 1910), 192 pages.

A detailed account of the great earthquake and its seismic record. Part 1 describes the shock, its time and origin, permanent ground displacements, the nature of the acting forces, shearing movements in the fault zone, vibratory movements and their effects, and the influence of the foundation on the apparent intensity. Part 2 summarizes instrumental records of the earthquake from observatories around the world. There is an appendix on the theory of the seismograph.

Richter, C. F., *Elementary Seismology* (W.H. Freeman, San Francisco, 1958), 768 pages.

A fundamental, relatively nonmathematical textbook on seismology written for the geologist and engineer. Part 1 describes the nature of earthquakes and observations of them. The major topics include the character of earthquake motion, descriptions of important earthquakes, foreshocks, aftershocks, earthquake swarms, earthquake effects on buildings, effects on ground and surface water, intensity, volcanic earthquakes, tectonic earthquakes, faulting, seismograph theory, elasticity, seismic waves, deep-focus earthquakes, magnitude, statistics, energy, microseisms, and earthquake risk. Part 2 discusses earthquakes in California, New Zealand, Japan, Taiwan, and other regions. An extensive set of tables gives mathematical details and useful seismic data.

Scholz, C. H., *The Mechanics of Earthquakes and Faulting* (Cambridge University Press, Cambridge, 1990), 439 pages.

This is an excellent intermediate level treatment of rock mechanics and earthquakes. Subjects include approaches to brittle failure, friction, structure of faults, quantification of earthquakes, the seismic cycle, and earthquake prediction.

NINE

Flows in Porous Media

9-1 Introduction

Fluids such as water, steam, petroleum, and natural gas often migrate through the Earth's crust. If these flows occur through open fractures, they can be studied using the channel flow theory developed in Chapter 6. In many cases, however, flows in the Earth's crust occur through a matrix of interconnecting passages provided by large numbers of small fractures or through the voids of naturally porous rocks. If the scale of the flow system is large compared with the scale of the interconnected passages, it is often appropriate to consider flow through a uniform *porous medium*.

Sand is an example of a naturally porous material. Because of the irregular shapes of sand particles there is a considerable void or pore space between them. The fraction of the volume made up of pore space is known as the *porosity* ϕ. Loose sand is particularly porous, $\phi \approx 40\%$, while oil sands have porosities in the range of 10 to 20%. Values of ϕ for some porous rocks, for example, sandstone and limestone, are listed in Table 2–2. As noted before, rocks that are not naturally porous can still be approximated as porous media if they are extensively fractured. The distribution of fractures must be reasonably uniform and the separation of fractures small compared with the scale of the overall flow.

Fluids can flow through a porous medium under the influence of an applied pressure gradient. In Chapter 6 we showed that, for laminar flow, the flow rate in channels and pipes is linearly proportional to the pressure gradient and inversely proportional to the viscosity. This is also the result obtained for many porous medium flows. The resistance of a porous medium to flow depends on the size, number, and "tortuosity" of the fluid pathways through the solid matrix. A measure of this resistance is the *permeability k* of the medium. The viscous flow theory in Chapter 6 can be used to determine the permeability for idealized models of porous media. However, the permeability of actual rocks must be determined in the laboratory or by field experiments.

In this chapter we develop the theory of flow through porous media and consider a number of geological applications. Calculations for the flow of groundwater have many important implications. In the immediate vicinity of a cooling intrusion the groundwater may boil, leading to hot springs, geysers, and geothermal reservoirs. These processes are responsible for many mineral deposits because minerals dissolve freely in hot groundwater and precipitate when the temperature drops or boiling occurs. Models of flow in porous media can also be applied to problems involving the migration of magma.

9-2 Darcy's Law

In many applications the flow through a porous medium is linearly proportional to the applied pressure gradient and inversely proportional to the viscosity of the fluid. This behavior is known as *Darcy's law*, an empirical relationship credited to Henry Darcy, who carried out experiments on the flow of water through vertical homogeneous sand filters in 1856. For a one-dimensional geometry in which the volumetric flow rate per unit area u is driven by the applied pressure gradient dp/dx, Darcy's law takes the form

$$u = -\frac{k}{\mu}\frac{dp}{dx}, \tag{9–1}$$

where k is the permeability of the medium and μ is the dynamic viscosity of the fluid.

The volumetric flow rate per unit area u has the dimensions of velocity, and it is referred to as the *Darcy velocity*. However, because it represents volume flow rate per unit area of the medium, and the pores or cracks occupy only a small fraction of this area, u is not the actual velocity of the fluid in the small channels. It is the average velocity per unit area.

The permeability characterizes the resistance of the porous medium to flow through it. The more permeable the medium is, that is, the larger k is, the smaller is the pressure gradient required to drive a given flow. The SI unit for permeability is square meter, m^2.

TABLE 9–1 **Typical Values for the Natural Permeabilities of Geologic Materials**

Permeability k (m²)	Classification	Material
10^{-7}		
10^{-8}		Gravel
10^{-9}	Pervious	
10^{-10}		
10^{-11}		Sand
10^{-12}		
10^{-13}	Semipervious	
10^{-14}		Sandstone
10^{-15}		
10^{-16}		
10^{-17}		Limestone
10^{-18}	Impervious	
10^{-19}		
10^{-20}		Granite

Typical values for the natural permeabilities of some geologic materials are given in Table 9–1. Coarse gravel is highly permeable, while unfractured homogeneous granite is virtually impermeable. Darcy's law is sometimes written in terms of the hydraulic head H defined in Equation (6–9)

$$u = -\frac{k\rho g}{\mu}\frac{dH}{dx} = -K\frac{dH}{dx}, \tag{9–2}$$

where $K \equiv k\rho g/\mu$ is known as the *hydraulic conductivity*. The form of Darcy's law given in Equation (9–2) is used frequently in applications to groundwater flow.

Darcy's law is applicable to flow through a solid matrix only if several conditions are satisfied. First, the scale of the porosity must be small compared with the other characteristic dimensions of the flow situation. For example, if a sedimentary layer of thickness h is made up of particles with a mean diameter b, the condition $b \ll h$ is required for Darcy's law to be valid. Alternatively, if the porosity of the layer is due to interconnected fractures, the dimensions and spacing of the fractures must be small compared with the thickness h. The second condition is that the flow in the individual channels must be laminar. This condition, discussed in Section 6–4, places an upper limit on the dimensions of the porosity and the flow rate of the fluid. Although Darcy's law is an empirical statement, it can be derived

theoretically for several simple models of the channel configuration within the solid matrix.

PROBLEM 9–1 To derive an upward flow in a porous medium, it is clear that pressure must increase more rapidly with depth y than it does when the fluid is motionless. Use this idea to justify writing Darcy's law for vertical flow in a porous medium in the form

$$v = -\frac{k}{\mu}\left(\frac{dp}{dy} - \rho g\right), \tag{9–3}$$

where v is the vertical Darcy velocity (positive in the direction of increasing depth), ρ is the fluid density, and g is the acceleration of gravity. Consider a porous medium lying on an impermeable surface inclined at an angle θ to the horizontal. Show that Darcy's law for the downslope volumetric flow rate per unit area q is

$$q = -\frac{k}{\mu}\left(\frac{dp}{ds} - \rho g \sin\theta\right), \tag{9–4}$$

where s is the downslope distance and q is positive in the direction of s.

PROBLEM 9–2 Consider an unconsolidated (uncemented) layer of soil completely saturated with groundwater; the water table is coincident with the surface. Show that the upward Darcy velocity $|v|$ required to *fluidize* the bed is

$$|v| = \frac{(1 - \phi)\,kg(\rho_s - \rho_w)}{\mu}, \tag{9–5}$$

where ϕ is the porosity, ρ_s is the density of the soil particles, and ρ_w is the water density. The condition of a fluidized bed occurs when the pressure at depth in the soil is sufficient to completely support the weight of the overburden. If the pressure exceeds this critical value, the flow can lift the soil layer.

9-3 Permeability Models

Idealized models for the geometrical configuration of the channels in a porous medium allow us to derive Darcy's law using viscous flow theory. Explicit relationships for the permeability are obtained. A variety of models can be used to approximate the structures of different materials. A sandstone may be approximated

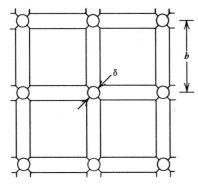

9–1 An idealized model of a porous medium. Circular tubes of diameter δ form a cubical matrix with dimensions b.

by a matrix of closely packed spheres, while fractured volcanic rock may be modeled with a regular (or random) matrix of thin channels. We do not consider alternative microscopic models in detail because they differ by geometrical factors that are of order one.

A typical model for the microscopic structure of a porous material is a cubic matrix of circular tubes. The matrix has a dimension b, and the tubes have a diameter δ, as illustrated in Figure 9–1. Each cube has a tube on each of its 12 edges; 1/4 of a tube of diameter δ and length b lies within the cube on each of these edges. Thus the equivalent of three tubes of diameter δ and length b lie within the cube. The porosity is therefore given by

$$\phi = 3\pi \left(\frac{\delta}{2}\right)^2 \frac{b}{b^3} = \frac{3\pi}{4}\frac{\delta^2}{b^2}. \tag{9–6}$$

We assume that the sides of the cubes lie in the x, y, and z directions. Under an applied pressure gradient, dp/dx, fluid flows through the tubes that are parallel to the x axis. If the flow through the tubes is laminar, the mean velocity in the tubes $\bar{u}_c$ is given by Equation (6–37)

$$\bar{u}_c = -\frac{\delta^2}{32\mu}\frac{dp}{dx}. \tag{9–7}$$

To obtain the mean velocity per unit area, the Darcy velocity u, consider a square with dimensions b lying perpendicular to the x direction with corners lying on the axes of four tubes. One-fourth of the cross-sectional area of each tube lies within the square. Thus the equivalent of the volumetric flow rate through a single tube $\pi(\delta^2/4)\bar{u}_c$ flows across the area b^2. The Darcy velocity

is therefore given by

$$u = \frac{\pi\delta^2}{4b^2}\bar{u}_c = \frac{\phi\bar{u}_c}{3}, \tag{9–8}$$

where Equation (9–6) has been used to eliminate δ^2/b^2. By combining Equations (9–7) and (9–8), we obtain

$$u = -\frac{b^2\phi^2}{72\pi\mu}\frac{dp}{dx}. \tag{9–9}$$

Our simple model predicts a linear relationship between the Darcy velocity and the pressure gradient in accordance with Darcy's law. A comparison of Equations (9–1) and (9–9) shows that the permeability k is given by

$$k = \frac{b^2\phi^2}{72\pi} = \frac{\pi}{128}\frac{\delta^4}{b^2}. \tag{9–10}$$

The permeability is a function of the geometry of the connected porosity. Taking $\delta = 1$ mm and $b = 0.1$ m, we find that $k = 2.45 \times 10^{-12}$ m^2. The porosity is only 2.36×10^{-2}%; the mean velocity in a single tube is 1.27×10^4 times larger than the Darcy velocity.

PROBLEM 9–3 Assume that a porous medium can be modeled as a cubic matrix with a dimension b; the walls of each cube are channels of thickness δ. (a) Determine expressions for the porosity and permeability in terms of b and δ. (b) What is the permeability if $b = 0.1$ m and $\delta = 1$ mm?

9–4 Flow in Confined Aquifers

Groundwater flow often takes place in a layer of porous sedimentary rock bounded above and below by impermeable strata. The channel formed by this layering is known as a *confined porous aquifer*. Consider one-dimensional flow through a confined aquifer with a constant cross-sectional area A, as sketched in Figure 9–2. For this flow the pressure p is a constant over the

9–2 Horizontal one-dimensional flow in a confined porous aquifer. The flow is driven by the indicated drop in head.

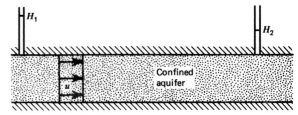

area $p = p(x)$. The Darcy velocity u is given by Equation (9–1); a uniform velocity profile, $u = u(x)$, is obtained. There is no requirement that the Darcy velocity in a porous medium satisfy a no-slip condition at the boundaries. The no-slip condition is valid on the microscopic scale of individual channels, but not on the macroscale of the aquifer.

The volumetric flow rate Q through the aquifer is the product of the Darcy velocity with the cross-sectional area

$$Q = uA = -\frac{kA}{\mu}\frac{dp}{dx}. \tag{9–11}$$

Because the flow rate Q is a constant independent of x, the pressure gradient is a negative constant; thus the pressure decreases linearly with x. For a porous aquifer with the circular cross-section and semicircular geometry previously considered in Section 6–5, the pressure gradient is given by Equation (6–43), and the volumetric flow rate through the aquifer can be written

$$Q = \frac{kR^2\rho g b}{\mu R'}, \tag{9–12}$$

where $\pi R'$ is the length of the aquifer and b is the elevation of the entrance of the aquifer relative to the exit.

PROBLEM 9–4 Assume that the model just described is applicable to an artesian spring (a spring driven by a topographic head). If a topographic head of 60 m drives water through the aquifer at the rate 8.3×10^{-5} m^3 s^{-1}, determine the permeability of the aquifer if its area is 2 m^2 and the distance from source to exit is 2 km.

PROBLEM 9–5 Consider one-dimensional flow through a confined porous aquifer of total thickness b and cross-sectional area A. Suppose the aquifer consists of N layers, each of thickness $b_i(i = 1,\ldots,N)$ and permeability $k_i(i = 1,\ldots,N)$. Determine the total flow rate through the aquifer if all the layers are subjected to the same driving pressure gradient. What is the uniform permeability of an aquifer of thickness b that delivers the same flow rate as the layered aquifer when the two are subjected to the same pressure gradient?

We next consider the drawdown of water into a well that completely penetrates a confined aquifer, as illus-

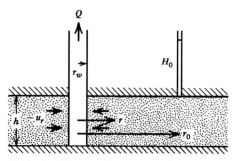

9–3 A model for the withdrawal of water from a well penetrating a confined aquifer.

trated in Figure 9–3. The well is considered to be a sink for the flow in the aquifer that is uniform, horizontal, and radially inward toward the well. The radial Darcy velocity in this cylindrically symmetric flow is u_r. The radial distance from the well is r, and u_r is positive in the direction of increasing r. Darcy's law for one-dimensional flow, Equation (9–1), can be generalized to apply to the radial flow of this problem if we replace the pressure gradient in Equation (9–1) by the radial pressure gradient dp/dr. We can therefore write

$$u_r = -\frac{k}{\mu}\frac{dp}{dr} \tag{9–13}$$

or, in terms of the hydraulic head H,

$$u_r = -\frac{k\rho g}{\mu}\frac{dH}{dr}. \tag{9–14}$$

If the thickness of the aquifer is h, the volume flow rate of water across a cylindrical surface of radius r is

$$Q_r = 2\pi r h u_r. \tag{9–15}$$

Note that Q_r is negative for radially inward flow. The substitution of Equation (9–14) into Equation (9–15) gives

$$Q_r = \frac{-2\pi h k \rho g r}{\mu}\frac{dH}{dr}. \tag{9–16}$$

Conservation of mass requires that Q_r be independent of r; the same amount of fluid must cross each cylindrical surface per unit time. Thus $|Q_r|$ is the volumetric flow rate into the well, and Equation (9–16) is a first-order, total differential equation relating the hydraulic head H to the radial position r. We integrate this equation to give

$$H - H_0 = \frac{-\mu Q_r}{2\pi h k \rho g}\ln\frac{r}{r_0}, \tag{9–17}$$

where H_0 is a prescribed hydraulic head at a radial position r_0($H = H_0$ at $r = r_0$). It is not possible to prescribe the ambient hydraulic head as $r \to \infty$ because Equation (9–17) exhibits a logarithmic singularity. The radial position r_0 is taken to be the distance to where the aquifer is being charged. The hydraulic head at the well H_w is obtained by setting $r = r_w$ (the well radius) in Equation (9–17) with the result

$$H_w = H_0 - \frac{\mu Q_r}{2\pi hk\rho g}\ln\frac{r_w}{r_0}. \tag{9–18}$$

We can rearrange this equation to yield an expression for Q_r in terms of the hydraulic heads H_0 and H_w

$$Q_r = \frac{2\pi hk\rho g(H_0 - H_w)}{\mu \ln\left(\dfrac{r_w}{r_0}\right)}. \tag{9–19}$$

As a typical numerical example we take $H_0 - H_w = 10\,\text{m}, h = 1\,\text{m}, r_w = 0.1\,\text{m}, r_0 = 1\,\text{km}, k = 10^{-11}\,\text{m}^2$, the viscosity of water $\mu = 10^{-3}$ Pa s, the density of water $\rho = 10^3\,\text{kg m}^{-3}$, and $g = 10\,\text{m s}^{-2}$. The flow rate to the well, from Equation (9–19), is $|Q_r| = 6.8 \times 10^{-4}\,\text{m}^3\,\text{s}^{-1}$.

PROBLEM 9-6 Consider the drawdown of a well penetrating a confined porous layer saturated with gas. The geometry is identical with that of Figure 9–3. Darcy's law in the form of Equation (9–13) is valid for the uniform flow of gas toward the well. Because the gas is compressible, both its density ρ and pressure p vary with radial distance r from the well. The mass flow rate of gas M_r crossing a cylindrical surface of radius r and height h must be constant to conserve mass. Assume that the perfect gas equation of state $p = \rho RT$ applies, where T is the gas temperature and R is the gas constant, and that the gas flow is isothermal. Show that the pressure distribution in the gas reservoir is

$$p^2 - p_0^2 = \frac{-M_r\mu RT}{\pi hk}\ln\frac{r}{r_0} \tag{9–20}$$

and that the mass flow rate into the well is

$$M_r = \frac{(p_0^2 - p_w^2)\pi hk}{\mu RT \ln\frac{r_w}{r_0}}. \tag{9–21}$$

Assume that the pressures p_0 at $r = r_0$ and p_w at $r = r_w$ are maintained constant during exploitation of the reservoir. Note that M_r is negative for flow toward the well.

9-5 Flow in Unconfined Aquifers

In many cases the flow in an aquifer with an impermeable lower boundary has a free upper surface. This upper surface is the *water table* and is often referred to as a *phreatic surface*. An aquifer in which the fluid has a free surface is known as an *unconfined aquifer*.

Let us consider a one-dimensional flow in an unconfined aquifer with a horizontal lower boundary, as shown in Figure 9–4. Below the phreatic surface the porous medium is saturated with water; above the free surface there is no groundwater. We denote the height of the phreatic surface above the lower boundary by $h(x)$ and assume that it is a slowly varying function of x; that is, we assume $dh/dx \ll 1$. In this case we can relate the horizontal pressure gradient in the saturated portion of the aquifer to the variation of h by

$$\frac{dp}{dx} = \rho g \frac{dh}{dx}. \tag{9–22}$$

This is known as the *Dupuit approximation*. The hydraulic head is the thickness of the water layer. This approximation breaks down when two-dimensional flow develops in the saturated part of the layer. Two-dimensional flow will occur if the slope of the free surface, dh/dx, becomes of order unity.

The Darcy velocity in the water-saturated region is found by substituting Equation (9–22) into Equation (9–1)

$$u = -\frac{k\rho g}{\mu}\frac{dh}{dx}. \tag{9–23}$$

The Darcy velocity depends on x because h is a function of x, but it is uniform over the thickness of the saturated zone. The total rate of fluid flow Q per unit width at position x in the aquifer is

$$Q = u(x)h(x) = -\frac{k\rho g}{\mu}h\frac{dh}{dx}. \tag{9–24}$$

Conservation of mass requires that the flow through the aquifer Q be constant. Thus Equation (9–24) can be

9-4 One-dimensional flow through an unconfined aquifer.

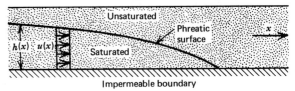

Unsaturated

Phreatic surface

x

$h(x)$ $u(x)$

Saturated

Impermeable boundary

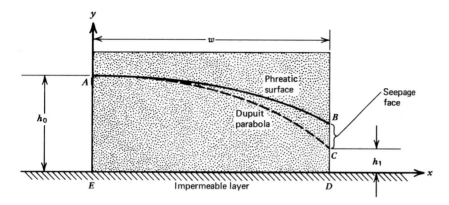

9–5 Unconfined flow through a porous dam. The Dupuit parabola *AC* is the solution if $(h_0 - h_1)/h_0 \ll 1$. The actual phreatic surface *AB* lies above the Dupuit parabola resulting in a seepage face *BC*.

considered as a differential equation for h as a function of x. Assuming that $h = h_0$ at $x = 0$, we integrate this equation to give

$$h = \left(h_0^2 - \frac{2Q\mu x}{k\rho g}\right)^{1/2}. \qquad (9\text{–}25)$$

The phreatic surface in the Dupuit approximation is a parabola. The free surface height is zero at the downstream distance

$$x_0 = \frac{k\rho g h_0^2}{2\mu Q}. \qquad (9\text{–}26)$$

However, the Dupuit approximation breaks down as $x \to x_0$ because $dh/dx \to \infty$.

As a specific example, consider the flow through the porous rectangular dam illustrated in Figure 9–5. The dam is constructed from material with a permeability k. The width of the dam is w, the height of the water behind the dam is h_0, and the height of the water in front of the dam is h_1. The dam and the reservoirs behind and in front of the dam are assumed to lie on an impermeable stratum. From Equation (9–25) the flow of water Q through the dam is

$$Q = \frac{k\rho g}{2\mu w}(h_0^2 - h_1^2). \qquad (9\text{–}27)$$

This is known as the *Dupuit–Fuchheimer discharge formula*. Since the Dupuit approximation has been used in the derivation of Equation (9–27), it might be expected to be valid only if $(h_0 - h_1) \ll h_0$; however, we

will show that this result is valid for an arbitrary value of h_1.

If h_1 is a small fraction of h_0, the actual phreatic surface *AB* lies above the Dupuit parabola, as illustrated in Figure 9–5. This results in a *seepage front BC* on the downstream face of the dam. If h_1 is a small fraction of h_0, the Darcy velocity components u_x and u_y are of the same magnitude in the porous dam. However, Darcy's law is still valid for the individual velocity components, and the horizontal component u_x is given by

$$u_x = -\frac{k}{\mu}\frac{\partial p}{\partial x}, \qquad (9\text{–}28)$$

where $p = p(x, y)$. We integrate this equation over the rectangular cross section of the dam, $0 < x < w$ and $0 < y < h_0$, with the result

$$\int_0^w \int_0^{h_0} u_x \, dy \, dx = -\frac{k}{\mu} \int_0^{h_0} \int_0^w \frac{\partial p}{\partial x} \, dx \, dy. \qquad (9\text{–}29)$$

The orders of integration on the two sides of the equation can be interchanged without affecting the result. However, the flow of water at a value of x is

$$Q = \int_0^{h_0} u_x \, dy, \qquad (9\text{–}30)$$

and Q is a constant independent of x. We also note that

$$\int_0^w \frac{\partial p}{\partial x} \, dx = p_1(y) - p_0(y), \qquad (9\text{–}31)$$

where $p_1(y)$ is the pressure distribution at $x = w$ and $p_0(y)$ is the pressure distribution at $x = 0$. Substitution of Equations (9–30) and (9–31) into Equation (9–29) yields

$$Q = \frac{k}{\mu w}\left(\int_0^{h_0} p_0 \, dy - \int_0^{h_0} p_1 \, dy\right). \qquad (9\text{–}32)$$

The pressure distributions are given by

$$p_0 = \rho g(h_0 - y) \tag{9-33}$$

$$p_1 = \rho g(h_1 - y) \qquad 0 < y < h_1$$

$$= 0 \qquad h_1 < y < h_0 \tag{9-34}$$

so that the integrals appearing in Equation (9–32) are

$$\int_0^{h_0} p_0 \, dy = \frac{1}{2} \rho g h_0^2 \tag{9-35}$$

$$\int_0^{h_0} p_1 \, dy = \frac{1}{2} \rho g h_1^2. \tag{9-36}$$

The substitution of Equations (9–35) and (9–36) into (9–32) once again yields Equation (9–27). Thus the Dupuit–Fuchheimer discharge formula is valid for all values of the ratio h_1/h_0.

PROBLEM 9–7 The base of an unconfined aquifer is inclined at an angle θ to the horizontal. Use Equation (9–4) and the Dupuit approximation to derive an expression for the flow rate Q if the free surface is at a constant height h above the base.

PROBLEM 9–8 Consider the unconfined flow through the stratified porous dam sketched in Figure 9–6. Assume that the height of the water in front of the dam h_1 exceeds the thickness a of the layer with permeability k_2 and that the phreatic surface lies totally in material with permeability k_1. Show that the flow rate through the dam is given by

$$Q = \frac{k_1 \rho g}{2\mu w}(h_0^2 - h_1^2)\left\{1 + \frac{2a}{(h_0 + h_1)}\left(\frac{k_2}{k_1} - 1\right)\right\}. \tag{9-37}$$

Do not use the Dupuit approximation.

The Dupuit approximation can be used to consider the drawdown of a well penetrating an unconfined

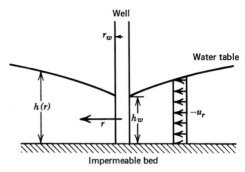

9–7 Draw down of a well penetrating an unconfined aquifer.

aquifer, as sketched in Figure 9–7. The radial Darcy velocity u_r in the Dupuit approximation can be obtained from a generalization of Equation (9–23)

$$u_r = \frac{-k\rho g}{\mu}\frac{dh}{dr}. \tag{9-38}$$

The flow rate through a cylindrical surface of height $h(r)$ and radius r is

$$Q_r = 2\pi r h u_r = \frac{-2\pi k \rho g}{\mu}\left(r h \frac{dh}{dr}\right), \tag{9-39}$$

where Q_r is positive if the flow is radially outward. Conservation of mass requires Q_r to be constant. We integrate Equation (9–39) with the condition that $h = h_0$ at $r = r_0$ and obtain

$$h^2 - h_0^2 = \frac{-\mu Q_r}{\pi k \rho g}\ln\frac{r}{r_0}. \tag{9-40}$$

If the well has a radius r_w and the height of the phreatic surface at the well is h_w, we find from Equation (9–40) that the flow to the well is

$$Q_r = \frac{\pi k \rho g(h_0^2 - h_w^2)}{\mu \ln(r_w/r_0)}. \tag{9-41}$$

Again r_0 is the distance to where the aquifer is being recharged.

PROBLEM 9–9 Determine the flow rate into the well for the case $h_0 = 10$ m, $h_w = 1$ m, $k = 10^{-12}$ m^2, $\mu = 10^{-3}$ Pa s, $r_0 = 5$ km, and $r_w = 0.1$ m. Calculate the shape of the phreatic surface.

So far we have limited our discussion to steady flows. To deal with flows in which the height of the phreatic surface and the flow rate through the aquifer vary with time t, we need to develop a continuity or conservation of mass equation. We continue to assume the

9–6 Unconfined flow through a stratified porous dam.

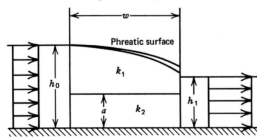

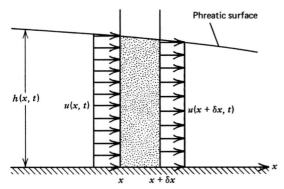

9–8 One-dimensional flow through an element of an unconfined aquifer.

validity of the Dupuit approximation and consider one-dimensional flow through an element of the unconfined aquifer between x and $x + \delta x$, as shown in Figure 9–8. The flow rate into the element per unit distance perpendicular to the plane in Figure 9–8 is $u(x, t)h(x, t)$. Similarly, the flow rate out of the element is $u(x + \delta x, t)h(x + \delta x, t)$. The net rate of flow out of the element is

$$u(x + \delta x, t)h(x + \delta x, t) - u(x, t)h(x, t)$$
$$\approx \frac{\partial}{\partial x}(uh)\, \delta x.$$

If the flow out of the element is not equal to the flow into the element, the height of the free surface must change, since the fluid is assumed to be incompressible. If the height of the phreatic surface changes from $h(t, x)$ to $h(t + \delta t, x)$ in time δt, the change in the volume of fluid in the element is

$$\phi[h(t + \delta t, x) - h(t, x)]\, \delta x \approx \phi \frac{\partial h}{\partial t}\, \delta x\, \delta t.$$

The factor ϕ is required because the fluid only fills the pore space in the matrix. Conservation of fluid requires that the net flow out of the element in time δt equals the decrease in fluid volume in the element so that

$$\phi \frac{\partial h}{\partial t} + \frac{\partial}{\partial x}(uh) = 0. \qquad (9\text{–}42)$$

Substitution of the Darcy velocity from Equation (9–23) yields

$$\frac{\partial h}{\partial t} = \frac{k\rho g}{\mu \phi} \frac{\partial}{\partial x}\left(h \frac{\partial h}{\partial x}\right). \qquad (9\text{–}43)$$

This nonlinear diffusion equation is often referred to as the *Boussinesq equation*.

If the variation in h is small, we can write

$$h = h_0 + h', \qquad (9\text{–}44)$$

where h_0 is constant and $|h'| \ll h_0$. Substitution of Equation (9–44) into (9–43) and neglecting the term that is quadratic in h' yields

$$\frac{\partial h'}{\partial t} = \frac{k\rho g h_0}{\mu \phi} \frac{\partial^2 h'}{\partial x^2}. \qquad (9\text{–}45)$$

This process of linearization is identical with that previously carried out in Section 6–19. Equation (9–45) is identical in form with the governing equation for the one-dimensional, unsteady conduction of heat given in Equation (4–68). A variety of solutions of this diffusion equation can be obtained using the methods introduced in Chapter 4.

As an example, consider how a specified periodic variation of h' at $x = 0$,

$$h' = h'_s \cos \omega t, \qquad (9\text{–}46)$$

causes the water table to fluctuate at distances $x > 0$ in a semi-infinite unconfined aquifer. The variation in h' could be due to the influence of annual runoff from a mountain range on the water table in an adjacent arid region. Noting that $k\rho g h_0/\phi\mu$ in the porous flow problem is equivalent to κ in the heat conduction problem, the solution given in Equation (4–89) can be appropriately modified to yield

$$h' = h'_s \exp\left\{-x\left(\frac{\omega\mu\phi}{2k\rho g h_0}\right)\right\}$$
$$\times \cos\left\{\omega t - x\left(\frac{\omega\mu\phi}{2k\rho g h_0}\right)\right\}. \qquad (9\text{–}47)$$

The amplitude of the periodic fluctuation in the phreatic surface decreases to $1/e$ of the applied value in a distance

$$x_e = \left(\frac{2k\rho g h_0}{\omega\mu\phi}\right)^{1/2}. \qquad (9\text{–}48)$$

Taking $\phi = 0.1$, $k = 10^{-11}$ m^2, $\omega = 2 \times 10^{-7}$ s^{-1} (a period of 1 year), $\mu = 10^{-3}$ Pa s, and $h_0 = 100$ m, we find that $x_e = 1$ km. We see that the influence of a change of hydraulic head propagates over a relatively short distance.

PROBLEM 9–10 The hydraulic head at $x = 0$ is increased from h_0 to $h_0 + \Delta h$ at $t = 0(\Delta h \ll h_0)$. Determine the flow into the unconfined aquifer that occupies the region $x > 0$.

PROBLEM 9–11 At $t = 0$ the phreatic surface in an unconfined aquifer in the region $x > 0$ has a height h_0. For $t > 0$ there is a constant discharge Q_0 (per unit length) from the aquifer at $x = 0$. Assuming that $|\Delta h| \ll h_0$ show that the height of the phreatic surface as a function of t and x is given by

$$h = h_0 - 2Q_0\left(\frac{k\rho g h_0 \phi}{\mu t}\right)^{1/2}\left(\xi \operatorname{erf}\xi - \frac{1}{\sqrt{\pi}}e^{-\xi^2}\right),$$

$$(9\text{–}49)$$

where

$$\xi = \frac{x}{2}\left(\frac{\mu\phi}{k\rho g h_0 t}\right)^{1/2}.$$

PROBLEM 9–12 At $t = 0$ the height of the phreatic surface in an unconfined aquifer is $h = h_1$ for $-L \le x \le L$ and $h = h_0$ for $|x| > L$ ($h_1 - h_0 \ll h_0$). Show that the height of the phreatic surface as a function of x and t for $t > 0$ is

$$h - h_0 = \frac{(h_1 - h_0)}{2}\left\{\operatorname{erf}\left[\frac{(L-x)}{2}\left(\frac{\mu\phi}{k\rho g h_0 t}\right)^{1/2}\right]\right.$$
$$\left. + \operatorname{erf}\left[\frac{(L+x)}{2}\left(\frac{\mu\phi}{k\rho g h_0 t}\right)^{1/2}\right]\right\}. \quad (9\text{–}50)$$

9–9 Drainage of water out of a bank due to the sudden change in the water level in the channel.

PROBLEM 9–13 For $t \le 0$ there is a steady flow Q_0 in an unconfined aquifer in the region $x > 0$. From Equation (9–25) the height of the phreatic surface is

$$h^2 = h_0^2 - \frac{2\mu Q_0 x}{k\rho g}, \quad (9\text{–}51)$$

where h_0 is the height of the surface at $x = 0$. At time $t = 0$ the height at $x = 0$ is changed from h_0 to h_1, $|h_0 - h_1| \ll h_0$. Show that the height of the phreatic surface at subsequent times is

$$h = h_1^2 + \left(h_0^2 - h_1^2\right)\operatorname{erf}\xi - \frac{2\mu Q_0 x}{k\rho g}, \quad (9\text{–}52)$$

where $\xi = (1/2)x(\mu\phi/k\rho g\bar{h}t)^{1/2}$ and $\bar{h}$ is a suitably defined average height. Proceed by recognizing that h^2 also satisfies a linearized diffusion equation when there are small changes in the square of the water table height. It is advantageous to solve for h^2 rather than h because the initial condition, Equation (9–51), is given in terms of h^2. Show that the flow into the channel at $x = 0$, is

$$Q = Q_0 - \frac{1}{2}\left(h_0^2 - h_1^2\right)\left(\frac{k\rho g\phi}{\pi\mu\bar{h}t}\right)^{1/2}. \quad (9\text{–}53)$$

A number of solutions of the nonlinear Boussinesq equation (9–43) can also be found. First consider the problem illustrated in Figure 9–9. At time $t = 0-$, Figure 9–9a, the water table in a channel and the adjacent porous bank is at a height h_0 above a horizontal, impermeable bed. At time $t = 0+$, Figure 9–9b, there is a sudden drop in the water level in the channel to a height h_1, and it remains at this value for $t > 0$. Water

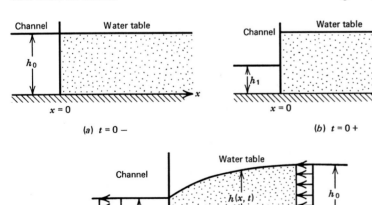

(a) $t = 0-$

(b) $t = 0+$

(c) $t > 0$

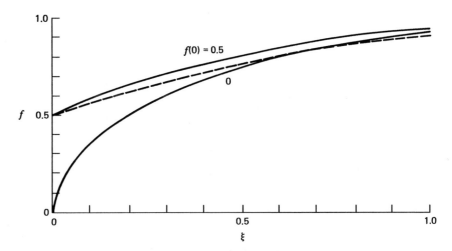

drains from the aquifer into the channel, and the height of the phreatic surface drops, Figure 9–9c. The required initial and boundary conditions for this problem are

$$h(0, t) = h_1 \qquad h(\infty, t) = h_0 \qquad h(x, 0) = h_0.$$
$$(9\text{--}54)$$

We noted before that the linearized form of the Boussinesq equation is identical with the equation governing the one-dimensional, unsteady conduction of heat. The boundary conditions for this problem are identical with those used for the instantaneous heating or cooling of a semi-infinite half-space considered in Section 4–15. Therefore it is not surprising that the similarity variable used for that problem – see Equation (4–96) – can also be used for this problem (noting the correspondence between κ and $k\rho g h_0/\mu\phi$)

$$\xi = \left(\frac{\mu\phi}{k\rho g h_0 t} \right)^{1/2} \frac{x}{2}.$$
$$(9\text{--}55)$$

In addition it is convenient to introduce the dimensionless water surface height

$$f = \frac{h}{h_0}.$$
$$(9\text{--}56)$$

In terms of f and ξ the Boussinesq equation (9–43) becomes

$$-2\xi \frac{df}{d\xi} = \frac{d}{d\xi}\left(f \frac{df}{d\xi} \right).$$
$$(9\text{--}57)$$

From Equations (9–54) and (9–55) the necessary boundary conditions are

$$f(0) = \frac{h_1}{h_0} \qquad f(\infty) = 1.$$
$$(9\text{--}58)$$

9–10 Numerical similarity solution of the nonlinear Boussinesq equation for water seepage into a channel from an adjacent bank after a sudden decrease in water level in the channel. The dimensionless water surface height is shown as a function of the similarity variable for $f(0) = 0$ and 0.5 (solid lines). The numerical solution is compared with the approximate linearized solution [Equation (9–59)] (dashed curve) for $f(0) = 0.5$.

While it is not possible to obtain an analytic solution of Equation (9–57), a solution can be obtained by numerical integration. The results of such an integration are given in Figure 9–10 for $f(0) = 0$ and 0.5.

If Equation (9–57) is linearized as discussed before, the solution is

$$f = f(0)\, \text{erfc}\, \xi.$$
$$(9\text{--}59)$$

This result is valid in the limit $f(0) \to 1$ (see Problem 9–10). The approximate result from Equation (9–59) is compared with the numerical integration of Equation (9–57) in Figure 9–10 for $f(0) = 0.5$. Reasonably good agreement is obtained, even though the linearization condition is not satisfied.

The fact that we have obtained a similarity solution shows that the shape of the phreatic surface varies with position and time in a self-similar manner; that is, h depends only on $x/t^{1/2}$. It must be noted, however, that this is only true for sufficiently large times. For small times dh/dx is of order unity, and the Dupuit approximation used in deriving Equation (9–43) is violated.

PROBLEM 9–14 For the problem in Figure 9–9 show that the rate at which water seeps into the channel Q

per unit distance along the channel is

$$Q = \frac{-\phi h_1}{2t^{1/2}} \left(\frac{k\rho g h_0}{\mu \phi} \right)^{1/2} f'(0), \qquad (9\text{–}60)$$

where $f'(0)$ is $df/d\xi$ at $\xi = 0$.

A similarity solution of the Boussinesq equation can also be obtained for the lateral spreading of a linear mound of groundwater. The mound of groundwater will diffuse outward in much the same way that a pulse of heat diffuses. The latter problem was studied in Section 4–21. Just as the total amount of heat was conserved in Equation (4–159), the total amount of fluid must be conserved in this problem. If the initial volume of water per unit distance parallel to the linear mound is V_1, conservation of fluid requires that

$$\int_0^{x_0} h \, dx = \frac{V_1}{2\phi}, \qquad (9\text{–}61)$$

where $h = h(x, t)$ is the height of the mound above an impermeable plane and $x_0(t)$ is the half-width of the mound at its base. We show that the spreading fluid mound has a well-defined front, which we denote by x_0; that is, $h \equiv 0$ for $|x| > x_0$. The fluid mound spreads symmetrically so that half the fluid is in the region $0 < x < x_0$.

Just as we introduced similarity variables to solve the thermal diffusion problem, we can also introduce similarity variables for this problem. The similarity variables relevant to the spreading of the groundwater mound are

$$f = \left(\frac{12k\rho g \phi t}{\mu V_1^2} \right)^{1/3} h \qquad (9\text{–}62)$$

$$\xi = \left(\frac{2\mu \phi^2}{3k\rho g V_1 t} \right)^{1/3} x. \qquad (9\text{–}63)$$

The Boussinesq equation (9–43) must be rewritten in terms of these variables. The required partial derivatives are

$$\frac{\partial h}{\partial t} = -\frac{1}{3t} \left(\frac{\mu V_1^2}{12k\rho g \phi t} \right)^{1/3} \left(\xi \frac{df}{d\xi} + f \right) \qquad (9\text{–}64)$$

$$\frac{\partial}{\partial x} \left(h \frac{\partial h}{\partial x} \right) = \left(\frac{\mu V_1^2}{12k\rho g \phi t} \right)^{2/3} \left(\frac{2}{3} \frac{\mu \phi^2}{k\rho g V_1 t} \right)^{2/3}$$
$$\times \frac{d}{d\xi} \left(f \frac{df}{d\xi} \right). \qquad (9\text{–}65)$$

Substitution of these expressions into Equation (9–43) gives

$$f \frac{d^2 f}{d\xi^2} + \left(\frac{df}{d\xi} \right)^2 + \xi \frac{df}{d\xi} + f = 0. \qquad (9\text{–}66)$$

Substitution of the nondimensional variables into the conservation of fluid condition, Equation (9–61), yields

$$\int_0^{\xi_0} f \, d\xi = 1, \qquad (9\text{–}67)$$

where ξ_0 is given by Equation (9–63) with $x = x_0$. An analytic solution that is symmetric about $\xi = 0$ and that satisfies Equations (9–66) and (9–67) is

$$f = \frac{3^{2/3}}{2} \left(1 - \frac{\xi^2}{3^{2/3}} \right) \qquad |\xi| \le \xi_0 = 3^{1/3}. \qquad (9\text{–}68)$$

The shape of the phreatic surface at any given time is parabolic. Equation (9–68) shows that $f = 0$ for $|\xi| = \xi_0 = 3^{1/3}$. Thus $\xi = \xi_0$ defines the fluid front or maximum half-width of the spreading mound. There is no fluid in the region $|\xi| > \xi_0$. The actual position of the fluid front as a function of time follows from Equation (9–63)

$$x_0 = \left(\frac{9k\rho g V_1 t}{2\mu \phi^2} \right)^{1/3}. \qquad (9\text{–}69)$$

Upon combining Equations (9–62), (9–63), and (9–68), we obtain the height of the phreatic surface as a function of position and time:

$$h = \left(\frac{3\mu V_1^2}{32k\rho g \phi t} \right)^{1/3} \left\{ 1 - \left(\frac{2\mu \phi^2}{9k\rho g V_1 t} \right)^{2/3} x^2 \right\}. \qquad (9\text{–}70)$$

At $t = 0$ the fluid mound has zero thickness and infinite height. However, the total volume of water is finite and equal to V_1 per unit distance along the mound.

PROBLEM 9–15 Show that the maximum height of the phreatic surface as a function of x is

$$h_{\max} = \frac{1}{2\sqrt{3}} \frac{V_1}{\phi x} \qquad (9\text{–}71)$$

and that this occurs at time

$$t_{\max} = \frac{2\mu \phi^2 x^3}{\sqrt{3} k\rho g V_1}. \qquad (9\text{–}72)$$

PROBLEM 9–16 Let h_0 be the height of the laterally spreading groundwater mound at $x = 0$ and $t = t_0$. Let the half-width of the mound at its base be l_0 at $t = t_0$. Show that the height of the mound at $x = 0$ and $t = t_0 + t'$ is given by

$$h_0 \left(1 + \frac{6k\rho g h_0 t'}{\mu \phi l_0^2} \right)^{-1/3}.$$

In addition, demonstrate that the half-width of the mound at its base at time $t = t_0 + t'$ is

$$l_0 \left(1 + \frac{6k\rho g h_0 t'}{\mu \phi l_0^2} \right)^{1/3}.$$

We next determine the height of the phreatic surface h as a function of x and t when water is introduced at $x = 0$ at a constant volumetric rate Q_1 per unit width. For $t < 0$, h is zero; for $t > 0$, there is a constant input of water at $x = 0$. Half of the fluid flows to the right into the region $x > 0$, and half flows to the left. From Equation (9–24) we can write the flow rate to the right at $x = 0+$ as

$$\frac{-k\rho g}{\mu} \left(h \frac{\partial h}{\partial x} \right)_{x=0+} = \frac{1}{2} Q_1. \tag{9–73}$$

The water table height $h(x, t)$ is the solution of the Boussinesq equation (9–43) that satisfies condition (9–73).

Once again we introduce similarity variables. The appropriate similarity variables for this problem are

$$f = \left(\frac{k\rho g \phi}{Q_1^2 \mu t} \right)^{1/3} h \tag{9–74}$$

$$\xi = \left(\frac{\phi^2 \mu}{k\rho g Q_1 t^2} \right)^{1/3} x. \tag{9–75}$$

Aside from numerical factors these variables are the same as the ones in Equations (9–62) and (9–63) if we replace V_1/t in those equations by Q_1. The introduction of these similarity variables into the Boussinesq equation yields

$$f \frac{d^2 f}{d\xi^2} + \left(\frac{df}{d\xi} \right)^2 + \frac{2}{3} \xi \frac{df}{d\xi} - \frac{1}{3} f = 0. \tag{9–76}$$

The boundary condition at $x = 0+$ given in Equation (9–73) becomes

$$\left(f \frac{df}{d\xi} \right)_{\xi=0+} = -\frac{1}{2}. \tag{9–77}$$

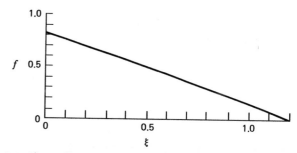

9–11 The nondimensional height of the phreatic surface f as a function of the similarity variable ξ for fluid injection at a constant rate from the plane $x = 0$.

The solution of this problem, unlike that of the previous one, requires a numerical integration. As was the case for the spreading mound of groundwater, there is a fluid front, and $f = 0$ for $\xi > \xi_0$. At the flow front Equation (9–76) yields

$$\left(\frac{df}{d\xi} \right)_{\xi=\xi_0} = -\frac{2}{3} \xi_0. \tag{9–78}$$

The numerical solution of Equation (9–76) subject to conditions (9–77) and (9–78) is given in Figure 9–11. The value of ξ_0 is 1.18, and the actual position of the fluid front from Equation (9–75) is

$$x_0 = 1.18 \left(\frac{k\rho g Q_1 t^2}{\phi^2 \mu} \right)^{1/3}. \tag{9–79}$$

The height of the phreatic surface at $x = 0$ is

$$h_{x=0} = 0.82 \left(\frac{Q_1^2 \mu t}{k\rho g \phi} \right)^{1/3}. \tag{9–80}$$

PROBLEM 9–17 If fluid is injected along a plane at $x = 0$ at a rate of 0.1 m^2 s^{-1}, how high is the phreatic surface at the point of injection and how far has the fluid migrated if $\mu = 10^{-3}$ Pa s, $\phi = 0.1$, $k = 10^{-11}$ m^2, $\rho = 1000$ kg m^{-3}, and $t = 10^5$ s?

To solve problems involving unsteady horizontal radial flow toward or away from a vertical line, we need to generalize the Boussinesq equation to cylindrical or polar coordinates. We do this by first deriving a fluid conservation equation for radial flow. The change in volume of fluid between r and $r + \delta r$ due to a change in the height of the phreatic surface occurring in the time interval t to $t + \delta t$ is

$$2\pi r\phi\{h(t + \delta t) - h(t)\}\,\delta r = 2\pi r\phi\frac{\partial h}{\partial t}\,\delta t\,\delta r.$$

The net flow out of the cylindrical element in time δt is

$$\{Q_r(r + \delta r) - Q_r(r)\}\,\delta t = \frac{\partial Q_r}{\partial r}\,\delta r\,\delta t,$$

where Q_r is the rate at which fluid flows across a cylindrical surface of height h and radius r. Conservation of fluid requires that any net outflow from the cylindrical element be balanced by a drop in height of the phreatic surface in the element. This enables us to write

$$\frac{\partial Q_r}{\partial r} + 2\pi r\phi\frac{\partial h}{\partial t} = 0. \tag{9–81}$$

By substituting for Q_r in terms of h from Equation (9–39), we obtain

$$\frac{\partial h}{\partial t} = \frac{k\rho g}{\mu\phi r}\frac{\partial}{\partial r}\left(rh\frac{\partial h}{\partial r}\right). \tag{9–82}$$

We first derive a similarity solution of this equation for the radial spreading of a mound of water. A volume of water V_0 is introduced at time $t = 0$ along the vertical line at $r = 0$. The fluid spreads radially outward; at any time $t > 0$, the region $r > 0$ must contain the original amount of fluid. Thus we can write the fluid conservation condition

$$2\pi\phi\int_0^{r_0} rh\,dr = V_0. \tag{9–83}$$

The integration extends only to $r = r_0$ because, as before, the water mound has a fluid front at $r = r_0$; that is, $h = 0$ for $r > r_0$.

The appropriate similarity variables for this problem are

$$f = \left(\frac{4k\rho gt}{\mu V_0}\right)^{1/2} h \tag{9–84}$$

$$\eta = \left(\frac{\mu\phi^2}{4k\rho g V_0 t}\right)^{1/4} r. \tag{9–85}$$

The introduction of these variables into Equation (9–82) yields

$$f\frac{d^2 f}{d\eta^2} + \left(\frac{df}{d\eta}\right)^2 + \frac{f}{\eta}\frac{df}{d\eta} + \eta\frac{df}{d\eta} + 2f = 0. \tag{9–86}$$

The nondimensional fluid conservation equation becomes

$$\int_0^{\eta_0} \eta f\,d\eta = \frac{1}{2\pi}, \tag{9–87}$$

where η_0 is the value of η corresponding to $r = r_0$. We can find an analytic solution for the axisymmetric spreading of a groundwater mound, just as we did for the spreading of the linear groundwater mound. The solution of Equation (9–86) that satisfies Equation (9–87) is

$$f = \frac{1}{\sqrt{\pi}}\left(1 - \frac{\eta^2\sqrt{\pi}}{2}\right) \qquad |\eta| \le \eta_0 = \left(\frac{2}{\sqrt{\pi}}\right)^{1/2}. \tag{9–88}$$

The actual position of the fluid front as a function of time follows from the value of η_0 and Equation (9–85)

$$r_0 = \left(\frac{16k\rho g V_0 t}{\pi\mu\phi^2}\right)^{1/4}. \tag{9–89}$$

By combining Equations (9–84), (9–85), and (9–88), we obtain the height of the phreatic surface as a function of position and time:

$$h = \left(\frac{\mu V_0}{4\pi k\rho gt}\right)^{1/2}\left\{1 - \left(\frac{\pi\mu\phi^2}{16k\rho g V_0 t}\right)^{1/2} r^2\right\}. \tag{9–90}$$

PROBLEM 9–18 Show that the maximum height of the phreatic surface as a function of r is

$$h_{\max} = \frac{V_0}{2\pi\phi r^2} \tag{9–91}$$

and that this occurs at time

$$t_{\max} = \frac{\pi\mu\phi^2 r^4}{4k\rho g V_0}. \tag{9–92}$$

PROBLEM 9–19 Let h_0 be the height of the spreading axisymmetric groundwater mound at $r = 0$ and $t = t_0$. Let the maximum radius of the mound at its base be b_0 at $t = t_0$. Show that the height of the mound at $r = 0$ and $t = t_0 + t'$ is given by

$$h_0\left(1 + \frac{8k\rho gh_0 t'}{\mu\phi b_0^2}\right)^{-1/2}.$$

In addition, demonstrate that the maximum radius of

the mound at its base at time $t = t_0 + t'$ is

$$b_0\left(1 + \frac{8k\rho g h_0 t'}{\mu\phi b_0^2}\right)^{1/4}.$$

An exact solution of the nonlinear diffusion equation for the height of the phreatic surface as a function of position and time is also possible when water is introduced at $r = 0$ at a constant volumetric rate Q_0. For $t < 0$, h is zero for all r; for $t > 0$, there is a constant input of water at $r = 0$. The boundary condition on the flow rate from Equation (9–39) is

$$-\frac{2\pi k\rho g r h}{\mu}\frac{\partial h}{\partial r} \to Q_0 \qquad \text{as } r \to 0. \qquad (9\text{–}93)$$

The appropriate similarity variables for this problem are

$$f = \left(\frac{k\rho g}{\mu Q_0}\right)^{1/2} h \qquad (9\text{–}94)$$

$$\eta = \left(\frac{\mu\phi^2}{k\rho g Q_0 t^2}\right)^{1/4} r. \qquad (9\text{–}95)$$

Aside from numerical factors, these variables can be obtained from the ones in Equations (9–84) and (9–85) by replacing V_0/t in those equations with Q_0. The introduction of these variables into Equation (9–82) yields

$$f\frac{d^2 f}{d\eta^2} + \left(\frac{df}{d\eta}\right)^2 + \frac{f}{\eta}\frac{df}{d\eta} + \frac{\eta}{2}\frac{df}{d\eta} = 0, \qquad (9\text{–}96)$$

while the boundary condition at $r = 0$ from Equation (9–93) becomes

$$\eta f\frac{df}{d\eta} \to -\frac{1}{2\pi} \qquad \text{as } \eta \to 0. \qquad (9\text{–}97)$$

A numerical solution is required for this problem. There is a fluid front at $\eta = \eta_0$ and $f = 0$ for $\eta > \eta_0$. From Equation (9–96) we can rewrite the condition $f = 0$ at the fluid front as

$$\left(\frac{df}{d\eta}\right)_{\eta=\eta_0} = -\frac{1}{2}\eta_0. \qquad (9\text{–}98)$$

The numerical solution of Equation (9–96) subject to the conditions given in Equations (9–97) and (9–98) is given in Figure 9–12. The value of η_0 is 1.16, and the position of the fluid front from Equation (9–95) is

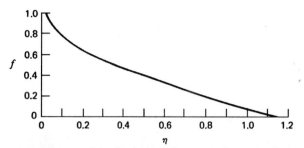

9-12 The nondimensional height of the phreatic surface f as a function of the similarity variable η for a line source releasing fluid at a constant rate at $r = 0$.

therefore

$$r_0 = 1.16\left(\frac{k\rho g Q_0 t^2}{\mu\phi^2}\right)^{1/4}. \qquad (9\text{–}99)$$

The height of the phreatic surface is logarithmically singular as $r \to 0$. This is the same singularity as we found for the steady flow problem in Equation (9–40).

9-6 Geometrical Form of Volcanoes

The shapes of many volcanoes are remarkably axisymmetric and similar. A large fraction of the composite volcanoes that make up the volcanic lines adjacent to ocean trenches have near-constant flank slopes and are concave upward near their summits. An example, Mount Fuji, in Japan, is shown in Figure 9–13. There are, however, a number of phenomena that can produce nonsymmetrical edifices. Examples include parasitic centers of volcanism on the flanks of a volcano, glacial and other types of erosion, and explosive eruptions.

One model for the geometrical form of volcanoes assumes that the volcanic edifice is a uniform porous medium and that the surface of the volcano is a surface of constant hydraulic head. The volcano is the constructional sum of many small lava flows. Each flow passes through the interior porosity of the edifice, flows onto the surface, and extends the porous matrix of the edifice as it solidifies.

The way in which these flows extend the edifice is illustrated qualitatively in Figure 9–14. At the beginning of an eruption, magma reaches the center of the base of the edifice through a volcanic pipe. Studies of groundwater migration indicate that volcanoes are permeated by fractures; presumably these are thermal contraction

9–13 Mount Fuji in Japan.

cracks formed during the solidification of individual flows. In some volcanoes the permeability is dominated by radial rift zones from which most eruptions emanate.

We assume that the magma is driven through the preexisting matrix of channels in search of the least resistant path to the surface of the volcano. This is illustrated in Figure 9–14a. In Figure 9–14b the magma

9–14 The mechanism by which a surface flow extends a volcanic edifice. (a) Magma penetrates the permeable edifice searching for the path of least resistance to the surface. (b) The magma reaches the surface at the point of least resistance. (c) The surface flow extends the edifice.

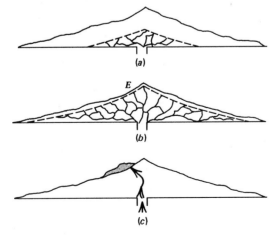

reaches the surface following the path of minimum hydraulic resistance. The magma continues to flow to the surface following this path creating a surface flow that extends the volcanic edifice (Figure 9–14c). Upon solidification, there will be an increase in the resistance to flow along this path, and the next eruption will follow a different path and occur at another point on the surface. If a volcano grows too tall, flank eruptions will widen it; if a volcano grows too wide, summit eruptions will increase its elevation. The equal resistance to flow requires that the volcano grow axisymmetrically.

The flow of magma through the volcanic edifice is essentially the same as the flow of groundwater through an unconfined aquifer. We assume that the surface of the volcano is a surface of constant hydraulic head, just as the phreatic surface in an unconfined aquifer is. We further assume that the slope of the volcano is small so that the Dupuit approximation can be made. When the magma reaches the surface, it extends the matrix instead of filling the pore space. Thus it is necessary to set $\phi = 1$ in the mass conservation equation (9–81) and in the nonlinear diffusion equation (9–82).

The similarity solution for the shape of the phreatic surface when fluid is introduced at $r = 0$ at a constant rate Q_0 is directly applicable to this problem. The similarity profile for the phreatic surface given in Figure 9–12 is therefore also the predicted geometrical form for volcanoes. The similarity profile yields an actual

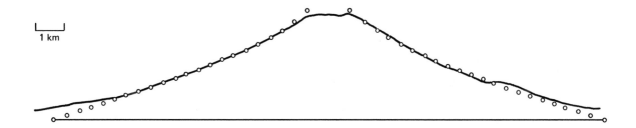

volcano shape $h = h(r)$ through the scaling factors in Equations (9–94) and (9–95). These scaling factors can be adjusted to give the best fits possible to actual volcanoes. The predicted form is compared with a cross section of Mount Fuji in Figure 9–15. In general the agreement is satisfactory. Near the base of the volcano the observed profile is more rounded; this can be attributed to deposits of alluvium. The theoretical profile is not expected to be appropriate near the summit where it is singular and where the Dupuit approximation is not valid.

The radius of the volcano is equivalent to the radial position of the fluid front given by Equation (9–99) with $\phi = 1$. From Equations (9–94), (9–95), (9–98), and (9–99) with $\phi = 1$, the flank slope of the volcano at its base is

$$\left(\frac{\partial h}{\partial r}\right)_{r=r_0} = -0.58\left(\frac{\mu}{k\rho g}\right)^{3/4}\frac{Q_0^{1/4}}{t^{1/2}}. \qquad (9\text{--}100)$$

9–16 Topographic profiles across seven volcanoes from the volcanic line in the western United States; R–Rainier, S–Shasta, A–Adams, B–Baker, H–Hood, L–Lassen, St–St. Helens. Four theoretical profiles are also included.

9–15 Comparison of the similarity solution for the constant hydraulic head volcanic surface (circles) with a cross section of Mount Fuji, Japan (solid line).

The negative product of Equations (9–99) and (9–100) defines a reference height for the volcano equal to

$$h_r = 0.673\left(\frac{\mu Q_0}{k\rho g}\right)^{1/2}. \qquad (9\text{--}101)$$

Because the reference height is independent of time, the theory predicts that volcanoes grow primarily by increasing their radii. A series of predicted volcanic profiles are compared with the cross sections of six volcanoes from the volcanic line in the western United States in Figure 9–16. Good agreement is obtained. It appears that volcanoes do, in fact, grow mainly by increasing their radii.

Just as we have applied the similarity solution for a line source of fluid in an unconfined aquifer to the determination of volcano shapes, we can also apply the similarity solution for a planar source of fluid to the determination of the geometrical form of volcanic ridges. The similarity form of the cross section of a volcanic

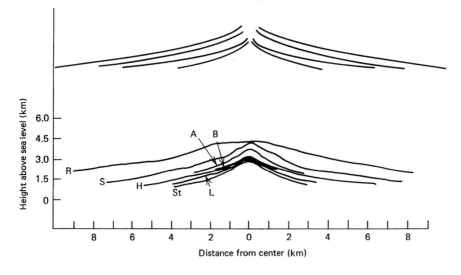

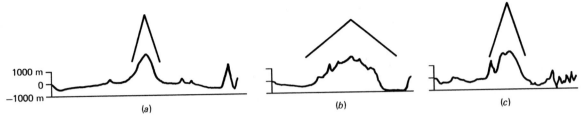

1000 m
0
−1000 m

(a)   (b)   (c)

9–17 Bathymetric profiles across the (a) Hawaiian, (b) Walvis, and (c) Ninety East ridges. Theoretical profiles predicted by the similarity solution are also shown.

ridge is therefore given in Figure 9–17. The transformation equations (9–74) and (9–75) can be used to convert this similarity profile into actual profiles of $h = h(x)$. The scale factors in these equations can be adjusted to give the best possible fits to observed ridge cross sections. A number of linear volcanic ridges rise above the seafloor. Typical bathymetric profiles across the Hawaiian, Walvis, and Ninety East Ridges are compared with predicted profiles from the similarity solution in Figure 9–17. Reasonably good agreement is obtained.

9–7 Equations of Conservation of Mass, Momentum, and Energy for Flow in Porous Media

So far we have considered one-dimensional or axisymmetric flows that only require Darcy's law and a simple mass balance equation for their solution. In this and subsequent sections we are concerned with both one- and two-dimensional flows in which there is also a transport of heat. Thus, in addition to Darcy's law, we require differential equations for conservation of mass and energy in two dimensions. We previously derived the relevant equations for a viscous incompressible fluid in Chapter 6; these equations can be applied to flows in porous media with minor modifications.

Conservation of mass for the flow of a viscous incompressible fluid in two dimensions requires that Equation (6–53) be satisfied. This equation can also apply to flows in porous media if the solid matrix cannot deform and if the fluid is incompressible

$$\frac{\partial u}{\partial x} + \frac{\partial v}{\partial y} = 0. \tag{9-102}$$

The velocity components in this equation are the Darcy velocities. Although the Darcy velocity components are not the actual fluid velocities in the microscopic channels of the porous medium, they are equivalent to the velocity components of an ordinary viscous fluid insofar as their transport of such quantities as mass and heat are concerned. This is apparent from the definition of the Darcy velocity as the volumetric flow rate per unit area of the entire medium. The Darcy velocity is an average velocity over an area element in a porous medium. The average is defined in such a way that it accounts for the transport of heat across the area element as well as the transport of mass.

The energy equation for a two-dimensional flow of an incompressible fluid in a porous medium can be written

$$\rho_m c_{p_m} \frac{\partial T}{\partial t} + \rho_f c_{p_f} \left(u \frac{\partial T}{\partial x} + v \frac{\partial T}{\partial y} \right)$$
$$= \lambda_m \left(\frac{\partial^2 T}{\partial x^2} + \frac{\partial^2 T}{\partial y^2} \right), \tag{9-103}$$

which is a generalization of Equation (6–293). The fluid and the solid matrix are assumed to have the same temperature T. The diffusion of heat by conduction occurs through the entire medium, and the appropriate value of the thermal conductivity appearing on the right side of Equation (9–103) must be a volumetric average over the fluid-filled pores and the solid matrix. We have used the symbol λ_m for this average thermal conductivity of the medium to avoid confusion with the symbol for permeability. Because a substantial fraction of the medium is made up of the solid matrix, which is usually a better conductor of heat than the fluid, it is generally a good approximation to assume that λ_m is the thermal conductivity of the solid matrix. Thermal energy is stored in both the fluid-filled pores and the solid matrix. Therefore the thermal inertia term on the left side of Equation (9–103) is also a volumetric average. The advection terms on the left side of the equation use the fluid density ρ_f and the fluid specific heat c_{p_f} because only the fluid transports heat.

To complete the formulation of the problem, we require Darcy's equations for the horizontal and vertical components of the Darcy velocity. These were given in Equations (9–1) and (9–3).

9-8 One-Dimensional Advection of Heat in a Porous Medium

Considerable observational evidence indicates that magma bodies induce large-scale motions of groundwater in the surrounding rocks. A substantial fraction of the hot springs with exit temperatures greater than about $50°C$ is believed to be the direct result of this type of hydrothermal circulation. The intrusion heats the groundwater, which becomes less dense and rises. Near the Earth's surface the water cools and becomes more dense. It can then sink and recharge the aquifers and porous rock in the vicinity of the intrusion. The water is reheated, and the cycle repeats. An analysis of the complete hydrothermal convection system requires the solution of a coupled set of nonlinear differential equations in at least two dimensions. This problem is considered in subsequent sections. Here, however, we study only the upwelling flow above the intrusion. A one-dimensional solution is obtained for the dependence of temperature on depth, and this is compared with measurements of the subsurface temperature in the Steamboat Springs, Nevada, hydrothermal system.

We simplify Equations (9–102) and (9–103) for steady one-dimensional upflow and obtain

$$\frac{dv}{dy} = 0 \qquad (9\text{–}104)$$

$$\rho_f c_{p_f} v \frac{dT}{dy} = \lambda_m \frac{d^2 T}{dy^2}. \qquad (9\text{–}105)$$

From the first of these equations v is a constant, and Equation (9–105) can be immediately integrated to give

$$\rho_f c_{p_f} v T = \lambda_m \frac{dT}{dy} + c_1. \qquad (9\text{–}106)$$

The constant of integration c_1 can be determined from the conditions at great depth where upwelling fluid has the uniform reservoir temperature T_r. Therefore as $y \to \infty$, we must have $dT/dy \to 0$ and $T \to T_r$. This gives

$$c_1 = \rho_f c_{p_f} v T_r \qquad (9\text{–}107)$$

and

$$\rho_f c_{p_f} v (T - T_r) = \lambda_m \frac{d}{dy}(T - T_r). \qquad (9\text{–}108)$$

The rearrangement of Equation (9–108) in the form

$$\frac{d(T - T_r)}{(T - T_r)} = \frac{\rho_f c_{p_f} v}{\lambda_m} \, dy \qquad (9\text{–}109)$$

and the integration of Equation (9–109) gives the result

$$\ln \frac{T - T_r}{c_2} = \frac{\rho_f c_{p_f} v}{\lambda_m} y \qquad (9\text{–}110)$$

or

$$T - T_r = c_2 \exp\left(\frac{\rho_f c_{p_f} v}{\lambda_m} y\right). \qquad (9\text{–}111)$$

As $y \to \infty$, the right side of this equation approaches zero because v is negative for upflow and $T - T_r$. To evaluate the integration constant c_2, we set $T = T_0$ at the surface $y = 0$ and find

$$c_2 = T_0 - T_r. \qquad (9\text{–}112)$$

The temperature as a function of depth is therefore given by

$$T = T_r - (T_r - T_0) \exp\left(\frac{\rho_f c_{p_f} v}{\lambda_m} y\right). \qquad (9\text{–}113)$$

We now apply this result to measurements of temperature versus depth in the Steamboat Springs, Nevada, hydrothermal system.

Steamboat Springs in southern Washoe County, Nevada, is an area of hot springs and some geysering. Extensive recent volcanics in the immediate vicinity suggest that this thermal system is associated with a solidifying magma body, although there is no direct evidence of the presence of the magma body. Probably the best information on the horizontal extent of the hydrothermal system comes from measurements of the groundwater temperature at the water table. A contour map of this temperature for the Steamboat Springs area is given in Figure 9–18. A well-defined thermal anomaly exists with a horizontal extent of 5 to 10 km. The dark areas are regions where boiling hot springs occur.

The bottom temperatures logged during the drilling of a well adjacent to the hot springs are shown in Figure 9–19. This well was located about 60 m from the nearest boiling spring. Also shown in Figure 9–19 is the temperature profile given by Equation (9–113). In making this comparison, we have taken $T_0 = 10°C$, $T_r = 165°C$, $\lambda_m = 3.35 \text{ W m}^{-1}\text{ K}^{-1}$, $\rho_f = 1000 \text{ kg m}^{-3}$, $c_{p_f} = 4.185 \times 10^3 \text{ J kg}^{-1}\text{ K}^{-1}$, and $v = -6.7 \times 10^{-8} \text{ m s}^{-1}$.

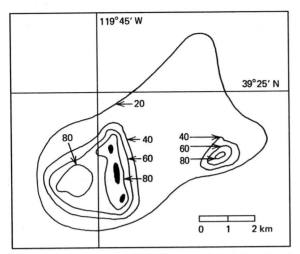

9–18 Groundwater temperature (°C) in the Steamboat Springs, Nevada, hydrothermal system. The solid areas are the regions of boiling hot springs and geysers.

This value of the Darcy velocity gives a predicted near-surface thermal structure that is in good agreement with observations. It is of interest to see how this velocity compares with the observed discharge of the thermal system. The approximately 70 hot springs associated with the Steamboat Springs thermal system discharge water at a rate of about $3.33 \times 10^{-3}\,\mathrm{m^3\,s^{-1}}$. How-

9–19 Temperature as a function of depth in the Steamboat Springs hydrothermal system. The data are from the GS-3 well (White, 1968) and the solid line is from Equation (9–113).

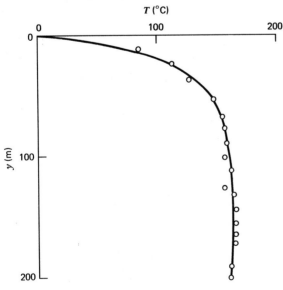

ever, geochemical studies indicate that a large fraction of the discharge of the system reaches Steamboat Creek directly without being fed through the hot springs. Based on these geochemical studies the total discharge of the system is estimated to be $7.12 \times 10^{-2}\,\mathrm{m^3\,s^{-1}}$. Assuming that our derived Darcy velocity of magnitude $6.7 \times 10^{-8}\,\mathrm{m\,s^{-1}}$ is correct, the total area required to discharge $7.12 \times 10^{-2}\,\mathrm{m^3\,s^{-1}}$ is 1.1 km². This indicates that ascending flow is confined to relatively restricted thermal plumes near the regions of boiling hot springs.

If the flow is driven by the buoyancy of the hot water, we can use this Darcy velocity to estimate the permeability of the system. Recall that the buoyancy force results from the small decrease in density that occurs upon heating

$$\rho_f = \rho_{f_0} - \alpha_f \rho_{f_0}(T_r - T_0), \qquad (9\text{–}114)$$

where ρ_{f_0} is the density of the water at temperature T_0 and α_f is the volume coefficient of thermal expansion of water. Upon substituting Equation (9–114) into Darcy's law, Equation (9–3), we obtain

$$v = -\frac{k}{\mu}\left(\frac{dp}{dy} - \rho_{f_0}g\right) - \frac{k}{\mu}\alpha_f \rho_{f_0}g(T_r - T_0). \qquad (9\text{–}115)$$

If we assume that the pressure gradient in excess of the hydrostatic value is negligible in the upwelling flow, we obtain

$$v = -\frac{k}{\mu}\alpha_f \rho_{f_0}g(T_r - T_0). \qquad (9\text{–}116)$$

Taking $v = -6.7 \times 10^{-8}\,\mathrm{m\,s^{-1}}$ as obtained above, $T_r - T_0 = 155\,\mathrm{K}$, $\alpha_f = 10^{-3}\,\mathrm{K^{-1}}$, and $\mu = 1.33 \times 10^{-4}\,\mathrm{Pa\,s}$, we find that the required permeability is $k = 5.75 \times 10^{-15}\,\mathrm{m^2}$, a low value. This calculation shows that the buoyancy of the hot water can easily drive the flow.

It should be emphasized that these calculations are only approximately valid. Several important fault zones are associated with the Steamboat Springs system, so the validity of the porous flow model is questionable. Also, only the ascending flow has been considered. This is only part of the hydrothermal system. In the next section we will consider the complete hydrothermal circulation pattern for convection in a fluid-saturated porous layer heated from below.

PROBLEM 9–20 Calculate the surface heat flux for the steady one-dimensional upwelling of fluid through a porous medium. Assume that temperature is uniform at great depth. How does the surface heat flow compare with the heat carried upward by the fluid at great depth? Use the parameter values given previously to estimate the total rate at which heat is being brought to the surface by the Steamboat Springs geothermal system.

PROBLEM 9–21 Consider the upwelling of a mixture of water and steam in a porous medium. Because of the cold temperatures near the surface, the mixture will reach a level where all the steam must abruptly condense. There will be a phase charge interface with upwelling water just above the boundary and upwelling steam and water just below it. Show that the temperature gradient immediately above the interface $(dT/dy)_2$ is larger than the temperature gradient just below the interface $(dT/dy)_1$ by the amount $-L\rho_s v_s$, where L is the latent heat of the steam–water phase change, ρ_s is the density of the steam, and $-v_s$ is the upwelling Darcy velocity of the steam.

9–9 Thermal Convection in a Porous Layer

In Section 6–19 we considered the onset of thermal convection in a fluid layer heated from below. Here we treat the analogous problem for a layer of fluid-saturated porous material contained between impermeable isothermal boundaries. The upper boundary, $y = 0$, is maintained at temperature T_0, and the lower boundary, $y = b$, is kept at temperature $T_1 (T_1 > T_0)$. The temperature gradient across the layer must exceed a critical value before convection will occur. Prior to the onset of convection the temperature distribution is given by the conduction solution (Equation (6–295))

$$T_c = T_0 + \left(\frac{T_1 - T_0}{b} \right) y. \tag{9–117}$$

At the onset of convection the temperature difference $T' \equiv T - T_c$ is arbitrarily small. The components of the Darcy velocity u', v' are similarly infinitesimal when motion first takes place.

As in Section 6–19 we adopt the Boussinesq approximation and consider the fluid to be incompressible except for the buoyancy term in Darcy's law for the

vertical Darcy velocity component. Thus the energy equation (9–103) can be written in terms of T' according to

$$\rho_m c_{p_m} \frac{\partial T'}{\partial t} + \rho_f c_{p_f} \left(u' \frac{\partial T'}{\partial x} + v' \frac{\partial T'}{\partial y} \right)$$
$$+ \rho_f c_{p_f} v' \frac{(T_1 - T_0)}{b} = \lambda_m \left(\frac{\partial^2 T'}{\partial x^2} + \frac{\partial^2 T'}{\partial y^2} \right). \tag{9–118}$$

Since T', u', and v' are small quantities, the nonlinear terms $u'\partial T'/\partial x$ and $v'\partial T'/\partial y$ on the left side of Equation (9–118) can be neglected. The appropriate forms of Equations (9–1), (9–3), (9–102), and (9–118) for the small perturbations of temperature T', velocity u', v', and pressure p' can be written

$$\frac{\partial u'}{\partial x} + \frac{\partial v'}{\partial y} = 0 \tag{9–119}$$

$$\rho_m c_{p_m} \frac{\partial T'}{\partial t} + \rho_f c_{p_f} v' \frac{(T_1 - T_0)}{b} = \lambda_m \left(\frac{\partial^2 T'}{\partial x^2} + \frac{\partial^2 T'}{\partial y^2} \right) \tag{9–120}$$

$$u' = -\frac{k}{\mu} \frac{\partial p'}{\partial x} \tag{9–121}$$

$$v' = -\frac{k}{\mu} \left(\frac{\partial p'}{\partial y} + \alpha_f \rho_f g T' \right). \tag{9–122}$$

These equations must be solved subject to the boundary conditions $v' = T' = 0$ at $y = 0, b$ because the boundaries are isothermal and impermeable.

As was shown in Section 6–19, the critical condition for the onset of convection can be obtained by setting $\partial/\partial t = 0$. Thus Equation (9–120) can be written

$$\rho_f c_{p_f} v' \frac{(T_1 - T_0)}{b} = \lambda_m \left(\frac{\partial^2 T'}{\partial x^2} + \frac{\partial^2 T'}{\partial y^2} \right). \tag{9–123}$$

The pressure perturbation can be eliminated from these equations by differentiating Equation (9–121) with respect to y and Equation (9–122) with respect to x and subtracting. The result is

$$\frac{\partial u'}{\partial y} - \frac{\partial v'}{\partial x} = \frac{k \alpha_f \rho_f g}{\mu} \frac{\partial T'}{\partial x}. \tag{9–124}$$

We can eliminate u' between Equations (9–119) and (9–124) by the same procedure of cross differentiation and subtraction to get

$$\frac{\partial^2 v'}{\partial x^2} + \frac{\partial^2 v'}{\partial y^2} = \frac{-k \alpha_f \rho_f g}{\mu} \frac{\partial^2 T'}{\partial x^2}. \tag{9–125}$$

A single equation for T' can be found by solving Equation (9–123) for v' and substituting into Equation (9–125) with the result

$$
\frac{\partial^4 T'}{\partial x^4} + 2\frac{\partial^4 T'}{\partial x^2 \partial y^2} + \frac{\partial^4 T'}{\partial y^4}
$$
$$
= \frac{-k\alpha_f \rho_f^2 g c_{p_f}(T_1 - T_0)}{\mu \lambda_m b}\frac{\partial^2 T'}{\partial x^2}. \qquad (9\text{–}126)
$$

The boundary conditions must also be written in terms of T'. Because $T' = 0$ on $y = 0, b$, $\partial^2 T'/\partial x^2$ is also zero on these boundaries. With $v' = 0$ and $\partial^2 T'/\partial x^2 = 0$ on $y = 0, b$, Equation (9–118) gives $\partial^2 T'/\partial y^2 = 0$ on the boundaries. Thus, the complete set of boundary conditions for the fourth-order differential equation for T' is $T' = \partial^2 T'/\partial y^2 = 0$ on $y = 0, b$.

The elementary solution for T' that will satisfy both the differential equation and the boundary conditions is

$$
T' = T_0' \sin\frac{\pi y}{b}\sin\frac{2\pi x}{\lambda}, \qquad (9\text{–}127)
$$

where T_0' is the amplitude of the temperature perturbation and λ is its wavelength. This form of the solution automatically satisfies all the boundary conditions. Its amplitude is indeterminate from a linear analysis, but

its wavelength can be found by substituting Equation (9–127) into Equation (9–126) with the result

$$
\frac{\left\{\left(\frac{2\pi b}{\lambda}\right)^2 + \pi^2\right\}^2}{\left(\frac{2\pi b}{\lambda}\right)^2} = \frac{\alpha_f g \rho_f^2 c_{p_f} k b(T_1 - T_0)}{\mu \lambda_m}. \qquad (9\text{–}128)
$$

The dimensionless combination of parameters on the right side of Equation (9–128) is the appropriate Rayleigh number for thermal convection in a layer of porous material heated from below

$$
\mathrm{Ra} \equiv \frac{\alpha_f g \rho_f^2 c_{p_f} k b(T_1 - T_0)}{\mu \lambda_m}. \qquad (9\text{–}129)
$$

In terms of this Rayleigh number Equation (9–128) becomes

$$
\frac{\left\{\left(\frac{2\pi b}{\lambda}\right)^2 + \pi^2\right\}^2}{\left(\frac{2\pi b}{\lambda}\right)^2} = \mathrm{Ra} = \mathrm{Ra_{cr}}. \qquad (9\text{–}130)
$$

The Rayleigh numbers given in Equation (9–130) are the critical Rayleigh numbers $\mathrm{Ra_{cr}}$ for the onset of convection with wavelength λ; see Equation (6–319). The dependence of $\mathrm{Ra_{cr}}$ on $2\pi b/\lambda$ is given in Figure 9–20. There is a minimum value of $\mathrm{Ra_{cr}}$ which is the lowest value of the Rayleigh number at which convection

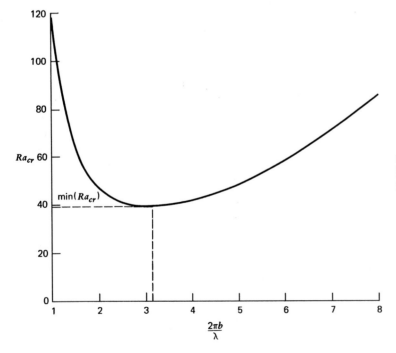

9–20 The Rayleigh number for the onset of convection in a layer of porous material heated from below as a function of the wavelength parameter $2\pi b/\lambda$.

can occur. The value of wavelength corresponding to min (Ra$_{cr}$) is obtained by differentiating the left side of Equation (9–130) with respect to $2\pi b/\lambda$ and setting the result equal to zero. When this is done, we obtain

$$\lambda = 2b. \tag{9–131}$$

The minimum value of Ra$_{cr}$ is found by substituting this result into Equation (9–130)

$$\min(\text{Ra}_{cr}) = 4\pi^2 = 39.4784. \tag{9–132}$$

We can now determine under what conditions thermal convection of groundwater will occur in a uniform permeable layer. Taking Ra $= 4\pi^2$, $\rho_f = 1000$ kg m^{-3}, $\alpha_f = 10^{-3}$ K^{-1}, $\mu = 1.33 \times 10^{-4}$ Pa s, $c_{p_f} = 4.2 \times 10^3$ J kg^{-1} K^{-1}, $\lambda_m = 3.3$ W m^{-1} K^{-1}, and $g = 10$ m s^{-2}, we can rewrite Equation (9–129) as

$$\frac{dT_c}{dy} = \frac{4.2 \times 10^{-10}}{kb^2}, \tag{9–133}$$

where all quantities are in SI units. Equation (9–133) gives the minimum value of the temperature gradient required for convection to occur in a porous layer of permeability k and thickness b. Figure 9–21 shows this relation in the form of a plot of dT_c/dy versus k for several values of b. This figure shows that for a typical geothermal gradient of 20 K km^{-1} and a layer thickness of 1 km a permeability greater than 2.1×10^{-14} m^2 is required for thermal convection. High geothermal gradients and large permeabilities favor the

occurrence of hydrothermal convection. Conditions in geothermal areas on the continents and in the oceanic crust near ocean ridges generally meet the minimum requirements for hydrothermal convection. Extensive fractures or fault zones usually provide the required permeability, and near-surface magma bodies usually provide the high thermal gradients.

PROBLEM 9–22 Determine the minimum critical Rayleigh number for the onset of convection in a layer of porous material heated from below with an isothermal and impermeable lower boundary and an isothermal constant pressure upper boundary. This boundary condition corresponds to a permeable boundary between a saturated porous layer and an overlying fluid. What is the horizontal wavelength that corresponds to the minimum value of Ra$_{cr}$? Take the layer thickness to be b, and let the upper boundary, $y = 0$, have temperature $T = T_0$ and the lower boundary, $y = b$, have temperature $T = T_1$. Assume that at the onset of convection T' has the form

$$T' = T_0' \sin \frac{2\pi X}{\lambda} Y(y) \tag{9–134}$$

and show that $Y(y)$ is a solution of

$$\frac{d^4 Y}{d\bar{y}^4} - 2a^2 \frac{d^2 Y}{d\bar{y}^2} + Y(a^4 - a^2\text{Ra}) = 0, \tag{9–135}$$

where

$$a \equiv \frac{2\pi b}{\lambda} \qquad \bar{y} \equiv \frac{y}{b}. \tag{9–136}$$

Show that the general solution of Equation (9–135)

9-21 The thermal gradient required for hydrothermal convection in a porous layer as a function of the permeability of the layer for several layer thicknesses.

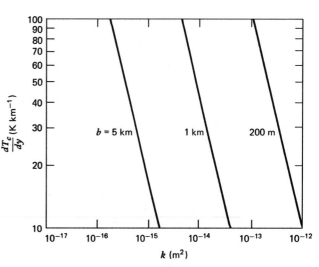

can be written as

$$Y = c_1 e^{\gamma \bar{y}} + c_2 e^{-\gamma \bar{y}} + c_3 \sin \delta \bar{y} + c_4 \cos \delta \bar{y}, \quad (9\text{--}137)$$

where c_1, c_2, c_3, and c_4 are constants of integration and

$$\gamma^2 = a^2 + a\sqrt{\text{Ra}} \quad (9\text{--}138)$$

$$\delta^2 = a\sqrt{\text{Ra}} - a^2. \quad (9\text{--}139)$$

Show that the boundary conditions are

$$Y = 0 \quad \text{on } \bar{y} = 0 \text{ and } 1 \quad (9\text{--}140)$$

$$\frac{d^2 Y}{d\bar{y}^2} = 0 \quad \text{on } \bar{y} = 1 \quad (9\text{--}141)$$

$$\frac{d}{d\bar{y}}\left(\frac{d^2 y}{d\bar{y}^2} - a^2 Y\right) = 0 \quad \text{on } \bar{y} = 0. \quad (9\text{--}142)$$

Substitute Equation (9–137) into each of these boundary conditions to obtain four homogeneous equations for the four unknown constants c_1, c_2, c_3, and c_4. Show that a nontrivial solution of these equations requires

$$\gamma \tan \delta + \delta \tanh \gamma = 0. \quad (9\text{--}143)$$

This transcendental equation is an eigenvalue equation that implicitly gives Ra_{cr} as a function of a, since both γ and δ are defined in terms of Ra and a in Equations (9–138) and (9–139). The critical Rayleigh number can be found by numerically solving Equations (9–138), (9–139), and (9–143). The value of min (Ra_{cr}) turns out to be 27.1. One way of proceeding is to choose a value of a (there exists an Ra_{cr} for each a). Then try a value of δ. Compute γ from $\gamma^2 = 2a^2 + \delta^2$. Then compute $\tan \delta/\delta$ and $-\tanh \gamma/\gamma$. Iterate on δ until these ratios are equal. With δ determined Ra_{cr} follows from Equation (9–139). Repeat the process for different values of a until min (Ra_{cr}) is found.

PROBLEM 9–23 Heat flow measurements as a function of distance from the Galapagos spreading center show an approximate periodic spatial variation with a wavelength of about 7 km. If these data are interpreted in terms of hydrothermal convection in the oceanic crust, what is the approximate depth of circulation? If the Rayleigh number for the convecting system is about 100 and the temperature rise across the layer is several hundred °C, estimate the permeability of the crustal rocks. Oceanic basalts are estimated to have permeabilities of about 10^{-16} m^2.

9–10 Thermal Plumes in Fluid-Saturated Porous Media

If an intrusion is of limited spatial extent, the heated buoyant groundwater in its vicinity rises in the form of a plume. Such a narrow plume resembles smoke rising from a chimney on a calm day. Figure 9–22 illustrates the two-dimensional plume above an intrusion of small cross-sectional area extending infinitely far in the z direction. For our analysis we approximate the intrusion as a line source of heat emitting Q units of energy per unit time and per unit distance in the z direction. The intrusion is embedded in a porous medium of permeability k completely saturated with groundwater. The vertical coordinate $-y$ is measured upward from the intrusion or line source of heat located at the origin of the coordinate system.

The equations governing the structure of the plume are Equations (9–1), (9–3), (9–102), and (9–103). Because the flow is steady, we set $\partial/\partial t = 0$ in Equation (9–103) and obtain

$$\rho_f c_{p_f}\left(u\frac{\partial T}{\partial x} + v\frac{\partial T}{\partial y}\right) = \lambda_m\left(\frac{\partial^2 T}{\partial x^2} + \frac{\partial^2 T}{\partial y^2}\right). \quad (9\text{--}144)$$

As before we adopt the Boussinesq approximation. The

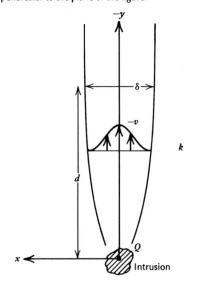

9–22 A two-dimensional plume of hot groundwater rising above an intrusion of small cross-sectional area emitting heat at the rate Q per unit distance perpendicular to the plane of the figure.

Darcy equations can be written

$$u = -\frac{k}{\mu}\frac{\partial p}{\partial x} \tag{9–145}$$

$$v = -\frac{k}{\mu}\left\{\frac{\partial p}{\partial y} + \rho_f \alpha_f g(T - T_0)\right\}, \tag{9–146}$$

where the hydrostatic pressure has been eliminated using Equation (6–284) and ρ_f and T_0 are the constant density and temperature of the ambient groundwater.

These equations can be simplified using a boundary-layer approximation if the plume remains thin as it moves upward. If δ is the width of the plume a distance h above the intrusion, the requirement that the plume be thin is equivalent to $\delta \ll h$. The narrowness of the plume depends on the rate at which the line source emits heat. For Q sufficiently large, the groundwater in the plume will be very buoyant and will rise at a relatively rapid rate. As a consequence, plume material will move upward quite far before it has an opportunity to spread laterally. The result will be a narrow plume. We will be able to state quantitatively just how large the heat source must be for the plume to remain thin because the analysis to follow will yield an equation for plume width as a function of distance above the intrusion.

If the plume is thin, quantities such as velocity and temperature will vary rapidly with distance across the plume compared with how they will vary with distance along the plume. This characteristic allows the problem to be simplified using boundary-layer approximations similar to the ones we have already used in our studies of the structure of the thermal lithosphere or boundary layer in Section 4–16. To determine the approximations appropriate to the present problem, we estimate the relative sizes of the terms in the governing equations. We will then neglect the terms that are demonstrably small. Let U and V represent the magnitudes of the horizontal and vertical Darcy velocities in the plume at a distance h above the origin where the plume thickness is δ. The gradient of u across the plume $\partial u/\partial x$ is approximately

$$\frac{\partial u}{\partial x} \approx \frac{U}{\delta}, \tag{9–147}$$

while the gradient of v along the plume $\partial v/\partial y$ is

approximately

$$\frac{\partial v}{\partial y} \approx \frac{V}{h}. \tag{9–148}$$

The continuity equation (9–102) requires that these two terms balance, which is only possible if

$$\frac{U}{\delta} \approx \frac{V}{h} \quad \text{or} \quad U \approx V\left(\frac{\delta}{h}\right). \tag{9–149}$$

The magnitude of the horizontal velocity in the plume is very small compared with the magnitude of the vertical velocity if $\delta \ll h$.

Darcy's law for the horizontal flow can now be used to relate the magnitude of the flow pressure P to the magnitude of the upward flow. According to Equation (9–145)

$$U \approx \frac{k}{\mu}\frac{P}{\delta}, \tag{9–150}$$

or, with the help of Equation (9–149),

$$P \approx \frac{\mu}{k}U\delta \approx \frac{\mu}{k}\frac{\delta^2}{h}V. \tag{9–151}$$

The pressure gradient term in Darcy's equation for the vertical flow $-(k/\mu)(\partial p/\partial y)$ thus has magnitude $V(\delta^2/h^2)$. The term is much smaller than the vertical flow itself if $\delta \ll h$; that is, the upward motion of the plume is driven by the buoyancy force. The pressure gradient term can thus be neglected in Equation (9–146), which becomes

$$v = -\frac{k\rho_f \alpha_f g}{\mu}(T - T_0). \tag{9–152}$$

The vertical velocity in the plume is directly proportional to the temperature excess.

The terms on the left side of the energy equation are comparable because

$$u\frac{\partial T}{\partial x} \approx \frac{UT}{\delta} \approx \frac{VT}{h} \approx v\frac{\partial T}{\partial y}. \tag{9–153}$$

The terms on the right side of the temperature equation are unequal, however, because

$$\frac{\partial^2 T/\partial y^2}{\partial^2 T/\partial x^2} \approx \frac{T/h^2}{T/\delta^2} = \frac{\delta^2}{h^2} \ll 1. \tag{9–154}$$

Thus heat conduction along the plume can be neglected compared with heat conduction across the plume, and we can write

$$u \frac{\partial T}{\partial x} + v \frac{\partial T}{\partial y} = \frac{\lambda_m}{\rho_f c_{p_f}} \frac{\partial^2 T}{\partial x^2}. \tag{9-155}$$

The boundary-layer equations for the plume structure are Equations (9–102), (9–152), and (9–155). Darcy's law for the horizontal velocity is not required because the effects of the flow pressure are negligible.

As in the case of the thermal plume considered in Section 6–21, the vertical flux of heat at any value of y must be equal to the heat input Q. Because the plume is symmetric about its centerline, this condition can be written

$$Q = -2 \int_0^\infty \rho_f c_{p_f} v (T - T_0) \, dx, \tag{9-156}$$

where the minus sign is required because the velocity v is negative. The symmetry of the plume also requires

$$u = \frac{\partial v}{\partial x} = 0 \qquad \text{at } x = 0. \tag{9-157}$$

At large distances from the plume the ambient temperature is T_0, and the fluid is motionless so that

$$T \to T_0, \qquad v \to 0 \qquad \text{as } x \to \infty. \tag{9-158}$$

It is once again appropriate to introduce the stream function defined in Equations (6–69) and (6–70) to satisfy the conservation of mass equation (9–102). In terms of the stream function, Equation (9–152) can be written

$$T - T_0 = -\frac{\mu}{k \rho_f \alpha_f g} \frac{\partial \psi}{\partial x}. \tag{9-159}$$

Upon substituting Equation (6–69), (6–70), and (9–159) into Equation (9–155), we obtain an equation for ψ:

$$\frac{\partial \psi}{\partial y} \frac{\partial^2 \psi}{\partial x^2} - \frac{\partial \psi}{\partial x} \frac{\partial^2 \psi}{\partial x \partial y} = -\frac{\lambda_m}{\rho_f c_{p_f}} \frac{\partial^3 \psi}{\partial x^3}. \tag{9-160}$$

The integral condition, Equation (9–156), becomes

$$\frac{k \alpha_f g Q}{2 \mu c_{p_f}} = \int_0^\infty \left(\frac{\partial \psi}{\partial x} \right)^2 dx \tag{9-161}$$

and the boundary conditions, (9–157) and (9–158), become

$$\frac{\partial \psi}{\partial y} = \frac{\partial^2 \psi}{\partial x^2} = 0 \qquad \text{at } x = 0 \tag{9-162}$$

$$\frac{\partial \psi}{\partial x} \to 0 \qquad \text{as } x \to \infty. \tag{9-163}$$

We must obtain a solution to Equation (9–160) subject to the conditions given in Equations (9–161) to (9–163). Once again we can use similarity variables. The appropriate variables are

$$f = \left(\frac{\mu c_{p_f}^2 \rho_f}{k \alpha_f g \lambda_m Q y} \right)^{1/3} \psi \tag{9-164}$$

$$\eta = \left(\frac{k \alpha_f g \rho_f^2 c_{p_f} Q}{\mu \lambda_m^2 y^2} \right)^{1/3} x. \tag{9-165}$$

Substitution of these variables into Equations (9–160) to (9–163) yields

$$f \frac{d^2 f}{d\eta^2} + \left(\frac{df}{d\eta} \right)^2 + 3 \frac{d^3 f}{d\eta^3} = 0 \tag{9-166}$$

$$\frac{1}{2} = \int_0^\infty \left(\frac{df}{d\eta} \right)^2 d\eta \tag{9-167}$$

$$f = \frac{d^2 f}{d\eta^2} = 0 \qquad \text{at } \eta = 0 \tag{9-168}$$

$$\frac{df}{d\eta} \to 0 \qquad \text{as } \eta \to \infty. \tag{9-169}$$

Equation (9–166) can be immediately integrated to give

$$f \frac{df}{d\eta} + 3 \frac{d^2 f}{d\eta^2} = 0. \tag{9-170}$$

The constant of integration is zero because the boundary conditions (9–168) require both f and its second derivative to vanish at $\eta = 0$. Another integration of Equation (9–170) gives

$$f^2 + 6 \frac{df}{d\eta} = c_1^2 \tag{9-171}$$

or

$$\frac{df}{c_1^2 - f^2} = \frac{1}{6} d\eta, \tag{9-172}$$

where c_1^2 is the constant of integration. The integral of this equation is

$$f = c_1 \tanh \left(\frac{c_1 \eta}{6} \right), \tag{9-173}$$

where the additional constant of integration is zero, since $f = 0$ at $\eta = 0$. Note that the boundary condition (9–169) is automatically satisfied by this form of f.

The constant of integration c_1 is determined by substituting Equation (9–173) into Equation (9–167) with the result

$$1 = \frac{c_1^4}{18}\int_0^\infty \operatorname{sech}^4\left(\frac{c_1\eta}{6}\right)d\eta = \frac{c_1^3}{3}\int_0^\infty \operatorname{sech}^4 s\, ds$$

$$= \frac{c_1^3}{3}\left[\tanh s - \frac{1}{3}\tanh^3 s\right]_0^\infty = \frac{2}{9}c_1^3 \qquad (9\text{--}174)$$

or

$$c_1 = \left(\frac{9}{2}\right)^{1/3}. \qquad (9\text{--}175)$$

Upon substituting this value of c_1 into Equation (9–173), we obtain

$$f = \left(\frac{9}{2}\right)^{1/3}\tanh\left(\frac{\eta}{48^{1/3}}\right). \qquad (9\text{--}176)$$

By combining Equations (6–70), (9–159), (9–164), (9–165), and (9–176), we find that the vertical Darcy velocity and temperature in a two-dimensional plume of groundwater is

$$v = \frac{-k\rho_f\alpha_f g}{\mu}(T - T_0)$$

$$= \left(\frac{k\alpha_f g Q}{\mu c_{p_f}}\right)^{2/3}\left(\frac{3\rho_f c_{p_f}}{32 y\lambda_m}\right)^{1/3}\operatorname{sech}^2\left(\frac{\eta}{48^{1/3}}\right). \qquad (9\text{--}177)$$

In applying this equation, recall that the upward coordinate is $-y$ and that $y = 0$ at the location of the line source of heat. Thus v is negative, as it should be for an upward flow, and $T - T_0$ is positive, consistent with a hot plume. The velocity and temperature structures are shown in Figure 9–23.

9–23 Velocity and temperature distributions in a two-dimensional plume.

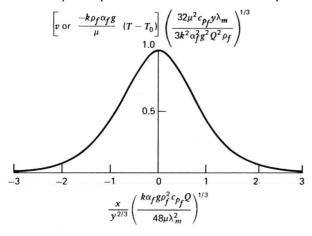

PROBLEM 9–24 Show that the width of the plume δ, defined as the region where $(T - T_0)/(T - T_0)_{\max} > 0.1$, is given by

$$\delta = 13.2\left(\frac{\mu\lambda_m^2 y^2}{k\alpha_f g\rho_f^2 c_{p_f} Q}\right)^{1/3}. \qquad (9\text{--}178)$$

What condition must Q satisfy for the boundary-layer approximations to be valid?

We next consider the axisymmetric plume that rises above a small cooling igneous body approximated by a point source of heat emitting Q units of energy per unit time. The heat source is located at the origin of the coordinate system with $-y$ vertically upward and r the radial distance from the plume centerline, as sketched in Figure 9–24. The boundary-layer approximations also apply to the axisymmetric plume, and we need only modify Equations (9–102), (9–152), and (9–155) for the effects of cylindrical geometry. Equation (9–152), Darcy's law for the vertical flow, does not require any change. The right side of Equation (9–155) gives the result of heat conduction normal to the plume centerline in the two-dimensional case. It needs to be replaced by the appropriate form for heat conduction in the radial direction that appears on the right side of Equation (6–253). In addition, the horizontal advection term $u\,\partial T/\partial x$ on the left side of Equation (9–155) must be replaced by the radial advection term $u_r\,\partial T/\partial r$, where u_r is the radial Darcy velocity. Thus the energy equation

9–24 An axisymmetric plume rising above a small igneous body emitting heat at the rate Q as it cools.

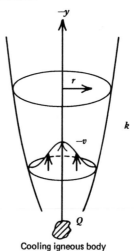

Cooling igneous body

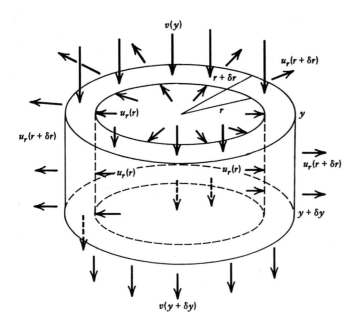

9–25 Flows into and out of an infinitesimal tubular cylindrical volume in a porous medium.

for the axisymmetric plume is

$$u_r \frac{\partial T}{\partial r} + v \frac{\partial T}{\partial y} = \frac{\lambda_m}{\rho_f c_{p_f}} \left(\frac{\partial^2 T}{\partial r^2} + \frac{1}{r} \frac{\partial T}{\partial r} \right). \qquad (9\text{–}179)$$

The appropriate form of the steady incompressible continuity equation in cylindrical geometry can be derived by carrying out a balance of fluid on the tubular cylindrical element in Figure 9–25. The rate of fluid flow into the element is $v(y)$ times the area $2\pi r\,\delta r$ plus $u_r(r)$ times the area $2\pi r\,\delta y$ or

$$v(y)2\pi r\,\delta r + u_r(r)2\pi r\,\delta y.$$

The rate of fluid flow out of the element is

$$v(y + \delta y)2\pi r\,\delta r + u_r(r + \delta r)2\pi(r + \delta r)\,\delta y.$$

These rates must balance for steady incompressible flow resulting in

$$0 = \frac{v(y + \delta y) - v(y)}{\delta y}$$
$$+ \frac{1}{r} \left\{ \frac{(r + \delta r)u_r(r + \delta r) - ru_r(r)}{\delta r} \right\}. \qquad (9\text{–}180)$$

In the limit $\delta y, \delta r \to 0$ Equation (9–180) gives the incompressible continuity equation in cylindrical geometry

$$\frac{1}{r} \frac{\partial}{\partial r}(ru_r) + \frac{\partial v}{\partial y} = 0. \qquad (9\text{–}181)$$

The equations governing the axisymmetric plume are Equations (9–152), (9–179), and (9–181).

The vertical flux of heat at any value of y must be equal to the heat input Q

$$Q = -2\pi \int_0^\infty \rho_f c_{p_f} rv(T - T_0)\,dr. \qquad (9\text{–}182)$$

The minus sign is required because Q and $T - T_0$ are positive whereas v is negative. The symmetry of the plume at the plume centerline requires

$$u_r = \frac{\partial v}{\partial r} = 0 \qquad \text{at } r = 0. \qquad (9\text{–}183)$$

At large distances from the plume the ambient temperature is T_0, and the fluid is motionless so that

$$T \to T_0, \qquad v \to 0 \qquad \text{as } r \to \infty. \qquad (9\text{–}184)$$

We proceed as we did in analyzing the two-dimensional plume by introducing a stream function ψ appropriate to axisymmetric incompressible flow

$$v = \frac{1}{r} \frac{\partial \psi}{\partial r} \qquad (9\text{–}185)$$

$$u_r = -\frac{1}{r} \frac{\partial \psi}{\partial y}. \qquad (9\text{–}186)$$

It can be verified by direct substitution that the continuity equation (9–181) is identically satisfied. The temperature in the plume is related to ψ through Equations

(9–152) and (9–185)

$$T - T_0 = \frac{-\mu}{k\rho_f \alpha_f gr} \frac{\partial \psi}{\partial r}. \tag{9–187}$$

A single equation for ψ is obtained by substituting Equations (9–185) to (9–187) into Equation (9–179)

$$\frac{1}{r^2} \frac{\partial \psi}{\partial y} \frac{\partial \psi}{\partial r} - \frac{1}{r} \frac{\partial \psi}{\partial y} \frac{\partial^2 \psi}{\partial r^2} + \frac{1}{r} \frac{\partial \psi}{\partial r} \frac{\partial^2 \psi}{\partial r \partial y}$$
$$= \frac{\lambda_m}{\rho_f c_{p_f}} \left\{ \frac{1}{r^2} \frac{\partial \psi}{\partial r} - \frac{1}{r} \frac{\partial^2 \psi}{\partial r^2} + \frac{\partial^3 \psi}{\partial r^3} \right\}. \tag{9–188}$$

The integral condition, Equation (9–182), becomes

$$Q = \frac{2\pi c_{p_f} \mu}{k\alpha_f g} \int_0^\infty \frac{1}{r} \left(\frac{\partial \psi}{\partial r} \right)^2 dr \tag{9–189}$$

and the boundary conditions, Equations (9–183) and (9–184), become

$$-\frac{1}{r} \frac{\partial \psi}{\partial y} \to 0, \qquad \frac{\partial}{\partial r} \left(\frac{1}{r} \frac{\partial \psi}{\partial r} \right) \to 0 \qquad \text{as } r \to 0 \tag{9–190}$$

$$\frac{1}{r} \frac{\partial \psi}{\partial y} \to 0 \qquad \text{as } r \to \infty. \tag{9–191}$$

We again find a solution by introducing similarity variables; the appropriate variables for the axially symmetric plume are

$$f = \frac{\rho_f c_{p_f} \psi}{\lambda_m y} \tag{9–192}$$

$$\eta = -\left(\frac{kc_{p_f} \alpha_f g Q}{\mu} \right)^{1/2} \frac{\rho_f r}{\lambda_m y}. \tag{9–193}$$

The minus sign is inserted into Equation (9–193) to make η a positive variable. Substitution of these variables into Equations (9–188) to (9–191) gives

$$\eta f \frac{d^2 f}{d\eta^2} - f \frac{df}{d\eta} + \eta \left(\frac{df}{d\eta} \right)^2 = -\frac{df}{d\eta} + \eta \frac{d^2 f}{d\eta^2} - \eta^2 \frac{d^3 f}{d\eta^3} \tag{9–194}$$

$$1 = 2\pi \int_0^\infty \left(\frac{df}{d\eta} \right)^2 \frac{d\eta}{\eta} \tag{9–195}$$

$$\frac{f}{\eta} - \frac{df}{d\eta} \to 0, \qquad \frac{1}{\eta} \frac{d^2 f}{d\eta^2} - \frac{1}{\eta^2} \frac{df}{d\eta} \to 0 \qquad \text{as } \eta \to 0 \tag{9–196}$$

$$\frac{1}{\eta} \frac{df}{d\eta} \to 0 \qquad \text{as } \eta \to \infty. \tag{9–197}$$

Equation (9–194) can be integrated to give

$$\frac{f}{\eta} \frac{df}{d\eta} = \frac{1}{\eta} \frac{df}{d\eta} - \frac{d^2 f}{d\eta^2} + c_1. \tag{9–198}$$

The boundary conditions (9–196) require that the constant of integration c_1 be given by

$$c_1 = \lim_{\eta \to 0} \left(\frac{df}{d\eta} \right)^2. \tag{9–199}$$

However, Equations (9–185), (9–192), and (9–193) show that df/dy is proportional to rv. Because v is finite at $r = 0$, $rv \to 0$ as $r \to 0$ and $df/d\eta \to 0$ as $\eta \to 0$. Thus $c_1 = 0$ and Equation (9–198) becomes

$$\frac{f}{\eta} \frac{df}{d\eta} = \frac{1}{\eta} \frac{df}{d\eta} - \frac{d^2 f}{d\eta^2}. \tag{9–200}$$

A solution of this equation that satisfies the boundary conditions (9–196) and (9–197) is

$$f = \frac{4c_2 \eta^2}{1 + c_2 \eta^2}. \tag{9–201}$$

The constant of integration c_2 is obtained by substituting Equation (9–201) into (9–195) with the result

$$c_2 = \frac{3}{64\pi}. \tag{9–202}$$

Equation (9–201) becomes

$$f = \frac{3}{16\pi} \frac{\eta^2}{\left(1 + \frac{3\eta^2}{64\pi} \right)}. \tag{9–203}$$

The vertical velocity and temperature distributions in the plume are obtained by combining Equations (9–185), (9–187), (9–192), (9–193), and (9–203)

$$v = \frac{-k\rho_f \alpha_f g}{\mu} (T - T_0)$$
$$= \frac{3}{8\pi y} \frac{k\alpha_f g \rho_f Q}{\mu \lambda_m} \left\{ 1 + \frac{3}{64\pi} \frac{r^2}{y^2} \frac{k\alpha_f g \rho_f^2 c_{p_f} Q}{\mu \lambda_m^2} \right\}^{-2}. \tag{9–204}$$

The dimensionless velocity and temperature profiles for the axisymmetric plume are shown in Figure 9–26.

PROBLEM 9–25 Show that the diameter of the axisymmetric plume δ, defined as the region where $(T - T_0)/(T - T_0)_{max} > 0.1$, is given by

$$\delta = \frac{24|y|\lambda_m}{\rho_f} \left(\frac{\mu}{k\alpha_f g c_{p_f} Q} \right)^{1/2} \tag{9–205}$$

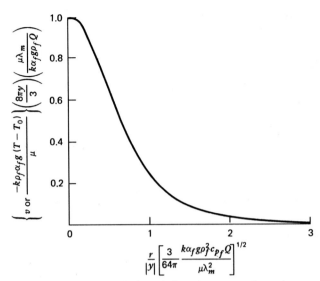

9–26 Profiles of the dimensionless velocity and temperature in an axisymmetric plume.

What condition must Q satisfy for the boundary layer approximations to be valid?

9–11 Porous Flow Model for Magma Migration

A large fraction of the Earth's volcanism occurs beneath ocean ridges. As mantle rock rises beneath a ridge partial melting occurs because of the decrease in pressure (see Figure 1–4). The resulting magma migrates upward through the mantle to form the basaltic oceanic crust. Although pressure-release melting explains why partial melting occurs at depth beneath an ocean ridge, it does not explain how the magma ascends through the mantle to form the overlying oceanic crust.

When partial melting occurs, the first magma produced collects along grain boundary intersections, as illustrated in Figure 9–27. When sufficient melting has occurred, the magma coalesces to form a network of

9–27 The formation of magma at grain intersections during the first stages of partial melting.

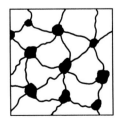

interconnected channels. The channels provide permeability for the migration of magma through the matrix of unmelted crystalline grains. Since the magma is lighter than the remaining crystalline rock, the gravitational body force drives the magma upward along the vertically connected channels.

A porous flow model can be used to quantitatively treat this upward migration of magma. The magma fills the porosity, and the solid crystals make up the matrix. As a reasonable approximation we assume that the microscopic porous flow model given in Section 9–3 and illustrated in Figure 9–1 is applicable to this problem. The differential buoyancy of the magma relative to the solid crystals is the pressure gradient that drives the magma upward, that is,

$$\frac{dp}{dy} = (\rho_s - \rho_l)g, \tag{9–206}$$

where ρ_l is the magma density and ρ_s is the density of the solid. In writing Equation (9–206), we assume that the pressures in the liquid and the matrix are equal. For this to be the case, the matrix must be able to deform and collapse as the magma migrates upward. At the high temperatures associated with partial melting and on the relevant time scales, solid-state creep processes are sufficiently rapid to provide this deformation.

The relative velocity between the magma in the vertically connected channels and the deformable matrix is given by substituting Equations (9–6) and (9–206) into (9–7)

$$v_l - v_s = -\frac{b^2\phi(\rho_s - \rho_l)g}{24\pi\mu}. \tag{9–207}$$

In this equation b is the grain size, v_l is the actual velocity of the magma in the vertically connected channels, v_s is the actual velocity of the solid matrix (upward velocities are negative), ϕ is the volume fraction of magma, and μ is the magma viscosity. Magma in the horizontally connected channels moves upward with the velocity v_s of the solid matrix. The quantities v_l, v_s, and ϕ are functions of depth y in the melt zone. Figure 9–28 shows the magma migration velocity relative to the solid matrix $v_l - v_s$ as a function of the volume fraction of magma for $b = 2$ mm, $\rho_s - \rho_l = 600$ kg m^{-3}, $g = 10$ m s^{-2}, and $\mu = 10$ Pa s. The viscosities of magmas are strong functions of temperature and vary considerably with magma composition. A typical viscosity

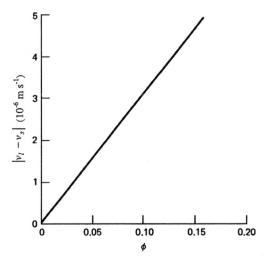

9–28 Magma migration velocity relative to the solid matrix $v_l - v_s$ as a function of the volume fraction of magma ϕ.

for a basaltic magma is 1 Pa s; andesitic magmas have viscosities of about 100 Pa s.

We now apply the magma migration model to the production and ascent of magma beneath an ocean ridge. We assume that mantle rock is rising vertically at a velocity $-v_0$ that is sufficiently large that prior to the onset of melting, heat conduction is negligible and the temperature of the rock is constant. The rock begins to melt when it reaches the depth y_0 at which its temperature profile intersects its melting temperature T_m profile. The dependence of the melting temperature on pressure is given by the slope of the Clapeyron curve

$$\gamma_m = \left(\frac{dp}{dT}\right)_m, \tag{9–208}$$

which is assumed to be constant. Since $dp/dy = \rho_s g$, the melting temperature gradient dT_m/dy is $\rho_s g / \gamma_m$ and T_m as a function of depth is given by

$$T_m = T_{m0} + \frac{\rho_s g}{\gamma_m} y, \tag{9–209}$$

where T_{m0} is the value of the melting temperature at the surface. The melting temperature profile and the isotherm of the ascending unmelted mantle rock are shown in Figure 9–29a for $y_0 = 50$ km, $\gamma_m = 7.5$ MPa K^{-1}, $g = 10$ m s^{-2}, $\rho_s = 3300$ kg m^{-3}, and $T_{m0} = 1400$ K.

9–29 Dependence of (a) temperature T, (b) melt fraction f, (c) upward velocity of the solid matrix $-v_s$, (d) upward velocity of magma in vertically connected channels $-v_l$, and (e) volume fraction of magma ϕ on depth y in the melt zone beneath an ocean ridge.

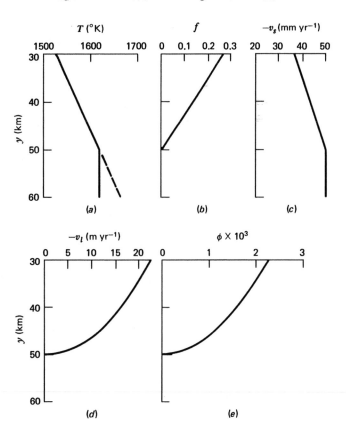

The temperature of the rising mantle rock prior to the onset of melting is 1620 K in this example.

Once melting commences, the temperature profiles of the ascending mantle rock and magma coincide with the melting temperature profile. As the rock and magma move upward, their temperatures decrease along the melting curve, and internal energy is made available to melt an increasing fraction of the rock. We define the melt fraction f to be the ratio of the total upward mass flow rate of magma to the upward mass flux of rock $-\rho_s v_0$ prior to the onset of melting. From Equation (9–8), the upward mass flow rate of magma in the vertically connected channels is $-\phi\rho_l v_l/3$, and the upward mass flow rate of magma in the horizontally connected channels is $-2\phi\rho_l v_s/3$. Thus the melt fraction is given by

$$f \equiv \frac{\phi\rho_l v_l + 2\phi\rho_l v_s}{3\rho_s v_0}. \tag{9–210}$$

If the upward mass flow rate of magma changes by

$$\frac{d}{dy}\left\{\frac{\phi\rho_l(v_l + 2v_s)}{3}\right\}$$

between y and $y - dy$, and the latent heat of fusion is L, an amount of energy

$$L\frac{d}{dy}\left\{\frac{\phi\rho_l(v_l + 2v_s)}{3}\right\}$$

must be extracted from the internal energy of both the rock and magma per unit time and per unit area over this same depth interval. Thus, the appropriate energy balance is

$$(-\rho_s v_0)c_p\frac{dT}{dy} = L\frac{d}{dy}\left\{\frac{\phi\rho_l(v_l + 2v_s)}{3}\right\}, \tag{9–211}$$

where we have assumed that the specific heats of the rock and magma, c_p, are equal. The coefficient of $c_p(dT/dy)$ on the left side of Equation (9–211) is the total upward mass flow rate of rock and magma in the melt zone; conservation of mass requires that this equal the mass flow rate of rock $(-\rho_s v_0)$ prior to the onset of melting. Substitution of Equation (9–210) into (9–211) yields

$$c_p\frac{dT}{dy} = -L\frac{df}{dy}. \tag{9–212}$$

In the melt zone $dT/dy = \rho_s g/\gamma_m$. We use this value of the temperature gradient and integrate Equation (9–212) with the boundary condition $f = 0$ at $y = y_0$ to obtain

$$f = \frac{c_p\rho_s g}{\gamma_m L}(y_0 - y). \tag{9–213}$$

Melt fraction increases linearly with decreasing depth in the melt zone as illustrated in Figure 9–29b for $c_p = 1\,\text{kJ}\,\text{kg}^{-1}\,\text{K}^{-1}$, $L = 320\,\text{kJ}\,\text{kg}^{-1}$, and the other quantities as given before. A 25% melt fraction is produced over a depth range of 18 km.

The velocity v_s can be determined as a function of depth by combining Equation (9–210) with the conservation of mass equation

$$\rho_s v_0 = \frac{\rho_l\phi(v_l + 2v_s)}{3} + \rho_s v_s(1 - \phi). \tag{9–214}$$

If we divide Equation (9–214) by $\rho_s v_0$ and subtract (9–210), we obtain

$$1 - f = \frac{v_s}{v_0}(1 - \phi). \tag{9–215}$$

Since ϕ is generally much smaller than 1 (we will see in an example below that ϕ is smaller than about 10^{-3}), we can neglect ϕ in Equation (9–215) and solve for v_s

$$\begin{aligned}v_s &= v_0(1 - f) \\ &= v_0\left\{1 - \frac{c_p\rho_s g}{\gamma_m L}(y_0 - y)\right\}. \end{aligned} \tag{9–216}$$

At $y = y_0$, v_s equals v_0; as y decreases, the upward velocity of the solid matrix $-v_s$ also decreases. Figure 9–29c shows $-v_s$ as a function of depth for the parameter values already given.

To determine v_l, we eliminate ϕ between Equations (9–207) and (9–210) and find

$$v_l^2 + v_l v_s - 2v_s^2 + \frac{3f\rho_s g(\rho_s - \rho_l)b^2 v_0}{24\pi\mu\rho_l} = 0. \tag{9–217}$$

The solution of this quadratic equation for v_l is

$$|v_l| = \frac{-|v_s|}{2} + \left\{\frac{9v_s^2}{4} + \frac{f\rho_s g(\rho_s - \rho_l)b^2|v_0|}{8\pi\mu\rho_l}\right\}^{1/2}, \tag{9–218}$$

which becomes, upon substituting for f and v_s from

Equations (9–213) and (9–216),

$$|v_l| = \frac{-|v_0|}{2}\left\{1 - \frac{c_p\rho_s g}{\gamma_m L}(y_0 - y)\right\}$$

$$+ \left[\frac{9v_0^2}{4}\left\{1 - \frac{c_p\rho_s g}{\gamma_m L}(y_0 - y)\right\}^2\right.$$

$$\left. + \frac{\rho_s^2 g^2(\rho_s - \rho_l)b^2|v_0|c_p(y_0 - y)}{8\pi\mu\rho_l\gamma_m L}\right]^{1/2}. \quad (9\text{–}219)$$

At $y = y_0$, v_l equals v_0; as y decreases, $|v_l|$ increases. Figure 9–29d shows the depth dependence of the upward velocity of the magma in the vertically connected channels. In addition to parameter values already given, we used $v_0 = -50\,\text{mm yr}^{-1}$, $\rho_l = 2700\,\text{kg m}^{-3}$, $b = 2\,\text{mm}$, and $\mu = 1\,\text{Pa s}$ to calculate v_l. The velocity of the magma in the vertically connected channels is tens of meters per year. Thus the magma flows freely upward along grain boundaries as soon as the melt is produced. Melts from different depths mix to form the magma reaching the surface.

The volume fraction of magma can be found by solving Equation (9–207) for ϕ and substituting for v_l and v_s from Equations (9–216) and (9–218) with the result

$$\phi = \frac{24\pi\mu}{b^2(\rho_s - \rho_l)g}\left[\frac{-3|v_0|}{2}\left\{1 - \frac{c_p\rho_s g}{\gamma_m L}(y_0 - y)\right\}\right.$$

$$+ \left\{\frac{9v_0^2}{4}\left(1 - \frac{c_p\rho_s g}{\gamma_m L}(y_0 - y)\right)^2\right.$$

$$\left.\left. + \frac{c_p\rho_s^2 g^2(y_0 - y)(\rho_s - \rho_l)b^2|v_0|}{8\pi\gamma_m L\mu\rho_l}\right\}^{1/2}\right]. \quad (9\text{–}220)$$

At $y = y_0$, ϕ is zero; as y decreases, ϕ increases. However, as can be seen in Figure 9–29e, for the parameter values given above, ϕ remains less than a few tenths of a percent. Although the volume fraction of magma is small in the melt zone, the magma in the vertically connected channels is rising so fast that the mass flow rate of this magma is a substantial fraction f of the ascending mantle rock.

PROBLEM 9–26 Melting in a layer with a thickness h is caused by a uniform rate of heat generation H per unit mass.

(a) Show that the rate at which magma migrates out of the top of the layer is given by hH/L.

(b) Derive an expression for the volume fraction of magma as a function of depth in the layer in terms of the grain size b, the magma viscosity μ, and the densities ρ_s and ρ_l.

9–12 Two-Phase Convection

If groundwater is heated sufficiently, boiling will occur. Geysers are evidence of boiling at depth. If water and steam coexist, the temperature of the mixture is at the boiling temperature, and the steam is said to be *wet*. If all the water is converted to steam, the temperature may exceed the boiling temperature and the steam is said to be *dry*. Wells of the geothermal power stations at the Geysers north of San Francisco and at Lardarello in Italy discharge dry steam. However, in the main reservoirs of these *vapor-dominated* systems, the steam is wet.

To better understand the behavior of such geothermal reservoirs, let us again consider a horizontal layer of a permeable medium saturated with water. The impermeable upper boundary at $y = 0$ is maintained at a temperature T_0, and the impermeable lower boundary at $y = b$ is maintained at a temperature T_1, $T_1 > T_0$. The stability of this layer with regard to thermal convection was considered in Section 9–9. Here we assume that the Rayleigh number for the layer is less than the minimum critical value of $4\pi^2$ given in Equation (9–132).

As long as the temperature T_1 of the lower boundary is less than the boiling temperature T_b, heat is transferred across the layer by conduction, and the temperature is given by Equation (9–117). When the temperature of the lower boundary reaches the boiling temperature, a two-phase, essentially isothermal zone develops adjacent to the lower boundary, as illustrated in Figure 9–30. In the two-phase zone the light steam rises through the denser water because of buoyancy. Boiling takes place at the lower boundary of the layer, and steam condenses at the boundary between the upper water layer and the lower two-phase zone. The depth of this boundary is y_b. An isothermal region in which heat is transferred by the counterflow of the two phases is known as a *heat pipe*. Measurements of temperature and pressure in drill holes in vapor-dominated geothermal systems show that their structures indeed resemble the model in Figure 9–30; they consist of

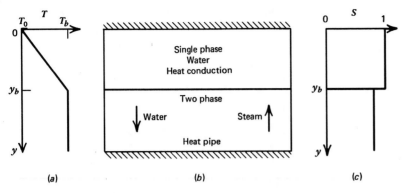

9–30 The heat pipe mechanism (*b*) for the vertical transport of heat in a two-phase fluid layer with the temperature profile (*a*) and the saturation profile (*c*).

near-surface water layers several hundred meters thick overlying the main vapor-dominated two-phase reservoirs.

Some laboratory measurements of heat transport in a counterflowing steam–water layer are given in Figure 9–31. The total thickness of the layer in which these data were acquired is $b = 0.159$ m, its porosity ϕ equals 0.37, its permeability $k = 8.5 \times 10^{-12}$ m^2, and the thermal conductivity of the saturated medium $\lambda_m = 0.92$ W m^{-1} K^{-1}. Figure 9–31*a* shows the temperature profiles in the layer for various values of the heat flux q.

9–31 (*a*) Temperature distribution in a porous layer saturated with water and heated from below for several values of the heat flux (Bau, 1980). (*b*) Measured dependence of the ratio of the depth of the two-phase zone to the layer thickness on the heat flux compared with Equation (9–221).

The isothermal lower zone and the linear temperature profile in the upper conduction-dominated region are clearly illustrated. The depth to the upper boundary of the two-phase zone y_b is easily obtained in terms of the heat flux by considering Fourier's law of heat conduction in the upper layer

$$y_b = \frac{-\lambda_m(T_b - T_0)}{q}, \qquad (9\text{–}221)$$

where it will be recalled that an upward heat flux is negative. Figure 9–31*b* shows that the predicted values of y_b/b are in good agreement with the observations.

We next consider the heat pipe mechanism for the transport of heat in the two-phase zone. Steam is produced at the lower boundary and flows upward at a mass flow rate per unit horizontal area dm_v/dt (upward velocities and mass flow rates are negative); water is condensed at the upper boundary of the two-phase zone and flows downward with a mass flow rate

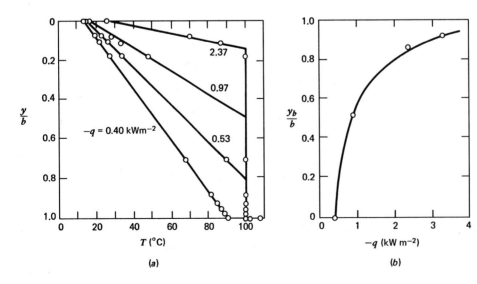

(a)   (b)

dm_w/dt that just balances the upward steam flow. In this steady state one-dimensional counterflow, mass conservation requires

$$\frac{dm_v}{dt} + \frac{dm_w}{dt} = 0. \qquad (9\text{–}222)$$

No condensation or boiling can occur within the two-phase zone because it is isothermal and there is no heat conduction. The enthalpy of the rising steam exceeds that of the descending water by an amount equal to the latent heat of vaporization of water. Thus there is a net upward advection of heat at the rate

$$q = L\frac{dm_v}{dt} = L\rho_v v_v = -L\rho_w v_w = -L\frac{dm_w}{dt}, \qquad (9\text{–}223)$$

where L is the latent heat of vaporization, ρ_v and ρ_w are the densities of steam and water, and v_v and v_w are the Darcy velocities of steam and water. In deriving Equation (9–223), we have used the relations

$$\frac{dm_v}{dt} = \rho_v v_v \qquad \frac{dm_w}{dt} = \rho_w v_w \qquad (9\text{–}224)$$

between the mass flow rates and the Darcy velocities.

For two-phase flow in a porous medium Darcy's law for vertical flow, Equation (9–3), can be modified to give

$$v_v = -\frac{k(1-S)}{\mu_v}\left(\frac{dp}{dy} - \rho_v g\right) \qquad (9\text{–}225)$$

$$v_w = -\frac{kS}{\mu_w}\left(\frac{dp}{dy} - \rho_w g\right), \qquad (9\text{–}226)$$

where μ_v and μ_w are the dynamic viscosities of steam and water and S is the *saturation*, the fraction of the porosity filled with water. The fraction of the porosity filled with steam is $1 - S$. Equations (9–225) and (9–226) can be derived by assuming a parallel model in which the water flows in a fraction S of the horizontal area occupied by the interconnected porosity, and steam flows in the remaining fraction $1 - S$.

The combination of Equations (9–222) to (9–226) yields a formula for the heat flux that is written in dimensionless form as

$$\Gamma = \frac{q\mu_v}{kLg\rho_v(\rho_v - \rho_w)} = \frac{S(1-S)}{\left\{(1-S)\frac{\mu_w\rho_v}{\mu_v\rho_w} + S\right\}}. \qquad (9\text{–}227)$$

The dimensionless heat flux Γ is a function of the

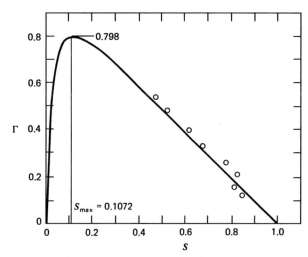

9–32 Dependence of the water saturation S on the dimensionless heat flux Γ from laboratory measurements (circles) and Equation (9–227) (curve).

saturation S. Since S is constant in the two-phase region, see Figure 9–30c, Γ is also constant. The dependence of Γ on S from Equation (9–227) is given in Figure 9–32 for the atmospheric pressure value of the ratio $\mu_v\rho_w/\mu_w\rho_v = 70$. The figure also includes data from the laboratory experiment described above; the agreement between theory and experiment is quite good.

The magnitude of the heat flux across the layer is a maximum $|q|_{max}$ at the value of the saturation S_{max} shown in Figure 9–32. The value of S_{max} can be obtained by differentiating Equation (9–227) and setting $d\Gamma/dS$ equal to zero

$$S_{max} = \frac{\left\{\left(\frac{\mu_w\rho_v}{\rho_w\mu_v}\right)^{1/2} - \frac{\mu_w\rho_v}{\rho_w\mu_v}\right\}}{\left\{1 - \frac{\mu_w\rho_v}{\rho_w\mu_v}\right\}}. \qquad (9\text{–}228)$$

For water and steam with $\mu_v\rho_w/\mu_w\rho_v = 70$, Equation (9–228) gives $S_{max} = 0.1072$. The maximum heat flux Γ_{max} is obtained by substituting Equation (9–228) into Equation (9–227)

$$\Gamma_{max} = \left[1 + \left(\frac{\mu_w\rho_v}{\mu_v\rho_w}\right)^{1/2}\right]^{-2}. \qquad (9\text{–}229)$$

For $\mu_v\rho_w/\mu_w\rho_v = 70$, Γ_{max} is 0.798. If the heat flux exceeds this value, burnout occurs. A large fraction of the layer is saturated with dry steam and heat is transported by the convection of the dry steam and conduction in the matrix. The temperature increases by a large amount in order to transport a small amount of additional heat. The maximum heat transport by the heat pipe mechanism is independent of the layer thickness – see Equations (9–227) and (9–229). With the properties appropriate to the boiling of water at atmospheric pressure $L = 2500$ kJ kg^{-1}, $\rho_w = 1000$ kg m^{-3}, $\rho_v = 0.598$ kg m^{-3}, $\mu_w = 0.284 \times 10^{-3}$ Pa s, $\mu_v = 1.25 \times 10^{-5}$ Pa s, we find that the heat pipe mechanism accommodates a heat flux $q = 0.95$ W m^{-2} if the permeability $k = 10^{-15}$ m^2.

PROBLEM 9–27 Consider a porous layer saturated with water that is at the boiling temperature at all depths. Show that the temperature–depth profile is given by

$$\frac{1}{T_{b0}} - \frac{1}{T} = \frac{R_v}{L} \ln\left(1 + \frac{\rho_l g y}{p_0}\right), \tag{9–230}$$

where T_{b0} is the boiling temperature of water at atmospheric pressure p_0, ρ_l is the density of liquid water which is assumed constant, and R_v is the gas constant for water vapor. Start with the hydrostatic equation for the pressure and derive an equation for dT/dy by using the formula for the slope of the Clapeyron curve between water and steam

$$\frac{dp}{dT} = \frac{L\rho_l\rho_v}{T(\rho_l - \rho_v)} \approx \frac{L\rho_v}{T}, \tag{9–231}$$

where ρ_v is the density of water vapor. Assume that steam is a perfect gas so that

$$\rho_v = \frac{p}{R_v T}. \tag{9–232}$$

Finally, note that $p = p_0 + \rho_l g y$ if ρ_l is assumed constant. What is the temperature at a depth of 1 km? Take $R_v = 0.462$ kJ kg^{-1} K^{-1}, $L = 2500$ kJ kg^{-1}, $T_{b0} = 373$ K, $p_0 = 10^5$ Pa, $\rho_l = 1000$ kg m^{-3}, $g = 10$ m s^{-2}.

PROBLEM 9–28 Calculate pressure as a function of depth in a vapor-dominated geothermal system consisting of a near-surface liquid layer 400 m thick overlying a wet steam reservoir in which the pressure-controlling phase is vapor. Assume that the hydrostatic law is applicable and that the liquid layer is at the boiling temperature throughout. Assume also that the steam reservoir is isothermal.

REFERENCES

Bau, H. H. (1980), Experimental and theoretical studies of natural convection in laboratory-scale models of geothermal systems, Ph.D. Thesis, Cornell University, Ithaca, N.Y.

White, D. E. (1968), Hydrology, activity and heat flow of the Steamboat Springs thermal system, Washoe County, Nevada, U.S. Geological Survey, Professional Paper 458-C.

COLLATERAL READING

Bear, J., *Dynamics of Fluids in Porous Media* (American Elsevier, New York, 1972), 764 pages.

A definitive textbook on the theory of the dynamics of fluids in porous media for advanced undergraduate and graduate students in the fields of groundwater hydrology, soil mechanics, soil physics, drainage and irrigation engineering, sanitary engineering, and petroleum and chemical engineering. A good background in advanced engineering mathematics is required including such subjects as vector analysis, cartesian tensors, and partial differential equations. Chapter 1 is an introductory chapter describing aquifers, groundwater, and oil reservoirs, the porous medium, and the continuum approach to porous media. Chapter 2 discusses fluid and porous matrix properties. In Chapter 3 the concepts of pressure and piezometric head are introduced. Succeeding chapters deal with the fundamental fluid transport equations in porous media, constitutive equations, Darcy's law, hydraulic conductivity, layered media, anisotropic permeability, derivations of Darcy's law, methods of solution of boundary and initial value problems, unconfined flow and the Dupuit approximation, flow of immiscible fluids, hydrodynamic dispersion, and models and analogs including the Hele–Shaw cell. There are problems at the end of each chapter and an appendix with answers.

Domenico, P. A., and F. W. Schwartz, *Physical and Chemical Hydrogeology* (John Wiley, New York, 1990), 824 pages.

This is a comprehensive textbook that covers a broad range of topics concerning both the physical and chemical aspects of hydrogeology. Concepts of porosity and permeability, Darcy's law, aquifers, and the governing equations are introduced. Methods of hydraulic testing, transport of heat and particulate matter, aqueous geochemistry, solution and precipitation, contaminant transport, and remediation are covered.

Freeze, R. A., and J. A. Cherry, *Groundwater* (Prentice Hall, Englewood Cliffs, N.J., 1979), 604 pages.

This undergraduate textbook covers virtually all aspects of the flow of groundwater. Both physical and chemical processes are considered along with engineering applications.

Fyfe, W. S., N. J. Price, and A. B. Thompson, *Fluids in the Earth's Crust* (Elsevier, Amsterdam, 1978), 383 pages.

A textbook for advanced undergraduate and graduate students on the generation and migration of fluids in the crust, their influence on structures, and their collection and concentration into commercially viable reservoirs. Major chapter headings include an introduction to fluid involvement in geochemical and tectonic processes, chemistry of natural fluids, volatiles in minerals, mineral solubility and solution chemistry, rates of metamorphic reactions, release of fluids from rocks during metamorphism, controls of fluid composition, melting, experimental rock deformation, crustal conditions of temperature, pressure, and strain rate, permeability, hydraulic fracture, elasticity, dewatering of the crust, diapirs and diapirism, and fluids, tectonics, and chemical transport.

Goguel, J., *Geothermics* (McGraw-Hill, New York, 1976), 200 pages.

This is an English translation of the author's book "La Géothermie." The book begins with a discussion of the thermal regime near the Earth's surface and how it is affected by groundwater circulation. The problem of extracting usable heat from geothermal resources is then addressed. The operation of wet and dry steam geothermal power plants is described and methods are given to estimate the power output and useful lifetime of a geothermal resource. The final few chapters consider more general problems of interest to students of Earth science: thermal effects of igneous intrusions and extrusions, heat produced by deformation and faulting, thermal stresses in rocks, the source of the heat-producing metamorphism, cooling of the Earth, thermal effects of radioactive decay, and convection in the mantle. Aside from the solutions of a few differential equations the book is nonmathematical in nature.

Muskat, M., *The Flow of Homogeneous Fluids Through Porous Media* (J. W. Edwards, Ann Arbor, Michigan, 1946), 763 pages.

A classic textbook on the theory of flow through porous media. Part 1 is an essentially nonmathematical treatment of the foundations of porous medium flow theory. It includes an extended introduction about the physical situations to which the theory is relevant – groundwater flow and the migration of oil and gas, and chapters on Darcy's law, the measurement of permeability, and the hydrodynamical equations. Part 2 discusses the steady-state flow of liquids stressing two-dimensional problems and potential theory methods; three-dimensional problems, gravity-flow

systems, nonuniform permeability, two-fluid systems, and multiple-well systems are also considered. Part 3 deals with the flow of compressible liquids in porous media and Part 5 treats the flow of gases.

Phillips, O. M., *Flow and Reaction in Permeable Rocks* (Cambridge University Press, Cambridge, 1991), 277 pages.

This is a graduate level textbook on flow through porous media. The governing principles, patterns of flow, reactions, instabilities, and thermal convection are considered.

Polubarinova-Kochina, P. Ya., *Theory of Ground Water Movement* (Princeton University Press, Princeton, New Jersey, 1962), 613 pages.

A classic textbook, translated from the Russian edition, on the mathematical theory of the flow of groundwater. The book is intended primarily for hydraulic engineers and the level of mathematical sophistication is high. There are two major parts on steady and unsteady flows. Major chapters include physical and mathematical fundamentals, two-dimensional flows in a vertical plane, confined seepage under hydraulic structures, the method of inversion and its applications, seepage in heterogeneous and anisotropic soils, natural and manmade wells, three-dimensional problems in seepage, inertia effects in unsteady flows, nonlinear unsteady flow with a free surface, linear equations of unsteady groundwater flow, two-dimensional unsteady flow, and numerical and graphical methods in steady and unsteady flows.

Scheidegger, A. E., *The Physics of Flow through Porous Media* (University of Toronto Press, Toronto, Canada, 1960), 313 pages.

A fundamental textbook on the physical principles of hydrodynamics in porous media. Topics covered are description and characterization of porous materials, physical properties of fluids, equations of fluid flow, hydrostatics in porous media, Darcy's law and solutions of Darcy's equation, physical aspects of permeability, and multiple-phase flow in porous media.

Waring, G. A., Thermal springs of the United States and other countries of the world—A summary, *U. S. Geological Survey Professional Paper, No. 492*, 1965, 383 pages.

An extensive review of the information on the location of the springs, the temperature of the water, flow rate, chemistry, gas content, and practical uses. The facts are tabulated by country or geographical area. The data are accompanied by a brief description of the geology and a map showing the location of the springs. There is a long annotated list of references to the literature on thermal springs.

TEN

Chemical Geodynamics

10-1 Introduction

Radioactive heating of the mantle and crust plays a key role in geodynamics as discussed in Section 4–5. The heat generated by the decay of the uranium isotopes ^{238}U and ^{235}U, the thorium isotope ^{232}Th, and the potassium isotope ^{40}K is the primary source of the energy that drives mantle convection and generates earthquakes and volcanic eruptions. Radiogenic isotopes play other key roles in the Earth sciences. Isotope ratios can be used to date the "ages" of rocks.

The science of dating rocks by radioisotopic techniques is known as *geochronology*. In many cases a rock that solidifies from a melt becomes a closed isotopic system. Measurements of isotope ratios and parent–daughter ratios can be used to determine how long ago the rock solidified from a magma and this defines the age of the rock. These techniques provide the only basis for absolute dating of geological processes. Age dating of meteorites has provided an age of the solar system of 4.55 Ga. The oldest rocks on the Earth were found in West Greenland and have an age of 3.65 Ga. Lunar samples returned by the Apollo missions have ages of over 4 Ga.

Quantitative measurements of the concentrations of radioactive isotopes and their daughter products in rocks form the basis for *chemical geodynamics*. Essentially all rocks found on the surface of the Earth have been through one or more melting episodes and many have experienced high temperature metamorphism. These episodes have clouded the chemistry of the source rocks from which the surface rocks have been derived. For example, the partial melting of mantle rocks concentrates incompatible elements into the resulting magma, but isotope ratios generally remain unaffected. Thus isotope systematics can place quantitative constraints on the evolution of the mantle as well as provide an absolute geological time scale.

Isotope studies of mid-ocean ridge basalts (MORB) show that they are remarkably uniform in their isotopic signatures. This uniformity is evidence that the upper mantle reservoir from which they are extracted is a well-mixed geochemical reservoir. These systematics also show that the reservoir is not pristine, but is systematically depleted in incompatible elements relative to the reference bulk silicate Earth. The relative concentrations of incompatible elements for the bulk silicate Earth are inferred from values found in chondritic meteorites. If the upper-mantle MORB source reservoir is depleted in the incompatible elements, then there must be a complementary enriched reservoir; this is the continental crust. When partial melting of the mantle occurs beneath a mid-ocean ridge the incompatible elements are systematically fractionated into the melt. Thus the basaltic oceanic crust is enriched in the incompatible elements. The complex magmatic processes responsible for the formation of the continental crust further concentrate these incompatible elements. This phenomenon was illustrated by the typical concentrations of the incompatible heat-producing elements given in Table 4–2. We will show that the isotope systematics of MORB can be used to determine the mean age of the continents and to estimate the size of the mantle reservoir from which the continental crust has been extracted.

Isotope studies of ocean island basalts (OIB) show systematic differences from MORB. The OIB, e.g., Hawaii, tend to be enriched in incompatible elements relative to MORB. Because OIB are generally associated with mantle plumes, we can conclude that plumes do not originate from the well-mixed upper mantle reservoir from which MORB is extracted. One hypothesis for the enriched OIB is that plumes sample a near pristine lower mantle reservoir. Mixing between a pristine lower mantle and a depleted upper mantle can explain some, but not all, of the isotope systematics of OIB. A second hypothesis is that the isotope heterogeneities associated with OIB developed in the continental crust and mantle and that plumes contain subducted marine sediments and delaminated continental lithosphere that have not been completely mixed into the mantle reservoir.

Geodynamic constraints can also be obtained from the concentrations of the rare gases helium and argon in the atmosphere and their fluxes out of the crust and mantle. As an example, all ^{40}Ar found in the atmosphere is the result of the decay of ^{40}K within the Earth's interior. Thus a balance can be made between the heat generated from potassium in the mantle and the mass of argon in the atmosphere.

10–2 Radioactivity and Geochronology

Lord Rutherford first pointed out the potential of radioactive isotopes for heating the interior of the Earth. In 1905 he proposed that uranium minerals could be dated by determining the amount of radiogenic helium in them. Boltwood (1907) published the first age determinations of uranite based on uranium–lead (U/Pb) ratios. His dates of 410–535 Ma are generally consistent with modern measurements on these rocks. Early studies of age dating were summarized and the first geological time scale was proposed by Holmes (1913).

Geochronology is based on the decay of a radioactive parent isotope with a mole density j (moles per unit mass) to a radiogenic daughter isotope with a mole density i^* and on a nonradiogenic reference isotope of the daughter element with a mole density i. The isotope ratio α is defined by

$$\alpha = \frac{i^*}{i}, \tag{10–1}$$

and the parent–daughter composition ratio μ is defined by

$$\mu = \frac{j}{i}. \tag{10–2}$$

As a specific example, consider the rubidium–strontium isotope system. The radiogenic parent rubidium isotope is ^{87}Rb, the radiogenic daughter strontium isotope is ^{87}Sr, and the nonradiogenic reference strontium isotope is ^{86}Sr. The concentrations of the radioactive parent isotope j and the radiogenic daughter isotope i^* vary with time t according to the principle of radioactive decay

$$\frac{dj}{dt} = -\lambda j \tag{10–3}$$

$$\frac{di^*}{dt} = \lambda j, \tag{10–4}$$

where λ is the decay constant and time t is measured forward. The concentration of a radioactive parent isotope decreases in time at a rate proportional to the concentration of the parent isotope, while the concentration of the radiogenic daughter isotope increases in time at the same rate. The integrals of Equations (10–3) and (10–4) are

$$j = j_0 e^{-\lambda t} \tag{10–5}$$

$$i^* = i_0^* + j_0(1 - e^{-\lambda t}), \tag{10–6}$$

where subscript zero refers to the concentrations at $t = 0$. The half-life $t_{1/2}$ of a radioactive parent isotope is defined to be the time required for one-half of the atoms present at $t = 0$ to decay. By putting $j = j_0/2$ in Equation (10–5) we obtain

$$0.5 = e^{-\lambda t_{1/2}} \tag{10–7}$$

or

$$t_{1/2} = \frac{\ln 2}{\lambda} = \frac{0.69315}{\lambda}. \tag{10–8}$$

Some of the more widely studied isotopic systems together with their decay constants and half-lives are given in Table 10–1.

The isotope and composition ratios α and μ can be determined using Equations (10–1), (10–2), (10–5) and (10–6). If the initial isotope and composition ratios at time $t = 0$, α_0 and μ_0 respectively, are specified, the subsequent time evolution of a closed system is given by

$$\alpha = \mu_0(1 - e^{-\lambda t}) + \alpha_0 \tag{10–9}$$

$$\mu = \mu_0 e^{-\lambda t}. \tag{10–10}$$

Eliminating the initial composition ratio μ_0 from this pair of equations gives

$$\alpha = \alpha_0 + \mu(e^{\lambda t} - 1). \tag{10–11}$$

This relation may be used to determine the "age" of a rock. The age refers to the time when the relevant elements became "frozen" into the rock. Under ideal conditions no further gain or loss of these elements would occur until the rocks are studied in the laboratory. Thus a measured date may represent the time since the crystallization of the rock or the time since a metamorphic event when the rock was heated to sufficiently high temperatures for chemical changes to occur.

TABLE 10-1 Isotope Systems Commonly Used in Chemical Geodynamics and Their Properties

Isotope System	Radioactive Parent Isotope	Radiogenic Daughter Isotope	Stable Reference Isotope	Decay Constant of Radioactive Parent λ (Gyr^{-1})	Half-Life of Radioactive Parent $\tau_{1/2}$ (Gyr)
Rubidium–Strontium	^{87}Rb	^{87}Sr	^{86}Sr	1.42×10^{-2}	48.8
Samarium–Neodymium	^{147}Sm	^{143}Nd	^{144}Nd	6.54×10^{-3}	106
Uranium–Lead	^{238}U	^{206}Pb	^{204}Pb	1.551×10^{-1}	4.469
Uranium–Lead	^{235}U	^{207}Pb	^{204}Pb	9.848×10^{-1}	0.704

For a number of isotopic systems it is appropriate to assume that $\lambda t \ll 1$. When this approximation is applied to Equations (10–9) and (10–10) we obtain

$$\alpha = \alpha_0 + \lambda t \mu_0 \qquad (10\text{–}12)$$

$$\mu = \mu_0. \qquad (10\text{–}13)$$

From Table 10–1 we see that this will be a good approximation for both the rubidium–strontium and samarium–neodymium systems. As a specific example of age dating, consider a rock that crystallized from a melt at time $t = 0$. We assume that the isotope ratio in the melt α_0 is a constant. The crystallized rock will have a variety of minerals in it. As these minerals form, fractionation of the parent and daughter isotopes occurs. In some minerals the parent isotope is enriched relative to the daughter isotope. In these minerals μ is large and the isotope ratio α becomes progressively larger over time. If α_0 was a constant and if the rock was not subsequently altered chemically, then measurements of α versus μ for different minerals in the rock should lie on a straight line known as the *whole-rock isochron*. The age is proportional to the slope of this line according to Equation (10–12).

Radiometric dating techniques presently in wide use include the decay of ^{87}Rb to ^{87}Sr, ^{147}Sm to ^{143}Nd, ^{40}K to ^{40}Ar, ^{235}U to ^{207}Pb, and ^{238}U to ^{206}Pb. We first consider the Rb–Sr dating method. From Table 10–1 the half-life for the system is 48.8 Gyr, thus the linear approximation given in Equation (10–12) is applicable. Rubidium is an alkali metal that substitutes for potassium in micas and K-feldspar. Strontium is an alkaline earth metal that substitutes for calcium in minerals such as plagioclase and apatite. Naturally occurring

rubidium typically contains 72.2% ^{85}Rb and 28.8% ^{87}Rb. Naturally occurring strontium typically contains 82.5% ^{88}Sr, 7.0% ^{87}Sr, 9.9% ^{86}Sr, and 0.6% ^{84}Sr. Radiogenic ^{87}Rb decays to the stable daughter ^{87}Sr by the emission of a beta particle and an antineutrino.

An example of a Rb–Sr whole-rock isochron is given in Figure 10–1. The rock is an Amitsoq gneiss from West Greenland and it is one of the oldest terrestrial rocks. Present values of isotope ratios α are plotted against present values of parent–daughter ratios μ for several minerals in this rock. The straight line is the best fit

10–1 Rubidium–strontium isochron for the Amitsoq gneiss from the Godthaab district of southwestern Greenland. The isotope ratio α is given as a function of the composition ratio μ for various minerals. (Data from Moorbath et al., 1972). This is one of the oldest terrestrial rocks. The correlation with Equation (10–12) gives an age $\tau = 3.65$ Ga.

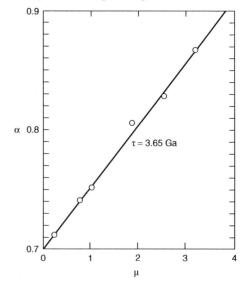

of Equation (10–12) to the data. To obtain this fit, we require that the age of the rock be $\tau = 3.65$ Ga and that the isotope ratio of the rock be $\alpha_0 = 0.70$. (We use the symbols t and unit yr when time is measured forward from the past and the symbols τ and unit a when time is measured backward from the present.)

Radiometric dating of rocks is not always as simple as this example. First, because decay products are isotopes of elements such as strontium, lead, and argon, there is uncertainty in the amount of the daughter element present at $t = 0$. In addition, rocks are not perfectly closed systems; there can be some exchange of both parent and daughter atoms with surrounding material. This is of particular concern when the decay product is a gas such as argon.

PROBLEM 10–1 Show that the mean life of the atoms of a radioactive isotope with decay constant λ is λ^{-1}.

PROBLEM 10–2 Four minerals in a rock were found to have the following $^{87}Sr/^{86}Sr$ and $^{87}Rb/^{86}Sr$ ratios: (1) 0.797 and 12.5, (2) 0.790 and 11.2, (3) 0.764 and 7.0, (4) 0.742 and 4.2. What is the age of the rock?

PROBLEM 10–3 Five minerals in a shale were found to have the following $^{87}Sr/^{86}Sr$ and $^{87}Rb/^{86}Sr$ ratios: (1) 0.784 and 18.0, (2) 0.769 and 14.0, (3) 0.750 and 9.4, (4) 0.733 and 5.7, (5) 0.716 and 2.0. What is the age of the rock?

PROBLEM 10–4 Two minerals, A and B, from a rock have strontium isotope ratios of 0.79 and 0.77 and rubidium–strontium composition ratios of 5.1 and 2.1, respectively. To understand these ratios, assume that the rock underwent a metamorphic alteration at some time after its formation. Assume that during the metamorphism ^{87}Sr was completely mixed but was not lost from the rock. Deduce the original age of the rock and the age of the metamorphic event. Assume that the mineral A is 8% of the rock and that mineral B is 18%. Take the ratio of the number of common ^{87}Sr atoms to the number of ^{86}Sr atoms to be 0.7.

A second important isotope system in chemical geodynamics is the samarium–neodymium system. Samarium and neodymium are rare earth elements that occur in many silicate and carbonate minerals. The radiogenic parent samarium isotope is ^{147}Sm, the radiogenic daughter neodymium is ^{143}Nd, and the non-radiogenic reference neodymium isotope is ^{144}Nd. Radiogenic ^{147}Sm decays to the stable daughter ^{143}Nd by alpha particle emission. The Sm–Nd system can be used for geochronology in exactly the same way as the Rb–Sr system. The primary advantage of both systems is that the parent and daughter elements are solids and therefore relatively secure against gain or loss during the life of the rock. The main disadvantages are the low concentrations of the elements and the relatively long half-lives, which make it difficult to date young rocks.

Another important isotope system for geochronology and chemical geodynamics is the uranium–lead system. The two principal isotopes of uranium are ^{238}U and ^{235}U with concentrations of 99.27% and 0.72%, respectively. Both isotopes are radiogenic and decay to lead isotopes through the emission of α and β particles: ^{238}U decays to ^{206}Pb and ^{235}Ur decays to ^{207}Pb. Other lead isotopes are ^{204}Pb which is not a radiogenic product and ^{208}Pb which results from the decay of the radiogenic isotope of thorium ^{232}Th. The uranium–lead system has the advantage that two different uranium isotopes decay to two different lead isotopes with different decay constants (Table 10–1). The system has the disadvantage that the elements are active chemically and the parent and daughter elements behave quite differently. In particular, uranium is very soluble in water under oxidizing conditions.

For the uranium–lead isotopic system, the radioactive parent isotopes ^{238}U and ^{235}U have mole densities j and j'. The radioactive daughter isotopes ^{206}Pb and ^{207}Pb have mole densities i^* and $i^{*\prime}$. The nonradiogenic reference isotope is ^{204}Pb with a mole density i. The decay constants are denoted by λ and λ'.

The uranium–lead system is commonly discussed in terms of the composition ratios

$$\mu = \frac{j}{i} \qquad \nu = \frac{j'}{j}, \tag{10–14}$$

where μ is the uranium–lead ratio and ν is the ratio of the uranium parent isotopes, which is a universal constant for the Earth. In addition, the lead isotope ratios

$$\alpha = \frac{i^*}{i} \qquad \beta = \frac{i^{*\prime}}{i} \tag{10–15}$$

are introduced. The time evolution of a closed system

is given by

$$\alpha = \mu_0(1 - e^{-\lambda t}) + \alpha_0 \qquad (10\text{–}16)$$

$$\beta = \nu_0\mu_0(1 - e^{-\lambda' t}) + \beta_0 \qquad (10\text{–}17)$$

$$\mu = \mu_0 e^{-\lambda t} \qquad (10\text{–}18)$$

$$\nu = \nu_0 e^{(\lambda - \lambda')t}, \qquad (10\text{–}19)$$

where α_0 and β_0 are initial values of the isotope ratios and μ_0 and ν_0 are initial values of the composition ratios. It is not appropriate to use the linear approximation of the exponentials for the decay of uranium isotopes because of the relatively large values of the decay constants (Table 10–1).

Initial values for the lead isotope ratios at the time the Earth formed have been obtained from studies of iron meteorites. Virtually no uranium or thorium is present in these meteorites so that the measured values are taken to be primordial; these are $\alpha_0 = 9.307$ and $\beta_0 = 10.294$. Initial values of the composition ratios μ_0 and ν_0 at the time the Earth formed are related to the composition ratios of the present bulk silicate Earth μ_{sp} and ν_{sp} (assumed to have evolved as a closed system) by

$$\mu_{sp} = \mu_0 e^{-\lambda \tau_e} \qquad (10\text{–}20)$$

$$\nu_{sp} = \nu_0 e^{(\lambda - \lambda')\tau_e} = \frac{1}{137.8}, \qquad (10\text{–}21)$$

where τ_e is the age of the Earth. (Note that subscript p is used to indicate present values.) The present uranium isotope ratio is $\nu_{sp} = j'_{sp}/j_{sp} = 1/137.8$ ($\nu_0 = 1/3.16$). The present lead isotope ratios for the bulk silicate Earth are

$$\alpha_{sp} = \mu_{sp}(e^{\lambda \tau_e} - 1) + 9.307 \qquad (10\text{–}22)$$

$$\beta_{sp} = \frac{\mu_{sp}}{137.8}(e^{\lambda' \tau_e} - 1) + 10.294. \qquad (10\text{–}23)$$

This is known as the Holmes–Houtermans model based on independent derivations by Holmes (1946) and Houtermans (1946).

Elimination of μ_{sp} from Equations (10–22) and (10–23) gives

$$\frac{\beta_{sp} - 10.294}{\alpha_{sp} - 9.307} = \frac{1}{137.8}\left(\frac{e^{\lambda' \tau_e} - 1}{e^{\lambda \tau_e} - 1}\right). \qquad (10\text{–}24)$$

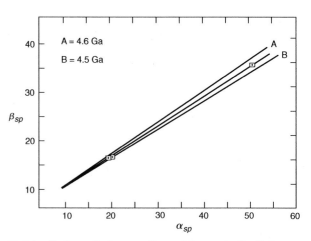

10–2 Lead isotope ratios for several iron and stony meteorites (Patterson, 1956). The ^{207}Pb ratios β_{sp} are given as a function of the ^{206}Pb ratios α_{sp}. Three isochrons are given from Equation (10–24); the best agreement is with an age $\tau_e = 4.55$ Ga, this is the age of the solar system.

The dependence of β_{sp} on α_{sp} defines an *isochron* for the bulk silicate Earth. Meteorites are a source of minerals that follow an isochron. The meteorites have been isolated closed systems since early in the evolution of the solar system. Some meteorites have considerably higher uranium–lead composition ratios μ than other meteorites. The high μ meteorites generate more radiogenic lead (^{206}Pb, ^{207}Pb) and thus have higher isotope ratios α_{sp} and β_{sp}. Data for α_{sp} and β_{sp} from several meteorites are given in Figure 10–2. The data correlate with the isochron given by Equation (10–24) taking the age $\tau_e = 4.55$ Ga, which is believed to be the age of the solar system.

The time evolutions of the isotope ratios in the bulk silicate Earth are given by

$$\alpha_s = \mu_s(e^{\lambda t} - 1) + 9.307 \qquad (10\text{–}25)$$

$$\beta_s = \frac{\mu_s}{137.8}(e^{\lambda' t} - 1) + 10.294 \qquad (10\text{–}26)$$

$$\mu_s = \mu_{sp} e^{\lambda(\tau_e - t)}, \qquad (10\text{–}27)$$

where α_s and β_s are the lead isotope ratios at a time t after the formation of the Earth and μ_s is the uranium–lead ratio at a time t after the formation of the Earth. The evolution of primeval lead in a closed system is illustrated in Figure 10–3. The isotope ratios β_s and α_s follow the growth curves given by Equations (10–25), (10–26), and (10–27); results are given for present

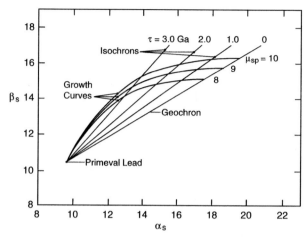

10-3 Growth curves for the lead isotope ratios in a closed system of primeval lead, α_s and β_s, as given by Equations (10–25), (10–26), and (10–27). Results are given for present uranium-lead ratios $\mu_{sp} = 8$, 9, and 10. The present values of α_s and β_s, α_{sp} and β_{sp}, for all values of the present uranium-lead ratio μ_{sp} define the geochron. The values of α_s and β_s at a time τ in the past for various values of μ_{sp} define an isochron. Isochrons are given for $\tau = 1, 2, 3$ Ga.

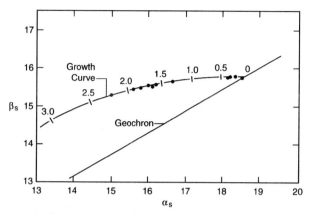

10-4 Lead isotope ratios α_s and β_s are given by the solid circles for several conformable lead deposits (Kanasewich, 1968). The data points correlate well with the growth curve from Equations (10–25), (10–26), and (10–27) taking $\mu_{sp} = 9$. The numbers on the growth curve represent ages τ in Ga. The oldest lead deposit has an age of about 2.2 Ga. The geochron is also shown.

uranium–lead ratios $\mu_{sp} = 8, 9, 10$. The values of β_s and α_s at the present time, β_{sp} and α_{sp}, for various values of μ_{sp} define the *geochron*. The intersections of the geochron with the growth curves give the values of β_{sp} and α_{sp} for a closed system with the three values $\mu_{sp} = 8, 9, 10$. The values of β_s and α_s at a time τ in the past for various values of μ_{sp} define an isochron; isochrons are given in Figure 10–3 for primeval lead with $\tau = 1, 2, 3$ Ga.

For lead to lie on a growth curve it must have been a closed system for the age of the Earth. Some galena (PbS) deposits that satisfy this condition are associated with sediments and volcanics in greenstone belts and island arcs that were conformable with the host rocks (in contrast to cross-cutting veins). Thus they are known as *conformable* lead deposits.

Lead isotope data for several conformable lead deposits are given in Figure 10–4. The β_s–α_s data in Figure 10–4 correlate well with the growth curve corresponding to $\mu_{sp} = 9$. It should be emphasized that conformable lead deposits are those that lie on a growth curve. Most lead deposits appear to have been derived from source regions that have generated excess radiogenic lead. This is also the case for the lead isotopes in both MORB and OIB as will be shown.

10-3 Geochemical Reservoirs

The Earth is subdivided into well-defined physical units. At the center is the solid inner core surrounded by the liquid outer core. The core is primarily iron and the core's formation occurred very early in the evolution of the Earth. The largest unit in the Earth is the mantle. In some cases it is appropriate to divide the mantle into two units, the upper mantle and the lower mantle. The near-surface layer on the Earth is divided into two units, the oceanic crust and the continental crust. We finally have the oceans and the atmosphere. In terms of geochemistry we define each of these units to be a geochemical reservoir. Some of these reservoirs are quite homogeneous chemically whereas others are quite heterogeneous. The chemical composition of the atmosphere is homogenous whereas the chemical composition of the continental crust is quite heterogeneous.

In terms of chemical geodynamics, we utilize a box model (Figure 10–5) in which each of these units is treated as a geochemical reservoir. The focus of our attention will be the geochemical cycle associated with plate tectonics and mantle convection. The primary processes are the generation of the oceanic crust at ocean ridges and its elimination at subduction zones, the generation of continental crust at subduction zones, the creation of oceanic and continental crust at hotspots, and the loss of continental crust by delamination

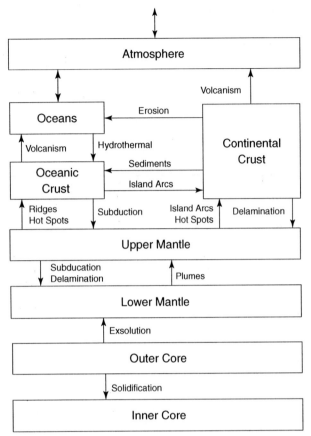

10–5 Schematic diagram of the geochemical reservoirs and interactions involved in the chemical geodynamic behavior of the Earth.

and sediment subduction. There are also important interactions with the core, oceans, and atmosphere.

The creation of the oceanic crust at mid-ocean ridges leads to the strong concentration of incompatible elements into the basaltic oceanic crust from the upper mantle through the partial melting process. Gases and fluids generated by this volcanism transfer incompatible and volatile elements to the oceans and atmosphere. Hydrothermal circulations also exchange material between the oceanic crust and the oceans. The oceanic crust is coated with sediments that are primarily derived from the continents.

At ocean trenches the altered oceanic crust is cycled back into the Earth's interior. Along with the descending lithospheric plate, some continental material is recycled into the mantle at subduction zones; this material includes chemically altered oceanic crust and entrained sediments. At a depth of about 100 km, the upper part of the oceanic crust melts. Partial melting also

occurs in the overlying mantle wedge. These processes further concentrate the incompatible elements. The result is island-arc volcanism. This volcanism along with continental flood basalts and hotspot volcanism forms new continental crust. However, all of these sources have compositions that are considerably more mafic (basaltic) than the present silicic composition of the continental crust. Further differentiation of the continental crust is attributed to remelting events and delamination of the mafic lower continental crust. The magmas from the mantle associated with subduction zone volcanics, flood basalts, and hotspot volcanism intrude the continental crust and in the presence of water produce silicic (granitic) magmas. These magmas rise into the upper crust making the upper crust more silicic and the lower crust more mafic. Subsequently the mafic dense rocks of the lower crust are returned to the mantle by delamination. The net result is that continental crust becomes more silicic and it becomes a reservoir for the incompatible elements, including the radiogenic elements U, Th, and K and the light rare Earth elements.

Although the continental crust is small in volume, its enrichment in incompatible elements is so large that it constitutes a significant global reservoir for these elements. The atmosphere constitutes an important reservoir for the radiogenic gases helium and argon. While the oceanic crust plays a critical role in chemical geodynamics, its volume is so small that it can be neglected in isotopic mass balances.

Isotopic studies of basalts provide important insights into the mantle reservoir or reservoirs from which they are derived. Basalts formed at mid-ocean ridges (MORB) and at oceanic islands (OIB) generally have the same major element composition. Both types are generated by pressure-release partial melting of mantle rock. But when considered in terms of trace element and isotopic compositions, the MORB and OIB can be quite different. Isotope ratios and the distributions of rare earth elements show that normal mid-ocean ridge basalts (n-MORB) are nearly uniformly depleted in incompatible elements. A normal MORB is defined in this context as the oceanic crust of that portion of the ridge system with bathymetric depths greater than 2 km. This excludes, for example, sections near Iceland and the Azores. The uniformity of n-MORB is evidence that the upper mantle from which it is derived is nearly homogeneous as discussed in Section 6–24. The depletion of the n-MORB source region is complementary to the enrichment of the continental crust. Thus, it is necessary that the rocks from which the continental crust has been extracted be uniformly mixed back into the upper mantle reservoir.

Although n-MORB is remarkably uniform, other mantle-generated basalts are more heterogeneous. Shallow parts of the mid-ocean ridge system generally have enriched mid-ocean ridge basalts (e-MORB) and OIB have a wide variety of isotopic signatures. These signatures range from depleted n-MORB values, to bulk silicate Earth values, to enriched continental crust values. Therefore, the mantle cannot be a single homogeneous chemical reservoir.

Some e-MORB and OIB appear to lie on a mixing line between n-MORB and a primordial bulk-Earth reservoir. For such a chemical reservoir to have survived over some 4.5 Gyr it must have remained essentially isolated over this period. One hypothesis that explains the mixing line is a two-layer mantle. Within the lower mantle there is an isolated region with primordial mantle rock. The primordial lower mantle region is separated from the depleted upper mantle by a compositional boundary. The density differences associated with this compositional boundary prevent subducted and delaminated lithosphere from entering the lower mantle. When plumes rise from the thermal boundary layer above this compositional boundary, they entrain primordial lower mantle material. Thus the primordial lower mantle region is decreasing in size with time. The mixing of primordial lower mantle rock with depleted upper mantle rock can explain the principal isotopic characteristics of the basalts generated by the major Hawaiian and Icelandic plumes. An alternative explanation for these geochemical characteristics is that their source regions are "blobs" of primitive mantle scattered throughout the lower mantle. Mantle plumes sample these blobs but mid-ocean ridge volcanism does not.

Although the mixing of primitive and depleted mantle reservoirs can explain some of the geochemical characteristics of e-MORB and OIB, other characteristics require an alternative explanation. The anomalous isotopic signatures of OIBs from ocean islands such as Tristan, Gough, Kerguelen, St. Helena, Azores, and the Society Islands can be attributed to the presence in the depleted mantle reservoir of incompletely homogenized subducted oceanic crust and entrained sediments

and delaminated continental crust and lithosphere. The observed isotopic anomalies must have developed in old continental crust and lithosphere because chemical heterogeneities would not have persisted for the required length of time in the convecting mantle.

Because of the kinematics of plate tectonics, the mid-ocean system randomly migrates over the upper mantle. The position of ridges is specified by the symmetry of the seafloor spreading process. The ascending flow beneath an ocean ridge is generally a passive process and the volcanism that generates n-MORB randomly samples the upper mantle reservoir. The near uniformity of n-MORB in terms of rare earth distributions and isotope ratios is evidence that the upper mantle is a nearly uniform depleted reservoir. Exceptions are ridge segments that interact with mantle plumes, for example Iceland and the Azores. In these regions the ridges are anomalously shallow and e-MORB is produced.

Now let us briefly consider the core reservoir. In Figure 10–5 the core reservoir is divided into two parts in accordance with our knowledge of the core structure. Cooling of the Earth through geologic time has resulted in the growth of a solid inner core. In this process, light alloying elements such as silicon are concentrated into the liquid outer core which becomes progressively richer in the light elements with time. Of particular importance are the possible modes of interaction of the lower mantle and outer core including chemical reactions at the core–mantle boundary, exsolution of light elements from the outer core into the lower mantle, and dissolution of heavy elements from the lower mantle into the outer core. While there is much speculation about core–mantle mass exchange, there is no strong evidence for significant transport between these reservoirs so in the following discussions we will assume such transport is negligible and neglect the core reservoir.

10–4 A Two-Reservoir Model with Instantaneous Crustal Differentiation

The simplest model for the geochemical evolution of the mantle and the continental crust is a two-reservoir model consisting of the continental crust with a mass M_c that was instantaneously separated from a mantle reservoir at a time τ_c before the present (τ_c is the age of crustal separation), the mass of the complementary mantle reservoir is M_m. This complementary mantle reservoir may be the entire mantle, in which case M_m is the mass of the mantle, or there may be an isolated pristine mantle reservoir in the deep mantle, in this case M_m is less than the mass of the mantle. The mass of the mantle reservoir and the time of crustal separation can be constrained by the measured values of the isotope ratios and composition ratios.

When considering reservoir models it is standard practice to express isotope ratios in terms of a normalized isotope ratio defined by

$$\varepsilon = \left(\frac{\alpha}{\alpha_s} - 1\right) \times 10^4 = \left[\left(\frac{i^*/i}{i_s^*/i_s}\right) - 1\right] \times 10^4,$$

$$(10\text{--}28)$$

where the subscript s refers to the bulk silicate Earth, and to express concentration ratios in terms of a fractionation factor defined by

$$f = \frac{\mu}{\mu_s} - 1 = \left(\frac{j/i}{j_s/i_s}\right) - 1.$$

$$(10\text{--}29)$$

If the parent isotope j is more incompatible than the daughter isotope i, then $f > 0$ in the enriched reservoir and $f < 0$ in the depleted reservoir. If the parent isotope j is less incompatible than the daughter isotope i then $f < 0$ in the enriched reservoir and $f > 0$ in the depleted reservoir. If $f > 0$ then the isotope ratio becomes more radiogenic and $\varepsilon > 0$; if $f < 0$ then the isotope ratio is less radiogenic and $\varepsilon < 0$. For the rubidium–strontium system the parent isotope ^{87}Rb is more incompatible than the daughter isotope ^{87}Sr and $f > 0$ and $\varepsilon > 0$ in the enriched crustal reservoir and $f < 0$ and $\varepsilon < 0$ in the depleted mantle reservoir. For the samarium–neodymium system the parent isotope ^{147}Sm is less incompatible than the daughter isotope ^{143}Nd and $f < 0$ and $\varepsilon < 0$ in the enriched crustal reservoir and $f > 0$ and $\varepsilon > 0$ in the depleted mantle reservoir.

To compare model results with observed values of isotopic ratios and fractionation factors, we need to develop expressions for the model quantities f_c, f_m, ε_c, and ε_m, where subscripts c and m refer to the crustal and mantle reservoirs, respectively. From the definitions of ε and f in Equations (10–28) and (10–29) it is clear that we need to derive formulas for $i_c^*(t)$, $i_m^*(t)$, $i_s^*(t)$, $j_c(t)$, $j_m(t)$, and $j_s(t)$. This can be done using Equations (10–5) and (10–6) provided we are careful to identify the amounts of the parent and daughter isotopes in

the initial state of each reservoir and to measure time forward from the instant of creation of each reservoir. Prior to crustal separation, all material evolves as bulk silicate Earth with an initial time $t = 0$ and the initial amounts of the parent and daughter isotopes are j_{s0} and i_{s0}^*.

For both the rubidium–strontium system and the samarium–neodymium system, it is a good approximation to assume that $\lambda t \ll 1$. With this approximation the isotope and composition ratios for the bulk silicate Earth from Equations (10–12) and (10–13) are given by

$$\mu_s = \mu_0 \tag{10--30}$$

$$\alpha_s = \alpha_0 + \lambda t \mu_0. \tag{10--31}$$

The composition ratio μ_s can be taken to be constant and the isotope ratio α_s increases linearly in time.

We assume that the continental crust is instantaneously removed from the mantle source reservoir at $t = \tau_e - \tau_c$. Subsequent to this separation the two reservoirs – the enriched continental crust and the depleted mantle – evolve as closed geochemical systems. Although the continental crustal reservoir is enriched in both the parent and daughter isotopes, the enrichment factors for the two differ. For the crustal and mantle source reservoirs, the initial time is $\tau_e - \tau_c$ (τ_e is the age of the Earth) and the initial amounts of the parent and daughter isotopes are written $\langle j_c \rangle$, $\langle j_m \rangle$, $\langle i_c^* \rangle$, and $\langle i_m^* \rangle$, where the brackets indicate that a quantity is evaluated at $t = \tau_e - \tau_c$. The initial values of $\langle j_c \rangle$ and $\langle i_c^* \rangle$ in the crustal reservoir are represented by enrichment factors D_{sj} and D_{si} relative to the bulk silicate Earth at the time of separation

$$D_{si} \equiv \frac{\langle i_c^* \rangle}{\langle i_s^* \rangle} = \frac{\langle i_c \rangle}{\langle i_{s0} \rangle} \tag{10--32}$$

$$D_{sj} \equiv \frac{\langle j_c \rangle}{\langle j_s \rangle}. \tag{10--33}$$

In writing Equation (10–32) we assumed that the radiogenic and nonradiogenic daughter isotopes i and i^* have the same enrichment factors. If the parent isotope is more incompatible than the daughter isotope then $D_{sj} > D_{si}$, if the daughter isotope is more incompatible than the parent isotope, then $D_{si} > D_{sj}$. The initial values of i^* and j in the mantle reservoir at $t = \tau_e - \tau_c$, $\langle i_m^* \rangle$ and $\langle j_m \rangle$, can be related to D_{si} and D_{sj} by the application of mass conservation for the parent radionuclide and daughter species at the time of crustal separation:

$$\langle i_s^* \rangle (M_c + M_m) = \langle i_c^* \rangle M_c + \langle i_m^* \rangle M_m \qquad (10\text{–}34)$$

$$\langle j_s \rangle (M_c + M_m) = \langle j_c \rangle M_c + \langle j_m \rangle M_m. \qquad (10\text{–}35)$$

A similar equation applies to the nonradiogenic isotope. Equations (10–34) and (10–35) can be rearranged to give

$$\frac{\langle i_m^* \rangle}{\langle i_s^* \rangle} = \frac{\langle i_m \rangle}{\langle i_{s0} \rangle} = 1 - \frac{M_c}{M_m}(D_{si} - 1) \qquad (10\text{–}36)$$

$$\frac{\langle j_m \rangle}{\langle j_s \rangle} = 1 - \frac{M_c}{M_m}(D_{sj} - 1), \qquad (10\text{–}37)$$

where, consistent with Equation (10–32), we have assumed the equality of the enrichment factors of the radiogenic and nonradiogenic daughter isotopes in the mantle source reservoir at the time of crustal formation.

After crustal separation the isotope and composition ratios for the two reservoirs are obtained using Equations (10–1), (10–2), (10–30) to (10–33), (10–36), and (10–37) with the result

$$\mu_c = \mu_0 \left(\frac{D_{sj}}{D_{si}} \right) \qquad (10\text{–}38)$$

$$\mu_m = \mu_0 \left[\frac{1 - \frac{M_c}{M_m}(D_{sj} - 1)}{1 - \frac{M_c}{M_m}(D_{si} - 1)} \right] \qquad (10\text{–}39)$$

$$\alpha_c = \alpha_0 + \lambda(\tau_e - \tau_c)\mu_0 + \lambda(\tau_c - \tau)\mu_c \qquad (10\text{–}40)$$

$$\alpha_m = \alpha_0 + \lambda(\tau_e - \tau_c)\mu_0 + \lambda(\tau_c - \tau)\mu_m. \qquad (10\text{–}41)$$

Using Equations (10–28), (10–29), and (10–38) to (10–41) the fractionation factors and normalized isotope ratios for the continental crustal and depleted mantle reservoirs are given by

$$f_c = \frac{D_{sj}}{D_{si}} - 1 \qquad (10\text{–}42)$$

$$f_m = \left[\frac{1 - \frac{M_c}{M_m}\{D_{sj} - 1\}}{1 - \frac{M_c}{M_m}\{D_{si} - 1\}} \right] - 1 \qquad (10\text{–}43)$$

$$\varepsilon_c = Q f_c (\tau_c - \tau) \qquad (10\text{–}44)$$

$$\varepsilon_m = Q f_m (\tau_c - \tau) \qquad (10\text{–}45)$$

where

$$Q = 10^4 \frac{\dot{j}_{s0}}{i_{s0}^*} \lambda = 10^4 \frac{\mu_{s0}}{\alpha_{s0}} \lambda \qquad (10\text{–}46)$$

and τ is the age

$$\tau \equiv \tau_e - t. \qquad (10\text{–}47)$$

In writing Equations (10–44) and (10–45), we assumed that $\lambda \tau_e \mu_0 / \alpha_0 \ll 1$ and that $\lambda \tau_e \ll 1$.

Upon evaluating Equations (10–42) to (10–46) at the present time $\tau = 0$, expressions for τ_c and M_c/M_m can be obtained in terms of the measurable quantities ε_{mp}, f_{cp}, and f_{mp}

$$\tau_c = \frac{\varepsilon_{mp}}{Q f_{mp}} \qquad (10\text{–}48)$$

$$\frac{M_c}{M_m} = \left\{ D_{si} \left(1 - \frac{f_{cp}}{f_{mp}} \right) - 1 \right\}^{-1}. \qquad (10\text{–}49)$$

These results are also valid for the gradual formation of the continental crust if τ_c is interpreted as the mean age at which the crust was extracted from the mantle reservoir. We next evaluate these formulas for τ_c and M_c/M_m using data from the widely investigated Sm–Nd and Rb–Sr systems.

To employ Equations (10–48) and (10–49) to estimate τ_c and M_c/M_m we must specify the numerical values of the composition ratios, isotope ratios, and other parameters that enter these equations. The values we need are given in Table 10–2 with error estimates. Reference values of the samarium–neodymium system for the bulk silicate Earth are based on data from chondritic meteorites (this is commonly known as a chondritic uniform reservoir, or CHUR). Meteorite data do not yield a bulk silicate Earth value for the rubidium–strontium composition ratio because the Earth is significantly depleted in these elements relative to chondritic meteorites.

The value of μ_s for the Rb–Sr system given in Table 10–2 has been inferred from Rb–Sr versus Sm–Nd systematics. Values for all relevant quantities are given except for the enrichment factor D_{si} for the Rb–Sr system. This is because of the extreme variability of the concentrations of rubidium and strontium in the crust. In some cases parameter values are better constrained for the Sm–Nd system and in other cases they are better constrained for the Rb–Sr system.

TABLE 10-2 Present Parameters for the Sm–Nd and Rb–Sr Isotope Systems

Isotopic System	Sm–Nd	Rb–Sr
Composition ratio, bulk silicate Earth μ_{sp}	0.1967 ± 0.0030	(0.0892 ± 0.0073)
Isotope ratio, bulk silicate Earth α_{sp}	0.51262 ± 0.00011	0.70476 ± 0.00044
Composition ratio, mantle μ_{mp}	0.22 ± 0.04	0.020 ± 0.020
Isotope ratio, mantle α_{mp}	0.51315 ± 0.00015	0.70271 ± 0.00058
Fractionation factor, mantle f_{mp}	0.118 ± 0.22	−0.78 ± 0.24
Normalized isotope ratio, mantle ε_{mp}	10 ± 5	−29 ± 15
Composition ratio, crust μ_{cp}	0.108 ± 0.012	0.9 ± 0.9
Isotope ratio, crust α_{cp}	0.5114 ± 0.001	0.718 ± 0.02
Fractionation factor, crust f_{cp}	−0.451 ± 0.07	9.1 ± 10
Normalized isotope ratio, crust ε_{cp}	−23.8 ± 22	190 ± 300
Enrichment factor D_{si}	29 ± 9	—
Q, Gyr^{-1}	25.3 ± 0.02	17.7 ± 1.5

Source: Allègre et al., 1983.

From Equation (10–48) and Table 10–2 we can obtain the mean age of the crustal reservoir. Substitution of values gives $\tau_c = 3.3$ Ga for the Sm–Nd system and $\tau_c = 2.1$ Ga for the Rb–Sr system. This difference in ages can be attributed either to uncertainties in the isotope parameters or to differential crustal recycling. The uncertainties in the isotope parameters, particularly f_{mp}, are sufficiently large to explain the difference in ages. The values for mantle fractionation are better constrained for the Rb–Sr system because the degree of fractionation is larger. Thus the lower age of 2.1 Ga is favored by most geochemists. Isotope parameters consistent with this age are given in Table 10–3. The time evolution of the Sm–Nd system consistent

TABLE 10-3 Parameter Values for the Sm–Nd and Rb–Sr Isotope Systems Used in the Two-Reservoir Model*

Isotope System	Sm–Nd	Rb–Sr
$Q(\text{Gyr})^{-1}$	25.3	17.7
ε_{mp}	10	−29
f_{mp}	(0.188)	−0.78
f_{cp}	−0.451	—
D_{si}	29	—
ε_{cp}	−23.8	—

* Values are either from Table 10–2 or are calculated (shown in parentheses). Calculated values are based on a separation age $\tau_c = 2.1$ Ga and a mass ratio $M_c/M_m = 0.010$.

with the values in Table 10–3 is given in Figures 10–6 and 10–7. Composition ratios μ and isotope ratios α are given as functions of time t and age τ in Figure 10–6 for the bulk silicate Earth, depleted mantle, and enriched continental crustal reservoirs. The continental crust is separated from the depleted mantle reservoir at an age $\tau_c = 2.1$ Ga. The mantle is enriched in the parent isotope ^{147}Sm and the crust is enriched in the daughter isotope ^{143}Nd relative to the bulk silicate Earth, thus $\mu_m > \mu_s > \mu_c$ in Figure 10–6a. Because of its enrichment in the parent isotope, the mantle becomes more radiogenic than the bulk silicate Earth. Because of its depletion in the radiogenic isotope, the crust becomes less radiogenic than the bulk silicate Earth, thus $\alpha_m > \alpha_s > \alpha_c$ in Figure 10–6b.

Fractionation factors f and normalized isotope ratios ε for the Sm–Nd system are given as functions of time t and age τ in Figure 10–7. Values are given for the bulk silicate Earth, depleted mantle, and enriched continental crustal reservoirs. The behavior of the fractionation factors in Figure 10–7a is essentially similar to the behavior of the composition ratios in Figure 10–6a. The positive values of ε for the depleted mantle reservoir in Figure 10–7b indicate relative enrichment in the parent isotope. Similarly, the negative values of ε for the enriched continental crustal reservoir indicate relative depletion in the parent isotope.

We will now determine the constraints on reservoir masses. If $\tau_c = 2.1$ Ga and $(\varepsilon_{mp})_{Nd} = 10$ we find from Equation (10–48) that $(f_{mp})_{Nd} = 0.188$, a value that is

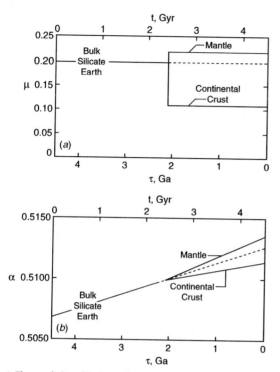

10–6 Time evolution of the Sm–Nd system consistent with the values given in Table 10–3. (*a*) Composition ratios μ are given as a function of time t and age τ for the bulk silicate Earth, the depleted mantle, and the enriched continental crustal reservoirs. (*b*) Isotope ratios α are given as a function of time t and age τ for the bulk silicate Earth, depleted mantle, and enriched continental crustal reservoirs.

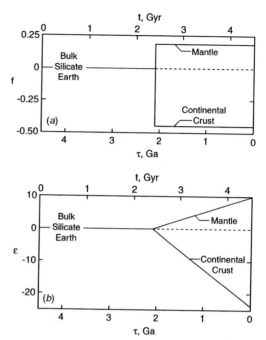

10–7 Time evolution of the Sm–Nd system consistent with the values given in Table 10–3. (*a*) Fractionation factors f are given as a function of time t and age τ for the bulk silicate Earth, the depleted mantle, and the enriched continental crustal reservoirs. (*b*) Normalized isotope ratios ε are given as a function of time t and age τ for the bulk silicate Earth, depleted mantle, and enriched continental crustal reservoirs.

within the uncertainties given in Table 10–2. To obtain the ratio of the mass of the continental crust to the mass of the depleted mantle using Equation (10–49) the Sm–Nd system must be used since $(f_{cp})_{Sr}$ is poorly constrained. With the assumption $(f_{mp})_{Nd} = 0.188$ and the use of other values from Table 10–2, Equation (10–49) gives $M_c/M_m = 0.010$. For comparison, the ratio of the mass of the crust to the mass of the entire mantle is 0.0050 and the ratio of the mass of the crust to the mass of the mantle above the 660-km seismic discontinuity is 0.0180. This mass balance suggests that the depleted upper mantle reservoir constitutes about one-half of the mantle and that the remainder of the mantle is a near-primordial reservoir in the lower mantle. There are certainly major uncertainties in making this determination of the masses of the mantle reservoirs. The uncertainty in the value of $(f_{mp})_{Nd}$ is the most important. However, as we will discuss, other geochemical observations support the presence of a near-primordial reservoir of approximately this size in the lower mantle.

The U–Pb system can also be studied in the context of the instantaneous crustal differentiation model. For the uranium–lead system, the linear approximation for radioactive decay cannot be used and the full exponential relation is required; also, the composition ratios are not constant. At the time of crustal separation $t = \tau_e - \tau_c$, the applicable bulk silicate Earth values of the isotope and composition ratios α_{sc}, β_{sc}, μ_{sc}, and ν_{sc} are obtained from Equations (10–16) to (10–19) with the result

$$\alpha_{sc} = \mu_0\left(1 - e^{-\lambda(\tau_e-\tau_c)}\right) + \alpha_0 \tag{10–50}$$

$$\beta_{sc} = \mu_0\nu_0\left(1 - e^{-\lambda'(\tau_e-\tau_c)}\right) + \beta_0 \tag{10–51}$$

$$\mu_{sc} = \mu_0 e^{-\lambda(\tau_e-\tau_c)} \tag{10–52}$$

$$\nu_{sc} = \nu_0 e^{(\lambda-\lambda')(\tau_e-\tau_c)}. \tag{10–53}$$

Only the composition ratio μ changes when the crust is formed because α_{sc}, β_{sc}, and ν_{sc} are isotope ratios. From Equations (10–14), (10–32), and (10–33), the changes in μ are related to the mean enrichment factors by

$$\frac{\mu_{c0}}{\mu_{sc}} = \frac{j_{c0} i_{sc}}{i_{c0} j_{sc}} = \frac{D_{sj}}{D_{si}} \tag{10–54}$$

$$\frac{\mu_{m0}}{\mu_{sc}} = \frac{1 - (M_c/M_m)(D_{sj} - 1)}{1 - (M_c/M_m)(D_{si} - 1)} \equiv \xi, \tag{10–55}$$

where μ_{c0} and μ_{m0} are the composition ratios for the enriched crust and depleted mantle, respectively, at the time of formation of the crust. The parameter ξ is a measure of the fractionation at this time. Formulas for the present lead isotope ratios of the crust and mantle can be found by the manipulation of the preceding equations

$$\alpha_{cp} = \mu_{sp}\left[e^{\lambda \tau_e} - e^{\lambda \tau_c} + \frac{D_{sj}}{D_{si}}(e^{\lambda \tau_c} - 1)\right] + \alpha_0 \tag{10–56}$$

$$\beta_{cp} = \frac{\mu_{sp}}{137.8}\left[e^{\lambda' \tau_e} - e^{\lambda' \tau_c} + \frac{D_{sj}}{D_{si}}(e^{\lambda' \tau_c} - 1)\right] + \beta_0 \tag{10–57}$$

$$\alpha_{mp} = \mu_{sp}[e^{\lambda \tau_e} - e^{\lambda \tau_c} + \xi(e^{\lambda \tau_c} - 1)] + \alpha_0 \tag{10–58}$$

$$\beta_{mp} = \frac{\mu_{sp}}{137.8}[e^{\lambda' \tau_e} - e^{\lambda' \tau_c} + \xi(e^{\lambda' \tau_c} - 1)] + \beta_0, \tag{10–59}$$

where Equations (10–20) and (10–21) have been used to introduce present bulk silicate Earth values.

Let us compare these lead isotope results for instantaneous crustal differentiation with MORB data. The MORB isotope data are taken to be representative of the depleted mantle reservoir. The envelope for the correlation between β_{mp} and α_{mp} for the lead isotope data from MORB is shown in Figure 10–8. The geochron from Equations (10–22) and (10–23) is shown for $\tau_e = 4.55$ Ga. For the bulk silicate Earth, $\mu_{sp} = 8.05$ is a preferred value. The point on the geochron corresponding to this uranium–lead ratio is indicated in Figure 10–8. This model assumes that lead was preferentially segregated into the continental crust at a time τ_c ago. Thus ξ, defined by Equation (10–55), is greater than 1 since D_{si} is larger than D_{sj}. The magnitude of ξ is a measure of the degree of preferential segregation of uranium into the mantle reservoir.

We assume the lead isotope data for MORB reflect the isotopic state of the depleted mantle reservoir. The dependence of β_{mp} on α_{mp} for $\tau_c = 1, 2,$ and 3 Ga is given

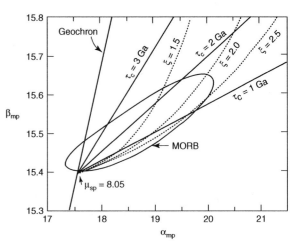

10–8 Predicted values of the lead isotope ratios for the depleted mantle based on a model of instantaneous crustal formation. The geochron for the present bulk silicate Earth from Equations (10–22) and (10–23) is shown. The uranium–lead composition ratio for the bulk silicate Earth is assumed to be $\mu_{sp} = 8.05$. The isotope ratios β_{mp} and α_{mp} for the depleted mantle from Equations (10–58) and (10–59) are given for various ages τ_c of crustal formation by the solid lines. Values corresponding to $\xi = 1.5, 2.0,$ and 2.5, as defined in Equation (10–55), are shown by the dotted lines. The distance from the geochron increases with increased removal of lead from the mantle, i.e., larger ξ. The intersections of the dotted lines with the solid lines give the required values of ξ. The field for lead data obtained from MORB (White, 1985) is also shown.

by the solid lines in Figure 10–8. The distance from the geochron increases as ξ increases. Values corresponding to $\xi = 1.5, 2.0,$ and 2.5 are illustrated by the dotted lines. The MORB field can be explained by a crustal segregation age τ_c between about 1.5 and 2.0 Ga and values of ξ between 1 and 2. The mean age of the continental crust inferred from the model and the MORB lead isotope data is about 1.7 Ga. With $\xi = 1.5$, $D_{sj} = 32$, and $M_c/M_m = 0.010$, Equation (10–46) gives $D_{si} = 55$; thus, lead must be strongly fractionated into the continental crust.

The model for MORB lead based on instantaneous crustal formation requires the removal of lead. The upper continental crust is not enriched in lead relative to uranium, however, and thus there is no direct evidence that the continental crust serves as the complementary lead-enriched reservoir. This lack of an obvious lead-enriched reservoir has become known as the missing lead paradox. The missing lead might reside in the lower continental crust or in the core. Because the mean age of extraction from the mantle of the missing lead is between 1.5 and 2.0 Ga and the age of core formation is about 4.5 Ga, the core is unlikely to be the reservoir of the missing lead.

PROBLEM 10–5 Isotope studies show that the uranium–thorium ratio is 4.0 for the bulk silicate Earth, 5.5 for the continental crust, and 2.5 for the MORB source reservoir. What fraction of the mantle is the MORB source reservoir?

10–5 Noble Gas Systems

In this section we consider the two isotopic systems (listed in Table 10–4) that produce the noble gases helium and argon. The radiogenic parent isotopes of these gases are the principal heat-producing isotopes in the mantle. The noble gas systems have both advantages and disadvantages when used in reservoir modeling. The gases (particularly helium) have high diffusion rates in the mantle and therefore migrate readily. Thus, measured concentrations must be treated with considerable care; however, the high mobility leads to high rates of mantle degassing, and measurements on concentrations and fluxes in the oceans and the atmosphere can be interpreted to provide insights into rates of production and transport in the mantle. The noble gas observations on MORB and OIB can also be interpreted in terms of a "primitive" signature. The Earth's mantle (and presumably the core) was extensively outgassed during accretion. High concentrations of nonradiogenic noble gas isotopes ^{3}He and ^{36}Ar are evidence for a primitive mantle (or core) reservoir that was less extensively outgassed.

There is essentially no primordial ^{40}Ar in the Earth; for practical purposes all ^{40}Ar found in the atmosphere

is the result of the radioactive decay of ^{40}K within the Earth's interior. The present isotope ratio for the atmosphere is ^{40}Ar/^{36}Ar $= 295.5$. The mass of ^{40}Ar in the atmosphere is $M_{^{40}Ar} = 6.60 \times 10^{16}$ kg. Because of the heavy atomic mass of argon, significant quantities of the argon isotopes do not escape from the atmosphere into space.

The ^{40}Ar in the Earth's atmosphere must have been transported from the Earth's interior to the atmosphere. Transport processes include volcanism, hydrothermal circulations through the continental and oceanic crusts, and erosion. As we discussed earlier, the fundamental hypothesis of chemical geodynamics is that the enriched continental crust has been separated from a depleted mantle reservoir. This depleted mantle reservoir may be all or part of the mantle. In the latter case there will be a second buried, near-pristine mantle reservoir. The near-pristine reservoir would have elemental concentrations close to those of the bulk silicate Earth. Similarly, the sum of the enriched continental crustal reservoir and the depleted mantle reservoir will also have elemental concentrations close to those of the bulk silicate Earth. We assume that a large fraction of the argon produced by the radioactive decay of potassium to argon has escaped from the crust and upper mantle reservoir to the atmosphere. The mass of ^{40}Ar produced over the age of the Earth $\tau_e = 4.55$ Ga in a reservoir of mass M is related to the mean concentration of potassium in the reservoir C_K by

$$M_{^{40}Ar} = 1.19 \times 10^{-4} \frac{\lambda_{^{40}Ar}}{\lambda_{^{40}K}} [\exp(\lambda_{^{40}K} \tau_e) - 1] M C_K,$$

(10–60)

where $\lambda_{^{40}Ar} = 5.81 \times 10^{-2}$ Gyr^{-1} is the decay constant of ^{40}K to ^{40}Ar and $\lambda_{^{40}K} = 5.543 \times 10^{-1}$ Gyr is the decay constant of ^{40}K to both ^{40}Ar and ^{40}Ca (see Table 10–4). The constant 1.19×10^{-4} is the fraction of potassium that is the isotope ^{40}K. From Table 4–1 we take $C_K = 31 \times 10^{-5}$ and with $M_{^{40}Ar} = 6.60 \times 10^{16}$ kg from above and $\tau_e = 4.55$ Ga we find from Equation (10–60) that $M = 1.38 \times 10^{24}$ kg. This is 34% of the mass of the entire mantle. The mass of ^{40}Ar in the atmosphere is 34% of the mass of ^{40}Ar that has been produced by the decay of ^{40}K in the crust and mantle over the past 4.55 Ga. One explanation is that there is a primordial reservoir in the mantle from which the argon has not

TABLE 10–4 Isotope Decay Reactions That Produce Rare Gases and the Associated Decay Constants

Isotopic Reaction	Decay Constant λ (Gyr^{-1})	Half-Life $\tau_{1/2}$ (Gyr)
^{238}U $\rightarrow$ ^{206}Pb + 8 ^{4}He	1.551×10^{-1}	4.469
^{235}U $\rightarrow$ ^{207}Pb + 7 ^{4}He	9.849×10^{-1}	0.7038
^{232}Th $\rightarrow$ ^{208}Pb + 6 ^{4}He	4.948×10^{-2}	14.009
^{40}K $\rightarrow$ ^{40}Ar	5.81×10^{-2}	11.93
(^{40}K $\rightarrow$ ^{40}Ar, ^{40}Ca)	(5.543×10^{-1})	(1.2505)

Source: Allègre et al., 1987.

escaped to the atmosphere with a mass that is 66% of the entire mantle. This is consistent with the 50% mass for a primordial reservoir deduced on the basis of Nd–Sm and Rb–Sr systematics in Section 10–4.

While global balances of ^{40}Ar can be carried out, it is not possible to do the same for ^4He because of the relatively rapid loss of this light constituent from the atmosphere. The isotope composition ratio (molal) of primordial helium from carbonaceous chondrites enriched in gases is $^4He/^3He = 1500$–3500. The present isotope ratio for the atmosphere is $^4He/^3He = 7.2 \times 10^5$. A large fraction of the ^4He in the atmosphere is attributed to the decay of ^{235}U, ^{238}U, and ^{232}Th in the mantle and crust (Table 10–4). This radiogenic helium plus primordial helium migrate from the mantle and crust to the oceans and atmosphere. A small amount of ^3He is also produced in the Earth's interior by nuclear reactions involving ^6Li. The helium isotopes escape from the atmosphere because of their relatively low masses. The residence time of helium in the atmosphere is estimated to be 500,000 years. ^3He is also generated in the upper atmosphere by cosmic ray bombardment.

Measurements of the $^4He/^3He$ ratios for gases trapped in MORB have relatively little scatter with a mean value of about $^4He/^3He \times 86,000$. The values of $^4He/^3He$ for MORB are a factor of eight smaller than the atmospheric value. The $^4He/^3He$ ratio from the mantle is smaller than the $^4He/^3He$ ratio in the atmosphere because the helium flux into the atmosphere from the continents is highly radiogenic, i.e., high $^4He/^3He$.

In direct contrast to MORB, measurements of the $^4He/^3He$ ratios for gases trapped in basalts from the Hawaiian Islands show significantly lower $^4He/^3He$ ratios. Samples from Loihi seamount give $^4He/^3He = 30,000 \pm 10,000$ with samples from other Hawaiian volcanoes giving somewhat higher values. The interpretation of the high ^3He values from the Hawaiian volcanoes is that these volcanoes are sampling a primitive reservoir. High ^3He values are also found in Iceland.

10–6 Isotope Systematics of OIB

So far we have concentrated on the crustal and depleted mantle reservoirs. We now turn to the isotope systematics of OIB. Unlike MORB, the OIB have considerable

isotopic variability. Interpretations of this variability require an identification of OIB sources. It is clear that OIB cannot come entirely from the near-homogeneous upper mantle reservoir that is the source of MORB. The ocean islands where basalts are found are hotspots attributed to partial melting in mantle plumes. In turn, the likely source of the mantle plumes is the instability of the hot thermal boundary layer at the base of the mantle reservoir that is the source of MORB. This boundary layer could lie at the core–mantle boundary or it could be at the upper boundary of a primordial reservoir in the lower mantle.

Pressure-release melting occurs in the ascending plume material resulting in OIB. The isotopic signatures of OIB can be attributed to the source region in the thermal boundary layer and to material entrained in the plume as it ascends through the mantle. If mantle convection is layered, then at least part of the signature can be attributed to a pristine or near-pristine layer in the lower mantle. A second possible source of anomalous isotopic signatures is subducted oceanic lithosphere and delaminated continental lithosphere. These units sink through the mantle and are gradually heated by adiabatic compression and by heat conduction from the hot surrounding mantle. After heating, this material can become entrained in the general mantle circulation and eventually into upwelling plumes, or the sinking material can descend to the bottom of the convecting region, be heated within the lower thermal boundary layer, and then swept into the plumes rising from the boundary layer. As part of the ascending mantle plumes, the formerly subducted and delaminated material can contribute to the isotopic heterogeneity associated with OIB. Thus, some of the isotopic heterogenity of OIB could have developed within the aging continental crust and mantle lithosphere.

The isotope systematics of OIB are illustrated in Figures 10–9 to 10–11. Normalized isotope ratios for the samarium–neodymium system $\varepsilon_p(Nd)$ are plotted against the normalized isotope ratios for the rubidium-strontium system $\varepsilon_p(Sr)$ in Figure 10–9. By definition, the bulk silicate Earth (BSE) lies at $\varepsilon_p(Nd) = \varepsilon_p(Sr) = 0$. As discussed in Section 10–4, our preferred values for the depleted mantle reservoir (DMR) are $\varepsilon_p(Nd) = 10$ and $\varepsilon_p(Sr) = -29$ (Table 10–3). The actual measurements for MORB lie within the envelopes given by Ar (mid-Atlantic ridge), PR (east Pacific rise), and

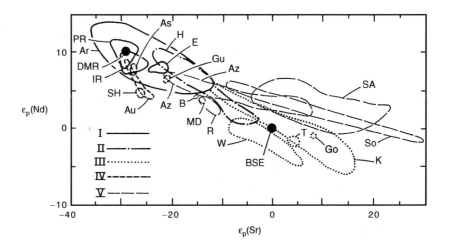

IR (the southwest section of the Indian Ridge). The classification of the OIB will be discussed later. The normalized isotope ratios for the rubidium–strontium system $\varepsilon_p(Sr)$ are plotted against the lead $^{206}Pb/^{204}Pb$ isotope ratios $\alpha_p(Pb)$ in Figure 10–10. The BSE lies at $\varepsilon_p(Sr) = 0$ and $\alpha_p(Pb) = 17.55$. The lead $^{207}Pb/^{204}Pb$ isotope ratios $\beta_p(Pb)$ are plotted against the lead $^{206}Pb/^{204}Pb$ isotope ratios $\alpha_p(Pb)$ in Figure 10–11. This is an extension of the lead–lead correlation given in Figure 10–8 to include the OIB. The BSE lies at $\beta_p(Pb) = 15.4$ and $\alpha_p(Pb) = 17.6$ and the geochron for the BSE from Equations (10–22) and (10–23) is given for $\mu_{sp} = 8.05$. To systematize these results we consider four MORB and OIB classifications as follows:

MORB (I). In addition to normal mid-ocean ridge basalts (n-MORB), this group includes a substantial fraction of the OIB from Iceland, the Galapagos, and

10–9 Neodymium–strontium isotope correlations for MORB and OIB. Normalized isotope ratios for the samarium–neodymium system $\varepsilon_p(Nd)$ are plotted against the normalized isotope ratios for the rubidium–strontium system $\varepsilon_p(Sr)$. Observations lie within the specified envelopes. The data for the basalts are divided into five groups: I. MORB; includes data from the Mid-Atlantic Ridge (Ar), East Pacific Rise (PR), the southwest section of the Indian Ridge (IR), and Easter Island (E). II. Hawaii (H). III. Kerguelen (K); also includes data from Gough (Go), Tristan da Cunha (T), and the Walvis Ridge (W). IV. St. Helena (SH); also includes data from Ascension (As), the Australs (Au), and Guadalupe (Gu). V. Society (So); also includes data from Samoa (SA), the Azores (Az), Rapa Ridge (R), MacDonald (MD), and Bouvet (B). Solid circles indicate bulk silicate Earth (BSE) values, $\varepsilon_p(Nd) = \varepsilon_p(Sr) = 0$, and depleted mantle reservoir (DMR) values, $\varepsilon_p(Nd) = 10$ and $\varepsilon_p(Sr) = -29$.

10–10 Strontium–lead isotope correlations for MORB and OIB. Normalized isotope ratios for the rubidium–strontium system $\varepsilon_p(Sr)$ are plotted against the $^{206}Pb/^{204}Pb$ isotope ratios $\alpha_p(Pb)$. The data are divided into five groups as described in Figure 10–9. The solid circle indicates BSE values, $\varepsilon_p(Sr) = 0$ and $\alpha_p(Pb) = 17.6$.

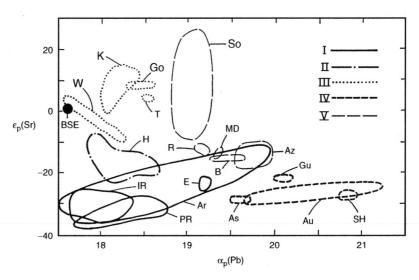

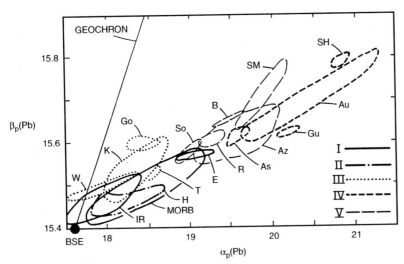

10–11 Lead-lead isotope correlations for MORB and OIB. The $^{207}Pb/^{204}Pb$ isotope ratios $\beta_p(Pb)$ are plotted against the $^{206}Pb/^{204}Pb$ isotope ratios $\alpha_p(Pb)$. The data for basalts are divided into five groups as described in Figure 10–9. The solid circle indicates BSE values, $\beta_p(Pb) = 15.4$ and $\alpha_p(Pb) = 17.6$.

Easter Island. The source is the well-mixed, depleted upper mantle reservoir, as discussed before. Mantle plumes contribute to the basaltic volcanism on Iceland, Galapagos, and Easter Island, but much of this basaltic volcanism comes from the same depleted mantle reservoir as n-MORB. The Nd–Sr correlation for MORB from the East Pacific Rise in Figure 10–9 tightly clusters near the depleted mantle value, $(\varepsilon_{mp})_{Nd} = 10$ and $(\varepsilon_{mp})_{Sr} = -29$. This can be attributed to the large quantities of basalt produced by this rapidly spreading ridge.

Hawaii (II). The Hawaiian hotspot is the most voluminous source of OIB. Hawaiian basalts appear to be a mixture between basalts from the depleted mantle reservoir and basalts from a relatively pristine reservoir. The positions of Hawaii in the isotope correlation plots of Figures 10–9 to 10–11 confirm this association. In particular, the Hawaiian basalts in the Nd–Sr correlation plot given in Figure 10–9 lie on a mixing line trend between the DMR and the BSE. Both Hawaii and Iceland can be associated with strong mantle plumes that contain considerable mantle rock from the depleted n-MORB source reservoir but also contain entrained pristine mantle rock. The association of pristine mantle rock is consistent with primordial noble gas signatures discussed in the previous section.

Kerguelen (III). The OIB from Kerguelen have a distinctive enriched isotopic signature relative to MORB. This signature is also found in OIB from Gough, Tristan da Cunha, and the Walvis Ridge. This group is referred to as enriched mantle member number one (EMI). These islands have a nearly pristine signature for Sr, Nd, and Pb but not for ^{3}He. The pristine signature is particularly striking in the Sr–Pb correlation plot in Figure 10–10. The $(\beta_p)_{Pb}-(\alpha_p)_{Pb}$ correlation in Figure 10–11 shows relatively little lead loss but an old model age. One interpretation is that the source is recently delaminated old continental lithosphere. Such a delamination would not be expected to include the radiogenic upper continental crust. Its Sr, Nd, and Pb isotope signatures would be nearly pristine but ^{3}He would be low because this mantle is outgassed.

St. Helena (IV). The OIB from St. Helena is distinctive in having suffered a very large lead loss relative to MORB and other OIB. This signature is also found in OIB from Ascension, the Australs, and Guadalupe. The large lead loss associated with this group implies high values for the uranium–lead ratio μ; for this reason they are referred to as HIMU. The source of the large values of μ is the radiogenic continental crust. This OIB group can be attributed to mixing between depleted mantle and delaminated old continental crust.

Society (V). The OIB from the Society Islands also have a distinctive enriched isotopic signature. It is distinct from the Kerguelen group in the strontium isotopic signature. The end member of this group is denoted as the enriched mantle member number two

(EMII). This group includes the Marquesas, Samoa, and the Azores including Sao Miguel, Rapa Ridge, MacDonald, and Bouvet. The OIB from the Society Islands and Samoa are particularly anomalous in Nd–Sr systematics as shown in Figure 10–9. The end member of this group lies on a mixing line between the depleted mantle reservoir and marine sediments. The Azores have a large spread that can be associated with a mixing line between the depleted mantle and subducted young continental crust. The large lead loss shown in the $(\beta_p)_{Pb}$–$(\alpha_p)_{Pb}$ correlation (Figure 10–11) can be associated with subducted radiogenic sediments from the upper continental crust.

The Kerguelen, St. Helena, and Society types of OIB are associated with weak plumes that do not entrain pristine rock. However, all the OIB include substantial fractions of basalts from the depleted mantle reservoir. The neodymium–strontium correlation given in Figure 10–9 shows a strong mixing-line trend between depleted MORB and BSE. This trend is strong evidence for the existence of a near-pristine reservoir. Only the Society group shows a significant deviation from the trend. As shown in Figure 10–11 all OIB lie to the right of the geochron. This characteristic lead signature may have developed in the continental crust.

The OIB heterogeneities probably arise from two sources: (1) nearly pristine rock that is entrained from the lower mantle and (2) continental crust and mantle that have recently been subducted or delaminated. Hawaii and Iceland are examples of the entrainment of lower mantle material. The basalts from these islands lie on mixing lines between depleted and primitive isotopic ratios and have excess primordial rare gas isotopes. Other OIB are divided into three types: The Kerguelen group is characterized by an isotopic signature associated with a contribution from recently delaminated continental lithosphere. The St. Helena group is characterized by a very large lead loss attributed to old, radiogenic continental crust. The Society group has anomalous Nd–Sr systematics and a large lead loss that is attributed to young continental crust.

REFERENCES

Allègre, C. J., S. R. Hart, and J.-F. Minster (1983), Chemical structure and evolution of the mantle and continents determined by inversion of Nd and Sr isotopic data, II. Numerical experiments and discussion, *Earth Planet. Sci. Lett.* **66**, 191–213.

Allègre, C. J., T. Staudacher, and P. Sarda (1987), Rare gas systematics: Formation of the atmosphere, evolution and structure of the Earth's mantle. *Earth Planet. Sci. Lett.* **81**, 127–150.

Boltwood, B. B. (1907), On the ultimate disintegration products of the radioactive elements, *Am. J. Sci.* **4**, 77–88.

Holmes, A. (1913), *The Age of the Earth* (Harper and Brothers, London) 194 p.

Holmes, A. (1946), An estimate of the age of the Earth, *Nature* **157**, 680–684.

Houtermans, F. G. (1946), Die isotopenhäufigkeiten in natürlichen Blei und das Alter des Urans, *Naturweiss.* **33**, 185–186, 219.

Kanasewich, E. R. (1968), The interpretation of lead isotopes and their geological significance. In *Radiometric Dating for Geologists*, pp. 147–223, eds. E. I. Hamilton and R. M. Farquhar (Interscience Publishers, London).

Moorbath, S., R. K. O'Nions, R. J. Pankhurst, N. H. Gale, and V. R. McGregor (1972), Further rubidium–strontium age determinations on the very early Precambrian rocks of the Godthaab district, West Greenland, *Nature Phys. Sci.* **240**, 78–82.

Patterson, C. C. (1956), Age of meteorites and the Earth, *Geochim. Cosmochim. Acta* **10**, 230–237.

White, W. (1985), Sources of oceanic basalts: Radiogenic isotope data, *Geology* **13**, 115–118.

COLLATERAL READING

Dickin, A. P., *Radiogenic Isotope Geology* (Cambridge University Press, Cambridge, 1995).

An advanced and very complete treatment of isotope geochemistry. The isotope systematics of the Rb–Sr, Sm–Nd, U–Th–Pb, Re–Os, and Lu–Hf systems are given. Rare gas geochemistry and U-series geochemistry are treated. Experimental techniques, fission track dating, cosmogenic nuclides, and extinct radionuclides are covered along with a variety of applications.

Faure, G., *Principles of Isotope Geology*, 2nd edition (Wiley, New York, 1986).

A comprehensive treatment of the uses of isotopes in geology. Various age dating techniques are comprehensively reviewed. The radiogenic systematics of the Rb–Sr, Sm–Nd, Lu–Hf, Re–Os, and U–Th–Pb systems are given. The stable isotopic systems of oxygen, hydrogen, carbon, nitrogen, and sulfur are also treated.

APPENDIX ONE

Symbols and Units

A. SI Units

Quantity	Unit	Symbol	Equivalent
Basic Units			
Length	meter	m	
Time	second	s	
Mass	kilogram	kg	
Temperature	Kelvin	K	
Electrical current	ampere	A	
Derived Units			
Force	newton	N	$kg\ m\ s^{-2}$
Energy	joule	J	$kg\ m^2\ s^{-2}$
Power	watt	W	$kg\ m^2\ s^{-3}$
Pressure	pascal	Pa	$kg\ m^{-1}\ s^{-2}$
Frequency	hertz	Hz	s^{-1}
Charge	coulomb	C	$A\ s$
Capacitance	farad	F	$C^2\ s^2\ kg^{-1} m^{-2}$
Magnetic induction	tesla	T	$kg\ A^{-1}\ s^{-2}$
Multiples of Ten			
10^{-3}	milli	m	
10^{-6}	micro	μ	
10^{-9}	nano	n	
10^{-12}	pico	p	
10^{-15}	femto	f	
10^{-18}	atto	a	
10^{3}	kilo	k	
10^{6}	mega	M	
10^{9}	giga	G	
10^{12}	tera	T	
10^{15}	peta	P	
10^{18}	exa	E	

B. Conversion Factors

To Convert	To	Multiply by
acre	ft^2	43560
	m^2	4046.9
angstrom, Å	cm	10^{-8}
	nm	10
astronomical unit, AU	cm	1.496×10^{13}
	Tm	0.1496
bar, b	atm	0.98692
	$dyne\ cm^{-2}$	10^6
	$lb\ in^{-2}$	14.5038
	mm Hg	750.06
	MPa	10^{-1}
barrel, bbl (petroleum)	gal (US)	42
	liter	158.98
British thermal unit, (B.T.U.)	cal	251.996
	joule	1054.35
calorie (gm), cal	joule	4.184
centimeter, cm	inch	0.39370
	m	10^{-2}
darcy	m^2	9.8697×10^{-13}
dyne	$g\ cm\ s^{-2}$	1
	newton	10^{-5}
erg	cal	2.39006×10^{-8}
	dyne cm	1
	joule	10^{-7}
fathom, fath	ft	6
feet, ft	in	12
	m	0.3048
furlong	yd	220
gal	$cm\ s^{-2}$	1
	$m\ s^{-2}$	10^{-2}
gallon, gal (U.S. liq.)	in^3	231
	liter	3.78541
gallon, gal (Imp.)	in^3	277.419
	liters	4.54608
gamma	gauss	10^{-5}
	tesla	10^{-9}
gauss	tesla	10^{-4}
gram, g	pound	0.0022046
	kg	10^{-3}
heat flow unit (H.F.U.)	$cal\ cm^{-2}\ s^{-1}$	10^{-6}
	$mW\ m^{-2}$	41.84
hectare	acre	2.47105
	cm^2	10^8
horsepower	W	745.700

(cont.)

B. Conversion Factors (cont.)

To Convert	To	Multiply by
inch, in	cm	2.54
joule, J	erg	10^7
	cal	0.239006
kilogram, kg	g	10^3
	pound	2.20462
kilometer, km	m	10^3
	ft	3280.84
	mile	0.621371
kilometer hr^{-1}	mile hr^{-1}	0.621371
kilowatt	watt	10^3
	HP	1.34102
knot	mi hr^{-1}	1.150779
liter	cm^3	10^3
	gal (U.S.)	0.26417
	in^3	61.0237
meter	ft	3.28084
micron, μ	cm	10^{-4}
mile	ft	5280
	km	1.60934
mm Hg	dyne cm^{-2}	1333.22
newton	dyne	10^5
ounce	lb	0.0625
pint	gallon	0.125
poise	g cm^{-1} s^{-1}	1
	kg m^{-1} s^{-1}	0.1
pound	kg	0.453592
poundal	newton	0.138255
quart	gallon	0.25
stoke	cm^2 s^{-1}	1
slug	kg	14.594
tesla	gauss	10^4
ton (short)	lb	2000
	kg	907.185
watt	J s^{-1}	1
	HP	0.00134102
yard	in	36
	m	0.9144
year (cal)	days	365
	s	3.1536×10^7

C. List of Symbols

Symbol	Quantity	Equation Introduced	SI Unit
a	equatorial radius of the Earth	(1–6)	m
	radius	(6–213)	m
A	equatorial moment of inertia	(5–29)	kg m^2
	area	(3–36)	m^2
	Madelung constant	(7–10)	
b	distance	(2–2)	m
b^*	Burgers vector	(7–116)	m
B	magnetic field	(1–2)	T
	equatorial moment of inertia	(5–31)	kg m^2
	buoyancy flux	(6–244)	kg s^{-1}
Br	Brinkman number	(7–154)	
c	specific heat	(4–67)	J kg^{-1} K^{-1}
	polar radius of the Earth	(5–55)	m
c_D	drag coefficient	(6–226)	
C	degree of compensation	(3–116)	
	concentration	(4–6)	
	polar moment of inertia	(5–26)	kg m^2
d	distance	(2–141)	m
D	magnetic declination	(1–4)	
	flexural rigidity	(3–72)	N m
	drag	(6–218)	N
	pipe diameter	(6–266)	m
	diffusion coefficient	(7–62)	m^2 s^{-1}
	enrichment factor	(10–32)	
e	isotropic strain	(2–132)	
	eccentricity	(5–65)	
	charge on an electron	(7–10)	C
	specific internal energy	(7–27)	J kg^{-1}
E	Young's modulus	(3–4)	Pa
	Eckert number	(6–412)	
	energy	(8–72)	J
E_a	activation energy	(7–77)	J mole^{-1}
E_0	barrier energy	(7–76)	J mole^{-1}

Symbol	Quantity	Equation Introduced	SI Unit	Symbol	Quantity	Equation Introduced	SI Unit
f	coefficient of friction	(2–23)		m	magnetic dipole moment	(1–6)	$A\,m^2$
	frequency	(4–73)	s^{-1}		mass	(5–1)	kg
	force	(5–1)	N		earthquake magnitude	(8–72)	
	flattening of the Earth	(5–56)		M	mass	(2–66)	kg
	friction factor	(6–39)			bending moment per unit length	(3–58)	N
	fractionation factor	(10–29)			moment of an earthquake	(8–73)	J
F	force per unit width	(2–14)	$N\,m^{-1}$	M_a	atomic mass	(7–27)	$kg\,mole^{-1}$
	force	(8–49)	N	n	number of atoms per unit volume	(7–46)	m^{-3}
g	acceleration of gravity	(1–1)	$m\,s^{-2}$	ΔN	geoid anomaly	(5–69)	m
Δg	gravity anomaly	(5–100)	$m\,s^{-2}$	N_0	Avogadro's number	(7–27)	
G	gravitational constant	(2–66)	$m^3\,kg^{-1}\,s^{-2}$	Nu	Nusselt number	(6–267)	
	shear modulus	(3–1)	Pa	p	pressure	(1–1)	Pa
h	depth	(2–2)	m	P	horizontal force per unit length	(3–58)	$N\,m^{-1}$
	height of topography	(3–101)	m		flow pressure	(6–66)	Pa
	heat transfer coefficient	(6–264)	$W\,m^{-2}\,K^{-1}$	Pe	Peclet number	(6–274)	
	height of phreatic surface	(9–22)	m	Pr	Prandtl number	(6–3)	
H	heat generation rate	(4–5)	$W\,kg^{-1}$	q	load	(3–56)	Pa
	thickness of reference crust	(5–148)	m		heat flow per unit area	(4–1)	$W\,m^{-2}$
	hydraulic head	(6–9)	m	Q	heat flow	(4–5)	W
i	mole density	(10–1)			volumetric flow rate	(6–35)	$m^3\,s^{-1}$
I	magnetic inclination	(1–2)			integrated heat flux	(4–117)	$J\,m^{-2}$
j	mole density	(10–1)			heat source strength	(4–119)	$W\,m^{-1}$
J	flux of atoms	(7–58)	$m^{-2}\,s^{-1}$		heat content per unit area	(4–158)	$J\,m^{-2}$
J_2	ellipticity coefficient	(5–43)		r	radial coordinate	(2–65)	m
k	thermal conductivity	(4–1)	$W\,m^{-1}\,K^{-1}$	R	radius of the earth	(2–74)	m
	Boltzmann constant	(7–27)	$J\,{}^\circ K^{-1}$		pipe radius	(6–33)	m
	permeability	(9–1)	m^2		radius of curvature	(3–66)	m
$\bar{k}$	spring constant	(7–18)	$N\,m^{-1}$		universal gas constant	(7–27)	$J\,mole^{-1}\,K^{-1}$
K	bulk modulus	(3–50)	Pa	Ra	Rayleigh number	(6–316)	
	transport coefficient	(4–258)	$kg\,m^{-1}\,s^{-1}$	Re	Reynolds number	(6–40)	
	hydraulic conductivity	(9–2)	$m\,s^{-1}$	s	surface distance	(1–19)	m
					distance to rotation axis	(5–46)	m
l	length	(3–66)	m		specific entropy	(4–252)	$J\,kg^{-1}\,K^{-1}$
L	length	(2–22)	m	S	shear strength	(8–36)	Pa
	latent heat	(4–138)	$J\,kg^{-1}$				*(cont.)*

C. List of Symbols (Cont.)

Symbol	Quantity	Equation Introduced	SI Unit
t	time	(4–7)	s
T	temperature	(4–1)	K
u	horizontal velocity	(1–17)	m s^{-1}
U	velocity	(4–151)	m s^{-1}
	geopotential	(5–53)	m^2 s^{-2}
	lattice energy	(7–4)	J
v	vertical velocity	(6–52)	m s^{-1}
	specific volume	(4–173)	m^3 kg^{-1}
V	volume	(3–51)	m^3
	vertical force per unit length	(3–56)	N m^{-1}
	gravitational potential	(5–51)	m^2 s^{-2}
V_a	activation volume	(7–77)	m^3
w	width	(2–6)	m
	displacement	(2–75)	m
	vertical deflection	(3–58)	m
W	depth of compensation	(5–151)	m
x	horizontal coordinate	(2–75)	m
y	vertical coordinate	(1–1)	m
Z	complex distance	(8–88)	m
z	horizontal coordinate	(2–85)	m
	number of charges per ion	(7–10)	
α	angle	(1–22)	
	stretching factor	(2–6)	
	flexural parameter	(3–127)	m
	coefficient of thermal expansion	(4–176)	K^{-1}
	isotope ratio	(10–1)	
β	angle	(2–150)	
	compressibility	(3–50)	Pa^{-1}
	geothermal gradient	(4–61)	K m^{-1}
	isotope ratio	(10–1)	
γ	slope of Clapeyron curve	(4–256)	Pa K^{-1}
	Euler's constant	(4–120)	
Γ	jump frequency	(7–56)	s^{-1}
δ	tube diameter	(9–6)	m
Δ	subtended angle	(1–17)	
	dilatation	(2–86)	
ε	strain	(2–74)	
	small quantity	(5–22)	
	normalized isotope ratio	(10–28)	

Symbol	Quantity	Equation Introduced	SI Unit
$\dot\varepsilon$	rate of strain	(2–127)	s^{-1}
ε_0	permittivity of free space	(7–10)	F m^{-1}
ζ	distance	(4–19)	m
η	distance	(4–19)	m
	similarity variable	(4–96)	
θ	colatitude	(1–6)	
	angle	(2–26)	
	nondimensional temperature	(4–93)	
κ	thermal diffusivity	(4–68)	m^2 s^{-1}
λ	decay constant	(10–3)	s^{-1}
	Lamé parameter	(3–1)	Pa
	wavelength	(3–107)	m
	thermal conductivity	(9–103)	W m^{-1} K^{-1}
μ	viscosity	(6–1)	Pa s
	coefficient of internal friction	(8–36)	
	composition ratio	(10–2)	
μ_0	permeability of free space	(1–6)	T m A^{-1}
ν	Poisson's ratio	(3–4)	
	kinematic viscosity	(6–2)	m^2 s^{-1}
	isotope ratio	(10–14)	
ξ	distance	(4–151)	m
ρ	density	(1–1)	kg m^{-3}
σ	stress	(2–1)	Pa
	Stefan–Boltzmann constant	(4–242)	W m^{-2} K^{-4}
	surface mass density	(5–106)	kg m^{-2}
τ	shear stress	(4–243)	Pa
	period of oscillation	(4–74)	s
	relaxation time	(6–104)	s
	growth time	(6–157)	s
$\tau_{1/2}$	half-life	(4–7)	s
ϕ	latitude	(1–10)	
	angle	(2–87)	
	kinetic energy	(7–46)	J
	porosity	(9–5)	
ψ	longitude	(1–14)	
	stream function	(6–69)	m^2 s^{-1}
ω	angular velocity	(1–17)	s^{-1}
	solid body rotation	(2–93)	
	circular frequency	(4–72)	s^{-1}

APPENDIX TWO

Physical Constants and Properties

A. Physical Constants

Quantity	Symbol	Value
Speed of light	c	2.99792458×10^8 m s^{-1}
Electronic charge	e	$-1.60217733 \times 10^{-19}$ C
Permeability of vacuum	μ_0	$4\pi \times 10^{-7}$ T m A^{-1}
Permittivity of vacuum	ε_0	$8.8541878 \times 10^{-12}$ F m^{-1}
Planck constant	h	$6.6260755 \times 10^{-34}$ J s
Boltzmann constant	k	1.380658×10^{-23} J K^{-1}
Stefan–Boltzmann constant	σ	5.67051×10^{-8} W m^{-2} K^{-4}
Gravitational constant	G	6.6726×10^{-11} N m^2 kg^{-2}
Electron rest mass	m_e	$0.91093897 \times 10^{-30}$ kg
Atomic mass unit		$1.6605402 \times 10^{-27}$ kg
Avogadro's number	N_A	6.0221367×10^{23} mol^{-1}
Universal gas constant	R	8.314510 J mol^{-1} K^{-1}

B. Properties of the Earth

Quantity	Symbol	Value
Equatorial radius	a	6.378137×10^6 m
Polar radius	c	6.356752×10^6 m
Volume	V	1.0832×10^{21} m^3
Volume of core	V_c	1.77×10^{20} m^3
Volume of mantle	V_m	9.06×10^{20} m^3
Radius of sphere of equal volume		6.3708×10^6 m
Radius of core	r_c	3.480×10^6 m
Radius of inner core	r_{ic}	1.215×10^6 m
Mass	M	5.9736×10^{24} kg
Mean density	ρ	5.515×10^3 kg m^{-3}
Mass of core	M_c	1.883×10^{24} kg
Mass of mantle	M_m	4.043×10^{24} kg
Mass of crust	M_{cr}	2.36×10^{22} kg
Equatorial surface gravity	g_e	9.7803267715 m s^{-2}
Polar surface gravity	g_p	9.8321863685 m s^{-2}
Area	A	5.10×10^{14} m^2
Land area		1.48×10^{14} m^2
Continental area including margins	A_c	2.0×10^{14} m^2
Water area		3.62×10^{14} m^2
Oceans excluding continental margins	A_o	3.1×10^{14} m^2
Mean land elevation	$\bar{h}$	875 m
Mean ocean depth	$\bar{w}$	3794 m
Mean thickness of continental crust	$\bar{h}_{cc}$	40 km
Mean thickness of oceanic crust	$\bar{h}_{oc}$	6 km
Mean surface heat flow	$\bar{q}_s$	87 mW m^{-2}
Total geothermal flux	Q_s	44.3 TW
Mean continental heat flow	$\bar{q}_c$	65 mW m^{-2}
Mean oceanic heat flow	$\bar{q}_o$	101 mW m^{-2}
Solar constant		1373 W m^{-2}
Angular velocity	w	7.292115×10^{-5} rad s^{-1}
Ellipticity coefficient	J_2	1.08263×10^{-3}
Flattening	f	$3.35281068118 \times 10^{-3}$
Polar moment of inertia	C	8.0358×10^{37} kg m^2
Equatorial moment of inertia	A	8.0095×10^{37} kg m^2
Age of Earth	τ_e	4.55 Ga

C. Properties of the Planets

	Mercury	Venus	Earth	Mars	Jupiter	Saturn	Uranus	Neptune	Pluto
Semimajor axis of orbit, 10^6 km	57.9	108.2	149.6	227.9	778.2	1431	2887	4529	5936
Sidereal period of revolution, days	87.969	224.701	365.25636	686.986	4332.660	10,759.43	30,688.9	60,189	90,465
Eccentricity	0.205614	0.006821	0.016721	0.093313	0.0481	0.051	0.047	0.007	0.253
Inclination to ecliptic, deg	7.005	3.394	—	1.850	1.305	2.49	0.773	1.770	17.13
Equatorial radius, km	2439	6052	6378	3394	71,398	60,330	26,200	25,225	1157
Ellipticity	0	0	0.0034	0.0059	0.0637	0.102	0.024	0.0266	—
Mass 10^{24} kg	0.3303	4.869	5.9736	0.64185	1899	568.5	86.83	102.4	0.015
Density, kg m^{-3}	5427	5204	5515	3933	1326	687	1318	1638	1100
Equatorial gravity, m s^{-2}	3.701	8.870	9.7803	3.690	23.1	8.96	8.69	11.0	0.72
Sidereal period of rotation, days	58.65	−243.0	0.997270	1.025956	0.413538	0.4375	−0.65	0.768	−6.387
C/MR^2	—	0.34	0.3335	0.365	0.26	0.25	0.23	0.23	—
Surface temperature, K	440	730	288	218	129	97	58	56	50

D. Properties of the Satellites

Planet	Satellite	Distance from Planet 10^3 km	Sidereal Period Days	Radius Km	Mass 10^{21} kg
Earth	Moon	384.4	27.322	1737.103	73.483
Mars	Phobos	9.38	0.3189	11	1.08×10^{-3}
	Deimos	23.48	1.26244	6	1.80×10^{-4}
Jupiter	Amalthea	181.3	0.49818	120	—
	Io	422	1.76914	1821	89.33
	Europa	670.9	3.5512	1569	48.0
	Ganymede	1070	7.155	2634	148
	Callisto	1883	16.689	2403	107.6
	Leda	11,094	238.7	8	—
	Himalia	11,480	250.6	93	—
	Lysithea	11,720	259.2	18	—
	Elara	11,737	259.6	38	—
	Ananke	21,200	631 R	15	—
	Carme	22,600	692 R	20	—
	Pasiphae	23,500	735 R	25	—
	Sinope	23,700	758 R	18	—
Saturn	Mimas	185	0.942	199	0.0375
	Enceladus	238	1.370	250	0.074
	Tethys	295	1.888	530	0.622

Planet	Satellite	Distance from Planet 10^3 km	Sidereal Period Days	Radius km	Mass 10^{21} kg
	Dione	377	2.737	560	1.05
	Rhea	527	4.518	765	2.28
	Titan	1222	15.945	2575	134.5
	Hyperion	1481	21.277	145	0.1
	Iapetus	3561	79.331	718	1.59
	Phoebe	12,952	550.4 R	110	—
Uranus	Miranda	130	1.413	1235	0.066
	Ariel	191	2.520	579	1.35
	Umbriel	266	4.144	585	1.17
	Titania	436	8.706	789	3.53
	Oberon	583	13.463	761	3.01
Neptune	Triton	355	5.877 R	1353	21.5
	Nereid	5513	360.14	170	—
Pluto	Charon	19.4	6.387	—	—

E. Properties of Rock

	Density kg m^{-3}	E 10^{11} Pa	G 10^{11} Pa	ν	k W m^{-1} K^{-1}	α 10^{-5} K^{-1}
Sedimentary						
Shale	2100–2700	0.1–0.7	0.1–0.3	0.1–0.2	1.2–3	
Sandstone	1900–2500	0.1–0.6	0.04–0.2	0.1–0.3	1.5–4.2	3
Limestone	1600–2700	0.5–0.8	0.2–0.3	0.15–0.3	2–3.4	2.4
Dolomite	2700–2850	0.5–0.9	0.2–6.4	0.1–0.4	3.2–5	
Metamorphic						
Gneiss	2600–2850	0.4–0.6	0.2–0.3	0.15–0.25	2.1–4.2	
Amphibole	2800–3150		0.5–1.0	0.4	2.1–3.8	
Marble	2670–2750	0.3–0.8	0.2–0.35	0.2–0.3	2.5–3	
Igneous						
Basalt	2950	0.6–0.8	0.25–0.35	0.2–0.25	1.3–2.9	
Granite	2650	0.4–0.7	0.2–0.3	0.2–0.25	2.4–3.8	2.4
Diabase	2900	0.8–1.1	0.3–0.45	0.25	2–4	
Gabbro	2950	0.6–1.0	0.2–0.35	0.15–0.2	1.9–4.0	1.6
Diorite	2800	0.6–0.8	0.3–0.35	0.25–0.3	2.8–3.6	
Pyroxenite	3250	1.0	0.4		4.1–5	
Anorthosite	2640–2920	0.83	0.35	0.25	1.7–2.1	
Granodiorite	2700	0.7	0.3	0.25	2.0–3.5	
Mantle						
Peridotite	3250				3–4.5	2.4
Dunite	3000–3700	1.4–1.6	0.6–0.7		3.7–4.6	
Miscellaneous						
Ice	917		0.092	0.31–0.36	2.2	5

F. Mantle Properties

Radius r (km)	Depth y (km)	Density ρ (kg m⁻³)	Pressure P (GPa)	Gravity g (m s⁻²)	Bulk Modulus K (GPa)	Shear Modulus G (GPa)	Poisson's Ratio ν
3480.0	2891.0	5566.5	135.8	10.69	655.6	293.8	0.31
3500.0	2871.0	5556.4	134.6	10.66	653.7	293.3	0.30
3600.0	2771.0	5506.4	128.8	10.52	644.0	290.7	0.30
3700.0	2671.0	5456.6	123.0	10.41	627.9	285.5	0.30
3800.0	2571.0	5406.8	117.4	10.31	609.5	279.4	0.30
3900.0	2471.0	5357.1	111.9	10.23	591.7	273.4	0.30
4000.0	2371.0	5307.3	106.4	10.16	574.4	267.5	0.30
4100.0	2271.0	5257.3	101.1	10.10	557.5	261.7	0.30
4200.0	2171.0	5207.2	95.8	10.06	540.9	255.9	0.30
4300.0	2071.0	5156.7	90.6	10.02	524.6	250.2	0.29
4400.0	1971.0	5105.9	85.5	9.99	508.5	244.5	0.29
4500.0	1871.0	5054.7	80.4	9.97	492.5	238.8	0.29
4600.0	1771.0	5003.0	75.4	9.95	476.6	233.0	0.29
4700.0	1671.0	4950.8	70.4	9.94	460.7	227.3	0.29
4800.0	1571.0	4897.9	65.5	9.93	444.8	221.5	0.29
4900.0	1471.0	4844.3	60.7	9.93	428.8	215.7	0.28
5000.0	1371.0	4789.9	55.9	9.94	412.8	209.8	0.28
5100.0	1271.0	4734.6	51.2	9.94	396.6	203.9	0.28
5200.0	1171.0	4678.5	46.5	9.95	380.3	197.9	0.28
5300.0	1071.0	4621.3	41.9	9.96	363.8	191.8	0.28
5400.0	971.0	4563.1	37.3	9.97	347.1	185.6	0.27
5500.0	871.0	4503.8	32.8	9.99	330.3	179.4	0.27
5600.0	771.0	4443.2	28.3	10.00	313.4	173.0	0.27
5701.0	670.0	4380.7	23.8	10.02	299.9	154.8	0.28
5701.0	670.0	3992.1	23.8	10.02	255.6	123.9	0.29
5800.0	571.0	3939.3	19.9	10.00	239.7	116.2	0.29
5900.0	471.0	3813.2	16.0	9.99	209.7	100.7	0.29
5971.0	400.0	3723.7	13.4	9.97	189.9	90.6	0.29
5971.0	400.0	3543.3	13.4	9.97	173.5	80.6	0.30
6000.0	371.0	3525.9	12.3	9.96	170.1	79.5	0.30
6100.0	271.0	3466.2	8.9	9.93	158.6	75.9	0.29
6151.0	220.0	3435.8	7.1	9.91	152.9	74.1	0.29
6151.0	220.0	3359.5	7.1	9.91	127.0	65.6	0.28
6200.0	171.0	3364.8	5.5	9.89	128.1	66.2	0.28
6300.0	71.0	3375.7	2.2	9.86	130.4	67.6	0.28
6346.6	24.4	3380.7	0.6	9.84	131.5	68.2	0.28
6346.6	24.4	2900.0	0.6	9.84	75.3	44.1	0.25
6356.0	15.0	2900.0	0.3	9.84	75.3	44.1	0.25
6356.0	15.0	2600.0	0.3	9.84	52.0	26.6	0.28
6368.0	3.0	2600.0	0.0	9.83	52.0	26.6	0.28
6368.0	3.0	1020.0	0.0	9.83	2.1	0.0	0.50
6371.0	0.0	1020.0	0.0	9.82	2.1	0.0	0.50

Answers to Selected Problems

1–1	57 Myr
1–2	25 km
1–3	439 m
1–4	3.74×10^{-4} T
1–5	1660 km
1–6	51°N, 78°E
1–7	75°N, 123°E
1–8	70°N, 186°E
1–11	0.3°, 70°
1–12	5.5°, 69°
1–13	1886 km, 2441 km
1–14	1028 km, 2080 km
1–17	9.1 mm yr^{-1}
1–18	75 mm yr^{-1}
1–19	47 mm yr^{-1}
1–20	210°, 120 mm yr^{-1}; 0°, 104 mm yr^{-1}
1–22	260°, 43 mm yr^{-1}
1–23	200°, 83 mm yr^{-1}; 180°, 78 mm yr^{-1}
1–24	250°, 92.4 mm yr^{-1}; 50 mm yr^{-1}
1–25	91°, 51 mm yr^{-1}
2–1	220 MPa
2–2	68 km
2–3	290 m
2–4	1.18
2–5	1.40
2–6	5.3 km, 30 km
2–7	−150 MPa
2–9	4.1×10^{11} N, 41 MPa
2–10	8.6×10^{4} N
2–14	166.5 MPa, −23.5 MPa
2–17	4.6 GPa
2–18	13,200 kg m^{-3}, 433 GPa, 129 GPa, 12.9 m s^{-2}

2–19	47 and 63 MPa, 75° and 165°
2–20	0.5, 3, 5, 5.5 MPa
2–26	$x_c - x_a = [(x_b - x_a)^2$ $+ (z_b - z_a)^2]^{1/2} \dfrac{\sin\theta_2}{\sin(\theta_1 + \theta_2)}$ $\times \left\{ -\cos\left[\theta_1 + \tan^{-1}\left(\dfrac{z_b - z_a}{x_b - x_a}\right)\right]\right\}$ $z_c - z_a = [(x_b - x_a)^2 + (z_b - z_a)]^{1/2} \dfrac{\sin\theta_2}{\sin(\theta_1 + \theta_2)}$ $\times \sin\left[\theta_1 + \tan^{-1}\left(\dfrac{z_b - z_a}{x_b - x_a}\right)\right]$
2–27	45 mm yr^{-1}
2–29	$\dot{\epsilon}_{xx} = 10^{-7}$ yr^{-1}, $\dot{\epsilon}_{yy} = 2.1 \times 10^{-7}$ yr^{-1}, $\dot{\epsilon}_{xy} = 10^{-7}$ yr^{-1}
2–30	20.2 mm/yr ($\Delta_{PG} = 75.3°$, $\Delta_{PW} = 59.7°$, $\Delta_{GW} = 61.4°$, $\beta = 89.1°$)
2–31	−85 mm/yr ($\Delta_{PS} = 79.5°$, $\Delta_{PM} = 62.6°$, $\Delta_{SM} = 60.6°$, $\beta = 64.4°$)
2–32	−73.8 mm/yr ($\Delta_{PE} = 84.1°$, $\Delta_{PA} = 74.9°$, $\Delta_{EA} = 36.6°$, $\beta = 72.5°$)
2–33	34.5 mm/yr^{-1}, 35.7° E of S
2–34	35.8 mm/yr^{-1}, 41.9° E of S
3–1	180 MPa
3–2	42 MPa, 125 MPa
3–3	$\nu\sigma_1$, $-\nu(1 + \nu)\sigma_1/E$
3–4	$(2\nu^2 + \nu - 1)\rho g y/E$, $(1 - 2\nu)\rho g y$
3–7	$x(L - x)M_0/2D$
3–8	$x(3L^2 - 4x^2)V_a/48D$
3–9	$q(L - x)$, $-q(L - x)^2/2$
3–10	32.4 MPa, 0.27 MPa
3–11	$-x^3 L \times q/12D + 3x^2 L^2 q/16$, $0 < x < L/2$ $(16x^4 - 64x^3 L + 96x^2 L^2 - 8x L^3 + L^4)$ $\times q/384 D$ $L/2 < x < L$
3–12	$(9 Lx^2 - 4x^3)V_a/12D$ $0 < x < L/2$ $(24 Lx^2 - 8x^3 + 6 L^2 x - L^3)V_a/48D$ $L/2 < x < L$
3–13	$(x^4 - 2x^3 L + x L^3)q/D$, $L/2$, $3L^2 q/4h^2$
3–14	0.405 m
3–15	$(L^4 q_0/\pi^4 D) \sin(\pi x/L)$
3–17	$(p - \rho g h)(L^2 - 12x^2)/24$, $\pm L/2$
3–18	$(p - \rho g h)(12x^2 - L^2)/4h^2$, $\pm L/2$
3–20b	1.0 GPa
3–21	18.1 km
3–22	16.5 km

4–3   18 ppb, 72 ppb, 0.108%

4–4 7.4×10^{-12} W kg^{-1}, 3.1×10^{-13} W kg^{-1}, 1.5×10^{-11} W kg^{-1}, 1.1×10^{-9} W kg^{-1}, 7.7×10^{-10} W kg^{-1}, 3.4×10^{-10} W kg^{-1}, 3.5×10^{-12} W kg^{-1}

4–5 9×10^{-11} W kg^{-1}

4–7   16 mW m^{-2}, 10 km

4–8 $q^* = q_m + \rho h_r H^*$

4–9   137 km, $0.52 \ \mu$W m^{-3}

4–10 $q_s = q_m + \dfrac{1}{2} \rho H_s b$

$T = T_0 + \dfrac{1}{k} \left(q_m y + \dfrac{1}{2} \rho H_s b y - \dfrac{1}{2} \rho H_s y^2 + \dfrac{1}{6} \dfrac{\rho H_s y^3}{b} \right)$

4–11   55.5 mW m^{-2}, 608.5°C

4–14   58,000 K

4–15 $T_0 + \dfrac{\rho H}{6k}(a^2 - b^2) + \Gamma b$

4–16 $T_0 + \dfrac{q_s a}{k} \left(\dfrac{3}{2} \dfrac{a}{b} - 1 \right)$

4–17 $T_0 + \dfrac{\rho H}{6k}(a^2 - r^2) + \dfrac{1}{3} \dfrac{\rho H b^3}{k} \left(\dfrac{1}{a} - \dfrac{1}{r} \right)$, $a > r > b$

4–19   733 m

4–20   43.8 mW m^{-2}

4–23 1.3×10^{12} yr, 10^{11}, 3.6×10^{11}, 1.2×10^{12}, 1.9×10^{11}

4–24 7.4×10^{-15} K s^{-1}

4–25 9.3×10^{-15} K s^{-1}

4–26   3.16 m, 9.94 m

4–27   950 m

4–28   1.14 m

4–29 5.2×10^{-3} K

4–30 $\dfrac{\pi}{4} + n\pi, n = 1, 2, 3 \ldots$

4–31 $3.64(\kappa t)^{1/2}$

4–32 3×10^5 sec

4–34 $T_0 + \beta y - \Delta T_0 \mathrm{erfc} \dfrac{y}{2\sqrt{\kappa(t + \tau)}}$, $-\tau < t < 0$

$T_0 + \beta y - \Delta T_0 \mathrm{erfc} \dfrac{y}{2\sqrt{\kappa(t + \tau)}}$
$+ \Delta T_0 \mathrm{erfc} \dfrac{y}{2\sqrt{\kappa t}}, t > 0$

4–37   0.32 m

4–38   26 Myr

4–39   65.9 Myr, 92 km

4–40 $\dfrac{\kappa(T_m - T_0)}{(\pi \kappa t)^{1/2} \mathrm{erf} \lambda_1}, e^{\lambda_1^2}$

4–41 $T_0 + \dfrac{(T_m - T_0)}{\mathrm{erf} \lambda_3} \mathrm{erf} \left(\dfrac{y}{2\sqrt{\kappa t}} \right), 0 < y < y_s$

$T_v - \dfrac{(T_v - T_m)}{\mathrm{erfc} \lambda_3} \mathrm{erfc} \left(\dfrac{y - y_s}{2\sqrt{\kappa t}} \right), y > y_s$

$\dfrac{(T_m - T_v)}{\mathrm{erfc} \lambda_3} - \dfrac{(T_v - T_m)}{\mathrm{erfc} \lambda_3} = \dfrac{\sqrt{\pi} \ L \lambda_3 e^{\lambda_3^2}}{c}$

4–42   32 km

4–43   140 km

4–44   1.29 days

4–46 $T_0 + (T_\infty - T_0) \exp \left(-\dfrac{U\xi}{\kappa} \right)$

4–47 $\dfrac{T_0 - T_\infty \mathrm{erf}(\alpha/2\sqrt{\kappa})}{1 - \mathrm{erf}(\alpha/2\sqrt{\kappa})} +$

$\left[\dfrac{T_\infty - T_0}{1 - \mathrm{erf}(\alpha/2\sqrt{\kappa})} \right] \mathrm{erf} \dfrac{y}{2\sqrt{\kappa t}}$

4–48 $\dfrac{q_m}{\rho[L + c(T_m - T_0)]}$

4–49   16.5 MPa

4–50   6.4 km

4–52   500 m, 200 m

4–53   0.04 mm yr^{-1}

4–54 $\dfrac{\chi(\rho_m - \rho_l) y_L}{(\rho_m - \rho_w)}, 0.6$

4–55   320 m

4–56   4.2 km

4–57   2–3 km

4–58   7.6 km, 12.8 km

4–59   10.6 km, 18.5 km

4–60   2.46 km

4–61   68 K

4–62   17,000 K

4–63   2.7 K

4–67   7.1 m^2 s^{-1}

4–68   2.3 m^2 s^{-1}

5–1 1.67×10^{-3}

5–2 3.4621×10^{-3}

5–3   (a) 6.25×10^7 J, (b) 1.12×10^4 m s^{-1}, (c) 62,200 K

5–4   No

5–5   9.78031846 m s^{-2}, 9.832177 m s^{-2}, 9.7804629 m s^{-2}, 9.832011 m s^{-2}

5–7 $g_e + 2Gb/a^3$

5–8   4140 kg m^{-3}, 12,410 kg m^{-3}

5–9   4460 kg m^{-3}

5–10   3.74 days

5–11   21 hours

5–12 13 million tons

5–16 3.98 mm s^{-2}

5–18 8.1 km

5–19 4.3×10^6 kg m^{-2}

5–20 0.044978 mm s^{-2}, -0.56184 mm s^{-2}

5–22 $\dfrac{\pi G}{g} \rho_{cu} \left\{ 2h \left[H + \left(\dfrac{\rho_m - \rho_{cl}}{\rho_m - \rho_{cu}} \right) b_L \right] + \dfrac{h^2 \rho_m}{(\rho_m - \rho_{cu})} \right\}$

5–23 14.4 km

6–2 $\dfrac{1}{2}\dfrac{dp}{dx}(2y - h) - \mu\dfrac{u_0}{h}, -\mu\dfrac{u_0}{h}, \dfrac{1}{2}\dfrac{dp}{dx}(2y - h)$

6–3 $\dfrac{h}{2} + \dfrac{\mu_0 u_0}{h(dp/dx)}$

6–4 $\dfrac{\rho g h^2 \sin\alpha}{3\mu}, \left(\dfrac{3\mu Q}{\rho g \sin\alpha} \right)^{1/3}$

6–5 0.317 MPa

6–6 19.2 MPa

6–7 1.27×10^{-20}

6–8 0.025 m, 0.84 m s^{-1}, 4×10^4

6–9 $(\rho_s - \rho_l)gd^3/12\mu$

6–10 $-\dfrac{1}{2\mu}\dfrac{dp}{dx}\left(\dfrac{y^3}{3} - \dfrac{hy^2}{2} \right) + \dfrac{u_0 y^2}{2h} - u_0 y,$

$-u_0 \left[y - \dfrac{y^2}{2h} + 6\left(\dfrac{h_L}{h} + \dfrac{1}{2} \right)\left(\dfrac{y^3}{3h^2} - \dfrac{y^2}{2h} \right) \right]$

6–12 0.61 km, 0.22 km, 4×10^{21} Pa s

6–13 $u = \dfrac{-\sqrt{2}U}{(2 - \frac{\pi^2}{4})}\left[\dfrac{\pi}{2} + \left(\dfrac{\pi}{2} - 2 \right)\tan^{-1}\dfrac{y}{x} \right.$

$\left. - \left(\dfrac{\pi x}{2} + \left[2 - \dfrac{\pi}{2} \right]y \right)\left(\dfrac{x}{x^2 + y^2} \right) \right]$

$v = \dfrac{-\sqrt{2}U}{(2 - \frac{\pi^2}{4})}\left[\dfrac{\pi}{2}\tan^{-1}\dfrac{y}{x} \right.$

$\left. - \left(\dfrac{\pi x}{2} + \left[2 - \dfrac{\pi}{2} \right]y \right)\left(\dfrac{y}{x^2 + y^2} \right) \right]$

$u = \dfrac{U}{(\frac{9\pi^2}{4} - 2)}\left\{ \left(\dfrac{9\pi^2}{4} - 2 \right) + \pi(2\sqrt{2} - 3\pi) \right.$

$- \left[\sqrt{2}\left(2 + \dfrac{3\pi}{2} \right) - 2\left(1 + \dfrac{3\pi}{2} \right) \right]\tan^{-1}\dfrac{x}{y}$

$- \left[\left(2 - \sqrt{2}\dfrac{3\pi}{2} \right)x \right.$

$+ \left(\sqrt{2}\left[2 + \dfrac{3\pi}{2} \right] - 2\left[1 + \dfrac{3\pi}{2} \right] \right)y \right]$

$\left. \times \dfrac{x}{x^2 + y^2} \right\}$

$v = \dfrac{U}{(\frac{9\pi^2}{4} - 2)}$

$\times \left\{ \left(2 - \sqrt{2}\dfrac{3\pi}{2} \right)\pi + \left(2 - \sqrt{2}\dfrac{3\pi}{2} \right)\tan^{-1}\dfrac{y}{x} \right.$

$- \left[\left(2 - \sqrt{2}\dfrac{3\pi}{2} \right)x + \left(\sqrt{2}\left(2 + \dfrac{3\pi}{2} \right) \right. \right.$

$\left. \left. -2\left[1 + \dfrac{3\pi}{2} \right] \right)y \right]\dfrac{x}{x^2 + y^2} \right\}$

6–15 4.6×10^{18} Pa s

6–16 195,000 yr

6–20 3.3×10^{18} Pa s

6–21 8.46×10^{20} Pa s

6–24 52 km, 55 m^3 s^{-1}, 4.6×10^{10} W, 0.21 Myr^{-1}, 1.38×10^8 km^3

6–25 74 km, 232 m^3 s^{-1}, 1.9×10^{11} W, 0.42 myr^{-1}, 2.22×10^9 km^3

6–26 $T = C_1 x + C_2 + \dfrac{C_1}{4\kappa\mu}\left(\dfrac{1}{6}y^4 - \dfrac{1}{4}d^2 y^2 + \dfrac{5}{96}d^4 \right)$

$q = \dfrac{C_1 k d^3}{24\kappa\mu}, h = \dfrac{70}{17}\dfrac{k}{d}, \text{Nu} = \dfrac{70}{17}$

6–29 $2.4 \times 10^5, 2.9 \times 10^9, 4.2 \times 10^7, 1.2 \times 10^6$

6–30 1.16 b, 7.85 b

6–33 14.4 km

6–35 0.32

6–38 $T_0 + \dfrac{\mu u_0^2}{kh^2}\left(hy - \dfrac{1}{2}y^2 \right), \dfrac{\mu u_0^2}{h}, \dfrac{\mu u_0^2}{2, k}$

7–1 547 kJ mole^{-1}

7–3 1.6×10^8 Pa

7–4 1.24×10^3 J kg^{-1} K^{-1}

7–11 -72 bars

7–13 137 MPa, 860 K

7–21 $\sigma^*, \sigma^*/\sqrt{6}$

7–22 $\sigma_0/[(1 - \nu)^2 + 1]^{1/2},$
$(1 - \nu^2)\sigma_0/E[(1 - \nu)^2 + \nu]^{1/2}$

7–23 $2\tau, \sqrt{3}\tau$

7–24 $\dfrac{a^2 \sigma_0}{2}\left(1 - \dfrac{c^2}{3a^2} \right), \dfrac{\sigma_0 a^2}{3}, \dfrac{\sigma_0 a^2}{2}$

Index